Modelling Behaviour

Mette Ramsgaard Thomsen
Martin Tamke · Christoph Gengnagel
Billie Faircloth · Fabian Scheurer
Editors

Modelling Behaviour

Design Modelling Symposium 2015

Springer

Editors
Mette Ramsgaard Thomsen
Schools of Architecture, Design and Conservation
The Royal Danish Academy of Fine Arts
Copenhagen
Denmark

and

Schools of Architecture, Design and Conservation, Centre for Information Technology and Architecture (CITA)
The Royal Danish Academy of Fine Arts
Copenhagen
Denmark

Martin Tamke
Schools of Architecture, Centre for IT and Architecture
Royal Danish Academy of Fine Art
Copenhagen
Denmark

Christoph Gengnagel
University of Arts
Berlin
Germany

Billie Faircloth
Kieran Timberlake
Philadelphia, PA
USA

Fabian Scheurer
designtoproduction GmbH
Erlenbach, Zürich
Switzerland

ISBN 978-3-319-24206-4 ISBN 978-3-319-24208-8 (eBook)
DOI 10.1007/978-3-319-24208-8

Library of Congress Control Number: 2015949452

Springer Cham Heidelberg New York Dordrecht London

Cover and chapter graphics: Olga Krukovskaya

Printed on acid-free paper

Springer International Publishing AG Switzerland is part of Springer Science+Business Media (www.springer.com)

Scientific Committee

We would like to thank the members of the Scientific Committee of the Design Modelling Symposium Copenhagen 2015 for their efforts in reviewing and guidance to the authors of the published contributions.

Al Fischer, Buro Happold, Bath
Areti Markopoulou, IAAC, Barcelona
Axel Kilian, Princeton University
Billie Faircloth, Kieran Timberlake
Bob Sheil, UCL, London
Brady Peters, Toronto
Branko Kolarevic, UCA
Christian Derix, Woods Bagots, London
Christoph Gengnagel, UdK, Berlin
Daniel Davis, CASE, New York
David Gerber, USC, Los Angeles
David Stasiuk, CITA, Copenhagen
Fabian Scheurer, designtoproduction
Günther Filz, Universität Innsbruck
Jane Burry, RMIT, Melbourne
Jonas Runberger, White Architects
Jonathan Rabagliati, Foster & Partners
Julian Lienhard, Universität Stuttgart
Mark Pauly, EPFL, Lausanne
Martin Tamke, CITA, Copenhagen
Mette Ramsgaard Thomsen, CITA, Copenhagen
Michael Hensel, Oslo School of Architecture and Design
Neil Burford, University of Dundee
Norbert Palz, UdK Berlin
Oliver Tessmann, KTH, Stockholm
Olivier Bavarel, UR Navier, Université Paris-Est
Paul Nicolas, CITA, Copenhagen
Paul Shepherd, University of Bath
Peter v. Bülow, University of Michigan
Phil Ayres, CITA, Copenhagen
Philipp Block, ETH Zürich
Raoul Wessel, University of Bonn
Roland Wüchner, TU München
Sean Ahlquist, Taubmann College

Contents

Contributors

Sean Ahlquist Taubman College of Architecture and Urban Planning, University of Michigan, Ann Arbor, USA

Francis Aish Foster + Partners, City Hall, London

Desislava Angelova Institute for Computational Design, University of Stuttgart, Stuttgart, Germany

Gideon D.P.A. Aschwanden Faculty of Architecture Building and Planning, University of Melbourne, Melbourne, Australia

Nicolas Azel Cornell University, Ithaca, USA

Roderick Bates KieranTimberlake, Philadelphia, USA

David Benjamin The Living an Autodesk Studio, New York, Brooklyn, USA

Christopher Beorkrem School of Architecture, University of North Carolina Charlotte, Charlotte, USA

Uta Berger Forest Biometrics/Systems Analysis, Dresden University of Technology, Dresden, Germany

Kai-Uwe Bletzinger Structural Analysis, Technische Universität München, Munich, Germany

Philippe Block Block Research Group, ETH Zurich Institute of Technology in Architecture, Zürich, Switzerland

Sofien Bouaziz École Polytechnique Fédérale de Lausanne, Lausanne, Switzerland

Jeffrey L. Boyer University of Michigan and dbHMS, Chicago, USA

Milan Brandt Advanced Manufacturing Precinct (AMP), RMIT University, Melbourne, Australia

Michael Breitenberger Structural Analysis, Technische Universität München, Munich, Germany

Jane Burry Spatial Information Architecture Laboratory (SIAL), School of Architecture and Design, RMIT University, Melbourne, Australia

Mark Burry Faculty of Architecture Building and Planning, University of Melbourne, Melbourne, Australia

Brandon Clifford Massachusetts Institute of Technology (MIT)/Matter Design, Boston, USA

Christopher Connock KieranTimberlake, Philadelphia, USA

Kristof Crolla Chinese University of Hong Kong, Hong Kong, People's Republic of China

Dana Cupkova Carnegie Mellon School of Architecture, Pittsburgh, USA

Steve Danilowicz School of Architecture, University of North Carolina Charlotte, Charlotte, USA

Enrico De Angelis Politecnico di Milano, DABC, Milan, Italy

Anders Holden Deleuran The Royal Danish Academy of Fine Arts Schools of Architecture Design and Conservation, Centre for Information Technology and Architecture (CITA), Copenhagen, Denmark

Bailin Deng École Polytechnique Fédérale de Lausanne, Lausanne, Switzerland

Mario Deuss École Polytechnique Fédérale de Lausanne, Lausanne, Switzerland

Karola Dierichs Institute for Computational Design, University of Stuttgart, Stuttgart, Germany

Romain Duballet Ponts ParisTech, ENSA Paris Malaquais, Paris, France

Moritz Dörstelmann Institute for Computational Design (ICD), Stuttgart, Germany

Eric Eisele KieranTimberlake, Philadelphia, USA

Khaled El-Ashry Foster + Partners, City Hall, London

Henrik Leander Evers The Royal Danish Academy of Fine Arts Schools of Architecture Design and Conservation, Centre for Information Technology and Architecture (CITA), Copenhagen, Denmark

Günther H. Filz Faculty of Architecture, Institute of Design/unit koge. Structure and Design, University of Innsbruck, Innsbruck, Austria

Corentin Fivet Massachusetts Institute of Technology, Cambridge, USA

Isak Worre Foged Department of Architecture and Media Technology, Aalborg University, Aalborg, DenmarkAREA, Copenhagen, Denmark

Ursula Frick Block Research Group, ETH Zurich Institute of Technology in Architecture, Zurich, Switzerland

Salomé Galjaard Arup, Amsterdam, The Netherlands

Christoph Gengnagel Department of Structural Design and Technology (KET), University of Arts Berlin, Berlin, Germany

Michail Georgiou HUB Design + Engineering Platform, ARC, University of Nicosia, Nicosia, Cyprus

Odysseas Georgiou HUB Design + Engineering Platform, ARC, University of Nicosia, Nicosia, Cyprus

Clément Gosselin ENSA Paris Malaquais, Paris, France

Kostas Grigoriadis School of Architecture, Royal College of Art, London, UK

Nicola Haberbosch Institute of Building Structures and Structural Design (ITKE), University of Stuttgart, Stuttgart, Germany

Dominik Holzer Faculty of Architecture Building and Planning, University of Melbourne, Melbourne, Australia

Christopher Hutchinson Department of Materials Science and Engineering, Monash University, Melbourne, Australia

Sam Joyce Foster + Partners, City Hall, London

Stefan Kainzwaldner Faculty of Architecture, Institute of Design/unit koge. Structure and Design, University of Innsbruck, Innsbruck, Austria

Justyna Anna Karakiewicz Faculty of Architecture Building and Planning, University of Melbourne, Melbourne, Australia

Markus Kayser Department of Architecture and Urban Planning (SA+P), Media Lab, Mediated Matter Group, Massachusetts Institute of Technology (MIT), Cambridge, MA, USA

Jan Knippers Institute of Building Structures and Structural Design (ITKE), University of Stuttgart, Stuttgart, Germany

Jakub Kolarik Technical University of Denmark, Kongens Lyngby, Denmark

Valentin Koslowski Institute of Building Structures and Structural Design (ITKE), Stuttgart, Germany

Thomas Krijnen Department of the Built Environment, Eindhoven University of Technology, Eindhoven, The Netherlands

Tom Kvan Faculty of Architecture Building and Planning, University of Melbourne, Melbourne, Australia

Riccardo La Magna Institute of Building Structures and Structural Design (ITKE), University of Stuttgart, Stuttgart, Germany

Donna Lanclos University Library, University of North Carolina Charlotte, Charlotte, USA

Martin Leary Advanced Manufacturing Precinct (AMP), RMIT University, Melbourne, Australia

Juney Lee Massachusetts Institute of Technology, Cambridge, USA

John Locke The Living an Autodesk Studio, New York, Brooklyn, USA

Frans Magnusson White arkitekter AB/Chalmers, Gothenburg, Sweden

Henrik Malm Foster + Partners, City Hall, London

Massimiliano Manfren Politecnico di Milano, DABC, Milan, Italy

Jacob Mans GSD, Harvard University, Cambridge, USA

Wes McGee University of Michigan/Matter Design, Ann Arbor, USA

Achim Menges Institute for Computational Design (ICD), Stuttgart, Germany

Achim Menges Institute for Computational Design, University of Stuttgart, Stuttgart, Germany

Christine Mondor Carnegie Mellon School of Architecture, Pittsburgh, USA

Philippe Morel ENSA Paris-Malaquais/EZCT Architecture and Design Research, Paris, France

Caitlin T. Mueller Department of Architecture, Massachusetts Institute of Technology, Cambridge, USA

Danil Nagy The Living an Autodesk Studio, New York, Brooklyn, USA

Ajit Naik dbHMS, Chicago, USA

Kristoffer Negendahl Technical University of Denmark, Kongens Lyngby, Denmark

Paul Nicholas Centre for IT and Architecture, Royal Danish Academy of Fine Art, School of Architecture, Copenhagen, Denmark

Esben Clausen Nørgaard Centre for IT and Architecture, Royal Danish Academy of Fine Art, School of Architecture, Copenhagen, Denmark

Patrick Ole Ohlbrock ETH Zurich, Zurich, Switzerland

Neri Oxman Department of Architecture and Urban Planning (SA+P), Media Lab, Mediated Matter Group, Massachusetts Institute of Technology (MIT), Cambridge, MA, USA

Anke Pasold AREA, Copenhagen, Denmark

Sayjel V. Patel Massachusetts Institute of Technology, MIT-SUTD Collaboration, Cambridge, USA

Mark Pauly École Polytechnique Fédérale de Lausanne, Lausanne, Switzerland

Thomas Perkov Technical University of Denmark, Kongens Lyngby, Denmark

Benedikt Philipp Structural Analysis, Technische Universität München, Munich, Germany

Daniel Piker Robert McNeel and Associates, London, UK

Eva Pirker Institute of Structural Design, Graz University of Technology, Graz, Austria

Marshall Prado Institute for Computational Design (ICD), Stuttgart, Germany

Daniel Prohasky Spatial Information Architecture Laboratory (SIAL), School of Architecture and Design, RMIT University, Melbourne, Australia

Gregory Quinn Department of Structural Design and Technology (KET), University of Arts Berlin, Berlin, Germany

Felix Raspall Harvard University Graduate School of Design, Cambridge, MA, USA

Andrew Rastetter College of Environmental Design (CED), University of California, Berkeley, USA

Shibo Ren Arup, Amsterdam, The Netherlands

Philippe Roux Arts et Métiers ParisTech, ENSA Paris Malaquais, Paris, France

Jorge Duro Royo Department of Architecture and Urban Planning (SA+P), Media Lab, Mediated Matter Group, Massachusetts Institute of Technology (MIT), Cambridge, MA, USA

Jonas Runberger White arkitekter AB/KTH, Stockholm, Sweden

Yasushi Sakai Nikken Sekkei, Tokyo, Japan

Eric Sauda School of Architecture, University of North Carolina Charlotte, Charlotte, USA

Gundula Schieber Institute of Building Structures and Structural Design (ITKE), Stuttgart, Germany

Simon Schleicher College of Environmental Design (CED), University of California, Berkeley, USA

Michel Schmeck Department for Structural Design and Technology (KET), University of Arts Berlin, Berlin, Germany

Thibault Schwartz HAL Robotics Ltd, London, UK

Andreas Schönbrunner Institute of Building Structures and Structural Design (ITKE), University of Stuttgart, Stuttgart, Germany

Hamed Seifi Centre for Innovative Structures and Materials, RMIT University, Melbourne, Australia

Laia Mogas Soldevila Department of Architecture and Urban Planning (SA+P), Media Lab, Mediated Matter Group, Massachusetts Institute of Technology (MIT), Cambridge, MA, USA

Richard Souvenir Department of Computer Science, University of North Carolina Charlotte, Charlotte, USA

Scott Spurlock Department of Computer Science, University of North Carolina Charlotte, Charlotte, USA

David Stasiuk Centre for IT and Architecture, Royal Danish Academy of Fine Art, School of Architecture, Copenhagen, Denmark

Djordje Stojanovic Faculty of Architecture, University of Belgrade, Belgrade, Serbia

Lavinia Chiara Tagliabue Angelo Luigi Camillo Ciribini, University of Brescia, DICATAM, Brescia, Italy

Martin Tamke Centre for IT and Architecture, Royal Danish Academy of Fine Art, School of Architecture, Copenhagen, Denmark

Mette Ramsgaard Thomsen Centre for IT and Architecture, Royal Danish Academy of Fine Art, School of Architecture, Copenhagen, Denmark; Centre for Information Technology and Architecture (CITA), The Royal Danish Academy of Fine Art, School of Architecture, Design and Conservation, Copenhagen, Denmark

Ida Katrine Friis Tinning Centre for IT and Architecture, Royal Danish Academy of Fine Art, School of Architecture, Copenhagen, Denmark

Martha Tsigkari Foster + Partners, City Hall, London

Daisuke Tsunoda Nikken Sekkei, Tokyo, Japan

Tom Van Mele Block Research Group, ETH Zurich Institute of Technology in Architecture, Zurich, Switzerland

Keith Van de Riet School of Architecture, Florida Atlantic University, Fort Lauderdale, USA

Lauren Vasey Institute for Computational Design (ICD), Stuttgart, Germany

Robert Vierlinger Institute of Architecture, Institute for Structural Design, University of Applied Arts Vienna, Vienna, Austria

Frédéric Waimer Institute of Building Structures and Structural Design (itke), Stuttgart, Germany

Ryan Welch KieranTimberlake, Philadelphia, USA

Marcus White Faculty of Architecture Building and Planning, University of Melbourne, Melbourne, Australia

Chris Williams University of Bath, Bath, UK

Nicholas Williams Spatial Information Architecture Laboratory (SIAL), School of Architecture and Design, RMIT University, Melbourne, Australia

Roland Wuchner Structural Analysis, Technische Universität München, Munich, Germany

Mike Xie Centre for Innovative Structures and Materials, RMIT University, Melbourne, Australia

James Yamada MetaLAB, Harvard University, Cambridge, USA

Yao Yu dbHMS, Chicago, USA

Malgorzata A. Zboinska Chalmers University of Technology, Gothenburg, Sweden

Introduction

The 5th Design Modelling Symposium moves to Copenhagen and to CITA (Centre for Information Technology and Architecture) at The Royal Danish Academy of Fine Arts, Schools of Architecture, Design and Conservation. The Conference maintains its focus on the creation of a shared platform in which architects and engineers, and those in associated fields such as materials science, computer science, ecology, physics or urban design, can exchange research questions, methods and results. The foundational motivation for the Design Modelling Symposium continues to be the need to investigate the practices that arise when computational design is dispersed across the knowledge fields that make up building design. By assembling leading professional and academic researchers biennially, the Symposium queries how our practices change, expand and crossover as methods emerge to assemble design, analysis, simulation, communication, specification, fabrication, construction and post-occupancy evaluation into an integrated sequence. The Symposium asks what the nature of these new practices will be and how promised ideals will form; these are ideals of information flowing uninterrupted between design phases, supporting strategic and creative feedback between stakeholders and enabling collaboration.

This broad vision contains a multitude of contributions to the field—from the advanced design of City Information Models and Geographic Information Systems that integrate Open Data and analyse spatial information, across the exploratory development of methods for design integrating simulation, to the intricate experiments in materials deposition and assembly using industrial robots. The emerging field is heterogeneous and often contradictory: Are we aiming to devise a unified method by which all project information can be assembled into one integrated model? Or, Is the distributed nature of a networked model, one that is project specific and parses only the information it needs, a more holistic vision for the future? Can architecture extend its relevance as we become increasingly aware of Big Data methods for analysing and integrating data into meaningful design decisions? What are the means by which we can simulate and analyse time-based phenomena in our design representations? And, What are the real consequences of the experiments in advanced robotic fabrication on building culture and the conception of its underlying material practices?

The Design Modelling Symposium Copenhagen 2015 brings these divergent conversations together through the concept of *behaviour* and the *behaving model.*

The Behaving Model

Models that integrate computational logic into design permit us to discover, predict or orchestrate the nuanced behaviour of architecture. Such models seek to emulate phenomena at scale and in-time, attempting a congruous simulation of the potential relationship between architecture and its contexts. We are entering a new practice in which a rich array of extended information models fundamentally enrich our field. Here, a mashup of diverse inputs may be pursued: environmental, social, structural or material inputs formerly particularised by discrete disciplines now assembled, associated and synthesised in the model. By integrating the simulation of events, whether occupancy, environmental impact or the dynamic behaviour of structures and materials under stress, our models are radically changing from static description of intent to dynamic analyses of event.

This shift in focus from 3-dimensional geometric extension to a n-dimensional field of divergent, heterogeneous and conflicting information triggers deep changes in the way we work as architects. It necessitates new methods by which the parsing of information, its analysis and its calculation can become part of our representations. We need to develop modelling paradigms by which these information flows can be critically assessed and employed in ways that support creative practice. As such, we need to learn from neighbouring fields in which parallel efforts for information modelling have longer and more established traditions. Rather than reinventing our own practices for design modelling, we need to consolidate new hybrid practices that transfer core concepts as well as techniques by which information can be made practical in design execution. Modelling behaviour thus affects the boundaries of our disciplines. Becoming the shared interface for multiple disciplines—architects, engineers, planners and fabricators as well as material scientists, ecologists and physicists—such models interface the breadth of design practice expanding the concerns of architectural design while simultaneously infiltrating our thinking with proprietary design methods and traditions for analysis and representation.

A Need for Validation

The behaving model expands architecture's address and defines new methods for designing to the extent that it is imperative for us to devise ways to validate and verify our new modelling practices. As the field of information modelling matures, we need to discuss and query the solidity of our modelling conventions. The processes of validation necessarily vary. The validation of an urban model necessitates other means of testing than those that engage directly with structure and material. In some parts of our practice, the role of the prototype has taken on a central role by which design simulations are not only tested but also calibrated and informed. Here, the empirical analysis and evaluation of structural and material behaviour becomes central for a new modelling practice. Jointly, these new modelling practices are poised for robust querying asking how we develop a shared criticality across the field.

Across Scale

A central focus of the Design Modelling Symposium Copenhagen—Modelling Behaviour is the inter-scalar nature of contemporary design practice. Building design practices have traditionally been organised hierarchically into a sequenced set of subsystems—foundation, substructure, superstructure, envelope, and roof—each one with its own team of specialised professions. The dogma of systems obfuscates our ability to see systems, whole buildings and hence cities as (forms of and forming) matter, energy, and ecology across miniscule and monumental scales. New design-based information modelling methods are increasingly focussed on investigating inter-scalar interdependencies: urban morphology may shape regional weather patterns, the presence or absence of vegetation or habitat. A single building's form—its height and solar orientation—contributes to the creation of local microclimates, while its materials absorb or reflect sunlight, instantiating the overall increase in the city's temperature when compared to rural surroundings. These observations expand the field of architecture and urban planning, and hence the agency of design, to include the dynamic and temporal aspects of social and ecological infrastructure. The agency of design is further expanded to the design of materials for architecture, ones which based on advanced fabrication tools can lead to materials with bespoke performances directly engineered at the molecular scale in response to local site or use scenarios. These emergent practices expand the architectural design space necessitating the development of new modelling and validation methods that can simultaneously engage design activity at the material scale—interfacing fabrication protocols and controlling manufacture—and design activity at building scale allowing meaningful and operational feedback between scales.

Design Modelling Symposium 2015 aims to foster a discussion on the future of these practices. By building on the understanding that current tools for architectural representation are impeding design innovation, we ask what new modelling paradigms and infrastructures we need to develop and to support real progress in our field. The conference broadly seeks to ascertain ways in which advanced information models delimit system boundaries, capture information and integrate feedback in future design scenarios across the expanded digital design process. Here, the digital design process is inclusive of structures, envelopes and materials—from their analysis, simulation and specification, to their fabrication and inhabitation.

The conference engages this breadth of concern through 3 trajectories. These form the three parts of this book.

Modelling and Design of Behaviour

As the profession is moving into performance-based practices, buildings are increasingly evaluated through their observed behaviour. This can be seen in the strengthening of a collective academic research interest in questioning how structural or thermal behaviour of architectural objects can be designed, or in professional practice, where clients and society rightfully ask that a building's projected costs or energy consumption are retained during building lifespan. Over the last decade, the profession has made great leaps in the modelling and simulation of behaviour. In thermal simulation, tools and methods have been created that provide feedback in early design stages and large-scale industry efforts are set in place to verify the built outcome, while in structural design the tradition of almost linear design with defined mechanical models has been transformed into the modelling of interactive systems including multi-perspective models. But what is the identity and purpose of the behavioural model for other relevant areas of building practice? And how are behaviours defined, represented and simulated, which cross domains? Contemporary design research is linking a wide array of simulation practices. Here, researchers and industry are investigating project-specific design of the thermal heat conductivity of materials, the use of material behaviour for kinetic building parts and the linkage of occupancy monitoring and simulation for design. Through these emergent practices, it becomes evident that design models are growing in complexity and have to predict behaviour in nonlinear and multi-scalar relations. This challenges the current practices of modelling in the profession, as models of a diverse range of disciplines have to be linked and modellers are facing the problem to do this in the context of a volatile design process. And while the sciences have developed modelling paradigms, such as discreet event simulation, to codify the behaviour of complex system into an ordered sequence of well-defined events and behaviours, building design resists a direct transfer of approaches. Simultaneously, new means for measuring are emerging, which use statistical methods and the datasets, that are provided through the ubiquity of the net and the interconnectedness of data. How do we create new design methods for engaging with these new information flows while retaining the intuitive, creative and communicable dimensions of architectural design practice?

Modelling and Design of Processes

Where CAD was focused on defining the end result of a design process, the profession is venturing into new areas in which the design and description of the processes itself is central. On one side, the digital models become dynamic by containing parametric descriptions that algorithmically define a path to the end result instead of an explicit solution. On the other side, the processes of building and operation are moving back into the focus of the design. Robotic fabrication and assembly, the planning of large-scale construction, life cycle management or facility management all necessitate new dynamic representations in which processes can be designed, simulated, coordinated and communicated within large teams of distinct disciplinary backgrounds. Contemporary architectural research is increasingly concerned with the exploration of new digital fabrication technologies employing a host of advanced CNC tools including the industrial robot. Where these lead to the invention of novel material and structural systems, they remain largely speculative and foreign to building production. However, in parallel industries such as the automotive industry and industrial design, new concepts such as Industry 4.0 are challenging the way we understand production. Here, computational technologies including CNC fabrication, sensing, cloud computing and virtual prototyping are fundamentally innovating industry. The conference aims to discuss how we bridge to these innovations in parallel industries while retaining our own advances within the field? How do the more speculative research led design experiments contribute to this development? What are the methods of dynamic modelling, how do we critically assess and understand input, and how can the model integrate into a heterogeneous workflow?

Modelling and Design of Information

What defines the underlying infrastructures of the behavioural model? For the model to become a true analytical tool, we need to develop our own critical practices by which to discuss and evaluate how information is defined, stored and retrieved as well as computed, handled and scaled. As architects and engineers become increasingly involved in research into new structural morphologies and hybrid structures, it becomes central to understand the data employed by the model as something generated as part of the design project rather than being predefined. As material design and fabrication similarly becomes a new part of an extended digital chain, it becomes fundamental that we find ways by which these materials can be formalised and parameterised so as to control material composition, grading and performance. At the same time, the field is positioned within a societal sea change in which data are being generated, analysed and exploited. The global push towards Open Data is creating new foundations as cities, municipalities and private shareholders are sharing data on their infrastructures, energy consumption and events. This new platform allows architects and urban designers to engage a breadth of relevant data into their models. The challenge remains how these data streams are parsed, interfaced and importantly how they are compounded within the design model.

Mette Ramsgaard Thomsen
Martin Tamke
Christoph Gengnagel
Billie Faircloth
Fabian Scheurer

The Design Modelling Symposium 2015: Modelling Behaviour was held in Copenhagen, Denmark, at CITA (Centre for Information Technology and Architecture, Royal Danish Academy of Fine Arts, Schools of Architecture, Design and Conservation).

Part I

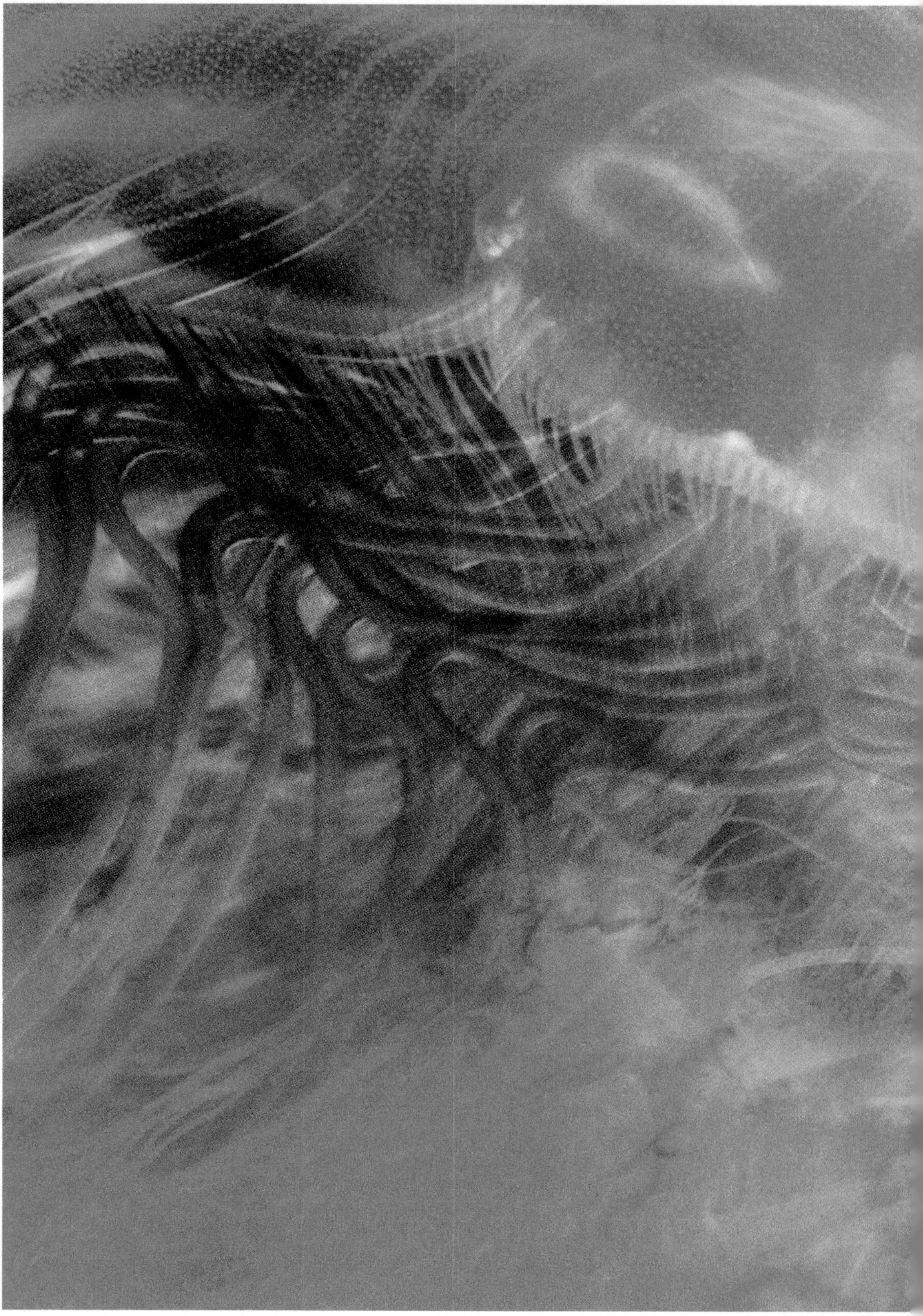

Modelling and Design of Behaviour

Modelling Aggregate Behaviour

Karola Dierichs, Desislava Angelova and Achim Menges

Abstract

Two Case Studies are presented, that investigate different methods of modelling aggregate behaviour in an architectural design application. Aggregates are a novel branch of architectural material systems and are defined as systems of large numbers of loose elements in frictional contact. Especially if the individual grain is custom-designed, the range of possible behaviours is widened—the granular system becomes programmable matter. As opposed to conventional construction systems in architecture, where local and global geometries are precisely described, aggregate systems require new modes of designing which are based on their emergent behaviour. The paper presents two cases both from the realm of numerical simulation and from material-based robotic experimentation. Case Study 1 uses Distinct-Element Modelling for the predictive and analytic modelling of an arch made of designed particles. Case Study 2 introduces an online robotic process that uses luminance values in an aggregate structure to control a six-axis robot using a feedback loop. As an outlook strategies for developing the numerical and the material model both individually and in combination are proposed.

Aim and Context—Behavioural Modelling for Aggregate Architectures

Aggregates are large numbers of individual particles lying only in loose contact with each other, such as sand, gravel or snow (Jaeger et al. 1996). Especially if the individual grain is custom-designed these granulates can perform as architectural material-systems requiring no

K. Dierichs (✉) · D. Angelova · A. Menges
Institute for Computational Design, University of Stuttgart, Stuttgart, Germany
e-mail: karola.dierichs@icd.uni-stuttgart.de

M.R. Thomsen et al. (eds.), *Modelling Behaviour*, DOI 10.1007/978-3-319-24208-8_1

additional binding agent (Hensel and Menges 2006a, b; Dierichs and Menges 2012). Such a designed aggregate system is considered to be a —programmable—material in itself, which through geometry and physical properties of the individual particles is calibrated to have specific behaviours.

As compared to conventional architectural assembly systems—in which both the geometry of the individual element and that of the overall formation are exactly known—the behavioural nature of aggregates requires the development and interrelation of an entirely different set of digital tools, methods and material processes (Dierichs and Menges 2012). In a conventional design process the architect is able to fully control both the geometry of individual parts and the overall structure. However if designing with a material system that has a high degree of indeterminacy, the designing architect needs to shift his attitude to design from controlling a geometry to observing and interacting with a material's behaviour (Hensel and Menges 2008). Developing suitable tools of observation then goes beyond being merely a design method but rather introducing a novel approach towards designing itself. Modelling the behaviour of aggregate formations in architecture can be conducted both through numerical and physical computational processes. The research presented here introduces two applications from both these computational realms—a Distinct-Element Simulation model and an online-controlled robotic process—and evaluates them with regards to their application in a full-scale architectural project. Material behaviour is in both cases introduced as an active driver of architectural design. In this context, the focus of the research presented here is the introduction of both mathematical simulation models and behaviour-based online-

controlled robotic processes for designing with custom-made granulates (Fig. 1).

State of the Art

Designed granulates have initially been developed at the Architectural Association and Rice University (Hensel and Menges 2006a, b). This research into artificial granulates has been further developed in recent years at the ICD Stuttgart and complemented by a parallel branch in Granular Physics (Dierichs and Menges 2012; Gravish et al. 2012; Jaeger 2015).

There is a wide range of mathematical models for simulating granulates depending on the problem at hand (Poeschel and Schwager 2005). The Distinct-Element Method (DEM) is the most widely used one and has developed into a very versatile digital tool for particle simulations

Fig. 1 Full-scale aggregate structures such as vaults can be poured from designed granulates. Varying layer thickness allows for functionally grading their architectural performance, such as the luminance values across a dome

(Cundall 1971; Cundall and Strack 1979; Poeschel and Schwager 2005; Parteli 2013). One of the core challenges within DEM models is the predictive and analytic modelling of granulates with non-spherical particles.

Online-controlled robotic processes are increasingly explored in architectural research (Doerfler et al. 2012; Johns 2014; DFAB 2015). The application of these processes in conjunction with designed granulates however is relatively unexplored (Dierichs et al. 2012).

In this context, the two proposed Case Studies aim to contribute both in the realm of numerical modelling of non-convex aggregates as well as the online-controlled robotic aggregation and disaggregation of designed granular matter. In the first case recent developments have been made with regards to the modelling of non-convex, designed aggregates, however the simulation of a full-scale architectural structure with a significantly large amount of particles has not been implemented so far (Dierichs and Menges 2015). In the second case, online-controlled processes for highly porous aggregates have been simulated, but not implemented in a real-life robotic system, the presented project is thus the first application to realize a feedback loop between a six-axis robot and designed aggregates.

Methods

Computational Modelling—Mathematical and Material Model

Computation is in principle the processing of information. Given this, both the mathematical-numerical and the material-physical realm can be understood as producing computational processes. In the mathematical-numerical realm information is based on the existence of a verified theoretical model. In the material-physical realm information is gathered from the material substrate itself using analytical tools to visualize and read data-sets (Dierichs and Menges 2012).

Mathematical Model—DEM

A state-of-the art software package based on the Distinct Element Method (DEM) has been used. The Distinct-Element Method has first been developed by Cundall and Strack (Cundall 1971; Cundall and Strack 1979). It is the most widely-spread numerical method for computing particle interactions. The forces and torques acting on the particles are computed and their positions consequently updated in a continuous loop. It is based on time-stepping model and allows for soft particle collisions (Poeschel and Schwager 2005). The software package which has been used in the context of this research project can model both polyhedral particles and spherical ones, which in both cases are rigidly-joined together to form a non-convex particle from convex blocks or spheres. If particle numbers exceed a thousand, the bonded sphere—or 'clump'—model is faster to compute and has been applied in this project (Dierichs and Menges 2013).

Material Model—Online-Controlled Robotics

A photogrammetric method is developed that allows for analysing luminance values in a granulate (Hiscocks 2013). It is linked to a robotic sensory interface thus enabling a six-axis robot to modulate an aggregate formation with regards to a target luminance model. The information feedback is realized through a visual programming interface, which translates picture values into data points and on that basis produces a KRL out-put to the robot updating its next point of interacting with the granulate.

Results

The two Case Studies presented here show computational models from both the numerical and the physical realm. They are conducted within a larger framework of research into Aggregate Architectures aiming at the development of full-scale architectural structures made from designed granules. In this context they are developing and deploying tools from both the digital and physical realm for a problem-specific prototypical Aggregate Structure.

Case Study One: Simulation of an Aggregate Arch

A numerical model based on the Distinct-Element Method has been developed in order to both predict and analyse the full-scale construction of an arch made of designed granules. The basic principle of the arch formation is a so-called clogging effect: Aggregates flowing through a funnel form stable arches over the outflow hole through clogging (Duran 2000; Gass and Otto 1990). This effect—well-known as a problem in sand processing plants—is utilized as a formative model for an aggregate arch: An elevated tank with removable bottom lids is modelled, non-convex designed aggregates are filled in and the lids are removed to let the granulate flow out and eventually form a stable arch over the outflow hole, in some of the models unloaded particles are removed after the arch has settled. The focus of the research presented in this context is on the simulations and their innate capacity to analyse both the macro- and the micro-behaviour of the evolving granular formation.

In the applied Distinct-Element Model, the six-armed non-convex particles are built from rigidly bound spheres, so-called clumps. The particles are modelled having a density of 950 kg/m^3, joint normal stiffness of 10^4 N/m, a joint shear stiffness of 5.0×10^3 N/m and a size of 100 mm diameter. The stiffness values are set relatively low, an increase will lead to a more accurate wall-clump interaction. The size of the arch is 2500 mm in width and 750 mm in depth, the height varies depending on how the aggregates settle. The model is used to analyse deformation under self-weight as well as surface speeds and directions of the particles.

All clumps are distributed evenly in a box prior to the actual initiation of the calculation. Subsequently inner and outer formwork are removed and the model is time-cycled to observe how the granulate reacts and behaves under a change in boundary conditions. Various construction techniques were simulated in order to predict how a specific construction-sequence might affect the overall aggregation process of the arch. Initially the basic clogging-effect was simulated with 25,000 particles. The following steps are implemented: clump generation in a box, removal of the outer boundary walls and removal of the bottom lid. Through the computation of both the contact forces and the clump surface velocities it can be observed that after removal of the lid, a stable arch is forming in the granular mass. Only a small portion of the granules are actually load-bearing, the vast amount around them are loose particles (Fig. 2).

Consequently in a next step, the removal of particles is introduced in the model after the removal of the bottom lids, the entire simulation sequence thus consists of altogether five steps: clump generation in a box with a filtered distribution adding more particles in the middle than on both ends, removal of the outer boundary walls, removal of the bottom lid, removal of particles within a cylindrical filter, removal of particles within a filter that is matching unloaded particles. After initially displaying the same behaviour as the first model, the removal of mainly unloaded particles within a cylindrical filter triggers a new force-distribution within the system. After this formation has settled a second filter is implemented, whose geometry is based on the load distribution within the aggregate and allows to remove loose particles from the top and bottom layers. After cycling the arch settles into a stable state again, in this case with the majority of the particles being loadbearing (Fig. 3).

Especially relevant is the analysis of detail force-distributions and motion vectors within the system. Front views show the increase of load

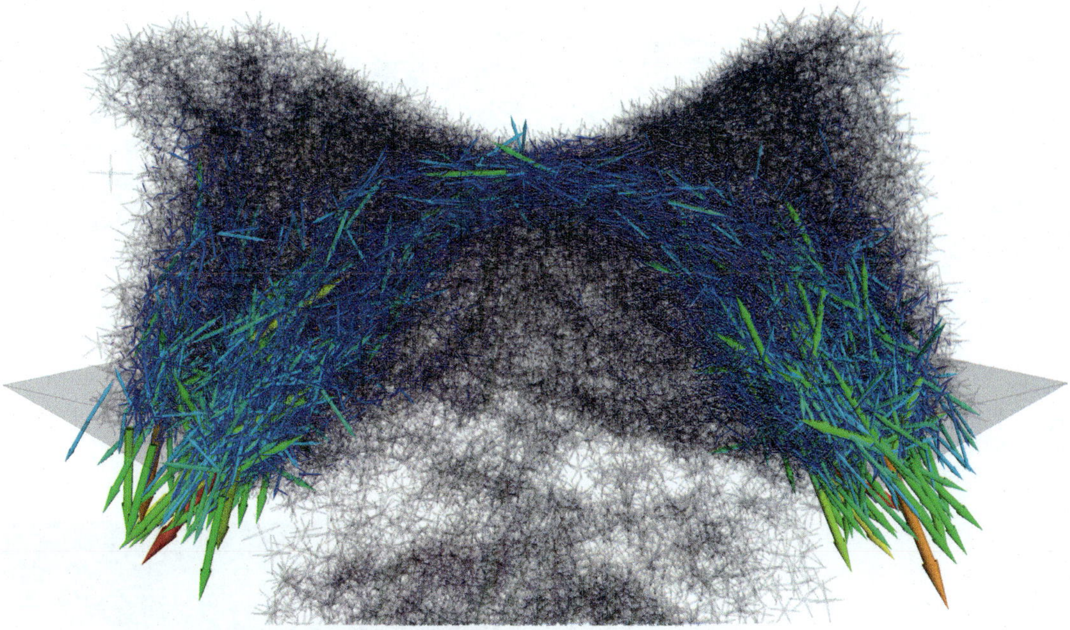

Fig. 2 The DEM simulation shows how an arch forms through a probabilistic effect within a large granular mass that flows out of a removable *bottom* lid or funnel, this clogging effect is enhanced when non-convex granules are used. Only a small portion of the entire granulate is load-bearing as indicated by the contact forces

Fig. 3 The DEM simulation models how an arch comes to rest when non-loadbearing particles are removed after an initial arch has formed over a funnel through clogging

from the top of the arch to both supports through a force-transfer from grain to grain. In the top views quasi-ribs, or force-lines can be observed to form across the depth of the arch. Within the redundancy of the material, force-lines thus form and re-form according to changing load-conditions (Figs. 3, 4 and 5).

An analysis plot of the clump speeds show that the aggregate system has come to rest as indicated by the green coloration of the arrow vectors, the directionality of the clump arrows throughout the structure gives an indication as to the vectorial motion particles or groups of particles are in at this given point in time: Whereas in the front view clumps direct symmetrically down on the left and right side of the structure except for a disturbance in the top right part, in the top view the vectors more clearly indicate this disturbance and how it percolates across the depth of the arch. Speed vector analyses like this

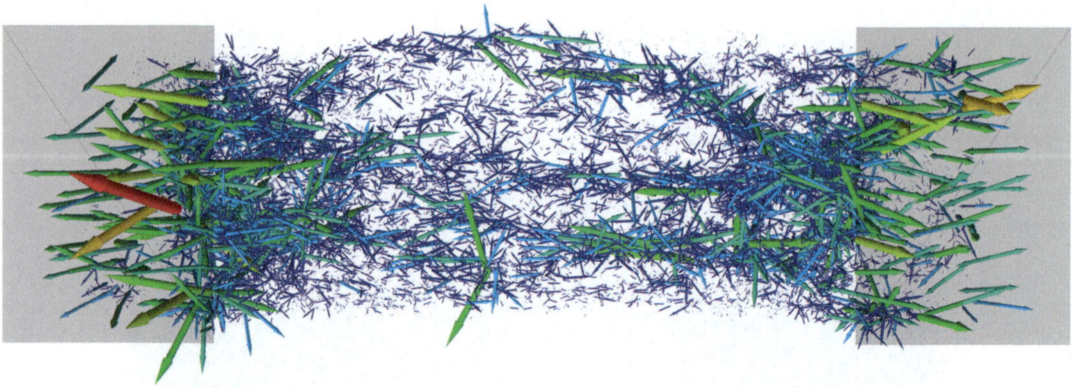

Fig. 4 *Top* view of the contact forces in an aggregate arch show how forces are distributed in the system from *low* at the *top* to *higher* at the supports forming quasi-ribs across the depth of the structure

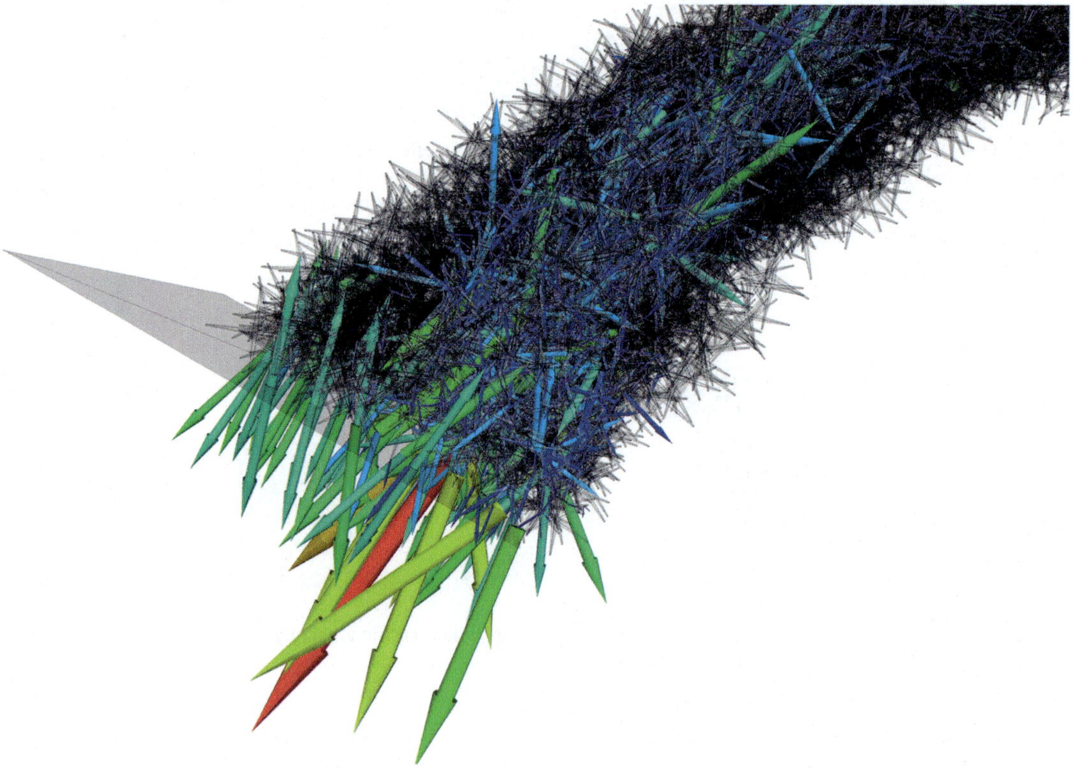

Fig. 5 A detail view of the *left* support shows the increase in overall contact force from *top* to *bottom* as well as the micromechanical loading-behaviour within the aggregate structure

are mainly used to understand if a structure actually has reached a stable state as well as to give one possible indicator about irregularities within the system (Fig. 6).

Case Study 1 has focused on simulating granular behaviour—i.e. contact forces and clump surface speeds—in the formation of an aggregate arch. In the first step the basic

Fig. 6 *Top* and *front* view of surface speeds and directions within an aggregate arch with most clumps having a velocity of around 1E−10 mm/s at 100,100 cycles after the final filtered deletion of particles

clogging-effect known from aggregate processing was verified for designed granulates, based on that the second model focused on extracting unloaded particles within the arch and observing the behaviour of contact forces and surface speeds. The simulated behaviour thus becomes informative to the construction of the overall computational model and eventually crucial for physical construction techniques. The computational model is both a predictive and an analytical tool. Further research will aim to directly stream the simulation to a robotically manufactured aggregate system.

Case Study Two: Online-Controlled Robotic Grading of an Aggregate Wall

The second Case Study is based on the physical modelling of an aggregate behaviour—in this case its light filtering capacity. The aim is to achieve a defined target luminance distribution in a curved aggregate wall through particle subtraction using an online-controlled robotic feedback process.

The mechanisms of the online feedback model are essentially based on the evaluation of photogrammetric luminance data which are obtained from a physical aggregate structure. The entire feedback-loop is managed through a parametric modelling environment in seven steps: A photogrammetric survey of a backlit aggregate structure is taken with a web-camera, the photogrammetric survey is analysed and the luminance values for each pixel are calculated (Hiscocks 2013), these values are projected on a digital representation of the physical structure, pixels are assigned to a grid of points, which are holding information for both the luminance and the location of the pixels on the physical model based on a previously defined interaction model, consequently subtraction points are defined for the robot and the robot code is fed through the robot sensory interface, based on that the robot subtracts aggregates from the structure with a gripper. To initiate the next loop a new photogrammetric survey is taken, streamed to the parametric modelling environment, luminance values are updated (Fig. 7).

The implemented behavioural model works with a finite target luminance distribution. Once the goal is been achieved, the feedback loop is meant to be terminated. Initial manual studies with grading light in an aggregate structure were

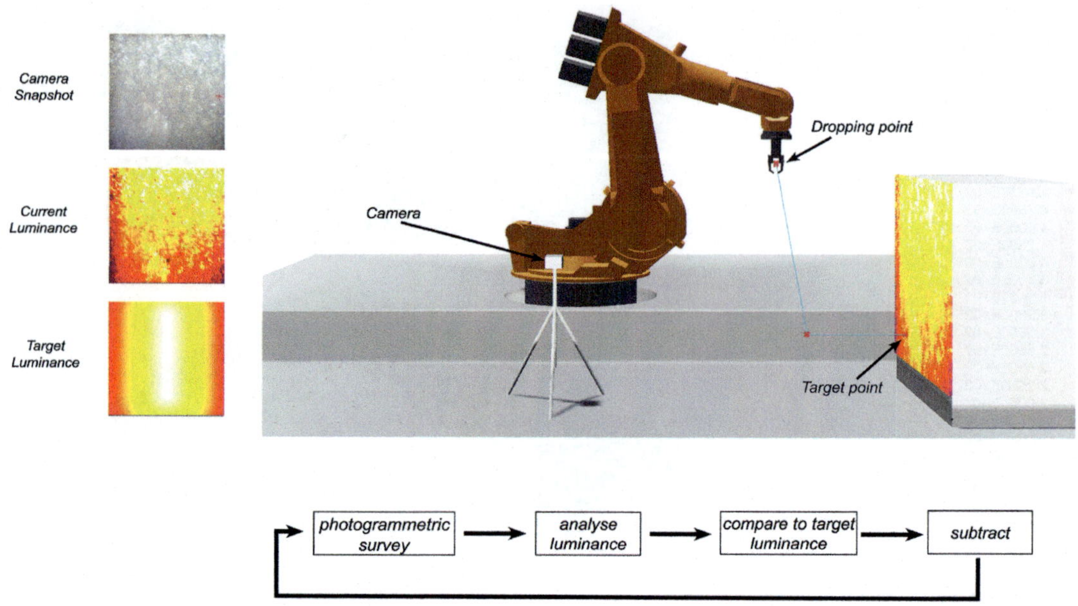

Fig. 7 An online-controlled feedback loop using luminance values has been implemented. The robot reacts to an aggregate structure based on sensory data input from a camera

conducted, testing linear-, centric- and point-distribution patterns of granular layer-thickness and light transmission. In parallel various illumination models were tested and analysed both under even and changing daylight conditions (Fig. 8).

For the prototype model a curved wall with a graded target light-distribution under even lighting conditions was implemented, allowing for the investigation of relatively subtle variations in layer thickness in a three-dimensional structure. The interaction model is consequently a goal model, in which the current luminance is compared to the target luminance model. The point value with the maximum difference in luminance value and its location are calculated as the next point for the robot arm to subtract particles from the structure. The process is repeated until the physical aggregate structure's luminance reaches the target value within a previously defined threshold of 15–20 units difference between target luminance and actual luminance. In the Case Study presented here, the target luminance distribution was approximated in about 100 steps, lasting eight hours. A simulation of the robot's kinematics runs in parallel to the real-time implementation (Fig. 9).

The goal-model was not achieved yet approximated until the structure disaggregated prior to completion of the graded luminance model. Consequently in a next development step, the online-sensing tool implemented in this Case Study technique also needs to be linked to other behavioural criteria based on suitable sensory input into the parametric control system, such as deformation of the aggregates or load-distributions within the system. Equally the digital model in the parametric environment can be updated using 3D scanning as a sensor in order to render results more accurate. Another future development step on both a technical and a conceptual level is the integration of both construction and reconstruction into the online loop: the model presented here has been constructed offline and only the subtraction process has been implemented as an online-controlled sensing process. The extension of the behaviour-based construction process to encompass both

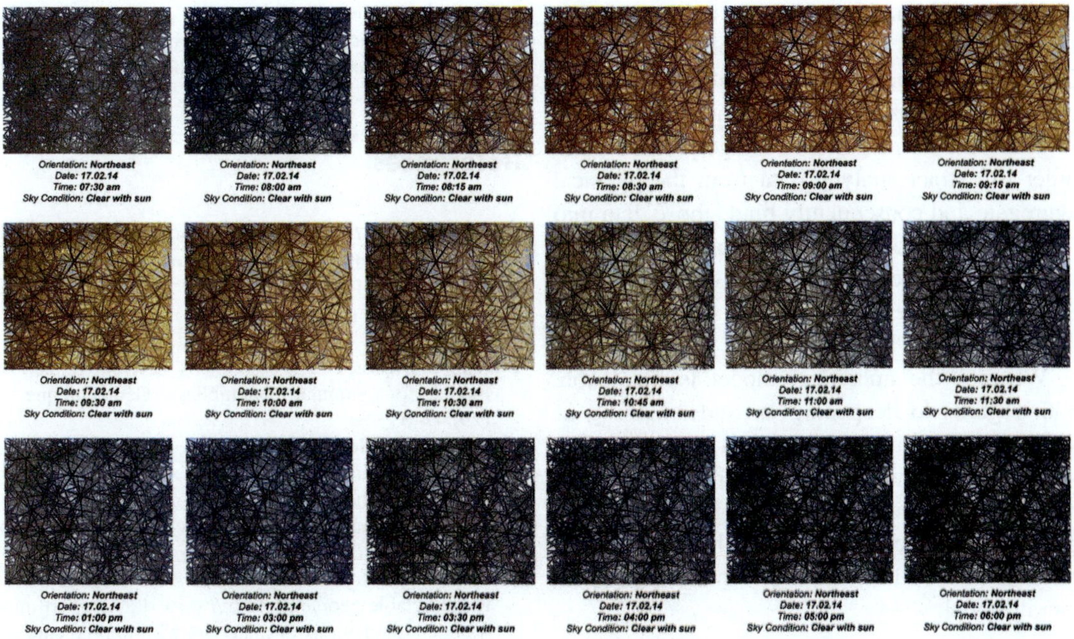

Fig. 8 Experiment series have been conducted under changing daylight condition in order to establish a catalogue of possible luminance values

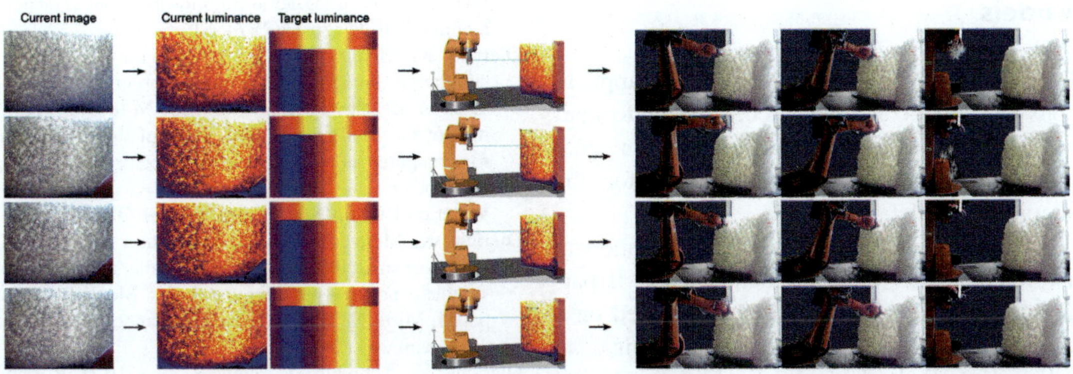

Fig. 9 The current luminance, the target luminance, the robot simulation and the actual robot positions and tool actions have been observed in a continuous loop during the construction process

aggregation and disaggregation of the structure also widens the conceptual scope of the project: The structures are constantly evolving, theoretically being never finite in time and only reaching temporary stable states. Based on that, a third research trajectory will focus on the interaction model itself, moving from goal-oriented systems to emergent ones (De Wolf and Holvoet 2005).

Discussion—Modelling Aggregate Behaviour

Two models have been presented that allow for observing and analysing aggregate behaviour in order to make it an active design agent. The numerical simulation model enables initially the prediction of a specific aggregate behaviour under given boundary conditions. More importantly though, it reveals information on behavioural patterns such as force-distributions that are impossible to observe using a material substrate.

The material model uses image processing in order to extract analytic data from the physical aggregate and consequently binds those data into a design-process using an online-controlled feedback loop between material behaviour, digitally analysed sensor data and a six-axis robotic arm.

Whereas the numerical model is invaluable with regards to deepening an understanding of the underlying micro-mechanics in a granular structure, the advantage of the online-controlled robotic process lies in making a given aggregate behaviour a direct design driver through filtering material data and integrating them into a behavioural model.

Outlook—Synthesizing Behavioural Models

Especially with regards to larger scale applications an online-controlled process needs to integrate multiple parameters as well as numerical and material modelling as an input for the interaction model.

In this respect, synthesizing both numerical and material-analytic behavioural models is paramount. Initially this can only be achieved on a comparative level through simulating an online-controlled robotic process and overlaying the resultant data sets. Ultimately though the aim is to merge these two realms of behavioural computing through binding the numerical simulation into the design model, which is driving the online-controlled feedback process. The behavioural model then spans both numerical and material methods into a synthesized computational model.

Acknowledgements The authors would like to thank Sacha Emam, Matthew Purvance and Reza Taghavi of ITASCA Consulting Group, Inc. for their support within the IEP Program, Tobias Schwinn and Julian Höll of the Institute for Computational Design (ICD) for sharing their expertise in online-controlled robotic processes as well as Matthias Helmreich and Ondřej Kyjánek (ICD) for their support in conducting large-scale experiments.

References

Cundall PA (1971) A computer model for simulating progressive large-scale movements in blocky rock systems. In: Proceedings of the symposium of the international society of rock mechanics, Nancy, France, 1971, 1: paper no II-8

Cundall PA, Strack ODL (1979) A discrete numerical model for granular assemblies. Géotechnique 1 (29):47–65

De Wolf T, Holvoet T (2005) Emergence versus self-organisation: different concepts but promising when combined. In: Engineering self organising systems: methodologies and applications, lecture notes in computer science, vol 3464, pp 1–15

DFAB (2015) Robotic aggregations of materials with unpredictable geometry. http://www.dfab.ch/portfolio/robotic-aggregations-of-materials-with-unpredictable-geometry/. Accessed 17 Jan 2015

Dierichs K, Menges A (2012) Aggregate structures: material and machine computation of designed granular substances. In: Menges A (ed) Material computation: higher integration in morphogenetic design, architectural design 82(2):74–81

Dierichs K, Menges A (2013) Aggregate architecture: simulation models for synthetic non-convex granulates. In: Beesley P, Khan O, Stacey M (eds) Proceedings of the 33rd annual conference of the association for computer aided design in architecture (ACADIA)—adaptive architecture, Waterloo/Buffalo/Nottingham, pp 301–310, 2013

Dierichs K, Menges A (2015) Simulation of aggregate structures in architecture: distinct-element modeling of synthetic non-convex granulates. In: Block P, Knippers J, Mitra N, Wang W (eds) Advances in architectural geometry, London, pp 1–13

Dierichs K, Schwinn T, Menges A (2012) Robotic pouring of aggregate structures: responsive motion planning strategies for online robot control of granular pouring processes with synthetic macro-scale particles. In: Brell Çokcan S, Braumann J (eds) Proceedings of the robots in architecture conference 2012, Vienna, pp 196–205, 2012

Doerfler K, Rist F, Rust R (2012) Interlacing: an experimental approach to integrating digital and physical design methods. In: Brell Çokcan S, Braumann J (eds) Proceedings of the robots in architecture conference 2012, Vienna, pp 82–91, 2012

Duran J (2000) Sands, powders and grains: an introduction to the physics of granular materials. Springer, New York

Gass S, Otto F (1990) Experimente/Experiments: form-Kraft-Masse 5/form-force-mass 5. Mitteilungen des Instituts für leichte Flächentragwerke (IL) Universität Stuttgart Nr. 25/Information of the Institute for Lightweight Structures (IL) University of Stuttgart Nr. 25. Karl Krämer Verlag, Stuttgart

Gravish N, Franklin SV, Hu DL, Goldman DI (2012) Entangled granular media. Phys Rev Lett 108, 208001—Published 17 May 2012

Hensel M, Menges A (2006a) Eiichi Matsuda—Aggregates 01 2003–2004. In: Hensel M, Menges A (eds) Morpho-ecologies. AA Publications, London, pp 262–271

Hensel M, Menges A (2006b) Anne Hawkins and Catie Newell—Aggregates 02 2004. In: Hensel M, Menges A (eds) Morpho-ecologies. AA Publications, London, pp 274–283

Hensel M, Menges A (2008) Aggregates. In: Hensel M, Menges A (eds) Versatility and vicissitude—performance in morpho-ecological design, architectural design, vol 78(2). pp 80–87

Hiscocks PD (2013) Measuring luminance with a digital camera: case history, Syscomp Electronic Design Limited. http://www.ee.ryerson.ca/~phiscock/astronomy/light-pollution/luminance-case-history.pdf. Accessed 14 Apr 2015

Jaeger H (2015) Toward jamming by design. Soft Matter 11(1):12–27. doi:10.1039/c4sm01923g—Published 7 Jan 2015

Jaeger H, Nagel S, Behringer R (1996) Granular solids, liquids and gases. Rev Mod Phys 68:1259—Published 1 Oct 1996

Johns RL (2014) Augmented materiality: modelling with material indeterminacy. In: Gramazio F, Kohler M, Langenberg S (eds) Fabricate: negotiating designing and making. gta Verlag, Zuerich, pp 216–223

Parteli EJR (2013) DEM simulation of particles of complex shapes using the multisphere method: application for additive manufacturing. AIP Conf Proc 1542:185–188

Poeschel T, Schwager T (2005) Computational granular dynamics: models and algorithms. Springer, Berlin

A Multiscale Adaptive Mesh Refinement Approach to Architectured Steel Specification in the Design of a Frameless Stressed Skin Structure

Paul Nicholas, David Stasiuk, Esben Clausen Nørgaard, Christopher Hutchinson and Mette Ramsgaard Thomsen

Abstract

This paper describes the development of a modelling approach for the design and fabrication of an incrementally formed, stressed skin metal structure. The term incremental forming refers to a progression of localised plastic deformation to impart 3D form onto a 2D metal sheet, directly from 3D design data. A brief introduction presents this fabrication concept, as well as the context of structures whose skin plays a significant structural role. Existing research into ISF privileges either the control of forming parameters to minimise geometric deviation, or the more accurate measurement of the impact of the forming process at the scale of the grain. But to enhance structural performance for architectural applications requires that both aspects are considered synthetically. We demonstrate a mesh-based approach that incorporates critical parameters at the scales of structure, element and material. Adaptive mesh refinement is used to support localised variance in resolution and information flow across these scales. The adaptation of mesh resolution is linked to structural analysis, panelisation, local geometric formation, connectivity, and the calculation of forming strains and material thinning.

P. Nicholas (✉) · D. Stasiuk · E.C. Nørgaard
M.R. Thomsen
Centre for IT and Architecture, Royal Danish Academy of Fine Art, School of Architecture, Copenhagen, Denmark
e-mail: paul.nicholas@kadk.dk

C. Hutchinson
Department of Materials Science and Engineering, Monash University, Melbourne, Australia

M.R. Thomsen
Centre for Information Technology and Architecture (CITA), The Royal Danish Academy of Fine Art, School of Architecture, Design and Conservation, Copenhagen, Denmark

Introduction

The research structure *StressedSkins* investigates the highly integrated material and formal specification of an inexpensive, long-standing architectural material best known for its homogeneity-steel. In this paper, we describe the process of asymmetric incremental sheet forming (ISF). We link ISF process parameters to variable

M.R. Thomsen et al. (eds.), *Modelling Behaviour*, DOI 10.1007/978-3-319-24208-8_2

specification at three architectural scales: within the material, cold working increases the strength; within the panel, forming out of plane increases the local stiffness of the sheet; within the structure, overall rigidity is obtained and increased in relation to panel locations, and where geometric formations on one side of the stressed skin connects with geometric formations on the other. We introduce a computational modelling approach for operationalising these relations, based on the use of an adaptive, unstructured mesh that instrumentalises interscalar feedback during the simulation process.

ISF Process

Incremental sheet forming (ISF) is an innovative fabrication method for imparting 3D form on a 2D metal sheet, directly informed by a 3D CAD model. In the ISF process, a simple tool moves over the surface of a thin (0.5–1 mm) metal sheet so as to cause localised plastic deformation (Fig. 1) (Jeswiet et al. 2005). ISF is of interest for three major reasons: it avoids the need for a costly, die (negative forming), by instead directly machining semi-finished pieces of metal. Secondly, because forming is highly localized, the force required does not increase with scale, meaning that there is theoretically no limit to the size of the sheet that is formed (Tisza 2012). Lastly, ISF has been shown to extend the formability of metals beyond what is achievable via conventional forming via stamping or deep drawing (Bagudanch et al. 2013). New research in this field examines scaling up the process: Ford and Boeing aim towards a system with an effective work area of 2 × 1.5 m (Energy 2015).

Transferred into architecture, ISF moves from a prototyping technology to a production technology. Within the context of mass customisation, it provides an alternate technology through which to incorporate, exploit and vary material capacities within the elements that make up a building system. Potential architectural applications have been identified in folded plate thin metal sheet structures (Trautz and Herkrath 2009) and customised load-adapted architectural designs (Kalo and Newsum 2014; Brüninghaus et al. 2013). Recent research has established ISF as structurally feasible at this scale (Bailly et al. 2015), and explored the utilisation of forming cone geometries as means to reach from one skin to another (Kalo and Newsum 2014; Bailly et al. 2015).

Stressed Skin Structures

In this research, the ISF process is used to fabricate a stressed skin structure. Stressed skin structures are thin sheet structures in which the skin is structurally active, bearing a considerable part of the load and providing significant rigidity. They are an intermediate between monocoque

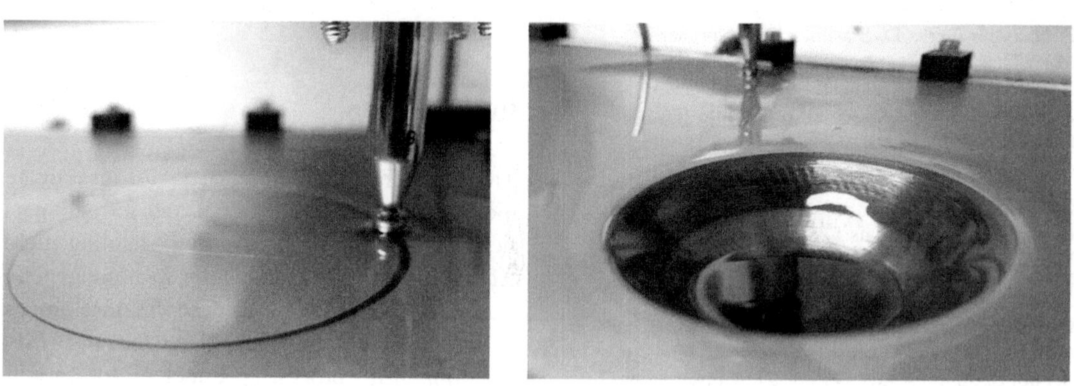

Fig. 1 Single point incremental forming

and rigid frame approaches, and have been particularly associated with light weight structures. In their design, rigidity is a central concern. One of the main problems is to ensure rigidity at multiple scales: against the instability of the whole structure and also the local buckling of the parts which have to carry compressive load. *StressedSkins* develops a structural approach in which the skin carries planar and shear forces, without an additional framing system, at the scale of a pavilion. Local corrugation avoids buckling through geometric stiffening of the skin, while shear connectors transfer loads between upper and lower skins to rigidize the entire structure (Fig. 2). These features, as well as all in-plane connections, and shear connections between the upper and lower skin, are achieved through the deformation of the skin and are outcomes of the computational modelling process.

Research Objective: A Mesh-Based Approach to Communication Across Scales

The object of this study is to explore how a mesh can work as a substrate for enacting and communicating various types of analysis across multiple scales to support the geometric specification, structural simulation and fabrication of a stressed skin structure. There are two established mesh-based methods for adapting resolution where required to capture complex dynamics, small scale geometry and scale sensitive calculations: the nesting of structured grids (multiple contiguous domains) and the adaptation of a non structured grid (a single continuous domain). The research begins with the aim to deploy a single, continuous-domain multi-scale mesh as an exclusive design medium for negotiating the form-finding and

Fig. 2 The *StressedSkins* installation at the Danish Design Museum

analysis, and producing all relevant outputs for fabrication and representation. This is ultimately achieved, however, via a hybrid approach that implements both contiguous and continuous approaches.

Modelling Framework

The modelling framework for *StressedSkins* considers macro, meso and micro scales as markers along a continuum describing variable, interdependent functionalities within the design system (Fig. 3).

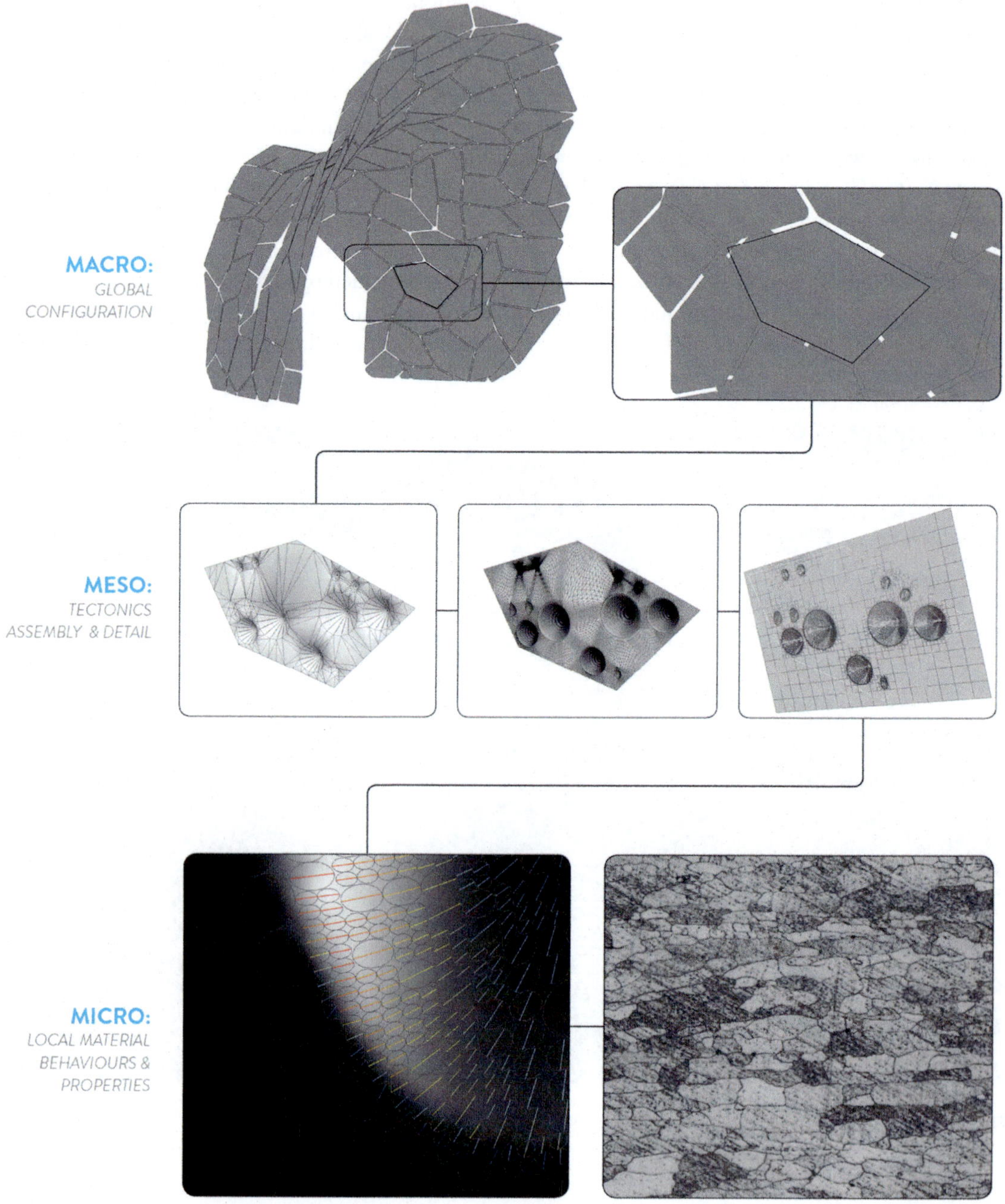

Fig. 3 Multiscale considerations in *StressedSkins*

In general, the macro scale encompasses the resolution of global design goals, overall geometric configurations, and a full-scale understanding of structural performance. The meso scale considers the project at an assembly and sub-assembly level, and is concerned with material behaviours tied to geometric transformation, detailing and component-level tectonic expression. The micro scale is concerned with relevant material characteristics at the most discretised level. The multi-scale modelling approach used here is then comprised of those techniques, which enable the information generated at each of these markers to flow both up and down the continuum.

Strategy and Computational Tooling

These modelling parameters are organised using a half-edge (or directed-edge) mesh data structure (Campagna et al. 1998). Half-edge meshes enable the deployment of n-gon faces (rather than more standard triangulated or quadrilateral faces). This opens up the possibility for designing with more complex topologies. In this case, a pentagonal tiling algorithm was used, resulting in a base mesh comprised of five-sided faces. The features of this mesh—such as its vertices, half-edges and faces—are coupled with a series of lists, dictionaries and Grasshopper data trees that effectively bundle within mesh elements critical design data related to: topology, form-finding and geometry; structural behaviour; material characteristics; connection detailing; and patterning and tectonic expression (Fig. 4).

The primary digital design instruments used are Rhino and Grasshopper as a base modelling environment, the Plankton library for scripting half-edge meshes, a beta library of the Kangaroo2 physics engine for form finding, and Karamba for FE analysis. A series of bespoke tools and implementations are created for managing and modifying the design meshes. These operate such that the meshes are transformed to operate at scales appropriate to particular function—or, when necessary, handshake with other meshes—while retaining the integrity of these key relationships and informing the design with new relevant data.

Macro Scale

Preconfiguration

The configuration of the overall form is developed through a multi-stage modelling process. The first stage entails the top-down drawing of a single target design surface in response to constraints introduced by the site and in pursuit of specific design ambitions. This small site—the foyer of the Danish Design Museum—requires the maintenance of adequate open circulation along its two primary axes and through two additional doors. The design ambitions lie in producing an asymmetrical, cantilevered form that will help test the forming limits of the steel, its capacity for geometric adaptation, and its structural performance (Fig. 5).

With this surface form established, a shell FE analysis is performed, loaded under self-weight. Utilisation of shell elements resulting from this analysis are used to drive locally varying offsets from this single surface into two discrete target surfaces, one for the upper and one for the lower skin. Higher utilisations result in greater target depth between the two skins.

Generative Model

The two skins are then grown on each of these offset target surfaces by sequentially locating pentagonal panels, spiralling outward from a seed tile on each surface. The Kangaroo2 physics engine is a constraint (or goal) based form-finding and simulation design system. In addition to pre-configured goals embedded in the library, the scripting interface enables the coding of custom goals. The recursive form-finding employed here includes edge length and angle goals, which seek to maintain the ideal geometry prescribed by the planar tiling rule set. Target mesh pulling goals draw the mesh vertices out of plane and to the target surfaces. As each new panel is located in the assembly, the solver reconciles these geometrically competing interests

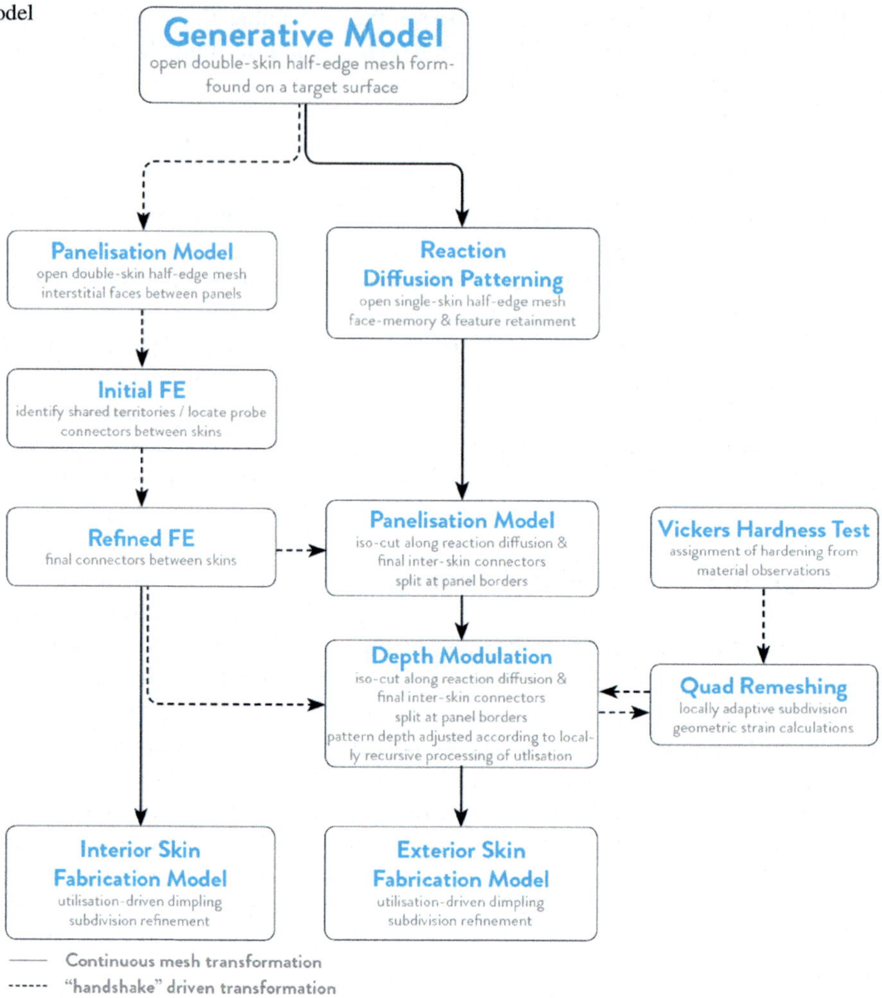

Fig. 4 Networked model dependencies

into a configuration that retains the topology of the tiling strategy but minimally adapts its emerging form to approximate the target surface. Additional goals include: vertex repelling goals between the upper and lower skins that increase diversity in connectivity between them; goals that neither skin collide with the other; and finally, after the target surfaces are fully tiled, a planarising goal is introduced to the panels. This last goal is essential to make fabrication viable, and further deforms each panel from its ideal planar geometry. Throughout this generative process, all relevant data between the form-finding solver and the mesh topology are coordinated as a unified data structure.

Panelisation Model

This pentagonal mesh then directly folds into a new, hybridised mesh that incorporates panel offsets for the geometric definition of boundaries, and specifies the connection detail between panels on the same skin. This consists of a male element on one panel and a female element on its adjacent panel. This male/female relationship is determined for each edge according to the generative sequence of each panel, such that older panels reach out with the male connection. This base model provides the specifications necessary for laser cutting the profiles—or dies—for each panel. A simple etched labelling strategy allows topological information to directly activate a

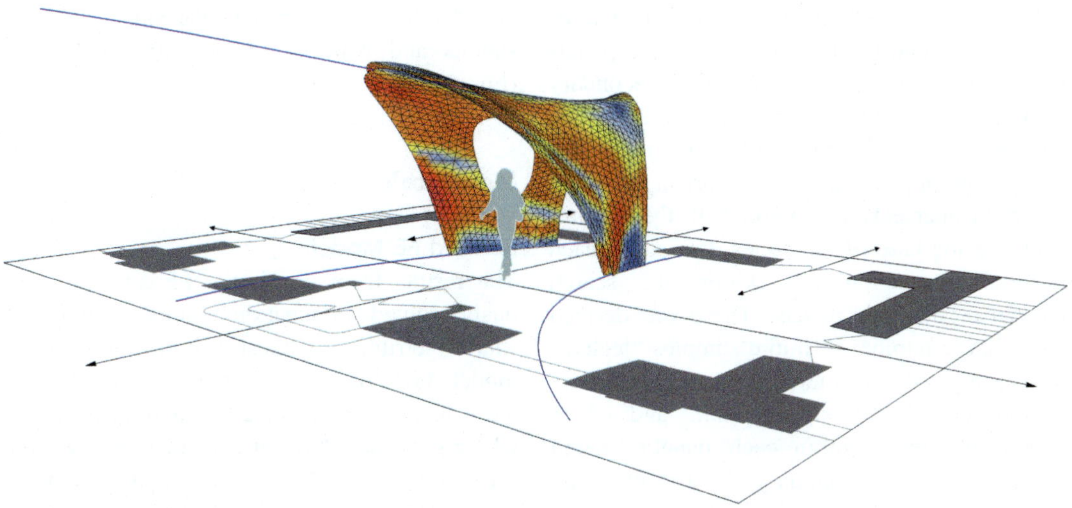

Fig. 5 Target design meshes for upper and lower skin, deployed in response to circulation constraints and with the ambition to produce an asymmetrical, cantilevered form. Variable thickness between the skins is determined by a simple FE analysis on the base mesh

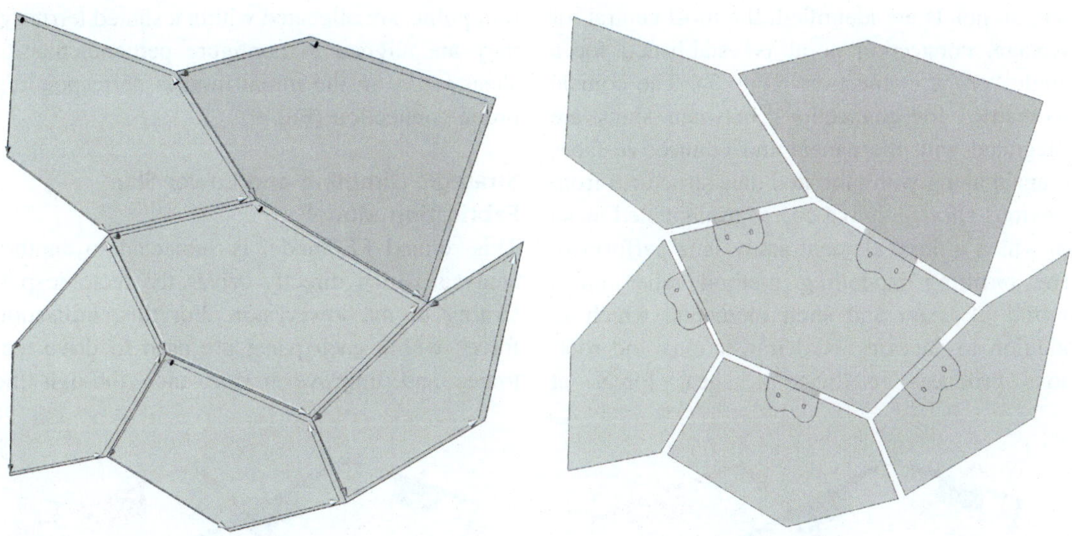

Fig. 6 Panelisation based on pentagonal tiling realised through a halfedge mesh. Planar geometry defined from generative form finding enables precise *offsets*, and order of formation and embedded topology simply organises male/female connection details

self-jigging assembly approach during installation (Fig. 6).

Simultaneously, connective faces within the design mesh are produced to bridge the gap between the newly offset panels. This supported topological consistency in the mesh and is subsequently key to the production of a finite element analysis mesh. Fabrication requires a minimum offset from the panel edges, as well as from the connection details. Based on the knowledge of these geometrical constraints, formable regions called "search boundaries" are then identified within each offset panel.

From these search boundaries, "shared territories" are identified between proximate panels on the lower and upper skins. This secondary form-finding procedure is comprised of a custom, iterative process whereby lines drawn normal to planes from opposite panels self-orient such that they each intersects at a point, with the requirement that the base of the cones defined by each potential connection fall within the search boundaries on both panels. These are derived from a large number of initial samples "feelers" from each panel which identify proximate feelers from panels on the opposite skin, and which dynamically relax within each panel's shared territory and search boundary. As a result many such many potential connection points are identified within each shared territory.

Initial FE Model

Once these territories and their constituent connection points are identified, the most central, or average, connection point is established for a single "probe connection" (Fig. 7). The conical geometries for connections between skins are integrated with the panels and connective faces —again along with inherited data structures from previous efforts—into a coarse triangulated mesh on which a finite element analysis is performed. The resulting modelling method relies on a hybrid of beam and shell elements, which in addition to data on nodal translations and rotations produces readings of shear forces at connection points between the skins, and utilisations and bending forces within the panels (Fig. 8).

Meso Scale

Refined FE Model

The results from this initial FE and the previously solved connection points within each shared territory synthesise into a refined FE model. For this it is asserted that each territory have at least one connection, and each panel—where possible—have at least three connections. Based on both of these interests, panels range in connection count from 0 to 5. The goal here is to maximise the diversity of panels connecting with each other across the skins in an effort to prevent possible hinging and maximise tri-angulation across skins each panel. When multiple connection points are allocated within a shared territory, they are aligned to configure perpendicular to shear forces in the initial models corresponding probe connection (Fig. 9).

Strategic Dimpling and Lower Skin Fabrication Model

This refined FE model is subjected to another analysis which directly drives the tectonic patterning of the lower skin. For this, utilisation forces within each panel are used to drive tectonics and improve performance through the

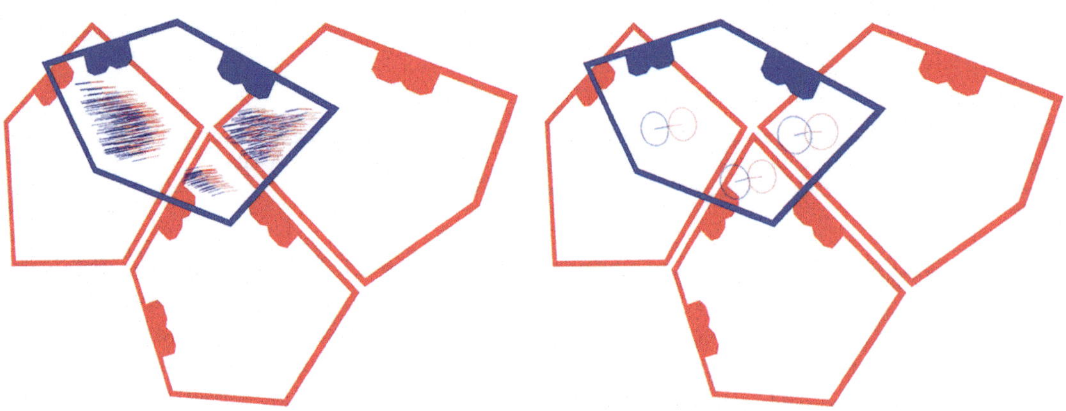

Fig. 7 Identification of "shared territories" between panels on the upper and lower skins, and the establishment of a single "probe connection" within each

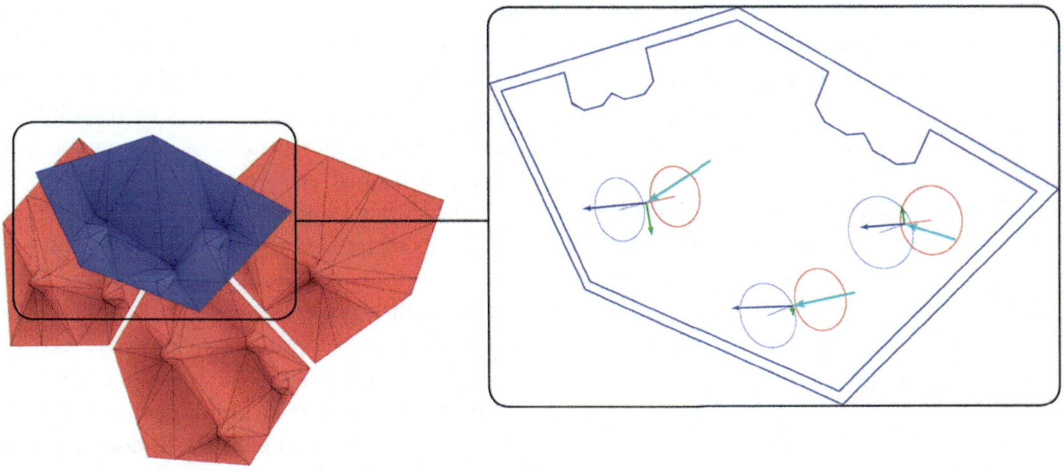

Fig. 8 Geometric definition of each probe connection, and extraction of shear forces at each connection point through an initial FE analysis

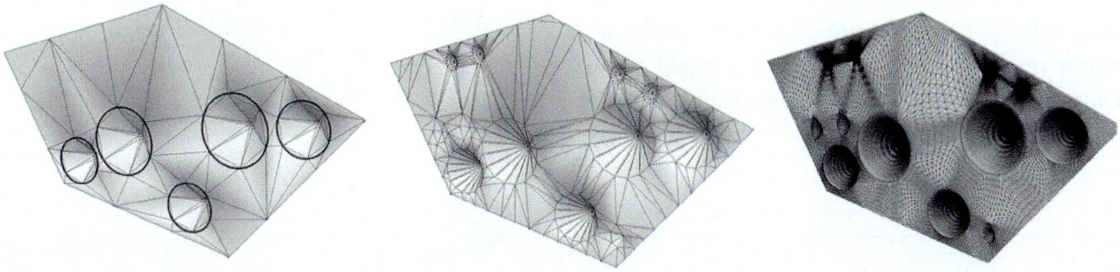

Fig. 9 (*Left* to *right*) **a** Redistribution of connections between skins based on improving resistance to shears identified in the initial FE analysis. **b** Strategic dimpling of panel as a tectonic response to in-panel utilisations. **c** Direct subdivision refinement for production of fabrication model

forming of a responsive pattern of oriented dimples within the structure. This pattern is located in areas of high utilisation and bending energy, the former of which is locally reduced through the strain hardening of the material and the latter better managed through the geometric stiffening resulting from the dimples. Shear forces within nearby connections orient the dimples, and enhance the structural expression of the pattern.

A final lower skin fabrication model is then synthesised from the refined FE model and the resulting dimple pattern, and subdivided to a higher resolution for fabrication, each panel systematically arrayed for extracting toolpaths.

Macro Scale

Reaction Diffusion

The foundation for the patterning strategy on the upper skin is the implementation of a Gray-Scott reaction-diffusion algorithm (comprised of the virtual ingredients U and V). This is executed on a higher-resolution, topologically persistent tri-angulated subdivision of the original generative pentagonal tile mesh, and informed with fixed vertex locations for connection elements that later enable the instantiation of precise connection geometries. The goal here is to produce a pattern on the upper surface whose isotropic nature assists in stiffening panels without favouring any particular directionality.

The subdivision technique used here allows for newly introduced features to inherit key data elements during both decimation and subdivision. It is an extension of a re-meshing script developed by Daniel Piker, which itself allows for the specification of fixed geometric features during these operations as well (Fig. 10). The mesh was then further discretised according to a surface-level iso cut, which split faces along edges according to the reaction-diffusion U and V ingredient parameters at each vertex. The areas inside the iso surface were considered formable (Fig. 11).

Meso to Micro Scale

Variable Resolution Quad Mesh

The meshing technique deployed here is an adaptive, variable resolution square mesh. This quad mesh is mapped onto an existing plane at a coarse resolution, starting from the panel plane. The vertices for each quad face are projected to the target geometry, and the face is tested for planarity. If the face is within a set tolerance, it remains at its current resolution. If it is not, it subdivides into four faces. This test is recursively applied to the mesh such that it locally adapts its resolution to relevant geometric features. After several iterations, each quad can be understood both in its initial unformed square state on the starting plane, and in its strained quad state resulting from geometric forming. A circle inscribed in the initial square is then projected onto the deformed quad, resulting in an ellipse whose primary and secondary axes produce both direction and lengths of strains resulting from the forming process (Fig. 12) (Emmens and Boogaard 2007). Strains are calculated as true strains (Eq. 1)

$$\varepsilon \text{ True } = \ln(L0 \times L1) \tag{1}$$

The resulting thickness strain (ε3) is determined by volume constancy (Eq. 2)

$$\varepsilon 1 + \varepsilon 2 + \varepsilon 3 = 0 \tag{2}$$

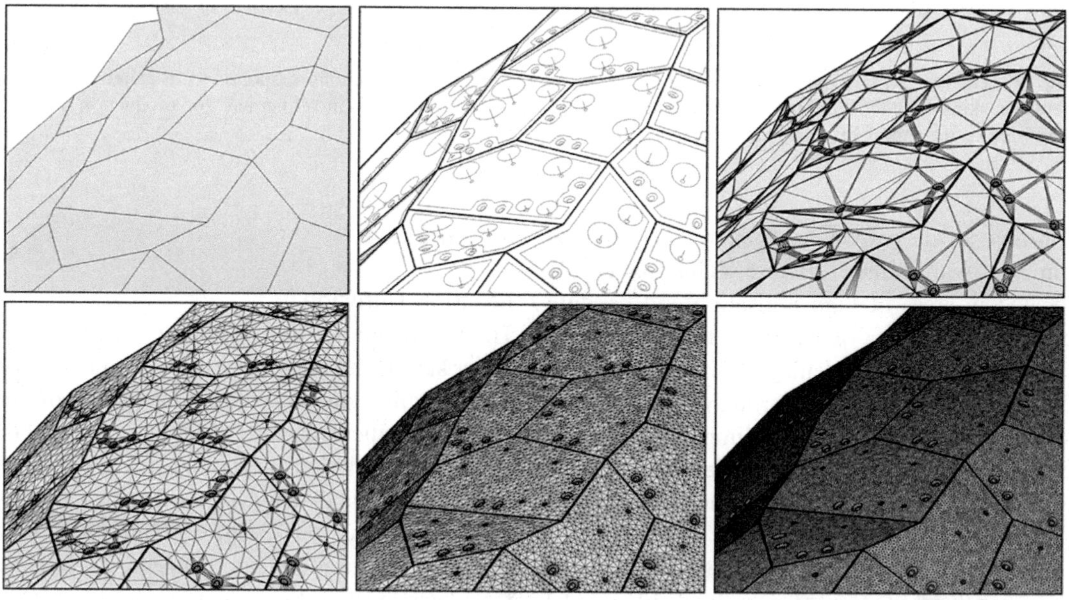

Fig. 10 (*Left* to *right*, *top* to *bottom*) **a** Unrefined upper skin pentagonal mesh. **b** Location of features for connections between panels within upper skin, as well as between upper and lower skin. **c** Introduction of features into pentagonal mesh as triangulated elements. **d** Through. **f** Subdivision of mesh to resolution required for executing reaction diffusion patterning scheme at desired resolution, while retaining key features both geometrically and topologically

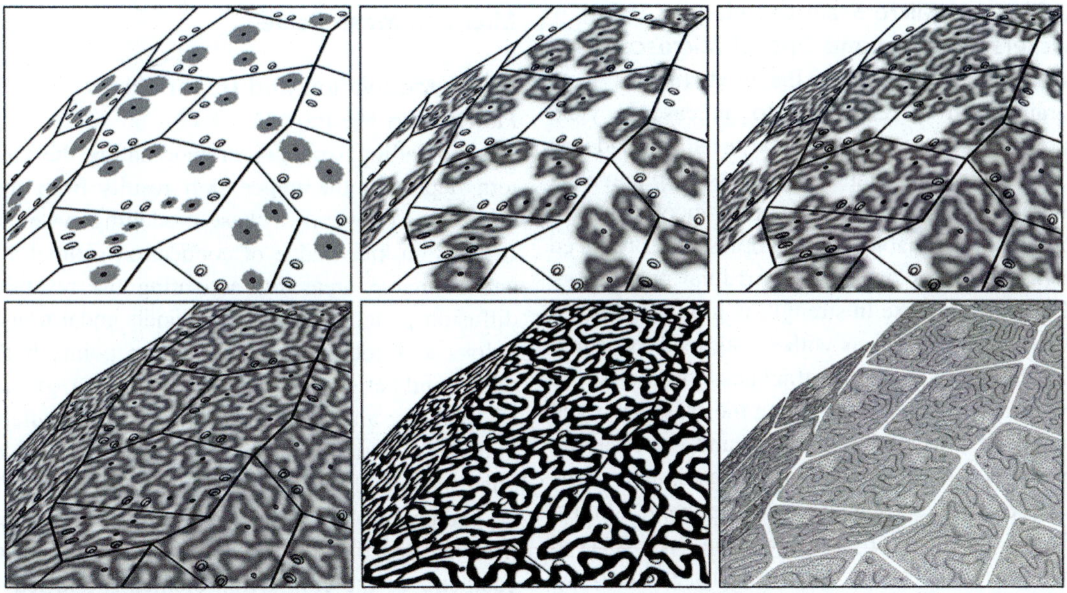

Fig. 11 (*Left* to *right*, *top* to *bottom*) **a** Initialisation of reaction diffusion algorithm on sub-divided mesh. **b** Through. **d** Pattern generation on the mesh, with values represented as colour gradient **e** Iso-splitting of pattern for sharp discretisation of formable areas. **f** Integration of connections between skins with pattern array

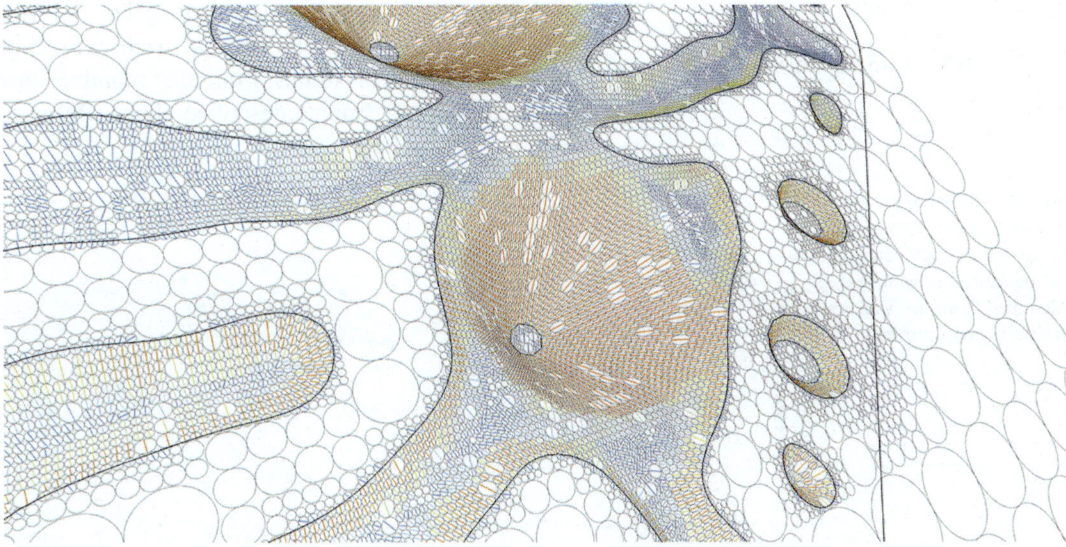

Fig. 12 Quad recursively remeshed to adapt through subdivision in order to more precisely describe geometric features. *Circles* inscribed in each quad are projected from the plane onto the mesh. Here deformation is read in the long axis of *ellipses* that have been deformed through projection. Flat areas on the mesh reflect unformed areas

Calibrating Micro Scale Calculations using Vickers Hardness and Optical Microscopy

To calibrate and verify the prediction of local strains, and to relate this to increase in yield strength, a series of empirical tests are made on material samples that vary processing parameters in a systematic fashion. To best incorporate all processing variables, all samples are produced using the same rig as used for final production. The local increase in strength is monitored using Vickers hardness tests with a 5 kg load measured on the cross sectional thickness of the sheet. Measurements are made along the cross-sectional length of a formed component and the resulting hardnesses are converted to estimated flow stresses and correlated with the local strains. Figure 13 shows a graph charting the observed relationship between local strains induced by forming and the measured flow stress calculated from the Vickers hardness measurements.

Equation 3 is derived from this curve, and is used to calculate, on individual mesh faces, local yield strength that was fed back into a Karamba finite element model.

$$\sigma Y = 43.407 \ln(\varepsilon) + 431.23 \quad (3)$$

Micro to Meso Scale

Depth Modulation and Upper Skin Fabrication Model

The modelling strategy for modulating the pattern depth on the upper skin results from the synthesis of several of these modelling systems. It relies on knowledge of connection geometries and formable areas for activating the reaction diffusion pattern within each panel; nodal translations and rotations at connection points both within and between skins from the refined, global FE analysis; and the capacity to locally calculate granular material properties through the adaptive quad mesh and the measured flow stress.

The process begins by discretising each panel from the reaction diffusion model, applying the geometry of the connection elements, and leaving the rest of the panel "un-formed". This minimum forming baseline geometry is then further subdivided into individual, triangulated elements, each of which is capable of having unique material properties assigned to it in the Karamba finite element modelling environment. An analysis is performed on each face, with extracted local strains from the adaptive quad remeshing technique described above. Resulting

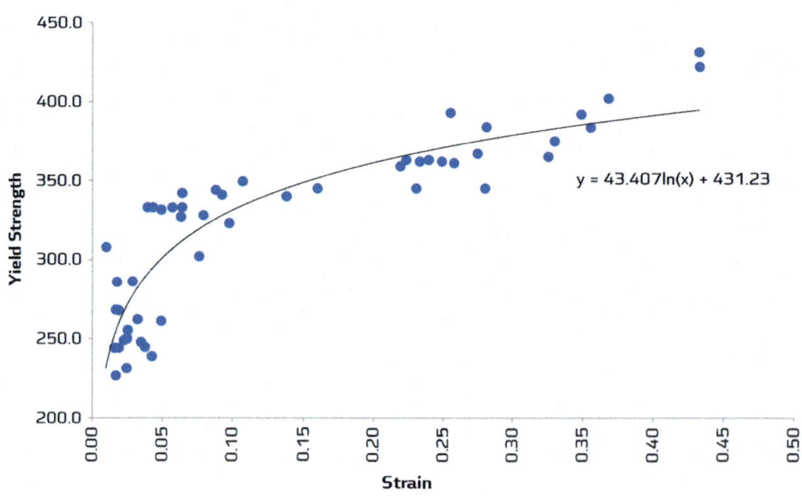

Fig. 13 Observed relationship between local strains induced by forming and measured flow stress calculated from the Vickers hardness measurements

yield strengths for each face are specified by the measured relationship between strain and yield strength.

Using Karamba's prescribed displacements enables the process of subjecting this locally informed mesh to corresponding nodal rotations and translations along its connection points, as extracted from the refined FE model. By virtually working the panel in this way, the depth of the each panel is incrementally and locally modulated. Here, vertices in the prescribed patterned area respond to proximity to faces that are heavily utilised when subjected to the prescribed displacement. Through an iterative accumulation of change in depth (Fig. 14)—and iterative analysis and application of resulting local changes in yield strength due to strain hardening (Figs. 15 and 16)—the patterning emerges as a

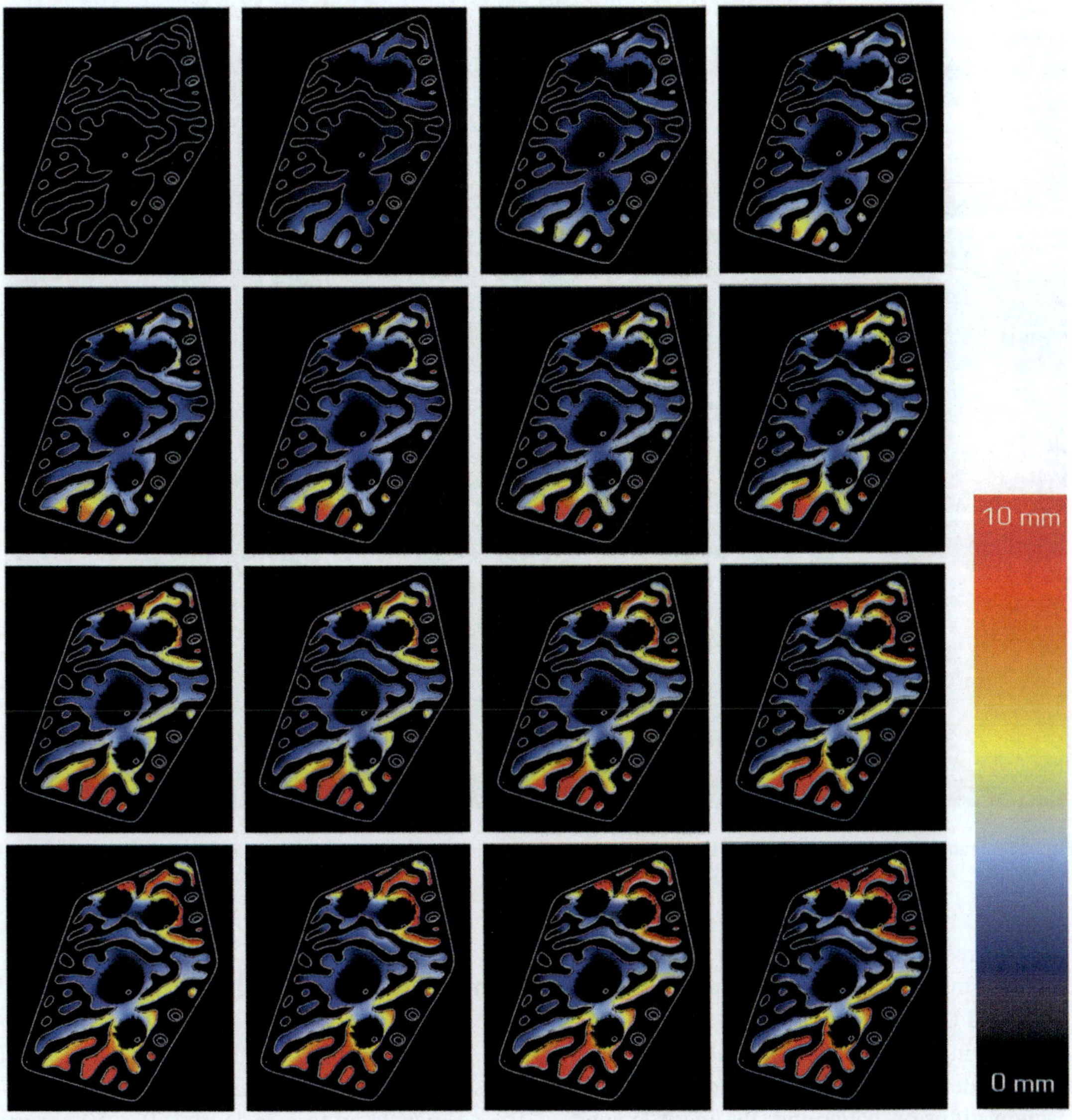

Fig. 14 Depth modulation of patterning introduced iteratively. The incremental increase in depth from a minimum *offset*. Here *black* represents areas that remain planar or are fixed at a depth required for connections, and *coloured areas* reflect formable areas iteratively deepening in response to local utilisations

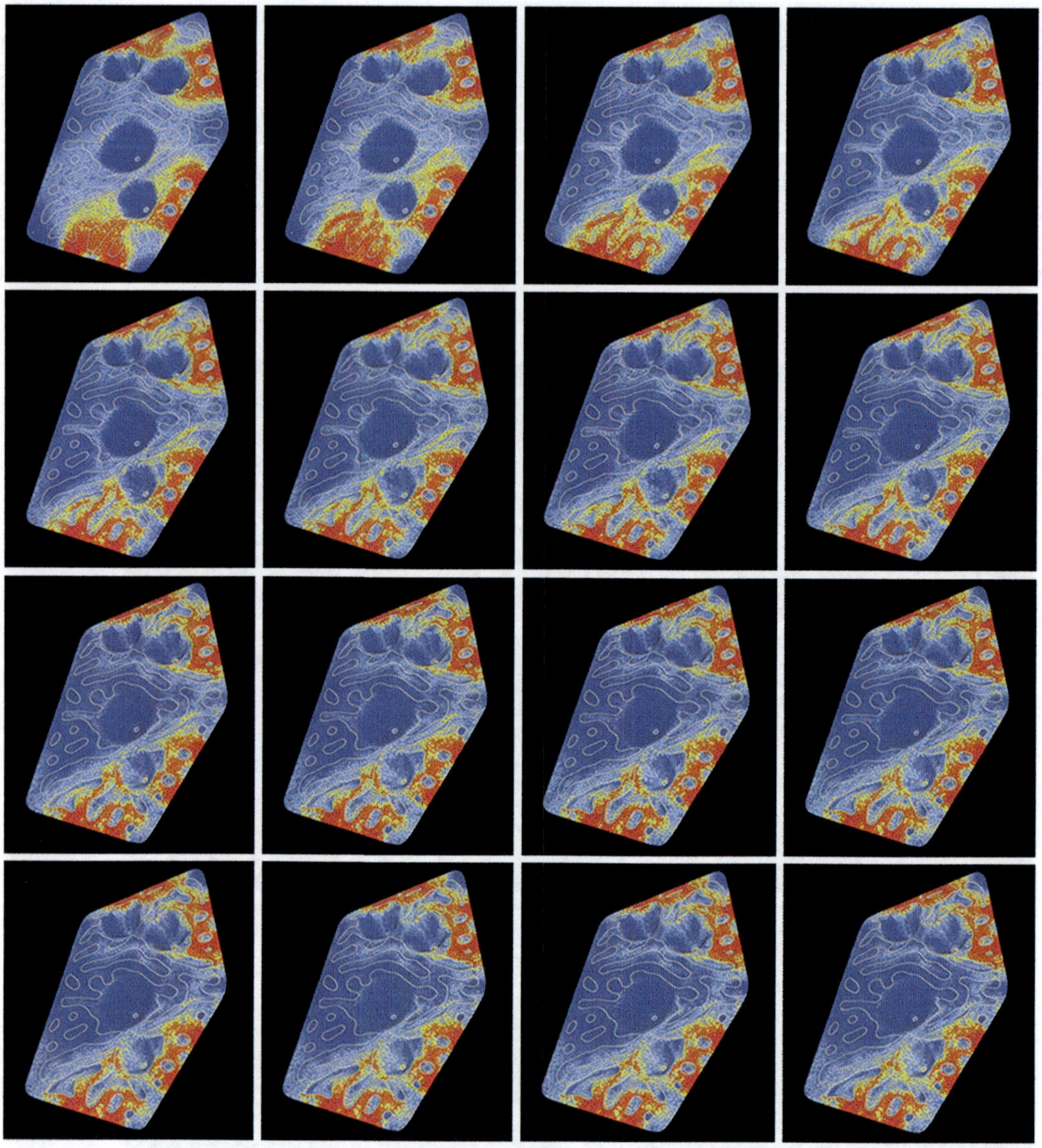

Fig. 15 Transformation in utilisation through local deepening of features and resultant strain hardening, exercised iteratively

tectonic response to utilisation and bending energies introduced by the global structural conditions (Fig. 17).

Following this iterative introduction of locally adaptive depth within channels defined by the reaction diffusion patterning, the mesh is subdivided to a higher resolution for fabrication, arrayed by panel toolpath extraction.

Fabrication and Toolpathing

An ABB IRB140 multi-purpose industrial robot is used to fabricate *StressedSkins*. During prototyping, panels up to a scale of 150 × 50 cm are produced; the working area for the final panels is approximately 50 × 100 cm. Conventional

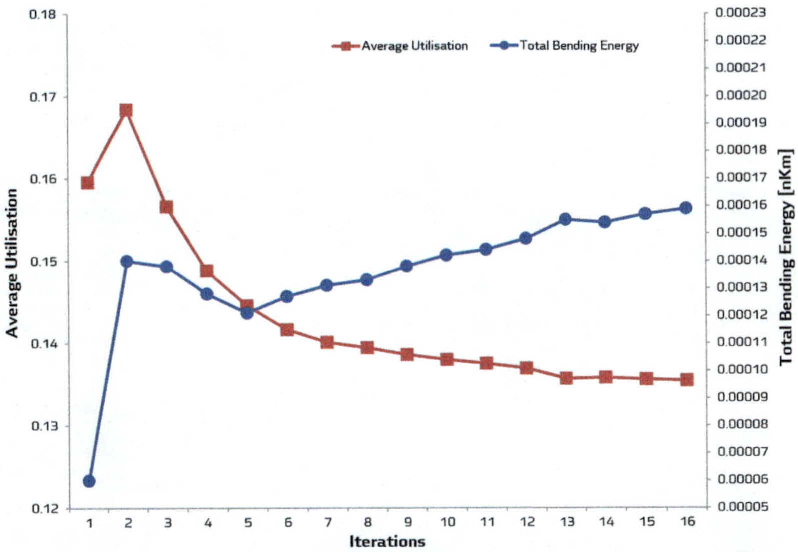

Fig. 16 Tracking of change in average utilisation for each mesh element, along with total bending energy captured within each panel as depth modulation is introduced and iterated through patterning and forming

Fig. 17 Variable depth patterning on an upper skin panel

toolpath generation algorithms generate simple sliced contours in the horizontal plane. However this strategy does not produce optimal results when applied to more complex geometries formed with ISF. To improve control over tooling time and surface quality, we develop a

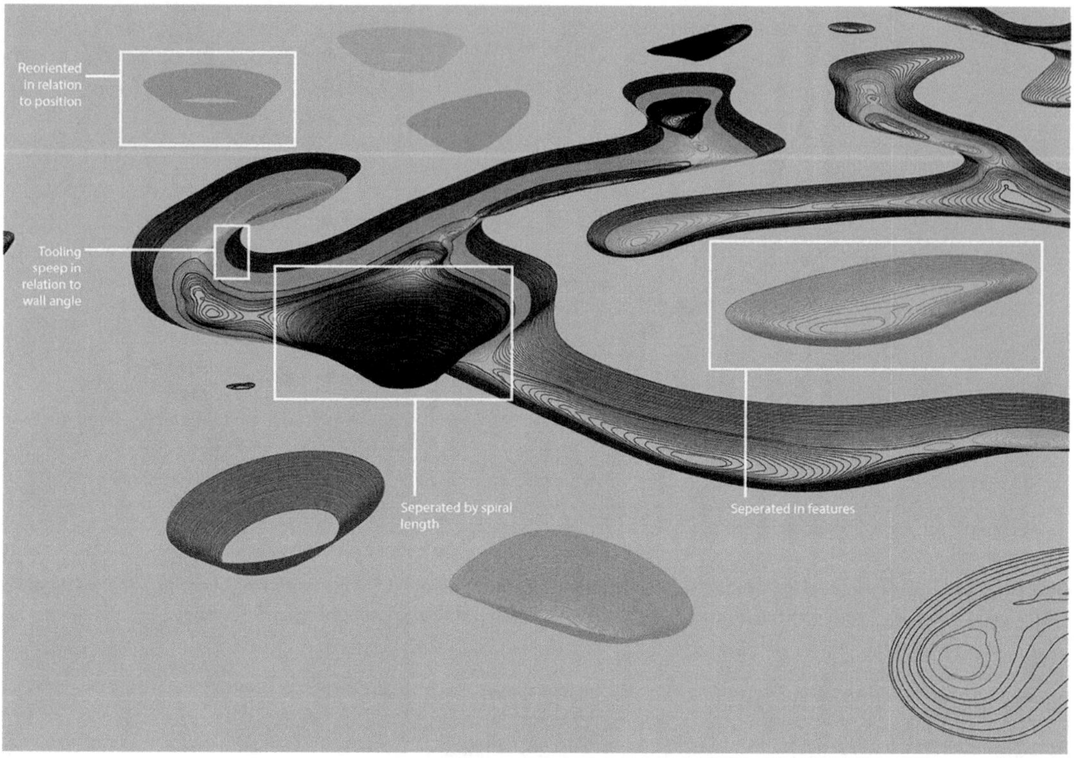

Fig. 18 Diagram of key drivers for toolpath extraction and organisation

toolpathing algorithm based on the established method of a spiral descent (Jeswiet et al. 2005). This algorithm integrates the grouping of features, the position of features, toolpath length and tooling speed in relation to wall angle (Fig. 18). Additionally, the algorithm is informed by knowledge gained directly through prototyping: this includes the observation of forming limits, optimal working areas, and tooling speed.

Discussion

This research demonstrates a mesh-based modelling method that synthesises structural performance interests with locally varying material properties, in the production of an architectural scale installation. The model is based around geometry, properties and mechanics at three scales. With the exception of two "handshakes", the model varies a single mesh topology to manage the complexity of bridging scales and functions while maintaining speed, flexibility and continuity of information flows up and down scales. Further work aims to incorporate the measure of strain into this continuous mesh topology by performing measures on triangular rather than quad faces (Fig. 19).

The model inputs include direct user inputs, geometric parameters, material and processing parameters, and feedback from finite element analysis. The micro scale calculations are calibrated through experimental testing of Vickers hardness, and are shown to be an accurate predictor. While at this stage the calculation of local yield strength is limited to informing finite element analyses, a next step will be to understand how parameters at this scale might also activate optimisation processes (Fig. 20).

Fig. 19 Connectivity between the upper and lower skins

Fig. 20 The *StressedSkins* installation

Acknowledgements This project was undertaken as part of the Sapere Aude Advanced Grant research project "Complex Modelling," supported by The Danish Council for Independent Research (DFF). The authors want to acknowledge the support of several collaborators: Clemens Preisinger and Robert Vierlinger of Bollinger Grohmann consulting engineers assisted in the forming of intuitions regarding structural behaviours and appropriate finite element modelling strategies to represent them; Daniel Piker and Will Pearson provided direct support with both the Kangaroo2 and Plankton libraries, and the development of computational tooling; the research departments DTU Mekanik supplied access to and assistance using their ISF-designated CNC rig, as well as insight into several ISF-related calculation techniques; robotic command and control was enabled through the software HAL; and introductory guidance regarding ISF operations was given from RWTH Aachen.

References

Bagudanch I et al (2013) Forming force in single point incremental forming under different bending conditions. Proc Eng 63:354–360

Bailly D et al (2015) Flexible manufacturing of double-curved sheet metal panels for the realization of self-supporting freeform structures. In: Key engineering materials 639, Trans Tech Publ. pp 41–48

Bouaziz O, Brechet Y, Embury JD (2008) Heterogeneous and architectured materials: a possible strategy for design of structural materials. Adv Eng Mater 10(1–2):24–36

Brüninghaus J, Krewet C, Kuhlenkötter B (2013) Robot assisted asymmetric incremental sheet forming. In: RobArch 2012. Springer, Heidelberg, pp 155–160

Campagna S, Kobbelt L, Seidel HP (1998) Directed edges—a scalable representation for triangle meshes. J Graph Tools 3(4):1–11

Danckert J, Wanheim T (1979) The use of a square grid as an alternative to a circular grid in the determination of strains. J Mech Working Technol 3(1):5–15

Echrif SBM, Hrairi M (2011) Research and progress in incremental sheet forming processes. Mater Manuf Process 26(11):1404–1414

Emmens WC, Van den Boogaard AH (2007) Strain in shear, and material behaviour in incremental forming. In: Key engineering materials 344. Trans Tech Publ. pp 519–526

Jackson K, Allwood J (2009) The mechanics of incremental sheet forming. J Mater Process Technol 209 (3):1158–1174

Jeswiet J et al (2005) Asymmetric single point incremental forming of sheet metal. CIRP Ann—Manuf Technology 54(2):88–114

Kalo A, Newsum MJ (2014) An investigation of robotic incremental sheet metal forming as a method for prototyping parametric architectural skins. In: Robotic fabrication in architecture, art and design 2014. Springer, Heidelberg, pp 33–49

Kobbelt L et al (1998) Interactive multi-resolution modeling on arbitrary meshes. In: Proceedings of the 25th annual conference on computer graphics and interactive techniques. ACM, New York, pp 105–114

Lu B et al (2013) Feature-based tool path generation approach for incremental sheet forming process. J Mater Process Technol 213(7):1221–1233

Rauch M et al (2009) Tool path programming optimization for incremental sheet forming applications. Comput Aided Des 41(12):877–885

Tisza M (2012) General overview of sheet incremental forming. Manuf Eng 55(1):113–120

Trautz M, Herkrath R (2009) The application of folded plate principles on spatial structures with regular, irregular and free-form geometries. In: Symposium of the International association for shell and spatial structures (50th. 2009. Valencia). Proceedings of the evolution and trends in design, analysis and construction of shell and spatial structures. Editorial Universitat Politecnica de Valencia

US Department of Energy (2015) Rapid freeform sheet metal. http://energy.gov/sites/prod/files/2015/03/f20/rapid_freeform_sheet_metal_forming_factsheet.pdf. Accessed 15 Jun 2015

Topology Optimisation for Steel Structural Design with Additive Manufacturing

Shibo Ren and Salomé Galjaard

Abstract

The application of steel based Additive Manufacturing (AM) processes on the design of a steel node for a tensegrity structure has been explored with the use of topology optimisation as the form finding method. AM bears the potential of increasing efficiency by reducing processing steps, material use and labor intensity, and offers a feasible manufacturing solution for structural building elements with complex geometry and function. Based on the redesign of an existing steel node, a number of design iterations have been developed and benchmarked using topology optimisation method, followed by a further rationalized design process for AM production. A comparison has also been performed between the new and the conventional design and resulted in insights into the new opportunities and the design freedom created by AM on production of metallic elements for the building industry.

Introduction

The contemporary digitalisation of the design process in the disciplines of architecture and engineering, coupled with the development of the digital fabrication technique, has resulted in an increased interest in the design integration of material, form and structure. Such a design process is informed by the material property and performance, and benefits from the availability of computational methods by merging the numerical evaluation with the optimization technique. Integrated with the digital production technique such as Additive Manufacturing (AM) it enables the design freedom and the possibility of fabrication for components with highly complex shape and various functional requirements. AM as a fabrication technique has been applied in industries such as aerospace and automotive for years, but the questions on whether AM can be implemented to manufacture parts for the built environment, and how the design process of architects or engineers can be best informed in such a process, are still to be understood. This paper looks into the design method and

S. Ren (✉) · S. Galjaard
Arup, Amsterdam, The Netherlands
e-mail: Shibo.Ren@arup.com

M.R. Thomsen et al. (eds.), *Modelling Behaviour*, DOI 10.1007/978-3-319-24208-8_3

fabrication process by redesign of a conventional steel node for a tensegrity structure using AM technique and the topology optimisation method.

Structural Design and Analysis

From the structural engineering perspective, the process of design is linear in most cases. Programmatic requirements define the initial design input. Geometric constraints are then placed upon the design in order to understand the structure of the form. With the structural logic in place, a material is then chosen which is most suitable for the structure. With each step completed, fabrication thinking then begins. This process, although most of the time includes a number of design iterations of intuitive interaction between architect and engineer, is limited on many projects when the interaction between architectural form and rational structural thinking is required. Empirically-based engineering intuition can easily find its limit when it comes to drive the complex and organic characteristic of a structural form.

In such a linear design process, it is unquestionable that the traditional engineering design and calculation method such as finite element analysis (FEA) is of great value in the evaluation and refinement of well-developed design proposals. From a design perspective, however, the usefulness of FEA at the decisive early phase of the design process has to be questioned (Kotnik and Acunto 2013). The transformation of a huge amount of numeric data into a coherent design-relevant feedback in the iterative design process requires an amount of knowledge and close interaction between the process of analysis and the process of design development. For instance, structural analysis can accurately predict the stress flows within the two given design proposals as indicated in Fig. 1, however, it lacks the ability to easily develop the design from its intimal from to the evolved optimal shape.

Topology Optimisation

In nature there is ample evidence of complete integration between form, material and structure. For instance, reaction wood in angiosperms reorients the trunk in response to external tensile forces. Osseous tissue reacts to applied load with its relatively hard and lightweight composite material. By observing the evolution of natural structures, it becomes obvious that the topology and the shape of such structures achieve their optimum over a long evolutionary period and adapt to whatever environment they find themselves in Jeronimidis 2004). Imitating this growing and adaptive process of biological structure allows for a design and modelling process that combines structure with geometry, material, function and manufacturing information to achieve the best performance through appropriate material distribution.

With the rapid development of computer aided design techniques, topology optimisation as a design approach has been widely implemented to optimise material layout when designing load-bearing structures with various techniques and theories (Michell 1904; Schmit 1960).

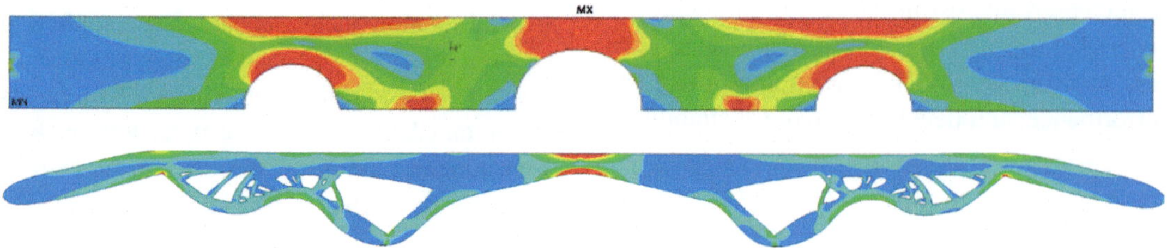

Fig. 1 FEA analysis for an initial shape (*top*) and an optimal shape (*bottom*), both were designed for the same loads and boundary condition

The establishment of a topology optimisation problem for structural design normally requires the definition of three statements: design variables, objective and the constraints. Design variables can be certain geometrical parameters, material properties, or the finite elements mesh that the optimisation algorithm can change to achieve different responses. Objective is a non-numerical indicator of the goal of the optimisation process, for instance, minimize the total structural weight or maximize the overall stiffness. Constraints are normally in the form of a numerical boundary limit applied to certain design response, for instance, the allowable level of stress for a steel part.

The well-developed gradient-based optimisation algorithm has been adopted by many optimisation packages (Schumacher 2005). By assuming a design domain subject to a set of boundary conditions, the optimal form can be predicted with iterative computation subject to the prescribed performance targets when all the design constraints are satisfied (Fig. 2). In this process, analysis is implemented at each generation to evaluate the results and provided design feedback for the new direction for the shape to develop towards. In this iterative way, topology optimisation allows for an integrated design and optimisation process that opens up the possibility of achieving an optimal design solution with organic and complex geometry.

Additive Manuacturing

Mostly used in aviation, aerospace and automotive industry, the term Additive Manufacturing refers to a whole set of different production techniques and process of joining materials to make objects from 3D model data, usually layer upon layer, as opposed to subtractive manufacturing methodologies, such as traditional machining (ASTM 2009). The process always starts with a 3D digital model which is then sliced into a stack of 2-dimensional planar layers (Fig. 3). These layers are built by the machine and stacked one after the other, whereby the contours and fillings of the part are cured from a feedstock.

AM technologies enable the creation of part of any arbitrarily individual geometry which fits well to the building industry's demand for geometrically complex design. The layer-based, additive nature means that virtually any shape can be fabricated without hard tooling. Geometries which are customized to individual solutions can be economically fabricated. For instance, the complicated but efficient shape from the topology optimisation process as described in previous section can be easily fabricated without the need for any conventional moulds, dies or fixtures. However, currently the architectural sector is only 3 % of the total AM business and those are not products used in actual buildings

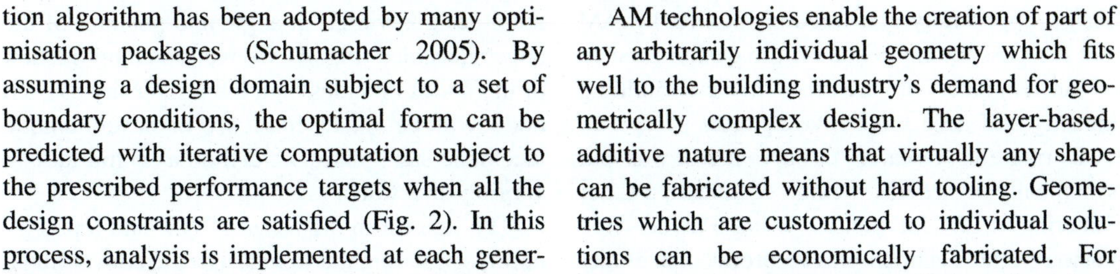

Fig. 2 Topology optimisation process of a beam structure: from the initial generation to the end results

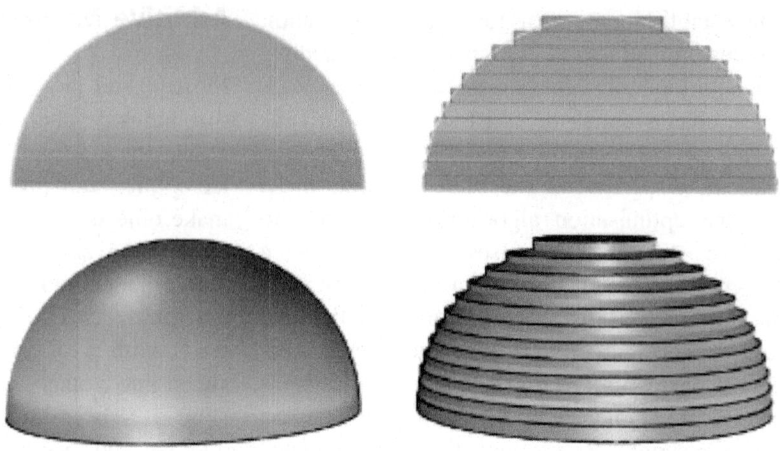

Fig. 3 Indication of the AM slicing process

but mostly scaled models and mock-ups. The design opportunity and freedom offered by AM and the possible influence on the design process of an engineer or architect in the building industry still need more investigation.

During this research project, focus was put specifically on the AM method of Selective Laser Sintering (SLS) for metals, using Powder Bed Fusion Metal Technologies to produce steel element as structural component. In the production process of SLS, powdered metal material is selectively melted layer by layer via lasers or electron beams until fully melted into a solid homogeneous mass. In this way, multiple parts that were conventionally manufactured and assembled as standard steel components can be replaced with a single part of steel that can be geometrically more complex and structurally more efficient, also with a numerous benefits for downstream activities.

Steel Node Design

The design of the steel node for a tensegrity structure has been used as a case study to explore the topology optimisation method and the opportunity provided by AM technology. Due to the irregular shape of the structure, the structural nodes, connecting the cables to the struts within the tensegrity, all have slightly different shapes. The structure as a whole asked for 1200 different node designs varied in angle and position of the attached cables (Fig. 4). Conventional production of the nodes includes high labor intensity as each node will be made from six or seven unique machined steel plates welded on a central tube in varying directions.

The interest in the usage of AM and topology optimisation method to redesign the conventional node relies on the fact that by employing such techniques, it is possible to consistently integrate customized node design, structural efficiency and manufacturing process within the design process. That is, structural topology optimisation allows the designers to define the direct correlation between the form and the internal force flow, and AM offers the possibility to produce the complex geometry physically without all the fabrication constraints as can be found in conventional fabrication method.

The design research on the AM steel node included three design phases:

- Traditional Node: Defined the functionality and structural requirements of the node with the traditional design approach. Designed and fabricate it with conventional fabrication technique.
- AM Node 1.0: Designed the first AM node with topology optimisation approach, within a similar design space of the traditional node, to satisfy the exactly same structural and

Fig. 4 Rendering of one of the original tensegrity structures in The Hague (*left*), *Architect* ELV Architecten © Studio i2, and model of the conventional structural node and light fixture, linking six cables and strut (*right*)

functionality requirements. Fabricated it with SLS technique.

- AM Node 2.0: Redefined and tuned the design space with the freedom allowed by AM, focusing on further integration on function and structural efficiency. Fabricated it with SLS technique.

Traditional Node

The structural function of the node was defined as to connect the structural elements (cables and struts) of the tensegrity structure, with an additional function in holding up the lighting fixture on top. The cables have a fork and pin that need enough space to be connected and therefore requires the positions of the connection openings to be some distance apart. Given the design forcers in the cable, the node could be designed with 6 machined steel plates which can be welded to a central tube manually (Fig. 5), with the total weight of 20 kg for a single node.

AM Node 1.0

Based on the design of the traditional node, the design domain for AM Node 1.0 was defined in such a way that all the original functional requirements were satisfied. The well-documented topology optimisation technique was used, starting with the assumption of a volume of material based on the original node and a set of boundary conditions. The analysis then

Fig. 5 Model of the tradition node, produced in the traditional way by cutting and welding

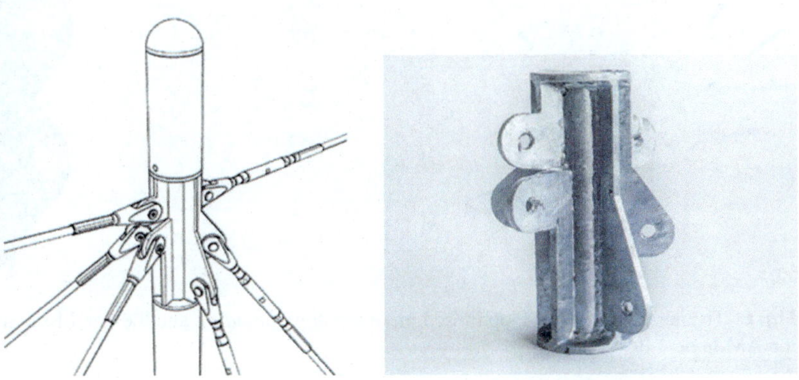

started and the density was reassigned in an iterative way in response to the loads to find the most optimal load path to transfer the internal forces. A maximum allowable von Mises stress was set as the design constraints in the process with the design objective of minimizing the total structural weight. In such a loop process, design inputs, geometry constraints, structural evaluation and the design optimisation are integrated to achieve the desired objective (Fig. 6).

Having obtained the initial optimal shape, a further rationalised design for production was followed that the AM fabrication constraints were taken as geometrical rules to refine the design. Products produced by SLS can have deep cavities or hidden channels, but sometimes need support structures due to the weight of the solidified material and heat dissipation. In order to simplify the fabrication process and eliminate the extra support material needed during the production process, the self-supporting features of the geometry were interactively evaluated and the shape was modified via parametric associative modeling to shape the geometry and to introduce parts which could support the overhangs. In that case, the geometry itself became the support structure (Fig. 7).

The optimised AM node driven by the computational process features an organic form with less material (14 kg, 70 % of the traditional node). The form itself is directly driven by the internal forces flow and the product functionality. In such a process, through optimal material distribution, the structural efficiency is maximized while still ensure all the original functions as a cable connector (Fig. 8).

AM Node 2.0

For the purpose of comparing the design results, the design starting point of AM Node 1.0 followed the traditional node where the cable connection points were some distance apart from each other to provide the space for the fork and pin of the cables. This is, however, not necessary given the design possibility of AM. The design could be further optimized with a different interface design for the cable-to-node connection, leading to a more compact and lighter node with better function integration.

As an alternative connecting method from the cables to the node, a cable with a standard threaded end to lead into the node and fix with a bolt was used. This allowed for an easy interface in the node in the shape of a circular plate with an opening. The interface to the strut was set as a short piece of tube with a diameter that would fit

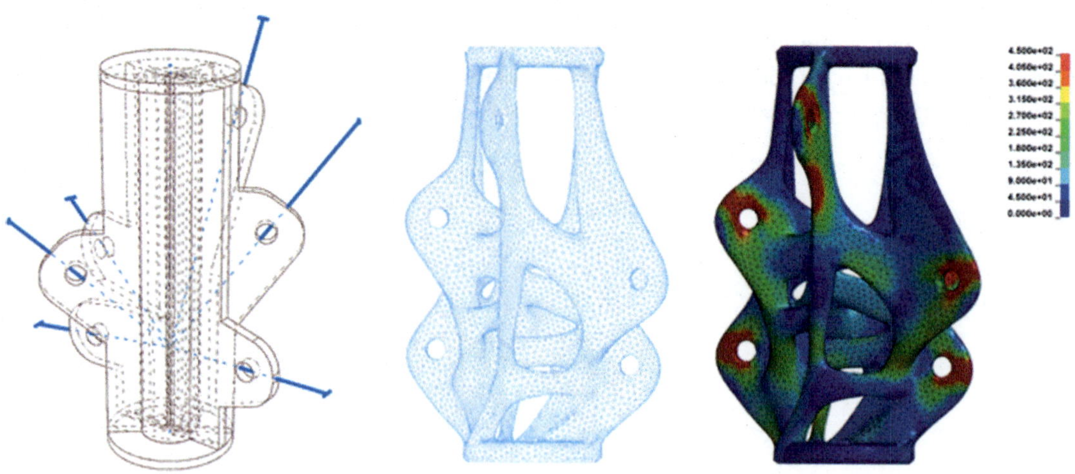

Fig. 6 The design space (*left*), optimised mesh topology (*middle*), and the von Mises stress analysis in [MPa] (*right*) of the AM node 1.0

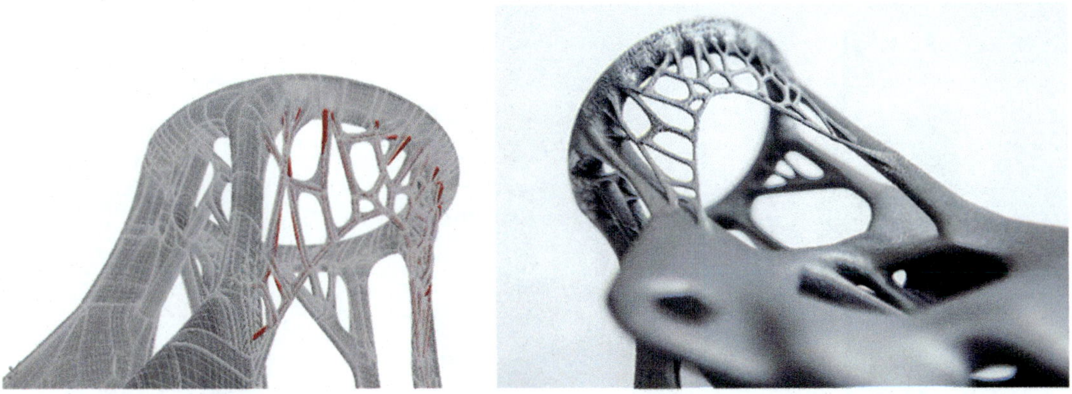

Fig. 7 Generative pattern as the support structure for the AM production: digital model in Rhino/Grasshopper (*left*) and the node produced by AM (*right*)

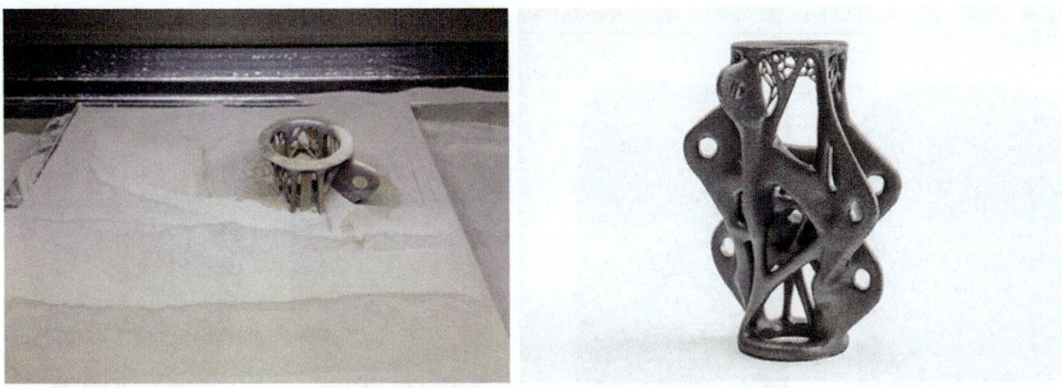

Fig. 8 The printed node emerges from the building chamber (*left*) and the final result (*right*) of AM Node 1.0

exactly in the inside of the tube. Following these, the new initial design space and boundary constraints for the optimisation process was refined in a Rhino/Grasshopper environment which allowed for setting the model parametrically. Two types of design areas were set: design space and non-design space. The central volume in the shape of a cylinder was defined as the main design space with a number of internal access openings of 50 mm as void areas within the design space. Material will be redistributed within this domain following the internal forces flow. The two types of ring elements as the contact interface to the cables and the contact interface to bottom strut were defined as non-design space. The material in this area will not be removed during the optimisation process to ensure the functionality (Fig. 9).

Having the design space and boundary conditions defined, topology optimisation was performed with the design objective of minimising the total structural weight. The material properties for stainless steel was defined and Von Mises stress was analysed as design constraints during the process to ensure the material will not yield. As shown in Fig. 10, starting with a solid mesh within the entire design space, element which was less loaded was eliminated as the generations extended. The node gradually evolved to an organic form with the most

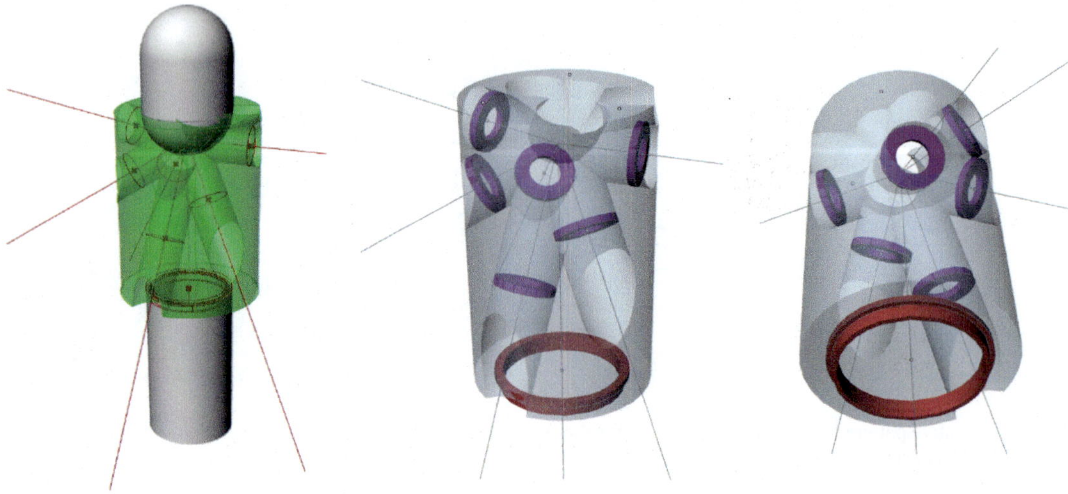

Fig. 9 Parametric definition of the design space for AM Node 2.0 with McNeel's Rhinoceros/Grasshopper

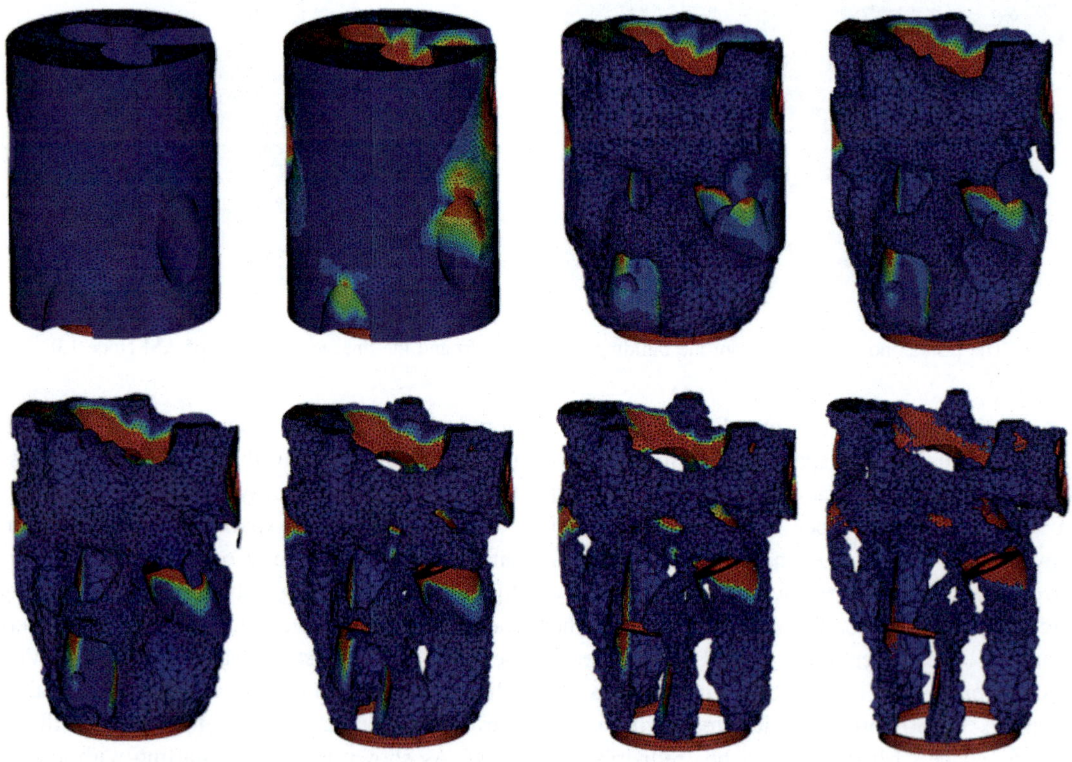

Fig. 10 Topology optimisation process of AM Node 2.0 (using OptiStruct)

efficient load path for the given design forces in the cables.

The output of the topology optimisation indicated the optimal shape for the node with the given boundary conditions and functional requirement. It was, however, still in the form of a rough solid mesh and needed another design process for design interpretation. The solid mesh

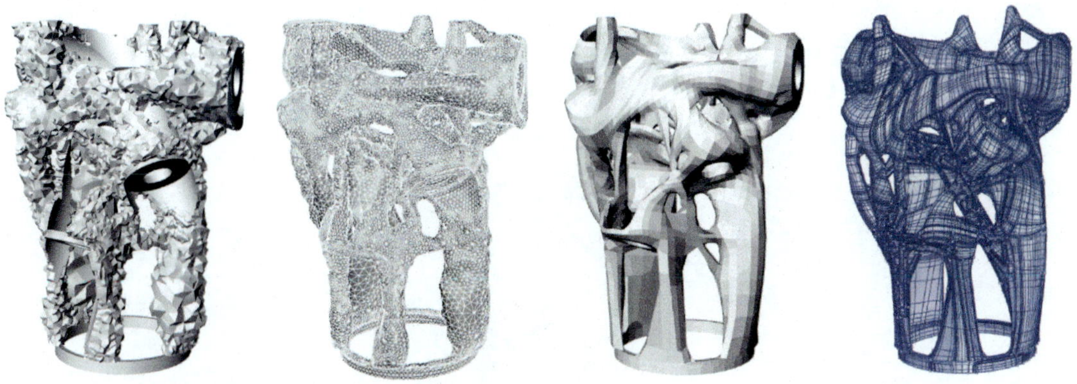

Fig. 11 Interpretation of the design results from a rough mesh to the final shape

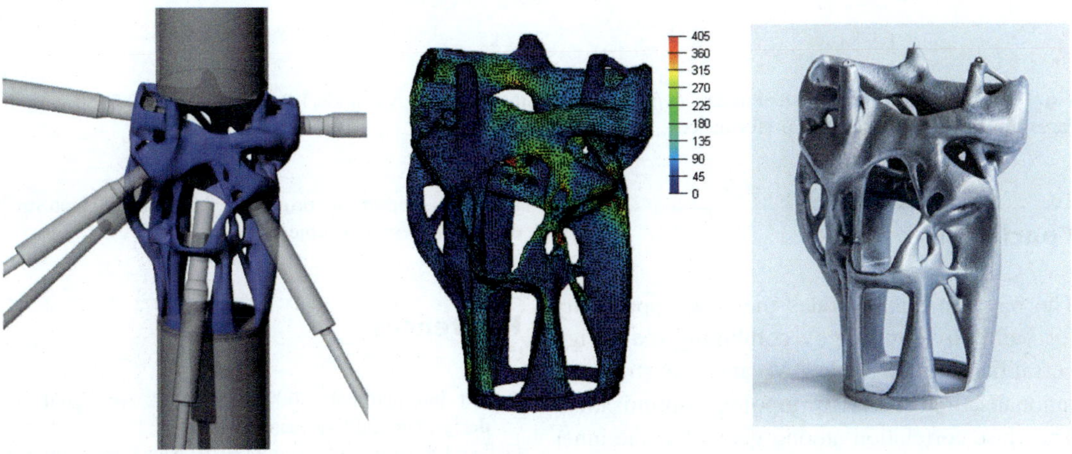

Fig. 12 Rendering of Node 2.0 with the cables and connectors (*left*), structural evaluation with von Mises stress plot in MPa (*middle*) and the node produced by AM (*right*)

was firstly converted to a triangular surface mesh and smoothed with an algorithm. A quad control mesh was then built up based on the triangular mesh for the generation of the final node geometry. The functionalities parts and interfaces elements was also further detailed and incorporated in the design in this process (Fig. 11).

A further structural analysis and comparison on the final shape of the nodes indicated that the optimal shape as in AM Node 2.0 had a similar structural behaviour as its precedents while the weight was dramatically reduced from 20 to 5 kg, with a structural height of 250 mm only (Figs. 12 and 13).

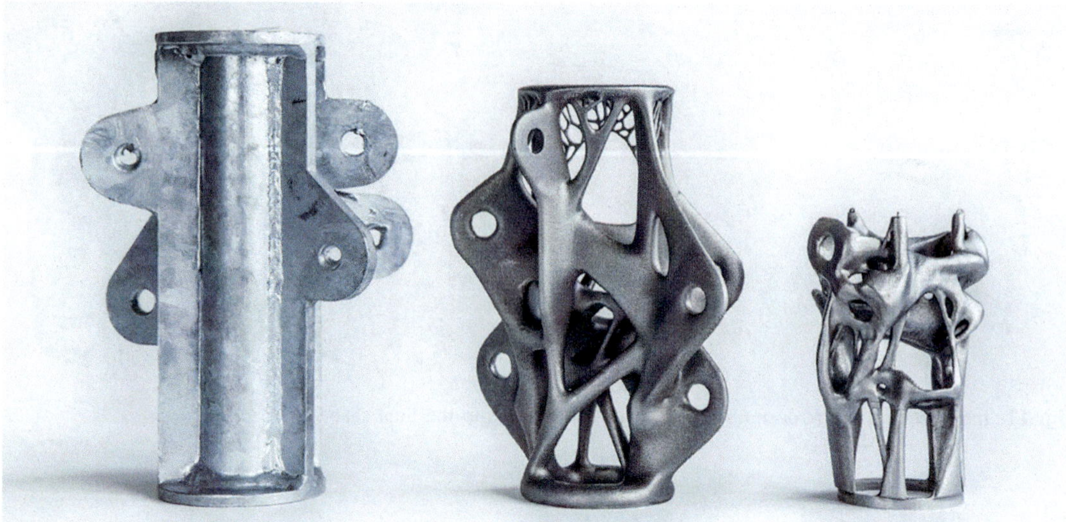

Fig. 13 Comparison of three steel nodes. *Left* to *right* Traditional node, AM node 1.0, and AM node 2.0, all with the same functional and structural performances

Conclusion

The research demonstrates the new opportunity for build environment by combining the design versatility provided by AM and the structural rationalisation driven by topology optimisation. The close correlation among geometry, the inner force flow, and the fabrication method underlines the integrated design process and offers new design freedom and possibility which go beyond the traditional design and fabrication method. This is, however, not a suggestion of relying only on the automated computational decision-making tool or the versatile AM machine that can produce almost any shape. These approaches do not attempt to replace the process of design thinking as a designer's fundamental ability but rather offer new operative medium to help the form and the structure converged.

Acknowledgements The author would like to express the gratitude to Luca Frattari and Michal Wanski of Altair for their support on providing the software OptiStruct during the research project.

References

ASTM International (2009) WK38342: new guide for design for additive manufacturing

Galjaard S, Hofman S, Ren S (2014) New opportunities to optimize structural designs in metal by using additive manufacturing. Advances in Architecture Geometry, London

Jeronimidis G (2004) Biodynamics, emergence: morphogenetic design strategies. Architectural Des 74(3)

Kotnik T, Acunto PD (2013) Operative diagramatology: structural folding for architectural design. In: Design modelling symposium 2013, Berlin

Michell AGM (1904) The limits of economy of material in framed structures. Phil Mag 6:589–597

Ren S, Tiliakos MA, Ureta PV (2011) Fibre morphologies. Thesis at Architectural Association, London

Schmit LA (1960) Structural design by systematic synthesis. In: Proceedings of 2nd ASCE conference on electronic computation. ASCE, New York

Schumacher A (2005) Optimierung mechanischer Strukturen. Springer, Berlin

Challenges of Scale Modelling Material Behaviour of Additive-Manufactured Nodes

Nicholas Williams, Daniel Prohasky, Jane Burry, Kristof Crolla, Martin Leary, Milan Brandt, Mike Xie and Hamed Seifi

Abstract

The application of Additive Manufacturing (AM) technologies promises much innovation across the manufacturing sector, and has generated great interest in the Architecture, Engineering and Construction (AEC) industry. This paper presents and reflects upon early prototypes for integrating AM in a construction application, through the design of a prototype frame structure with linear members connected by nodes of unique shapes. As an enabler for design, a system is developed and implemented to integrate expertise across architecture, structural engineering and advanced manufacturing in order to design and detail components for single layer canopies. This includes the topology optimisation and additive manufacture of nodes, both of which require the control of material behaviour at small scales. A scaled pavilion structure and full-scale node prototypes were successfully realised. However, this first stage of research presented a number of challenges to modelling material behaviour across scales—both the physical size and production volumes.

N. Williams (✉) · D. Prohasky · J. Burry
Spatial Information Architecture Laboratory (SIAL), School of Architecture and Design, RMIT University, Melbourne, Australia
e-mail: nicholas.williams@rmit.edu.au

K. Crolla
Chinese University of Hong Kong, Hong Kong, People's Republic of China

M. Leary · M. Brandt
Advanced Manufacturing Precinct (AMP), RMIT University, Melbourne, Australia

M. Xie · H. Seifi
Centre for Innovative Structures and Materials, RMIT University, Melbourne, Australia

Introduction

Amidst much excitement about the potentials of additive manufacturing to change manufacturing, the construction industry presents significant challenges of scale. These are both physical—related to the size of buildings—and in terms systems design, related to production volumes suitable to the project-by-project context of the AEC industry. The research presented here takes a clear position on each aspect of scale to explore a potential application for additive manufacturing in construction. To bypass limitations of physical

M.R. Thomsen et al. (eds.), *Modelling Behaviour*, DOI 10.1007/978-3-319-24208-8_4

size, additive manufacturing is targeted at the relatively small, connection components, so-called nodes as part of larger assembly. Embracing a manufacturing context of mass-customisation, each node is considered as a unique iteration derived from a common design system.

A team of experts from across architectural design, structural design, topology optimisation and AM was assembled to address the challenge of prototyping a design system and exemplar structure. The project aimed to integrate knowledge through connecting a series of existing technologies and techniques, namely Selective Laser Melting (SLM) for additive manufacture, Bi-directional Evolutionary Structural Optimisation (BESO) (Huang and Xie 2010) for topology optimisation of node design, and parametric models with lightweight analysis tools for architectural and structural design of building form and detailing.

A digital workflow was implemented, linking modules across a design and production process. With these tools, both full-scale exemplar nodes printed in stainless steel (Fig. 1), and a pavilion structure at scale 1:5, constructed with timber beams linked by ABS printed plastic nodes (Fig. 4), were realised. These early prototypes offer promise that such a system might be feasible in the construction industry as technologies, in particular SLM additive manufacture, improve in terms of speed and cost.

Despite this promise, however, this early research raises significant further design questions. The project demands that node design incorporates the modelling of material behaviour in multiple contexts. Firstly, current techniques abstract material behaviour in order to optimise topologies in response to given load cases. While this reduces weight and material volume, manufacturing parameters are not considered. Secondly, AM requires that material behaviour is controlled to ensure successful results. For further development, the system needs to respond to manufacturing parameters to improve the design of nodes.

Background

Additive Manufacturing in Metals

ASTM: F2792—12a (2012) defines AM as the "process of joining materials to make objects from 3D model data, usually layer upon layer, as opposed to subtractive manufacturing methodologies, such as traditional machining". AM requires a Computer Aided Design (CAD) representation of the intended geometry. This CAD data has traditionally been represented in Stereolithography (STL) format (Gibson et al. 2010). However, enhanced data representations are under development, such as the Additive Manufacturing Format (AMF) (ISO/ASTM 2014).

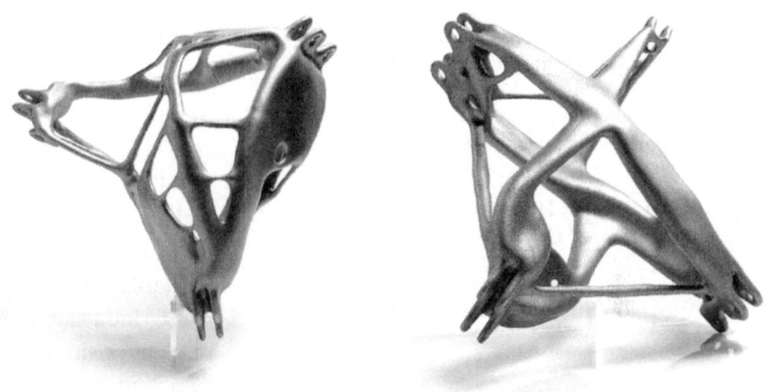

Fig. 1 Stainless steel prototypes of proposed construction nodes

This AM data is then converted to a sequence of two-dimensional scan paths (Zhai 2014) that are sequentially processed by the AM process.

AM is an enabling technology with economic and technical attributes that allow design innovation including: reduced cycle times and unit cost (Ruffo et al. 2006), repair of high-value components (Shepeleva et al. 2000), and manufacture of high efficiency structures not compatible with traditional methods (Brandt et al. 2013). Plastic parts manufactured through Fused Deposition Modelling (FDM) are nowadays common and accessible to non-expert users. Metal-based AM systems are less common but provide opportunity for commercial benefit (Atzeni and Salmi 2012; Atzeni et al. 2010).

Selective Laser Melting (SLM) applies a laser power source to selectively solidify powder in a powder-bed. SLM can process a range of useful engineering materials including: high-strength and alloy steels; light alloys, including aluminium, titanium and magnesium; and precious metals. Despite the associated design opportunities, SLM is subject to complex, dynamic manufacturing constraints. For example, SLM melt pool dynamics and is governed by over 130 parameters (Rehme and Emmelmann 2006). Furthermore, the associated heat transfer is both spatially and temporally transient, resulting in significant challenges to the manufacture of commercially robust components (Gu et al. 2012); especially associated with optimisation of orientation and distribution of support materials (Brandt et al. 2013; Leary et al. 2014).

Node Precedents

Many in the AEC industry have identified that relatively small connection elements are key to the design of larger assemblies of parts. This includes key examples of pre-fabricated, modular housing systems (Wachsmann 1946); space frames (Bardell et al. 1996); and gridshell frames (Carlow and Crolla 2013). This latter case presents particular challenges as nodes are highly varied to meet angles of adjacent members.

The design of connection parts has also received attention in high-value automotive and aerospace design. As SLM technologies have matured, designs have been proposed for key structural components in jet engine (Halchak et al. 2012), aerospace (Brandt et al. 2013), and automotive (Cooper et al. 2012) that are incompatible with traditional manufacturing methods. These are designed to replace existing components and exploit weight savings which can be achieved with optimised, additive manufactured parts. As though parts are produced for runs of vehicles, multiple versions and design prototypes can be produced and tested.

A first proposal for such an application of AM in construction has been proposed for application to a node in a tensegrity structure (Galjaard et al. 2015). This focusses on a case of a joint connecting a series of tension cables to a rigid member in compression. The design of the part employed optimisation techniques to reduce weight and a first prototype has been manufactured using SLM technology.

Design and Optimisation for Additive Manufacture

A number of preliminary studies have highlighted benefits of topology optimisation for additive manufacturing. Two techniques have received particular attention—BESO and SIMP (Aremu et al. 2010; Brackett et al. 2011).

BESO routines, used here, develop optimal results through the iterative removal of inefficient material. A given design space is divided into spatial volumes, typically a grid of voxels, and results can be determined for stress, stiffness and other factors (Yang et al. 2005, p. 957). This results in a minimisation of material volumes through specific and commonly freeform geometries. While these shapes present challenges to manufacture using conventional technologies, AM bypasses many common limitations.

The Smartnodes Design System and Prototypes

A design system was conceived to link modules across architectural design, structural analysis and design, topology optimisation of nodes and the generation of data for fabrication. Such a close linking a series of modules enables information to quickly flow both downstream and upstream, offering improved designs through iterative input across many project details. Key to these systems is consistent interfaces between modules, concisely defining data but with sufficient redundancy to enable flexibility (Marble 2012, p. 130).

The design system was implemented, firstly, around a series of parametric models in *McNeel Rhinoceros*. In parallel to this, modules were developed to undertake more detailed structural analysis and topology optimisation of nodes (Fig. 2). Agreed interfaces in a neutral csv format describe firstly, the setout of axes across the form allowing them to be imported into GSA for structural analysis, and later, results from this analysis of the model which were interpreted into Abaqus tools for BESO algorithms (Fig. 3).

Through a custom interface created for *Rhinoceros*, a pavilion form was designed with linear members meeting at nodes. In order to accommodate a freeform shape, each node was assumed to have a unique geometric set-out, connecting between three and five beams in directions unconstrained by local conditions. The geometric setout was then exported in a neutral format for use in each branch of the workflow.

Structural analysis in response to gravity loads was generated. The results produced by *Karamba,* a plugin for Grasshopper, were compared to the results generated in the commercial structural design software *Oasys GSA*. The results correlated closely, with the latter providing verification for results from the lightweight tools.

Using the lightweight tools, the 144 nodes of the structure were topology optimised using *millipede*, a plugin for *Grasshopper*. In parallel, example nodes were identified and optimised in more detail for full-scale production. These parallel processes involved significant differences in resolution. For lightweight analysis, starting forms were modelled as spheres of 35,000 voxels. BESO algorithms were then run for six iterations, providing relatively simple topologies with crude levels of detail. This could be processed in only a few minutes on a desktop computer. The resulting forms were suitable for additive manufacture as demonstrator components at scale 1:5, providing minimum material thicknesses of 0.5 mm. In some contrast to this, the BESO algorithms applied to the full scale prototypes used a starting mesh with greater than 500,000 voxels. The three node cases, one connecting each of 3, 4 and 5 members, each

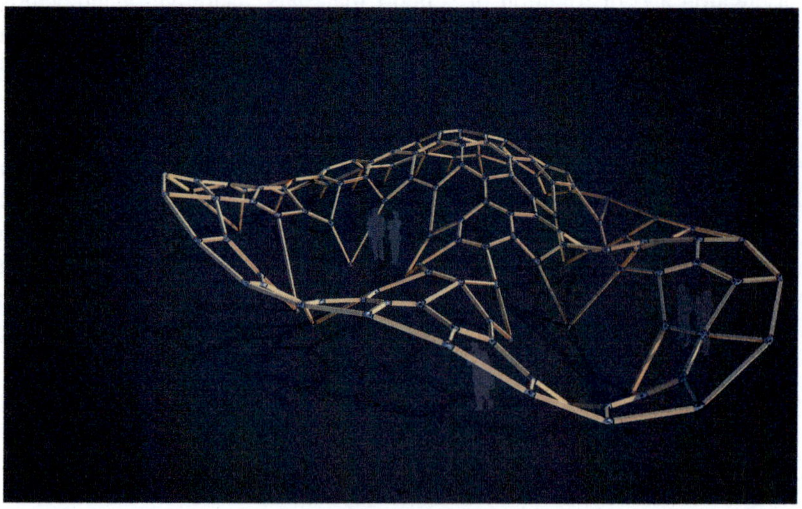

Fig. 2 Proposed design of a *single layer* canopy with standard beams connected to customised nodes

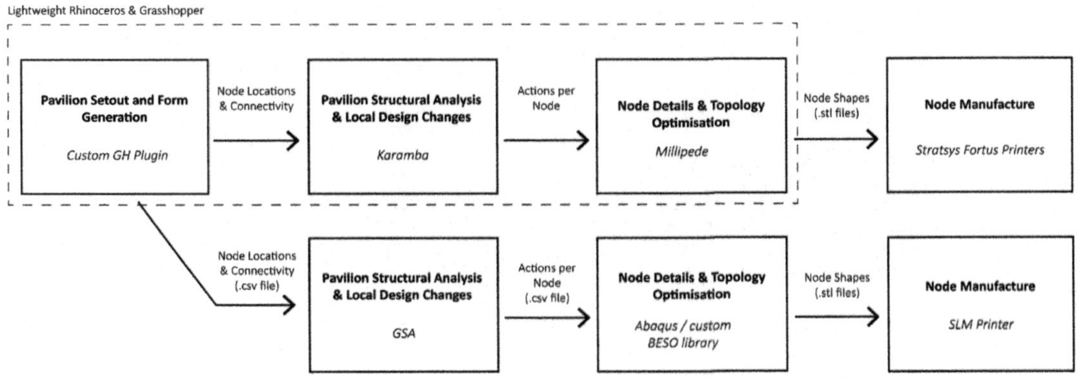

Fig. 3 Key workflow structure showing modules and interface formats

required over 48 h to process 100 iterations. Results were of a much finer grain and offer significantly better outcomes, in terms of volume and weight, than the scaled versions.

The manufacturing of nodes through each of the parallel processes was undertaken. The scaled demonstrator nodes at 1:5 were printed using a *Stratasys Fortus* printer. Support material was generated using proprietary *Stratasys* software and manufacture required between 2 and 2.5 h per part. At full scale, two of the three nodes were manufactured in stainless steel using an SLM printer. Print times of greater than 40 h were required for these. As well, further skilled labour was required to orient the part, design support material and subsequently to remove this material after manufacture. Parts were successfully manufactured (Fig. 1) however some minor problems were encountered.

Discussion

Flexibility for Customisation in Construction

The bifurcation of the project workflow into parallel streams was driven by pragmatic issues around managing levels of detail. As the project developed, large model sizes and lengthy processing times were identified as threats to both feasibility and flexibility for design. For example, taking the required BESO optimisation time of a full-scale node and applying it to the 144 node scenarios across the pavilion would require many days using current computers. Even assuming improvements in computer processing, a relatively lightweight process is required to be feasible for a volume of parts such as this.

The manufacture of 144 nodes with FDM machines was straightforward, with all jobs successfully completed and parts within tolerance (Fig. 4). This demonstrates 'file to factory' principles which promise customised manufacture and flexibility in design.

In comparison to this, the SLM technology applied to the full scale parts is relatively new and caution has been advised in drawing simple comparisons between the two (Brewster 2014). Given this context, it was unsurprising that some problems were experienced. Firstly, incorrect humidity levels caused failure through material slumping in one job. The second problem centres on difficulties in orienting the complex forms for printing. The design of an adequate support structure to ensure successful manufacture proved very challenging, and the proposals contained excessive amounts of material. As a result a prototype of a 5-sided node was not attempted.

These technical challenges to SLM demand further development of the technology for consideration in commercial construction. Much research is being pursued in this area and it is foreseeable that these issues will be largely overcome. Beyond this, however, human labour

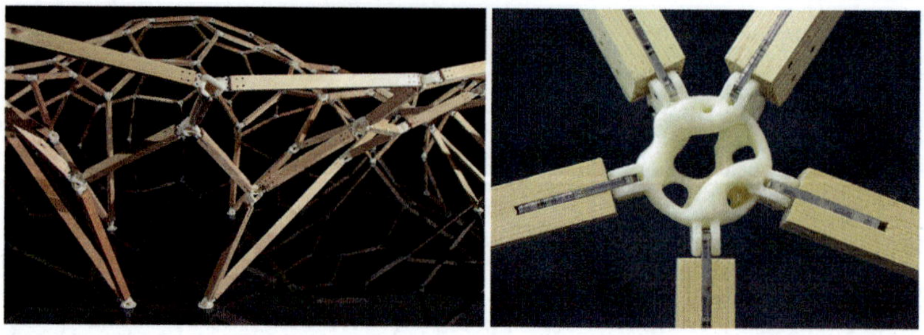

Fig. 4 The fabricated pavilion prototype at scale 1:5 showing (*left*) a network of beams and (*right*) a typical node manufactured by an FDM machine

required for designing and removing support material is a significant impost to SLM manufacture. To enable flexibility for customised manufacture, this labour must be significantly reduced through algorithms using rules of thumb to automate orientation and design of efficient support structures.

Optimisation for Broader Criteria

BESO optimisation techniques are conceived to address a given set of load cases. Simulations typically use stopping conditions related to material stress or volume. While such optimisation techniques are highly sophisticated, the optimisation criteria addressed here reflect only a fraction of broader design issues being addressed. Simply, weight and volume are not directly proportional to economic issues underlying manufacture.

The challenges in SLM manufacture encountered in this research suggest that the inclusion of manufacturing parameters in optimisation processes would be of benefit. It is conceivable that suitable orientation of parts could be automated in response to geometric analysis. Once a suitable orientation is identified, support material could be generated for required areas using rules of thumb. Algorithms for generating efficient support structures for FDM printing are available through softwares such as *Autodesk Meshmixer*. Future research will explore whether similar ideas can be applied to SLM.

In opposition to this, however, this research also highlights significant computation time. As a result, the simple addition of further optimisation criteria and routines is not straightforward. Trade-offs between computation time for optimisation routines, and the material, labour and time for manufacturing must be negotiated. The consideration of broader criteria might alternatively lead to exploration of other optimisation strategies.

Conclusion

Through the study of parallel workflows applying both BESO and AM to the design of components at different scales, this paper highlights challenges in modelling material behaviour at an architectural scale. Scale parts using relatively simple optimisation routines and mature FDM technology were rapidly realised. This offers significant customisation and flexibility in the design of built form. However, for full-scale components using relatively complex BESO optimisation and SLM manufacture, technical challenges were encountered and design and production times were excessive.

This suggests that recent claims that AM allows architects to design at the scale of matter are overly simplistic. While it is possible to design highly detailed parts, the sheer size of a building requires that designers are strategic in identifying key components for such a level of attention. Further to this, the short production volumes and high degrees of customisation typical to construction demand flexibility in manufacturing. This requires robust processes (Piller 2013) with many challenges still ahead for AM in metals to meet such a context.

References

ASTM F2792—12a (2012) Standard terminology for additive manufacturing technologies, West Conshohocken

ISO/ASTM 52915 (2014) 2013 Standard specification for additive manufacturing file format (AMF) Version 1.1

Aremu A et al (2010) Suitability of SIMP and BESO topology optimization algorithms for additive manufacture. In: 21st annual international solid freeform fabrication symposium (SFF)—an additive manufacturing conference, Austin, Texas, pp 679-692

Atzeni E, Salmi A (2012) Economics of additive manufacturing for end-usable metal parts. Int J Adv Manuf Technol 62(9–12):1147–1155

Atzeni E, Iuliano L, Minetola P, Salmi A (2010) Redesign and cost estimation of rapid manufactured plastic parts. Rapid Prototyp J 16(5):308–317

Bardell N et al (1996) The development of MURJ-3D: a modular, universal, reconfigurable joint for in 3D Space Frame Applications. Int J Space Struct 2(12): 89–107

Brewster S (2014) Why you won't see a laser sintering printer on your desk any time soon. https://gigaom.com/2014/04/25/why-you-wont-see-a-laser-sintering-3d-printer-on-your-desk-anytime-soon. Accessed 29 Apr 2014

Carlow J, Crolla K (2013) Shipping complexity: parametric design for remote communities. In: Zhang et al (eds) Global design and local materialization. In: 15th international conference, CAAD Futures 2103, Shanghai, China, 3–5 July 2013. Springer, New York

Galjaard S, Hoffman S, Ren S (2015) New Opportunities to optimize structural designs in metal by using additive manufacturing. In: Block P et al (eds) Advances in architectural geometry 2014. Springer, Switzerland, pp 79–92

Gibson I, Rosen D, Stucker B (2010) Additive manufacturing technologies. In: Additive manufacturing technologies rapid prototyping to direct digital manufacturing, Boston, MA. Springer, New York

Gu D, Meiners W, Wissenbach K, Poprawe R (2012) Laser additive manufacturing of metallic components: materials, processes and mechanisms. Int Mater Rev 57(3):133–164

Halchak J, Wooten J, McEnerney B (2012) Layer build of titanium alloy complex-geometry components for rocket engines. In: Ti 2011—proceedings of the 12th World conference on titanium 3, Beijing; China, pp 1715–1718

Huang X, Xie YM (2010) Evolutionary Topology optimization of continuum structures: methods and applications. Wiley, England

Leary M, Merli L, Torti F, Mazur M, Brandt M (2014) Optimal topology for additive manufacture: a method for enabling additive manufacture of support-free optimal structures. Mater Des 63:678–690

Marble S (ed) (2012) Digital workflows in architecture: designing design—designing assembly—designing industry. Birkhäuser, Basel

Piller F (2013) Three capabilities that make mass-customisation work. In: Piroozfar P, Piller FT (eds) Mass customisation and personalisation in architecture and construction. Routledge, New York, pp 17–30

Rehme O, Emmelman C (2006) Rapid manufacturing of lattice structures with selective laser melting. In: Solid freeform fabrication, vol 6107, p 61070K

Ruffo M, Tuck C, Hague R (2006) Cost estimation for rapid manufacturing-laser sintering production for low to medium volumes. Proc Inst Mech Eng Part B J Eng Manuf 220(9):1417–1427

Yang X, Xie YM, Steven G (2005) Evolutionary methods for topology optimisation of continuous structures with design dependent loads. Comput Struct 83: 956–963

Zhai Y, Lados D, LaGoy J (2014) Additive manufacturing: making imagination the major limitation. JOM 66(5):808–816

Form-Finding and Design Potentials of Bending-Active Plate Structures

Simon Schleicher, Andrew Rastetter, Riccardo La Magna, Andreas Schönbrunner, Nicola Haberbosch and Jan Knippers

Abstract

This work presented investigates the form-finding and design potentials of bending-active plate structures. Using two reference projects from the recent past, the authors present different design methodologies that either follow a geometry-based or integrated approach. A closer look at the newly accessible tools for digital form-finding and analysis reveals their increasing importance for the design process. In order to better demonstrate their potential, the authors present three case studies, which each separately enhances the integrated approach and in combination indicate the existence of a much larger design space of bending-active plate structures.

Introduction

Bending-active plate structures use the elastic deformation of planar, off-the-shelf building materials to generate curved surface structures (Knippers et al. 2011). While the traditional maxim in engineering is to limit the amount of bending in structures, this typology actually harnesses bending for the creation of complex and extremely lightweight designs. In the past, thin plates have rarely been used as primary structure in architecture because of their low bending stiffness. Many sheet materials like plywood, metals, plastics, and fibre-reinforced polymers, however, are not only flexible but also have high tensile strength. The two properties together are a perfect match for bending-active structures because they enable elements to undergo large elastic deformations and to resist high stresses before failure. This behaviour opens up new possibilities for the design of bent static and kinetic structures (Lienhard et al. 2014; Schleicher et al. 2015). The most significant advantage of these systems is that they can be constructed from simple planar parts, which can be fabricated with inexpensive, conventional flatbed processes. Additionally, the assembly of

S. Schleicher (✉) · A. Rastetter
College of Environmental Design (CED), University of California, Berkeley, USA
e-mail: simon-s@berkeley.edu

R. La Magna · A. Schönbrunner · N. Haberbosch · J. Knippers
Institute of Building Structures and Structural Design (ITKE), University of Stuttgart, Stuttgart, Germany

M.R. Thomsen et al. (eds.), *Modelling Behaviour*, DOI 10.1007/978-3-319-24208-8_5

these structures does not require skilled labour or auxiliary formwork. Despite these benefits, the design of bending-active plate structures is a major challenge. This is because it is difficult to assess their structural behaviour and to accurately anticipate their deformed geometry. Therefore, it is essential to develop new design approaches.

Design Approaches to Bending-Active Plate Structures

A recent study identified three main strategies for the design of bending-active structures. These are behaviour-based approaches, geometry-based approaches, or integrated approaches (Lienhard et al. 2013). Bent huts and tents of vernacular architecture, for example, fall into the first category. Here, bending is used rather intuitively during the construction process. While this first approach relies heavily on experience, the other two categories are more scientific and require experimental and analytical form-finding techniques. Their similarities and differences are illustrated by the examples of Buckminster Fuller's plydomes and the 2010 ICD/ITKE Research Pavilion (Figs. 1 and 2).

Fig. 1 Two-frequency geodesic plydome in Des Moines, Iowa, 1957. The hemisphere spans 7.3 m and is made out of marine plywood sheets with a thickness of 6.4 mm. (Marks 1973, p. 210)

Fig. 2 ICD/ITKE research pavilion 2010 spans 10 m and consists of 80 birch plywood strips with a thickness of 6.4 mm

Self-strutted Geodesic Plydomes

As part of his research on geodesic dome structures, Buckminster Fuller (1895–1983) experimented with various structural systems and materials. His motivation was to find an optimum balance between a structure's stability, weight, and cost (Marks et al. 1973). While the majority of his larger dome structures are triangulated lattice shells that use steel struts as structural framework and solid panels or textiles as cladding, some of his smaller projects investigated different structural typologies that further increased the efficiency of the systems. Of particular interest are the geodesic plydomes that he developed from the fifties on. Here, the key idea was to construct a dome-shaped enclosure through repetitive tiling of rectangular plywood sheets. Plywood sheets were used because they are mass-produced, easily obtainable, relatively inexpensive, and can be stacked compactly for shipment. Fuller used a geometry-based approach to fit the flat sheets to the doubly curved surface of a sphere. This approach is described in Fuller's 1959 patent for the system (Richard 1959). He first approximates the target geometry of a sphere with a regular polyhedron (Fig. 3a).

Next, he arranges individual plates along the edges of the polyhedron (Fig. 3b). Adjacent

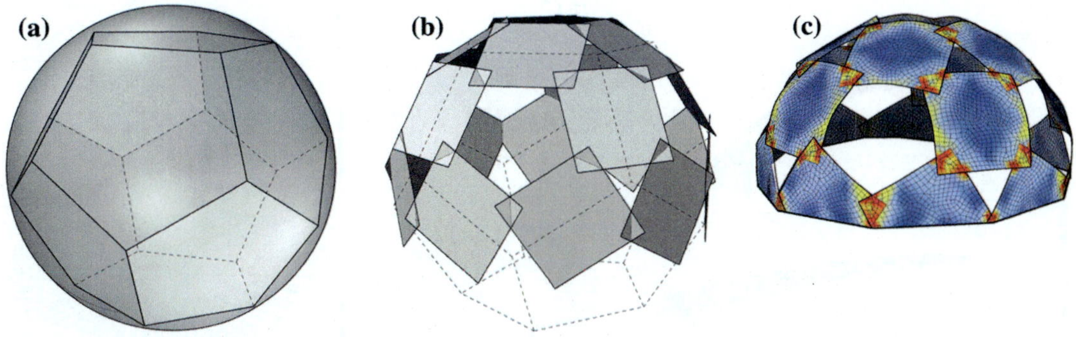

Fig. 3 Geometry-based approach of a plydome approximates a sphere with a polyhedron (**a**). Based on the polyhedron's edges multiple sheets get to arrange spatially (**b**) and then bent and fastened together (**c**)

plates are partially overlapped, pulled together, and fastened to one another with pre-drilled holes. This generates the double curvature of the global geometry even though the individual plates only experience single curvature at any given location (Fig. 3c). The greatest challenge in implementing Fuller's system is determining the amount of overlap between adjacent sheets and the location of the attachments points. Both are dictated by the deformed geometry of the plates and may also vary according to the sheet's position in the overall pattern. Without the digital tools that we have today, Fuller was forced to compute this information mathematically. Over time he calibrated these numerical results to the actual behaviour of the system.

ICD/ITKE Research Pavilion 2010

A good example of a more advanced approach to bending-active plate structures is the ICD/ITKE Research Pavilion 2010 (Fig. 2). This pavilion, which follows an integrated approach, was designed and built by students and teachers at the University of Stuttgart, as a collaborative effort between the Institute of Computational Design (ICD) and the Institute of Building Structures and Structural Design (ITKE) (Lienhard et al. 2011; Fleischmann et al. 2012). The project achieves a more complex curved geometry by alternating segmentally bent plywood strips that are linked together. The team began the design process by considering the limiting material characteristics of the plywood strips. The first step was to calibrate finite-element simulations with physical experiments. This ensured that the digital form-finding techniques provided an accurate description of the actual material behaviour while at the same time offering full control over the geometry. Of particular importance was the ratio between the pitch and span of a bent strip (Fig. 4a).

This ratio describes the maximum achievable deflection under a given safety factor. Once established, the ratio informed a parametric model that was used to design the global geometry of the pavilion (Fig. 4b). This model determined the dimensions and connection logic for each strip. The last step was to translate the geometry from the parametric model to a more advanced finite element analysis, which re-created the bending process under consideration of relevant material properties for birch plywood (Fig. 4c).

This last step was of key importance because it provided precise information about the pavilion's deformations and structural performance under different loading scenarios as well as essential data for the subsequent fabrication and assembly.

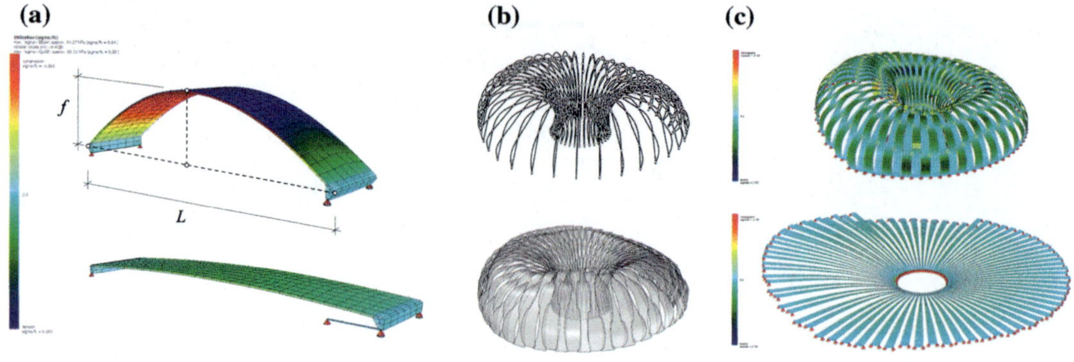

Fig. 4 ICD/ITKE research pavilion 2010 illustrates an integrated approach to the design of bending-active plate structures

Form-Finding and Analysis of Bending-Active Plate Structures

The biggest challenge regarding bending-active plate structures is the difficulty of predicting their deformed geometry and structural performance. As demonstrated by previous projects, options for their form-finding and analysis include physical modelling, mathematical calculations, and advanced digital simulations. These digital simulation tools, however, have only been available for a few years to a broader public. They can be subdivided into two categories:

The first method relates to real-time physics-based simulations. These are used extensively in the computer graphic community and are now also available for common CAD environments. The Rhinoceros® plugin Kangaroo Physics is a good example of this type of software. For the simulation of the bending behaviour of shells, it employs a discrete shell flexural energy model as described in Grinspun (2008) and Piker (2013). Mesh deformations are computed employing a dynamic relaxation scheme, requiring the introduction of lumped mass values and damping coefficients in the computational model. During the simulation process, the system converges to an equilibrium position that represents the final bent geometry. Although the definition of such additional parameters (mass, damping coefficients) is often arbitrary and not physically motivated, the strength of this computational scheme relies on the calculation speed and easiness of setup. This makes it ideally suited for iterative studies in early design stages or design explorations for structures with many elements.

The second method relies on finite element simulation (FEM). While this technique was originally used mainly for post-design analysis, non-linear FEM routines have advanced so much over the last few years that it has become practical to integrate them early in the design process (Lienhard et al. 2011). Programs like SOFiS-TiK®, for example, allow a designer to calculate the deformations and stresses of structures under large deformations and to predict complex equilibrium states. In doing so, the software simulates the bending of a structure by considering both external forces and internal material stresses. Considering both simultaneously is particularly important because the geometry of a deformed structure depends significantly on the balance of forces that are exerted on it. Unlike real-time physics simulations, FEM can be also used to visualize the evolution of stresses within the material during the form-finding process. FEM simulations offer the most complete and correct mechanical description of the behaviour of shell elements, representing an invaluable tool for the correct evaluation of the mechanical behaviour and structural capacity of bending—active structures. On the other hand, the completeness

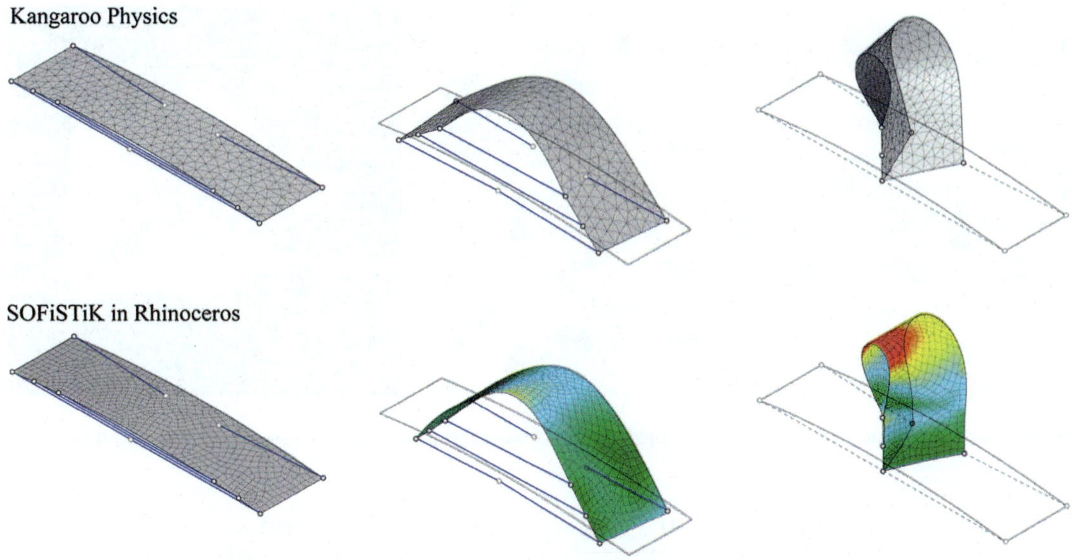

Fig. 5 Form-finding with contracting elastic cables using Kangaroo Physics and SOFiSTiK

of the mechanical model does not come for free, as it is computationally more intensive and can be rather slow for large models.

Contracting elastic cables provide a practical method to induce bending in both of these digital simulation techniques (Fig. 5). The cables are shortened through a reduction in stiffness and a simultaneously applied pre-stressing load (Lienhard et al. 2014). Each cable is attached to pairs of nodes on one or multiple meshes. These nodes are pulled together during the simulation process, which produces a controlled deformation of the attached meshes. This technique is very versatile and easy to use because it does not require the input of an explicit nodal displacement path.

Case Studies that Render New Potentials

The authors conducted a series of case studies to enhance the previously described integrated approach. These case studies separately explore new design potentials but also build upon each other to generate more complex bending-active plate structures.

Case Study 1—Effective Pinching

The first example explores the benefits and opportunities of the single-curvature that typically results from bending thin plates. Unlike the reference projects, which primarily use cylindrical shapes, this case study explores the potential of conical bending (Fig. 6). It conceptualizes a triangular facade module out of initially flat panels and challenges the design process by investigating shapes whose cutting patterns are not predefined from the start but needed to be determined through a series of simulations.

In this example, internal openings are pinched together to provoke global deformations in a plate. Material is strategically removed from the sheet's centre and the internal edges are forced together with contracting cables in Kangaroo Physics (Fig. 7a). This leads to a conical out-of-plane buckling that gives the plate a unique developable form. Each plate is then mirrored, merged, and trimmed with its counterpart, which brings about first changes to the initial cutting pattern (Fig. 7b). The result is a dual layer module with two plates that affect each other in form and significantly enhance the structural rigidity of the system. Trimming the

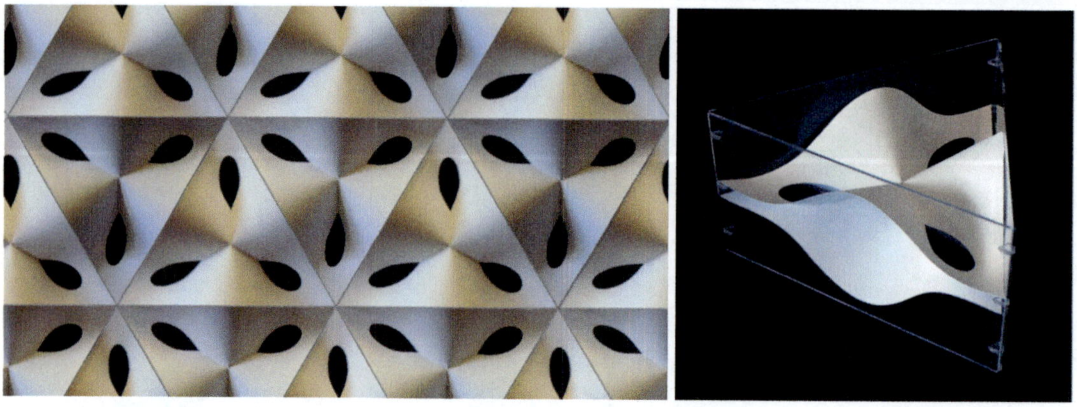

Fig. 6 Triangular facade tessellation with a module that is based on pinching and cross-connecting two thin sheets

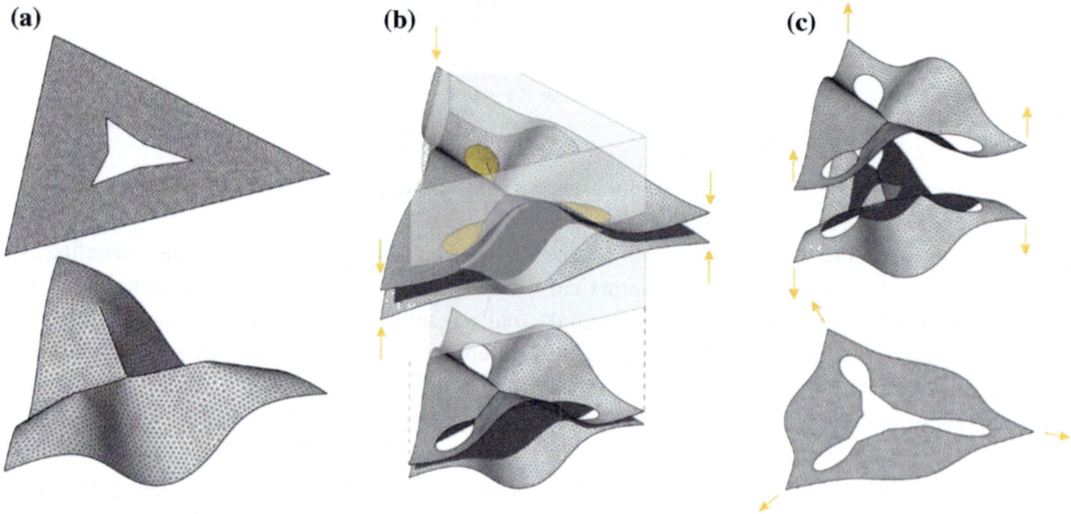

Fig. 7 Pinching a plate in the centre causes a conical deformation (**a**). Two plates can be merged with each other and trimmed (**b**) in order to derive with a special informed cutting pattern (**c**)

module's periphery with an extruded triangular profile produces the second substantial modification to the initial pattern. This step is necessary to guarantee that the module will fit into a symmetric facade tessellation. What makes this trimming action so special, however, is the fact that it is conducted on a plate that is already deformed, which results in a unique cutting pattern. The surprising complexity of the final pattern becomes particularly apparent in the last step of the form-finding process, where the layers of the module are separated and flattened by applying a uniformly distributed load that presses the mesh against a virtual floor (Fig. 7c).

Case Study 2—Mutual Reinforcement

The second case study aims to push the research one step further and focuses not only on an individual model but also on the complexity of the global system. It explores the question of how flat plates can be bent into a doubly curved, multi-layered structure that is extremely lightweight and has a high load-bearing capacity.

Similar to a piece of corrugated cardboard, this project uses the technique of pleating to cross-connect multiple thin sheets in order to form an assembly that is more rigid than the sum of its individual layers. What is most special about this project, however, is that it transfers the idea to a doubly curved global geometry (Fig. 8). This is not an easy task because a continuous flat plate can only be bent into single but not into double curvature. In order to achieve the global double curvature, the individual plates are perforated with holes and slits. These perforations divide the plates into individual zones that can be bent differentially. The role of the perforations, therefore, is to liberate individual plate segments from the necessity to take on double curvature. Instead, the Y-shaped subunits are only single curved and can conform to different synclastic and anticlastic target geometries. Multiple surfaces are stacked, offset, and then connected to one another in order to lock the shape of the global geometry and to provide a load path through the structure. In this way, every hole in the first layer gets filled and structurally supported by the second layer. This process is repeated by adding a third layer and more layers can be added to increase the stiffness of the structure.

A series of physical and digital studies were used to design and test the structural system. Material tests were synchronized with FEM simulations in order to determine the limit of feasible deformation, which is dictated by the local stress concentrations in the material (Fig. 9a). An additional factor that had to be considered was the amplification effect that results from the interaction between the interconnected layers. This interaction was used to determine the degree of variation between adjacent perforation patterns (Fig. 9b). Once a feasible deformation for a given sheet material was found, it was possible to develop an abstracted parametric model that preserved the internal relationships of the system and limited the design space to achievable solutions (Fig. 9c). This approach required less computing power than a comprehensive FEM analysis and allowed for the efficient tessellation of various doubly curved surfaces.

Case Study 3—Functionalized Instability

The third case study builds on the insights of the previous two examples and explores bending

Fig. 8 Mock-up of a 2 × 1 m plate structure uses the coupling of three PET-G layers of 1 mm thickness to obtain a freeform geometry with a high load-bearing capacity

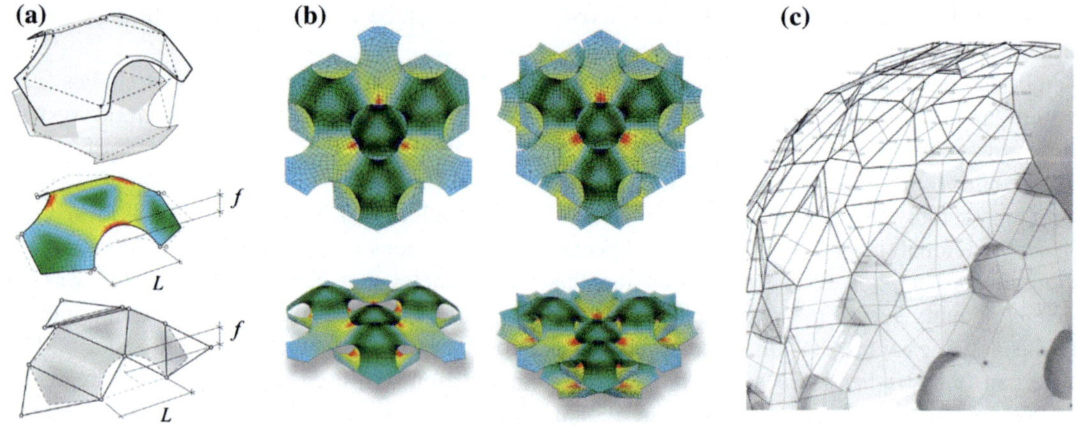

Fig. 9 Considering local stress concentrations in the individual unit as well as in the interacting layers, allowed to simplify the system to a parametric model with which to easily populate given freeform surfaces

both as a form-giving strategy and also as practical means to quickly assemble a larger structure out of multiple smaller subunits. It uses a snap-through instability mechanism to connect the components of the system.

Like the previous examples, this project restricts itself to elements that are constructed from flat sheet material. In this case, the structure is assembled from variations of only four basic shapes. These include two annular components and two longer undulating strips (Fig. 10). The components vary in size and edge conditions, and are fitted with mortise and tenon joints to connect adjacent pieces together. The annular components get pulled together at their ends to create conical frustums. These truncated cones fit precisely to the undulating strips and act as spacers between two layers in the system. Once all parts are loosely assembled, they are fastened together by pressing on the cones and deliberately causing them to buckle until they snap through to a new equilibrium position. This process is used to clamp adjacent parts and it is repeated three times to fully lock all of the structural components together. The result is a double-layered sandwich structure that is geometrically versatile and can be applied to both synclastic and anticlastic shapes (Fig. 11).

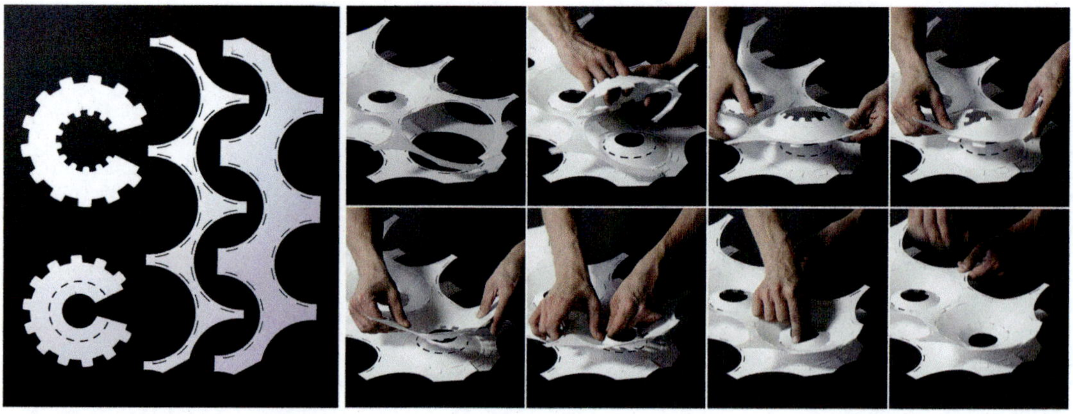

Fig. 10 The entire structure consists of four basic shapes that are clamped together by a local snap-through instability mechanism

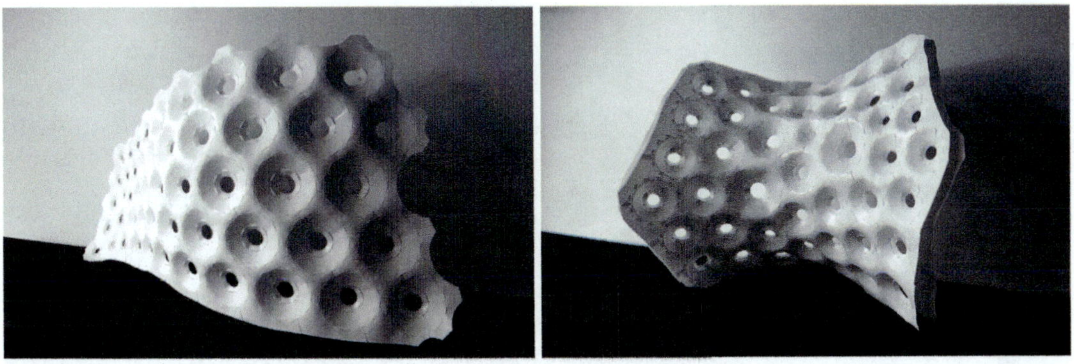

Fig. 11 This construction methodology can be applied to the design of synclastic and anticlastic surfaces

An important step in this project was to translate this construction method into a computational process that allows for morphological differentiation. In order to do so, an integrated approach was developed to incorporate both geometric and material constraints. FEM analyses were conducted at two different scales. At the level of the subunit, FEM simulations were used to determine suitable material characteristics and restricting geometric dimensions to ensure that the truncated cones can be snapped by hand. At the global level, FEM simulations were used to optimize the thickness of the sandwich structure depending on the stress distribution in the target surface (Fig. 12). Areas with larger bending moments have more distance between the two layers, while areas with higher shear forces have smaller cones and thus feature greater material density. This structural information was used to inform the packing pattern that was applied to the target surface. There is a direct relationship between cone radius, assembly thickness, and material density, which all act together on the final composition of the structure. Finally, the detailing and layout of the individual parts was automated in order to facilitate an easy translation to new global geometries. Each individual component is automatically unrolled and detailed with the mortise and tenon joints that are used to connect the parts together. As a proof of concept, a large-scale model was constructed using flatbed fabrication technologies and light PET-G plates of 0.5 and 0.75 mm thicknesses (Fig. 13).

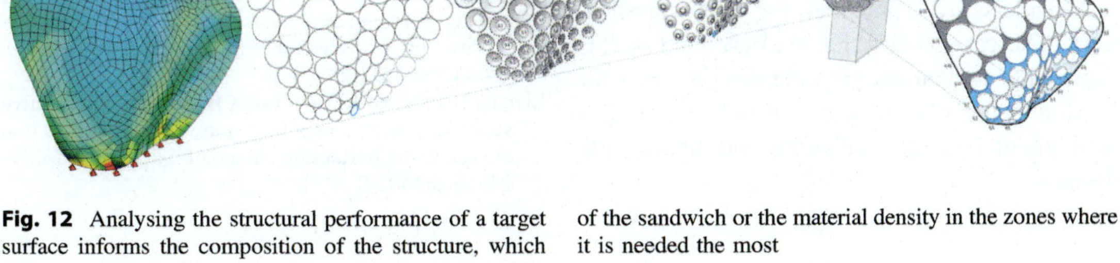

Fig. 12 Analysing the structural performance of a target surface informs the composition of the structure, which then gets optimized by either increasing the static height of the sandwich or the material density in the zones where it is needed the most

Fig. 13 The full-scale mock-up was built of PET-G plates with a thickness of only 0.5 and 0.75 mm

Conclusion

The evolution of bending-active plate structures and the associated form-finding and analysis techniques demonstrate the significant advancements that have occurred since Buckminster Fuller's first plydomes. The increasing availability of computing power and advancement of simulation tools have made it much easier to understand the complex interdependencies of bending-active structural systems as well as to master them for new designs. The integration of both geometric and material constraints from the start of the design process is a powerful approach that renders great potential for many applications. The case studies that are described in this paper indicate that the possible design space for bending-active plate structures is rich and offers a plethora of beautiful, efficient, and lightweight designs.

References

Fleischmann M et al (2012) Material behaviour: embedding physical properties in computational design processes. Archit Design 82:44–51. doi:10.1002/ad.1378

Grinspun E (2008) A discrete model of thin shells. Discrete Differ Geom 325–337. doi:10.1007/978-3-7643-8621-4_17

Knippers J et al (2010) Atlas Kunststoffe+Membranen: Werkstoffe und Halbzeuge, Formfindung und Konstruktion. Institut für Internationale Architektur-Dokumentation, München

Lienhard J et al (2011) Bending-active structures—research Pavilion ICD/ITKE. In: Proceedings of the international symposium of the IABSE-IASS symposium

Lienhard J et al (2013) Active bending—a review on structures where bending is used as a self-formation process. Int J Space Struct 28(3–4):187–196. doi:10.1260/0266-3511.28.3-4.187

Lienhard J et al (2014) Form-finding bending-active structures with temporary ultra-elastic contraction elements. Mob Rapidly Assembl Struct IV. doi:10.2495/mar140091

Lienhard J et al (2014) Bio-inspired, flexible structures and materials. Biotechnol Biomim Civil Eng 275–296. doi:10.1007/978-3-319-09287-4_12

Marks RW, Fuller RB (1973) Dymaxion world of Buckminster Fuller. Anchor Books, New York

Piker D (2013) Kangaroo: Form finding with computational physics. Archit Des 83(2):136–137. doi:10.1002/ad.1569

Richard BF (1959) Self-strutted geodesic plydome. US Patent 2,905,113, 22 Sept 1959

Schleicher S et al (2015) A methodology for architecture. Comput Aided Des 60:105–117. doi:10.1016/j.cad.2014.01.005

Form-Finding of Architectural Membranes in a CAD-Environment Using the AiCAD-Concept

Benedikt Philipp, Michael Breitenberger, Roland Wuchner and Kai-Uwe Bletzinger

Abstract

In the design process of membrane structures, the gap between design and analysis becomes one of the main challenges, since these structures require consistent interaction between both disciplines. This gap between engineering and design can often be identified between the environments in use, CAD and CAE. Various approaches to this problem have been proposed up to now, mainly focussing on easier and faster bridging of the gap. In the present contribution, a novel approach called *Analysis in CAD* (AiCAD) is proposed, which aims on omitting this gap instead of only reducing it. Therefore, all analyses are performed on the augmented CAD model without the need for a separate, specialized model for the structural analysis. In order to perform the analyses on this NURBS-based B-Rep model, the *Isogeometric B-Rep Analysis* (IBRA) has been developed. The application to the form-finding of architectural membranes demonstrates the concept and its implementation as a plugin to Rhinoceros 5. This implementation and the application to form-finding can be seen as prototypes, since the AiCAD approach is very general and can be applied to all fields of analysis and design.

Introduction and Motivation

Architectural membranes as tensile structures offer a unique language of shapes, mainly characterized by their curved silhouette, underlining their light-weight, efficient nature (Figs. 1 and 2). The shape of tensile structures is motivated by their core property to bear loads exclusively via tension, due to their prestressed state.

In the design process of architectural membranes three major steps can be identified: (i) the form-finding, (ii) the structural analysis, and (iii) the cutting pattern determination.

Form-finding has the task to determine the shape that fits the prescribed pre-stress state and given boundary conditions. Thus, membrane

B. Philipp (✉) · M. Breitenberger · R. Wuchner · K.-U. Bletzinger
Structural Analysis, Technische Universität München, Munich, Germany
e-mail: benedikt.philipp@tum.de

M.R. Thomsen et al. (eds.), *Modelling Behaviour*, DOI 10.1007/978-3-319-24208-8_6

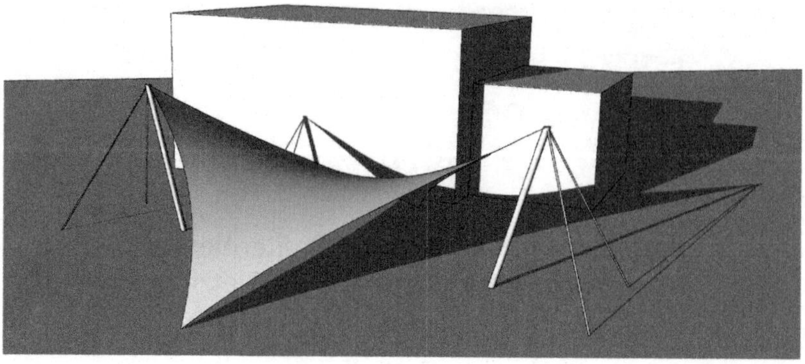

Fig. 1 Rendering of a form-found four-point sail with schematic surroundings and supporting structure (Philipp et al. 2014, 2015). The form-finding of this four-point sail has been performed within Rhinoceros using the presented plugin based on the AiCAD-approach (Breitenberger et al. 2015)

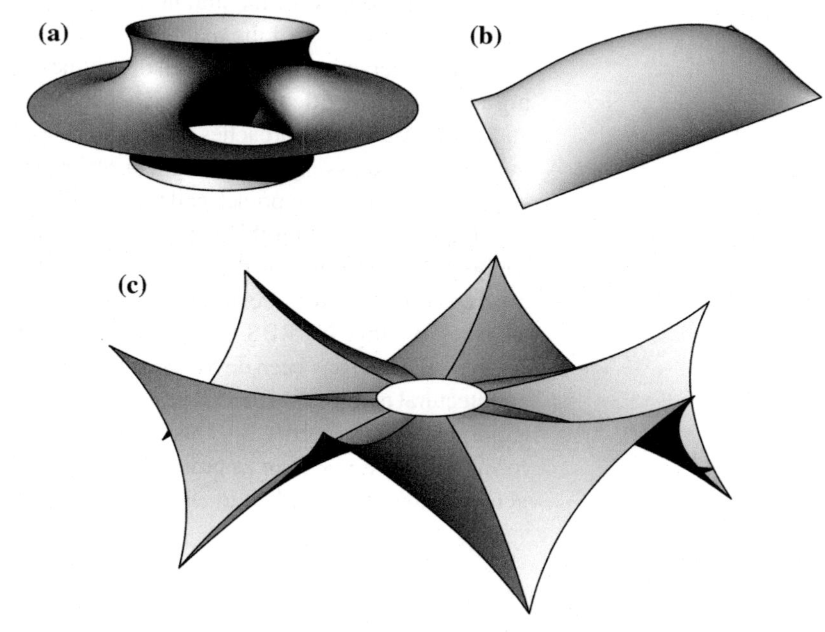

Fig. 2 Selection of shapes form-found with the presented framework: **a** Costa's minimal surface (slightly adapted) in the intermediate configuration. **b** Inflated cushion in a rectangular frame: Here internal pressure is considered during the form-finding. **c** Dancing fountain (Tanzbrunnen) by Frei Otto as an example for an actually built structure including edge, ridge and valley cables

structures are mechanically motivated surfaces (Bletzinger et al. 2010), in contrast to most other structures. The numerical determination of these surfaces (often free-form surfaces) encounters several challenges, mainly characterized by the non-unique parametrizations of the final surface. Various form-finding approaches have been developed to overcome this difficulty, like the dynamic relaxation (Wakefield 1999), adapted linearization techniques (Haug 1972) or the force density (Linkwitz 1999). In the present paper, the *Updated Reference Strategy* (URS) (Bletzinger and Ramm 1999) is used as form-finding approach.

Figures 1 and 2 show examples for shapes form-found with the present framework.

Form-finding usually is an iterative approach towards an architecturally desirable structure, altering between mechanical requirements and design aspects. The interlacing between design and structural engineering is extremely close. In this context, the gap between *Computer Aided Design* (CAD) and *Computer Aided Engineering* (CAE) presents a major obstacle: Multiple conversions and adaptations between different co-existing, specialized geometric models for these tasks require considerable amounts of resources and obviously are prone to errors and deviations (Hughes et al. 2005; Stimpfle 2013). Although these models are aimed to represent the same, "real" geometry, they may differ substantially.

The *Analysis in Computer Aided Design* (AiCAD) concept (Breitenberger et al. 2015) has been invented with the aim of consistently omitting this gap by using the same model for both CAD and CAE. The necessary mechanical analysis technique has been laid out with the *Isogeometric B-Rep Analysis* (IBRA) (Breitenberger et al. 2015). The present contribution shows the realization of the AiCAD concept for the form-finding as exemplary application. The paper is outlined as follows: "The AiCAD concept and its implementation" presents the AiCAD concept, reviews the IBRA as the necessary framework for structural analysis and describes the implementation in a CAD environment. "Form-finding as an application of AiCAD" shows the application of the AiCAD to the form-finding of structures and discusses its main features. Finally "Conclusions and outlook" gives conclusions and an outlook on future work.

The AiCAD Concept and Its Implementation

The *Analysis in CAD* (AiCAD) concept, initially proposed by Breitenberger et al. (2015) is based on using one common model for both the design as well as structural analysis, see "*The AiCAD concept*". In this case, the exemplary structural analysis application is the form-finding of architectural membranes. In order to provide the mechanical framework, the *Isogeometric B-Rep Analysis* (IBRA) has been developed, see "*The Isogeometric B-Rep Analysis (IBRA)*". The realization of the AiCAD concept within a commercial CAD environment is presented in "*Implementation in Rhinoceros*".

The AiCAD Concept

Most approaches towards a better CAD-CAE-integration aim on facilitating the automatic creation of a separate, specialized analysis model from the CAD model, with so-called "mesh-generation" as one major task. The AiCAD concept follows a fundamentally different philosophy: Unlike these mentioned approaches, the AiCAD concept follows the philosophy of using one single model that comprises the information for all needed purposes. Therefore the CAD model is enriched by the information necessary for the analysis without building up a separate model. It has to be pointed out that for performing the structural analyses all information is consistently stored within one single model, no separate finite element model is created. Figure 3 illustrates this idea of one model for various purposes.

The core of this single model is the CAD geometry. Since it is at the basis of all analyses, the CAD geometry provides the structure along which further information is added. In addition to geometry related information there might be abstract information that cannot be assigned to geometric entities but which are rather functional parameters like a required accuracy for structural analyses. Note the information which is not only related to one purpose, but used for different purposes: For instance the diameter of a cable is used in analysis, while it may also serve as basis for an automatic cost estimation based on the mass of steel applied.

As a well known example, the idea of AiCAD could be compared to the rendering functionality which is part of most common CAD environments: In addition to the pure geometry description, the CAD model also contains information that is related to the rendering. This information can be specifically related to geometric entities (e.g. a surface has some color or texture) as well as to the model in general (e.g. position and intensity of sources of light). When the rendering operation is launched, the related information from the model is "filtered" from the

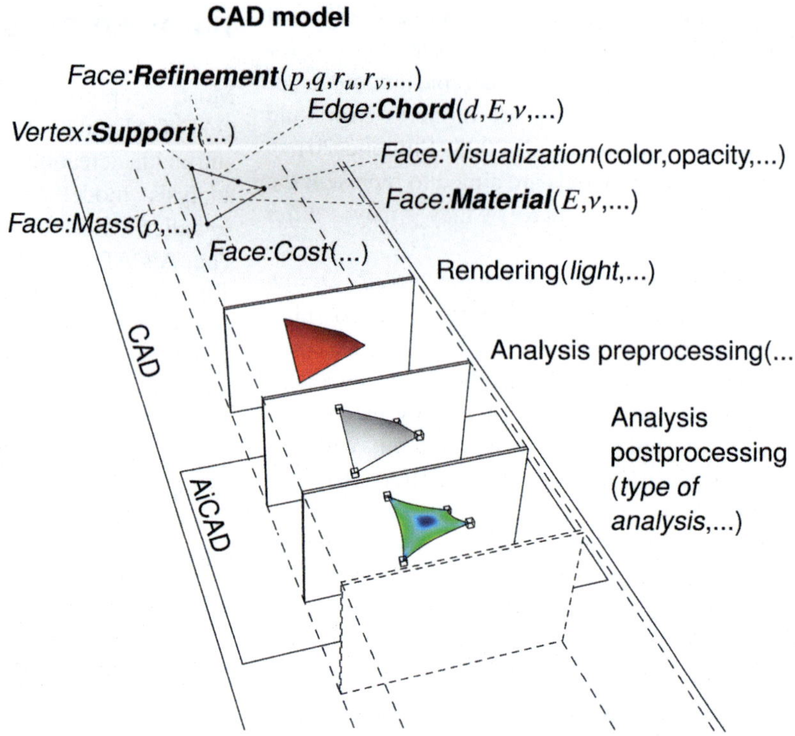

Fig. 3 Illustration (from Philipp et al. 2015) of the AiCAD-approach: The CAD model is enriched by information for various purposes (*top*). This information is structured along the B-Rep description of the CAD model. The AiCAD-approach (Breitenberger et al. 2015) interprets the analysis like "glasses" which filter out the required information for the structural analysis, in analogy to other operations like rendering

overall data of the model and the visualization is realized based on this information (Fig. 1). Figuratively, one could see the rendering as "glasses" which let pass the required information and transform it to the desired output. Nevertheless the underlying CAD model may contain other data as well, which eventually is not used in the current operation.

In analogy, in the AiCAD concept the CAD model is enriched by data related to structural analysis like support conditions, loads, Young's modulus E, a pre-stress σ0 or abstract information like a certain accuracy ε or load combinations. The operations of pre- and post-processing then refer to that information.

Together with the underlying CAD model this information is used in the analysis. In form-finding, the result of this analysis might be a displacement field **u** which leads from the initial geometry to the form-found shape, visualizing the "output" of the "analysis glasses".

At this point it is important to note that a broad variety of attempts to integrate structural analysis in CAD environments already exists, like *Kangoroo3D* (Piker et al.), *Karamba* (Preisinger et al.), *SmartForm* (BuroHappold) or *RhinoMembrane* (Bletzinger). However all of these approaches rely on an explicit FE adapted meshing of the CAD geometry. In contrast, the AiCAD concept directly performs structural analyses on the CAD model and again provides a CAD representation of the resulting shape, fully integrated in the CAD tool-chain.

A review of the necessary framework for the mechanical part of the analysis follows.

The Isogeometric B-Rep Analysis (IBRA)

The *Isogeometric B-Rep Analysis* (IBRA), introduced in (Breitenberger et al. 2015), is an analysis approach that enables structural analysis directly on the geometric representation of a B-Rep model. The displacements of the control-points of the geometry (Fig. 4) are taken as degrees of freedom for the mechanical system to be solved.

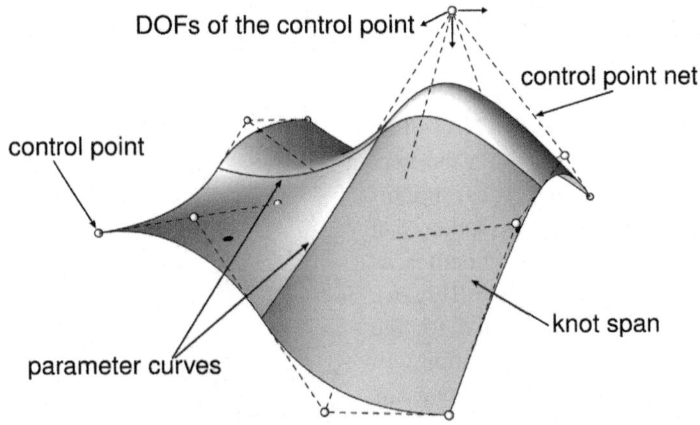

Fig. 4 Example for a NURBS-based surface for analysis with IBRA: The surface is controlled by the position of *control points*. The displacement of these control points is used as degrees of freedom (DOFs) in the structural analysis with IBRA. Thus, the isogeometric paradigm is followed, since the same philosophy of controlling a geometry is also used in CAD

B-Rep as Standard for Geometry Description

The *Boundary-Representation* (B-Rep) is a technique used in CAD to describe arbitrary geometrical entities by their boundaries. For a three dimensional object a set of adjacent bounded surface elements called faces describes the "skin" of the object and thus the object itself. The faces at their turn are bounded by sets of edges which are curves lying on the surface of the faces. Several edges meet in points called vertices. As basis functions, commonly non-uniform rational B-splines (NURBS) are used (Piegl and Tiller 1997). Hence using a NURBS-based B-Rep model, the description of arbitrarily trimmed geometries is possible.

IBRA as Generalization of the IGA Concept

The *Isogeometric B-Rep Analysis* (IBRA) (Breitenberger 2015) is a generalization of the *Isogeometric Analysis* (IGA) introduced by Hughes et al. (2005).

IGA uses the same basis functions for the computation of the discretized system that are used for the geometry representation in CAD (i.e. commonly NURBS). This approach was aimed at providing a first step to overcome the gap between CAD and CAE, introduced by separate specialized models for different purposes.

While IGA in its pure form is restricted to complete patches (Fig. 4), IBRA refers to the full B-Rep description from the CAD system. Therefore, IBRA provides the framework for a consistent mechanical interpretation of the geometrical trimming. Thus, also trimmed and coupled patches can be used. Moreover, the concept of IBRA allows enforcing various properties to the B-Rep entities. This can be seen as a new type of element, the so-called *B-Rep edge element*: Elements of this type permit enforcing e.g. coupling boundary conditions or supports, loads or even mechanical properties like cable elements with prestress. These elements are presented in detail in (Breitenberger et al. 2015; Philipp et al. 2015).

It should be pointed out that in order to perform a structural analysis with IBRA, only an augmented version of the underlying geometry model, enriched by the necessary information, is needed in order to provide an analysis suitable model. Thus, following the AiCAD concept, with the development of IBRA the gap between CAD and CAE can finally be omitted, not only reduced.

Implementation in Rhinoceros

The presented AiCAD approach has been realized as a plugin for *Rhinoceros 5* (McNeel et al. 2015) (Figs. 6 and 7). With this plugin the necessary properties like element properties, support conditions, loads or prestress information can directly be assigned to the CAD geometries as additional information in dialogue boxes (Fig. 6). All assigned data refers dynamically to the underlying geometry. Hence the information will also be kept for a modification of the geometry as it will usually be necessary within the design process. Figuratively spoken, a surface knows its texture or mechanical property, independent of eventual changes that are made to its geometrical description.

The analysis itself is performed based on the IBRA approach which is realized in CARAT++, the structural mechanics code at the authors' institute. CARAT++ is started from the plugin as an encapsulated process.

The visualization of the results of the analysis is also directly performed within the plugin. Thus the user stays within the CAD environment throughout all the design and analysis process. Following the AiCAD concept, the geometry stays an intact CAD geometry along all steps, just overlaid by additional information.

Form-Finding as an Application of AiCAD

For a better understanding of the AiCAD concept, the form-finding of a prototype five-point sail is demonstrated in detail "*Form-finding of a prototype five-point sail*" and a full-scale form-finding problem is shown in "*Form-finding of a full scale tent with integrated eye-cable*".

Form-Finding of a Prototype Five-Point Sail

For the form-finding of the five-point sail within Rhino, a starting geometry is modeled out of two NURBS surfaces, with one of the patches being trimmed (Fig. 5). The vertices (future high- and low-points) are defined at the desired positions. At this point it has to be emphasized that with IBRA the analysis can be performed directly on the trimmed NURBS patches.

The pure geometry model is augmented by the necessary mechanical information within the developed plugin (Fig. 6). The mechanical properties of the edge cables are assigned to the B-Rep edges and the surface is defined to be a membrane with isotropic prestress. Additionally the vertices are defined as fixed supports. Along

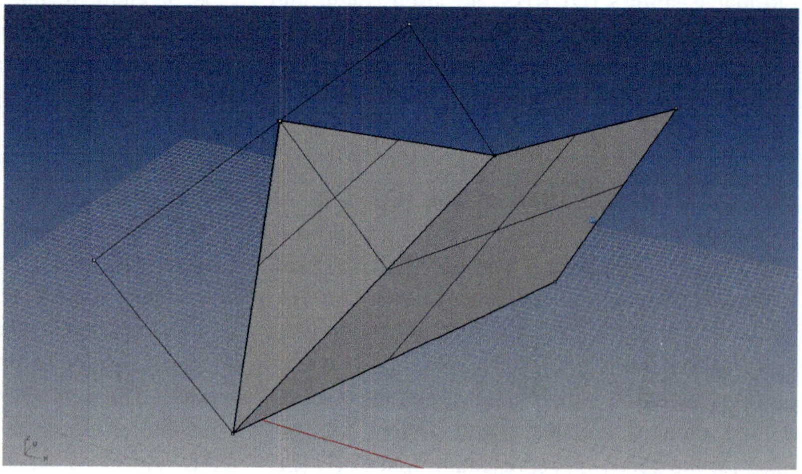

Fig. 5 Geometrical modeling of the initial geometry of the five-point sail out of two patches. Trimming of one edge yields the fifth edge. Along the common edge, the two patches have to be coupled

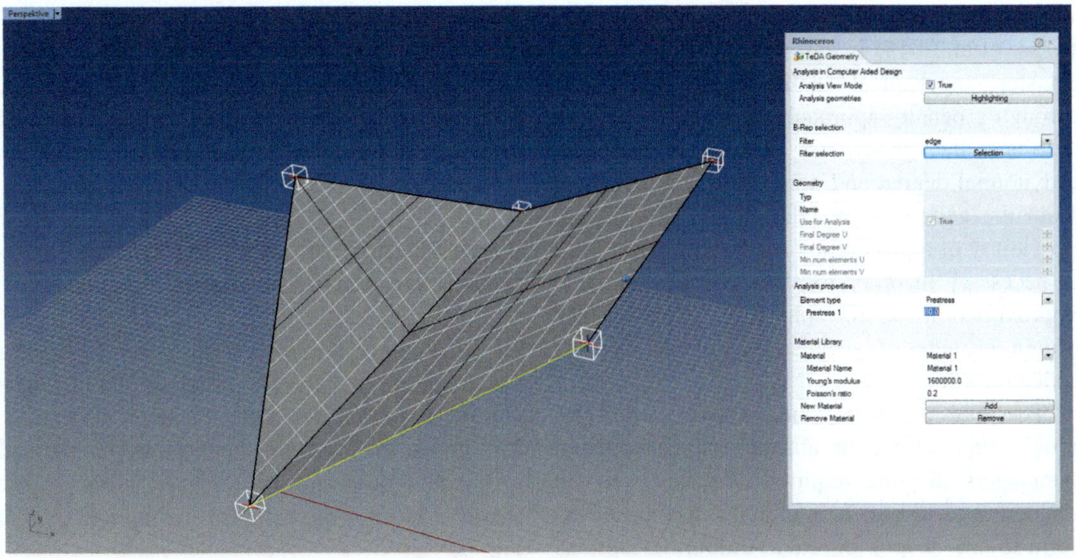

Fig. 6 Pre-processing for the form-finding of the five-point sail: Screen-shot of the analysis pre-processing in the developed Rhinoceros-plugin. Supports are applied at the edge points of the patches. The refinement that is necessary for a physically relevant analysis is visualized for the membrane. In the plugin's *dialogue box* the mechanical properties of the highlighted edge cable are specified

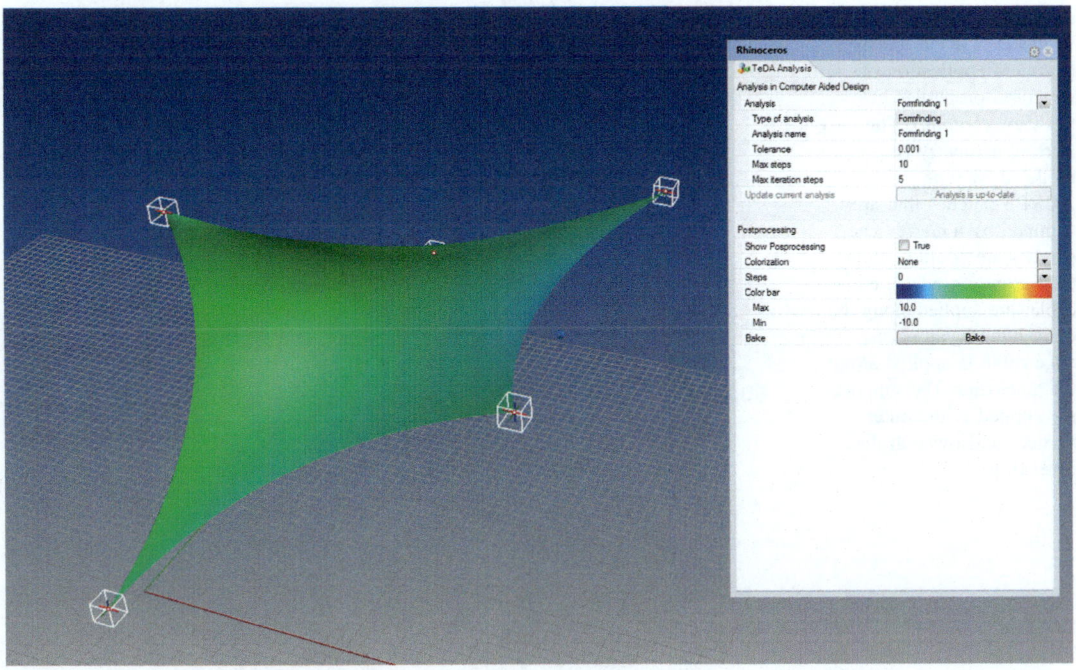

Fig. 7 Form-finding of a *five-point* sail: *Screen-shot* of the analysis post-processing in the Rhinoceros-plugin. The abstract analysis parameters are specified within the respective *dialogue box*, like the tolerance or number of form-finding steps. The surface is coloured with the principal stress

the coupling edge, "IBRA coupling" (Breitenberger et al. 2015) is enforced, which is able to transfer mechanical deflections and forces through a penalty approach.

Since the geometries with their standard polynomial degree and number of control points from the geometry modeling provide a limited possibility of representing geometries, refinement is necessary in order to allow for a reasonable approximation of the physical solution. The plugin provides a visualization of the applied refinement (Fig. 6).

The analysis itself is also started within the plugin after setting the abstract, analysis-related parameters like the required accuracy, see the dialogue box in Fig. 7. In the visualization of the form-found shape, different colouring options are available, e.g. principal stresses or displacements.

As a distinguishing feature of the AiCAD approach compared to other CADCAE-integration attempts, the result is still provided as a valid CAD geometry, all subsequent CAD operations can be performed with it.

Form-Finding of a Full Scale Tent with Integrated Eye-Cable

For the demonstration of the potential of the AiCAD approach, the form-finding of a full scale tent with an integrated eye-cable is performed. The geometric modeling and the application of mechanical properties is illustrated in Fig. 8. The shape resulting from the form-finding is shown in Fig. 9.

The example illustrates the applicability of the AiCAD approach for full scale structures.

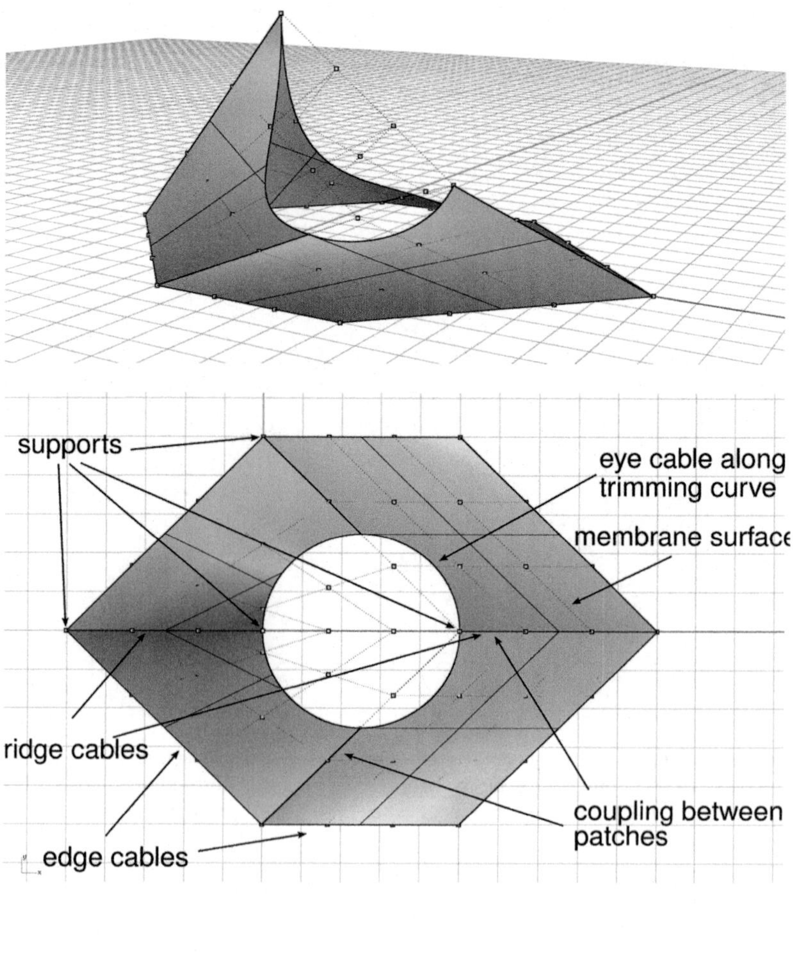

Fig. 8 Tent with an eye-cable: Geometry modeling within Rhinoceros in perspective (*top*) and application of structural properties in top-view (*bottom*). The starting geometry is modeled by four NURBS-patches that are trimmed by a *circle*. The patches are coupled along their common edges, edge cables are applied along the outer bounds, an elastic edge cable is applied along the inner ring. The supports are applied at the outer vertices and twice at the inner ring

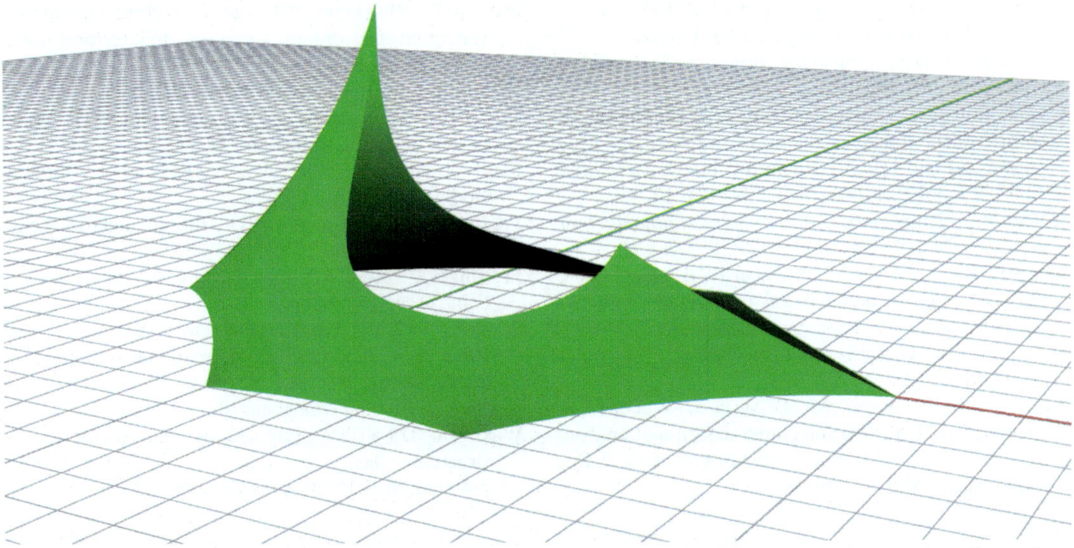

Fig. 9 Form-found tent with eye cable. The form-found shape is a CAD-geometry and can thus be consistently treated in further steps of design and manufacturing

Conclusions and Outlook

This contribution presents the realization of a membrane form-finding workflow using the Ai-CAD concept. For the analysis, the recently introduced IBRA-approach has been briefly reviewed. The presented implementation allows performing accurate membrane form-finding within a plugin to Rhinoceros, without the need for a separate simulation environment or even a separate FE mesh while consistently keeping the same NURBS-based B-Rep model.

The presented concept is very general and can be directly extended to other types of analysis and other types of elements. In the context of tensile structures, the developed framework shall be extended by structural analysis. Introducing elastic, bending active elements and a staging analysis would open the way towards hybrid structures which recently gained more attention (Lienhard 2014; Philipp and Bletzinger 2013; Philipp et al. 2015; Dieringer et al. 2013).

Acknowledgements The support of the *Deutsche Forschungsgemeinschaft* (DFG) under various grants is gratefully acknowledged.

References

Bletzinger KU, D'Anza G, Linhard J. RhinoMembranes. http://www.ixforten.com/rhinomembrane.htm. Accessed 24 June 2015

Bletzinger KU, Ramm E (1999) A general finite element approach to the form finding of tensile structures by the updated reference strategy. Int J Space Struct 14:131–145. doi:10.1260/0266351991494759

Bletzinger K-U, Firl M, Linhard J, Wuchner R (2010) Optimal shapes of mechanically motivated surfaces. CMAME 199:324–333. doi:10.1016/j.cma.2008.09.009

Breitenberger M, Apostolatos A, Philipp B, Wuchner R, Bletzinger KU (2015) Analysis in computer aided design: nonlinear isogeometric B-Rep analysis of shell structures. CMAME 284:401–457. doi:10.1016/j.cma.2014.09.033

BuroHappold. SmartForm http://www.smart-solutions-network.com/page/introduction-tosmartform. Accessed 24 June 2015

Dieringer F, Philipp B, Wuchner R, Bletzinger KU (2013) Numerical methods for the design and analysis of hybrid structures. Int J Space Struct 28:149–160. doi:10.1260/0266-3511.28.3-4.149

Haug E (1972) Finite element analysis of nonlinear membrane structures. Dissertation, University of California

Hughes TJR, Cottrell JA, Bazilevs Y (2005) Isogeometric analysis: CAD, finite elements, NURBS, exact geometry and mesh refinement. CMAME 194:4135–4195. doi:10.1016/10.1016/j.cma.2004.10.008

Lienhard J (2014) Bending-active structures. Dissertation, University of Stuttgart

Linkwitz K (1999) Formfinding by the "Direct Approach" and pertinent strategies for the conceptual design of prestressed and hanging structures. Int J Space Struct 14:73–87. doi:10.1260/0266351991494713

McNeel et al. Rhinoceros 5.0. http://www.rhino3d.com/. Accessed 24 June 2015

Philipp B, Breitenberger M, Wuchner R, Bletzinger K-U (2014) Design of architectural membranes with isogeometric elements. Paper presented at the 11th world congress on computational mechanics, Barcelona, 20–25 June 2014

Philipp B, Breitenberger M, D'Auria I, Wuchner R, Bletzinger K-U (2015) Integrated design and analysis of structural membranes using the isogeometric B-Rep analysis. CMAME (submitted)

Philipp B, Bletzinger K-U (2013) Hybrid structures—Enlarging the design space of architectural membranes. J IASS 54:35–49

Philipp B, Dieringer F, Wuchner R, Bletzinger K-U (2015) Form-finding and structural analysis of the design of hybrid structures. J IASS 56:17–24

Piegl L, Tiller W (1997) The NURBS book. Springer, Berlin

Piker D et al. Kangoroo3D. http://kangaroo3d.com. Accessed 24 June 2015

Preisinger C. et al. Karamba3D. http://www.karamba3d.com/. Accessed 24 June 2015

Stimpfle B (2013) From NURBS to textile architecture. Paper presented at the 6th international conference on textile composites and inflatable structures. Technische Universität Munchen, Munich, 9–11 Oct 2013

Wakefield DS (1999) Engineering analysis of tension structures: theory and practice. Eng Struct 21:680–690. doi:10.1016/S0141-0296(98)00023-6

Balancing Behaviours—Designing with Combinatorial Equilibrium Models

Patrick Ole Ohlbrock

Abstract

The present research tries to provide a theoretical framework (named Combinatorial Equilibrium Modelling), based on Graphic Statics and Graph Theory, which allows to shape and explore spatial equilibrated structures that are not restricted to typologies but are constituted by combinations of tension and compression forces. The focus of this paper is to demonstrate with the help of three different design scenarios how this novel interactive modelling approach can be used to negotiate between the structures different behavioural properties.

Introduction

An architectural design process is characterized by a constant negotiation of different physical and non-physical aspects. The goal of structural design is generally to harmonize the structures form and its inner stresses. Since the separation between the disciplines of architects and engineers, started during the industrial revolution in the nineteenth century, analytical methods became more and more the basis for the design and control of the balance between form and inner stresses (Schwartz 2011, p. 243). On the one side of the current practice, numerical methods such as FEM are used to analyse and evaluate the inner stresses for given forms. On the opposite side, numerical form-finding methods [for example the "Force-Density" method (Scheck 1974)], are used to find the form for given stress conditions. Both approaches share a one way—sequential process which generally makes a negotiation hardly to happen (Au 2012, p. 237). An exception is RhinoVault (Rippmann et al. 2012), which is based on the force density method respectively TNA (Block 2009). This tool allows to actively balance form and force simultaneously to a certain extent. A similar approach has been proposed by the Interactive Equilibrium Modelling (Lachauer 2015). A visualization of the relationship between form and inner stresses that also allows for an active manipulation is the key for a synthetic design process (Kotnik and D'Acunto 2013, p. 195). As observed by Akos Moravanszky: "*Like Culmann in his "graphic statics", Max Bill thinks that the*

P.O. Ohlbrock (✉)
ETH Zurich, Zurich, Switzerland
e-mail: Ohlbrock@arch.ethz.ch

M.R. Thomsen et al. (eds.), *Modelling Behaviour*, DOI 10.1007/978-3-319-24208-8_7

key to solving the problems of sciences lies in an adequate and precise visualization of abstract processes of thought." (Moravanszky 2008, p. 129).

Graphic Statics has demonstrated its significant potential for a visual control of the relationship between a structural shape and its inner stresses. This relationship, which describes the equilibrium state of structures, is expressed in a geometrical dependency between two diagrammatic models: the form diagram which describes the geometry of the inner forces (representing the resultants of the inner stresses) and the force diagram which illustrates the intensity of the corresponding forces. According to Jasienski, the main problem of the few available 3D design tools which are based on Graphic Statics is their limitation to typologies, mainly to shell structures that can be described by 3d curved surfaces. There is a need for a new development which also allows a geometry to emerge that is out of plane and is not restricted to any typology (Jasienski 2014, p. 1).

The present research tries to provide a theoretical framework (named Combinatorial Equilibrium Modelling), based on Graphic Statics and Graph Theory, which allows to shape and explore spatial equilibrated structures that are not restricted to typologies but are constituted by combinations of tension and compression forces (Ohlbrock 2015). The focus of the present paper is to demonstrate how this approach can be used to negotiate between the structures different behaviours. The paper is organized in four sections. The first section briefly summarizes different contemporary developments in the field of Graphic Statics, and briefly describe the technical background of the Combinatorial Equilibrium Modelling approach. This will be followed by a description of external boundary conditions and internal behaviour conditions that can be balanced in a structural design process based on this modelling approach. With the help of three different case studies, the subsequent section demonstrates the potential of this approach; and in the last section concludes the results of this paper.

Technical Framework

Graphic Statics

Graphic Statics has been formalized in the 19th century as a method for the analysis of structures by Culmann (1864). Based on concepts of vector equilibrium and reciprocal diagrams, the method allows to evaluate the inner forces of a structure with the help of geometrically constructed force diagrams.

Recently there has been renewed interest in Graphic Statics because of its power and simplicity. In the realm of planar systems, Gronewolds ActiveStatics (Greenwold and Allen 2003) and Equilibrium by Van Mele et al. (2009–2012) are illustrative examples for contemporary computer aided tools, in which the user can didactically experience the reciprocal relationship between form and force. Nevertheless they are restricted to relatively simple and predefined typologies (Mueller 2014, p. 30). For spatial systems, Block for instance developed Thrust Network Analysis (TNA), an approach in which an algorithm finds equilibrium for compression-only vaults based on the designer manipulations on the support location and on the planar projection of the force diagram (Block 2009). Laffranchi proposed a method for the determination of spatial funicular polygons for the design of curved bridges (Laffranchi 1999). Lachauers Interactive Equilibrium Modelling overcomes compression and tension-only typologies but has the limitation that the user needs to have a certain background in the field of equilibrium modelling, to ensure that the algorithm is able to find equilibrium (Lachauer 2015, p. 128). In general the developed spatial approaches are either fixed to specific typologies such as shells or bridges and/or they require a deep knowledge in the field of equilibrium modelling. It is the aim of the present research to overcome the described shortcomings and develop another proposal for the category of "full 3 D approaches" described by Jasienski (2014).

Representation of CEM

Combinatorial Equilibrium Modelling (CEM) is based on the Theory of Plasticity and its Lower Bound Theorem (Ohlbrock 2015). The latter allows to focus on equilibrium solutions and thereby makes it possible to visualize the force flow with the help of Graphic Statics (Muttoni et al. 1997). Additionally the design process is supported by representation techniques derived from the Graph Theory.

A character of CEM is that the visualization of the quantitative reciprocal relationship between form and force diagram (Fig. 1, right side), which is peculiar to Graphic Statics, is extended to the representation of the qualitative behaviour (Fig. 1, left side). This qualitative behaviour is characterized by a force flow diagram and a force state diagram. The force flow model, describing the topological connectivity and the flow direction of the inner forces, is visualized with a flow network with force connection nodes V and directed force edges e; the term "force flow" is here used not to represent a dynamic effect, but describes the sequence of forces from the source L_i (input of a load) to the sink S_i (support). The force state diagram, visualizes whether the force edges are in compression or in tension. Conventionally, the diagram is coloured red for tension, blue for compression and green for external forces (loads, support forces) (Muttoni et al. 1997).

Based on the Combinatorial Equilibrium Modelling approach, different possible applications for the conceptual design phase are conceivable.

Structural Design Process with CEM

A CEM-based negotiation process between form and force and its parameters depends on specific external conditions and desired internal behaviours imposed on the design. Those conditions can be subdivided in two groups (Fig. 2). On the one hand are the external boundary conditions, which comprise the structures boundary geometry (I), the location of the support (II) and the load scenario (III). An aspect introduced by this research, is to consider the support locations not

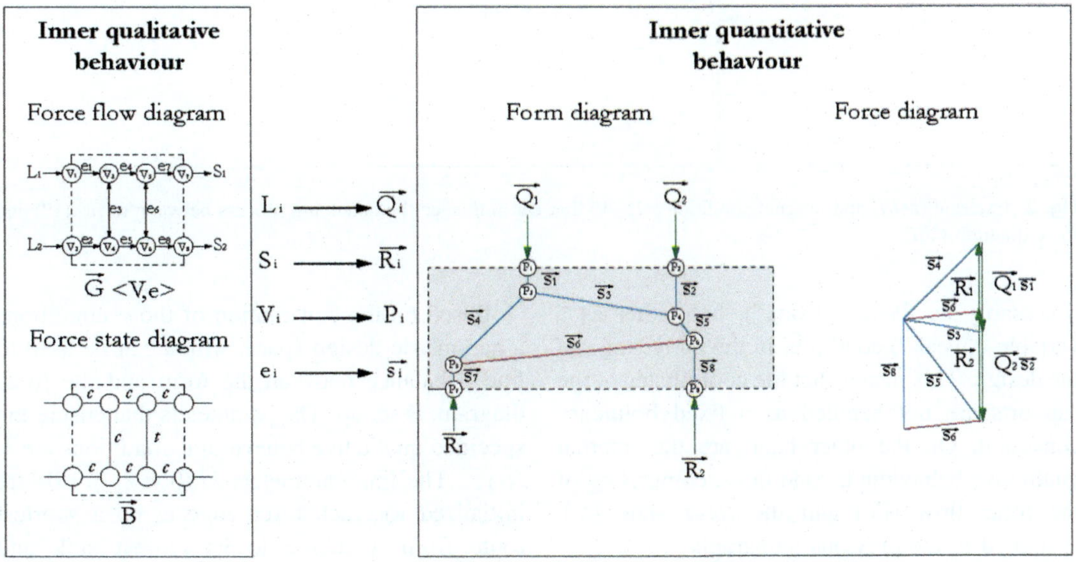

Fig. 1 Force flow-, force state-, form- and force-diagram representing the structures qualitative and quantitative behaviour. A force edge e_i in the qualitative diagram implies a force vector $\vec{s_i}$ in the quantitative diagrams. A load L_i in turn a load vector $\vec{Q_i}$ and a support S_i a vector for the reaction force $\vec{R_i}$

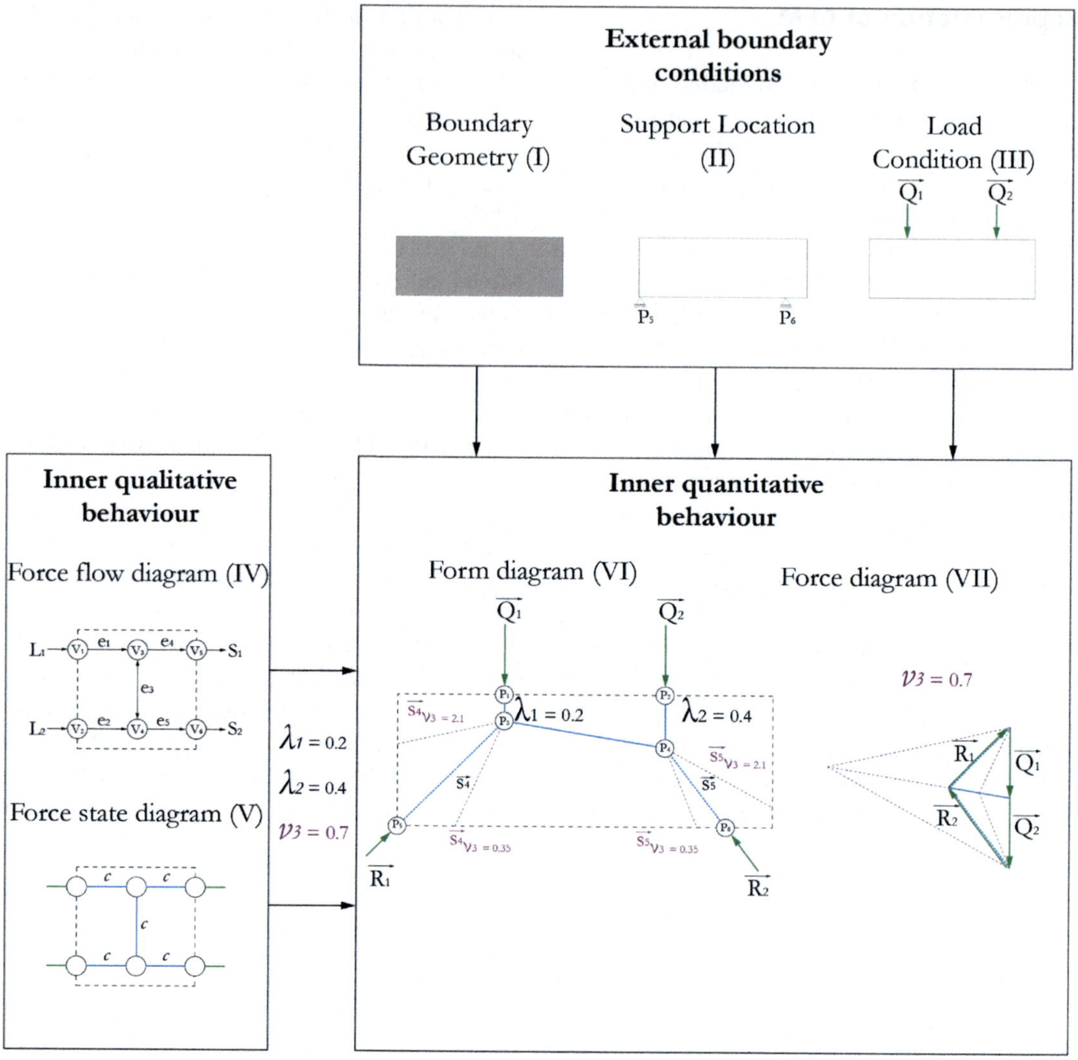

Fig. 2 External (*I–III*) and internal conditions (*IV–V*) that can influence the balancing process between form- (*VI*) and force-diagram (*VII*)

necessarily as fixed ("strong"), but rather as a variable ("weak") condition in the early stage of the design. This means that the coordinates of the supports are not handled as a fixed boundary constraint. On the other hand are the internal qualitative behavioural conditions, comprising of the force flow (IV) and the force state (V), described in the previous paragraph.

Based on the constitution of those conditions a parametric design space, which can be used to find a balance between the form and the force diagram, is set up. The parameters that ensure the specified qualitative behavioural conditions are λ_s & v_{ns}. The line parameters λ_s $(0 < \lambda_s < 1.0)$ are initialized for each force edge e_s on a shortest route from a source to its closest sink and describe the location of the consecutive node along the line of action from the current node ($\lambda_s = 0$) to the intersection point of the line of action of the input force and the boundary geometry of the structure ($\lambda_s = 1.0$). The force magnitude factors v_{ns} ($0 < v_{ns} < \infty$) are initialized for each force edge e_{ns} that is not on the shortest routes and describe the factors that are multiplied with the input forces. For the balancing process visualized in Fig. 2, two line parameters (λ_1 for e_1 and λ_2 for e_2) and one force magnitude parameters (v_3 for e_3) describe the parametric design space. A variation of v_3 and the consequences for the magnitude and the directions of the consecutive forces ($\overrightarrow{s_4}$, $\overrightarrow{s_5}$, $\overrightarrow{R_1}$ and $\overrightarrow{R_2}$) is illustrated in Fig. 2 in purple. The force state defines whether the forces in the form diagram are regarded as tensile or compressive forces. The 3-D case follows exactly the same rules as the described 2-D case in Fig. 2. Hence, within this approach it is possible to model 3-dimensional pin-jointed structures, resulting in solely axial forces in each member. The generated structures are in equilibrium for the specified load case.

To make the potential and the advantages of this balancing process visible, three different design scenarios are described in the next paragraph. They have been developed with the help of a computer-aided design tool within the software *Rhinoceros*. The required code for the tool is written in the scripting environment *RhinoPython*.

Case Studies

Stress Analysis

The first design scenario corresponds to a conventional stress-analysis procedure. In this design scenario, the topology of the force flow (IV) as well as the boundary geometry (I), the form of the force flow (VI), the load vector (III) and the supports locations (II) are fixed (grey shaded boxes). What is unknown and variable is the force state (V) and the magnitude (VII) of the inner forces, represented in the force diagram. Due to the pre-defined form of the force flow it is not really a negotiation, but rather a one way process from form to force (Fig. 3).

Based on this setup of conditions, the user can find out that this desired force flow geometry of the structure is realizable with a combinatorial force state sequence of e_1–e_4 = compression, e_5–e_{10} = tension and e_{11}–e_{16} = compr. (Fig. 4, upper part) in comparison with the "cantilever" combination of e_1–e_4 = compr., e_5–e_6 = tension and e_7–e_{16} = compr. as shown in the lower part of Fig. 4. In this scenario all the parameters in one layer are identical (e.g. $\lambda_1 = \lambda_2 = \lambda_3 = \lambda_4 = \lambda_{1-4}$).

By varying the four initialized parameters the user will furthermore realize that the structure will be balanced with the values set to $\lambda_{1-4} = 0.6$, $v_{5-6} = 0.9$, $\lambda_{7-10} = 0.8$ and $v_{11-12} = 0.67$ (Fig. 6) in comparison to the initial values represented in Fig. 5 where the last force magnitude factor v_{11-12}, which is responsible for the redirection, is too low.

Form-Finding

In this scenario, which can be regarded as a form-finding approach, the following conditions are defined by the designer; the boundary geometry (I), describing the volume of a roof design; the force flow (IV) and the force state (V) reflecting a compression-focused force flow with only six members in tension and a load vector (III) that characterises a vertical dead-load scenario. The peculiarity of this scenario is that the support location (II) is handled as a weak condition (Fig. 7, dashed red box). This means that the designer formulates desired support conditions, which are then tried to be fulfilled during the form-finding-process. For that purpose

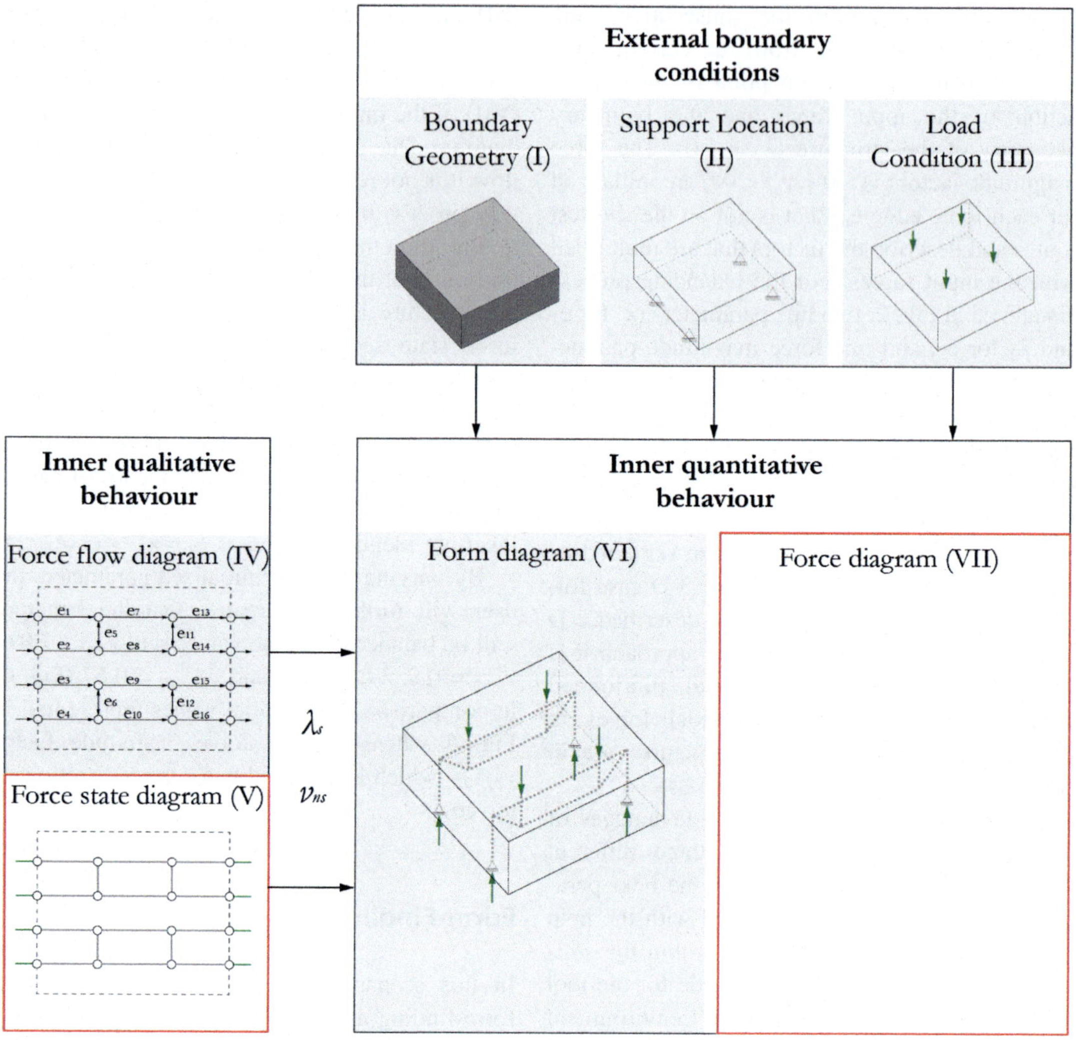

Fig. 3 Fixed conditions (*I*, *II*, *III*, *IV* and *VI*) and two variable conditions (*V* and *VII*) for a stress-analysis procedure

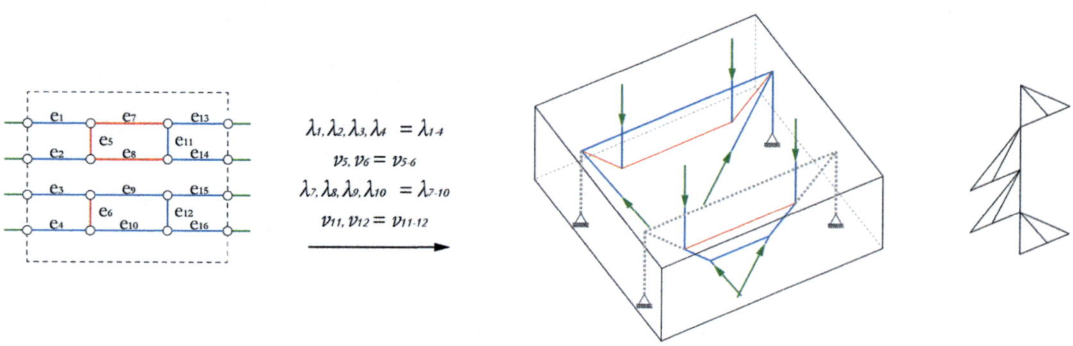

Fig. 4 Force state-, form- and force-diagram for two different behaviours and identical λ_s & ν_{ns} for each "layer"

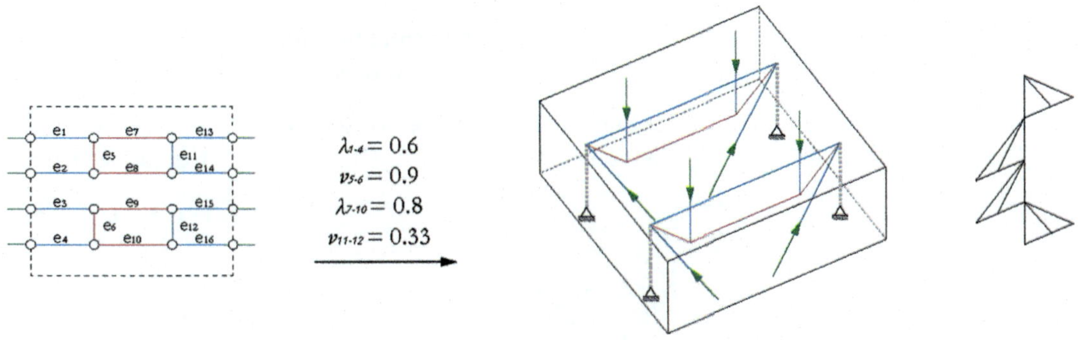

Fig. 5 Resulting force state-, form- and force-diagram for initial choices of λ_s & ν_{ns}

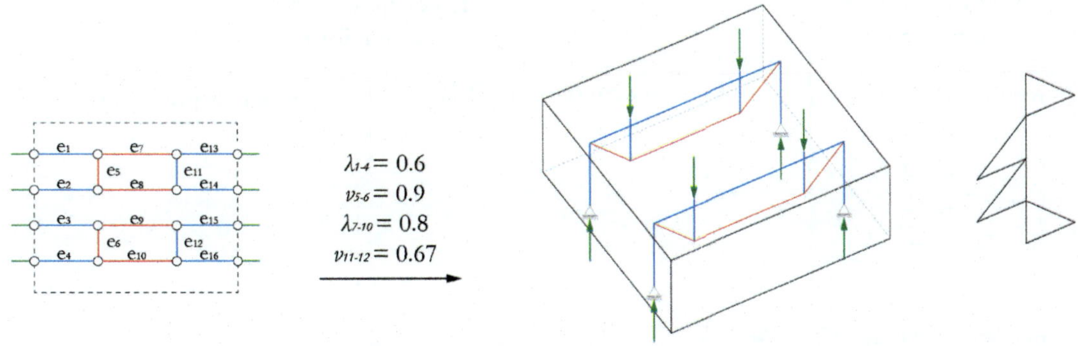

Fig. 6 Resulting force state-, form- and force-diagram for final choices of λ_s & ν_{ns}

an optimization process is used to minimize the *residual sum r* of the distances between the current support locations P_i^c to their desired support locations P_i^d. In this scenario the goal was formulated in such way that pairs of two adjacent vertices (e.g. V_{25} and V_{26}) in the force flow (IV) should coincide in their geometric support location (II) (e.g. P_{25-26}).

With the help of an evolutionary optimization algorithm (*Galapagos*), the balancing process between form and force is then carried out passively by the computer and stopped if the residual *r* is smaller than the tolerance value ε. The result of this process is the spatial form diagram below (Figs. 8 and 9), which could be easily used as the starting point to the design of a shell (Fig. 10).

Active Design

With this last balancing scenario, this paper demonstrates the full potential of the CEM approach. The boundaries of the geometry (I) reflect the designers aim to design a stadium roof. The force flow topology (IV) has more than 30 members. The force state (V) reflects a combinatorial transformation from a typical spoked-wheel system (Bergermann et al. 2000, p. 598). The difference to this often used structural system is that one of two outer compression rings is not supported by a column, but is pulled with tension force edges. The load vector (III) is again reflecting a vertical exposure for the structure. Unlike the previous balancing scenarios, no specific desired support location is

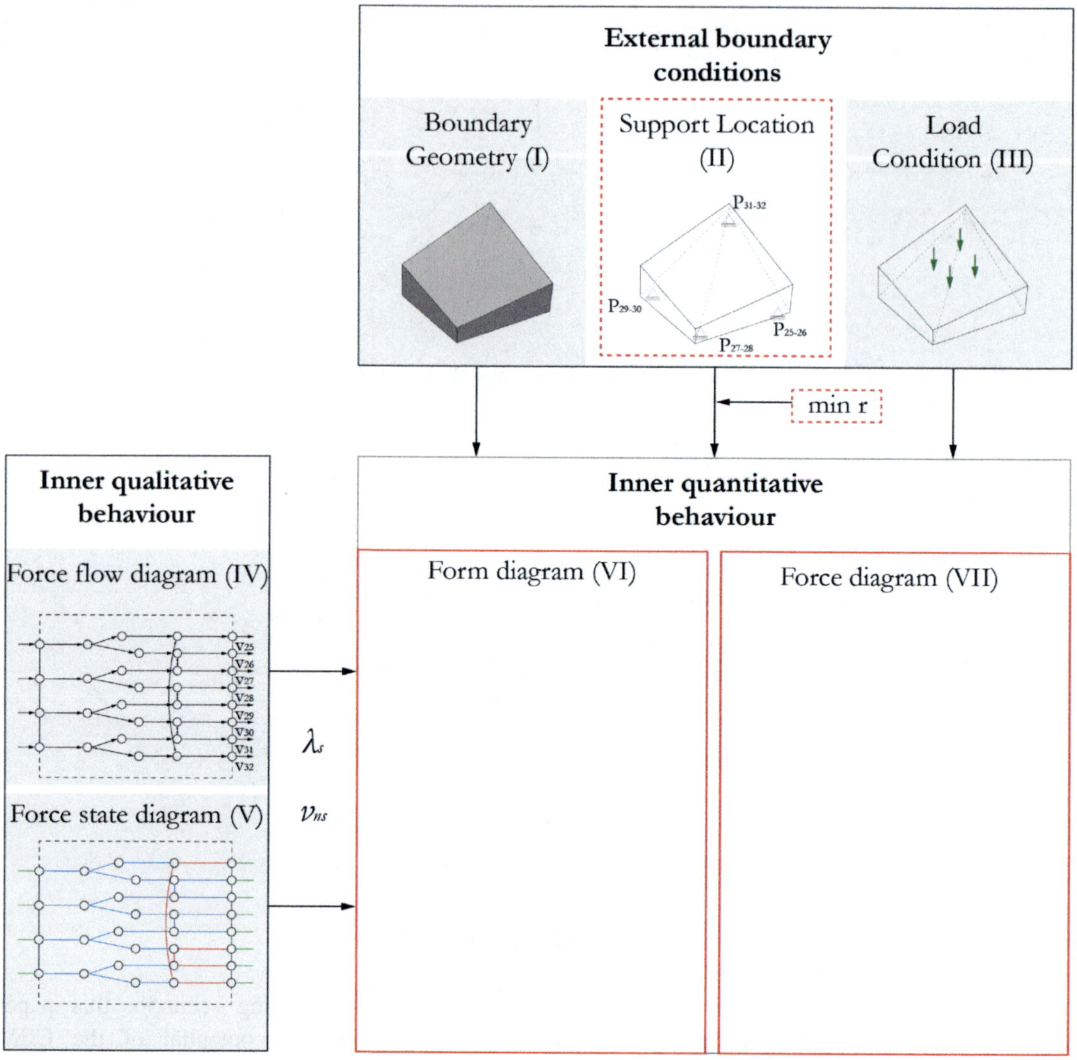

Fig. 7 Four fixed, "strong" conditions (*I*, *III*, *IV* and *V*) and one weak condition (*II*) for a form-finding process, carried out by a minimization of the residual objective r

defined. The only desired requirement is that the support is located on the ground of the boundary geometry (grey area in diagram II). Based on these user-defined conditions, the designer can then actively vary the arising metric parameters λ_s & ν_{ns} until the spatial form and force diagram in real-time reach a desired state. If both diagrams do not reach such a state for the set conditions, the designer can change the boundary geometry or the inner qualitative behaviour with ease and start the metric variation again. Hence, CEM supports a non-linear, chronology free and interactive balancing process between the seven external and internal conditions of a structure. Whatever conditions set by the designer, equilibrium is not a result of an analytical process, but from the beginning guaranteed. The presented formulation might therefore also allow

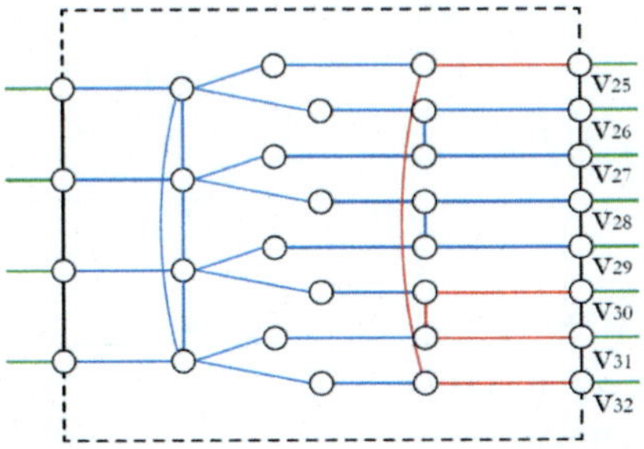

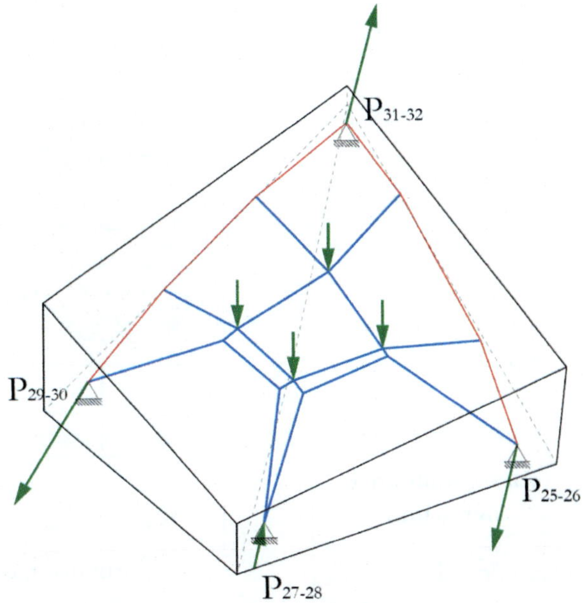

Fig. 8 Force state diagram (*left*) and an axonometry of resulting spatial form diagram (*right*)

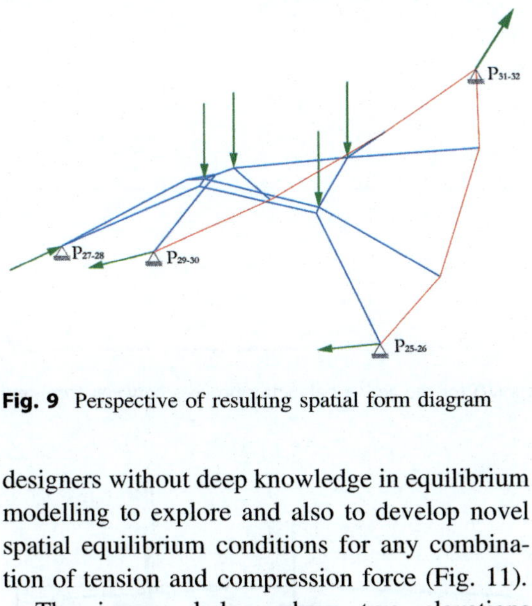

Fig. 9 Perspective of resulting spatial form diagram

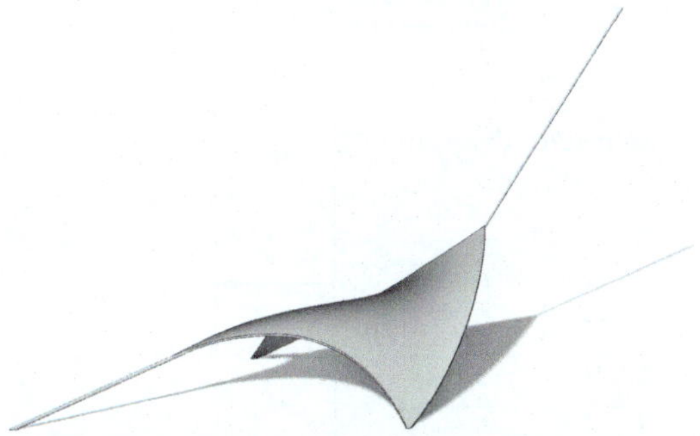

Fig. 10 Rendering of a shell structure that is designed according to the form diagram

designers without deep knowledge in equilibrium modelling to explore and also to develop novel spatial equilibrium conditions for any combination of tension and compression force (Fig. 11).

The images below show two elevations (Fig. 12) and an axonometry (Fig. 13) of a possible force flow result of this active balancing process. The metric parameters λ_s & ν_{ns} are kept identical at each layer of the force flow. The two tension forces at the longitudinal side of the stadium could be connected with an additional force member underneath the pitch. In an additional step the resolution (the amount of elements) of the force flow is increased to get a more detailed impression of the spatial force flow (Fig. 14). In a last step one of the many possible interpretations of this spatial force flow is visualized on the last page (Fig. 15).

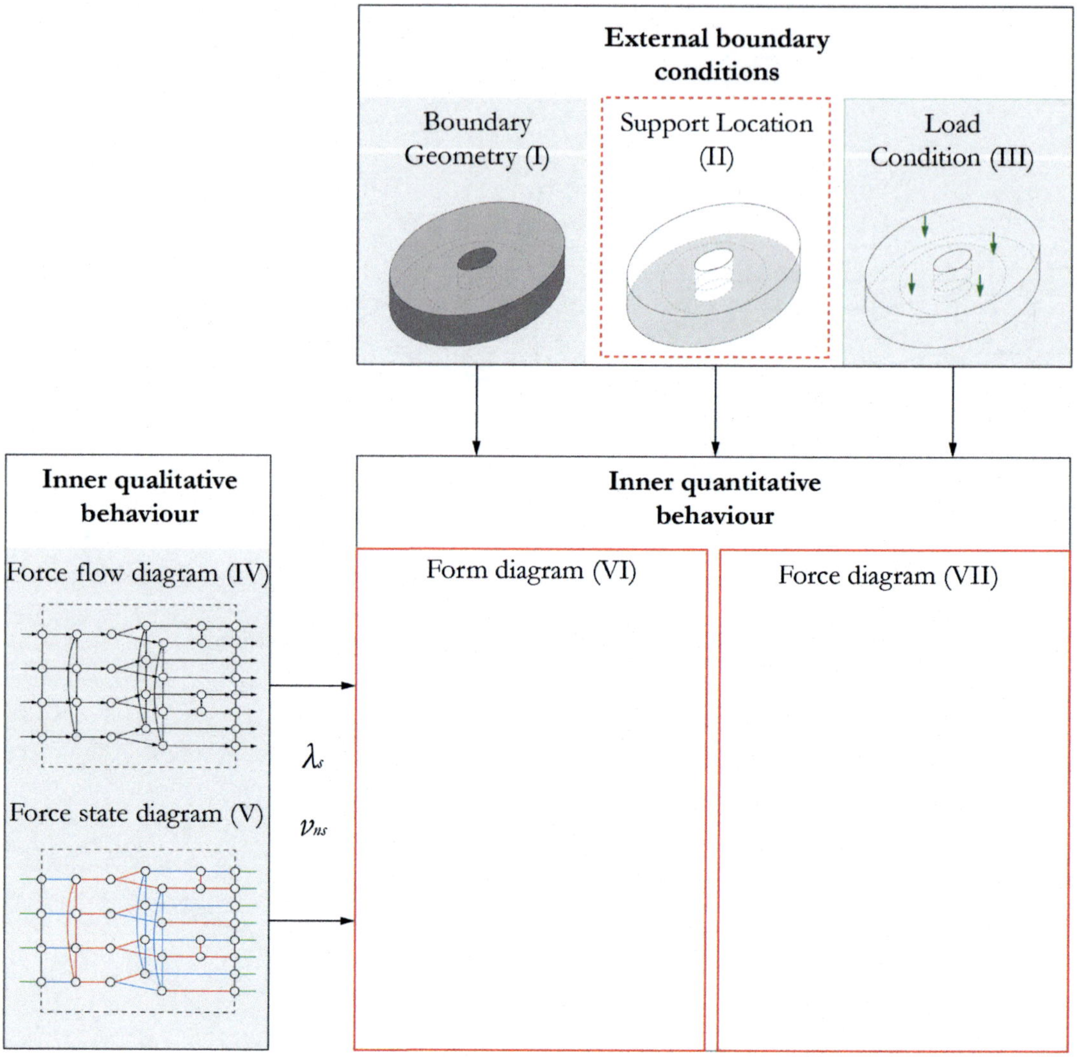

Fig. 11 Four fixed (*I*, *III*, *IV* and *V*) and one variable aspects (*II*) for the active balancing process between form and force diagram carried out by the designer

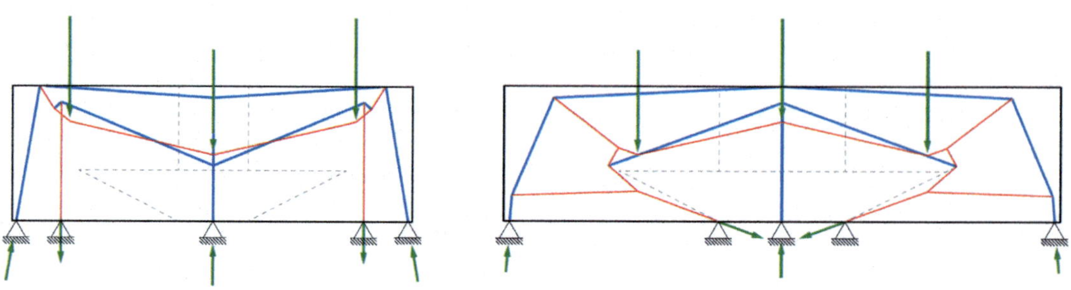

Fig. 12 Elevation of designed spatial form diagram

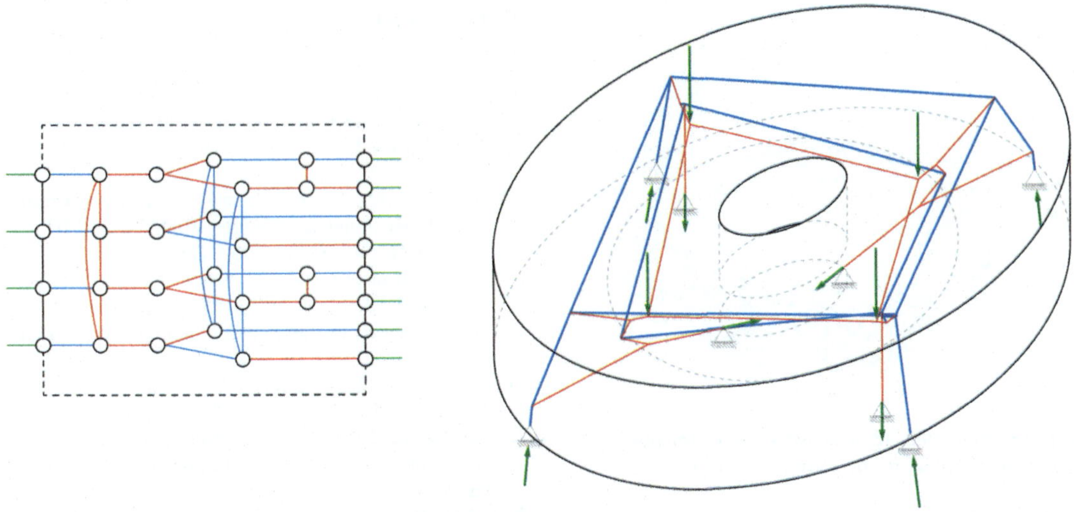

Fig. 13 Force state diagram (*left*) and corresponding form diagram (*right*)

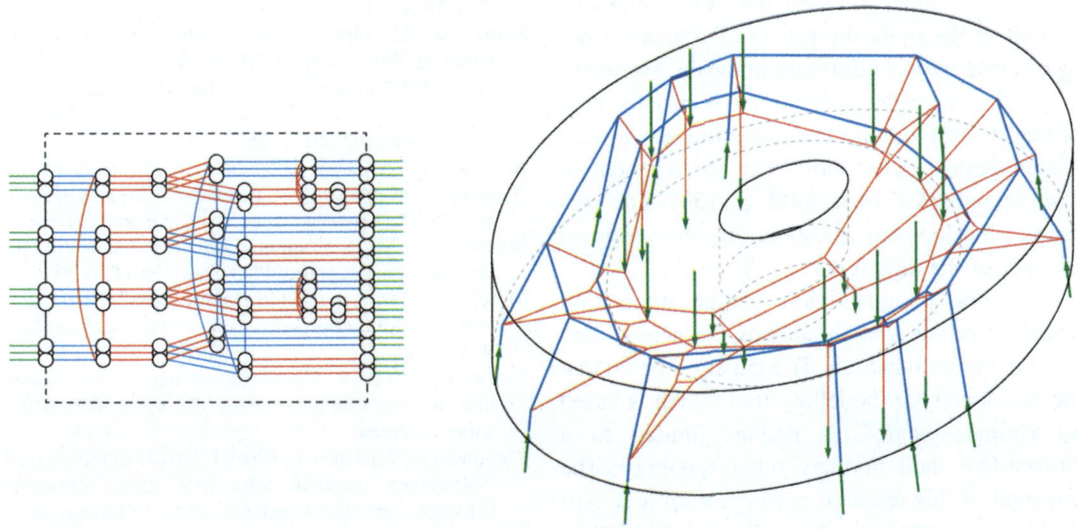

Fig. 14 Force state diagram (*left*) and corresponding form diagram (*right*) with higher resolution (more elements)

Conclusion

There are three main advantages of including the Combinatorial Equilibrium Modelling approach in the conceptual design phase. Firstly CEM allows an interactive spatial negotiation process between form and force to take place, based on robust geometric operations (by means of Graphic Statics) and visual representations only. Secondly it allows quickly to impose and thereby compare different internal qualitative behaviours for the same external design conditions, being at the same time not restricted to any formal nor structural typology (Ohlbrock 2015). The third and main difference to many other computer-aided design processes is that the designer can, to a certain extent, actively shape

Fig. 15 Rendering of a possible roof structure that is designed according to the form diagram in Fig. 14

form and force simultaneously on top of the two more conventional approaches of stress-analysis and form-finding. According to Lesek a design process is more influencing the cognitive awareness the more the process is characterized by errors, personal effort and intensity. However, in the digital realm, a computer is usually conceptually used to minimize this effort in "*a never disappointing expectation that he will do so without errors*" (Lesek 2008, p. 153). The presented development question this conventional concept of the digital era.

With the demonstration of the three case studies it is the goal to address the needs identified in the introduction. The paper rendered that the developed 3D modelling tool which is based on Graphic Statics, is neither limited to a surface-like shell nor any other typology. The potential of this research is to open up new replicable and synthetic ways to negotiate the formal and the structural behaviour in the early conceptual design phase in an active way.

Acknowledgments I would like to thank Pierluigi D'Acunto for his suggestions and critical comments during the development of this work and for his careful and patient editing.

References

Au F (2012) A necessary resistance within architect-engineer collaboration, in interdisciplinary design—new lessons from architecture and engineering. p 237

Block P (2009) Thrust network analysis, exploring three-dimensional equilibrium. PhD-Thesis, MIT

Bergermann R, Göppert K (2000) Das Speichenrad – Ein Konstruktionsprinzip für weitgespannte Dachkonstruktionen. Stahlbau 69(8):598

Culmann C (1864) Die graphische Statik. Meyer & Zeller, Zurich

Greenwold S, Allen E (2003) Active statics. http://acg.media.mit.edu/people/simong/statics/Start.html

Jasienski JP (2014) Various perspectives on the extension of graphic statics to the third dimension. In: Proceedings of the IASS 2014, p 1

Kotnik T, D'Acunto P (2013) Operative diagramatology: structural folding for architectural design. In: Proceedings of the design modelling symposium, Berlin, p 195

Lachauer L (2015) Interactive equilibrium modelling, PhD-Thesis, ETH Zurich, p 128

Laffranchi M (1999) Zur Konzeption gekrümmter Brücken, PhD-Thesis, ETH Zurich

Lesek F (2008) Precision as a virtue, as a necessity, and for its own sake. In: Precisions—architecture between sciences and the arts, p 153

Moravanszky A (2008) The utopia of zero tolerance: the question of precision in architecture. In: Precisions—architecture between sciences and the arts, p 129

Mueller C (2014) Computational exploration of the structural design space, PhD-Thesis, MIT, p 30

Muttoni A, Schwartz J, Thürlimann B (1997) Bemessung von Betontragwerken mit Spannungsfeldern, Birkhäuser

Ohlbrock PO (2015) Combinatorial equilibrium modelling. In: Proceedings of the IASS symposium 2015, paper accepted

Rippmann M, Lachauer L, Block P (2012) RhinoVAULT—designing funicular form with Rhino, computer software. http://block.arch.ethz.ch/tools/rhinovault/

Schek HJ (1974) The force density method for form finding and computation of general networks. Comput Methods Appl Mech Eng 3(1):115–134

Schwartz J (2011) Tragwerkswissenschaft und Tragwerkslehre. In: Kooperation. Zur Zusammenarbeit von Ingenieur und Architekt. pp 243

Van Mele T, Block P, Ernst C, Ballo L (2009–2012) eQUILIBRIUM. http://block.arch.ethz.ch/equilibrium/

Hybrid Tower, Designing Soft Structures

Mette Ramsgaard Thomsen, Martin Tamke, Anders Holden Deleuran, Ida Katrine Friis Tinning, Henrik Leander Evers, Christoph Gengnagel and Michel Schmeck

Abstract

This paper presents the research project Hybrid Tower, an interdisciplinary collaboration between CITA—Centre for IT and Architecture, KET—Department for Structural Design and Technology, Fibrenamics, Universidade do Minho Guimarães, AFF a. ferreira & filhos, sa, Caldas de Vizela, Portugal and Essener Labor für Leichte Flächentragwerke, Universität Duisburg-Essen. Hybrid Tower is a hybrid structural system combining bending active compression members and tensile members for architectural design. The paper presents two central investigations: (1) the creation of new design methods that embed predictions about the inherent interdependency and material dependent performance of the hybrid structure and (2) the inter-scalar design strategies for specification and fabrication. The first investigation focuses on the design pipelines developed between the implementation of realtime physics and constraint solvers and more rigorous Finite Element methods supporting respectively design analysis and form finding and performance evaluation and verification. The second investigation describes the inter-scalar feedback loops between design at the macro scale (overall structural behaviour), meso scale (membrane reinforcement strategy) and micro scale (design of bespoke textile membrane). The paper concludes with a post construction analysis. Comparing structural and environmental data, the predicted and the actual performance of tower are evaluated and discussed.

M. Tamke
Centre for IT and Architecture, Royal Danish Academy of Fine Art, School of Architecture, Copenhagen, Denmark

M.R. Thomsen (✉) · A.H. Deleuran · I.K.F. Tinning · H.L. Evers
The Royal Danish Academy of Fine Arts Schools of Architecture Design and Conservation, Centre for Information Technology and Architecture (CITA), Copenhagen, Denmark
e-mail: martin.tamke@kadk.dk

M.R. Thomsen
Centre for Information Technology and Architecture (CITA), The Royal Danish Academy of Fine Art, School of Architecture, Design and Conservation, Copenhagen, Denmark

C. Gengnagel · M. Schmeck
Department for Structural Design and Technology (KET), University of Arts Berlin, Berlin, Germany

I.K.F. Tinning
Centre for IT and Architecture, Royal Danish Academy ofFine Art, School of Architecture, Copenhagen, Denmark

M.R. Thomsen et al. (eds.), *Modelling Behaviour*, DOI 10.1007/978-3-319-24208-8_8

Introduction

Hybrid Tower examines the creation of hybrid structures that combine two or more structural performances to create a stronger whole (Gengnagel et al. 2013). By merging membrane performance with overlapping actively bent compression rods, the tower's load behaviour and structural integrity performs as a grid, whose stability can only be achieved through the interaction of the material systems. The project explores the utilisation of flexibility on the material and structural level. Where architecture is traditionally conceived as static, this approach positions adaptability as the idea of reducing stiffness, allowing for deformations and thus minimizing material use. Similar to the concept of resilience in nature, a system's survival is understood as a function of its ability to adjust to environmental change that allows the system to become robust and endure. Hybrid Tower employs its inherent 'softness'—its flexibility and bending—as resilient measures. Built for the outside courtyard of the Design Museum Denmark it engages environmental forces of wind and weather (Fig. 1).

Hybrid Tower is part of the Complex Modelling research project (Complex Modelling 2013–17). The overall aim is to develop new interdisciplinary design strategies that enable the integration of material performance as a particular challenge to architectural design, opening up new horizons for design practice. The ability to design *for* and *with* material performance is understood as a core resource for design innovation being closely tied to material optimisation. The objective of Hybrid Tower is to pilot the forming of these methods. Compared to static and homogenous systems this involves an increased level of complexity in terms of modelling, analysis, fabrication and construction. Based on a close collaboration between architecture and structural engineering, technical textiles design and textile evaluation, the project develops new form finding methods that combine tensile and compressive behaviours and realise

Fig. 1 The interior of the hybrid tower is characterised by the tensioning system and the resulting *inward* oriented cone-like membranes

these through the direct interfacing between design and CNC knit fabrication. In Hybrid Tower a central goal has been to prototype and understand how these pipelines between realtime physics/constraint solvers and FE modelling can be devised.

Structural Concept of the Hybrid Tower

Hybrid Tower is constructed from stacked overlapping glass fibre reinforced plastic (GFRP) rods, which are connected and braced by a bespoke knitted membrane made from high tenacity polyester yarn. The rods are bent into arch like shapes utilizing the material's ability to be deformed elastically. The main advantage of using this bending active approach is the simplicity of shaping elements and the possibility to explore a wide architectural language of shapes. Furthermore it allows the, minimisation of material and energy consumption (Gengnagel et al. 2013). As the residual stress from the shaping process grows in proportion to bending curvature and element thickness, actively bent rods have to be very slender and of a high strength to stand the high stresses. Glassfiber reinforced plastics (GFRP) combine high strength with a low Young's modulus and are like timber and similar material preferred materials for such systems. To stabilise these systems membranes have proven to be very efficient restraining the slender elements and increase the hybrid system's stiffness dramatically (Alpermann and Gengnagel 2012).

The concept of the textile hybrid tower requires a balance between stiffness and rigidity to withstand external impact and a certain flexibility and softness to allow the structure to adapt to impact, store energy by deforming elastically and releasing this energy upon recovery. The arrangement of the actively bent rods forms a grid shell like structure embedded in channels and pockets inside a bespoke knitted fabric. The structure forms an almost hyperbolic tower. Built for the outdoors courtyard of the Design Museum Denmark, the structural weakness of tall structures are on the global level: wind compresses the front and creates suction forces at the sides, which create deformations that lead eventually to buckling and collapse (Fig. 2). Hence, the system needs to be restrained. This is done through pulling the membrane between the rods of each layer with tension cables to the tower's central axis. This results in a spoke wheel effect which provides horizontal stiffness and braces the rods.

The system's overall stability relies on the interaction between the GFRP grid and the membrane behaviour. The scale and amount of detail necessary to realise the membrane system makes the use of woven PVC membranes, which requires pattern cutting, seeming and application of details, a highly laboursome process. Knit was therefore chosen as a particular textile technique that allows bespoke detailing and is structurally

Fig. 2 Wind distribution according to Danish building code (*left*) and the deformation of the hybrid tower under respective wind load (*middle*) and without (*right*)

flexible allowing for a high degree of 3 dimensionality in the membrane design. The integration of detailing into the textile fabrication alleviates the need for sewn on details such as pockets for receiving the rods, but also creates the opportunity to design and fabricate a continual connection between rod and membrane enabling the best possible performance of the hybrid system.

In Hybrid Tower, the micro scale design and specification of the membrane is integral part of the overall design project, and highly interwoven with the form finding and the analysis of the towers overall structural behaviour.

Computational Modelling Approach

Challenges

During the design process, extensive physical prototyping was used to ideate a hybrid system that would be structurally performing and could be resolved at micro, meso and macro scale (Fig. 3).

From a computational design modelling perspective the primary development challenge has been how to model the hybrid behaviour of bending active tensile membrane structures in an interactive form finding and design system.

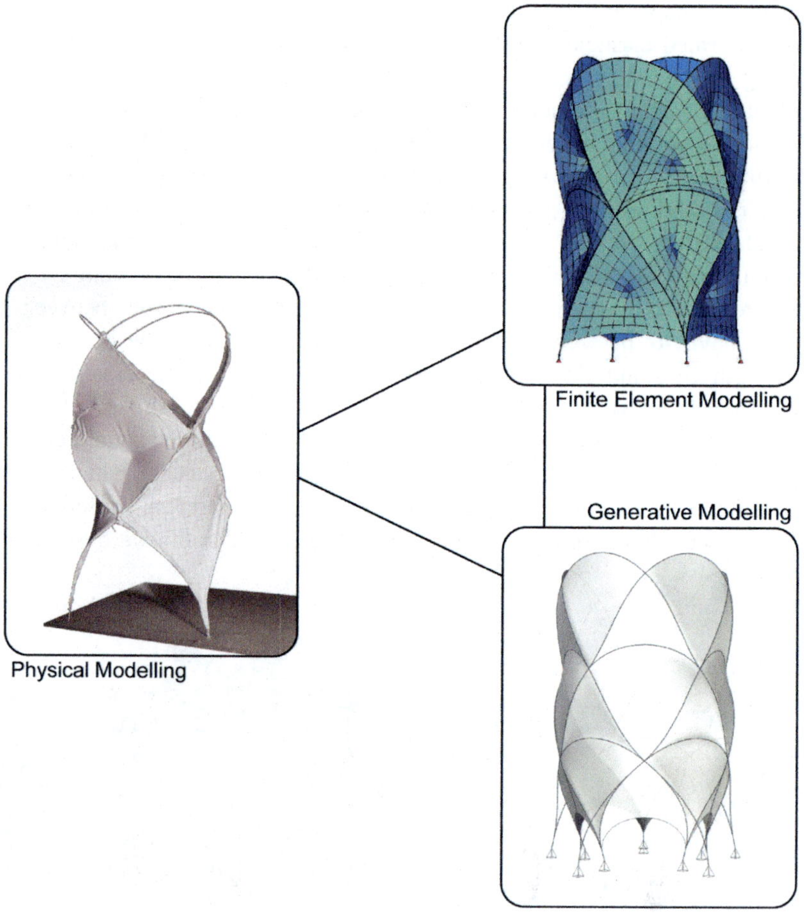

Fig. 3 The design process takes place in three interconnected types of models

This involved solving several subproblems including how to model the bending members and membranes in one integrated system, which physics solver to use and how to implement it. A large part of the project has therefore been focussed on how to generate and manipulate the tower topology and dimensions, which analysis models to implement as well as define processes by which the form found geometry can be further processed so as to develop the final fabrication data for the bespoke membranes.

Computational Workflow

To handle the challenges of the Hybrid Tower a computational workflow was developed consisting of a four-step-process (Fig. 4): Model Variables contain the definition for the tower design, constant and flexible geometric parameters and material properties, the Generative Model instance produces a sets of geometric possible results based on the parameters set before. The Analytical Model produces performance data of environmental impact, The Design Instance transfers the model data to the required production data. If the design doesn't match the desired requirements, the process is started again.

Generative Modelling Methods

Our generative design modelling approach extends previous work on actively bent structures (Adriaenssens and Barnes 2001; Deleuran et al. 2011; Quinn et al. 2013; Alpermann and Gengnagel 2012; Ahlquist and Menges 2013; Mele et al. 2013; Lienhard 2014) by implementing the new Kangaroo2 real-time physics and goal-driven constraint solver library. This library overcomes the stability and performance problems, which dynamic relaxation form finding methods have, as it employs methods which have been popularized as position based dynamics (PBD) (Bender et al. 2014), where instead of using summed forces to calculate accelerations, from which velocities, then positions are updated, the particles are simply to a position which satisfies the constraints, modifying the positions directly. Iterating repeatedly over all projected constraints lets the system converge fast, stable, allows it to include hard constraints and runtime user interaction. The Kangaroo2 library uses a

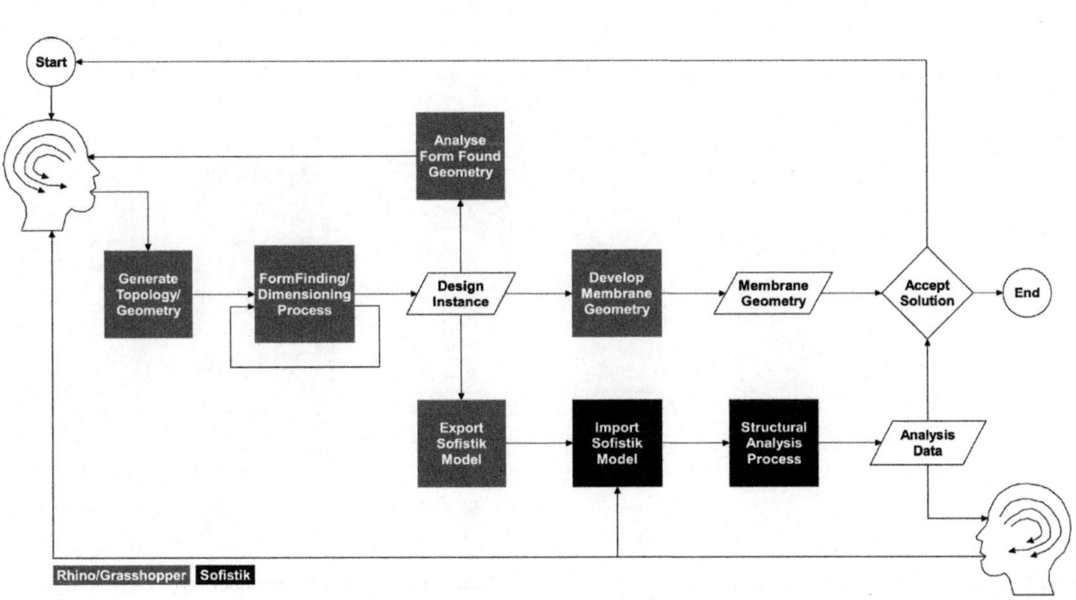

Fig. 4 Flowchart of the developed modelling pipeline

similar approach and treats herein geometric constraints, elastic materials, applied loads and other energies as different forms of the same type of object—encompassing them under the term goal.

Our modelling pipeline is based on the Rhino/Grasshopper CAD environment and implements the central algorithms as Python components with RhinoCommon and Kangaroo2 libraries. The pipeline is divided into four consecutive stages:

Topology and Geometry: The principle of stacking overlapping bending members around a central vertical axis is used to generate the fundamental tower topology. Bending members are abstracted to constituent polylines discretised and used to generate meshes representing the knit. This forms the base for the tension system (Fig. 5).

Form Finding and Dimensioning: Kangaroo2 goals are defined for each member representing its behavior and conditions, as the exact member dimensions per layer and possible flexibility. These are passed to the solver component which iteratively form finds the structure and allows the designer to interactively manipulate the system (Fig. 6).

Analysis of Form Found Geometry: A desired property in membrane design is high double curvature as this stabilises the membrane. For the bending members a key geometric property with structural implications is the local and maximum bending radii. These properties are analysed and visualised in the viewport in order to allow for informed design decisions (Fig. 7).

Unrolling Membranes: Developing the knit meshes in the plane for fabrication is problematic as they are double curved. A constraint-based approach was developed which forces a 2D mesh to have the same edge lengths as the 3D knit mesh. This in-plane mesh can constantly be checked against the maximum dimension of the fabrication machines (Fig. 8).

Analytical Modelling Methods

Bending stress utilization can only be passively determined in the generative model by evaluating beam bending curvature, unlike FE models which generate accurate and complex stress matrices describing all possible stress components (bending, torsion, shear).

The generative model can approximate realistic material behaviour for certain physical phenomena; for example beam bending is simulated accurately following the Barnes and Adriaenssens model (Adriaenssens 2001). The generative model currently falls short on its ability to accurately simulate phenomena such as beam torsion, and shell/plate elements. A validation in an advanced FE environment for

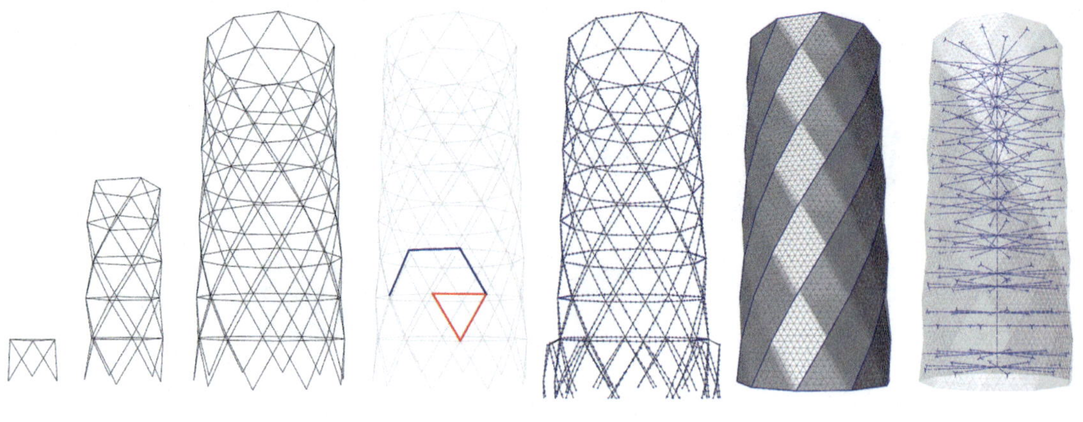

Fig. 5 Three instances of tower topologies with 3, 5, 10 sides and 1, 5, 9 stories. The local *polyline* members of the system are highlighted in the *fourth image*, followed by the three structural member types (bending rods, membranes, tension wires) and their geometric representations (*polylines*, *meshes lines*)

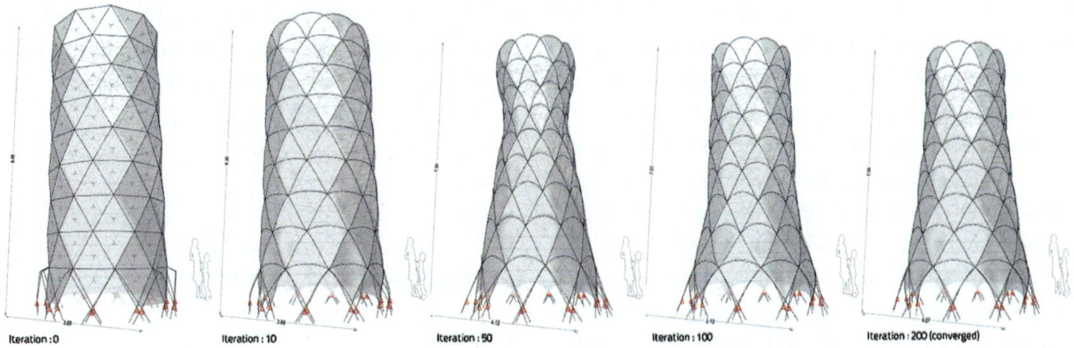

Fig. 6 Screenshots from the iterative and interactive form finding and dimensioning process, showing the initial shape (*left*), intermediate states after 10, 50, 100 iterations and the converged state after 200 iterations (*right*)

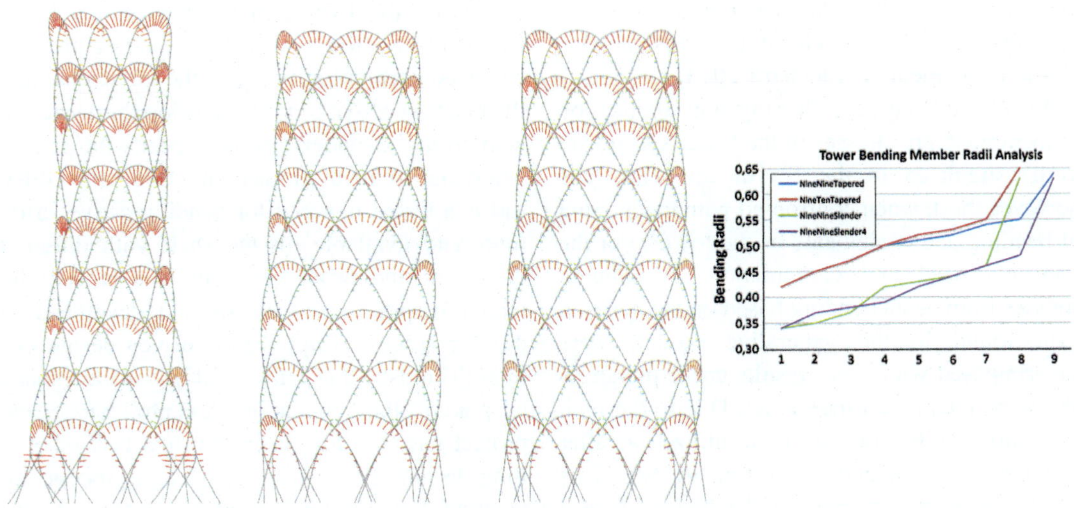

Fig. 7 Comparative bending radii analysis of differently dimensioned towers. Note the relationship between macro shape and bending radii

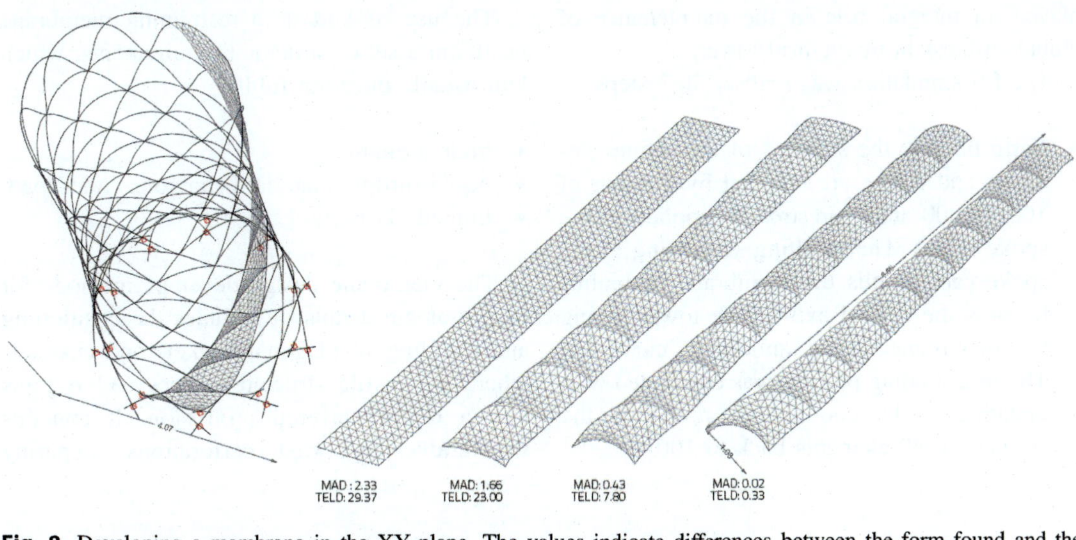

Fig. 8 Developing a membrane in the XY-plane. The values indicate differences between the form found and the in-plane meshes (*MAD* Mesh area difference, *TELD* Total edge length difference)

structural systems is hence necessary in order guarantee precision of physical behaviour in the generative model for the phenomena in question —bending, torsion and membrane behaviour.

Large deformations are generally unproblematic in mass spring system (MSS) environments, however in order to achieve the same in an FE environment, which is inherently founded on the concept of small deformations, extra measures are necessary (such as load step iterations, non-linear solvers and 3rd order differential equations).

In the SOFiSTiK model, beams are usually "pulled" to their target positions using single element cables with reduced stiffness and incrementally increasing levels of prestress, i.e. the elastic cable method (Lienhard 2014).

Ideally, all physical shaping and application of membrane pre-stresses in the tower system are to be simulated in the FE model. This falls however short, when confronted with the amount of bending active elements of the tower. For the sake of speed and computational simplicity, stresses from beam bending based on the bending radii found in the generative model, were superimposed with stress results under pre-stress and wind loading on the model. Due to the high curvatures of the beams, their utilization after forming alone ranged from (60–80 %). This meant the stress reserves in the beams were a precious resource under additional loading from membrane pre-stress and external loads. Subsequently form-stabilization from the spoke wheels played an integral role in the maintenance of global stiffness in this hybrid tower.

The FE simulation was realised in 3 steps:

- Form finding: the stiffness of membrane elements and cables are reduced by a factor of 100–10.000 and a prestress is applied to the spoke cables. The resulting shortening of the spoke cables pulls the membrane midpoints towards the central axis of the tower giving the membrane strong anticlastic curvature. The form finding process was concluded with updating node coordinates reverting the stiffness of all elements back to 100 %.
- Loading: wind loads were applied according to building codes and projected on to the local z-axis of the membrane quad elements.
- Analysis: stresses and deformation

Fabrication and Construction

Designing the Knitted Membrane Material

Designing the hybrid interaction of membrane and bending active system necessitates a finely to balanced correlation between the membrane and the bending active rods. To develop this system it is important to understand and specify the precise behaviour of the used materials. While the GFRP rod is a tested material for bending active structures and available off the shelf with different diameters and properties, knit is a process by which bespoke materials can be developed in direct response to application driven design criteria (Thomsen and Hicks 2008) is always made bespoke to the application. A central part of the project has therefore been the dual task of eliciting design criteria from the design process and creating methods for knitting the bespoke membranes, while at the same time developing means by which these can be formally tested so as to be simulated within the design system.

The use of knit as a restraining membrane results in a set of structural requirements, which knit usually does not fulfill:

- high strength
- near isotropic material behaviour under load
- limited elasticity.

The membrane design develops methods for four strategic details. It includes the positioning and detailing of diagonal pockets and channels where the textile structure splices and re-joins into a double layered protrusion. It contains structurally supported perforations preparing

Fig. 9 Defining three of the four details: the pocket, channel, the seam line and the reinforcement

seam lines. It is knitted to shape alleviating the need for pattern cutting. And it develops methods for local reinforcement strengthening edges as well as central radial pulling points (Fig. 9).

The development of knitting patterns and the choice of fibres took place in collaboration with Fibrenamics, Universidade do Minho, Guimarães, Portugal. Through an iterative process of prototyping small scale samples were developed and tested on a uni-axial force gage. Initial tests showed that 110 and 55tex High Tenacity Polymer in a Piquet-lacoste knit structure demonstrated high stability in structure, high strength and were relatively isotropic. In order to get insights into the material behaviour the Japanese MSAJ/M-02-1995 standard for the bi-axial testing of weave, was adopted to our knit by the University of Duisburg Essen. The tests measures in intervals in both material directions and the approach allows to determine fictitious material constants, suitable for application in FE, for highly non-linearly behaving materials, such as knit. In our knit the tensile stiffness is extremely low with Young's moduli of approximately $E = 5$ kN/m for wales and course for the full stress interval and $E_{wales} = 10$ kN/m and $E_{course} = 26$ kN/m in stress interval of 1–3 kN/m, which is the one determined relevant for the setup in Copenhagen. In contrast, the Poisson's ratios are very high with minor $n = 0.83$ and 0.66, respectively. Strains are large with up to 70 % under uniaxial loading.

The process of creating solid test data for simulation and design was however compromised during the production process. Moving from the production of samples to the final membrane patches necessitated a larger knitting machine in turn resulting in changing material properties. The final fabric was therefore more elastic than anticipated. This had a severe influence on the construction, geometry and behaviour of the tower.

Generating Fabrication Data

The fabrication data was directly derived from the generative model, after the design was verified by the FE model. The model generates the final length of the GFRP rods and the in-plane mesh of the membrane. This was further processed into the CNC knit pattern for the fabricator (Fig. 10). The process of developing the CNC knitting pattern necessitated the incorporation of the stretch of the material through the active bending system, as well as the anisotropy of the knit and the specifics of the fabrication process. The necessary compensation factors were established through thebi-axial testing of samples and some tests with the final assembly system.

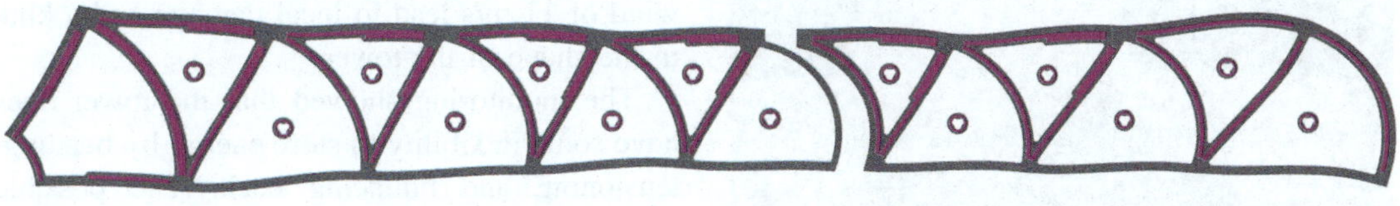

Fig. 10 A *pixel image* with details for pockets, channels, reinforcements and seam holes informs the CNC knitting machines on a loop by loop base. The image shows the pattern before the compensation of knit for the machines

Assembling the Tower

The assembly strategies for the tower were tested on 1:5 and 1:1 prototypes with membrane surfaces made from weave and partly with CNC knit. Though more prototyping would have been necessary these physical tests allowed us to assess the behaviour of the assembly and the functioning of inserting the rods in the textile through channels and pockets.

The aim for the assembly was to create a strategy, which would not necessitate scaffolding or other expensive supports. Assembly therefore took place on the ground, starting with the top layer, and lifting the tower up, constructing each consecutive 'storey' beneath it. The Tower was planned to be 8 storeys high. The low stiffness values of the bespoke knitted fabric resulted in an increase of material stretch and hence increase in size of the tower, so that the final height was already reached with 6 stories (Fig. 11). A simulation with the material properties of the finally produced knit, showed as well a 80 % decrease in load capacity under wind load.

Fig. 11 The tower in the courtyard of the Danish design museum—April 2015

Post Construction Analysis

During the exhibition we conducted a 10-day evaluation monitoring of the structure with two time lapse cameras to the front and side of the tower as well as a synchronised wind measurement device. These evaluation data were used to determine the differences between the behaviour of the built demonstrator and the predicted behaviour of our computational design models.

The simulation was conducted in Sofistik and uses the developed modelling pipeline, with the material dimensions and properties from the built tower. The rods are defined as beam elements with diameters of 12, 10 and 8 mm. Tension cables at the naked edges and inside the structure are defined as ideal cables elements with a diameter of 8 mm of Steel S355. Supports were fixed and the tower height was 6.90 m.

Simulating the behaviour of the tower under wind loads in the Sofistik FE environment was challenging, as the softness of the structure lead to numerical problems and diverging simulations. The knitted fabric was approximately 50 times softer than a usual pvc-type-I-membrane, and convergence could only achieved for 20 % of the initially predicted wind loads, which were taken from the Danish code for temporary structures.

The overlay of the videos with the weather data allows for correlation of wind (gust) speed and deflection of the tower. Up to wind speed of 6.5 m/s (which equals Beaufort 3) the tower stayed in the elastic domain. The measured peak wind of 11 m/s lead to local damage and a kink in the shape of the tower.

The monitoring showed that the tower does have some flexibility to store energy by bending, tensioning and bouncing back. The possible deformation is actually a lot higher than calculated (Table 1). This is due to differences between the real structure and the simulated model, especially in detailing:

Table 1 Wind speed used in the simulation of the tower

	Maximum wind speed		Beaufort	Wind pressure	Deflection	
	[km/h]	[m/s]		[kN/m^2]	Prototype	FE-Model
Temporary structure	107.8	30.0	11	0.54		
Temporary structure 20 %	48.2	13.4		0.108		
Monitored	23.4	6.5	3	0.0254	Ca. 200 mm	110 mm

Temporary structures shows the basic wind speed for temporary structures, below 20 % of that speed and in the bottom the monitored wind speed from the monitoring

- The connection between the rods and the membrane can not be modelled correctly, as the friction can not be considered in the simulation to sufficient precision.
- When under tension the rods can slide within the tunnels and pockets, increasing deformation and local stress.
- The ends of the rods fitted into pockets and induce point loads into the membrane that enlarge deformation even further and produce wrinkles in the membrane.

The evaluation of the collected weather data resulted in a maximum wind (gust) speed of 6.5 m/s. The associated deflection was measured from the video footage and compared to the deflection of a FE model simulating the same load level (Figs. 12 and 13).

As expected the biggest deformation occurs at the top of Tower. In the simulated model the deformation occurs only in a small region where compressive wind occurs, while the whole top is shifted more evenly to the side on the prototype. The maximum deflection of the model is 110 mm, while the prototype moves ca. 200 mm. These deviations can be explained by differences in the stiffness of the jointing. This observation is in line with earlier research (Quinn et al. 2013), which indicated, that the modelling of details is crucial for a realistic simulation. In the digital model, these are defined as rigid nodes with direct load transfer, in the prototype the forces are transferred from one rod to the other via the membrane, that stretches and slides to a certain amount. This local behavior of (elastic) stretch and the friction at the transition from rod to membrane to rod has not been considered in the simulation (Fig. 14).

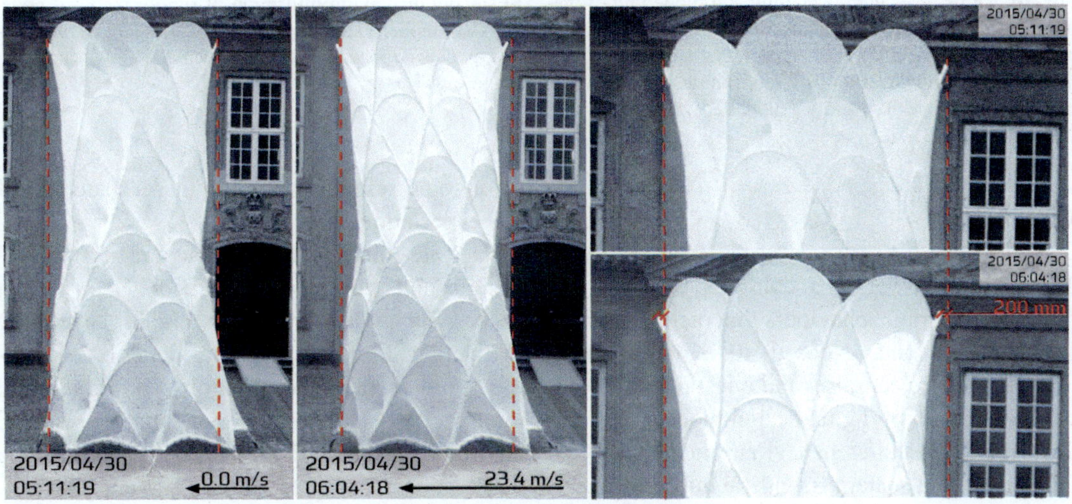

Fig. 12 Horizontal deflection observed in the video foo tage is approximately 200 mm (wind from *right*). Figures from *left* to *right*: undeformed tower, deformed tower, comparison of deformed/undeformed tower (*top* section)

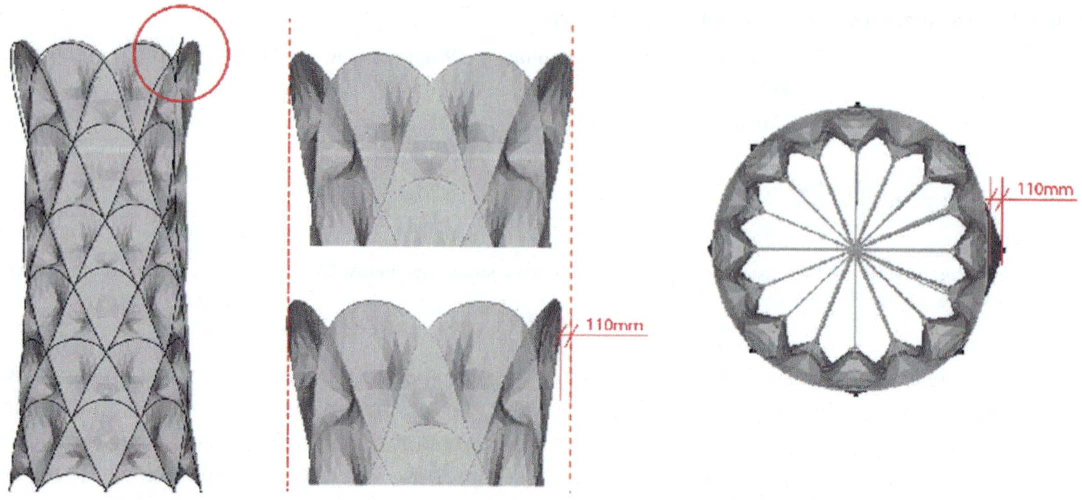

Fig. 13 Horizontal deflection in the simulated model occurred mostly local (*top right*) with a maximum of 110 mm (wind from *right*). Figures from *left* to *right*: undeformed tower with deformed grid (deformation scaled by factor 3.0), comparison of deformed/undeformed tower (*top* section) (deformation scale factor 1.0), deformation in plan (deformation scale factor 1.0)

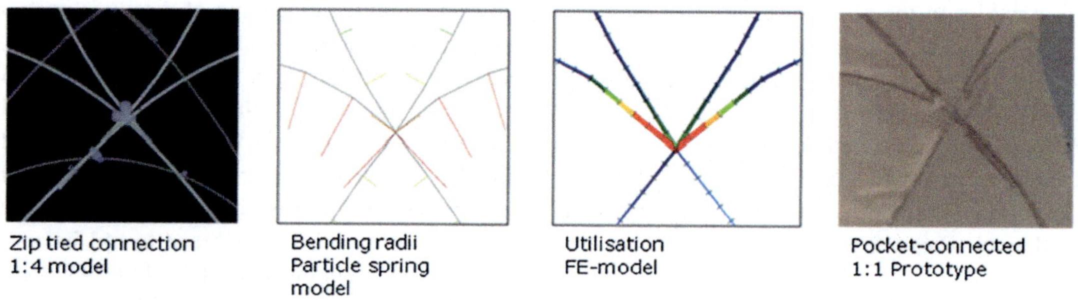

Fig. 14 Differences in the detailing and modelling of the intersecting and overlapping profiles. From *left* to *right*: physical prototype (1:4) with zip-tied connections, digital model in Rhino/Grasshopper environment displaying bending radii with rigidly modelled node, digital FE-model displaying utilisation of rigidly modeled node, physical model (1:1) displaying rods integrated in tubes and pockets—load transfer between the rods via membrane only

Conclusion

The Hybrid Tower is a complex construct of multiple material behaviours interacting at different scales and exposed to environmental impact. The project raises the question about the setup and depth of digital design models in order to predict a structures behaviour and specify the fabrication and materials with a sufficient accuracy. The Hybrid Tower is an intermediate stage of ongoing research and did not yet match all of the expectations.

Despite the challenges encountered in the design and development process the project successfully developed a digital workflow (Modelling Pipeline), creating a stable environment for generative constraint based form finding (Generative Model), processing data to an FE environment (Analytical Model) allowing for the necessary structural feedback. Though the feedback loop from the analytical FE model to the generative model was not automated the overall process was sufficient to drive the specification and digital production of materials and elements of the structure.

The deviation in the behaviours of the simulated model and the prototype confirm the impact of connections, intersections and joints on the structures behaviour, as the need for proper testing and evaluation of materials and assemblies. The question remains, how to improve the implementation of analysis, model building, detailing and testing in the design process, to deliver an increased level of accurate feedback in early design stage and to shorten the specification, development and evaluation cycles of material and systems.

Key is to embed the digital design pipeline in a dynamic process of constant calibration between the digital and physical instances. Where precise definitions are not yet possible sensible simplifications have to be estimated and tested for validity. The focus in the ongoing research will be to improve the exchange of data between the instances of the pipeline as well as to parallely investigate methods of implementing the analysis directly into the process.

Acknowledgement The project is funded by The Danish Council for Independent Research (DFF). Membranes were developed by Fibrenamics, Universidade do Minho, Guimarães, Portugal. Further development and fabrication took place with AFF a. ferreira & filhos, sa, Caldas de Vizela, Portugal. Mechanical testing of the knit was conducted by University Duisburg-Essen, Laboratory for Lightweight Structures. The Tower was exhibited at Designmuseum Danmark. We wish to thank: Daniel Piker for generously involving us in testing Kangaroo2. Dongil Kim, Esben Clausen Nørgaard and the students of CITA. studio for their tireless support. Photos by Anders Ingvartsen.

References

Adriaenssens SM, Barnes M (2001) Tensegrity spline beam and grid shell structures. Eng Struct 23(1):29–36

Ahlquist S, Menges A (2013) Frameworks for computational design of textile micro-architectures and material behavior in forming complex force-active structures. ACADIA 2013:281–292

Alpermann H, Gengnagel C (2012) Shaping actively-bent elements by restraining systems. In: Conference proceedings IASS- APCS symposium: from spatial structures to space structures, 2012

Bender J, Müller M, Otaduy MA, Teschner M, Macklin M (2014) A survey on position-based simulation methods in computer graphics. Comput Graph Forum 33:6

Complex modelling (2015). http://www.complexmodelling.dk/. Accessed 30 June 2015

Deleuran AH, Tamke M, Thomsen MR (2011) Designing with deformation—sketching material and aggregate behaviour of actively deforming structures. Symp Simul Architect Urban Des 2011:5–12

Gengnagel C, Alpermann H, Lafuente E (2013) Active bending in hybrid structures. Form—Rule | Rule—Form 2013:12–27

Lienhard J (2014) Bending-active structures—form-finding strategies using elastic deformation in static and kinetic systems and the structural potentials therein

MSAJ/M-02-1995 (1995) Testing method for elastic constants of membrane materials. Membrane Structures Association of Japan

Mele TV et al (2013) Shaping tension structures with actively bent linear elements. Int J Space Struct 28 (3):127–135

Quinn G et al (2013) Structural analysis and optimisation of a computationally designed plywood gridshell

Thomsen MR, Hicks T (2008) To build a Knitted Wall. In: Proceedings, ambience, smart textiles conference, Gøteborg, 2008

Integrating Differentiated Knit Logics and Pre-Stress in Textile Hybrid Structures

Sean Ahlquist

Abstract
This paper describes research in the use of machine knitting for manufacturing highly differentiated textiles and their implementation as the tensile component of a textile hybrid structural system. The fundamental concept of a textile hybrid structure is in generating form through the integration of bending- and form-active behaviours implemented in materials comprised, in some or all parts, of a fibrous nature. A prototype entitled *Mobius Rib-knit* explores the nature of a knitted textile as a part of such a system. Operating at the level of stitch structure, differentiated form-active properties and non-planar geometries are materialized within a seamless textile. Utilizing CNC machine knitting, a fundamental stitch structure, the *rib-knit*, is exploited for its elastic nature, while the ability to generate a shaped 3d textile allows for a seamless material to fit to an intensely contorted geometry. These characteristics are tailored to describe visual, spatial and tactile qualities; ones which are unique to the field of pre-stressed lightweight structures. While the rib knit is a conventional knit structure, its novel use is described in this paper as the articulator of surface dimensionality and patterning within an architectonic system.

Introduction

In a textile hybrid system, material behaviour is activated at the moment in which elements are composed into a particular topology, with their pre-stress in mutual exchange, ultimately resolving into a particular form (Ahlquist et al. 2013; Ahlquist and Menges 2013; Lienhard et al. 2013a, b). This paper looks to refine that definition of material behaviour, where the design and manufacturing of variegated textiles poses the consideration of behaviour existing at the moment of material formation, expanding upon previous research in exploring the nature of textile hybrids. Ultimately, this proposes a consideration of *material* behaviour as a process of manufacturing and *system* behaviour as the

S. Ahlquist (✉)
Taubman College of Architecture and Urban Planning, University of Michigan,
Ann Arbor, MI, USA
e-mail: ahlquist@umich.edu

M.R. Thomsen et al. (eds.), *Modelling Behaviour*, DOI 10.1007/978-3-319-24208-8_9

Fig. 1 Mobius Rib-knit prototype developed as a textile hybrid structural system, integrating glass-fibre reinforced rods and machine knitted textile; (Ahlquist 2014)

moment in which all material components are assembled and pre-stressed. By examining the fundamental nature of a rib-knit structure and its response to imposed tensile forces, the play between material behaviour and system behaviour is exhibited (Fig. 1). This builds upon research which exploits the use of bespoke manufactured materials, specifically CNC fabricated textiles, exemplifying a significant shift from the manipulation of standardised materials to explicit production of graded materials, highly tailored for their performative value (Thomsen and Karmon 2012).

Textile Hybrid Structural System

The definition of a hybrid structural system is provided by Heino Engel, stating that it constitutes a system which integrates multiple types of structural action to realize a stable structure (Engel 2007). Specific to research in textile hybrid systems, form is attained at the equilibrium of interconnected tensile (form-active) surfaces and arrangements of elastically bent (bending-active) elements. The term "bending-active" refers to the use of initially straight beam elements that have been elastically deformed into a curved beam, gaining stiffness geometrically and through residual stress. The most suitable materials are those which combine high strength with low bending stiffness. The specific classification of a *textile* hybrid emphasizes the priority of the fibrous make-up of the bending-active elements and of the textile membranes in forming an integrated architecture. Central to this research is the shift from a bi-axial uniformity in the fibre-structure of the textile to one of a variegated organization, exploring

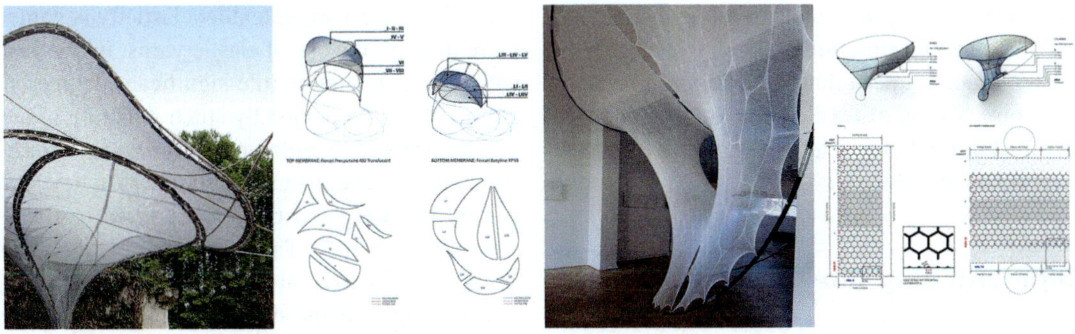

Fig. 2 Previous research in textile hybrid structures and related drawings indicating textile manufacturing data, including the multi-hierarchical system developed for the M1 at La Tour de l'Architecte project (*left*) and the semi-toroidal installation (*right*) integrating seamless differentiated machine knitted textiles; [Ahlquist and Lienhard 2012 (*left*), Ahlquist 2012 (*right*)]

differentiation in how it distributes the form-active pre-stress and affects the visual and spatial properties.

Two projects represent important developments in textile hybrid structures (Fig. 2). The M1 project, constructed in Monthoiron, France, is comprised of a multi-hierarchical system which balances scale, geometric variation and a minimal imposition to the surrounding site. The meta-scale bending-active structure, defining the overall form, utilizes a combination of gridshell-like arrangements and free spans supported by tensile membranes. A meso-scale system rests within the volumetric structures supporting the long-spanning regions. Conversely, the Semi-Toroidal textile hybrid installation greatly minimizes the topological complexity, while creating a more variegated surface and spatially complex architecture through the use of CNC machine knitting. In articulating the knit structure, a highly elastic textile is produced, enabling the entire structure to be formed of only two textile parts.

In comparison to these precedents, the Mobius Rib-knit prototype seeks to reduce topological complexity at the scale of the number of components which comprise the system, while advancing the potential for spatial and visual complexity. The M1 prototype requires an intense number of components, partially due to scale, but also driven by the use of standardized membrane materials. To accomplish the doubly-curved geometries which envelope the meta-scale structure, 15 separate parts are cut from pre-manufactured membrane material, and sewn together to create 8 components which are lashed to the GFRP rod structure, forming the non-planar geometries. This produces approximately 70 % material waste for the production of the individual parts, and significant post-production in sewing, edge detailing and assembly. In the M1, geometric and spatial differentiation is accomplished through the manipulation of pre-manufactured materials, shaping them into the specific profiles defined by pre-stress and geometry. Where the M1 accomplishes differentiation through a multi-hierarchical material assembly, the research related to the Mobius Rib-knit shifts the multi-hierarchical nature to the variegated construction of the textile itself. Some of this begins to be developed with the Semi-toroidal prototype, where CNC knitting allows for a hexagonal sub-structure to be tailored to the stresses and geometries of the overall system. In the Semi-toroidal prototype, developed in collaboration with a manufacturer for technical knitted textiles, the level of resolution to control the shaping of the textile is at the scale of the hexagonal pattern. The determination of the number of hexagonal units is largely done through full-scale mock-ups, requiring up to five iterations. Thus, material waste is still an issue, not in manipulation of the textile, rather in the number of full scale units needed to test and confirm the appropriate variegated structure. Research described in this paper looks to refine the method for dictating knit structure, producing a more fluid design process inter-relating CNC machine knitting, observation of textile behaviour, simulation of structural behaviour and material performance within a textile hybrid system.

Differentiating Rib-Knit Behaviour

This research utilizes the facilities at the University of Michigan's FabLab which features a large-scale CNC weft-knitting machine (Fig. 3). Weft-knitting forms a textile through the looping of yarn fibres in the horizontal (weft) direction, each loop or stitch defining a wale and each row of wales defining a course. This is in comparison to warp knitting which works in the vertical (warp) direction, and weaving which overlays straight fibres in the weft and warp directions (Fig. 4). In simple terms, a weft-knitted textile is least stable, meaning the structure of the stitch or the overlay of fibres is easily deformable, as the interconnection of stitches happens only in the horizontal direction. Certain stitches and increased knit densities can be utilized to improve the native stability of the textile. Used as a tensile surface, the stitch instability, and looped rather than straight fibers, is invaluable as it allows for great deformation or elasticity in multiple directions, as the stitches interact by being pulled taut against each other. With a woven or warp-knitted textile, elasticity is gained more through the use of elastic yarns. In contrast, with a weft-knitted textile significant degrees of elasticity can be gained by stitch structure alone.

The Mobius prototype utilizes a rib-knit structure, a fundamentally elastic structure driven by the weighted bias of a knitted textile to naturally "recoil". The direction of recoil is defined by the bed on which the stitch is made, determining how the head of the stitch overlays the legs of the stitch in the subsequent course (Fig. 5). A knitted textile manufactured only on a single needle bed, referred to as a single jersey structure, will exhibit the recoiling tendencies. A rib-knit is formed by stitching back and forth between the front and back needle bed (Fig. 6). Ultimately a balanced textile is generated where the recoiling tendencies are inverted between neighbouring ribs. As mentioned previously, elasticity is gained in a knitted textile by the ability to re-configure the geometry of the stitches when tensioned. Another scale of elasticity is activated by utilizing the rib knit. When tensioned, the textile will first uncoil the ribs and then the stitches themselves will be activated. A rib knit is the common technique used for cuffs and collars of a sweater given its natural ability to rebound without the need for an elastic yarn, which can quickly fatigue. A vertical rib is most often used, though in this research, the horizontal rib is utilized and at much greater scales than would be conventionally used in fashion.

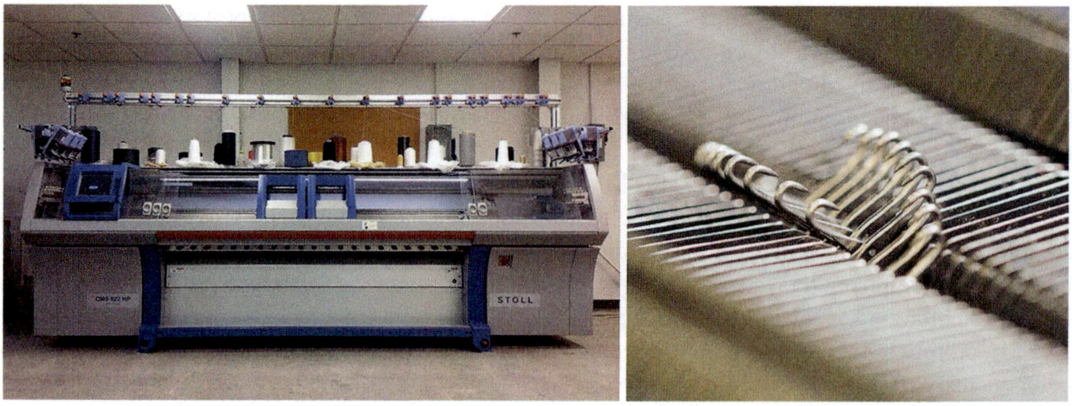

Fig. 3 Stoll v-bed knitting machine, 14 gauge with 82 needle bed, part of the FabLab facilities at the University of Michigan, Taubman College of Architecture and Urban Planning; (Ahlquist 2014)

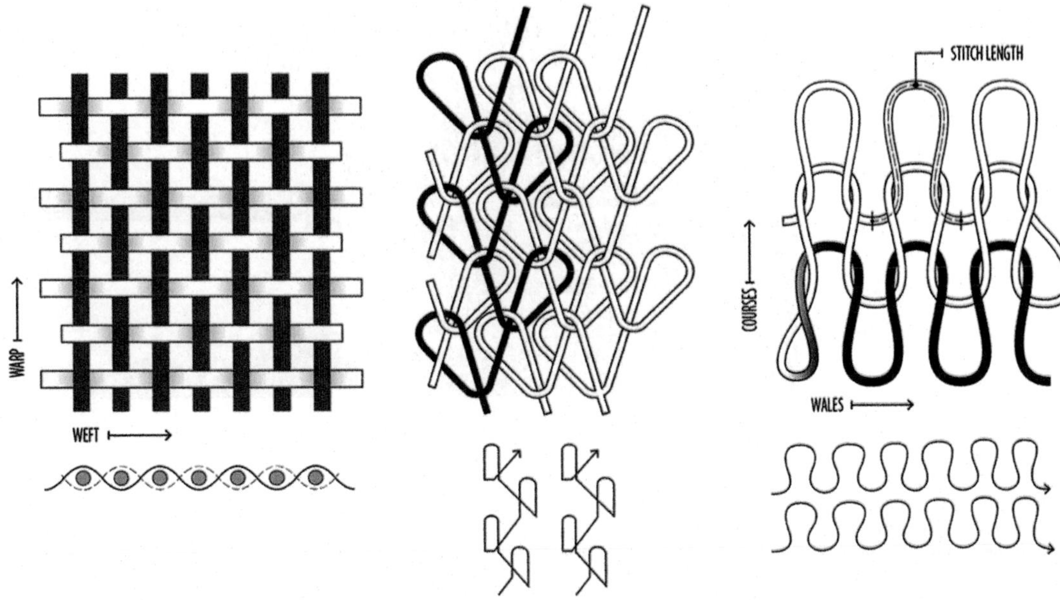

Fig. 4 Comparison of basic fiber logics between weaving, warp knitting and weft knitting; (Ahlquist 2015)

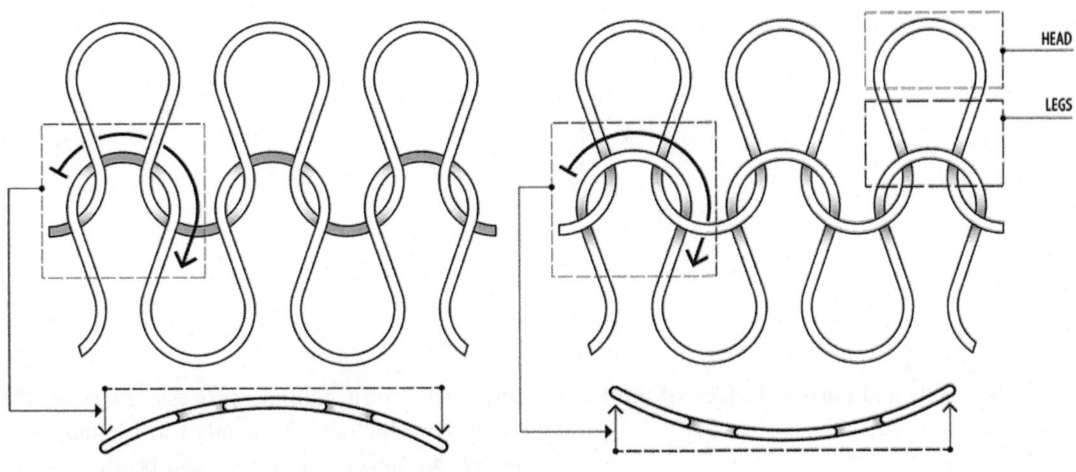

Fig. 5 How the "head" of one stitch overlays the "legs" of another stitch indicates which direction the textile will ultimately recoil, where a balanced structure is made by integrating both stitch types, vertically or horizontally; (Ahlquist 2015)

Mobius Rib-Knit Prototype

The topology of this prototype is simple in the numbers of components of which it is comprised: a single continuous GFRP rod boundary and a seamless four-sided knitted textile. It is in the manner of associating the components and articulating multiple aspects of the knitted textile in which the involuted surface and tectonic qualities emerge. Four "leaves" are generated within the continuous boundary element, and subsequently inverted on each other, upwards or downwards from a central ring (Fig. 7). In affixing the textile to the undulating boundary, the impression of a Mobius surface appears

Fig. 6 Detail of *vertical* rib-knit structure on knitting machine; (Ahlquist 2014)

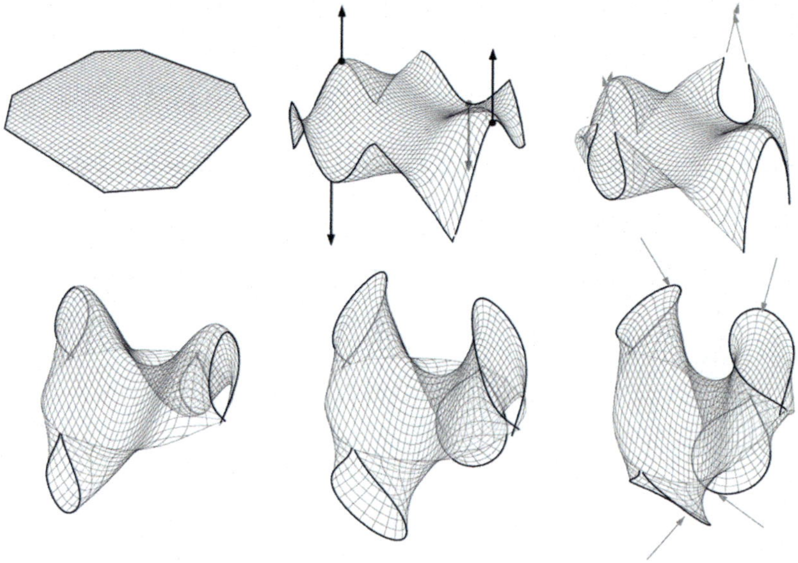

Fig. 7 Topology of the Mobius Rib-knit, creating an involuted surface from a single continuous boundary element; (Ahlquist 2014)

where the inside and outside surface of the form become continuous.

Relating Force Distribution and Knit Program

The form-finding of the relationship between the bending-active boundary and the tensile surface is accomplished in *springFORM*, a java-based software developed as a part of the author's doctoral research (Fig. 8). The software allows for the complex and inextricable structural actions to be actively computed and manipulated during the form-finding process, akin to the manner in which the physically form-found study models are developed. Form- and bending-active behaviours are computed through networks of springs. An important aspect of the form-finding process is determining the geometric relationship and topology of the bending-active boundary in relation to the geometry and form-active properties of the surface. Mimicking the exploration of such relationships through physical form-finding, *springFORM* is utilized to actively alter the boundary lengths, and reconfigure interconnection points and mesh topologies. Referred to as minimized "pre-planning" in previous research, active manipulation during

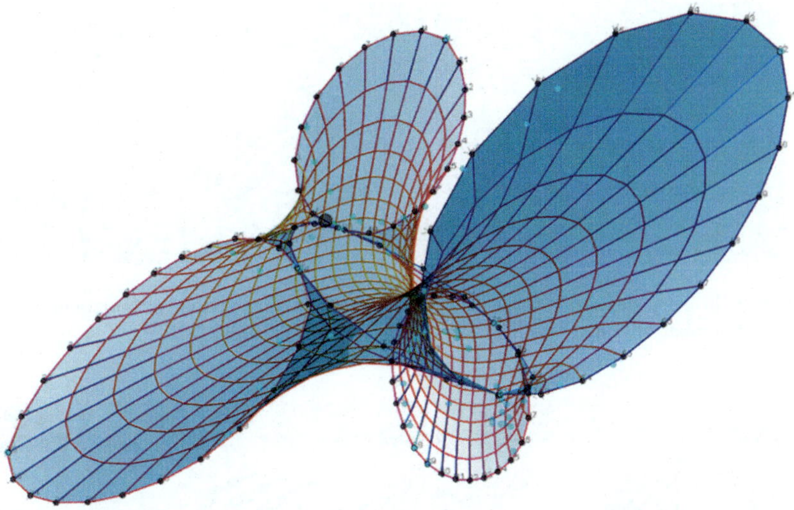

Fig. 8 Spring-based form-finding in *springFORM* to describe bending-active geometry and approximate tensile force distribution, used to define the "shaping" of the textile; (Ahlquist 2014)

form-finding allows for a generic model to be transformed into a specific design. This is often necessary for complex solutions where geometric and topological data cannot be determined prior to, or "pre-planned" before, the initiation of the form-finding process (Ahlquist et al. 2014). The output from *springFORM* provides specific dimensional data for the geometry and topology of the boundary element, along with a relative understanding of force distribution in the surface.

Simulating the behaviour of the knit structure requires consideration of the amount of pre-planning necessary to construct the computational model. The most accurate form of simulation necessitates yarn- and stitch-level models, with methods to specify the properties for a wide range of yarn types and fundamental knit structures. This also needs to reflect the particular structure and behaviour of weft-knitted textiles in comparison to methods utilized for simulating woven and warp-knitted textiles. Even in comparison to methods for finite element modelling of warp-knitted fabrics, an explicit model of yarn-loop structure is required, yet the configuration of the loops are approximated mathematically (Vassiliadis et al. 2007; Kurbak 2009). This is problematic for this research, where the geometries of the loops are heavily transformed when stressed under tensile loading. For a weft-knitted fabric, its mechanical properties are significantly related to yarn properties, local knit structure and global knit orientation. Therefore, in order to accurately compute the behaviour, the model has to operate at the scale of the individual stitch (Araujo et al. 2003). In the field of Computer Graphics, high resolution simulation has been accomplished through the modelling of yarns as rigid bodies. Utilizing a method where each face of a mesh is assigned a specific stitch type, a relaxation method calculated stitch deformation as well as "slip through" of yarns between stitches (Yuksel et al. 2012). While able to produce realistic behaviours with yarn level detail, the computationally heavy process demands intense pre-planning in providing a complete description of knit structure prior to the initiation of the simulation process. This is a valuable capability, yet it does not address the design process which needs to operate iteratively in order to explore and resolve the inter-relationships of material and system behaviour.

In developing the Mobius Rib-knit prototype, material behaviour and the innate properties of the rib-knit structure are established through observation of knit samples, examining the shaping of knit structure to produce variations in the rib-knit behaviour. The rib-knit structure is defined by knitting on the front needle bed with white yarn alternated with ribs knitted on the back bed with black yarn (Fig. 9). Because elasticity is gained with the rib-knit logic, the yarn is primarily inelastic; in this case an air-jet

Fig. 9 Primary samples for studying basic single jersey behaviour to horizontal rib-knit and horizontal rib-knit with varied number of courses within each rib

Fig. 10 Key parameters in designing textile behavior: directionality of stress along the length of the ribs, and varying width within each rib to shape the overall textile; (Ahlquist 2014)

textured polyester yarn. The patterning of successive ribs and the widths within individual ribs are varied in order to shape the textile to suit expanded or contracted regions within the bending-active rod boundary (Fig. 10). Shaping is defined by the ability to vary the number of courses within a specific area of the knitted textile. System behaviour, the impact of tensile loading on the textile, is first understood by calculating the ratio of forces between the longitudinal and latitudinal directions within the quad-mesh in *springFORM* (Fig. 11, left). Where the ratio is weighted towards the latitudinal direction of the mesh, a parallel-running standard rib-knit structure will maintain its dimensional corrugated nature. Where the forces are weighted towards the longitudinal direction, shaping of the ribs is necessary to ensure a certain degree of surface undulation. This concept for shaping the ribs is given dimensional properties by analysing the actual spring lengths in the longitudinal direction of the *springFORM* model. A map is then produced which plots an approximated number of ribs necessary at each point to realize a knit structure which maintains its undulating dimensionality when in tension (Fig. 11, right). The resulting knit program is based upon an approximation of the rib-count diagram (Fig. 12). The primary features in the knit structure is the narrowing of ribs towards the center, where the black rib maintains its width while the white rib is greatly reduced, and the introduction of new ribs at the edge, accounting for the increased linear dimension along the sides of the structure (Fig. 13).

Conclusion

In comparing to the Semi-Toroidal textile hybrid installation, the lowest level of resolution that could be studied was that of the hexagonal region (Fig. 14). Determining the number of rows and columns of hexagonal cells was the most refined level of control possible at the stage of research in implementing knitted textiles in textile hybrid structures. Advancing to the Mobius Rib-knit prototype, the operable level of resolution is at the

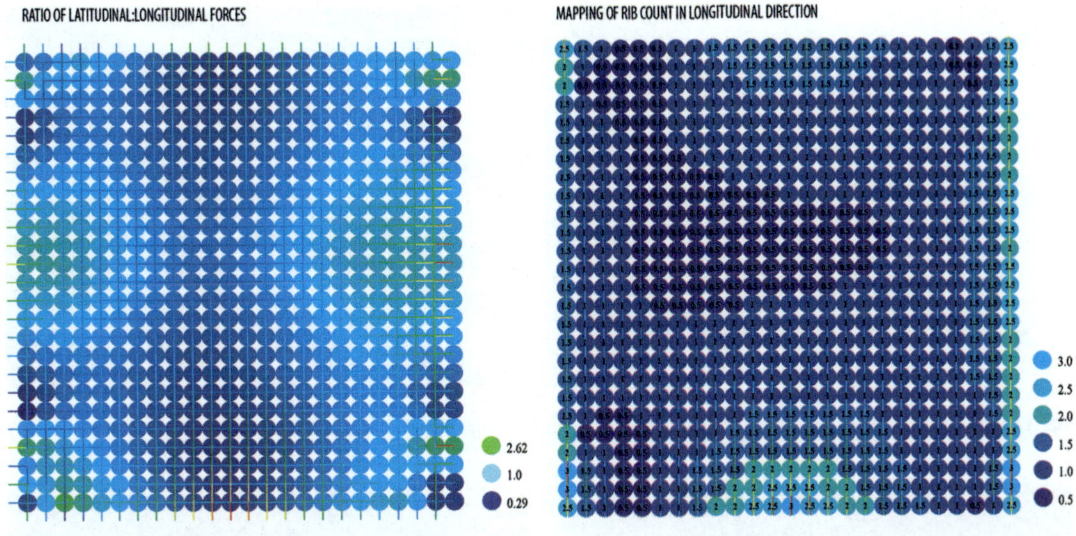

Fig. 11 Assessing the relative force distribution from the spring mesh, biaxially (*left*), and calculating the rib of ribs across the mesh based upon spring dimension in the longitudinal direction (*right*); (Ahlquist 2014)

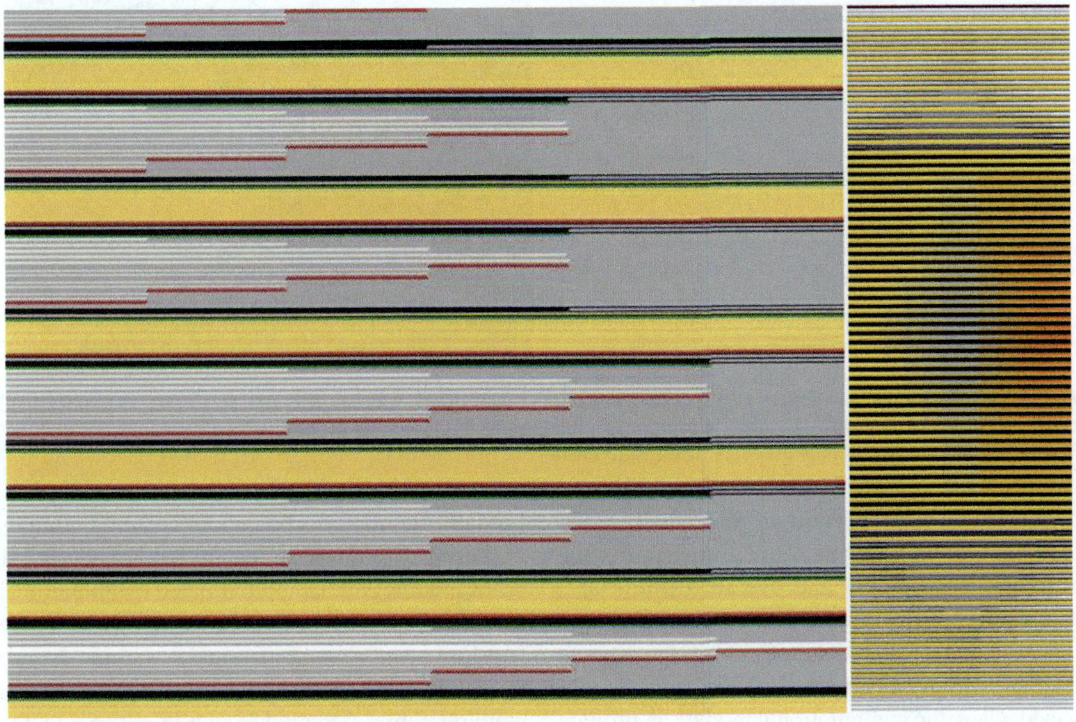

Fig. 12 Screenshot of the knit program, showing the shaping (*narrowing*) of a series of ribs where the *grey* area indicates where no stitches are made (*left*) and the overall structure where new ribs are introduced along the sides of the textile (*right*); (Ahlquist 2014)

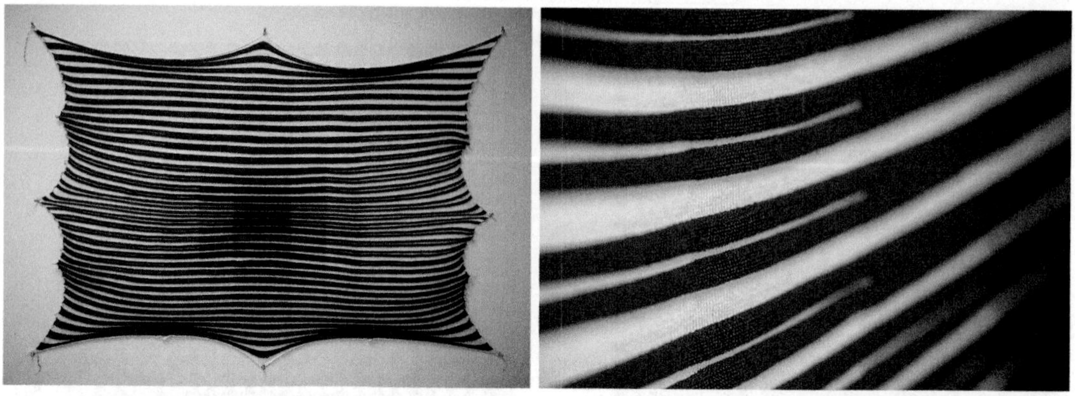

Fig. 13 Full scale knitted textile with bifurcating rib-knit structure for shaping; (Ahlquist 2014)

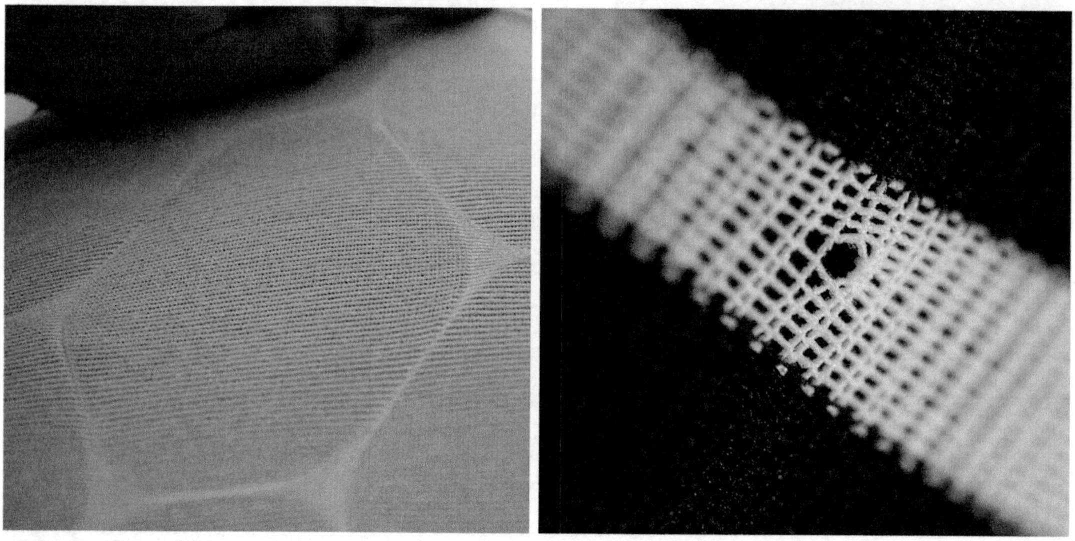

Fig. 14 Comparative resolution of designing textile behaviour, at the scale of the hexagon in the Semi-toroidal prototype versus the scale of the individual stitch in the Mobius Rib-knit prototype; [Ahlquist 2012 (*left*), Ahlquist 2014 (*right*)]

both the scale of rib as a non-uniform structure and the stitch as the ultimate vehicle for controlling both elastic behaviour and overall textile shape. The computational model does not serve to specifically define such parameters of the knitted textile. But a translational methodology is established where the computational model can provide information at the scale of rib width and count, locally within the seamless textile. With the Semi-Toroidal prototype, trial and error studies at full scale were the main vehicle for textile design. In comparison, the Mobius Rib-knit prototype offers a more fluid methodology for understanding fundamental material behaviour and projecting its influence, through studying knitted samples at 1:1 scale and integrating approximated computational models. Further research looks to maintain the freedom in computational (spring-based) form-finding as a method for integrating form- and bending-active behaviours, while advancing the methods of translation from the force diagrams to offer finer resolution of indicating knit structure. Nonetheless, the design process will always necessitate iterations of manufacturing, material study and simulation, as the spectrum of possibilities through machine knitting is immense

Fig. 15 Controlling the relationship of internal knit behaviour and global stress distribution, moments within the textile surface are allowed to maintain their dimensional rib-knit geometry; (Ahlquist 2014)

when the parameters of yarn, knit structure and textile behaviour are all exposed as design variables (Fig. 15).

Acknowledgements This research was developed through the support of the Research Through Making grant from the University of Michigan, Taubman College of Architecture and Urban Planning, as a part of the Knit Architectures project. The initial concepts for the Mobius Rib-knit prototype were developed with students Pandush Gaqi and Yi Yuan, in collaboration with Jane Scott at the University of Leeds. The prototype shown in this paper was developed with the assistance of Jacobo Mendoza.

References

Ahlquist S et al (2013) Physical and numerical prototyping for integrated bending and form-active textile hybrid structures. In: Gengnagel C et al (eds) Rethinking prototyping: proceedings of the design modelling symposium, Berlin, 2013

Ahlquist S, Menges A (2013) Frameworks for computational design of textile micro-architectures and material behavior in forming complex force-active structures. In: Beasley P et al (eds) ACADIA 2013 adaptive architecture: proceedings of the 33rd annual conference of the association for computer aided design in architecture, Cambridge, 2013

Ahlquist S, Kampowski T, Oliyan Torghabehi O et al (2014) Development of a digital framework for the computation of complex material and morphological behavior of biological and technological systems. Comput Aided Des Spec Issue Mater Ecol 60:84–104

Araujo M, Fangueiro R, Hong H (2003) Modelling and simulation of the mechanical behaviour of weft-knitted fabrics for technical applications part I: general considerations and experimental analyses. AUTEX Res J 3(3):111–123

Engel H (2007) Tragsysteme—structure systems, 4th edn. Hatje Cantz, Ostfildern

Kurbak A (2009) Geometrical models for balanced rib knitted fabrics part I: conventionally Knitted 1 × 1 rib fabrics. Text Res J 79(5):418–435

Lewis WJ (2003) Tension structures: form and behaviour. Thomas Telford Publishing, London

Lienhard J, Gengnagel C, Knippers J (2013a) Active bending, a review on structures where bending is used as a self formation process. Int J Space Struct 28 (3/4):187–196

Lienhard J, Ahlquist S, Knippers J, Menges A (2013b) Extending the functional and formal vocabulary of tensile membrane structures through the interaction with bending-active elements. In: Boegner-Balz H et al (eds) [Re]Thinking lightweight structures, Proceedings of Tensinet Symposium, Istanbul, 2013

Thomsen MR, Karmon A (2012) Listener: a probe into information based material specification. Stud Mater Thinking 7:1–10

Vassiliadis S, Blaga M, Provatidis C (2007) Finite element modelling of the warp knitted structure. RJTA 11(4):40–47

Yuksel C, Kaldor M, James D, Marschner S (2012) Stitch meshes for modelling knitted clothing with yarn-level detail. ACM Trans Graph 31(4):37:1–37:12

Thermal Responsive Envelope: Computational Assembling Behavioural Composites by Additive and Subtractive Processes

Isak Worre Foged and Anke Pasold

Abstract

The paper presents an architectural computational method and model, which, through additive and subtractive processes, create composite elements with bending behaviour based on thermal variations in the surrounding climatic environment. The present effort is focused on the manipulation of assembly composite layers and their relative layer lengths thereby embedding the merged material effect to create a responsive behavioural architectural envelope. Copper and polypropylene are used as base materials for the composite structure due to their high differences in thermal expansion, surface emissivity alterations, their respective durability and copper's architectural (visual and transformative) aesthetic qualities. Through the use of an evolutionary solver, the composite structure of the elements are organised to find the bending behaviour specified by and for the thermal environments. The entire model includes the calculation of bending behaviour, the calculation of perceived temperatures inside the envelope and the evolutionary module, which in a design process advance the composite structure in relation to the thermal environment desired. The research presents the methods used and developed, the way in which the behavioural composites are organised in modules and how they act and perform. Furthermore, a large full-scale prototype is made as a demonstrator and experimental setup for post-construct analysis and evaluation of the design research. The work finds that the presented method and model can create 'programmed' responsive composite architectural envelopes and that the organisational method of nested modular elements with nested responsive composites enables a modular building method with embedded dynamic responsive properties.

I.W. Foged (✉)
Department of Architecture and Media Technology,
Aalborg University, Aalborg, Denmark
e-mail: iwfo@create.aau.dk

I.W. Foged · A. Pasold
AREA, Copenhagen, Denmark

M.R. Thomsen et al. (eds.), *Modelling Behaviour*, DOI 10.1007/978-3-319-24208-8_10

Introduction

With an increasing need to address environmental thermal conditioning and energy use, previous research strongly indicates that static buildings do not have the capacity to meet governmental and intergovernmental requirements for energy usage (Winther et al. 2009) and are not meeting the explicit and diverse thermal conditions stated explicitly in building codes and implicitly by building occupants. Approached more ambitiously, architecture could turn this around and offer the construct of environments that seek to accommodate and enrich human life as its primary task beyond building codes, with the additional capacity to improve energy scores. In this way, environmental architecture becomes both the driving element for architectural articulation and a positive influence on sustainable assessment and constructs, currently understood as the task of the indoor climate engineer. The first problem that arises when buildings are environmentally based on pure technical approaches is the physical and mental separation of humans from their context, which has a direct negative physiological and psychological impact on humans (Ulrich 1984; Fich 2014). The second problem is the separation of building and context, limiting the potential for locally articulated architectures, which would aid a move towards a rich and enhanced building culture as a whole (Frampton 1995, 2011; Leatherbarrow 2009; Moe 2013; Mostafavi and Leatherbarrow 1993).

The background for previous solutions by architectural thermally responsive approaches is roughly divided into two orientations: one focused on the use of mechanical and electronic motor-based constructs, and another focused on the use of material assemblies with perceivable dynamic material behaviour. The former has seen industrial applications from automated blinds to complex mechanical systems, following the early work of Jean Nouvel's Institut du Monde Arabe in Paris. This orientation is typically guided by laboratory studies based on a sensor-microprocessor-material assembly-actuator setup, as can be seen in the work of Sterk (2003, 2006), among others (Biloria and Sumini 2009; Foged et al. 2010; Foged and Poulsen 2010; Hoberman and Schwitter 2008; Mossé 2011). Increasingly in recent projects, these efforts have been applied by large architectural offices such as Foster and Partners in the Central Market in Abu Dhabi (2010) and by Aedas Architects in the Al Bahar Towers façade in Abu Dhabi (2012).

The latter approach, material-based responsive systems, has predominantly been present in architectural and engineering academic laboratories. In particular, as discussed previously, numerous studies the response of wood to humidity changes have been heavily explored since the early studies by Hensel (2010), Menges and Reichert (2012). These efforts are currently being continued by the above authors and others.

The proposal here and the background for potential solutions are both contained in the previous studies (Pasold and Foged 2010), and the extended experiments with the organisation of isotropic materials into anisotropic behavioural material composites here. Instead of using a singular embedded material dynamic, such as the hygroscopic properties in wood, this study is based on exploring the dynamic environmental-material properties arising from metal-plastic material composite structures. The previous studies showed the capacities for this approach through the development of an architectural method that would search for these material assemblies in response to thermal agency.

What is done in the present effort is the manipulation of assembly composite layers, thereby 'programming' the merged material effect desired towards modifying thermal environmental conditions. Copper and polypropylene are used as base materials for the composite structure due to their high differences in thermal expansion, surface emissivity alterations, their respective durability, copper's architectural (visual and transformative) aesthetic qualities and their accessibility within the industry, making the study directly accessible to others.

The 'programming' of the combined material is approached by altering the relationship (lengths) between the two material layers (metals and plastics with isotropic thermal properties) into a variable composite structure. The research presents the methods used and developed, the way in which the behavioural composites act and perform, and a large full-scale prototype as a demonstrator and experimental setup for post-construct analysis and evaluation of the design research.

Methods

A set of computational methods and unique modules for the Rhinoceros/Grasshopper computational framework is developed to simulate material behaviour, thermal environmental behaviour and the derived occupancy sensation. These simulations are embedded into an architectural design process model based on evolutionary processes for the development of environmental-human-oriented design proposals and computational fabrication procedures for the making of the physical probe.

Parametric Model

The material behaviour of the two-layered composite is computationally simulated by the use of the Eqs. (1) and (2) (Kanthal 2008):

$$\frac{1}{R_T} - \frac{1}{R_0} = \frac{6(\alpha_2 - \alpha_1)(1+m)^2}{3(1+m)^2 + (1+m \cdot n)(m^2 + \frac{1}{m \cdot n})} \cdot \frac{T - T_0}{s}. \tag{1}$$

$$\left(R_T + \frac{s}{2}\right)^2 = \left(R_T + \frac{s}{2} - A\right)^2 + L^2 \tag{2}$$

Of specific importance are the material thermal expansion coefficients of the two materials (α_1 and α_2), the relationship between the two material elasticities (n, E1/E2, Young's Modulus), the material thickness of the combined composite (m = S1/S2) and material length (L). The former three values remain constant, assuming linear expansion, within the given temperature domain, while the latter, length, is applied as a variable in this specific study. By modifying the length of a three-layered composite (Fig. 1), a new two-layered composite is locally created, which in turn defines the bending behaviour. The bending is induced as one layer extends due to thermal variation more than the other layer. Furthermore, the temperature domain (T) at the surface of the material composite is a variable of the immediate thermal environment. The bending behaviour of the composite results from these interrelated constant and variable values. The format of the nested responsive elements are organised in a 1:3 module format. This is done to investigate the application of modular organisational system of responsive elements.

Environmental Simulation Model

The thermal environmental sensation is simulated with insolation and ambient temperature as input and perceived operative temperature as output. The simulation model is based on the mathematical models described by Fanger (1970), Höppe (1993), Khamporn and Chaiyapinunt (2013). The thermal simulation serves as an exploratory feedback mechanism by focusing on specific perceived temperatures to guide an evolutionary search. Thus, the method and model are open to architectural intent, rather than the search for generalised comfort temperatures.

Evolutionary Model

The open-source evolutionary solver 'Goat', developed by Rechenraum (2013), is used for

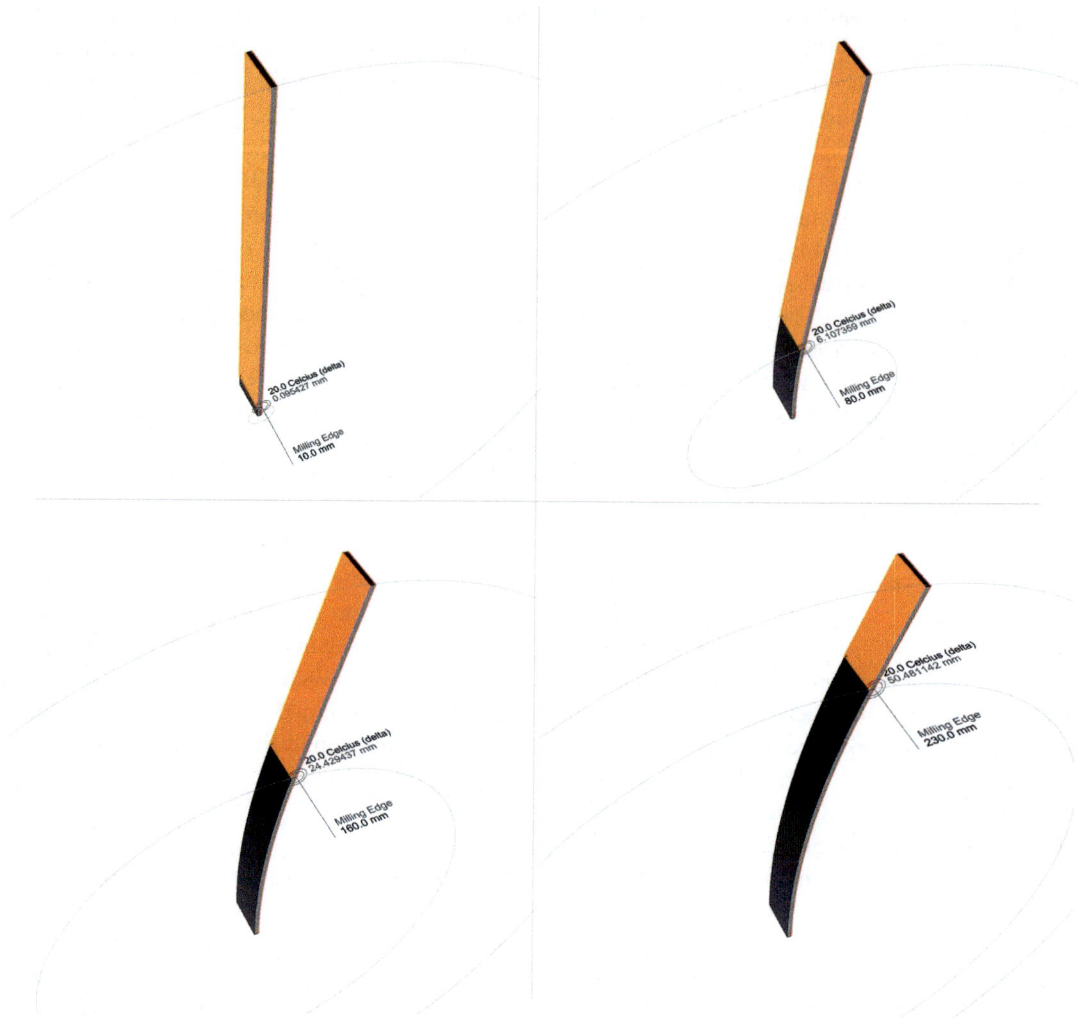

Fig. 1 Model of the bending behaviour (taken from the computational model) that results from applying temperature and one material layer length of the composite structure as variables. Notice the non-linear bending behaviour with linear modification of the variables

these studies. Goat is a solver, which offers a set of evolutionary computational processes with different progressive methods. The description of the evolutionary mechanics is outside the scope of this paper and can be found in the documentation of the solver. When the search algorithm is paired with the above simulations, it advances its architectural thermal proposals by modifying the material composite organisation. Specifically, by changing the length of one of the two layers, the responsive behaviour as a function of temperature variation is explored. This then initiates the development of an organisation of composites that is the makeup of an envelope oriented towards a desired thermal sensation on the *inner* side of the envelope.

By articulating dynamic matter properties through custom composites and integrating time-based thermal sensations (interrelations between material, environment and human occupancy), the method and model attempt to allow a deep time-based co-organisation of both immediate and offset thermal and visual sensations. This is caused by both instant and delayed conditions as a consequence of the dynamics within the model.

Design Experimentation

A computational design experiment is conducted to apply the developed architectural model to environmental search processes. The model enables computational testing of the search procedure, progressively subtracting material from the initial three-layered composite. This process creates the basis for a full-scale physical demonstrator. By using the above methods and models to define responsive behaviours, two milling patterns of the three-layered copper–polypropylene–copper (0.5–3–0.5 mm) composite material are developed. These include a milling pattern removing a series of areas of the 0.5 mm copper layer on one side (Fig. 2) and a milling pattern cutting through the entire composite along three of the four sides of the areas removed by the first milling pattern (Fig. 3). By this method, the new specific composites are 'released' from the base three-layered industrially produced composite into the new two-layered composites, while remaining in a structural lattice. As mentioned, the overall composite geometry is constructed in a 1:3 format, allowing the panels, as elements, to be arranged in a multitude of configurations as an additional instrumental variable when searching for thermal environmental conditions during the design process.

Fig. 2 First milling procedure as a result of the form-finding of behavioural characteristics

To monitor the thermal variations within the prototype, three enclosures are made, each with a set of temperature and light sensors installed at the location of the envelope and in the back of the enclosure (Fig. 5). These register a temperature (Celcius) and light level (Lux) every 60 s, creating a high frequency reading of the thermal environment (Fig. 4).

Three such enclosures are created, each with a different set of computed and fabricated behavioural panels. Furthermore, on one of the enclosures is an external transparent screen installed to encapsulate the thermal environment (Fig. 6). This is done to study how the composite behaviour and thermal environment may react in a potential double-layered envelope.

Results

By studying the organisation of isotropic materials by evolutionary processes, the presented method and model have shown the capacity to 'programme' material behaviour towards the creation of environmental sensations in architecture.

Specifically, the study finds that:

(1) The developed subtractive method can create a 'programmed' responsive composite architectural envelope. By the simple modification of the length of one layer in the composite, advanced behaviour and capacity for thermal articulation are identified and shown. The work finds that the composite

Fig. 3 Second milling pattern releases the behaviour of the new composite structure as a result of the first milling procedure

Fig. 4 Sensors and microprocessor installed for post-construct analysis just inside the membrane and in the back of the space. The prototype consists of 24 panels with the dimensions 490 × 1490 mm. Six clusters of four panels are organised with different milling patterns and orientations. Three clusters are spatially enclosed, with a single cluster being framed further by the external screen (Fig. 6)

Fig. 5 Full scale three-by-eight-meter prototype, including 24 composite behavioural panels and the embedded sensor network for post construct analysis

Fig. 6 A case with transparent acrylic plate installed on the external side of the membrane. It is possible to see the thermal and light sensors installed between the acrylic plate and the membrane just above and to the *right* of the image centre. Viewing the membrane through a transparent plate also modifies the visual appearance, as reflections of the context become visible

organisation (material difference and composite structure difference) has a significant influence on thermal sensations and the visual effect. This can be seen both in the plotted graphs (Fig. 8) based on installed thermal sensors and by direct observation in the full-scale prototype (Fig. 7). What has not been possible to compare is the relation between perceived temperature in the computational model and the physical prototype. The reason for this being that it has not been possible to install occupants in the built prototype.

(2) The organisational method of nested modular elements with nested responsive composites enables a modular building method with embedded transformative (responsive) properties. By nesting the composites in modular nested arrays, the method and model are further instrumental in their

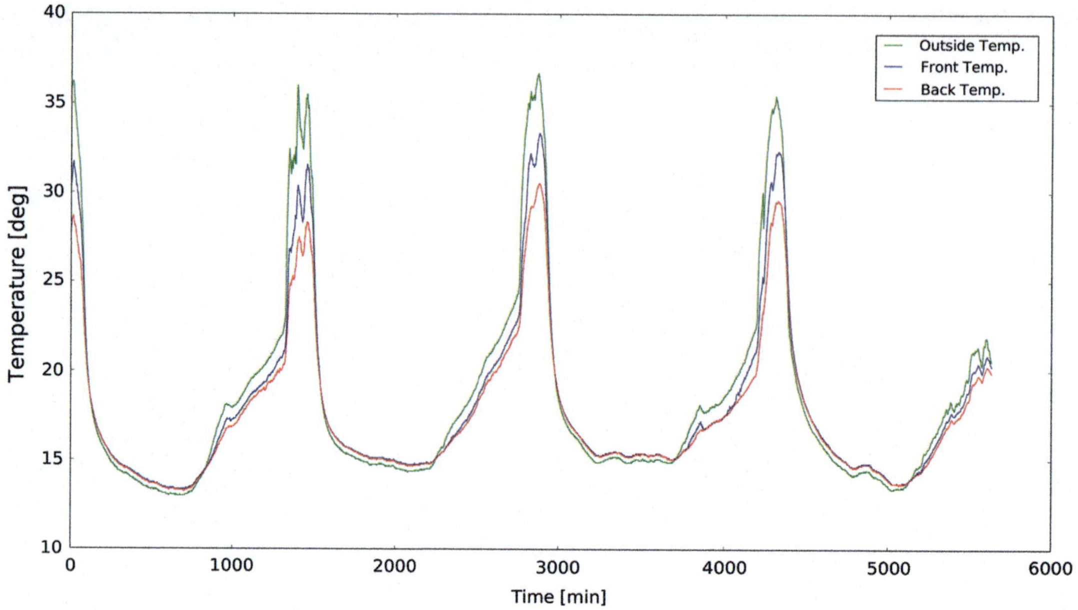

Fig. 7 Graph of plotted temperature values across four days in the cluster with an external screen attached. Visible, among other things, is the effect on the composite behaviour of the transparent screen as an increase in temperature and light is registered

Fig. 8 Full-scale prototype, measuring approximately 3 by 8 m as installed next to the Danish architecture centre for two months, for demonstration, measuring and observation. The image illustrates the articulation of the composite and its resultant bending behaviour

Fig. 9 Detail of behavioural composites created through the organisations of expanding and contracting composite layers

versatility of application and modification by simply changing the organisation of composite structure and array organisation. The simplicity of the modular element extends, in turn, the responsive capacity of the nested composites by allowing various configurations of the composites (Fig. 9).

Conclusions

The work contributes to the efforts to advance architectural behavioural methods and models towards an environmental-human-based architecture that is rooted in a search for environmentally perceived sensations. In contrast to mechanical electro-motor-based systems, environmental registration and actuation is created as an integrative instrumental composite structure. While it does not allow for a re-configuration of behaviour, as is possible with microprocessor based cybernetic models, it includes the capacity to 'programme' immediate and delayed response patterns that can be orchestrated with human behavioural patterns.

Additionally, a transient and dynamic architecture is supported by the environmentally open and responsive properties of the architectural envelope. Such an approach contrasts with the isolation principles described as the predominant approach today with its described problems. In advancing the proposed method and model, additional layers influencing the properties are proposed simply by adding the acrylic layer. This modifies the microclimatic environment of the composites and thus the subsequent modification of the thermal environment for humans. To extend such studies, more advanced modes of acting layers could be integrated, thereby merging several behavioural layers to increase the ability to function in relation to other input values besides that of temperature.

The combination of adding/subtracting composite layers appears to offer another level of design instrumentality, given that a method and model of additive principles alone logically will expand continuously in space. Within this study, the additive and subtractive processes are integrated on two levels: firstly, by the molecular transformative processes of the copper surface through its surface oxidation process; and, secondly, by the design process of adding and subtracting layers of the isotropic composites.

While the panels respond according to the computed temperature variations, there are also unforeseen variations in the bending behaviour detected in the physical prototype. Some elements bend more than others, despite the intention of creating the same behaviour. The reason for this is understood to be caused by three factors related to the manufacturing process. The first being that the first milling pattern, removing the surface layer, must be exact as to not remove too much material. Another aspect is the temperature control when the original three-layered panels are merged. If the temperature in this process is different to the information integrated into the design model or if there exist variations between panels, the base conditions for calculating the bending variations are flawed. Lastly, heat is created in the manufacturing process, which may influence the structural makeup of the polypropylene layer during the removal of the copper layer, effectively also modifying its expansion properties. If any of these factors are not as computed during the design process the bending behaviour will vary. This in turn means that a potential application needs to go through more studies of how to control the manufacturing process of the composites, beyond the method and model shown.

References

Biloria N, Sumini V (2009) Skin systems: a morphogenomic developing real-time adaptive building performative building. Int J Archi Comput 07(04):643–676

Fanger PO (1970) Thermal comfort. Danish Technical Press, Denmark

Foged IW, Poulsen EB (2010) Environmental feedback and spatial conditioning. In: ACADIA2010—Proceedings. ACADIA

Foged IW et al (2010) Shape control of responsive building envelopes. In: Proceedings of the international association for shell and spatial structures symposium 2010. IASS

Fich L (2014) Towards a neuroaffective approach to healing architecture. Aalborg University

Frampton K (1995) Studies in tectonic culture: the poetics of construction in nineteenth and twentieth century architecture. MIT Press, Boston

Frampton K (2011) Reflections on the state of things. Including the Membrane, Princeton

Hensel M (2010) Performance-oriented architecture: towards a biological paradigm for architectural design and the built environment. FORMakademisk 3(1):36–56

Hoberman C, Schwitter C (2008) Adaptive structures: building for performance and sustainability. Des Intell

Höppe PR (1993) Heat balance modelling. Experientia 49 (9):741–746

Kanthal (2008) Kanthal thermostatic bimetal handbook

Khamporn N, Chaiyapinunt S (2013) An investigation on the human thermal comfort from a glass window. Eng J 18(1):25–44

Leatherbarrow D (2009) Architecture oriented otherwise. Princeton Architectural Press, New York

Menges A, Reichert S (2012) Material capacity: embedded responsiveness. Archit Des 82(2):52–59

Moe K (2013) Convergence: an architectural agenda for energy. Routledge, New York

Mossé A (2011) Materializing a responsive interior: designing minimum energy structure based on dielectric elastomers. In: International adaptive architecture conference, pp 1–13

Mostafavi M, Leatherbarrow D (1993) On weathering—the life of building in time. The MIT Press, Massachusetts

Pasold A, Foged IW (2010) Performative responsive architecture powered by climate. In: Computer aided design in architecture, pp 1–14

Rechenraum (2013) Goat—evolutionary solver. http://www.rechenraum.com/en/

Sterk TE (2003) Using actuated tensegrity structures to produce a responsive architecture. In: Klinger K (ed) Connecting. ACADIA, pp 84–93

Sterk TE (2006) Shape control in responsive architectural structures—current reasons and challenges. In: 4th World conference on structural control and monitoring, pp 1–8

Ulrich RS (1984) View through a window may influence recovery from surgery. Science (New York) 224 (4647):420–421

Winther FV, Heiselberg P, Jensen RL (2009) Intelligent glazed facades for fulfilment of future energy regulations, pp 1–7

Formations of Energy: Modelling Toward an Understanding of Open Thermodynamic Systems

Jacob Mans and James Yamada

Abstract

This inquiry explores how cellular automations (CA) can be used to characterize open thermodynamic systems. Open thermodynamic systems describe the formation of buildings, landscapes, and cities; systems shaped by the exchange of energy and material flowing through the larger environments of which they are a part. Unlike models focused on the formation of objects, this inquiry focuses on cell-based modelling that brings us into contact with the forces that shape objects. This research uses CA to visualize the dynamic and constantly evolving exchange of mass and energy that characterize these systems. This methodology explores "non-expert" modelling, i.e. models generated across disciplinary boundaries, in an effort to yield trans-disciplinary insights. This model connects architectural practices to silviculture literature through a simulation that characterizes the role of disturbance on the spatial patterns of forest succession. While this model mimics forest development at the landscape scale, forests may also be modelled as a micro-scaled variable in a global carbon model or as macro-scaled variable in models that characterize thermodynamic performance at the building-scale.

Introduction

As digital modelling informs more of our design techniques we need to be aware of the habits of mind these workflows carry. This inquiry focuses on modelling that brings us into contact with the *forces* that shape objects in an effort to understand the flows of material and energy across our environments. These flows need to be defined thermodynamically, both in the context of conservation as well as the less often considered context of energy dissipation. As a framework for critique we consider modelling practices that produce fictions based on the accumulation of opaque numerical findings, predetermined inputs, and modelling practices themselves. Alternatively we explore open systems through the dynamic formations shaped by

J. Mans (✉)
GSD, Harvard University, Cambridge, USA
e-mail: jacobmans@gmail.com

J. Yamada
MetaLAB, Harvard University, Cambridge, USA
e-mail: jsyamada@gmail.com

M.R. Thomsen et al. (eds.), *Modelling Behaviour*, DOI 10.1007/978-3-319-24208-8_11

Fig. 1 Fire disturbance from forest simulation

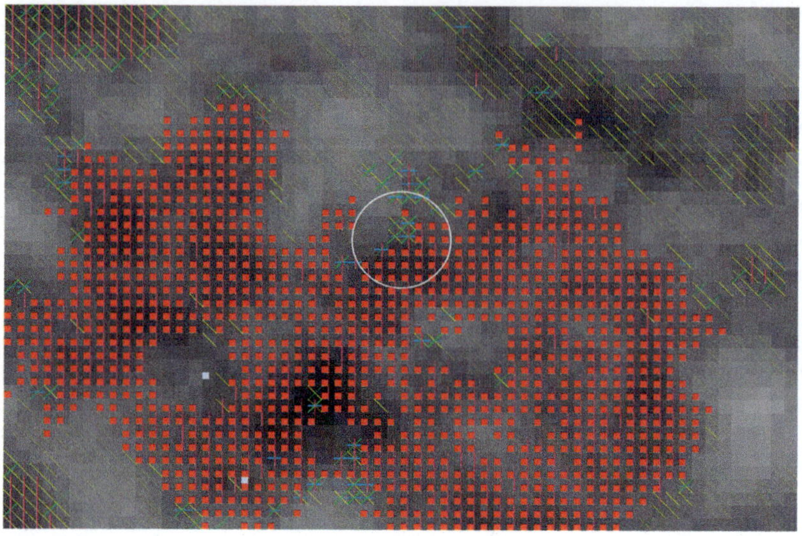

the neighbourhood exchanges and systems boundaries in non-expert CA models. We evaluate our methodology, which combines the use of system diagramming to synthesize reviewed literature and non-expert, but projective, CA modelling, to see what insights can be gained into the trans-scalar exchanges of open systems. We conclude with a discussion of boundaries, the limitations of our chosen modelling technique, and the goal of modelling toward an understanding of open-thermodynamic systems (Fig. 1).

Background

An Open World

Buildings, cities, and environments share an irrefutable similarity—they all behave as open thermodynamic systems. Moe, characterizes architecture's open system behavior as, "an absolute, irreducible, and uncontestable reality that must be the basis of any non-modern agenda for energy in architecture in the future" (Moe 2014, p. 20). Modern design practices, however, align with reductive methodologies that dissect and isolate objects from their larger systems. While these practices may occasionally be needed in order to simplify and understand those objects, they should remain temporary. An isolated analysis should be reintroduced into its proper non-isolated context. Not doing so risks misunderstanding of our energetic milieu (Moe 2014, p. 21).

Modelling Practices are Isolating

If we couple Moe's earlier characterization with a question posed by Schrödinger asking why should "laws and regularities… apply immediately to the behaviour of systems which do not exhibit the structure on which those laws and regularities are based" (Schrödinger 1956, pp. 2–3), then a fictions regarding our models appear. Simply put, our models are inherently isolated. There are, even in very complex models, a limited number of variables that the model can internalize. As a result an excessively large number of variables are externalized. For the variables our models internalize there *may* be logics that describe their phenomena individually, but the complexity of their interactions and the implications of their interlocking logics are not easily understood (Winsberg 2001, p. 442).

Furthermore, what is understood from the model is derived not only from these logics, but from the practice of modelling itself, which as Winsberg describes produces an additional set of fictions that must be assessed (Winsberg 2001, p. 445). If our models are isolated; if the implications of their interactions are not understood; if their results are influenced by the practice of modelling; how can we rely on them to describe what we attempt to model?

Toward an Understanding of Open Systems

Designers often rely on software that produces models that are closed, static or snapshots of dynamic and complex systems. As an experiment for breaking out of this habit of mind, a precedent for modelling complex, decentralized, dynamic systems was found in cellular automata, or CA. Using CA, these systems could consist of neighborhoods of cells and individual cells, and could seemingly mimic an open system up to the scale of the cell neighborhood and small collections of adjacent cell neighborhoods.

Furthermore, while it is impossible to establish universal laws that predict the behaviour of complex natural systems, it is possible to mimic wide-spread repeatable patterns in those systems (Lawton 1999), so while generalizations cannot project specific changes in complex natural systems over time, it may be possible to mimic patterns in nature using cellular automata (Rohde and Heikki 2005).

A design interface that utilized custom-scripted CA served as a point of departure for this effort and builds on its potential to express the complexity and self-organized propagation of natural systems through an iterative application of simple rulesets on a given condition. This lent itself to studying how temporary formations of energy in forests self-organize in space over time, particularly since other CA studies have been conducted for finding rule-sets that lead to the spatial complexity or apparent randomness found in natural systems (Wolfram 2002).

The interface developed in this effort is based in-part on a canonic form of CA: Conway's Game of Life, as developed in Processing by Daniel Shiffman. The Game of Life served as a structural precedent for this experiment to work from, and it was modified heavily from its topographically flat, environmentally static, binary form in order to produce a design interface for topographically varied forest landscapes, with stochastic exchanges between cells that are influenced by topography, and with environmental disruptions to those cells in the form of destructive natural events that have their own unique sets of spatial characteristics.

It could still be pushed further: CA practices exist for simulating the behaviour of natural systems over time (Rohde and Heikki 2005, p. 206), and have led to adjustments in how CA is organized. According to Dunn (2010), it was found that organizing a cellular automaton into a multi-resolution and multi-scale hierarchy helped to reflect spatial structures found in ecosystems, particularly when combined with isotropic models of propagation.

That said, it is important to note that this effort was a design experiment. Giving a user the ability to manipulate of the model with respect to its simplified emulation of the observed behavior for multiple distinct but interrelated phenomena with overlapping boundaries was the primary focus.

Methods

This study took the form of a CA-based design interface that permits users to actively experiment with encouraging, discouraging or preventing different forms of energy exchange across a fictional topographic landscape by manipulating the landscape's topography while the simulation runs across a contracted time scale, with each update corresponding to approximately 1 year. The user is only able to manipulate topography in order to influence the aforementioned energy exchanges—the other variables at work within the model are not directly manipulated by the user (Figs. 2 and 3).

Each cell represents a single large forest stand, and its neighborhood consists of its eight adjacent neighbors. The topographic elevation of the cell is assumed to determine its ability to support the growth of a forest stand. The higher the cell, the slower a forest stand will grow since it is assumed to be exposed to higher typical winds, and has both shallower and drier soil—if the cell is too high, nothing will grow. Conversely, the lower the cell, the faster a forest stand will grow.

Energy exchange occurs in two primary forms: "constructive" and "destructive" from the perspective of the forest. Both forms of energy exchange involve variables at the cell scale, the neighborhood scale and the global scale. Constructive energy exchange is expressed in the succession of a forest stand within each cell as influenced by factors internal and external to the cell. Destructive energy exchange is a set of disruptions to forest succession that occur with varying degrees of frequency and remove or redistribute the energy of cells that they impact across the landscape—they include lightning, fire, cyclones, downbursts, pests, and clear-cutting. Each of these disruptions have their own combination of spatial characteristics.

Global variables at work for both constructive and destructive energy exchanges consist of average annual high-air-temperature, humidity, rainfall, and the number of storms per annum. Wind direction is randomly updated for every

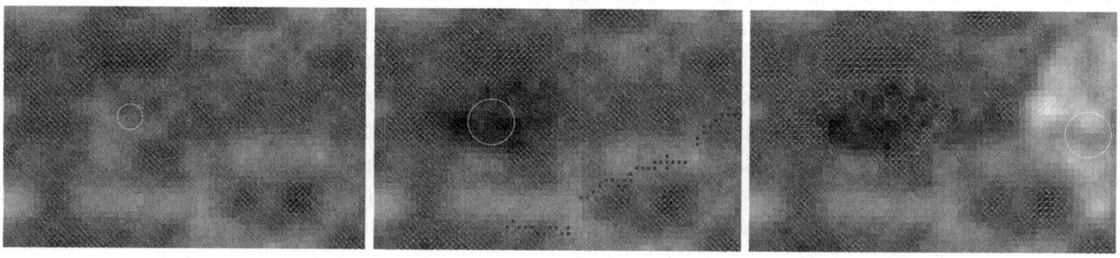

Fig. 2 Terrain modification time lapse

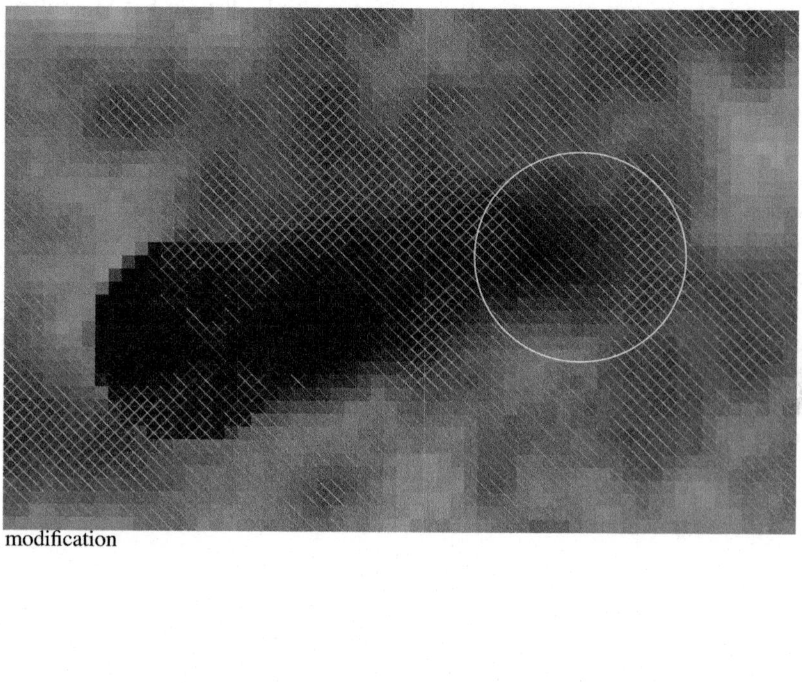

Fig. 3 Terrain modification

disruption in which it is a factor in order to account for its impact and variability across relatively short time spans.

The specific forms of energy exchange and the ways they are influenced by variables at different scales are described in more detail within a set of supporting systems diagrams. These diagrams are important on two fronts: first, they provide a visual framework to organize the information pulled from relevant scientific literature on each of the different disturbance types, and second, they document the system/boundary analysis concealed within the algorithm of the model. The diagrams below are a visual synthesis of the literature reviewed in an effort to understand the formations of energy in the landscape and were used to select variables that were parameterized to create the model.

Constructive Energy Exchanges and Their Boundaries—Forest Succession

As stated above, this study utilizes patterns of forest succession that are influenced by topography and external disturbances. Specifically the internal model represents a New England mixed hardwood forest where the cell growth of one phase impedes the growth of prior phases, with the exception of the old growth phase, which inhibits growth in phase besides those in the stand initiation phase, which then come to represent shade-tolerant tree species. The CA model progresses through 4 potential values for each cell in an automaton that corresponds to the phases of forest development: stand initiation (yellow), stem exclusion (green), understory re-initiation (blue), and old growth (red) (Oliver and Larson 1990, p. 467; Franklin et al. 2002, p. 405). Each colour was given its own alpha channel that corresponded to the amount of energy dissipated in each phase—i.e. the degree to which trees of a particular phase in each cell are working to resist energy gradients, or entropy (Kay and Schneider 1994).

As mentioned above, the elevation of the cell is assumed to determine its ability to support the growth of a forest stand. Perlin noise was used to produce initial variations in each cell's elevation. Cells at a lower elevation are darker, and those at a higher elevation are lighter.

While this fictional landscape begins with a distribution of stands in the stand initiation phase, is occasionally "seeded" in one or two random spots with every update, similar to seed transfer in natural systems. Cells that are seeded gradually develop over time through the phases of succession listed above. The more they develop, the more likely they are to seed adjacent cells. Once the value for one phase within a cell exceeds a certain threshold, the following phase of succession begins. The growth process continues this cycle up through the old growth phase. The model spatially responds to fire, wind, pest and clear-cutting disturbances (Figs. 4, 5, 6, 7, 8, 9, 10 and 11). When introduced, the disturbances alter the boundary conditions of the cells in the affected area, affecting future exchanges across affected cells.

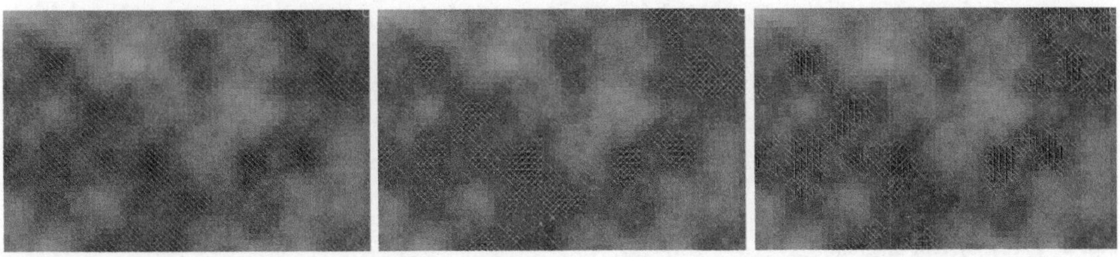

Fig. 4 Forest succession time lapse—internal model

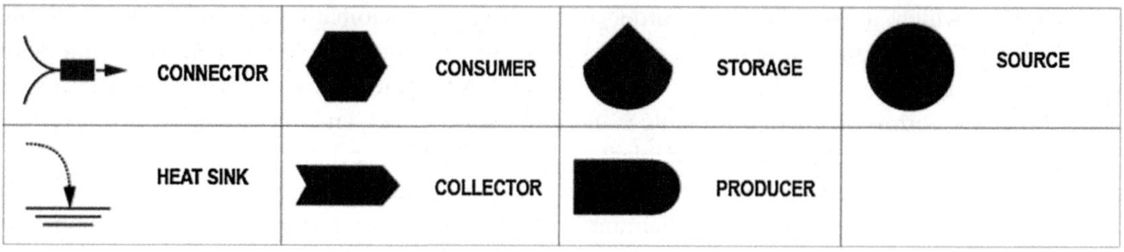

Fig. 5 System diagram key, based on Odum (2007) and Moe and Srinivasan (2015)

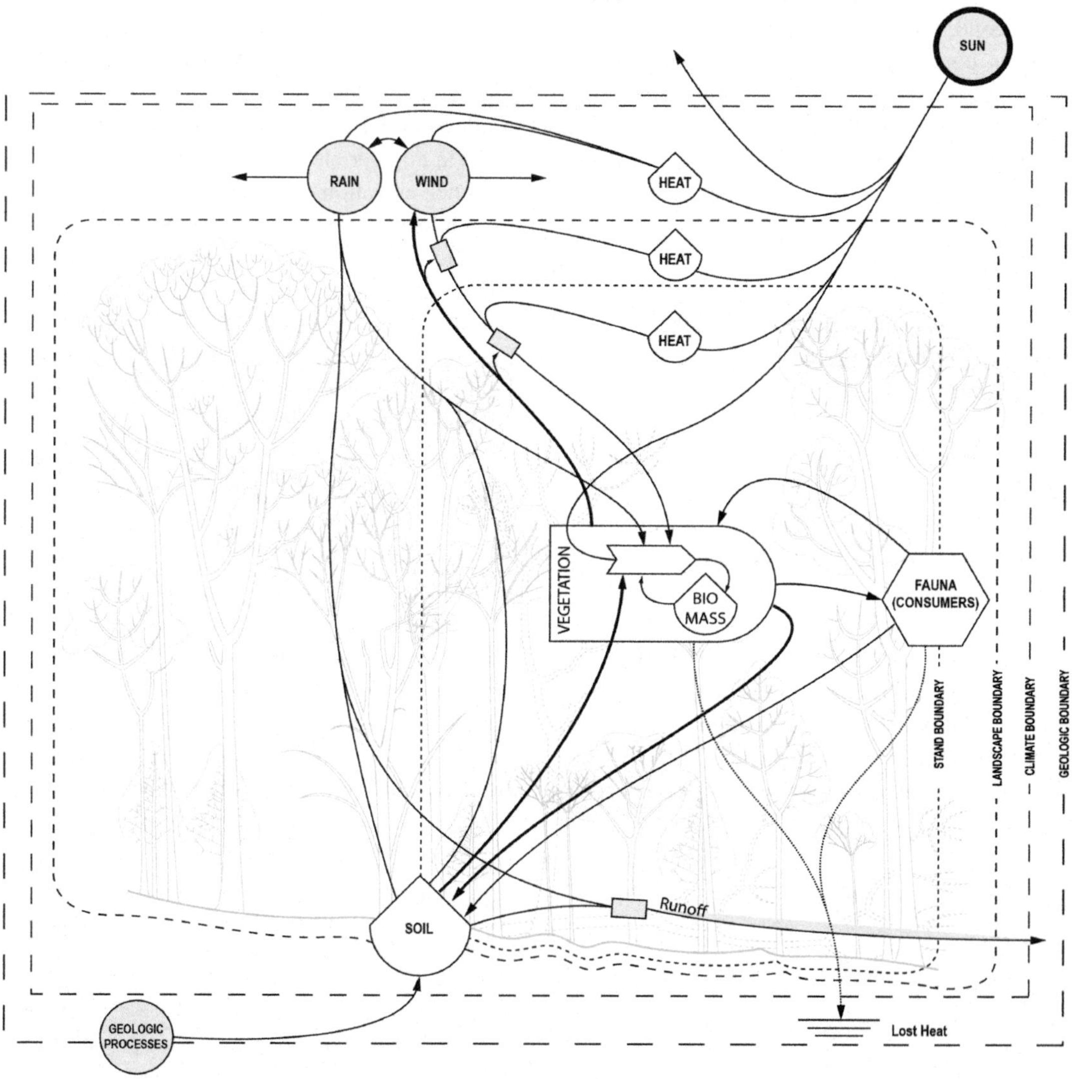

Fig. 6 Forest system diagram

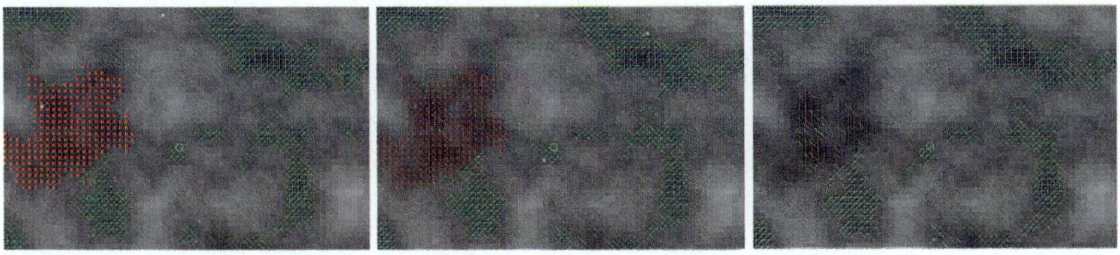

Fig. 7 Fire disturbance time lapse

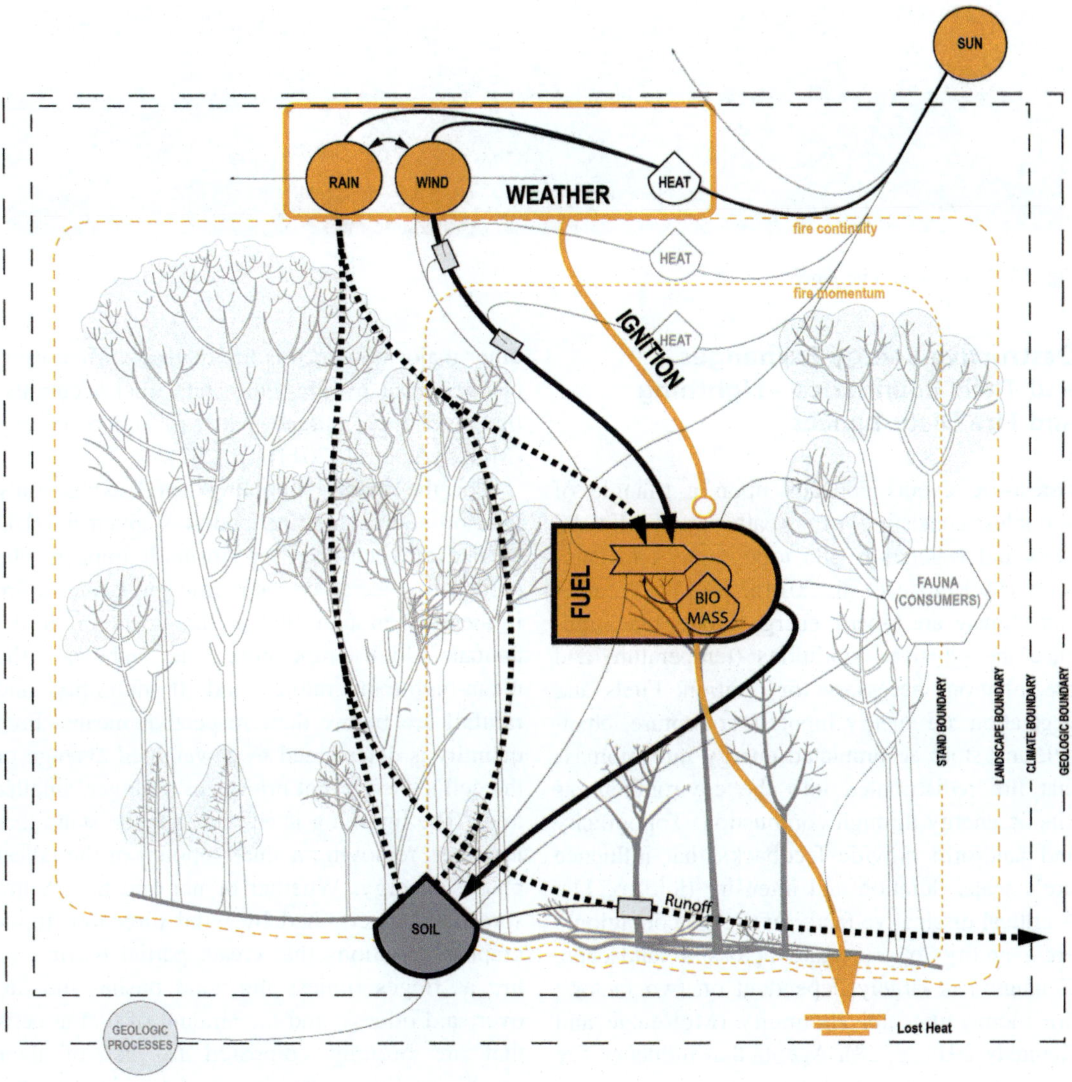

Fig. 8 System diagram of fire disturbance boundaries

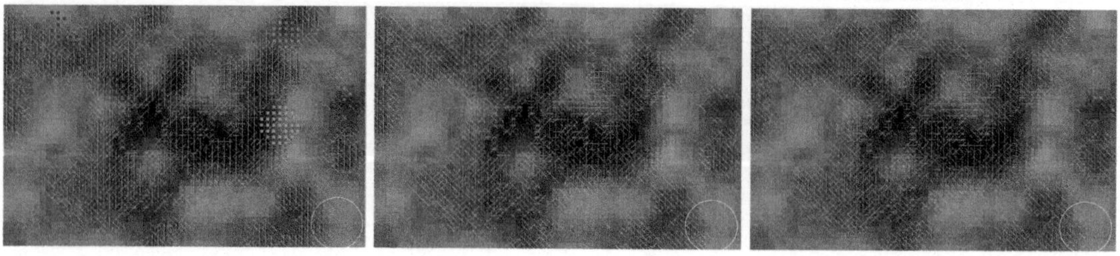

Fig. 9 Cyclone time lapse

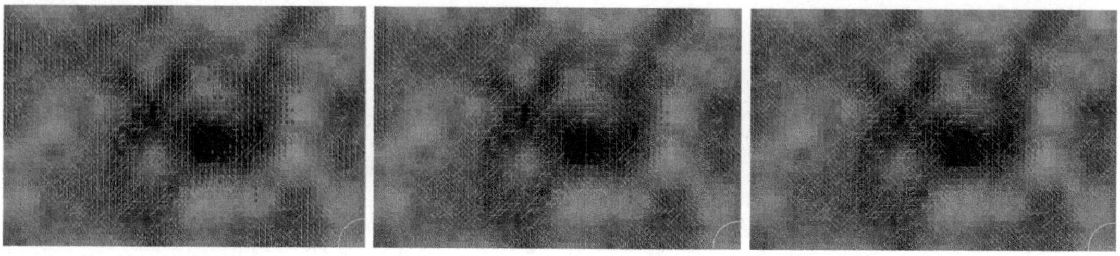

Fig. 10 Downburst time lapse

Destructive Energy Exchanges and Their Boundaries—Lightning and Fire Disturbances

Forest-fire events depends upon a triangle of variables consisting of: weather and climate, fuels and vegetation, and topography and landform (McKenzie et al. 2011a, p. 11). Weather and climate are kinetic energy inputs that influence the physical conditions (temperature and precipitation) necessary for ignition. Fuels and vegetation are energy inputs that capture, channel, and store accumulated energy into biomass that fire redistributed into the environment as kinetic energy through combustion. Topography and landform provide feedbacks that influence fire's scale, duration and intensity (ibid, p. 11). A critical dissipative factor of fire is "contagion"; fire's ability to propagate across a landscape. Contagion is largely dependent on two factors: fire momentum and continuity (McKenzie and Kennedy 2011, p. 28). Agents that influence fire momentum include fuel moisture and air movement patterns. Fire continuity is the product of the connectivity of fuel and topography. These influences interact with the landscape to determine the extent and intensity of the fire. One other major influence is fire memory. Memory is characterized by fire-return rate, fuel accumulation, fuel type, climate patterns, and continuity (McKenzie et al. 2011b, p. 297).

For the model, lightning acted as the only ignition source for fire, and was given a 1/100 chance of occurring per storm. It impacts one randomly selected cell at a time and ignites a fire if there is enough fuel at that location, if the annual high-temperature is above the mean-high-temperature, and if humidity and rainfall are below their respective means. Fuel quantity is determined by a weighted average of the cell's energy that prioritizes younger, smaller trees. The lightning itself damages the stand that it strikes, removing a small amount of the tallest phases' energy. Whether or not and how a fire spreads is determined by wind direction, topographic variations that create partial barriers to fire as ridges (unless the wind pushes the fire over said ridges), and the amount of fuel at cells that are burning compared to that of their neighbors. Fuel quantities are determined by the same weighted average as that for ignition by lightning. Due to these factors, fire is most severe in valleys with energy-rich forest stands, but does not usually spread very far if fuel near the

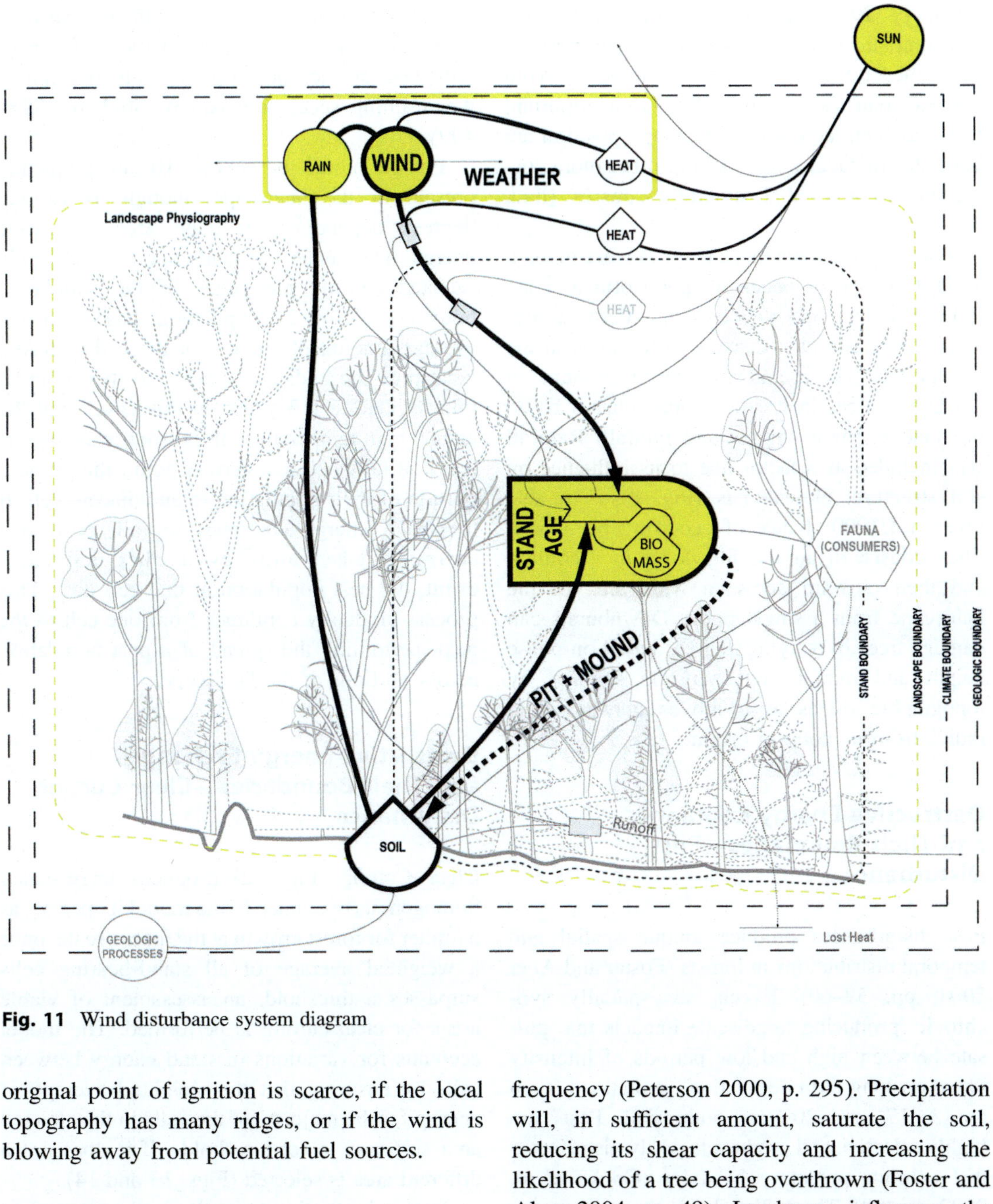

Fig. 11 Wind disturbance system diagram

original point of ignition is scarce, if the local topography has many ridges, or if the wind is blowing away from potential fuel sources.

Destructive Energy Exchanges and Their Boundaries—Wind Disturbances

The prime factors of wind events are weather, landscape, and forest vegetation. Disturbance patterns vary with weather type, intensity, and frequency (Peterson 2000, p. 295). Precipitation will, in sufficient amount, saturate the soil, reducing its shear capacity and increasing the likelihood of a tree being overthrown (Foster and Aber 2004, p. 49). Landscape influences the distribution, size and shape of areas damaged in an event. Physiography describes the forest's exposure to factors such as stand structure and topography. Soil structure varies in response to landscape conditions. Root structures develop in response hydrology producing soils with spatially distinct overturning capacities (Peterson

2000, p. 295). The most critical vegetative characteristic is size. It is generally understood that taller trees are more susceptible to wind damage than shorter trees. Age is an important factor as both large size and canopy position are attributes of older trees, making them more susceptible to non-lethal structural stresses (ibid, pp. 298–299).

For the model, cyclones (Fig. 7) were given a 1/1000 chance of occurring per storm and are influenced by wind direction and Perlin noise. They do catastrophic damage to the trees in the old growth phase and a reduced amount of damage to trees in less-developed phases based on relative phase heights and rainfall. Rainfall was included as a factor due to its influence on soil structure. Downbursts (Fig. 8) were also given a 1/1000 chance of occurring per storm. They consist of air moving down in a column and then outward across a wide area of the landscape from a single point. Downbursts can damage trees by varying degrees based on phase height and rainfall, but may be stymied by topographic ridges as it moves outward in a radius from its point of origin.

Destructive Energy Exchanges and Their Boundaries—Pest Disturbance

Pest disturbances produce unique spatial and temporal distributions in forests (Foster and Aber 2004, pp. 59–60). Events are spatially synchronic, producing large-scale impacts that pulsate between high and low periods of intensity based on long-term population oscillations (such as 13, 17, and 20 year outbreaks). Pests are highly selective and produce the only disturbance that exhibits predatory behavior (Cooke et al. 2007, p. 488). These disturbances are driven by complex biotic relationships between the host plant and the pest group including: pest's preference, pest population dynamics (reproductive and life-cycle dynamics), and host plant tolerance to herbivory. Abiotic factors include: seasonality, climate, and landscape/host species connectivity. Pest's select host species that provide a particular advantage to the pest species. This selection leads re-composition of forests with host species presence severely reduced in the resulting succession pattern (ibid, pp. 498–499).

For the model, pests (Fig. 10) are frequently present in the landscape, though to varying degrees depending on their access to food energy. They enter the landscape at a randomly selected cell on the edges of the simulation boundary, grow in population based on the average amount of energy for the cell's phases, what time the update is relative to their population oscillation (13 years, in this case), and the annual high temperature. If the energy average at that cell falls below a certain point, they send a portion of their population to an adjacent cell. If the cell's energy continues to fall, or if that energy is taken away by another destructive event, the pest population at that cell dies. This process gradually continues from one cell to the next, permitting the spread of a pest population across portions of the landscape.

Destructive Energy Exchanges and Their Boundaries—Clear Cutting Disturbance

Clear Cutting (Fig. 12) produces large-scaled homogenous openings. In the model it operates as a limiter for forest growth at the macro scale: once a weighted average of all stand-bearing cells surpasses a threshold, an assessment of viable areas for clear-cutting is performed. The model accounts for variations in stand energy between cells by ensuring that the weighted average of energy for the majority of the cells in the affected area surpasses a set threshold—if it does not a different area is selected (Figs. 13 and 14).

Timber harvesting in reality is also dependent on consumer demands. Furthermore, physiography is a significant factor that limits the access and extraction ability of harvesting operations. Season/climate have similar impacts and also factor into the severity of soil disturbance thus impacting the rate and species composition of subsequent stand initiations (Fig. 15).

Fig. 12 Pest disturbance time lapse. Pests are light cyan

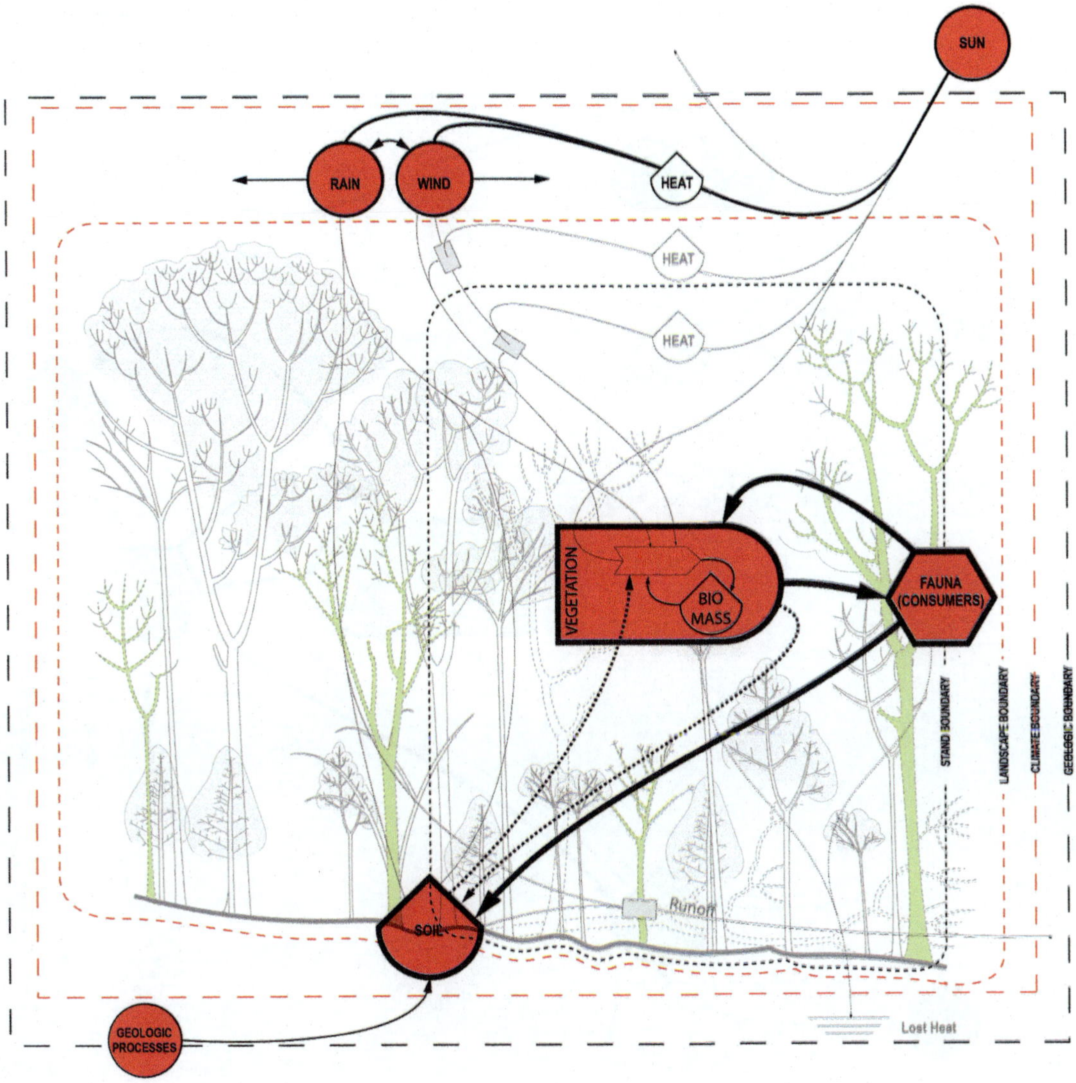

Fig. 13 Pest disturbance system diagram

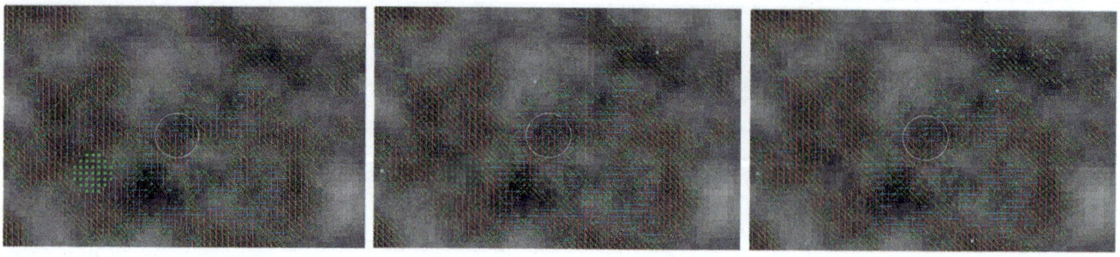

Fig. 14 Clear cutting disturbance time lapse

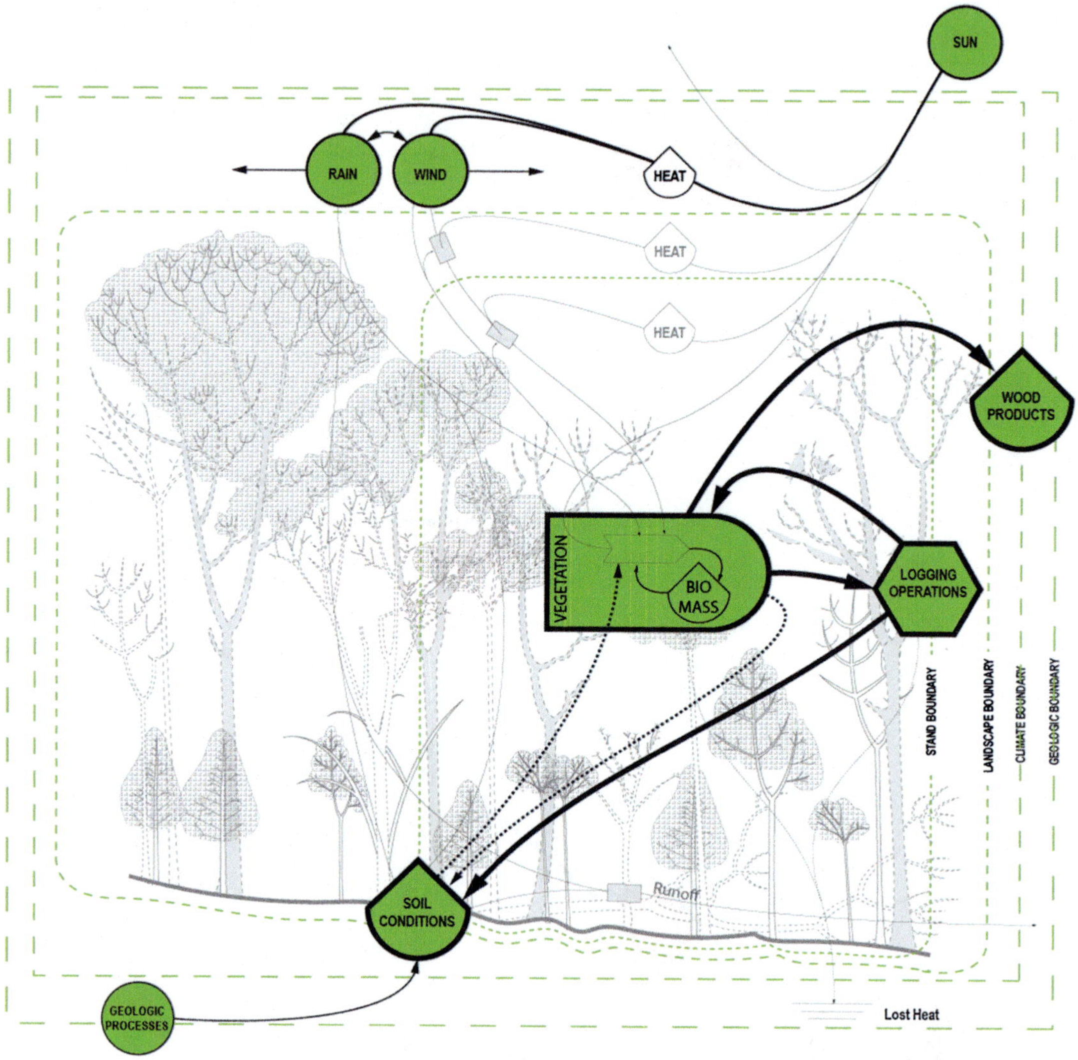

Fig. 15 Clear cutting system diagram

Discussion

Fictions are both intrinsic to the hypothetical aspects of modelling and are a practical necessity for computational tractability (Winsberg 2010, pp. 120–135)—however, with the need for fictions comes the need for their design and curation toward a valid model's construction.

Boundaries of Narrative

Our model included several disruptions to energy formation that were simplified representations of phenomena that are often described through narrative. While this could be more rigorous given further time for development, more thought must be given as to how they might work with or against each other, particularly in light of "agents" in the landscape. Repeatedly within the model, pest populations were wiped out by larger forest fires or were reduced by wind damage. However, while the weakening of a stand by pests should make it more susceptible to wind damage from a downburst or cyclone, it's unclear how badly wind damage would actually impact pest populations or if it might actually help them spread to other adjacent stands by carrying them aloft. Similarly, if enough plant matter were to be blown from once cell to another by a cyclone, would it in turn make the latter cell more vulnerable to fire? Would old growth in a stand that partially survived a fire but lost its canopy be less vulnerable to wind damage?

Boundaries of Computational Scope

Users were provided with a way to experiment with spatial boundaries for energy exchange through topography, and this impacted the spatial effect that every included form of energy exchange had on the model, either directly or indirectly by affecting forest stand growth. However, in many of the model's energy exchanges, climate is cited as a boundary and yet is left outside of the user's influence. In addition to being completely automated, climate is treated as a simple, noisy, cyclical set of global variables. There is a question as to whether or not the user should be given the ability to directly intervene in the model's approximation of a climate: is this effort ultimately about a user authoring systems that are open to each other but are all within an enclosed environment, or is it about authoring systems that are simply open to outside influences beyond the user's control?

Conclusion

Our non-expert CA forest succession model openly exhibits the isolated pitfalls of other models. However, as an open thermodynamic design tool our methodology presents several insights worthy of extended discourse. First, designers must consider more seriously the role of the non-expert in general, and the non-expert model in specific. Trans-scalar practices require us to selectively un-educate ourselves in order to see outside our increasingly narrow design specializations—hence, the architect mucking about in forest succession models seeking a better understanding of how the two may collaborate. Second, models that bridge disciplines will incorporate increasingly more disparate data inputs requiring rigorous workflows to identify internalities, externalities, biases and assumptions—hence, the inclusion of systems diagramming within a modelling methodology for transparency and as a provision of scale analysis applied to different trans-disciplinary inputs. Last is a consideration for the underlying model structure, in our particular case CA. The logic of CA relies on the boundary conditions between neighboring cells to dictate the type and quantity of data exchanged between cells. In this way, CA is a diagram of the open system being modeled—thus, CA provides an interesting connection between model theory and practical application.

References

Berlik M, Kittredge D, Foster D (2002) The illusion of preservation: a global environmental argument for the local production of natural resources (Harvard Forest paper; no. 26). Harvard Forest, Harvard University, Petersham, Mass

Bogost I (2006) Unit operations: an approach to videogame criticism. MIT Press, Cambridge

Cooke BJ, Nealis VG, Régnière J (2007) 15—insect defoliators as periodic disturbances in northern forest ecosystems. In: Johnson EA, Miyanishi K (eds) Plant disturbance ecology: the process and the response. Elsevier/AP, Amsterdam, Boston, pp 487–525

Dunn A (2010) Hierarchical cellular automata methods. In: Hoekstra AG, Kroc J, Sloot PMA (eds) Simulating complex systems by cellular automata. Springer, Heidelberg

Foster DR, Aber JD (2004) Forests in time: the environmental consequences of 1,000 years of change in New England. Yale University Press, New Haven

Lawton J (1999) Are there general laws in ecology? Oikos 84(2):177–192

McKenzie D, Kennedy MC (2011) Scaling laws and complexity in fire regimes. In: McKenzie D, Miller C, Falk DA (eds) The landscape ecology of fire. Springer, Dordrecht

McKenzie D, Miller C, Falk DA (2011a) Introduction. In: McKenzie D, Miller C, Falk DA (eds) The landscape ecology of fire. Springer, Dordrecht

McKenzie D, Miller C, Falk DA (2011b) Synthesis: landscape ecology and changing fire regimes. In: McKenzie D, Miller C, Falk DA (eds) The landscape ecology of fire. Springer, Dordrecht

Moe K (2014) Insulating modernism: isolated and non-isolated thermodynamics in architecture. Birkhäuser, Basel, Boston

Odum H (2007) Environment, power, and society for the twenty-first century: the hierarchy of energy. Columbia University Press, New York

Oliver CD, Larson BC (1990) Forest stand dynamics. McGraw-Hill, New York

Peterson CJ (2000) Catastrophic wind damage to north american forests and the potential impact of climate change. Sci Total Environ 262(3):287–311

Rohde K, Heikki S (2005) Cellular automata and ecology. Oikos 110(1):203–207

Schneider ED, Kay JJ (1994) Life as a manifestation of the second law of thermodynamics. Math Comput Model 19(6):25–48

Schrödinger E (1956) What is life? And other scientific essays. 1st edn. Doubleday, Garden City, NY

Shiffman D (2012) The nature of code. Creative Commons, Mountain View

Srinivasan R, Moe K (2015) The hierarchy of energy in architecture: emergy analysis. Routledge

Winsberg E (2001) Simulations, models, and theories: complex physical systems and their representations. Philos Sci 68(3):S442–S454

Winsberg E (2010) Science in the age of computer simulation. University of Chicago Press, Chicago

Wolfram S (2002) A new kind of science. Wolfram Media Inc, Champaign

Thinking Massively Parallel: Design Modelling Thermoactive Architecture

Jeffrey L. Boyer, Yao Yu and Ajit Naik

Abstract

Advancing the design of thermally activated surfaces in architecture requires modelling the complex interplay between form, materiality and building systems. In this paper we discuss accelerating numerical methods for modelling the specific multi-physics behaviour encountered by high performance computing (HPC) on massively parallel Graphics Processing Unit (GPU) hardware. Through two case examples: The Chicago Chinatown Public Library in collaboration with Skidmore, Owings and Merrill and a confidential project with Ateliers Jean Nouvel we illustrate how adapting simulation methods can enable lighting, thermal and energy solutions to co-exist and intensely interact within the design process. In both cases massively parallel serves as both technical approach and metaphor for a concurrent process of simulation, design and emergence.

Introduction

All building energy modeling software has been designed and optimized for sequential serial processing on CPU architecture. This runs counter to the computer graphics and gaming industries that have focused on massively parallel GPU computing. A consequence of legacy CPU computing is that substantial changes to an energy model, such as geometric manipulation, can have an unacceptably high latency cost constraining the reasonable design space for exploration. In contrast, parallel processing allows for multi-scalar, multi-parameter quantification, concurrently. This in turn can allow the simulation activity itself to run in parallel with the design process (Fig. 1).

To put the opportunity for parallelization into perspective, in 2001, when the Department of Energy released *EnergyPlus,* a combination of the legacy *DOE-2* and *BLAST* simulation programs, the fastest computer in the world, ASCI White, had a theoretical processing speed of 12.3 teraflops, cost 110 million US dollars and covered the space of two basketball courts. This year, four NVIDIA GTX 980 graphics cards can

J.L. Boyer (✉)
University of Michigan and dbHMS, Chicago, USA
e-mail: jboyer@dbhms.com

Y. Yu · A. Naik
dbHMS, Chicago, USA
e-mail: yyu@dbhms.com

A. Naik
e-mail: anaik@dbhms.com

M.R. Thomsen et al. (eds.), *Modelling Behaviour*, DOI 10.1007/978-3-319-24208-8_12

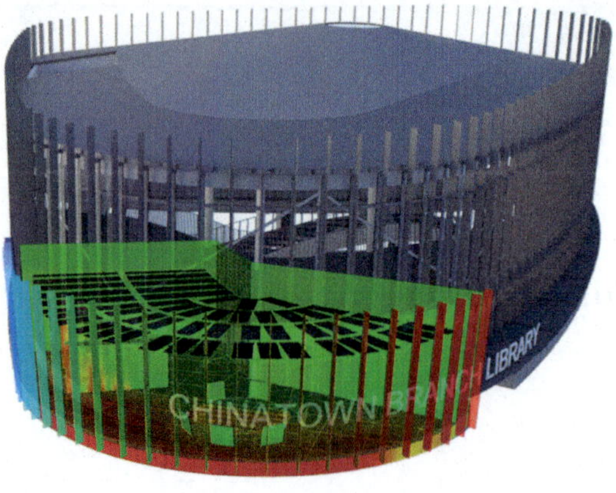

Fig. 1 The Chicago Chinatown public library

deliver around 24 TFLOPS of computing power for less than 3000 US dollars within a standard desktop machine.

We suggest that maximizing the potential of GPU acceleration requires a departure from traditional 8760-h serial simulation methods through statistical post-processing. This is possible as building performance is highly dependent on stochastic behavior: solar radiation, atmospheric conditions, diurnal temperature, occupancy and operation as but some examples. Typical meteorological year data (TMY) is already a statistical composite over decades (Wilcox and Marion 2008) where data is culled where there is weak agreement with the long-term cumulative distribution. By adapting simulation to parallel statistical methods, accuracy can in fact be further improved by allowing higher geometric resolution and enhanced multi-physics interactivity through computing power.

Introduction to Thermoactive Architecture

Thermoactive Architecture utilizes material and form through various modalities to heat and cool space, an approach predating modern HVAC systems by several thousands of years. The Chinese Kang bed-stove, the Korean Ondol and the much later Roman Hypocaust are all examples of direct radiant space tempering relying on indirect methods of energy transport, namely combustion exhaust (Bean et al. 2010). It was not until the early 1900s, with innovations attributed to Frank Lloyd Wright and Professor A.H. Barker that pipes were used to circulate warm fluid more efficiently and with less structural investment (Bean et al. 2010; Bine). Historically radiant heating gained greater acceptance because of a lack of understanding on the topics of humidity control and thermal comfort. Paradoxically, it was through scientific advancements in psychometrics that an era mostly ignoring radiant cooling was ushered in.

Specifically, two developments of consequence occurred in the early twenties: Willis Carrier's *Manufactured Weather* and Mies van der Rohe's daring use of an all glass skin for the *Friedirchstrasse Skyscraper* proposal. There is no mystery as to why the all-air HVAC paradigm supplanted the potential of radiant systems. The marginal cost of installing a single mechanical system that could provide heating, cooling and humidity control to any building was dramatically lower. Energy cost was also mostly irrelevant. Yet, in 2006, with heightened awareness of the adverse environmental impact of HVAC systems, the author was part of a team that renewed the radiant cooling concept. In conjunction with an interactive double-skin façade, The Pearl River Tower (Boyer and Dang 2007) is one of the most energy efficient skyscrapers in the world.

SOM's Pearl River Tower and more recently The Chinatown Public library in 2013 incorporates thermoactive design strategies for energy savings and improved indoor environmental quality. Radiant cooling and heating rely on water, significantly denser than air, to absorb, release and convey heat throughout the building. Comfort can therefore be achieved with a substantially lower parasitic energy cost with fan energy typically reduced by over two-thirds. Radiant systems also rely on mild 18–35 °C (65–95 °F) water temperatures as opposed to 6 and 60 °C (44 and 140 °F) for an all-air system. The lower exergy, or useful work, allows ground and water-source heat pumps to operate at high efficiency. Moreover, as operative temperature, an average of mean radiant and air temperature dictates human comfort; further savings can be realized through a broader range of air temperature set points in conjunction with heat recovery and economizer ventilation strategies.

Chinatown Branch Library and Simulation Challenges

Design of Chinatown Branch Library evolved the Pearl River Tower concept to incorporate a cost-effective capillary mat system in lieu of chilled metal panels. On its exterior, an array of vertical shading fins reduce peak solar gains and juxtapose the lattice-like network formed by the polypropylene mats. The integrated system provided designers the opportunity to realize a building projected to consume 42-percent less energy than code, while allowing for a 70-percent high performance glass skin.

At two stories and 1600 m^2, SOM designers worked closely with Chicago Public Library officials to identify opportunities for programmatic overlap, thereby increasing the projected utilization of the building's spaces. The radiant system aided in the quest for efficiency, allowing for a reduction in floor-to-floor heights by over 0.5-m while maintaining an equivalent ceiling height, as with the more traditional all-air approach. The combination of savings in facade and structure, sheet metal for ductwork, and interior finishing made high performance possible, at a comparable cost to the existing library prototype (Fig. 2)

In developing a model that was able to adequately capture the complex physics encountered with such as system, the design team encountered a number of simplifications embedded in industry-standard simulation tools that jeopardized accuracy or obscured results including:

1. Lack of direct indoor solar absorptivity
2. Perimeter zoning based on standard 15' while solar conditions and radiant response are dynamic
3. Assumed uniform surface temperatures

Fig. 2 Construction photo of Chicago Chinatown branch library and capillary radiant mat installation

4. Assumed uniform long- and shortwave radiation
5. Inability to account for specular reflection from complex fenestration
6. Assumed well-mixed, uniform air temperature
7. Prescribed uniform absorptivity and emissivity
8. One-dimensional heat conduction
9. Invalid assumption of lumped capacitance with slab systems (Biot number greater than one)
10. Inability to account for geometric complexities
11. Inability to account for composite or anisotropic materials

Moreover our team discovered rather dramatic differences in cooling load calculations between programs and calculation methodologies, even within the same tool, a finding corroborated in recent work by Feng (2013). Considering the ubiquitous use of TRACE700 and eQuest, tools that rely on the simplified Radiant Time Series (RTS) method, the capacity of these systems could be grossly under-predicted. The results of a single-zone cooling load calculation for various tools are shown: (Table 1)

Towards Massively Parallel Simulation

Many of the computational challenges encountered simulating radiant exchange by thermally activated surfaces are common to those in the computer graphics field today. With a touch of irony, the scientific foundation for some recent advances in computer gaming graphics, such as global illumination using progressive refinement radiosity, were first developed in the fifties within the engineering field of heat transfer (Goral et al. 1984). This realization led our team to begin investigating opportunities to buttress our annual energy simulation with more advanced physics simulation and in turn, opportunities for acceleration on the GPU due to the time consuming nature of these calculations (Fig. 3 and 4).

In the building simulation community the closest related efforts found were to accelerate daylight simulation through OpenCL (Zuo et al. 2011) or NVIDIA's proprietary CUDA based OptiX ray tracing engine for Radiance (Jones and Reinhart 2014). It is important to stress that the Radiosity method mentioned earlier differs from the better-known approach of Ray Tracing in that it is computed in *object-space* as opposed to *image-space* and is therefore view independent. Accurate thermal simulation requires the latter approach to understand complex radiant exchange between surfaces; however, the former could be refashioned to approximate the radiosity solution through progressive refinement.

To test whether there would be merit in this approach we began by running a 24 h simulation of a 140 square-meter south-western zone using *RadTherm*, a commercially available multi-dimensional transient heat transfer simulation engine, and *Radiance* through *DIVA* for *Rhino* with post-processing in *Matlab*. The *Radiance* simulation ran on two NVIDIA GTX 980 graphics cards using a beta version of *Accelerad*, a GPU-based lighting and daylighting analysis engine by Nathaniel Jones of MIT.

Utilizing CUDA processing architecture, tasks are executed in what NVIDIA terms a SIMT (single instruction, multiple thread) fashion.

Table 1 Comparison of cooling loads by tool and methodology

Tool—methodology	Design cooling load W/m^2
TRACE700—RTS (heat balance)—radiant time	151.4
TRACE700—CLTD-CLF (ASHRAE TFM)—exact transfer function method	123.0
TRACE700—CEC-DOE2—exact TFM method with CEC\DOE 2.1c constraints	119.8
TRNSYS—transfer function method	113.5

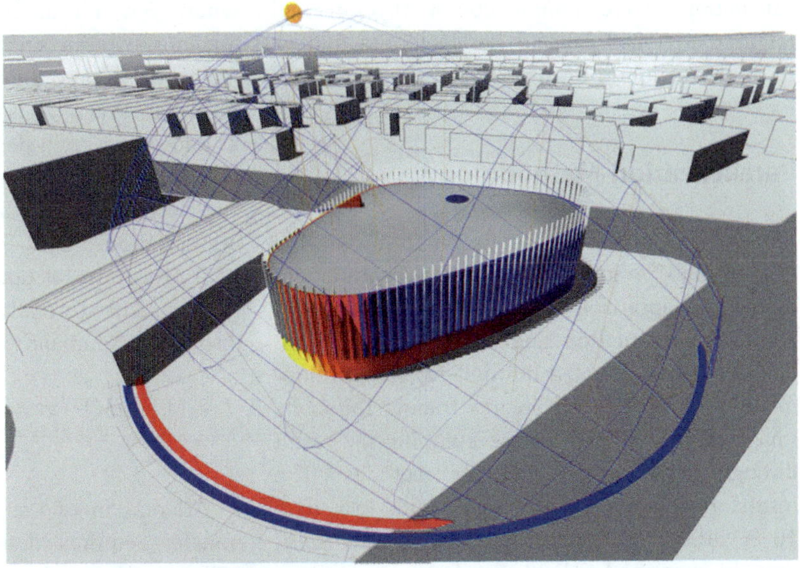

Fig. 3 Solar insolation map on the façade by Ecotect

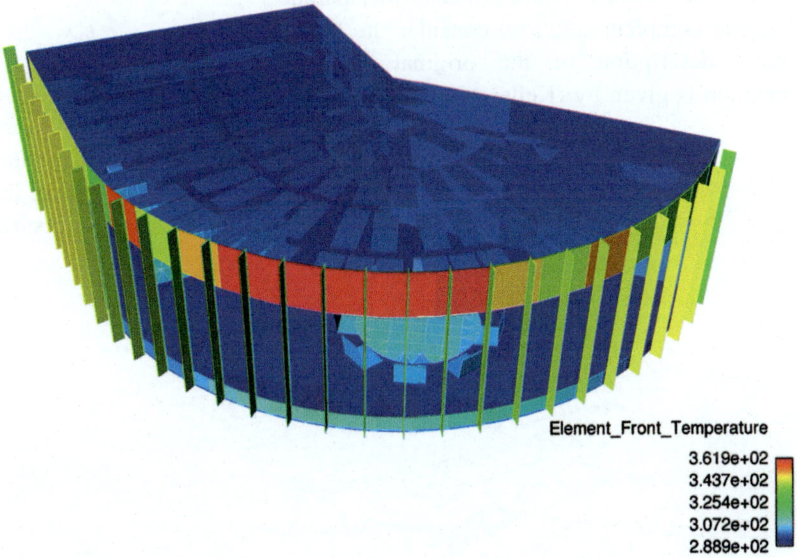

Fig. 4 RadTherm analysis of element surface temperatures at 2 pm July 19

Code is executed in groups of 32 threads, what NVIDIA terms a warp. In the case of the GTX 980 graphics card there are 16 streaming multi-processors (SMMs) featuring 128 CUDA cores each for a total of 2048 CUDA cores. The work distributor assigns each CUDA block to one SMM that divides the effort into warps. While each SMM operates independently, they have shared memory to quickly stitch and trim the parallel solution together. By adapting the instant radiosity method using virtual point lights by Keller (1997) we estimate heat transfer in parallel with virtual heat sinks or sources (VHS) as an analogue to artificial lighting in *Radiance*. Each VHS solution is adjusted to account for the fractional power density calculated in the visible spectrum relative to the entire transmission including UV and Infrared.

The resulting *RadTherm* solution converged in 0.38 h with a mesh resolution of 2581 Tris and 1379 quads on a 6-core Intel processor at 3.5 GHz at 5 min time-steps. *Radiance* calculations were completed in 0.76 h and required to 7.2e8 rays to iterate to the highest resolution solution on the GPU (Fig. 5).

Parallelization Methodology

At this point, while seemingly far from our goal of real-time progressive refinement simulation, the current approach of using *Radiance* requires additional calculation and translation overhead that could be mitigated through further development. Specifically, we are ray-tracing the same image multiple times, before combing the result through serial post-processing methods. The results however are promising, with less than 10 % difference between *RadTherm* and *Radiance* in quantified radiant transfer.

To understand how we could further parallelize the complete solution, consider the mathematic description of the original rendering equation is given by (Keller 1997)

$$I(x,x') = g(x,x')\left[\varepsilon(x,x') + \int_S \rho(x,x',x'')dx''\right] \tag{1}$$

where $I(x, x')$ describes the amount of light arriving at point x from x'; $g(x, x')$ is the form factor as a ratio of potential energy transport from 0 to 1; $\varepsilon(x, x')$ is the light emitted from x' to x; and $\rho(x, x', x'')$ describes the amount of light sent from x" to x by reflection from x'.

The common method for solving the rendering equation comes from rewriting the equation with M as a linear operator given by the integral in (1) and formally inverting the equation through a Neumann expansion method, thus

$$I = (1 - gM)^{-1}g\varepsilon = g\varepsilon + gMg\varepsilon + g(Mg)^N\varepsilon\ldots \tag{2}$$

Adaptation of the following method to heat transfer requires defining $\varepsilon(x, x')$ as thermal radiant exchange by

$$\varepsilon(x,x') = \varepsilon\sigma(T_{x'}^4 - T_x^4) \tag{3}$$

the temperature calculated by a thermal RC-network as prescribed by radiant system type, material conductivity and location of the heat sink or source in the assembly. In this case we are assuming surfaces as black body diffuse

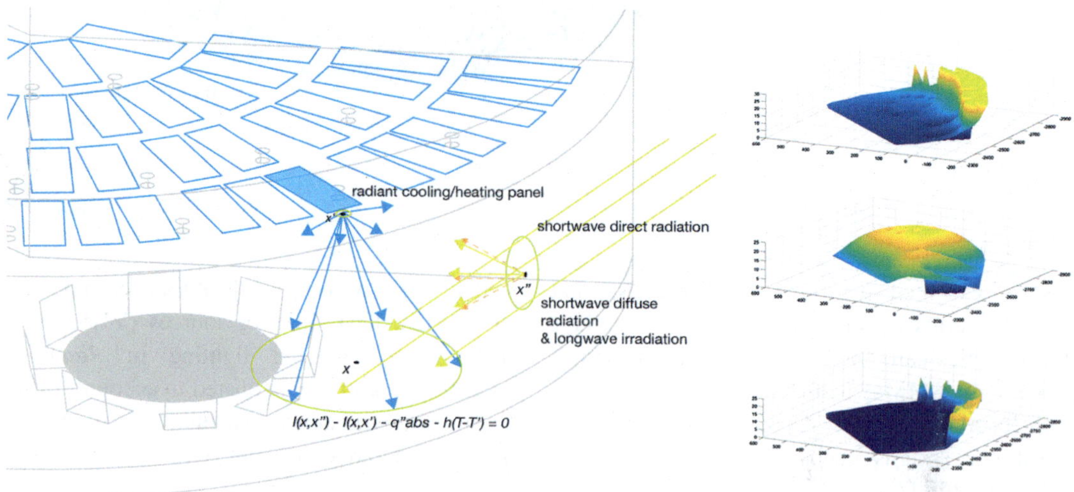

Fig. 5 *Left* Diagram of procedure for using radiance to quantify the interaction between chilled radiant panel and incoming natural light and heat at a specific node; *top right* insolation map to interior grid; *middle right* view factor and cooling capacity map to interior grid; *bottom right* resulting loads not addressed by radiant cooling assumed to go entirely to convection in this early development stage

Lambertian surfaces. Two methods for time series solutions to the conduction transfer functions, as used by *EnergyPlus,* are the Laplace Transform Formulation and the State Space Formulation (EnergyPlus 2014). The form of the conduction transfer function solution (CTF) is shown as

$$q''_{i,t} = \sum_{m=1}^{M} X_m T_{i,t-m+1} - \sum_{m=1}^{M} Y_m T_{o,t-m+1} + \sum_{m=1}^{k} F_m q''_{i,t-m} \quad (4)$$

where k is the order of the CTF, M is the number functions, and X, Y and F are the transfer functions.

Solving complex energy transport at multiple points across a spatial domain is where GPU acceleration excels, as we do not need to recalculate temperatures and flux with iteration. Following the CTF approach, temperatures are updated based upon the prior cached irradiance and temperature values with the latest rendered solution. This can be achieved through a Newton-Raphson method based upon core temperature and controlled to a maximum/minimum allowable surface temperature. Acceleration may also be possible by quickly testing the degree coherence between the current and prior rendered solution at each VHS prior to updating.

Further Acceleration

Raytracing is the most computationally expensive step, thus further acceleration is possible considering discritization of the Tregenza model (Perez et al. 1990; Robinson and Stone 2004). At 145 patches it is redundant to calculate the radiosity solution at a spatio-temporal resolution higher than this relative to solar-time. Less than one day, every 4–5 weeks, is all that would be necessary at this resolution, depending on location and time of year. Following Rienhart's extension of the Trezenga sky, each patch can additionally be subdivided into 16 patches, except at the zenith, for a total of 2305 patches (Jones and Reinhart 2014) requiring closer to a weekly interrogation. To account for stochastic variations in weather, multiple conditions are run in parallel for the same target day to determine the sensitivity of the system and then combined at the end in a statistical post-processing step. This approach could also be utilized to analyze defined energy performance measures, such as improvements to glazing solar heat gain coefficients (SHGC) as a convolution filter to the existing radiance solution (Fig. 6).

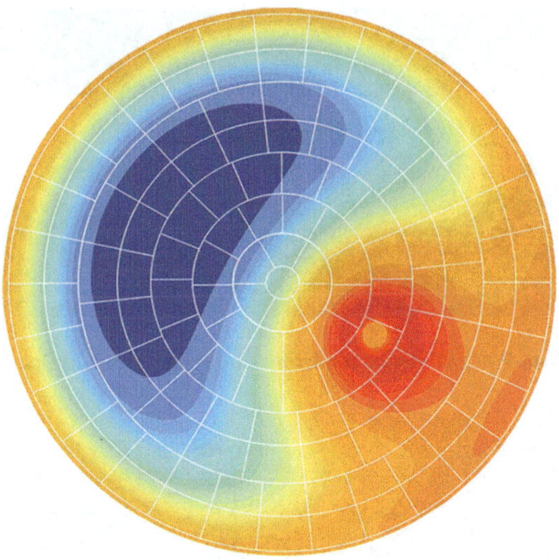

Fig. 6 Overlay of Tregenza sky model

Paralllel Daylight and Thermoactive Design Thinking

In practice it is not uncommon for daylight analysis to occur independent of building system design. When the two are finally brought together, through an annual energy model, it is often too late for any strong bi-directional influence. The result is that many buildings, particularly those designed in the past 10-years responding to sustainability rating systems, have pushed for unusably high vision glazing percentages and/or light transmission values. Not only can this lead to visual discomfort and glare, it can also dramatically increase the size, capital and operating cost of the building conditioning system (Fig. 7).

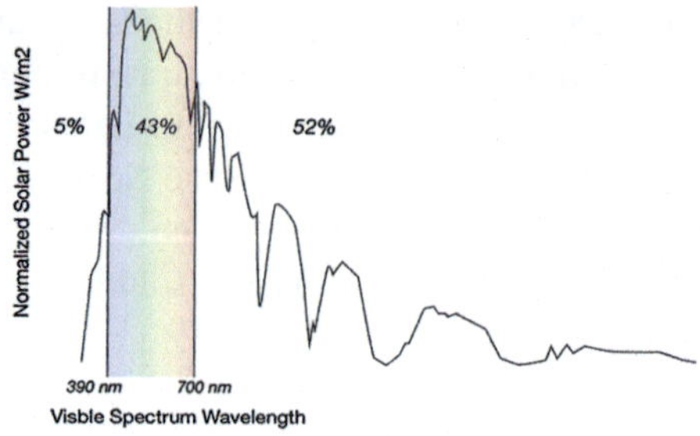

Fig. 7 *Left* Interior rendering of daylight diffusing panels. *Right* Normalized solar power spectral density

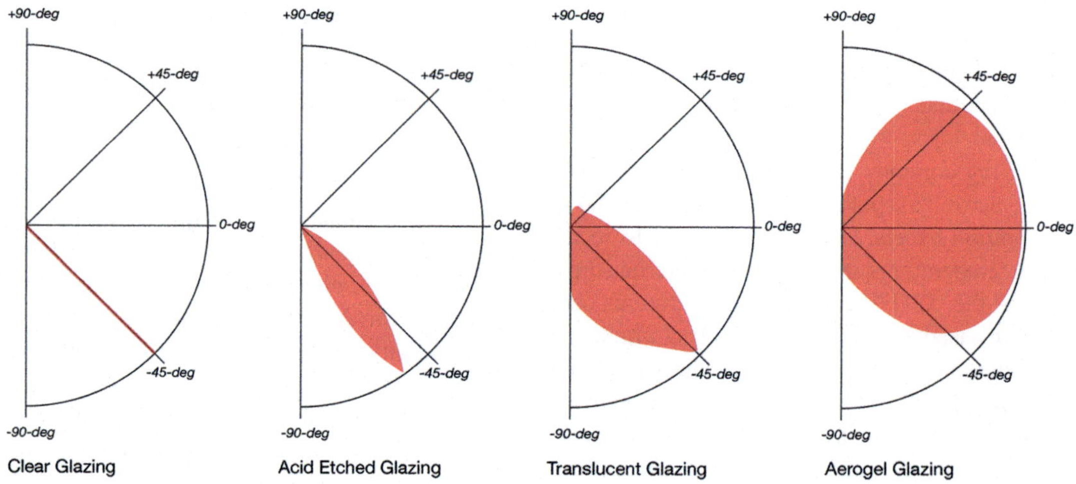

Fig. 8 Comparison of normalized daylight diffusion power (LDP) of clear glazing, acid etched, translucent and aerogel IGU to a common 45-degree angle of incidence

For Chinatown Branch Library external shading devices were incorporated to reduce peak solar gains, thereby allowing the radiant system to provide maximum benefit. The downside to this approach is a slight uptick in annual energy consumption in a climate such as Chicago, where passive solar heating of perimeter spaces in winter months is inhibited. On a recent competition, in collaboration with Ateliers Jean Nouvel, designers were able to overcome this effect whilst creating a stunning façade aesthetic by incorporating aerogel within a low-iron acid etched insulated glazing unit (IGU). The entirely glazed building not only achieves an overall U-value approaching 0.28 W/m^2 K, the facade also acts to evenly distribute light and heat. This effect allows the program to take advantage of daylight throughout the day, regardless of orientation, while space temperature is mediated through the integrated thermoactive slab system (Fig. 8, 9 and 10).

Even with a high performance aerogel façade, a typical radiant slab utilizing cross-linked polyethylene is deficient in meeting the total cooling demand. With a typical output of between 40–50 W/m^2, additional cooling airflow would be required, diminishing energy savings which was not acceptable to our client. We took this as an opportunity to build upon the analytical methodology developed on Chinatown Branch Library, focusing on the potential to increase the cooling capacity of an embedded thermoactive system through material, surface treatment, pipe placement and form.

Fig. 9 Daylight and glare analysis of clear glazing for west facing auditorium in late afternoon

Fig. 10 Daylight and glare analysis of aerogel glazing for west facing auditorium in late afternoon

After extensive modelling of multiple scenarios a 190 mm composite deck, in lieu of a more common 100 mm thickness, was proposed with polypropylene radiant tubing embedded within a layer of Ultra High Performance Concrete (UHPC). To prevent the exposed metal deck from reflecting radiation, a high emissivity dark coating would be applied to the underside of the deck. Through these interventions we were able to increase the peak cooling potential over a traditional radiant slab by over 50 % allowing us to meet the necessary cooling load. The high mass system, in contrast to the exposed capillary system for Chinatown, further allowed for a reduction in central plant demand by pre-cooling the building structure at night. Thus, the building structure provides heating, cooling, energy storage, and finished surface for the project (Fig. 11).

Quantifying the complex physical interaction of radiant heat transfer, transient

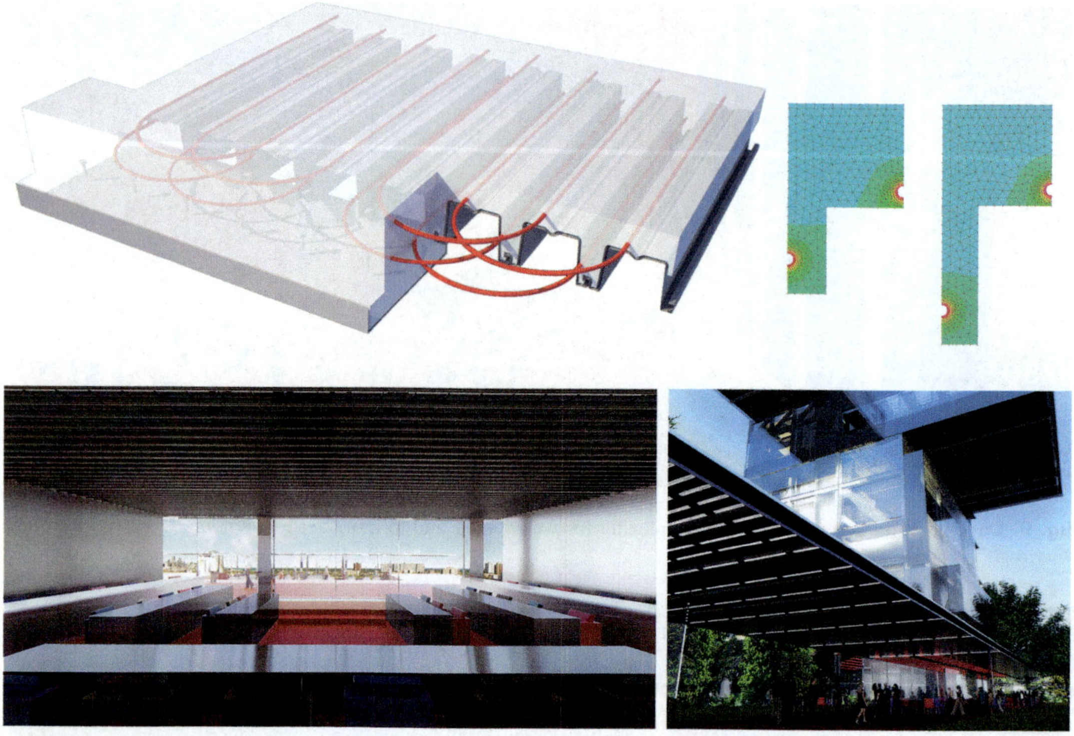

Fig. 11 *Top left* Diagram of deep radiant deck system; *top right* Steady-state thermal response of standard and deep deck radiant systems; *bottom* Renderings of thermoactive ceiling and aerogel façade integration

conduction and convection would have been time-consuming or simply impossible within industry standard whole building energy simulation tools. Some specific conditions encountered were:

1. The Solera Aerogel IGU diffuses both light and heat, thus the traditional 4–5 m perimeter zone used in modeling would suggest a substantially higher solar flux than actually encountered in real world operation
2. The deep 190 mm composite deck increases surface area over a flat slab by 230 %, increasing heat transfer efficacy
3. The ribbed profile of the deck further increases convection at the interface with the room and cooling load capacity
4. The composite beam and UHPC deep deck structure allows intermediate purlins to be removed, maximizing heat exchange area with substantially longer spans realizable with less structural material
5. UHPC has both a significantly higher thermal conductivity than reference concrete, 2.5 J/m/s/K at 2500 kg/m^3 density as compared to 1.5 J/m/s/K and 1700 kg/m^3 improving through slab conduction from radiant tubing to deck surface
6. Increased density at a comparable specific heat allows UHPC to store more energy per unit volume
7. UHPC also allowed for a tighter tubing spacing, closer to the deck surface, further improving through slab conduction
8. The low emissivity deck coating maximizes heat transfer by radiation allowing for greater passive solar heating in winter. The effect's biological analogue is that of polar bear or arctic fox fur, which only appears white due to light scattering (Fig. 12).

Fig. 12 *Left* Biomimetic design through advanced materials, as a polar bear's coat is clear yet appears white so does the skin of the building allowing for passive solar heating in winter; *top right* Scanning electron microscopy (SEM) image of polar bear fur; *bottom right* SEM image of aerogel micropores

Conclusions and Future Work

As illustrated through our work in collaboration with SOM and AJN, massively parallel processing offers more than a technical approach to accelerate design modelling. Low-latency data exchange and multi-physics simulation can enable the design process with real-time feedback providing a platform for informed adaptation and design emergence. Further development will focus expanding modes of heat transfer to conduction and convection, exploring the potential to adaptively tessellate NURBS geometry to reduce data exchange latency and increasing acceleration on a high performance visual computing appliance. An ultimate goal of this research is to enable top-down quantification and optimization of complex surface geometry, as might be possible with fabric formwork, in parallel with the bottom-up, nanoscale design of high performance composites utilizing novel materials such as Graphene for strength and conductivity.

References

ASHRAE (2013) Energy estimating and modelling method. In: Handbook of fundamentals

Bean R, Olesen B, Kim KW (2010) History of radiant heating and cooling systems. ASHRAE J

Bine "History" of thermally activated building systems. http://www.bine.info/en/publications/publikation/thermoaktive-bauteilsysteme/geschichte-thermisch-aktivierter-bauteilsysteme/

Boyer J, Dang A (2007) Designing for performance. In: Proceedings of the building simulation

Chantrasrisalai C, Ghatti V, Fisher DE, Scheatzle DG (2003) Experimental validation of the EnergyPlus low-temperature radiant simulation. ASHRAE Trans 109(2):614–623

Crawley DB et al (2001) EnergyPlus: creating a new-generation building energy simulation program. Energy Build 33(4):319–331

EnergyPlus Engineering Reference (2013) The reference to EnergyPlus calculations (in case you want or need to know)

Feng D (2013) Radiant cooling design: performance prediction and modeling methods. GGASHRAE Radiant Energy Seminar

Goral C, Torrance K, Greenberg D, Battaile B (1984) Modeling the Interaction of light between diffuse surfaces. Comput Graphics 18(3):213–222

Jones N, Reinhart C (2014) Irradiance caching for global illumination calculation on graphics hardware. In: IBPSA-USA building simulation conference, ASHRAE, Atlanta, GA

Keller A (1997) Instant radiosity. In: Computer graphics 31, annual conference series, pp 49–56

Moe K (2010) Thermally active surfaces in architecture, 1st edn. Princeton Architectural Press, New York

Perez R, Ineichen P, Seals R (1990) Modelling daylight availability and irradiance components from direct and global irradiance. Solar Energy 44:271–289

Robinson D, Stone A (2004) Irradiation modeling made simple: the cumulative sky approach and its applications. In Proceedings of the 21st conference on passive and low energy architecture
The Economist (2008) From blueprint to database. The Economist, London
Valkering E (2010) Raytracing of NURBS surfaces using CUDA. MS thesis at TU Delft
Wilcox S, Marion W (2008) Users manual for TMY3 data sets. NREL Technical Report
Zuo W, McNeil A, Wetter M, Lee E (2011) Acceleration of radiance for lighting simulation by using parallel computing with OpenCL. In Proceedings of the building simulation

The Architecture of the ILL-Tempered Environment

Djordje Stojanovic

Abstract

This study examines the role of a specific behavioural model in the design process, based on an on-going architectural project for redevelopment of a public space. The employed model may be used to register environmental conditions and process information to create visible effects within the given public space. The approach concentrates on overcoming the problem of insufficient contextual information provided in the design brief. The discussion focuses on advantages and shortcomings of a design strategy based on real time data collection, including vehicular and pedestrian traffic, temperature, relative air humidity, insulation, pollution etc. and on establishing relevance of such information for the proposed built form and design decision making.

Introduction

The project started as a winning entry at an open architectural competition and has gone through significant changes in the later stages of the design process, offering itself as a case study and an opportunity to probe into the relation between architectural research and practice (Fig. 1).

Currently, the design proposal is developing into a five-year research project funded by the National Ministry for Science and Technology, aiming at the development of a much larger system to be incorporated into the existing urban fabric. In accordance with the idea initiated through the architectural project, such system will be comprised of sensors, controllers and actuators in order to harness and analyse data from the surroundings, and produce intelligent and automated effects within the public space. Its purpose is to provide environmental information but also to visualize and present such information as an integral part of the built form. In addition to the commitment to making public areas in the city better, the principal objective of the research project is to structure a specific design model and thus enable architects to take part in a more comprehensive way of understanding and creating space. The interdisciplinary collaboration is observed as an essential aspect to the success of the research project.

D. Stojanovic (✉)
Faculty of Architecture, University of Belgrade, Belgrade, Serbia
e-mail: ds@4ofseven.com

M.R. Thomsen et al. (eds.), *Modelling Behaviour*, DOI 10.1007/978-3-319-24208-8_13

Fig. 1 An ongoing architectural project for redevelopment of a public square in Belgrade by 4 of 7 Architecture

At this point, the partnership between several institutions is established to tackle the problem. One of the key roles belongs to the City Institute of Public Health—Centre for Hygiene and Human Ecology which will provide meteorological and air pollution data through their network of monitoring stations operating in the city. Other important participants include institutions and researchers concerned with the modelling in mechanical engineering and mechatronics, interested to test and expand their knowledge of control systems in the domain of architecture. The focus of the research remains architectural and thus primarily concerned with spatial effects and dependencies. It is anticipated that the proposed system and the resulting architectural interventions could help achieve different ambient values in the public spaces, but also help manage traffic overcrowding, create microclimatic condition on the specific sites, relieve some pollution effects or notify people of the presence of bothersome allergens in the air.

The Design for ILL-Tempered Environment

Initially and within the framework of the architectural project, the design intent was to create a new public space at a busy and derelict traffic junction, currently underused by pedestrians. The given site is an important intersection of city's traffic routes but arguably, throughout the city's history, it has never been established as a successful public space, nor was it occupied by any buildings of architectural relevance. Interestingly, its identity may be perceived through changing intensities of many different flows of movement rather than through its physical structure and built form. At the same time, the existing heavy vehicular traffic and pedestrian overcrowding are principal contributors to the decaying image of the place and chief causes of the environmental degradation of the site. While it is true that city's traffic problems and environmental hazards could be truly remedied only as a part much larger planning initiative, this proposal is conceived as a dynamic and changing space, focused on the two specific goals: to create a new identity of the square; and to turn everyone's attention to the less visible but palpable relationship between the built form and the environment.

Many Versus One

The design is conceived as a new landmark, visible to drivers and passengers. The planned interventions consist of many elements. Rather than a single and central gesture (called for by the competition brief), 88 rotating lampposts will be installed in the same zone. Their movement is planned according to the changing conditions of their immediate surroundings. The overall form is less important while the relations between proposed elements and their surroundings are central to the project (Fig. 2).

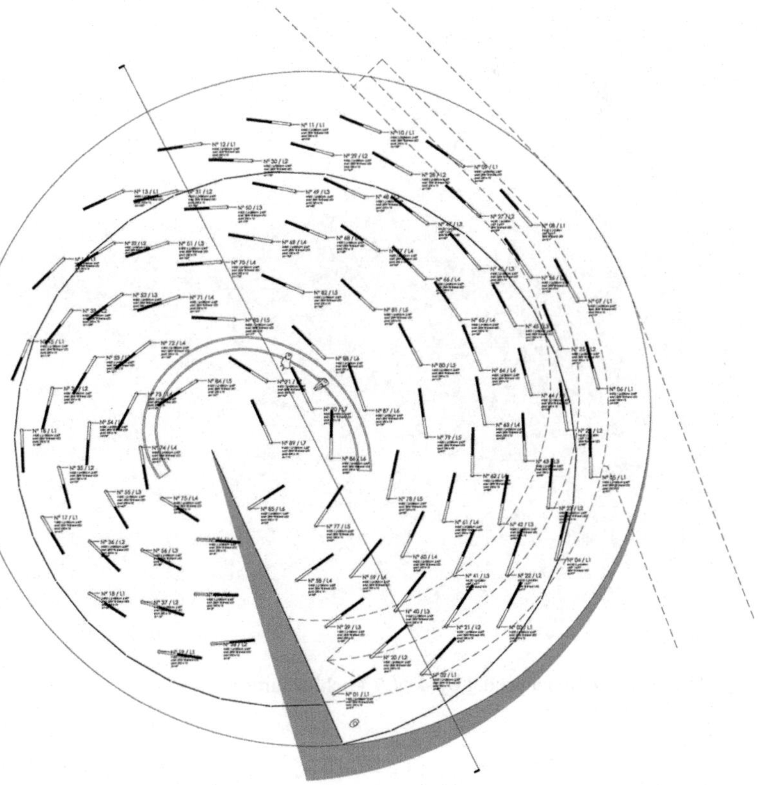

Fig. 2 Ground plan drawing, default set out and geometric relations between 88 lampposts

This approach is inspired by the idea presented in the book titled "The Geometry of Environment: an Introduction to Spatial Organization in Design" (March and Steadman 1971) in which relational reasoning is signalled out as a logical way of understanding of complex spatial organizations. The idea is developed further by Allen (1997) who named such spatial effects as Field Conditions and depicted them as bottom-up phenomena, defined not by overarching geometrical schemes but by intricate local connections. Accordingly, the behavioural capacity of the proposed design model for the public square in the city is based on the idea of locally regulated interdependencies between constitutional elements of the scheme. A specific behavioural model was employed at the core of the design workflow, in addition to the series of representation models and conventional drawings (Fig. 3).

Understanding the Given Site Through Movement and Change

The guiding idea of the project is to create a responsive spatial environment and to help alleviate some of the effects related to the troublesome character of the given site. The entire given site is understood through notions of movement and change, rather than its physical constitution. It is therefore assumed that the appropriate contextual analysis would consist of data, vectors and trajectories rather than of static representation of the urban morphology. At the very beginning of the design process, the notion of movement is understood through dynamics of numerous vehicular and pedestrian traffic routes intersecting at the square, while the notion of change is related to the weather conditions including oscillations in air quality, temperature, humidity and solar irradiation.

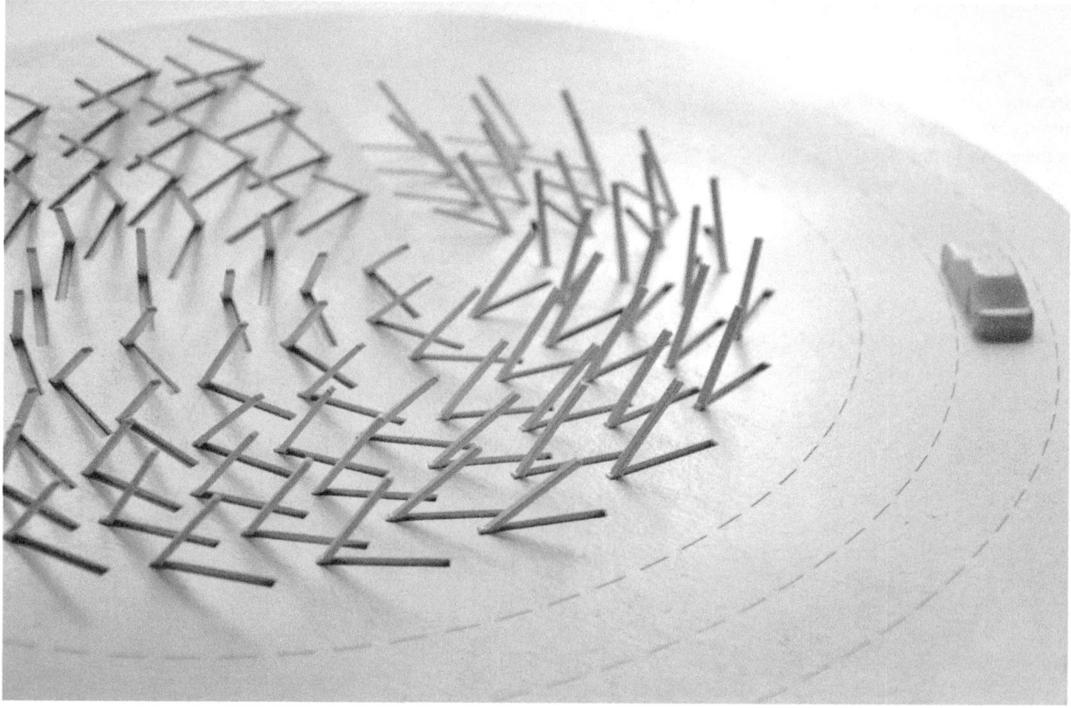

Fig. 3 Representational model by 4 of 7 Architecture

Although it has been well noted that the agenda of responsive environments, opens up an entirely new domain of design research (Schumacher 2004) and while the idea can occur on multiple levels and can involve readily available communication technologies comprised of sensors, actuators and controllers (Schnadelbach 2010) it remains seldom tested and resolved in large scale projects.

Introducing Actuators, Sensors and Controllers

At a certain scale, the complex physical interactions are already made possible by the creative fusion of embedded computation (intelligence) with a physical, tangible counterpart (kinetics) as noted and illustrated in the book "Interactive Architecture" by Fox and Kemp (2009). In reference, the feasibility of this proposal, concerned with a large scale problem, is based on readily available technologies comprised of actuators, sensors and controllers. The main feature of the design scheme consists of 88 revolving lampposts, put in motion by electrically powered actuators and controlled by custom developed software (Fig. 4).

The actual rotation of the lampposts is made possible via the electrically powered mechanism concealed in the concrete base and within the specifically designed base joint to allow planar rotation of the post around the central axis. The motor enabling controlled movement of the post is reminiscent of the motor used for parking barrier gates with an input voltage of AC220V/150w and work temperature from 35 to 80 °C. Lampposts are produced from round steel tubes (Ø152.4 mm, wall thickness 3 mm, total length of 5 m) and installed at an inclination of 60° relative to the floor surface. An acrylic light box holding a light emitting diode is inserted into the upper segment of each lamp post.

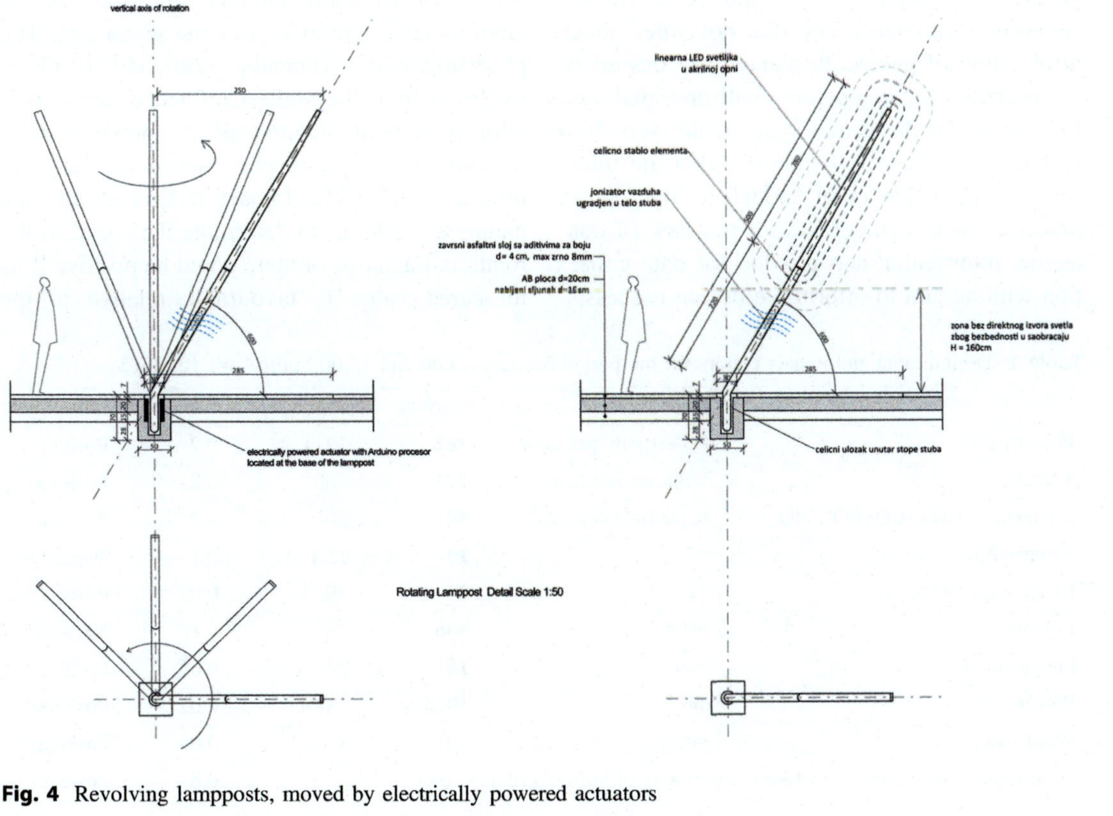

Fig. 4 Revolving lampposts, moved by electrically powered actuators

Surveillance and Real-Time Data Collection

The design proposal is not focused on a single and finite built form; instead it relies on a specific behavioural model and a system put into motion upon receiving information from the surroundings. Successful collection of such data is a prerequisite for the operation of the model but it also stands for the very first phase of the implementation of the proposal. A number of different sensors, including people/vehicle counters, thermostats, barometers, hygrometers, heliometers and electrostatic precipitators are to be strategically positioned across the square and its surroundings to collect real-time information and thus provide the context of the project. It is important to emphasise that the initial phase of the design process concentrates on gathering of data which was missing from the design brief. In 2012, city authorities, who are the project organizers, have commissioned a traffic study concerned solely with vehicular traffic. The study, generally not available to the public, confirmed congestion problems but did not offer much information of interest to planers and designers. Accordingly, the competition call provided very little input on the actual state of the site. It is therefore important to emphasise that the initial role of the behavioural model in the design process was to compensate for the lack of contextual information and provide for data collection with an aim to inform the design process.

Control and Information Processing

The design input is created by counting, recording and measuring, yet the plan and ways of data collection are reconsidered and significantly developed after the competition stage of the project. To illustrate initial thinking behind the design proposal, information for the specific date and time, consisting of vehicular and pedestrian numbers, temperature (°C), relative air humidity (%), insolation (W/m^2) etc. is first gathered, then analysed and finally used to determine the speed of rotation for each lamppost. Based on the data available from the local hydro-meteorological station and average weather conditions for the observed area found on several reputable weather portals, pedestrian and vehicle count for the hour between 13:00 and 14:00 h on April 12, 2015, design input could be structured, evaluated and processed to determine the speed of the lampposts rotation (Table 1).

Different impact factors are assigned for each of the measured parameters according to the subjective perception of their relevance for the environmental conditions at the given site. The pedestrian count, vehicular traffic and Air Quality Index hold the highest influence factor (0.2) while precipitation, atmospheric pressure, relative air humidity and wind speed are less significant (0.05). The Impact factors of all nine parameters add up to determine the speed of the rotation. Each one of them could be positive if its measured value is favourable relative to the

Table 1 Design input and impact factors for the period between 13:00 and 14:00 h, on April 12, 2015

	Units	Current	Average	Factor	Impact
Pedestrians	Pedestrians per hour	**182**	1211	0.2	Positive
Vehicles	Vehicles per hour	**221**	580	0.2	Positive
Air quality index (CO, SO_2, NO_2)	Scale 0–500 μg/m^3	**50**	50	0.2	Positive
Temperature	°C	**19**	15.1	0.1	Negative
Relative air humidity	%	**35**	49	0.05	Positive
Insolation	W/m^2	**836**	488	0.1	Negative
Precipitation	mm	**18**	54	0.05	Positive
Pressure	mb	**1025**	1.001	0.05	Positive
Wind speed	m/s	**2.5**	4	0.05	Positive
Total impact for the period between 13:00 and 14:00 h, April 12, 2015				**0.8**	Positive

median value of that parameter for the given time of day and year; or negative if its measured value is detrimental relative to the median value of that parameter for the given time of day and year. If all parameters are positive their impact factors add up to the value of 1.0 which is equivalent to the maximum rotation speed. In this case each lamp post will make a full rotation cycle per 1 h. In case that any one of the parameters is negative, its value is to be subtracted from the overall sum and the speed of rotation will decrease. A simple algorithm is employed to evaluate and assign the correct impact factor for each of the parameters according to the changing input received from sensors distributed across the square.

The rotation of each lamppost is correlated with the collective movement so that all lampposts progress in the same direction and according to synchronised velocities. The overall intensity of lamppost movement drops as environmental conditions at the square become less favourable. For example, if the car count increases, if temperature rises, and if the ambient air quality drops, lampposts will come to a standstill. On the contrary, when the sum of the environmental conditions in the square starts to improve and the space becomes inhabitable once again, the lampposts will come to life (Fig. 5).

This way, project delivers a design strategy whereby current environmental condition of the site is related to the behaviour of the proposed structure in real-time. Yet, such model is a classical fitness function that collapses all data into a single number. Even if the choreography of rotating lampposts would fulfil the promise of an intriguing architectural spectacle in the public space, it is reasonable to extend inquiry further by pondering why all environmental data has to map into a single parameter which is the speed of rotation. Since diverse information could be collected from the environment, more opportunities are opening up for research.

Further study will focus on finding more relevant data from the surroundings and will try to determine how such information can be used to animate different actuators in the built environment. For instance, preliminary data obtained from an onsite air quality measurement station, belonging to the City Institute of Public Health—Centre for Hygiene and Human Ecology, demonstrates that Air Quality Index is generally tolerable in this area. This is no surprise since Air Quality Index used by governmental agencies is predominantly based on the presence of gas pollutants, Carbon Monoxide, Sulphur Dioxide and Nitrogen Dioxide. However the same data shows frequent excesses of equally harmful Fine Suspended Particles. Could there be another actuator contained within lampposts or entirely different piece of equipment used in the public space, with the purpose to indicate presence of pollutants otherwise invisible to a human eye. If more strategically positioned sensors would be connected to such actuators, measurement and display of

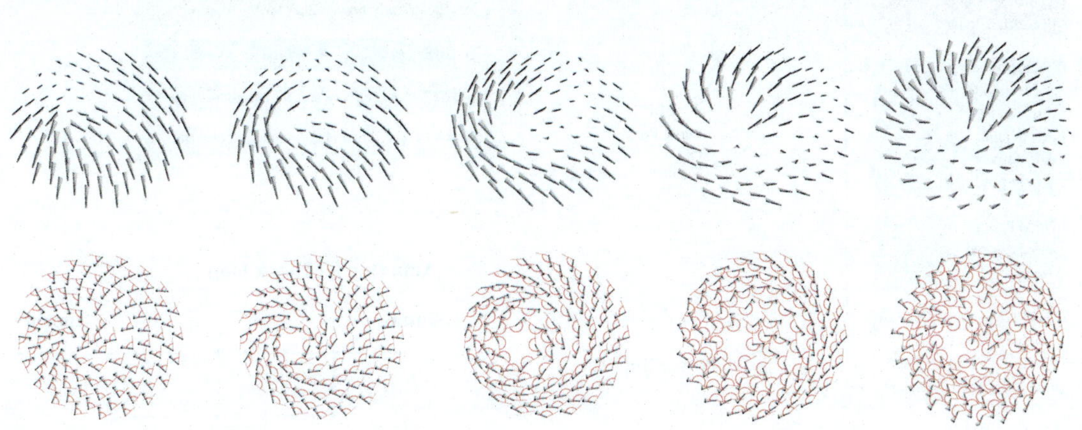

Fig. 5 Behavioural model output, 88 lamppost changing position during one hour cycle, according to environmental data

environmental data could be done more accurately. Should this be done for specific spots in the urban fabric to replace generalized understanding of environmental conditions based on median values across large areas or even entire cities? Equally useful information which could be released at specific micro sites could be the one of allergen levels. Other roles of the proposed system could be more proactive and could help control the microclimatic condition of the site, i.e. for the occasions when air temperature is too high; lampposts could be equipped with small fans and tanks for adiabatic cooling. Furthermore, the means of adjusting the way outdoor public space is lit according to environmental conditions could be another important topic for the future research. The change colour and intensity of street light in order to produce different ambient qualities could be linked to measurement of precipitation or air pressure. Similarly, the pulsation of emitting light could be related to the vehicular count with the purpose to complement the existing system of traffic signalisation. These possibilities constitute a list of tasks for the commencing research project. Departing from to the idea initiated in the architectural proposal, further study is aiming at the development of a much larger system to be incorporated into the existing urban fabric. Its purpose would be to collect environmental data but also to visualize and present such information

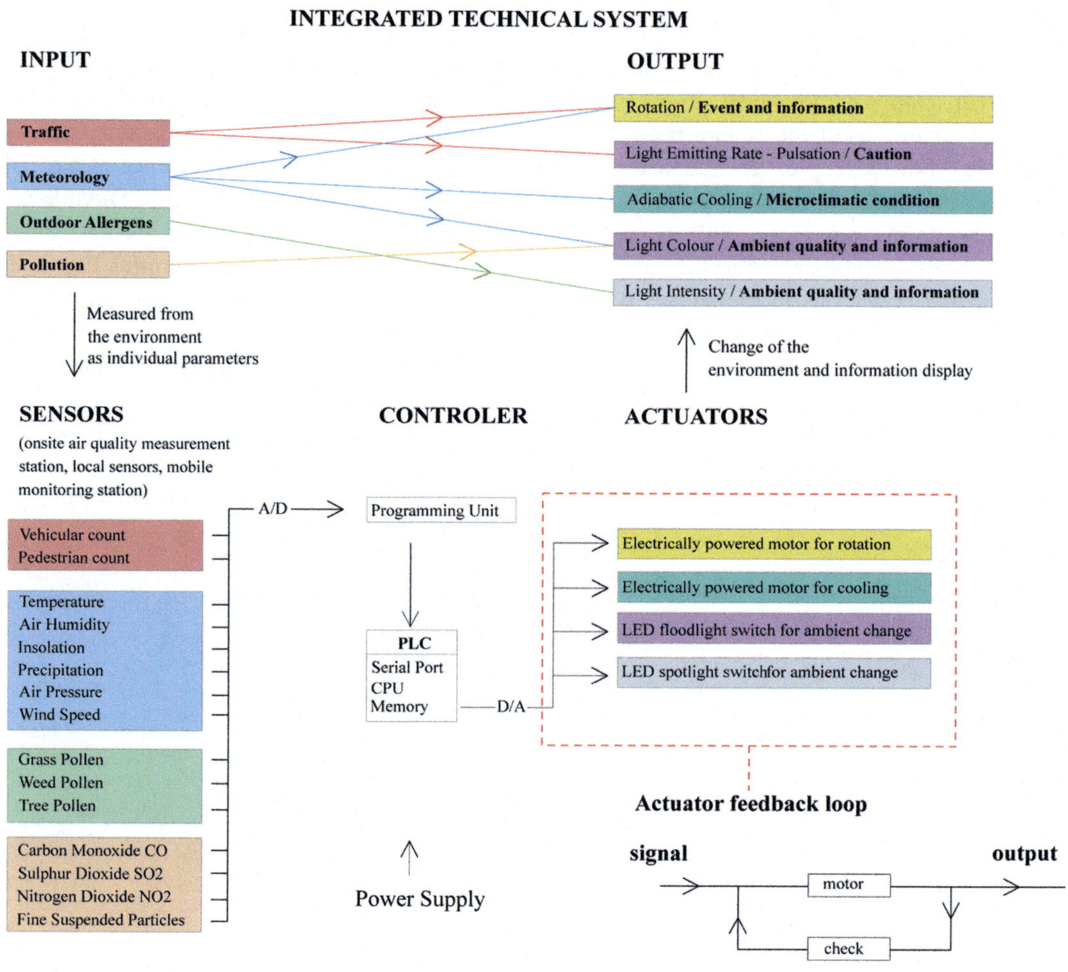

Fig. 6 Integrated technical system—behavioural mode: enables understanding of each individual parameter and shows how information will be interpreted and used in the control of the technologies

as an integral part of the built form. Coupling between an architectural intent and a technique of modelling originating in the field of mechatronics, may help articulate this idea toward development of an integrated technical system (Fig. 6).

Conclusions

The presented behavioural model is employed within the dual aim, to harness information from the environment and to test ways of using such information for the purpose of design decision making. As a consequence, design proposal is not focused on a single and finite built form, instead it relies on such behavioural model to study architectural qualities of the built system put into motion upon receiving information from the surroundings. Practical implications of the results of this study are in the development of the specific workflow and a model to help designers establish relations across several different scales and parameters of the given site, including movement intensities, weather conditions, pollution levels etc. The approach concentrates on overcoming the problem of insufficient contextual information provided in the design brief.

At the same time, the potential to establish a feedback, or reverse flow of information, from the proposed structure to the environment, was not achieved in this project. In the information theory, this is known as the "circular causal" relationship and architects and designers have a long lasting interest for such ways of modelling the built environment (Khan 2012). It could be intriguing to test in large scale projects if any such proposed system could cause a change in its environment and if that change could be fed back to the system via information that enables the system to change its behaviour. Future study will look into practical ways of structuring the entire design work flow around the "recursive interaction between the proposed system and the existing environment" (Khan 2012). In reference to this particular project, it is recognized that the proposed structure comprised of 88 lampposts will not send back any such information to its environment. It is understood that the rotation of lampposts will not immediately affect the city traffic or pollution problems; while it may help create a new identity of the square which would literally build upon its relationship to the surroundings via a number of measurable parameters.

The title of this paper echoes the title of Banham's seminal book "The Architecture of the Well-tempered Environment" (1969), concerned with the technology and its possible role in the reconciliation between human needs and environmental concerns. In a flash-forward, the agenda of responsive environments once again opens up the very same domain of design research and practice. The aim of this study is to expose some of the contemporary design workflows and methods through engaging with communicable dimensions of architectural design practice.

Acknowledgements The architectural project for redevelopment of Slavija square in Belgrade is made by 4of7 Architecture (Djordje Stojanovic, Milutin Cerovic, Andjela Karabasevic, Bogdan Obradovic and Milica Tasic). The author would like to thank Jelena Pejkovic for the last minute spelling and grammar check.

References

Allen S (1997) From object to field. AD 67:24–31

Banham R (1969) Architecture of the well-tempered environment. The Architectural Press, London

Fox M, Kemp M (2009) Interactive architecture. Princeton Architectural Press, New York

Khan O (2012) A communications primer revisited. In: Ayers P (ed) Persistent modelling. Routledge, London and New York, pp 51–61

March L, Steadman J (1971) The geometry of environment: an introduction to spatial organization in design. MIT Press, Cambridge

Schnädelbach H (2010) Adaptive architecture: a conceptual framework. http://www.cs.nott.ac.uk/~hms/pdfs/Schnadelbach_AdaptiveArchitectureConceptualFramework_MediaCity2010.pdf. Accessed 15 Mar 2015

Schumacher P (2004) Responsive environments: from drawing to scripting. http://www.patrikschumacher.com/Texts/AADRLDrawingScripting.html. Accessed 15 Mar 2015

Designing the Desert

Modelling the Walls of the UAE Pavilion for the 2015 World Expo

Henrik Malm, Sam Joyce, Martha Tsigkari,
Khaled El-Ashry and Francis Aish

Abstract

The design and modelling of the walls of the United Arab Emirates Pavilion for the 2015 World Expo in Milan was a rich and complex challenge in shape research, computational design, digital fabrication and considerations for logistics and construction. This paper will focus on how the shape and textures of the walls of the pavilion were developed and specifically presents a novel method for generating tileable, three-dimensional, sand ripple patterns based on the theory of reaction diffusion equations. The walls were to be panelised by glass fibre reinforced concrete panels and the paper also explains how a multi-objective optimisation approach was applied to maximise the randomness of the placement of the different textured panels under budget and fabrication constraints.

Introduction

The national pavilions at the World Exhibitions have often tried to express certain recognizable flavours of the specific countries and the United Arab Emirates pavilion for the 2015 World Expo in Milan follows in this tradition. Under this spectrum, the design brief was to create walls inspired by the stunning desert landscapes of the United Arab Emirates. The walls were to be panelised using Glass fibre Reinforced Concrete panels (GRC). The current paper will tell the story of the design and modelling of these walls and in the next section, "*The Shapes and Textures of the Desert*", we will describe the research conducted on the shape of sand dunes and how the walls found their main form based on this. Then, in "*Computational Generation of Ripples*", we will zoom in on the computational generation of the smaller scale ripple pattern that covers a large part of the panels. It will be explained how this pattern was generated in order to produce a tileable result and we will cover both how the 2D topology of the pattern was found and how this then was used to create the final 3D texture on the panels. Further, in "*Optimizing the Ripples*", a multi-objective

H. Malm (✉) · S. Joyce · M. Tsigkari · K. El-Ashry · F. Aish
Foster + Partners, City Hall, London

M.R. Thomsen et al. (eds.), *Modelling Behaviour*, DOI 10.1007/978-3-319-24208-8_14

optimisation approach will be outlined, which was used to find the optimal sequence of the panels along the walls, maximizing the randomness of the panels while minimising the total number of moulds. The paper is rounded off with "*Acknowledgements*" and "*Conclusions*".

The Shapes and Textures of the Desert

To find the form and textures of the pavilion walls, research on the shapes created by the natural elements in the sandy landscapes of the UAE deserts was conducted. In this process, it was early on observed that features on two main scales dominate the typical perception of sand dunes; At a large scale, the wave shaped dune formation dominates the picture, while at a smaller scale an elongated fine-grained sand ripple pattern is evident (Fig. 1). The physics of the formation of these dune and ripple structures has been extensively studied and simulated by other authors, c.f. (Nishimori and Ouchi 1993; Yizaq et al. 1991; Lamb et al. 2002; Tsoar 2001). Briefly, the small-scale ripple pattern is spontaneously formed by the saltation of sand grains when the wind force exceeds a critical value, while the large scale dune is gradually built up by the prevailing wind until it collapses on the leeward side.

Fig. 1 The two scales of sand dunes

In the current project, field trips were made in order to study and register real sand dune and ripple formations. The shape of several examples of sand ripples were 3D scanned and the resulting models 3D colour printed. This gave us the possibility to study the shape and proportions of the ripples, together with a true material and shadow perception (Fig. 2). Measurements on site in the desert outside Abu Dhabi were also made in order to get a first-hand feeling for the proportions (Fig. 3).

It was concluded that the features at the two main scales of the desert formations, the dunes and the ripples, should be expressed together in an important part of the building; the entrance. Following elongated site constraints, this part was designed as a canyon, with narrow passages

Fig. 2 3D prints of sand ripples

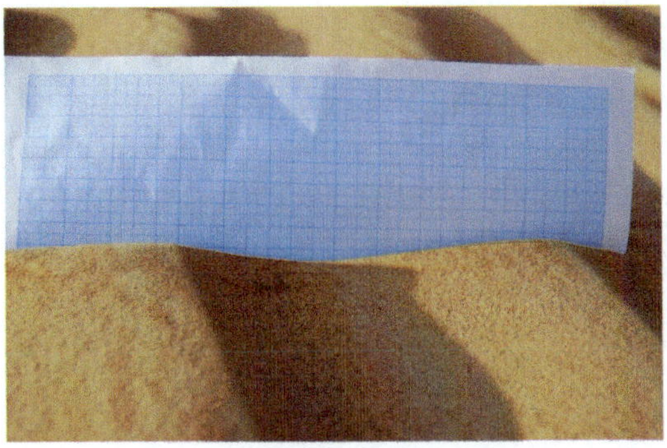

Fig. 3 Measurements on site outside Abu Dhabi

followed by more open areas. Here the visitors should be totally immersed in the experience of the dunes. The inner walls were designed with an undulating ridge on each side of the canyon reaching through the total length of the structure. On the upper part of these ridge walls, portraying the windward side of the dune, the sand ripple pattern was applied, while a smooth surface was modelled on the lower part, portraying the leeward side. See Fig. 4 for a visualization of the final canyon geometry at the entrance location and Fig. 6 for a plan view of the canyon.

There was also a wish to create some shelter from the sun or rain within the canyon and an array of different deployable structures for this were considered. However, in the end the undulating ridge of the walls was used for the purpose by allowing it to protrude further out into the canyon at the more open areas, in order for the structure itself to create an amount of

Fig. 4 Render of the entrance to the canyon

Fig. 5 The self-shading of the canyon

self-shading and shelter at those locations. In the narrower areas the ridges were instead modelled close to the baseline of the walls, creating a dynamic entrance sequence (Figs. 4, 5 and 6).

Fig. 6 Plan view of the entrance canyon. Figures 4 and 5 are rendered at the *marked* locations

Computational Generation of Ripples

Topology

In contrast to the canyon, the ripple pattern covers the whole 12 m height of the panels on the rest of the GRC walls. In order for these GRC panels to have some repetition, and to not all require unique moulds, a novel method for constructing a tileable ripple pattern was developed within the project. The repetition of these panels was also made possible by the fact that they are all straight extrusions from the base-line curves of the walls, c.f. "*Optimizing the Ripples*". In the initial design, these panels covered most of the exterior and interior walls, but in a later value-engineering stage only the exterior walls were clad with GRC. All of the panels of the canyon walls are still bespoke because of their free-form doubly curved shapes.

Some different approaches to 2D pattern generation were reviewed early in the process, such as work by Turner (2015), but it was quite soon observed that the solution space of a basic reaction-diffusion equation bore similarity to the branching structures of sand ripple formations. The theory of reaction-diffusion equations and their shape generative properties was first studied by Turing (1952), while more recently other authors have studied the texture and shape generating properties of these equations within the field of Computer Graphics, c.f. (Witkin 1991; Sanderson et al. 2006). In order to produce the elongated branching topology of the sand ripple pattern, an anisotropic reaction-diffusion equation was needed where the diffusion coefficient is replaced by a diffusion function which depends on the spatial directions (i.e. the amount of diffusion for "the chemical species" can be different in the x- and y-directions). To this end, we started looking at a solver implemented by Nervous System Inc (2015) in the Java-based coding environment Processing (Fry and Reas 2015), which we then modified as described below in order to meet our specific needs. Another implementation of a reaction-diffusion solver considered early in the process was the work by Schmidt in his toxiclibs libraries for Processing (Schmidt 2015).

The anisotropic reaction-diffusion equations used in this project follows the model of Gray-Scott:

$$\frac{\partial y}{\partial t} = r_u(\nabla u)\nabla^2 u + uv^2 + f(1-u) \qquad (1)$$

$$\frac{\partial v}{\partial t} = r_v(\nabla v)\nabla^2 v + uv^2 - (f+k)v \qquad (2)$$

where u and v represent the concentrations of two chemical species U and V. $r_u(u)$ and $r_v(v)$ are their diffusion rates, which in this anisotropic case depends on the direction of the gradients, ∇u and ∇n, of the concentrations in the current point. In the full explanation of the Gray-Scott model a third species P is also introduced and parameter k can then be interpreted as the rate of conversion of the species V to P and f as the amount of feeding of the species U and the draining of the species V and P.

By randomizing the initial conditions (seed) for the solver, new patterns were created for every simulation, giving an infinite range of branching patterns. As for all diffusion equations, this anisotropic reaction-diffusion equation also comes close to an equilibrium state for higher time values t, which with our settings meant a pattern of totally parallel lines (the diffusion of the dominating species is amplified in the x-direction, while the reaction and diffusion is set more equal in the y-direction). In order to instead get the desired amount of branching bifurcations in the pattern, the simulations needed to be supervised and the solver stopped at the right time giving the desired density of bifurcations.

However, for different seeds the above process outputs individual unrelated patterns that do not match when placed next to each other. The main challenge was now to create a process where the boundary conditions of the solver could be controlled in order to give a pattern that was continuous from tile to tile, both in position and in the tangents of the central curves of the ripple ridges. To this end, the first step in the process was to generate an initial base pattern. The boundary of this first pattern was stored in order to then be used as a part of the initial conditions of the rest of the ripple tile simulations, making all the generated patterns match. However, it was not enough to only store the very edge of the first generated pattern. In order to get the tangents of the ripples to match, which of course is very important for the flow of the pattern across panels, a wider edge band from the initial simulation needed to be stored. What was stored from the initial simulation was actually not the generated grey level of the pattern image, but the gradients of the intensity levels (i.e. the gradients of the relative amount of the chemical substances simulated, in the original interpretation of the equation). In the final simulation process, by weighting the gradients from the border conditions with a standard unbiased simulation, i.e. without border conditions, a satisfying flow of the ripples could be achieved, without any kinks or discontinuities.

This process gave us the possibility to simulate a theoretically infinite number of matching ripple textures. The process could be used to generate patterns matching on all four edges of a rectangular tile, however, in this project we finally chose to generate full height panels in one go, so that the patterns only needed to match along the horizontal direction. See Fig. 7 for a set of matching pattern tiles (cropped in height).

3D Structure

The process outlined above only gives the basic topology of the pattern. In order to actually achieve a 3D ripple pattern, with its wave like cross section, additional processing was required. To this end, the essential central line of the ripple curves was extracted. This was realised by initially thresholding the output images from the reaction-diffusion solver into binary black and white images. A process called skeletonization, c.f. (Gonzalez and Woods 1992), was then applied, resulting in only the central line of the pattern stripes being preserved. This method was implemented using Python and its OpenCV libraries (Itseez 2015). See Fig. 8, for a set of matching skeletonized patterns.

The next step was to replace these central line curves with a 3D geometry giving the correct cross section of the ripples. By studying our scanned ripple specimens and our on-site measurements, a number of options for the final cross section were designed, see Fig. 9 for a selection. The first method developed for generating the 3D geometry was based on 2D image filtering, where all the white "on" pixels in the binary skeleton images where replaced by a special filtering kernel. Interpreting the intensity of this kernel as a height measure, the kernel was created so that it had the correct 3D ripple section. Its pixel values were additionally chosen with a Gaussian fall-off to the outer edge of the kernel,

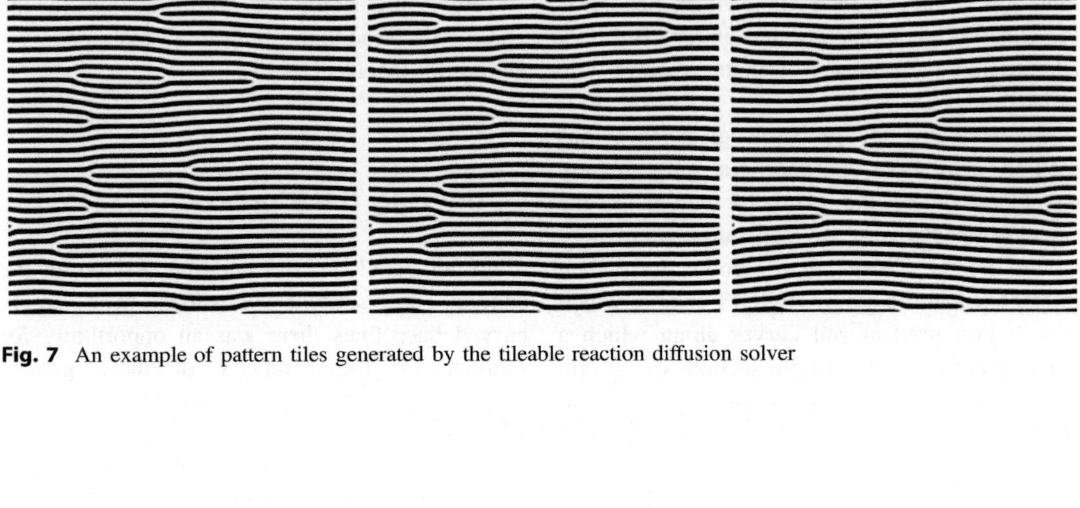

Fig. 7 An example of pattern tiles generated by the tileable reaction diffusion solver

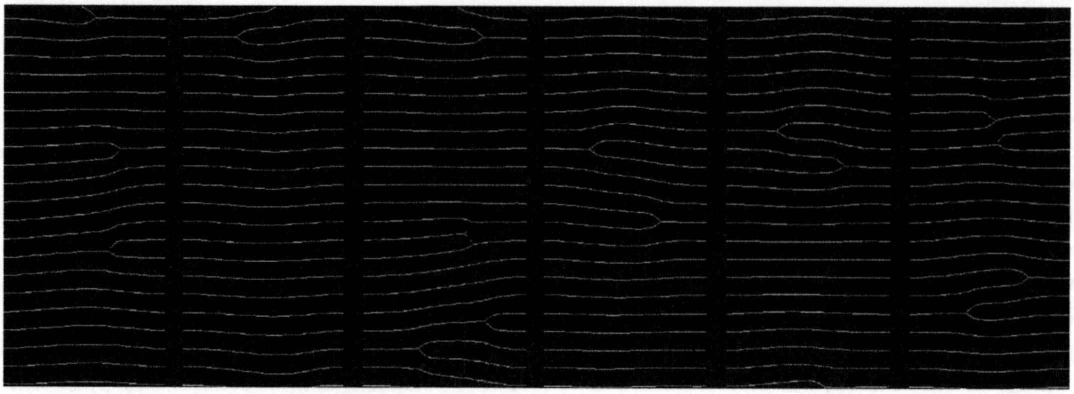

Fig. 8 Skeletonization of matching tiles of the reaction-diffusion pattern

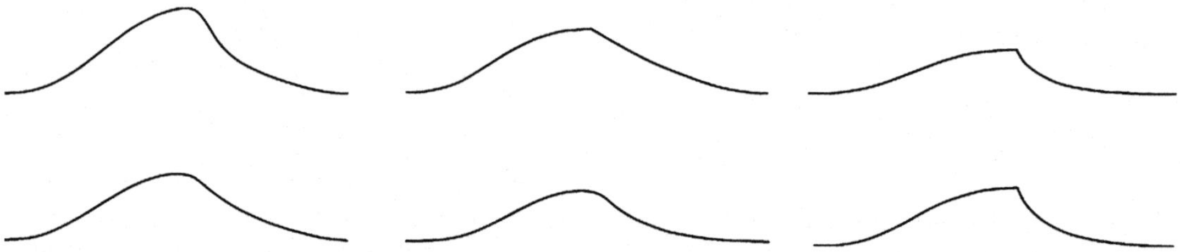

Fig. 9 Example of ripple sections. An option close to the lower right section was finally chosen

so that when applying the kernel the resulting output image had a smooth transition in every direction, except in the direction orthogonal to the ripple ridge. The resulting image was then used as 3D displacement data in order to obtain the actual 3D geometry. However, this approach had two major drawbacks. Firstly, despite the intrinsic smoothing in the above mentioned process, the image was still pixelated and gave a quite jittery look of each ripple ridge. Secondly, the resulting geometry became extremely heavy and would have been very difficult to work with in later stages.

To overcome the problems of the previous method, a different approach was subsequently developed where the pixelated curves in the skeleton images were divided into a number of control points. These control points were then interpolated into a set of NURBS curves, which was in turn used as rail curves along which a curve with the wanted ripple section was swept. This process was not as straightforward as it might seem, since some tricky special cases occurred, especially at the splitting of a rail curve (at the so-called bifurcations). However, the process could in the end be totally automated and resulted in a 3D ripple structure that is very eminiscent of the inspiration. See Fig. 10 for a comparison between a real sand ripple and an example of a computer generated bifurcation on one of the fabricated panels. The bespoke sweeping algorithm was implemented in the parametric modelling plug-in Grasshopper for Rhino 3D (McNeel 2015).

Optimizing the Ripples

Since all panels on the building, excluding the canyon, are straight extrusions from singly curved base lines there was an opportunity for rationalising the geometry of these panels. A "semi-facetted" approach was chosen for this

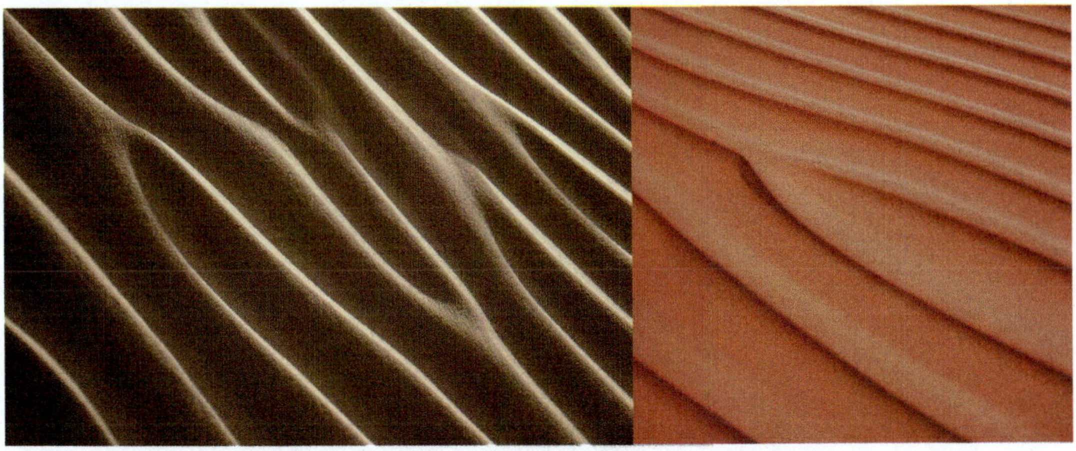

Fig. 10 Comparison of a real desert sand ripple and a detail from one of the final GRC panels

where the panels were divided into a family of arcs that fit the base lines as closely as possible. In the end, a family of 9 different arcs was chosen, where the radii was found through optimisation with the genetic solver Galapagos in Grasshopper. Additionally, 20 special panels were added, dealing with outliers with too large deviations from the baseline curves. We will not detail the geometry rationalization further here, but instead focus on the succeeding multi-objective optimization that relates to the randomisation of the ripple pattern.

Due to the GRC fabrication process, every combination of ripple pattern and panel arc radius demanded a unique mould to be made. The ripple generation technique outlined in Sect. 3 above could theoretically produces an infinite number of matching patterns. However, there would not be any benefit of the tiling if the same pattern was not reused at several locations. After visual testing, using rendered images, it was decided that 8 different patterns were sufficient to give a good overall feeling of randomness in the wall texture.

Let us say that all unique arcs in the panel family were only used 8 times or less on the building. Using as many different patterns (of the 8 available) as possible for each arc would then still demand a unique mould for every panel on the building. But, since each mould actually could be reused up to 10 times in the fabrication, there was a wish to get closer to this amount of the usages per mould, and minimize the number of total moulds to be milled and prepared.

To find a solution to the conflicting objectives of reducing the number of moulds while maximizing the randomness of the overall texture, a multi-objective optimisation approach was applied. The solver used was the Octopus plug-in for Grasshopper which implements SPEA2 (Strength Pareto Evolutionary Algorithm) by Zitzler and Thiele (1991). Two objective functions, f_1 and f_2, were used, where f_1 was based on an adjacency measure $a_{i,j}$ equivalent to the distance from panel i to another panel j along the wall, disregarding arc type. The inverse of $a_{i,j}$ was then summed over the 10 closest panels to panel i, counting only identically patterned panels. This was then summed over all N panels. Let P_i denote the pattern at panel i, then:

$$f_1 = \sum_{i=0}^{N} \sum_{j=i-5}^{i+5} b_{i,j}, \quad where \quad b_{i,j} \begin{cases} 0, & i = j \\ 0, & P_i \neq P_j \\ \frac{i}{a_{i,j}}, & P_i = P_j \end{cases} \tag{3}$$

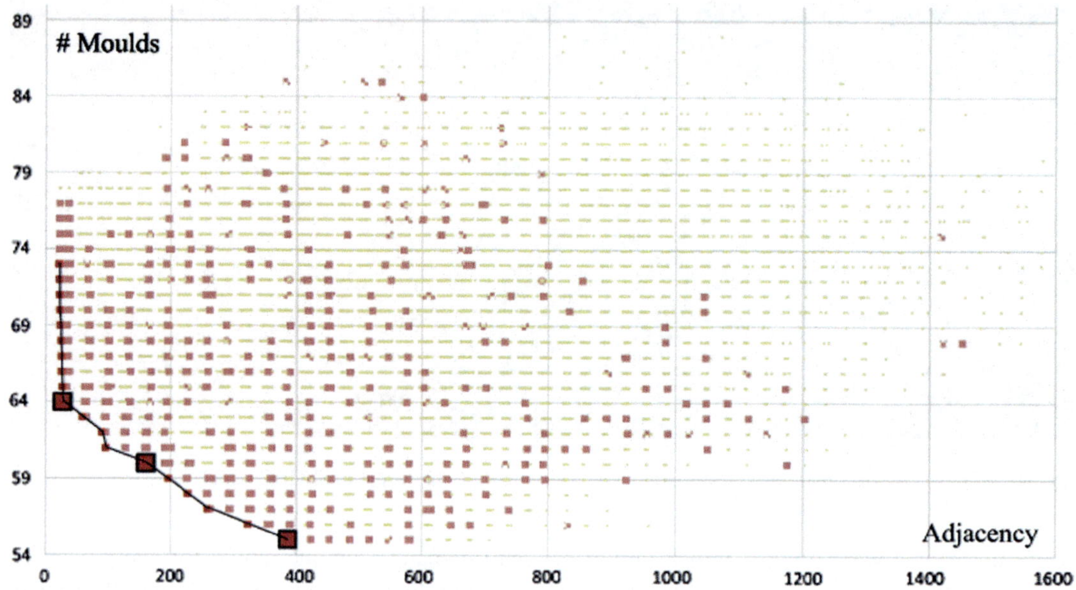

Fig. 11 The Pareto front of the multi-objective optimization after 1018 generations. *Dark red marks* points on the final Pareto front, *light red* marks previously non-dominated points and *light green* shows dominated evaluated points

The second objective function was just the total number of moulds. Since each mould could be reused a maximum of 10 times, this function can be written as:

$$f_2 = \sum_{i=0}^{n_A} \sum_{j=0}^{n_P} \frac{N_{i,j}}{10} \tag{4}$$

where $N_{i,j}$ is the number of panels with arc radius i and pattern j, n_A is the total number of arcs and n_P is the total number of patterns.

The genetic algortihm ran through 1018 generation, with a population size of 100. Figure 11 shows the Pareto front and evaluated panel configurations plotted in the objective space after the process had converged, with f_1 on the x-axis and f_2 on the y-axis. Three different non-dominated points on the Pareto front have been selected and Fig. 12 shows the corresponding panel configurations, in the same order, from left to right. The panels have been color-coded according to the level of adjacency: Red, orange, yellow, green and grey represents panels with a distance of zero, one, two, three and larger than three panels to a panel with the same ripple pattern. A point on the Pareto front,

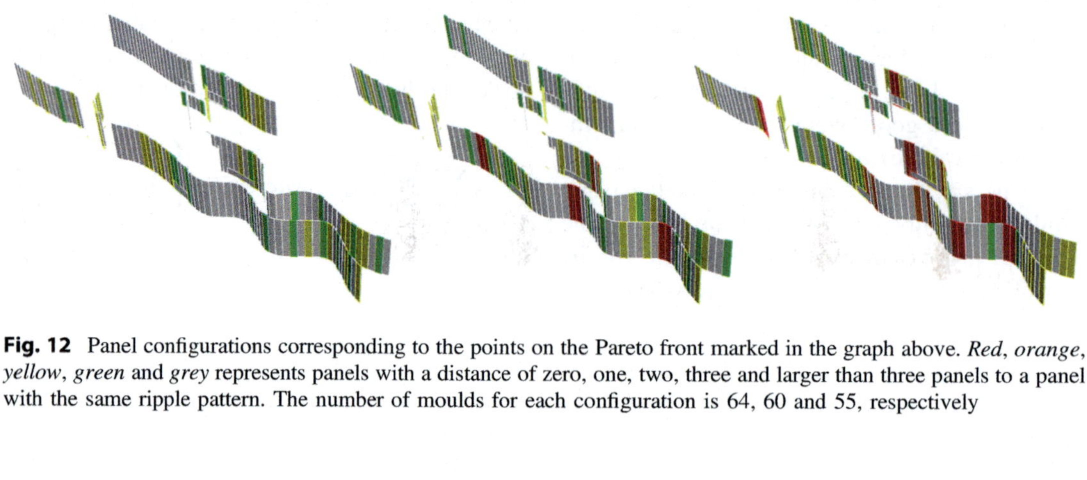

Fig. 12 Panel configurations corresponding to the points on the Pareto front marked in the graph above. *Red, orange, yellow, green* and *grey* represents panels with a distance of zero, one, two, three and larger than three panels to a panel with the same ripple pattern. The number of moulds for each configuration is 64, 60 and 55, respectively

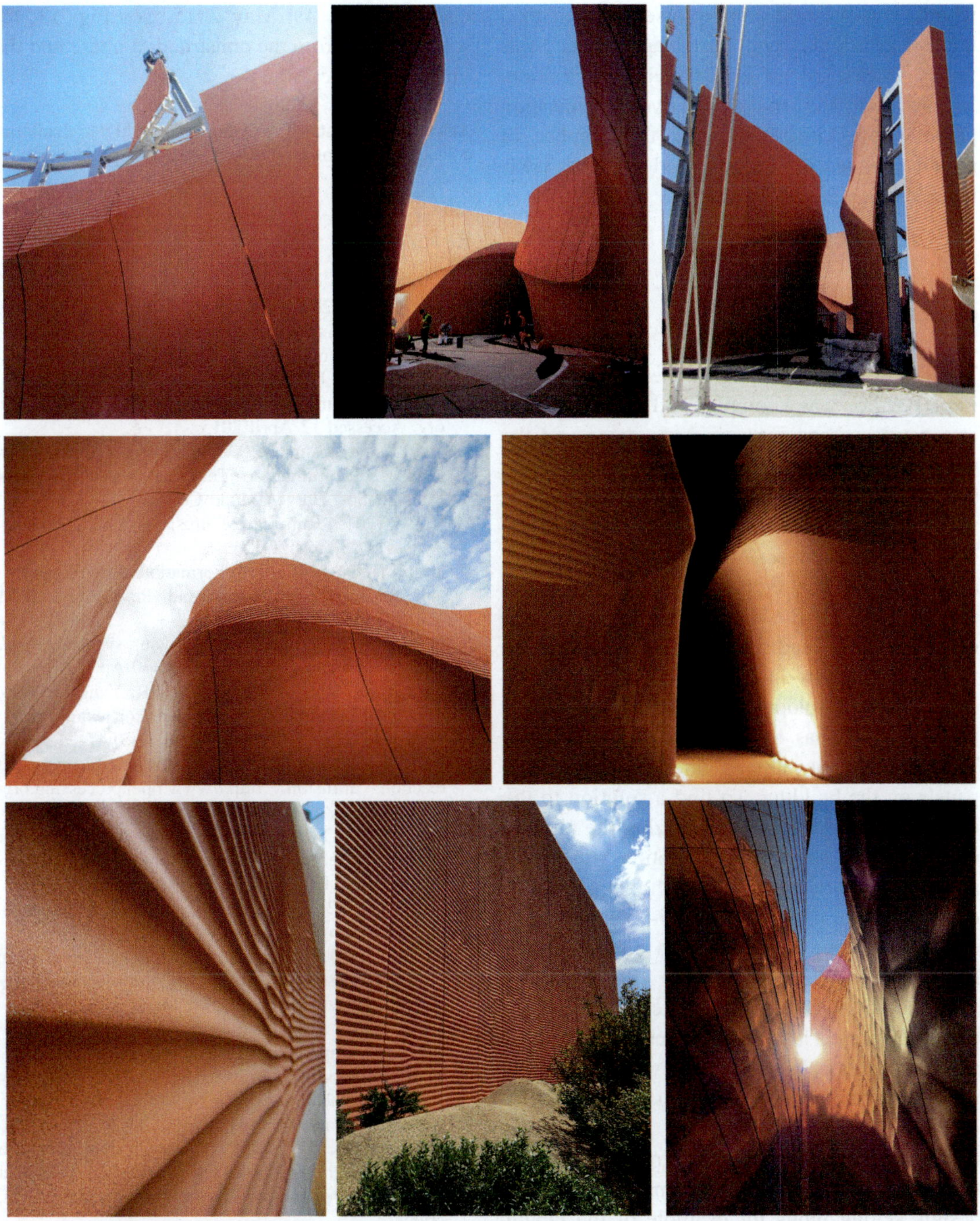

Fig. 13 Photos from the construction phase and the finished project

regarded as the best compromise between total number of moulds and visual appearance, was finally chosen as the final panel configuration. In the end, the multi-objective optimisation process approximately halved the number of moulds needed to be prepared for the GRC fabrication.

Conclusions

The design and modelling of the walls of UAE Expo Pavilion was a complex and inspiring study into natures shape formation processes and the usage of these studies to create an attractive space and building. The walls and entrance space was designed to reflect the perception of the UAE desert landscapes, with the undulating sand dune ridge of the entrance canyon creating a dynamic and self-shading structure. The creation of a tileable 3D sand ripple pattern, based on the theory of reaction-diffusion equations, was one of the major developments within this project, a process which has been a focus of this paper. This pattern was applied on all of the GRC panels on the building. To find a solution to the contrasting objectives of minimizing the number of moulds for the GRC fabrication while preserving a random appearance of the ripple pattern on the panelised walls, a multi-objective optimization strategy was applied, a process which has also been detailed here.

In addition to the subjects covered in this paper, the project also presented challenges when it came to rationalization of the wall geometries and the creation of an efficient pipeline from design to the GRC fabrication. Many aspects concerning the installation and transportation of the panels also had to be considered in the design, setting limits for example of the height and width of each individual panel.

The pavilion was constructed during the autumn of 2014 and the spring of 2015 and opened on the 1st May 2015. See Fig. 13, for some images from the construction phase and the finalized project.

Acknowledgments The authors would like to thank our colleagues and collaborators in the wider project team.

References

Fry B, Reas C (2015) Homepage of processing. https://www.processing.org. Accessed 15 April 2015

Gonzalez RC, Woods RE (1992) Digital image processing, 2nd edn. Prentice Hall, Boston

Itseez (2015) Homepage of opencv. https://www.opencv.org. Accessed 13 April 2015

Lamb P, Kwan J, Ahn S (2002) A computer simulation of sand ripple formation. Technical report, Dept of Mathematics, Harvey Mudd College

McNeel R (2015) Rhinoceros. https://www.rhino3D.com. Accessed 15 April 2015

Nishimori H, Ouchi N (1993) Formation of ripple patterns and dunes by wind-blown sand. Phys Rev Lett 71 (1):197

N. S. Inc (2015) Homepage of nervous system. https://www.n-e-r-v-o-u-s.com. Accessed 30 April 2015

Sanderson AR, Kirby RM, Johnson CR, Yang L (2006) Advanced reactiondiffusion models for texture analysis. J Graphics GPU Game Tools 11(3):47–71

Schmidt K (2015) Toxiclibs, simutils-0001, gray-scott reaction diffusion. http://www.openprocessing.org/sketch/2698. Accessed 01 May 2015

Tsoar H (2001) Types of aeolian sand dunes and their formation. Geomorphological Fluid Mechanics, Lecture Notes in Physics 582:403–429

Turing AM (1952) The chemical basis of morphogenesis. Phil Trans Roy Soc 237:37–72

Turner A (2015) Sand ripples. Processing sketch from open processing. http://www.openprocessing.org/sketch/2698. Accessed 01 May 2015

Witkin A (1991) Reaction-diffusion textures. ACM SIGGRAPH Comput Graph 25(4):299–308

Yizaq H, Balmforth N, Provenzale A (1991) Blown by wind: non-linear dynamics of aeolian sand ripples. Physica D Nonlinear Phenom 195:3–4

Zitzler E, Thiele MLL (1991) Spea2: improving the strength pareto evolutionary algorithm. In: Technical report 103, Department Electrical Engineering, ETH, Zrich

Part II

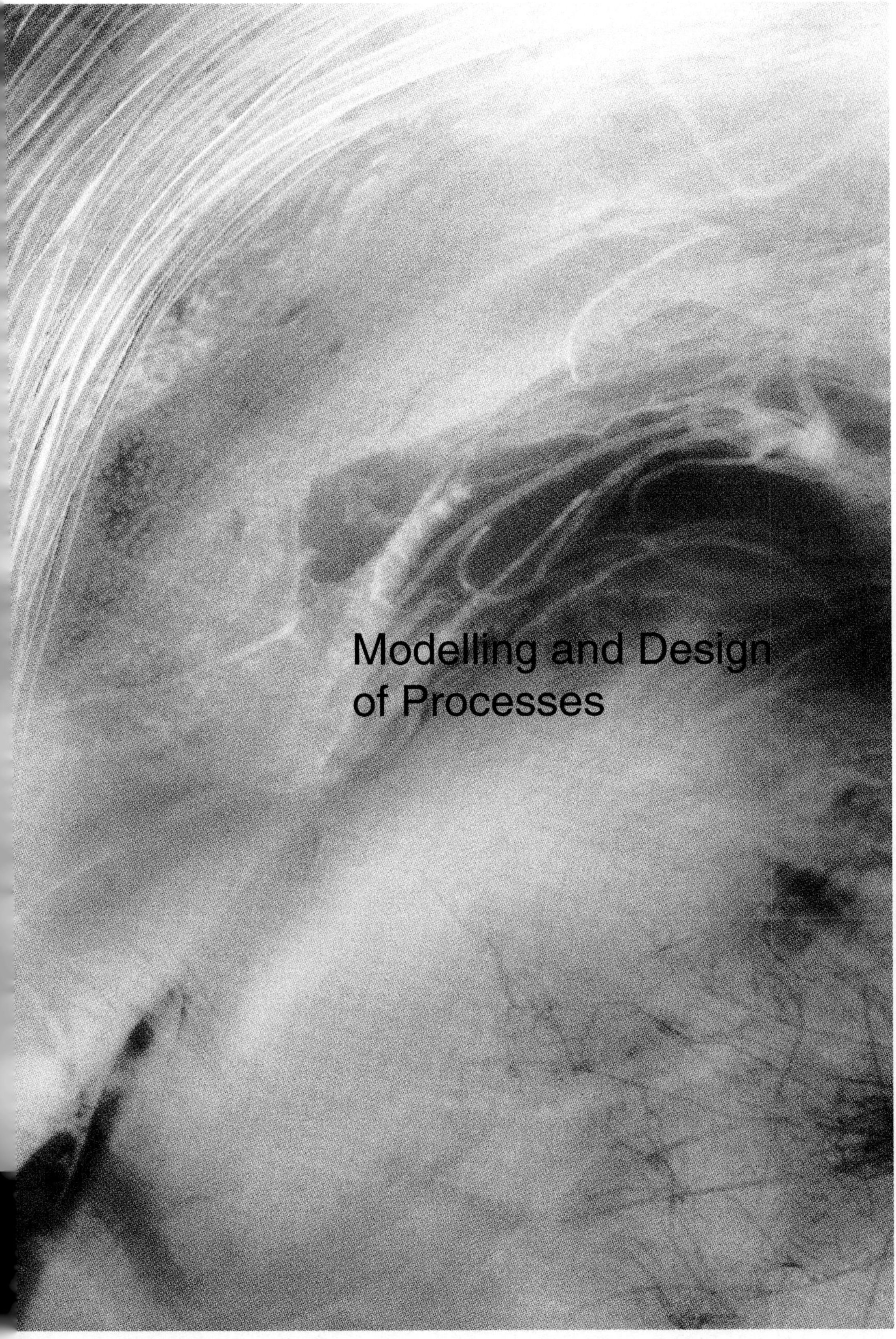

Modelling and Design of Processes

Digital Inca: An Assembly Method for Free-Form Geometries

Brandon Clifford and Wes McGee

Abstract

Inca masons mastered a unique procedure of stone construction resulting in complex and precise configurations of visually zero-tolerance alignment. Each stone is nibbled away by a gradual transition of hammerstones from course to fine as the carving approaches a precise alignment on each edge at the visible face. Beyond this shared front edge, a gap opens to allow mortar to be packed in from behind for structural and setting purposes. These formations are limited to vertical walls as a result of the use of plumb technology. With so much current attention being focused on free-form geometries and procedural methods for masonry vaulting (Andriaenssens et al. 2014), this paper proposes a translation of the Inca wedge method into a digital process whereby stones align with each other to produce a global figure without the use of templating, falsework, or formwork. This differs from most conventional masonry processes whereby mortar is applied to units before positioning, allowing alignment to templates through the malleability of the mortar. Instead, units are dry fit to each other and then set in place. This inversion of the alignment procedure relies on precision carving of voussoirs to align with each other at specific moments in order to ensure proper self-assembly while simultaneously accepting mortar for setting. The resulting prototype shell structure is composed of uniquely carved voussoirs of autoclave-aerated concrete set with plaster in a similar manner to the Catalan and Guastavino methods. Previous research (Ramage 2006) has demonstrated the advantage of this lightweight material for vaulting purposes. The combination of this material property with the Inca assembly method

B. Clifford (✉)
Massachusetts Institute of Technology (MIT)/Matter Design, Boston, USA
e-mail: bcliffor@mit.edu

W. McGee
University of Michigan/Matter Design, Ann Arbor, USA

M.R. Thomsen et al. (eds.), *Modelling Behaviour*, DOI 10.1007/978-3-319-24208-8_15

results in an expedient assembly of complex shell structures. The potentials of this proof-of-concept suggests through results that future research can advance the intelligence of error aggregation through the aid of measuring, correcting, and carving throughout the assembly process just as the Incas constructed.

Introduction

Inca stonemasons held tremendous knowledge in a particular method of making resulting in timeless stone constructions. This architecture emerged from procedural logics that are sequential, rule-based, and adaptable; therefore aligning with contemporary topics in the integration of computation with fabrication. In recent years, significant attention is being paid to the capacity of computation to aid in the production of free-form shell structures, both virtually and physically. In translating the conceptual principles of compression-only structures into physicality, a series of hurdles commonly emerge. These include determining how to intelligently carve volumetric material into unique geometries that precisely align with each other in space, how to determine the sequence of assembly to maintain structural integrity, and as a result, how to reduce the waste of falsework and formwork. These hurdles are not new. They have existed throughout the history of spatial masonry construction. A paper (Rippmann and Block 2011) demonstrates the advantages of re-learning ancient knowledge—in that case it was the sixteenth century technique stereotomy. This influential text contributes to a consortium of research that is attempting to re-engage the topic of volume, not only for the vacuum of current knowledge in the topic, but for the potentials it serves to define space structurally. Other contributions include 'Stereotomy' (Fallacara 2012), 'Volumetric Robotics' (Clifford 2014), as well as papers like Feringa and Sondergaard (2014) and McGee et al. (2012). This paper seeks to mine the potentials of Inca masonry construction and identify which knowledge is capable of contributing to the hurdles listed above. In doing so, it explains an Inca method and prototypes this procedure with contemporary tools and materials. It proposes a translated method that employs ancient knowledge into a process to precisely carve Autoclave Aerated Concrete (AAC) with self-inscribed information to assemble free-form geometry shell structures without the aid of templating, falsework, or formwork, while ensuring structural stability.

Inca Architecture

There is a great deal of speculation surrounding Inca stonework. How did such a primitive civilization produce these precise mortar-less alignments as shown in Fig. 1 at a time prior to mechanization? A paper (Protzen and Nair 1997) examines this conundrum. These voluptuous assemblies are indicative of Inca architecture; however, similar examples can be found in other cultures that employed hammerstones instead of metal tools such as Egypt and Rapa Nui (Easter Island). The most archetypal works of the Incas exhibit a number of inventions as a result of their own technology. From the unique process of quarrying, to the fitting, dressing, and assembly of stone, the Inca rarify their technology to produce incredible works of architecture. Significant research in the field of archaeology has contributed to a better understanding of how the Inca constructed, most notably 'Inca Architecture and Construction at Ollantaytambo' (Protzen 1993) and 'Inca Architecture' (Gasparini and Margolies 1980).

Fig. 1 Inca wall construction in Cusco Peru

Hammerstones

Perhaps the central technology that defines the works of the Incas is the hammerstone. Outwater stated in 1959 that "[v]ery few tools are in evidence at the site [Kechiqhata]. There were some hammerstones of diorite but very few picks or wedges." (Outwater 1959, p. 28) Protzen later suggests these picks and wedges must have been from a later era, as there is "only scant evidence that the Incas split rocks with the aid of wedges." (Protzen 1985, p. 166) Instead, the Incas were carving stones with other stones, requiring a paradigm shift in our conception of stone carving. As Dean explains, "the In[c]a referred to the working of finely joined masonry as *canincakuchini*, which is derived from the verb *kanini* (*canini*), meaning to bite or nibble." (Dean 2010) Hammerstones are employed in two manners—drafting and dressing. Drafting is a technique whereby the hammerstone strikes close to the edge and around the perimeter of a stone in order to remove large portions of the stone as they blow out from the side. The other technique of dressing involves hammerstone striking relatively perpendicular to the face of the stone to nibble away at a rounded figure. This technique removes less material, but allows for greater precision and is typically employed in a progressive manner from course to fine where greater definition is required. There is a further technique of polishing, though not employed in the prototypes of this paper.

Quarrying and Configuring

Because the Inca were using hammerstones instead of metal picks or wooden wedges, the process of quarrying in the proper sense was rare. The majority of stone acquisition was done through a process of selection from loose rockfalls. The approximate stone is then dressed "only minimally before it was sent on its way to the construction site." (Protzen 1985, p. 165) The configuration of these stones is not pre-determined, but the result of a sequential assembly from the bottom up of irregular forms that are scribed and minimally carved to align with their neighbors.

Fitting

Stones can be site-dressed to fit by either carving the existing assembly to accommodate the quarried stone, or vice versa. Occasionally both methods are applied to a single fit. Two common theories exist that explain how the Inca determined the geometries needed to carve these unique units of construction. The first theory employs templates and the second requires dry-fitting. This is not to suggest that the Inca only used one of these methods. Instead it is highly likely the Incas employed both.

Templating

This common theory suggests that a wooden template is constructed to scribe the existing condition of either the assembled wall or the quarried stone. This template is then translated onto the stone that will be dressed to fit and used as a guide to approach the desired profile geometry. This process is ideal for stones that are too large to maneuver over and over again, but does not produce the same precision of the dry fitting approach.

Dry Fitting

One byproduct of the carving method of nibbling with hammerstones is a great deal of stonedust. In the dry fitting theory, as described by Protzen (1985) this stonedust coats the set stones while the loose stone is set into place. This stone is then removed and where the stone is touching the existing condition, the dust is compacted and visualizes where material needs to be removed. This is a tedious process and limits the size of the stones to something easily maneuverable. While this fit is more precise than the templating approach, it is also more laborious.

Finishing and Dressing: The Wedge Method

The use of hammerstones to draft and dress the stones results in three geometric conditions—face, edge, and wedge (Figure 2). As previously described, the sides of the stone are drafted to produce the approximate profile of the stone. This act of drafting removes cleft chunks of stone from the sides at obtuse angles to the cut. When precision is required, the stone is flipped over and dressed from the fact, progressively moving closer and closer to the desired geometry. This face dressing explains the most characteristic geometry of the Inca stonework—the rounded face. This rounding is not just a result of the carving method, but also preforms an important task of providing the distinctive chiaroscuro that disguise subtle misalignments that invariably happen upon assembly. The drafting and dressing produce the geometries of the cleft sides and the rounded face. The condition between these two is the most important—the edge. The edge is rather well defined, but is always produced at an

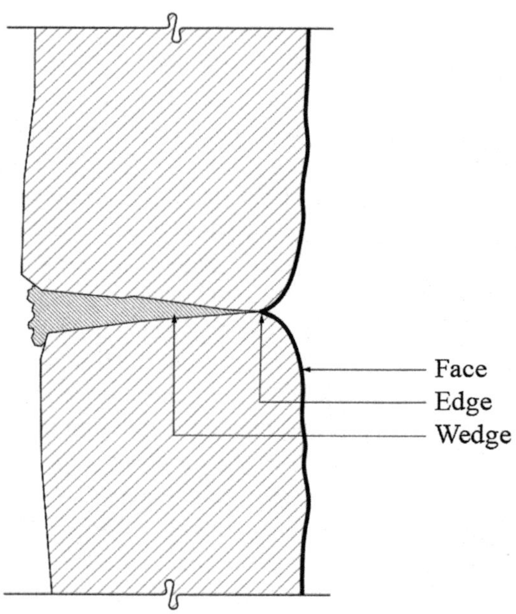

Fig. 2 Section through an Inca assembly revealing the three geometries

Fig. 3 A series of Inca constructions that reveal the voided wedge between stones

oblique angle. This edge is the only alignment between stones. The sides of the stone have been roughly drafted in order to not collide with their neighbors. This opening produces a negative wedge between stones that can be filled in with rubble from behind as shown in Fig. 3. Harth-terré introduces this *piedra-cuña* (wedge-stone) as the technique that allows the Inca masons to produce this illusion of apparently mortarless assembly (Harth-terré 1964).

Occasionally, the Inca did fit their stones with complete continuity from front to back, but these moments are rare. They are reserved for temple construction as well as the occasional corner condition. The rarity of these full-contact fits speaks to the difficulty of such an alignment. The wedge-method on the other hand takes less carving and fitting, but produces a similar visual result. It has one further advantage of privileging one side for visibility. In the case of the wedge construction, the back face of any stone is hidden inside the rubble, and therefore is not required to align to anything. It can be deep or shallow, depending n the needs of the visible face; a variable this paper employs.

Mortar Bed Versus Dry Fit

While it is commonly assumed that there is no mortar between the joints of these Inca walls, this is not true. This wedge simply disguises the mortar to the visible face. Even though the Inca constructions do contain mortar; the logic differs greatly from the conventional mortar bed approach. In conventional masonry, standard units are set into a mortar bed that employs viscosity in order to align the stone to a global goal—typically a horizontal string. This method uses the variability of the mortar joint to adjust for the imprecision of the unit. The Inca process shifts alignment from the process of setting to the process of fitting. Precision carving allows the stones to dry align with each other before setting.

Translation Prototypes

With this foundational understanding of how the Incas constructed, this paper seeks to find a viable translation of this technique into contemporary practice. It is not immediately feasible, or practical to directly carve stone with hammerstones and manual labor. The following two prototypes test a hypothesis that via subtractive machining of Autoclave Aerated Concrete (AAC) one can assemble a free-form geometry shell structure. Previous research by Ramage (2006) explored these unique material capacities toward the ability to erect compressive vault structures using the Guastavino method (Ochsendorf 2010), as the lower density of AAC allowed larger panels and thus a faster installation rate.

Fig. 4 Images of the units and assembly of prototype #2

Prototype 01

The first prototype distilled the Inca condition down to the wedge, neglecting the rounding of the face. This was intended to test that it could be possible to employ the edge as a spatial alignment from unit to unit. A free-form dome geometry serves as the global figure with three units applied to this figure. The front edges align precisely with an acute angle, and opens in the back to produce the wedge. The machining process revealed that this sharp front edge becomes fragile in such a brittle material, suggesting that the rounded pillowing has some material advantages. Upon assembly, two more issues emerge. The first being that with only one edge of alignment between parts, it is difficult to locate a unit because the edge acts as a hinge. Given there is adequate curvature to the figure, three parts better align in space. The second issue is that the assembly is difficult to get precise alignment and the sharp edge reveals the inaccuracies in the manual assembly process. Aside from these two concerns, the assembly process is rapid and stable.

Prototype 02

In order to combat the previous prototypes concerns, a second prototype produces a rounded geometry on the front face to protect the brittle edge of the material, as well as help disguise the mis-alignments upon assembly. This rounding is not done as a nostalgic replication of the Inca construction, but the result of understanding the role this rounding performs. It is important to round the front face geometry while maintaining a clear edge with an obtuse angle to aid in the alignment. The second advancement in this prototype is the nub that appears in the center of each side. This nub aligns between units as a stop that resists the hinging index issue. In the event this nub needs to be removed, it is minimal enough that it won't require the caving away of an entire face. The assembly of this ten-unit prototype emerged successfully and rapidly as seen in Fig. 4, serving as a proof-of-concept to move forward and test the proposed method against a larger assembly.

Method

A larger installation, titled 'Round Room' is produced in order to test these principles on a larger assembly. This production requires the use of complex geometry, structural calculations, fabrication limitations, and assembly procedures —each entangled with the rest. The following describes the process and techniques, many of which expand upon previous research by the authors (Clifford and McGee 2014) that manifest in the project 'La Voûte de LeFevre'.

Geometry

A free-form geometry is first composed to approximate a possible structural condition. A particle-spring system evenly distributes points across this surface with a dimensional variable set at 45 cm establishing the size of the resulting parts. A three-dimensional voronoi calculation breaks the single figure into discrete polygonal units of approximately 45 cm in diameter. Each central point then generates a vector normal to the figure and defines a plane on the back face for fabrication purposes. This plane moves backwards until it ensures all edges of the face geometry are at least one inch positive of the fabrication plane. The perimeter of the intersection between this plane and the voronoi calculation is offset inward to begin the wedge and nub geometry. Each edge is checked for length to make sure they are long enough to add a registration point, and if so, the offset polygon is pulled out to align with the initial intersection polygon precisely at the mid-point of the edge length as demonstrated in Fig. 5. A ruled surface geometry is generated between the front edge and the back offset, nub. This geometry is ruled in order to allow the parts to be swarf machined in a similar manner to the previous application. (Clifford and McGee 2014)

Structural Computation

After the geometric calculation is generated, a closed unit can define its own volume. This volume is used to inform another particle-spring system that ensures a thrust-network may fall within the thickness of the material. As opposed to a typical hanging chain model where the mass of each node is uniform producing a catenary geometry, this Thick-Funicular calculation (Clifford 2013) checks the volume of each unit and re-defines the particle of the unit with this number as the vertical thrust (Figs. 6, 7 and 8). This volumetric version of the Thick-Funicular calculation is identical to the one described in 'La Voûte de LeFevre' (Clifford and McGee 2014); however, the variable that allows the unit to be heavier or lighter relative to its neighbors is the depth dimension of the back face. While the calculation is running, if a particular node is higher than it should be, that back face is brought closer to the original one-inch offset thus reducing the volume and mass.

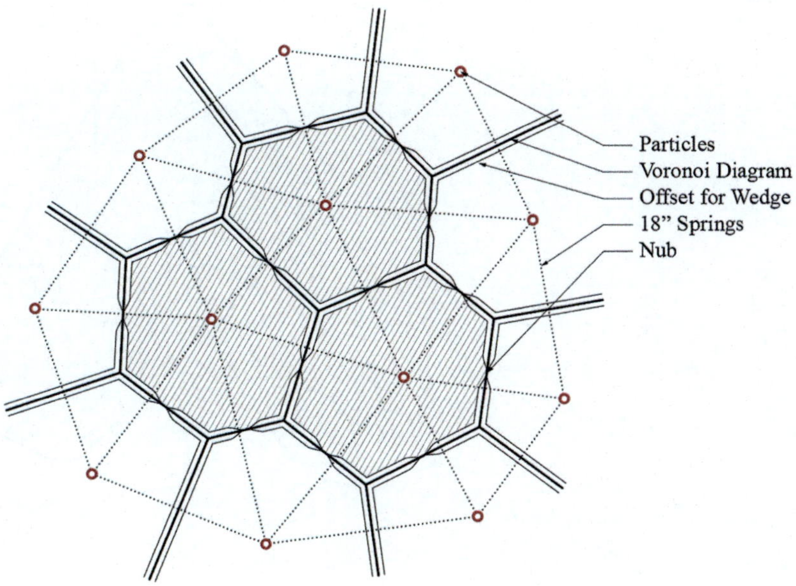

Fig. 5 A 2d diagram of the unit discretization, wedge offset, and indexing nub geometries

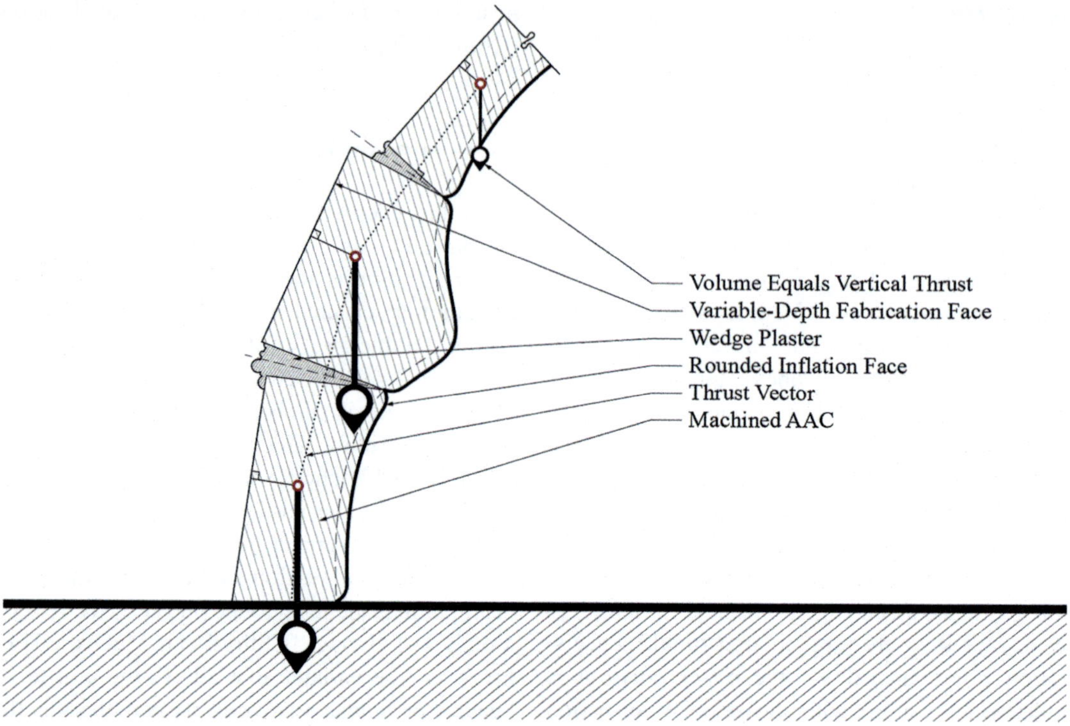

Fig. 6 A section revealing the wedge method as well as the geometries of the variable depth back face that redefined the volume in order to ensure the thrust network falls within the depth of material

Fig. 7 A 3d diagram explaining the geometries that discretize the units, determine the back face plane

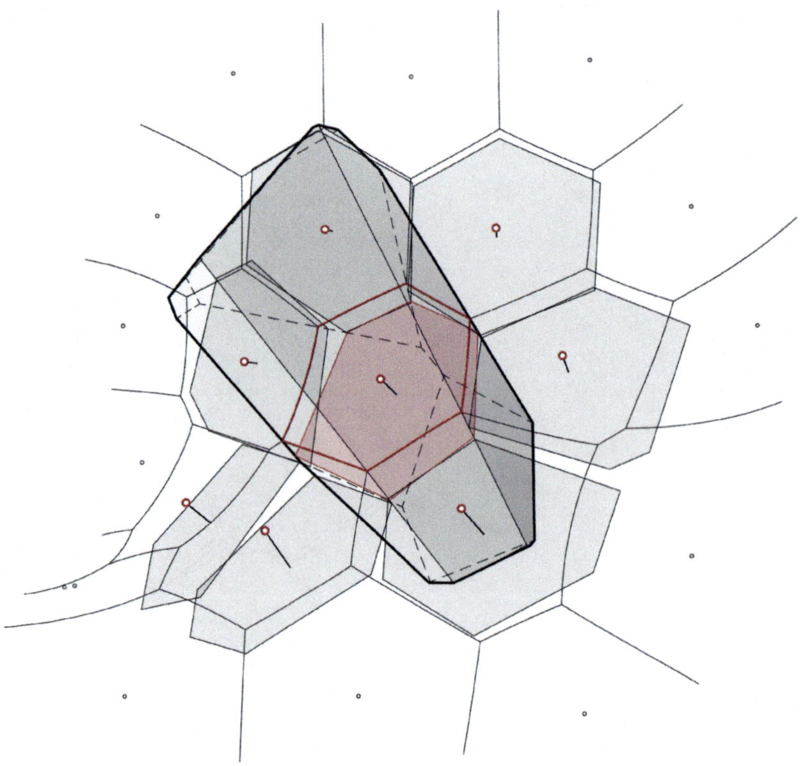

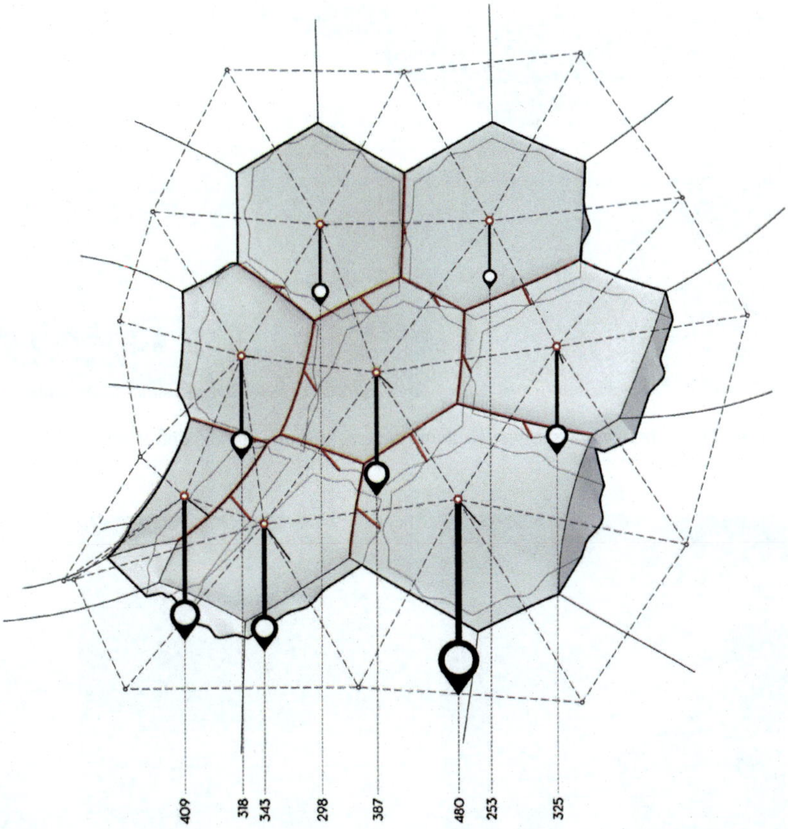

Fig. 8 A 3d diagram explaining the variable-volume particle-spring structural calculation. The *vertical pendants* represent the volume of each unit thus re-defining the vertical thrust of the particle in the system. The *red lines* represent the only coincidental geometries between units—the *front edge* as well as the *nub* defines a 'T' shaped intersection

Machining

The initial prototypes were machined on a typical five axis router, but the final installation components were wet machined with a Kuka KR500 at Quarra Stone, inc. AAC blanks were pre-sized using a slab sawing process, which allowed the final machining process to proceed as quickly as possible. The surface machining process used on the pillowed face of each unit is the primary time constraint, with the swarf or flank machining process used on the sides proceeding relatively quickly. An ongoing collaboration with Quarra Stone is investigating the development of new workflows, which translate directly from the generative model to machine code using the SuperMatterTools script library (McGee and Pigram 2011).

Assembly

The first course of units that align to the ground are wedge plastered from unit to unit, but glued to the base in order to act as a foundation and resist any horizontal thrust as seen in Fig. 9. In a manner similar to the Incas, each unit is dry fit to its neighbors prior to being mortared in place. The edges and nubs align the unit in space with the previous units as guides. Once fit, a batch of plaster is mixed and transferred into a piping bag in order to squeeze the plaster in from behind. A rather liquid mix is used in order to obtain the correct viscosity. Throughout this process, one installer is holding the unit in place while another is piping the plaster as demonstrated in Fig. 10. A single bead of plaster is applied to the backside of each wedge, drying in a time between one and

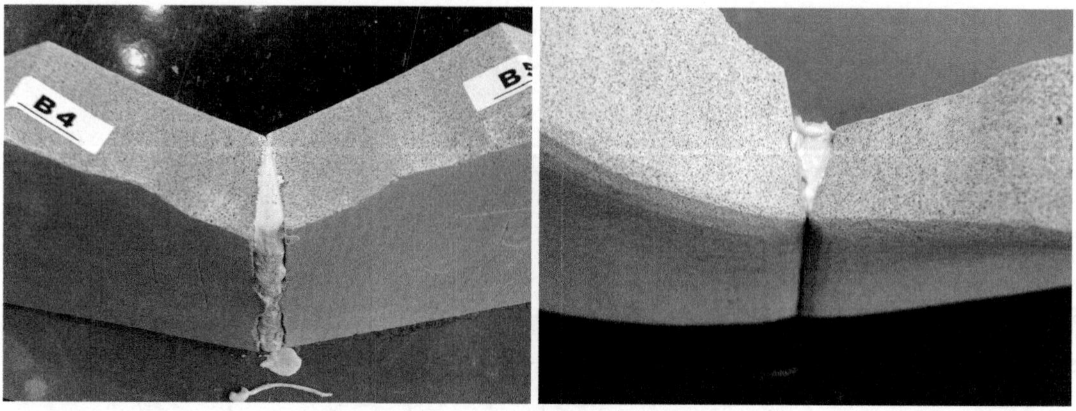

Fig. 9 Detail images of the wedge method from the non and visible sides

Fig. 10 Images of the assembly process as well as the undercarriage supports

three minutes. At that moment, the first installer is able to let go of the unit while the second installer continues to fill the voided wedge with plaster. The tolerance of misalignment from unit to unit was around 0.8 mm for the first few courses, shown in Fig. 11. This could be the result of tolerance issues in the machining process or assembly, but fell well within visible tolerance. During this period, no carving or fitting is employed. As predicted, this error aggregated as the assembly continued. When the errors became closer to 3 mm, they became more visible and a new process of the Inca dry fitting was employed. The first installer would set the stone where it was intended to be placed and immediately remove it. The collision between the geometries becomes visible as the AAC is compressed slightly producing a white powder on the surface. This white powder represents geometry that needs to be removed. The installer then shaves away at these moments in the collision and re-fits the parts until an entire edge is aligned and the assembly process can continue. The advantage of this shaving process is a precise fit

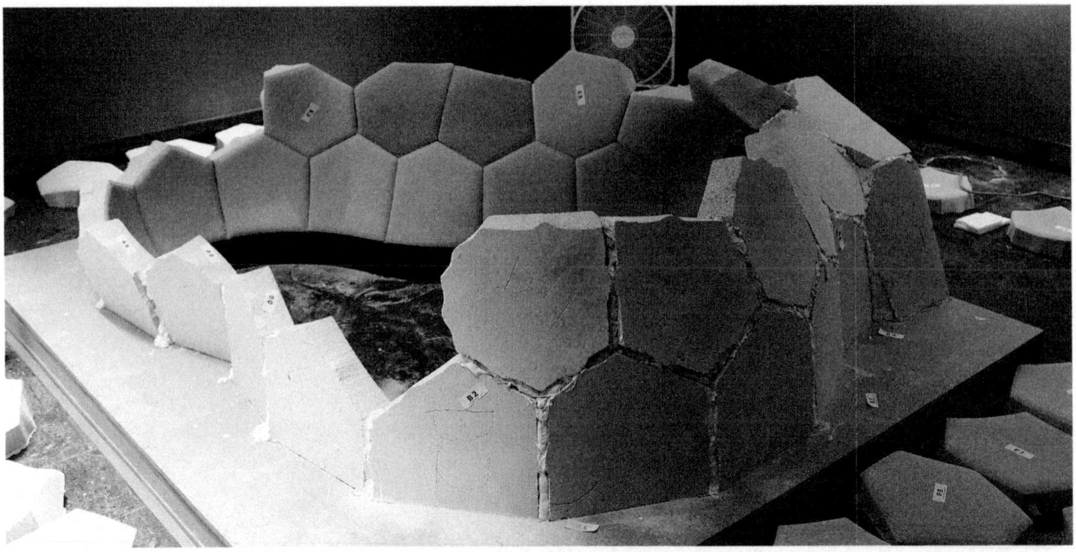

Fig. 11 The first couple courses of the 'round room' assembly process

Fig. 12 Image of the final visible assembly of the 'round room'

Fig. 13 The 'round room' by matter design and quarra stone

can be achieved without an ideal pre-existing condition. The dis-advantage being that is it ensures the following units will be further out of spec as the lower cells are now shorter on some edges. At a certain point, these errors aggregated so much that yet another process of Inca templating is employed. When a cell is too large to fit in a location, the surrounding cells are arched over the location leaving a void that is templated with a piece of paper and transferred onto the original unit. The unit is then manually cut to align with the template and face-placed into position. This manual process may be similar to the Inca method; however, undermines the computation and digital fabrication process. As the assembly continues and errors aggregate, the efficiency of the process progressively slows down to accommodate these manual fits, ultimately approaching the original manual process (Figs. 12 and 13).

Conclusions

This paper demonstrates the possibility of learning from the Inca in order to inform contemporary shell structure production. By employing this past knowledge of the wedge method into a way of carving AAC and assembling via the Guastavino approach, it is viable to construct a free-form shell structure without the aid of templating, falsework, or formwork. This method proves viable; however, will require future research to resolve a number of concerns —the primary being the aggregation of errors upon assembly. A number of techniques—such as the wedge and pillowing—carry over from the Inca method into this translation; however, there are a number that do not. One process that does not carry over and speaks to the difficulty of this prototype is the sequential carving of units with bed-joints to align a stone to an existing condition. In the Inca process, the stones are carved simultaneously to the assembly process in a sequential manner. This paper proposes a method whereby all the units are carved prior to the assembly in the hope that errors only aggregate within a tolerance. While the prototypes in this paper do employ the dry-fit and templating methods, these are employed in a manner that does not re-inform the initial calculations, or the subsequent units, but rather as a way to ensure pre-fabricated units may assemble together. In order to more intelligently respond to the shortcomings of this approach, future research can extend the computation into the process of assembly, carving and adjusting future units to account for assembly aggregation errors.

Acknowledgments The 'Round Room' is the result of a partnership between the authors (www.matterdesignstudio.com) and *Quarra Stone* (www.quarrastone.com) as fabricators with structural engineering by *Simpson Gumpertz and Heger* (www.sgh.com). Prototypes were produced at the *University of Michigan FABLab* and the installation was installed at the *MIT Keller Gallery*. This project is funded in part by the *Council for the Arts at MIT* and the *Belluschi Lectureship*. The thick-funicular calculation employs Kangaroo (www.grasshopper3d.com/group/kangaroo) as the physics engine solver for the particle-spring system developed by Daniel Piker to work inside Grasshopper (www.grasshopper3d.com), a plugin developed by David Rutten for Rhinoceros (www.rhino3d.com), a program developed by Robert McNeil.

References

Andriaenssens S, Block P, Veenendaal D, Williams C (eds) (2014) Shell structures for architecture: form finding and optimization. Routledge, London

Clifford B (2013) Thicker funicular: particle-spring systems for variable-depth form-responding compression-only structures. Struct Archit Concepts Appl Challenges 2:205–206

Clifford B (ed) (2014) Volumetric robotics: MIT architectural design workshop. Matter Design Press, Boston

Clifford B, McGee W (2014) La Voûte de LeFevre: a variable-volume compression-only vault. Fabricate: negotiating design & making, vol 2, pp 146–153

Dean C (2010) A culture of stone: inka perspectives on rock. Duke University Press, Durham

Fallacara G (2012) Stereotomy: stone architecture and new research. Presses des Ponts, Paris

Feringa J, Sondergaard A (2014) Fabricating architectural volume: stereotomic investigations in robotic craft. Fabricate: negotiating design & making. vol 2, pp 76–83

Gasparini G, Margolies L (1980) Inca architecture. Indiana University Press, Indiana

Harth-terré E (1964) Técnica y Arte de la Cantería Incaica. Garcilaso

McGee W, Ferringa J, Sondergaard A (2012) Processes for an architecture of volume: robotic wire cutting. Rob|Arch robotic fabrication in architecture, art, and design, pp 62–71

McGee W, Pigram D (2011) Formation embedded design: a methodology for the integration of fabrication constraints into architectural design. ACADIA 2011, pp 122–131

Ochsendorf J (2010) Guastavino vaulting: the art of structural tile. Princeton Architectural Press, Princeton

Outwater J (1959) Building the fortress of Olantaytambo. Archaeology. pp 12–28

Piker D (2014) Dynamic remeshing. www.grasshopper3d.com/profiles/blogs/dynamic-remeshing-now-with-feature-preservation-curvature. Accessed 19 May 2015

Protzen JP (1993) Inca architecture and construction at ollantaytambo. Oxford University Press, Oxford

Protzen JP (1985) Inca quarrying and stonecutting. J Soc Architectural Historians XLIV(2):161–182

Protzen JP, Nair S (1997) Who taught the inca stonemasons their skills?: a comparison of tiahuanaco and inca cut-stone masonry. J Soc Architectrual Historians LVI (2):146–167

Ramage M (2006) Structural vaulting built with aercrete masonry. Masonry Int 20:29–35

Rippmann M, Block P (2011) Digital stereotomy: voussoir geometry for freeform masonry-like vaults informed by structural and fabrication constraints. In: Proceedings of the IABSE-IASS symposium, London, UK

Decomposing Three-Dimensional Shapes into Self-supporting, Discrete-Element Assemblies

Ursula Frick, Tom Van Mele and Philippe Block

Abstract

This study investigates a computational design approach to generate volumetric decompositions of given, arbitrary, three-dimensional shapes into self supporting, discrete-element assemblies. These assemblies are structures formed by individual units that remain in equilibrium solely as a result of compressive and frictional contact forces between the elements. This paper presents a prototypical implementation of a decomposition tool into a CAD software, focusing on user-controlled design to generate such assemblies. The implementation provides an interactive design environment including real time visual feedback, in which the design space of self-supporting block assemblies can be explored and expanded. Some surprising results of such explorations are included and discussed.

Introduction

Volumetric decomposition as a means to reduce element size in assemblies is relevant to the building industry because it simplifies fabrication and transport. The connections between the individual units needed to establish equilibrium of the assembly are often problematic, material intensive or complicated. Especially tensile, mechanical connections often result in complicated detailing and can be expensive and intrusive. Glued connections are mainly simple, but typically difficult to adjust or remove.

This research presents a prototypical decomposition tool as a means to design volumetric decomposition of three-dimensional shapes into self-supporting, discrete-element assemblies. The generated structures, formed by individual, disjoint units, rely solely on spatial compression flows (arching), friction, balancing, and any combination of these actions to stand in equilibrium (Fig. 1). This means they are stable without additional mechanical or physical joinery between the blocks, which keeps the connections simple and adjustable.

Considering the structural integrity of such assemblies, the defining principles that must be evaluated are the assembly's overall stability and

U. Frick (✉) · T.V. Mele · P. Block
Block Research Group, ETH Zurich Institute of Technology in Architecture, Zurich, Switzerland
e-mail: frick@arch.ethz.ch

M.R. Thomsen et al. (eds.), *Modelling Behaviour*, DOI 10.1007/978-3-319-24208-8_16

Fig. 1 Photographs of a discrete element assembly in equilibrium as a result of arching, friction, and balancing, and without mechanical connections or glue

material failure at the scale of the individual unit. As stability is predominantly an issue of geometry and not of internal stress distributions, methods to address geometrical stability are embedded in the presented computational setup. Material failure of the discrete elements is not considered at this stage of the research.

Related Work

Discretization of architectural geometries and design of self-supporting structures are ongoing and active areas of research. Both topics play an important role in architectural design and associated fields and have a strong influence on manufacturing, assembly and building cost.

In recent years, various innovative computational techniques for topological and surface-based discretization of architectural geometries have been developed, such as Pottmann et al. (2008) and Eigensatz et al. (2010). However, few of these techniques consider structural and/or assembly constraints e.g. Rippmann et al. (2013), Panozzo et al. (2013) and Deuss et al. (2014). Furthermore, in most cases topological and surface based discretization approaches are not applicable to volumetric shapes.

Most existing volumetric discretization approaches, such as methods to decompose solids into parts optimized for layered fabrication (Hu et al. 2014), do not consider interactions that keep disjoint assemblies in equilibrium. Whiting et al. (2009), Whiting (2012) presented an approach for generating structurally sound masonry assemblies by refining coarse volumetric models with known/typical structural elements such as walls, arches, domes, etc.

Objectives and Outline

The presented approach focuses on interactive decomposition of given arbitrary shapes that would naturally not be considered as suitable shapes for self-supporting, discrete-element structures. By providing *real-time* visual feedback, it allows exploring and extending the design space of such assemblies. From this design perspective, the proposed approach offers a means of creating surprising equilibrium assemblies that go beyond the scope of known structurally sound configurations for unreinforced masonry and other discrete-element structures.

"*Decomposition process*" gives an overview of the decomposition process, the used structural analysis method and the computational implementation. In "*Results*", the results of explorations are presented with three case studies. The stability of the generated digital models are validated with 3D-printed physical models and illustrated with photographs. "Conclusion" discusses the presented approach and gives an outlook to future work.

Decomposition Process

This section gives an overview of a prototypical implementation of a decomposition tool in Rhinoceros (2014). Grasshopper (2009) was used to build up an interactive design environment. Equilibrium calculations were written in Python (2015) and solved with quadratic programming.

The decomposition process starts with an initial geometry, which is refined step by step until a satisfactory result is obtained. After every user-controlled refinement, interfaces between blocks, and between blocks and the surroundings are detected automatically. At every step, the discretization of the geometry can be changed or updated, whereafter no-tension equilibrium has to be (re-)established. An equilibrium solution can be found by changing the boundary conditions, modifying the location and orientation of the interfaces, changing the material properties, or any combination of these options. Providing intuitive, visual feedback on the current state of the model is clearly essential during this process.

Equilibrium Calculation

The equilibrium calculations are based on the method described in Whiting (2012) and Whiting et al. (2009, 2012).This method extends the Rigid Block Limit Equilibrium Analysis method by Livesley (1978, 1992) by including penalty forces to allow for configurations of discrete-element assemblies in which tension is required. This structural analysis method enables computing estimates of the occurring forces in a given structure that satisfy the equilibrium equations, including friction constraints, with quadratic programming.

Here, we briefly summarize the essential equations of the optimization problem. The static equilibrium equations (Whiting et al. 2009) can be set up in matrix form as follows:

$$A_{eq} * \mathrm{c} + \mathrm{b} = 0$$

The matrix A_{eq} contains the sub-matrices $\mathrm{A}_{\mathrm{j,k}}$ of size $6 \times 4\ \mathrm{v}_{\mathrm{k}}$, with v_{k} the number of vertices of interface k, representing the (global) xyz-components of the force and moment interactions between block *j* and interface *k* in the local coordinate system $(\hat{n}_k,\ \hat{u}_k,\ \hat{v}_k)$ of interface *k* (Fig. 2):

$$A_{j,k} * c_k + b_j = 0$$

which expands to:

$$\begin{bmatrix} f_{kx} & f_{kx} & & f_{kx} \\ f_{ky} & f_{ky} & & f_{ky} \\ f_{kz} & f_{kz} & & f_{kz} \\ m^1_{j,k_x} & m^2_{j,k_x} & \cdots & m^v_{j,k_x} \\ m^1_{j,k_y} & m^2_{j,k_y} & & m^v_{j,k_y} \\ m^1_{j,k_z} & m^2_{j,k_z} & & m^v_{j,k_z} \end{bmatrix} * \begin{bmatrix} c^1_k \\ c^2_k \\ \vdots \\ c^{v_k}_k \end{bmatrix} + \begin{bmatrix} F_{jx} \\ F_{jy} \\ F_{jz} \\ M_{jx} \\ M_{jy} \\ M_{jz} \end{bmatrix} = 0$$

The sub-vectors f_{kx}, f_{ky}, and f_{kz} of $\mathrm{A}_{\mathrm{j,k}}$ contain the xyz-components of the local coordinate system of interface *k*.

$$f_{kx} = \left[(\hat{n}_k)_x,\ -(\hat{n}_k)_x,\ (\hat{u}_k)_x,\ (\hat{v}_k)_x\right]$$

The sub-vectors m^i_{j,k_x}, m^i_{j,k_y} *and* m^i_{j,k_z} contain the xyz-components of the moment contributions of the interface forces of vertex *i* of interface *k*,

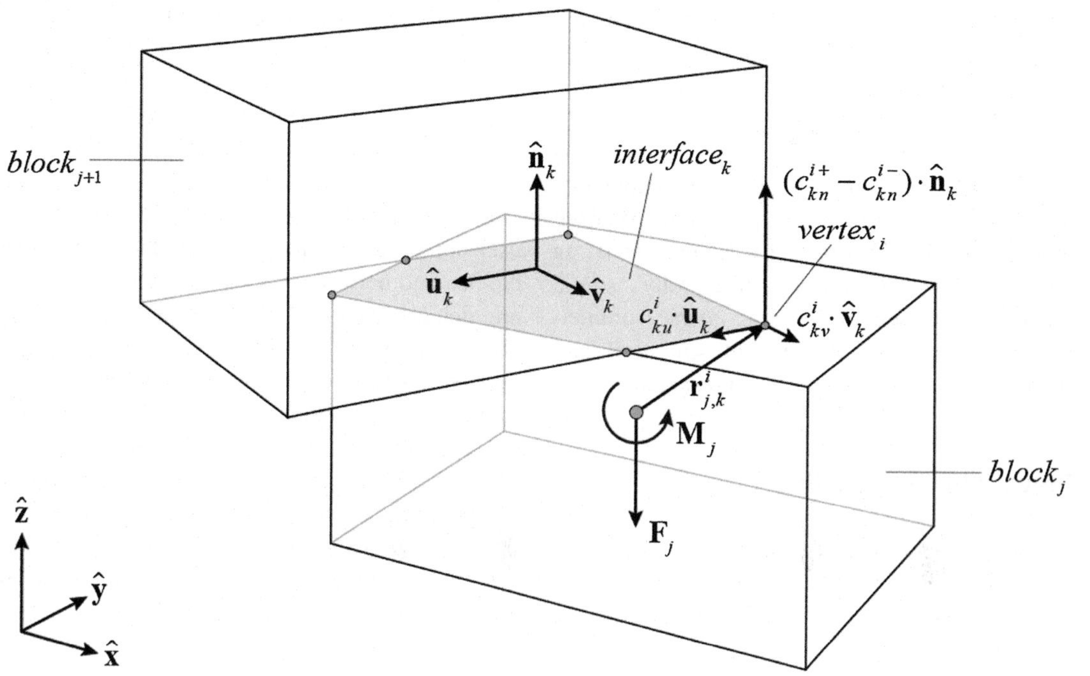

Fig. 2 Diagram of the interaction between block *j* and interface *k*

acting on block j. The vector $r^i_{j,k}$ defines the relative position of vertex i with respect to the mass centroid of block j.

$$m^i_{j,k_x} = \left[(r^i_{j,k} \times \hat{n}_k)_{x,} - (r^i_{j,k} \times \hat{n}_k)_{x,} (r^i_{j,k} \times \hat{u}_k)_{x,} (r^i_{j,k} \times \hat{v}_k)_x\right]$$

The $4_{vk} \times 1$ vector c_k, with v_k the number of vertices of interface k, contains the unknown normal and in-plane force coefficients (signed force magnitudes) for all vertices of interface k, including the penalty formulation as described in Whiting et al. (2009), Whiting (2012).

$$\begin{aligned} c_{normal} &= c_n^+ - c_n^- \\ c_{friction} &= c_u + c_v \end{aligned} \Rightarrow c^i_k = \begin{bmatrix} c^{i+}_{kn} \\ c^{i-}_{kn} \\ c^i_{ku} \\ c^i_{kv} \end{bmatrix}$$

The 6×1 sub-vector b_j contains the xyz-components of the body forces applied to the centroid of block j.

To improve stability of the calculation, an eight-sided friction pyramid (Livesley 1992) rather than the four-sided one used in Whiting (2012) has been used for friction constraints (μ is the static friction coefficient of the used material).

$$\left(|c_u|, |c_v|, \frac{1}{\sqrt{2}}|c_u| + \frac{1}{\sqrt{2}}|c_v|\right) \leq \mu * c_n^{c+}$$

The energy function, as described in Whiting (2012), has then been used to calculate the interface-forces.

$$\underset{c}{minimize} f(c) \; such \, that \begin{cases} A_{eq} * c = -b \\ c_{normal} \geq 0 \\ c_{friction} \leq \mu * c_n^+ \end{cases}$$

This structural analysis method enables the calculation of infeasible self-supporting structures. Infeasible self-supporting structures are those for which the no-tension equilibrium is violated. In comparison, stability simulations with physics engines, e.g. Bullet-Physics-Library (2012) and Nvidia physx library (2013), would typically result in a Yes/No answer. Furthermore, the equilibrium equations of the used method can be solved reasonably prompt with quadratic programming. Both of these points makes the approach particularly adequate to be implemented into a computational setup with emphasis on interactivity.

Interface-Force Diagrams

The models are visualized using interface-force diagrams, in which contact interfaces are represented by coloured surfaces. The colours provide different information about the forces at the interfaces depending on the selected feedback mode.

In compression-tension mode, blue indicates compression, and red tension. Colour gradients indicate variations in the distribution of forces over the interface and interfaces without compression are grey. Note, that the colour gradients represent the force distribution normalized per face and do not reflect the force magnitudes. Relative contact-force magnitudes are visualized by switching to vector mode.

In friction mode, interfaces without friction are also grey. Interfaces with friction have solid colours between yellow and red. All friction forces below a user-defined threshold are illustrated in yellow. Red indicates that the friction force exceeds the allowed maximum. Occurring friction forces between those bounds are illustrated with orange shades. For example, dark orange indicates that the resulting friction force is close to the allowed maximum.

For easier understanding of the assembly's equilibrium, mass-center locations are displayed and additional information of contact-force magnitudes in text form are optional. Figures 3, 4 and 5 illustrate the gradient and vector visualizations modes and the influence of user-controlled modifications.

Feedback and Equilibrium Modification

Figure 3 depicts different configurations of a solid object in the shape of the letters SC. The interfaces are shown in compression-tension mode. In Fig. 3a, the notension constraint is violated at the support interfaces. The left support is entirely in tension. At the right support interface, both tension and compression forces occur simultaneously. This equilibrium is the result of the location of the center of mass of the shape in relation to the supports. No-tension equilibrium can be established by adding an additional support interface (Fig. 3b), or by changing the overall geometry (Fig. 3c).

Figures 4 and 5 depict different configurations of a discretization of the letter “M”. Figure 4 shows the interfaces in compression-tension mode. In configuration a, no-tension equilibrium is violated due to the need for tension forces at the internal interface (between the top and bottom element). In configurations b and c, the violation is resolved by changing the location of the cut. The colour gradient in Fig. 4b indicates that, at this location, the compression forces are unevenly distributed over the interface. Moving the cut further down results in a more even distribution and thus a more robust equilibrium (Fig. 4c). Note that in all cases compression forces at the support interface are evenly distributed, as these forces depend only on global equilibrium, which remains unaltered in the three cases. Figure 5 shows the interfaces of a slightly different discretization of the M-shaped object in friction mode. In configuration a, the orientation of the cut is such that maximum friction is exceeded. By rotating the cut in configurations b and c, friction is reduced to allowable levels. Friction is lowest in configuration c. Note that in none of the configurations friction occurs at the support interface, since the applied loads are vertical and there is no arch action.

Results

This section presents the results of explorations of the no-tension discrete-element assembly design space with the decomposition tool discussed in “*Decomposition process*”. The equilibrium of the generated digital models has been verified with physical models. The physical models were 3D printed with a ZCORP ZPrinter 650, using a composite of zp150 powder and

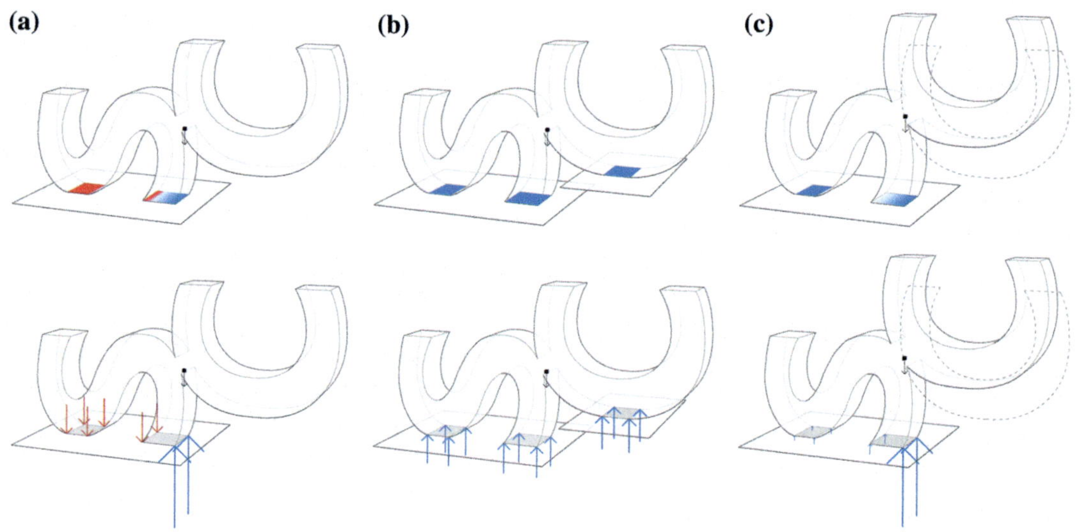

Fig. 3 Different configurations of a solid object in the shape of the letters SC. **a** Violation of no-tension equilibrium at the support. No-tension equilibrium established by (**b**) adding a support interface, or **c** changing the geometry

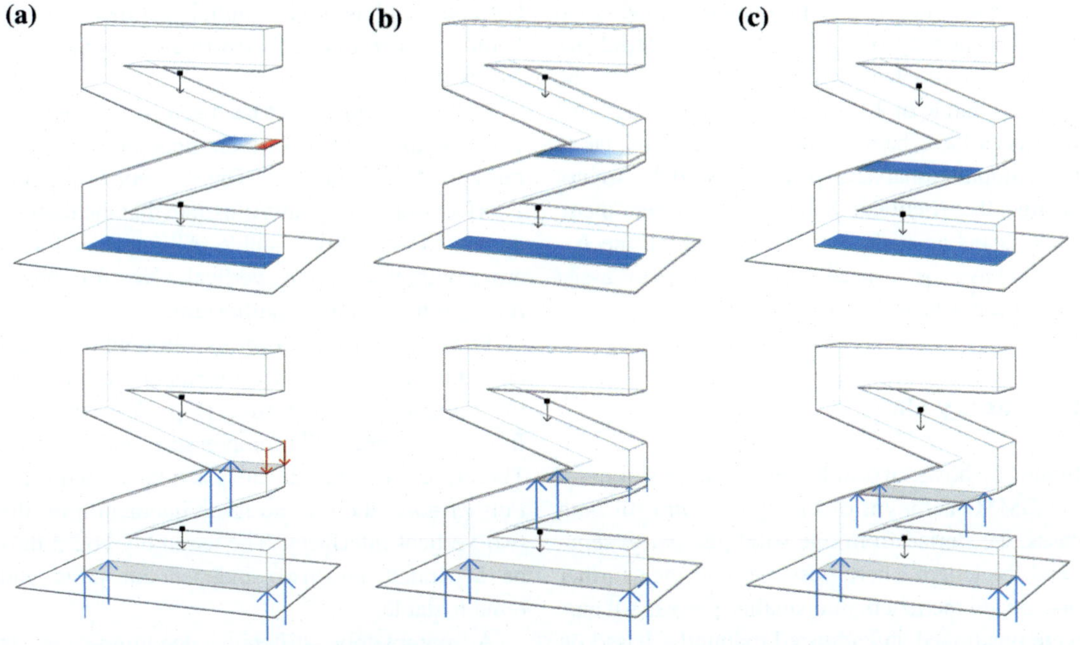

Fig. 4 Tension forces at the internal interfaces (**a**) can be removed by changing the location of the interface (**b**). More robust solutions are recognized by more uniform colouring of the interfaces (**c**)

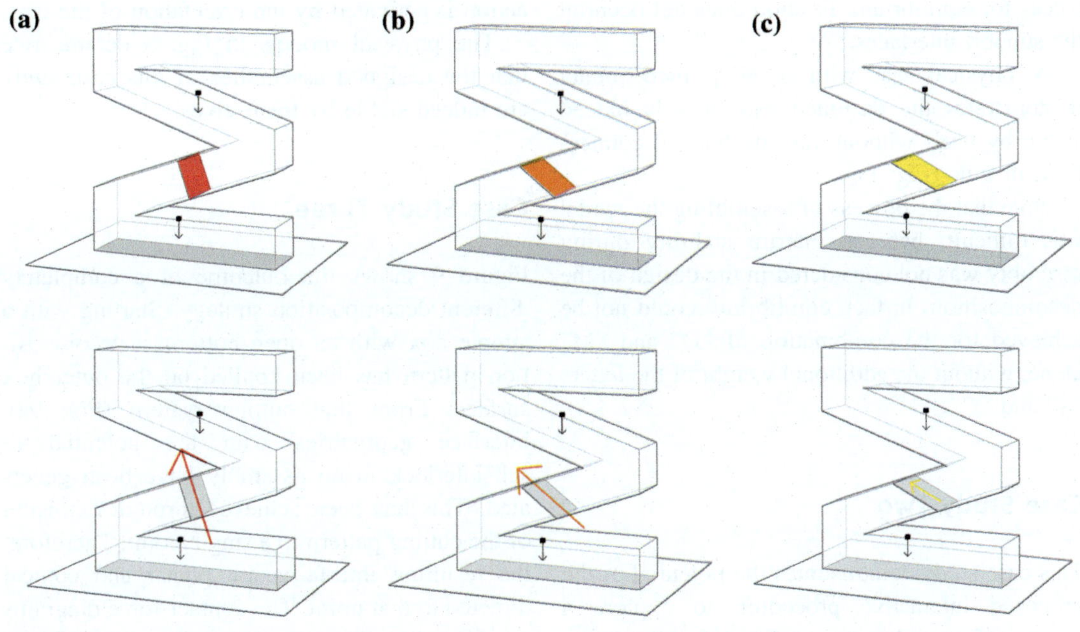

Fig. 5 Violation of friction limitations at an internal interface (**a**) can be resolved by, for example, rotation of that interface (**b**). The amount of friction is indicated by a colour ranging from *yellow* to *red*. *Red* indicates that maximum friction is exceeded (**a**). *Yellow* indicates that friction is below a user-defined threshold (**c**)

zb61 clear binder and impregnated with Z-Bond 101. The density of the composite material has been approximated with 0.60 g/cm^3, and the frictional angle of 40° (Van Mele et al. 2012). All interfaces have been modeled flat and planar in the computational and physical models, to not change the frictional behaviour between them. This means no male-female interlocking mechanisms have been used, for example to assist during the assembly process.

Case Study One

Figure 6 shows two different decompositions of the DMSC acronym of the symposium. In both cases, we started from one solid geometry, separated the letters where possible, and then further discretized the letters into smaller pieces. All cuts were positioned and adjusted manually, based on the visual feedback of the equilibrium calculations. Figure 6b, c depict the tension-compression and frictional contact forces at all interfaces. Both assemblies require compressive and frictional forces for equilibrium. Friction does not occur at the support interfaces.

A physical test with a 3D-printed model demonstrates the designed assembly is indeed stable by itself without any mechanical connections or glue (Fig. 1).

Note that the process of assembling the model was difficult, because interim stability during assembly was not considered in the design of the decomposition. In fact, equilibrium could not be achieved for the combination of "D" and "M" alone, without the additional weight of the letters "S" and "C".

Case Study Two

This case study demonstrates the potential of the proposed interactive procedure to explore a variety of equilibrium solutions for the same given initial shape (Fig. 7). As can be seen in Fig. 7a, the initial shape for this case study is a threedimensional, kinked loop positioned on a horizontal plane. The initial geometry is in equilibrium, with evenly distributed compression forces at the support.

In Fig. 7b, the object has been partitioned into two L-shaped and two cuboid geometries by four horizontal cuts. The corresponding interface-force diagrams illustrate that the thereby generated assembly is stable. All interface forces are compressive and vertical. No friction is required to establish equilibrium.

Even after further decomposition with vertical cuts through the L-shaped elements, the assembly remains self-supporting (Fig. 7c). All interface forces are still compressive and vertical. Therefore, as before, no friction is required. Furthermore, there is no force interaction on the two vertical interfaces. The assembly could thus be separated into two independent parts and remain stable.

A completely different decomposition is achieved with inclined cuts, as seen in Fig. 7d. In terms of force transfer, this configuration is more interesting. It requires both compressive and friction forces to be in equilibrium, because arch action is activated by the orientation of the cuts.

The physical models in Fig. 8 demonstrate that the designed assemblies of this case study are indeed stable by themselves.

Case Study Three

Figure 9 shows the outcome of a completely different decomposition strategy. Starting with a simple box with an open bottom, a decomposition pattern has been applied on the outer box surface. From that cutting pattern (Fig. 9a), interface geometries with the potential to self-interlock in an assembly have been generated. This has been achieved through extrusion of the cutting pattern to a single point. Therefore, the resulting interfaces are planar and conical directed to that point. The contact-force diagrams in Fig. 9a illustrate that the assembly is self-supporting. All contact forces are either compressive or frictional and the maximum friction is not exceeded.

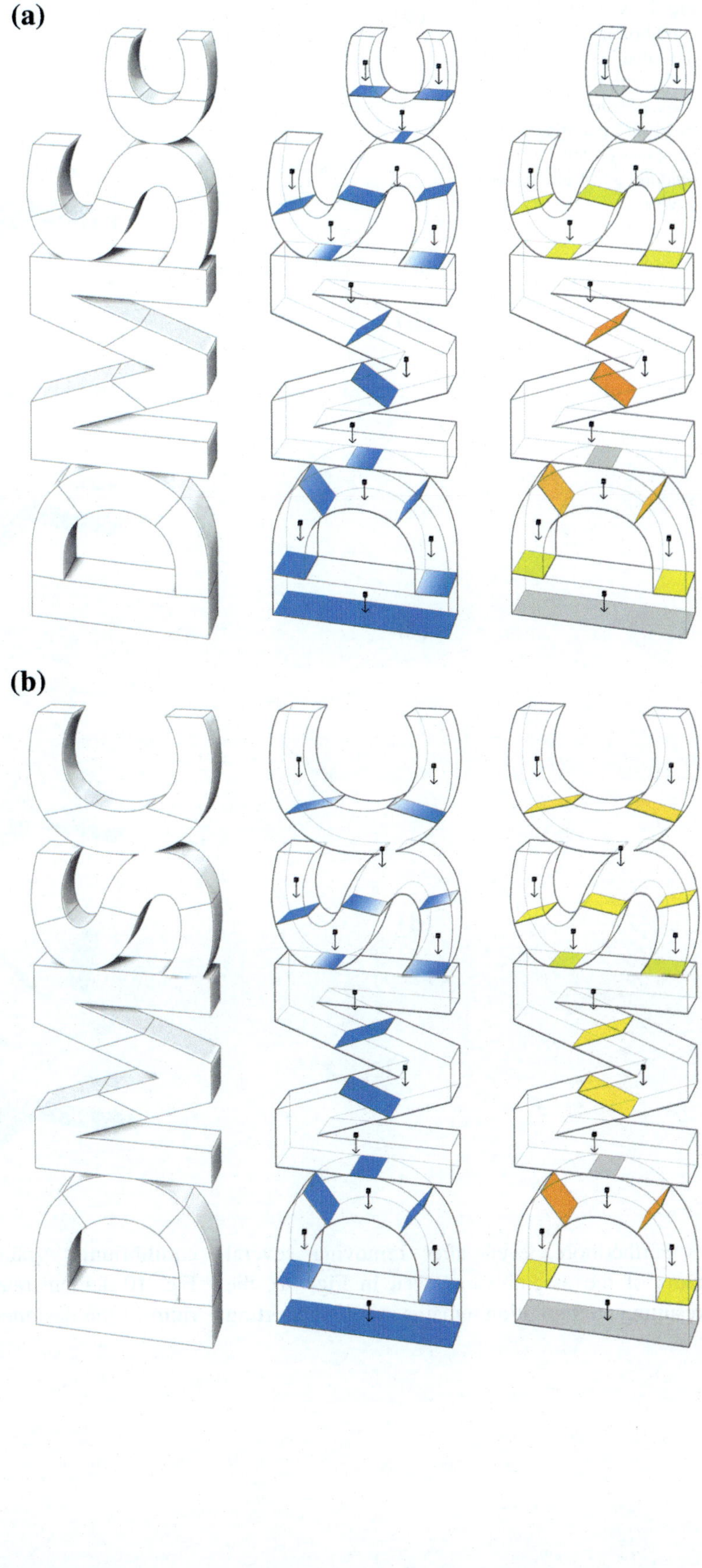

Fig. 6 Two different decompositions of the DMSC acronym (*left*), with the corresponding interface-force diagrams in compression-tension mode (*middle*) and friction mode (*right*)

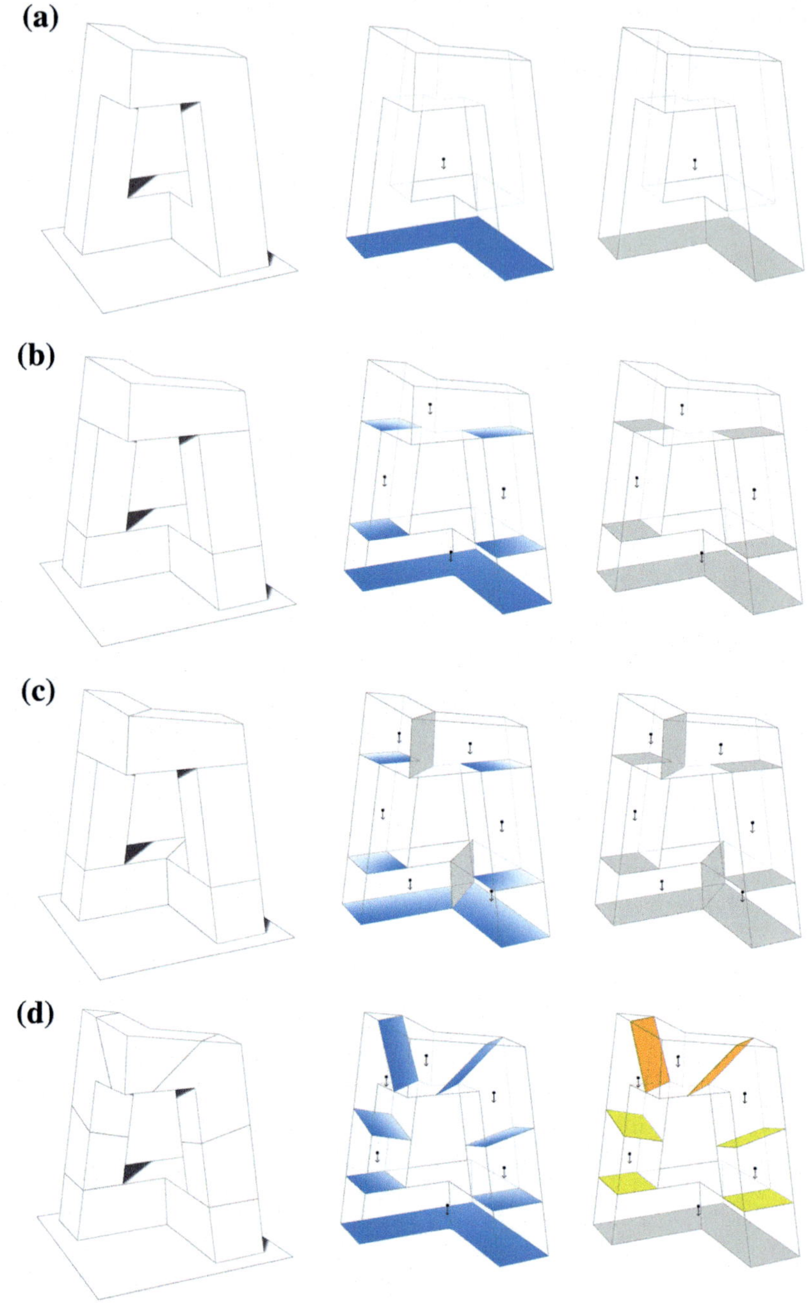

Fig. 7 Various equilibrium solutions for the same initial geometry (*left*), with the corresponding interface-force diagrams in compression-tension mode (*middle*) and friction mode (*right*)

Furthermore, even after removing several parts of the assembly, as seen in Fig. 9b, the resulting configuration remains in self-supporting equilibrium. Again, the physical models in Fig. 10 demonstrate the self-supporting equilibrium of the designed decompositions.

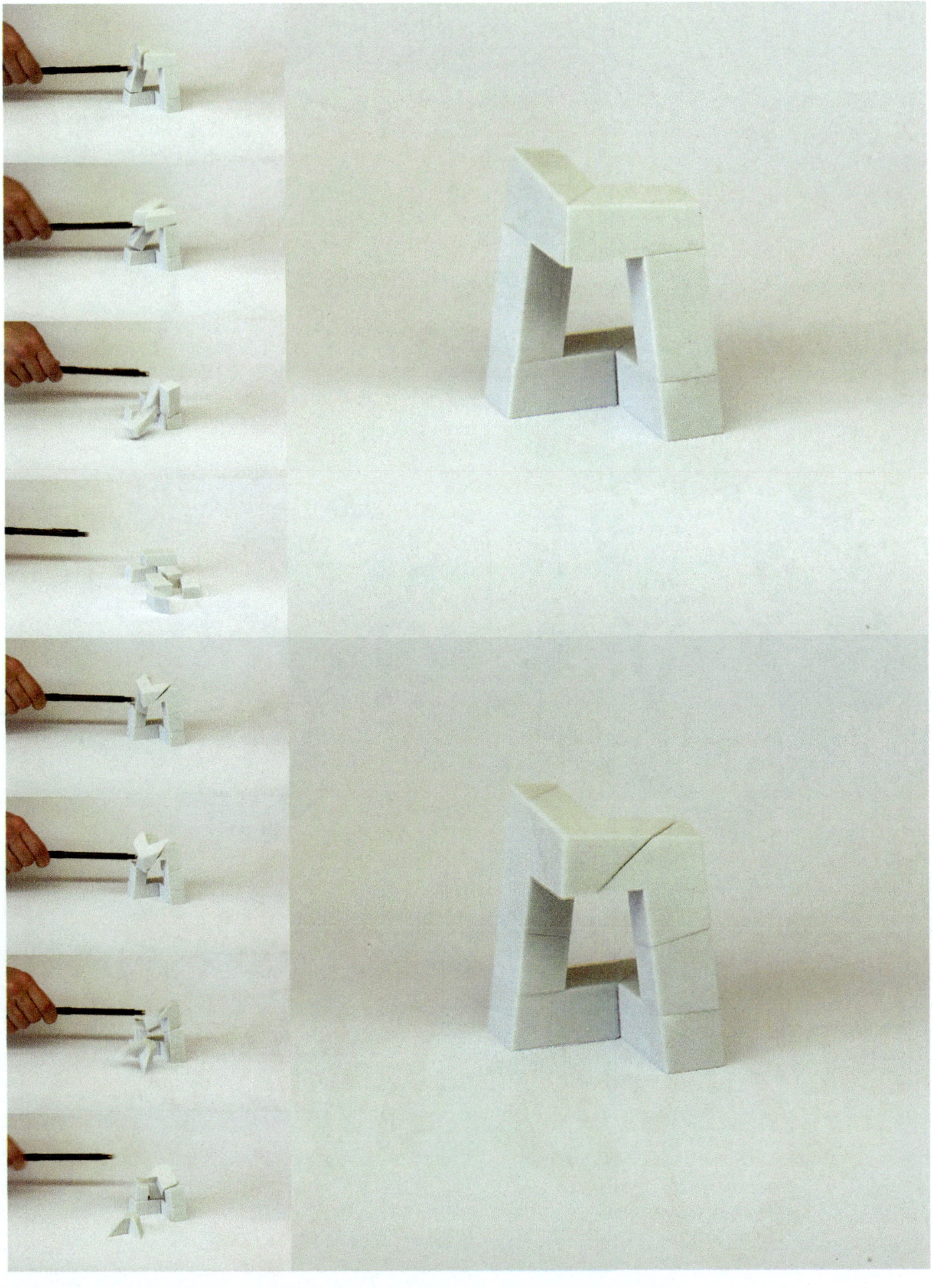

Fig. 8 Photographs of two different decompositions from the same initial geometry

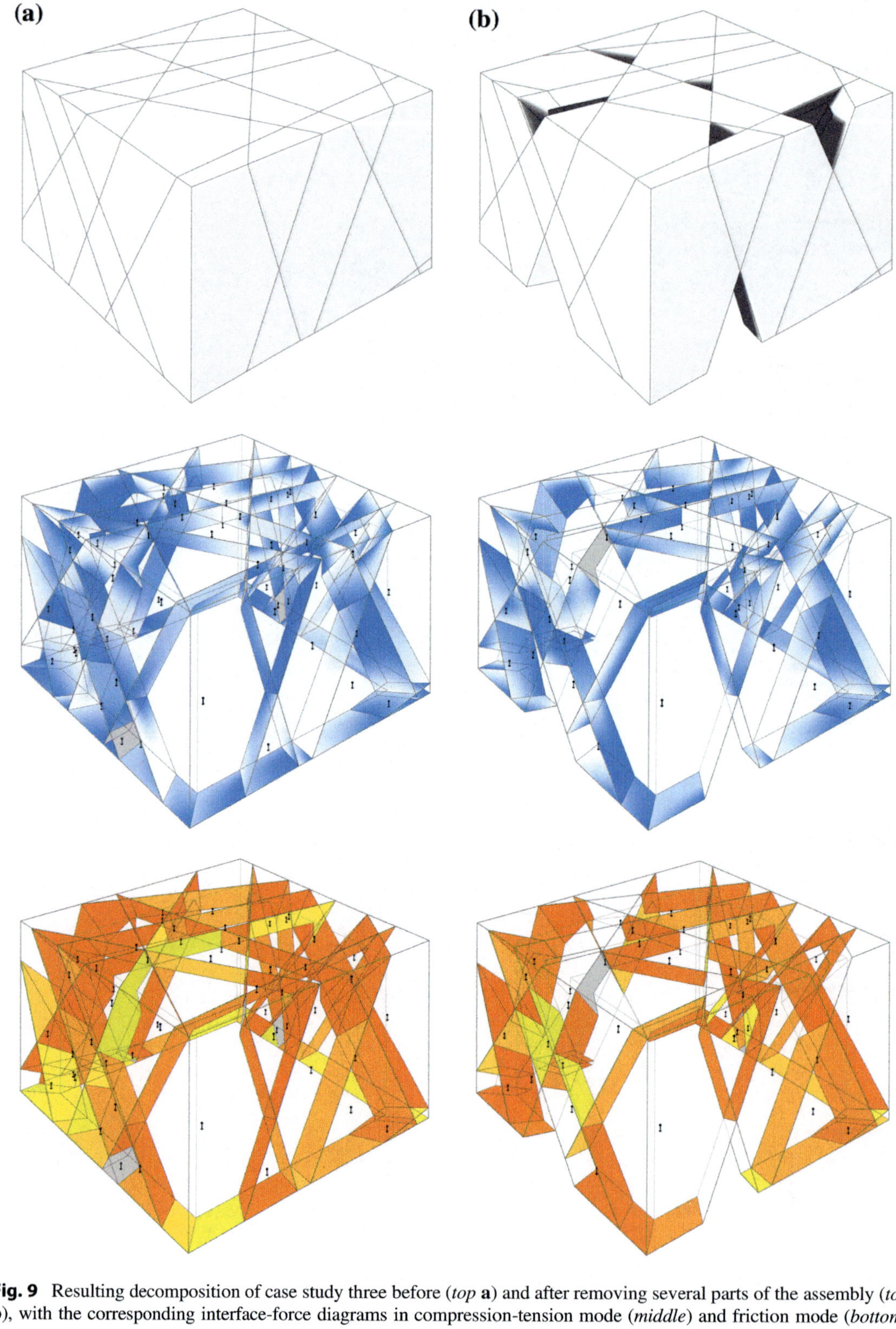

Fig. 9 Resulting decomposition of case study three before (*top* **a**) and after removing several parts of the assembly (*top* **b**), with the corresponding interface-force diagrams in compression-tension mode (*middle*) and friction mode (*bottom*)

Fig. 10 Photographs of case study three before (*top*) and after removing parts from the assembly (*bottom*)

Conclusion

In practice, volumetric decomposition into self-supporting assemblies has applications in areas ranging from the construction of large-scale, cut-stone masonry structures (Howeler et al. 2014) and prefabricated building assemblies, to smaller-scale prototyping. When applied to the design and production of prefabricated elements in the building industry, optimization of joints can be achieved as a means of limiting joinery to predominantly no-tension force transfer.

The prototypical decomposition tool presented in this paper is based on realtime, clear, visual feedback. The advantage of such an interactive process has been demonstrated through the surprising results shown in "*Results*". As can be seen in those case studies, the step-by-step, interactive discretization process has successfully been used to explore the design space of discrete-element assemblies and experiment with previously unseen forms.

Despite the promising results shown in "*Results*", the interactive decomposition procedure can be time consuming and the handling of large models can be problematic. Further research will be necessary to develop general, (semi-)automated decomposition strategies. A digital self-supporting equilibrium solution does not automatically guarantee a physical self-supporting application. As explained in previous research (Whiting 2012), an alternative equilibrium state could exits where friction constraints are violated. Furthermore, the implemented algorithms do not consider imperfections of element's geometry or assembly.

This research is part of a larger research project, which aims to develop better understanding and novel techniques for the equilibrium design of masonry and other discrete-element assemblies. Near-future goals related to the research presented are the implementation of additional methods for the equilibrium design, such as manipulation of weight distribution within the volume (Bacher et al. 2014), and the development of general, automatic decomposition strategies. Furthermore, the research will focus on the structural integrity and robustness by considering possible material failure at unit scale and interim stability during assembly.

Acknowledgments This research was supported by the NCCR Digital Fabrication, funded by the Swiss National Science Foundation (NCCR Digital Fabrication Agreement # 51NF40-141853).

References

Bächer M, Whiting E, Bickel B, Sorkine-Hornung O (2014) Spin-it: optimizing moment of inertia for spinnable objects. ACM Trans Graph 33(4):96:1–96:10

Bullet-Physics-Libary (Copyright (c) 2012) Bullet collision detection and physics library. http://bulletphysics.org

Deuss M, Panozzo D, Whiting E, Liu Y, Block P, Hornung-Sorkine O, Pauly M (2014) Assembling self-supporting structures. ACM Trans Graph 33 (6):214:1–214:10

Eigensatz M, Kilian M, Schiftner A, Mitra NJ, Pottmann H, Pauly M (2010) Paneling architectural freeform surfaces. ACM Trans Graph 29(4):45:1–45:10

Grasshopper (Copyright 2009) Grasshopper—algorithmic modeling for rhino version 27 August 2014. http://www.grasshopper3d.com

Höweler E, Yoon JM, Ochsendorf J, Block P, DeJong M (2014) Material computation—the collier memorial design using analog and digital tools. In: Gerber DJ, Ibaez M (eds) Paradigms in computing—making, machines and models for design agency in architecture, chapter 0. eVolo Press, Los Angeles

Hu R, Li H, Zhang H, Cohen-Or D (2014) Approximate pyramidal shape decomposition. ACM Trans Graph 33 (6):213:1–213:12

Livesley RK (1978) Limit analysis of structures formed from rigid blocks. Int J Numer Meth Eng 12:1853–1871

Livesley RK (1992) A computational model for the limit analysis of threedimensional masonry structures. Meccanica, 27(3):161–172. Nvidia-PhysX (Copyright 2013)

Nvidia physx library (2013). http://www.nvidia.com/object/physx-9.12.0213-driver.html

Panozzo D, Block P, Sorkine-Hornung O (2013) Designing unreinforced masonry models. ACM Trans Graph 32(4):91:1–91:12

Pottmann H, Schiftner A, Bo P, Schmiedhofer H, Wang W, Baldassini N, Wallner J (2008) Freeform surfaces from single curved panels. ACM Trans Graph 27(3):76:1–76:10

Python (Copyright 2001–2015) Python programming language. https://www.python.org

Rhinoceros (Copyright 1993–2014) Rhinoceros modeling tools for designers, version 5. https://www.rhino3d.com

Rippmann M, Curry J, Escobedo D, Block P (2013) Optimising stonecutting strategies for freeform masonry vaults. In: Proceedings of the international association for shell and spatial structures (IASS) symposium 2013. Wroclaw, Poland

Van Mele T, McInerney J, DeJong M, Block P (2012) Physical and computational discrete modeling of masonry vault collapse. In: Proceedings of the 8th international conference on structural analysis of historical constructions. Wroclaw, Poland

Whiting E, Ochsendorf J, Durand F (2009) Procedural modeling of structurally-sound masonry buildings. ACM Trans Graph 28(5):112

Whiting E, Shin H, Wang R, Ochsendorf J, Durand F (2012) Structural optimization of 3d masonry buildings. ACM Trans Graph 31(6):159:1–159:11

Whiting EJW (2012) Design of structurally-sound masonry buildings using 3d static analysis. PhD thesis, Department of Architecture, Massachusetts Institute of Technology

Computational Brick Stacking for Constructing Free-Form Structures

Danil Nagy, John Locke and David Benjamin

Abstract

Our work explores new design methods and workflows that operate at the intersection of emerging biological technologies and advanced computation and engineering. In 2014 we won a competition to construct a large scale temporary installation to host a series of weekend parties. Our proposal explores the future of architecture by using innovative computational tools to test a new material system at an architectural scale. This paper focuses on the design of a custom computational brick stacking logic that generated the structure's brick layout within the specific constraints of a new organic material. The resulting digital model was utilized directly to guide the structure's construction, creating a feedback loop where changes made on site could be fed back into the model to recalculate the layout in real time. While computation and technology play a crucial role in this proposal, the ultimate goal is to show how innovation in materials and methods can lead to a more responsive, intelligent, and sustainable architectural practice.

The Natural Ecosystem

Due to the installation's short two-month lifespan, we began by questioning the role of waste and permanence in architecture, and the way in which the building industry currently interfaces with the Earth's natural ecosystem. In nature, the carbon cycle provides an extremely efficient system for recycling matter and energy. In this system there is no true waste, and each organism's death creates food for new life. In contrast to this healthy cycle, typical construction projects follow a highly unsustainable linear approach, where productive materials are taken from the natural ecosystem, used in construction, and eventually end up as permanent waste (Fig. 1). The scale of this waste is enormous, with the construction and demolition industries accounting for as much as 40 % of the total trash entering landfills in the United States (US EPA 1995).

D. Nagy (✉) · J. Locke · D. Benjamin
The Living an Autodesk Studio, New York, Brooklyn, USA
e-mail: danil.nagy@autdodesk.com

M.R. Thomsen et al. (eds.), *Modelling Behaviour*, DOI 10.1007/978-3-319-24208-8_17

Not only are these materials non-organic, many of them are toxic and can harm the environment if not properly contained within a landfill.

In reaction to these issues, we proposed an alternative approach which uses a new class of building materials made only from organic and existing waste materials. Our vision was for a pavilion that would serve its purpose during the summer events, but could be quickly deconstructed and returned back to nature once its productive life came to an end. By looking not only at the construction of the structure but its entire life cycle, we proposed a building that was designed to disappear as much as it was designed to appear.

A New Material System

The key to our proposal was the use of a biological material, mycelium, which is the root structure of mushrooms. When combined with other plant material, the mycelium's filament-like micro root structures grow and bind the material to produce a solid mass in a very short time. Since it forms through a natural process, the material requires almost no energy or carbon emissions to produce. It is also completely organic, and can be quickly and easily broken down to create high quality soil through composting (Holt et al. 2012).

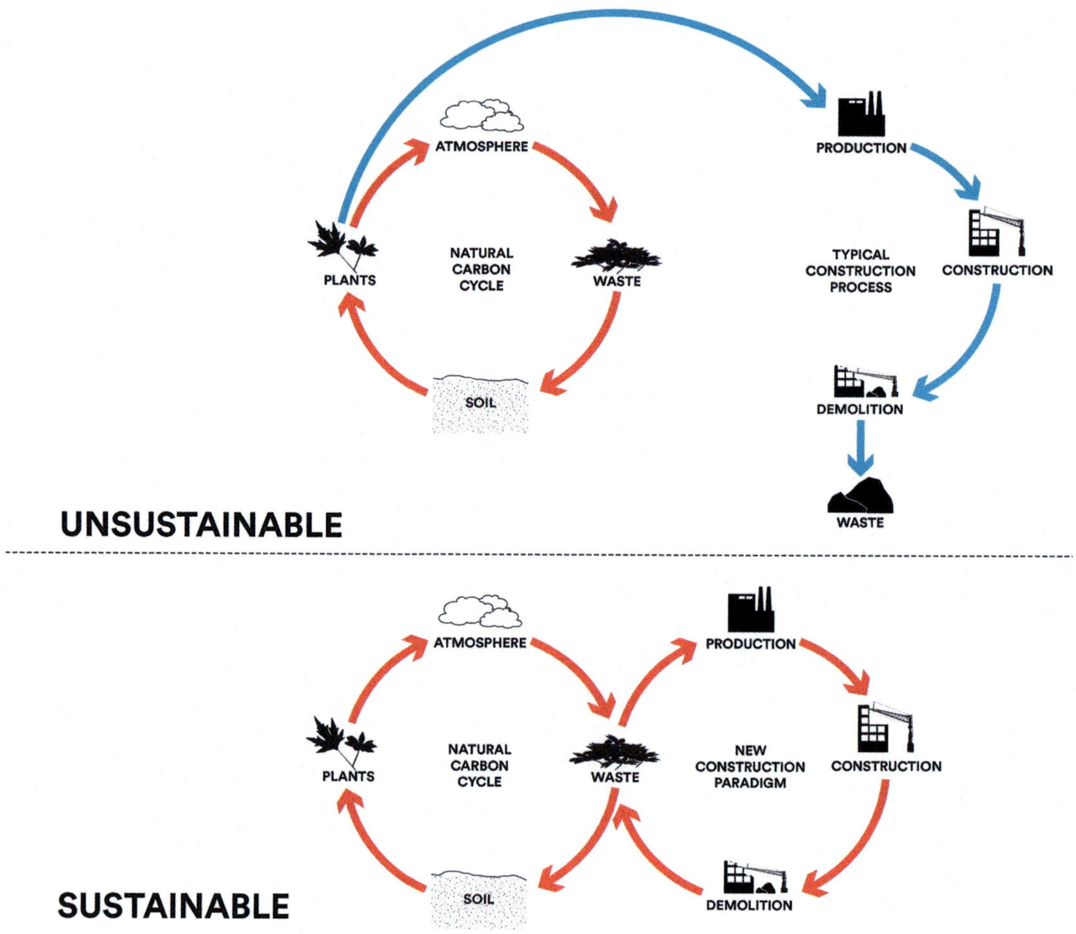

Fig. 1 Relationship of the building industry to the natural carbon cycle, current condition (*top*) and our proposal (*below*)

To produce the material we partnered with a local company called Ecovative that has been researching and developing mycelium-based materials for various uses over the last eight years. Ecovative's main product is a packing material with similar properties as Styrofoam, which they produce at a small factory in Green Island, located in upstate New York. The material is made from a combination of corn husks and hemp collected from local farms, and a custom engineered strain of mycelium. Besides producing a completely organic and compostable material, this process also taps into an existing agricultural waste stream with little or no inherent value. Thus it does not compete with other productive agricultural uses such as food and does not require additional investment in agriculture.

Ecovative was a key partner in this project because they shared our vision of creating a better and more sustainable future through new technology and materials. They also had a high level of expertise in the material, and had the technology to produce it and customize it to our needs. Most importantly, they had the industrial capacity to manufacture enough material to test it at a large scale. Working with this material, we designed a 40 foot tall branching tower made of 10,000 individual compostable bricks (Fig. 2). The structure's unique form would create a completely new kind of spatial experience for the museum's guests, while its large scale would force us to engage with the realities of the construction industry, and to test the material's capacity at a truly architectural scale.

The Design to Testing Workflow

In order to make this vision possible, we first had to ensure that the material could withstand the forces within the structure. This included both the weight of the structure under gravity, as well as potentially hurricane-level wind loads. Working with engineering firm ARUP, we created a simulation model based on a triangular finite element analysis (FEA) surface mesh. By routing forces through each triangle of the mesh, this model could predict how the structure would perform under various loading conditions and provide critical information about how the form would deflect, where the maximum stresses would collect, and whether these stresses were within the limits of the material.

In order to route forces through the model accurately, the simulation software needs a good definition of the structural properties of the material used. Most structural analysis software includes properties for a variety of typical

Fig. 2 Organic bricks on growing racks at Ecovative factory in Green Island, NY (*left*). Final brick bring removed from its mould (*right*)

Fig. 3 Wall assembly being tested for performance under compression (*left*). Single bricks and composites before and after testing (*right*)

construction materials such as wood, masonry, and metals. One of the difficulties of working with an entirely new material was that there was no built-in definition for a mushroom brick, meaning we had to derive these properties from scratch. To do this we partnered with the Carleton Strength of Materials Laboratory at Columbia University, who subjected our prototype brick modules to a series of tests including compression and bending (Fig. 3). Based on the results of these tests, ARUP reverse engineered a material definition which could be used with the FEA model to accurately predict how the 40 foot structure would behave under load.

The initial simulations showed a maximum displacement of 30 in. under maximum wind load, which exceeded the margin of safety and the limits of the material. To solve this issue we worked with our partners to develop a fast material innovation cycle which would take feedback from both physical tests and structural simulation to develop a stronger material that could meet the needs of the structure. Each iteration of the cycle began with Ecovative producing a series of bricks, based on varying properties such as grow time, the mix and composition of agricultural waste, the amount of mycelium, and the amount of extra nutrients. Each batch of bricks was then tested at Carleton Lab, and the data was used to update ARUP's material specifications. The simulation results were then sent back to Ecovative, who used them to develop the next batch of bricks. Each iteration of this cycle would take approximately one week, and within the time constraints of the project we were able to complete seven such iterations. By the seventh iteration the simulation showed a maximum displacement of 12 in., which was within the margin of safety for the structure (Fig. 4).

The Design Challenge

Although the FEA model showed that our design could work, its results were based on the assumption that the entire surface would act as a continuous membrane with homogeneous structural properties. By testing several brick assemblies and wall mockups, we determined that this assumption would hold as long as the bricks were arranged in a running bond pattern, with a minimum 2″ overlap between each brick and each of the two bricks below it (Fig. 5). To ensure that we could meet this requirement, we had to extend the basic FEA model to describe the actual placement of each brick in the structure. This model would not only validate that an optimal brick layout was possible, but could also be used to guide construction itself.

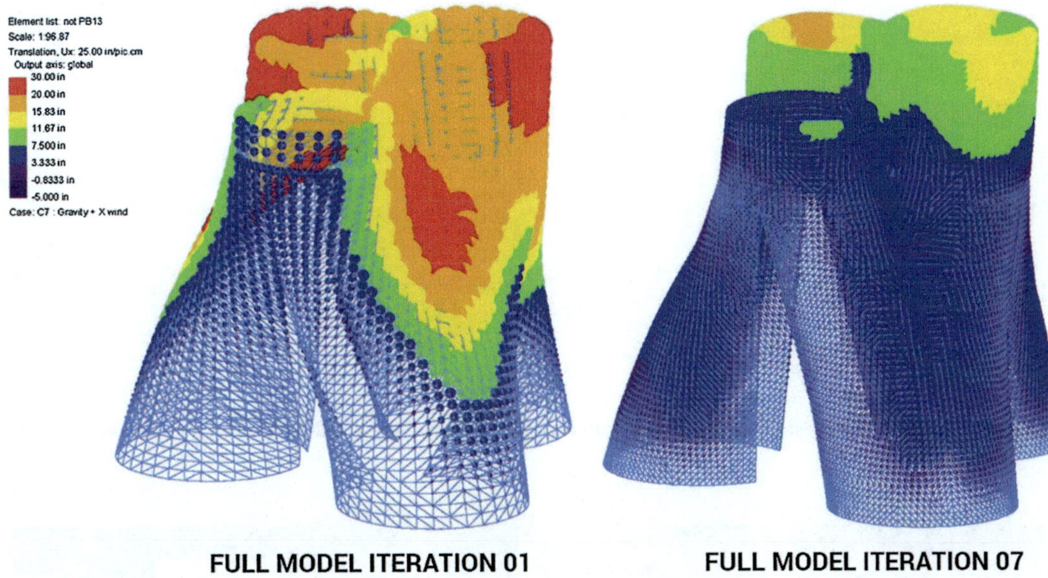

Fig. 4 Simulation of gravity and wind load on FEA model, showing performance with initial and final material

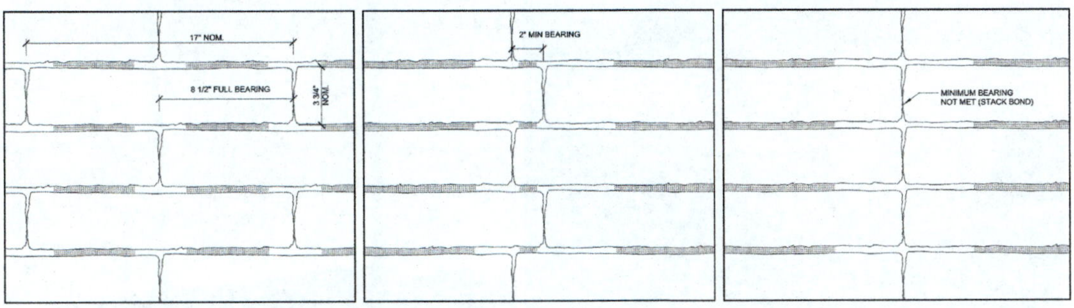

Fig. 5 Initial brick stacking tests, showing issues with description of double-curved surface: incomplete coursing (*left*) and improper bearing (*right*)

As we developed this model, we discovered a unique challenge based on the structure's complex, doubly-curved surface. As the surface curves in two directions, the length of each course is constantly changing. If the bricks are laid in a naïve way, the bricks will maintain a proper bearing, but eventually the final brick will not fit, or a gap will form at the end of the course. Alternatively, if this gap is distributed throughout the course, the running bond will be disrupted, leading to poor structural performance (Fig. 6). In traditional brick construction, masons can simply cut brick on site to fit the changing coursing or fill gaps. However, this was not a viable option for our material. As the bricks grow in moulds, they form a solid layer of mycelium at the exterior, which accounts for much of the brick's strength and resilience against water penetration. If a brick is cut, it reveals a more loose mix of mycelium and corn stalks which jeopardizes much of its durability and structural capacity (Fig. 7).

The Computational Model

To address this challenge we extended our digital model with a custom computational design tool which uses several discrete modules of brick

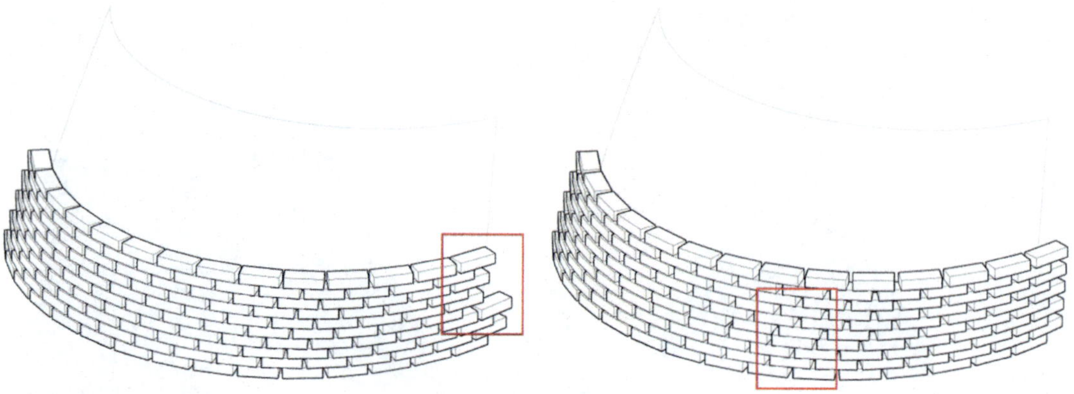

Fig. 6 Initial brick stacking tests, showing issues with description of double-curved surface: incomplete coursing (*left*) and improper bearing (*right*)

Fig. 7 Internal composition of bricks, consisting of a loose fill of crushed corn stalks and hemp

(full, half, and quarter size) to produce optimal layouts of bricks that describe any arbitrary surface. The goals of the program are to (1) fill each course as completely as possible and (2) make sure there are no bearing issues between each course. This tool was developed within Autodesk's Dynamo platform, a node-based graphic programming interface that runs on top of the Revit design software. It uses a combination of standard Dynamo nodes to handle data flow and geometric operations, as well as a custom Python script which describes its core logic.

To prepare the model, the surface was broken down along its seams into six vertical panels. Each panel was then contoured on a 4″ module to create curves that guide the brick laying for each course (Fig. 8). Each curve is then input into the main logic node which specifies how the bricks are laid according to a procedural, iterative logic (Fig. 9):

1. Starting at one end of the course, full size bricks are laid in an optimal running bond until there is no more room for a whole brick.
2. Based on how much space is left at the end of the course, smaller modules are inserted at a random location within the course until there is no more room left.

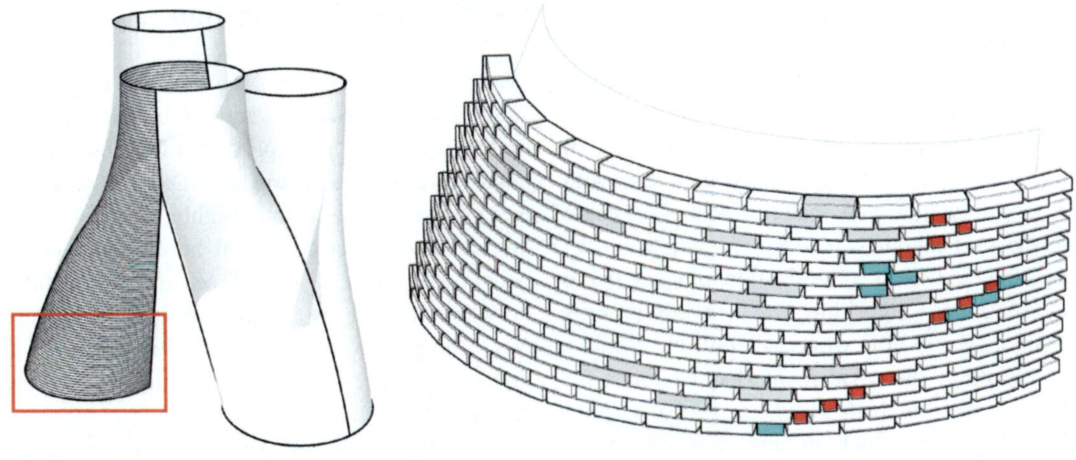

Fig. 8 Full surface model showing six vertical panels and course curves (*left*) and portion of surface described with final recursive brick stacking logic using three size modules (*right*)

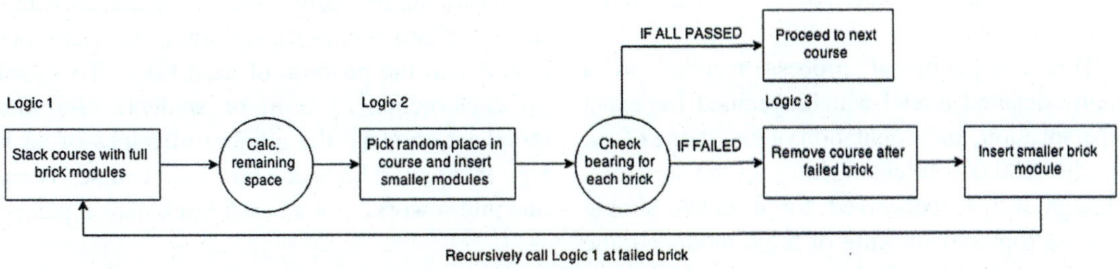

Fig. 9 Structure of brick stacking algorithm showing recursive logic

3. The remaining spacing is distributed throughout the course to even out the gaps between bricks.
4. Starting at the beginning of the course, each brick is checked to make sure it has the minimal 2″ bearing to the two bricks below.
5. If minimal bearing is not met, the failing brick is removed, a smaller unit is put in it's place, and the whole laying logic is called recursively starting at that unit.

A key feature of this logic is that it is not fully deterministic. Using a localized intelligence, the algorithm can recognize when bearing problems occur in individual bricks, but it cannot determine the most optimal layout from the very beginning. Instead, the algorithm relies on local testing and a recursive process to make multiple attempts and small adjustments until there are no bearing issues and the problem is finally solved for each course. This is in contrast to many previous digital architecture workflows, which use a top-down, deterministic process to describe complex surfaces with smaller modules (Anzelone et al. 2009). One limitation of such deterministic approaches is that they often result in modules of many varying sizes (often referred to as 'mass customization'), which was not possible in our case. Another limitation is that they cannot directly respond to local constraints within the material or construction system. Although somewhat less efficient, our approach is much more intuitive, and is derived directly from the constraints of the material and the construction process itself. Because all the agency is contained within the brick itself, this creates a more bottom-up approach and an emergent solution similar to how problems are solved in nature.

This emergent logic, along with the random placement of smaller modules in step 2, creates the possibility of multiple design solutions. Though each solution is guaranteed to be structurally sound, some solutions may use more of the smaller brick modules, which increases both the assembly time and material cost. To ensure that the best layout was used, the whole model was subjected to an optimization process using an evolutionary algorithm (EA). In this process the algorithm is able to create multiple designs by changing the placement factor for each course. It then evaluates each model based on a fitness function, which in this case was minimizing the number of half and quarter brick modules used. The algorithm can then create new designs by combining the parameters of the best performing designs, thereby 'evolving' more efficient layouts.

This computational process resulted in a highly detailed model which specified the exact type, location, and orientation of every one of the roughly 100,000 bricks making up the structure. Though it was conceived for a rather simple purpose (optimal stacking of brick modules), the process involved a combination of several computational techniques including generative design, structural simulation, and multi-objective optimization. The application of such sophisticated digital methods to brick stacking is especially interesting considering that brick construction is one of the most ancient forms of building, going back many thousands of years. Modern brick masons rely on this long history of accumulated knowledge and in-the-field experience to create solutions to many structural problems. However, the particularities of our new material, together with the unusual form of the structure, created a problem that would be extremely difficult or impossible to solve by humans alone. The interaction of traditional methods with cutting-edge computation became a major concept throughout the whole project, and extended into the construction phase, which was accomplished by a team of architecture design students working with a crew of New York City brick masons.

Feedback and Integration

While our computational model was extremely useful during the design and engineering phases of the project, it also proved invaluable as we began to plan its construction. Due to the structure's unique shape and highly specific brick layout, the project could not be documented or constructed with a typical set of architectural drawings. In fact, no construction drawings were produced prior to construction, with all structural details and specifications built directly into the digital project model (Fig. 10).

In collaboration with our contractor, we used the digital model to simulate the assembly process, creating targets for how many bricks would have to be laid each day. We also used the digital model to automatically create template drawings that specified the exact sequence of bricks per course and the position of each brick. To speed up construction, a team of students used the templates to stage the number of bricks of each type that would be needed for each course. Then, one intern working with each brick mason placed

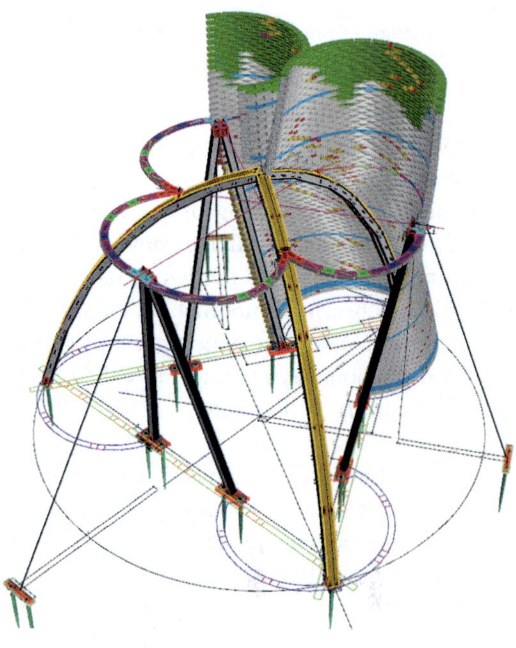

Fig. 10 Final design model integrating information about brick layout, structure, and construction

Fig. 11 On-site workshop used for prototyping, fabrication, and digital model maintenance

Fig. 12 Interior view of installation, showing material and spatial experience

the bricks in the right sequence on each course according to the templates. In this way, we were able to lay all 10,000 bricks within a span of less than three weeks with a minimum of corrections and delays. The model was also constantly referenced during construction to keep track of daily progress and to adjust daily targets accordingly to meet our final deadline (Fig. 11).

Conclusion

The digital model and the intelligence embedded within it created a type of feedback loop within the project. As new challenges arose, they could be incorporated directly into the logic of the model, leading to new design solutions. The nature of the model also speaks to the dual influences of the biological process on the project. On the one hand, the tower itself is composed of a novel biological material that offers a new paradigm for the future of manufacturing and design. At the same time, the computational tools we developed for the project are themselves inspired by the way that forms and solutions are developed in nature. Rather than designing a singular object, we designed a system that integrates all the complex logic and parameters of the design problem. This model is then allowed to evolve and adapt as the project evolves, leading to a design solution which would be impossible to conceive for a human designer alone (Fig. 12).

References

Anzelone P, Vidich J, Draper J (2009) Non-uniform assemblage: mass customization in digital fabrication. In: Clouston P, Schreiber S, Mann RK (eds) Without a hitch—new directions in prefabricated architecture. Lulu.com. pp 298–306

Holt GA, Mcintyre G, Bayer E (2012) Optimizing biomass blends for manufacturing molded packaging materials using mycelium. In: Proceedings of association for the advancement of industrial crops conference

US EPA (1995). Report on construction and demolition waste landfills. U.S. Environmental Protection Agency, Office of Solid Waste, Washington, DC, EPA/68-W3–0008

Automated Casting Systems for Spatial Concrete Lattices

Philippe Morel and Thibault Schwartz

Abstract

Today, concrete is (in weight) the most frequently-used material on the planet after water, with a quantity of one cubic metre per person per year. Every second, 126 tonnes of cement are poured across the world, amounting to some 3.4 billion tonnes per year, or the equivalent of just over 14,000 Empire State Buildings. Over the three years from 2011 to 2013, China consumed 6.6 billion tonnes of cement, in other words more than the 4.5 billion tonnes used by the USA during the entire 20th century (Sources: USGS, Cement Statistics 1900–2012; USGS, Mineral Industry of China, 1990–2013). If we add to these figures the fact that production of cement by the clinkerisation process involves firing at temperatures of around 1450–1500 °C, it is not difficult to understand why the cement industry is one of the most polluting industries. Just like the construction sector as a whole, this industry faces many challenges today. Amongst these challenges are optimising the cement production process in the face of the growing cost of more sophisticated "raw materials" and their proven or potential harmfulness, or rethinking the supply chains and the product lifecycle to reduce grey energy, and lastly developing new construction methods. This last problem, in which architects have the most influence, should be envisioned beyond the usual constraints associated with buildings' regulations as a great opportunity for the architectural discipline, especially through the (re)examination of new or deprecated concrete and/or cement production methods. This search for new approaches is at the heart of the works we are presenting in this paper, oriented towards integrative computational and fabrication methods for the design and realization of three-dimensional concrete-based spatial lattices.

P. Morel (✉)
ENSA Paris-Malaquais/EZCT Architecture and Design Research Co-Director, Paris, France
e-mail: philippe.morel@ezct.net

T. Schwartz
HAL Robotics Ltd, London, UK

M.R. Thomsen et al. (eds.), *Modelling Behaviour*, DOI 10.1007/978-3-319-24208-8_18

Introduction: Lightweight UHPFC Structures as a Potential Solution to Sustainable Concrete Construction

Background

Today, concrete is the most frequently-used material on the planet after water, with a quantity of one cubic metre per person per year. Every second, 126 tonnes of *cement* are poured across the world, amounting to some 3.4 billion tonnes per year, or the equivalent of just over 14,000 Empire State Buildings. Over the three years from 2011 to 2013, China consumed 6.6 billion tonnes of cement, in other words more than the 4.5 billion tonnes used by the USA during the entire 20th century (Sources: USGS, Cement Statistics 1900–2012; USGS, Mineral Industry of China, 1990–2013). If we add to these figures the fact that production of cement by the clinkerisation process involves firing at temperatures of around 1450–1500 °C, it is not difficult to understand why the cement industry is one of the most polluting industries. Just like the construction sector as a whole, this industry faces many challenges today. Amongst these challenges are optimising the cement production process in the face of the growing cost of more sophisticated "raw materials" and their proven or potential harmfulness, or rethinking the supply chains and the product lifecycle to reduce grey energy, and lastly developing new construction methods. This last problem, *in which architects have the most influence*, should be envisioned beyond the usual constraints associated with buildings' regulations as a great opportunity for the architectural discipline, especially through the (re) examination of new or deprecated concrete and/or cement production methods. This search for new approaches is at the heart of the works we are presenting in this paper, oriented towards integrative computational and fabrication methods for the design and realization of three-dimensional concrete-based spatial lattices. Most of the actual innovative concretes devoid of aggregates should be considered as *cement*, a fact especially valid for the kind of ultra-high performance fibre-reinforced concrete (UHFPC or UHPC when no fibre is present) we used for some of the work we are presenting, nevertheless we will refer in this paper to the various cementitious materials by using the generic word "concrete". After introducing a few contextual elements in the first part, the second part will present the results of research conducted at *EZCT Architecture & Design Research* as of 2010, dedicated to the design and realization of 3D spatial lattices made of concrete. This research was firstly based on the use of a smart assembly of "hollow voxels" and then oriented towards a different approach to the same problem—i.e. generating true three-dimensional ultra-light concrete lattices—based on the use of 3D-printed sand moulds. The third part presents another approach to the creation of spatial lattices, by showing the results of two experiments led at UCL Bartlett School of Architecture during the academic year 2013–2014. One approach was based on the use of "inflatable" but non-stretchable 3D fabric formworks and the other on fully reusable 3D printed clay moulds.

Overal Context

Mindful that buildings represent one of the main sources of energy consumption, occidental public authorities have dramatically increased the number of sustainability initiatives over the past 15 years. Although this new legislation means constraints for numerous architects, we are considering this change as a formidable evolution opportunity for the discipline, especially through the (re)examination of new or deprecated fabrication and assembly methodologies engaging economic considerations: it is the first time since the introduction of industrial prefabrication that we have observed such a need for change in the way we design and build. As the most widely used construction material on the planet after water (ISO 2005), concrete has been designated as a sensitive research topic in this quest for sustainability. Despite important evolutions towards better energy efficiency, the environmental performance of concrete is still considered as widely ameliorable and architects can play a significant role in this redefinition of the construction material catalogues by identifying appropriate uses of "high-tech" concrete. Multiple strategies to diminish the environmental impact of concrete structures have been developed since the 1960s. These strategies fall into three groups: (a) those aiming to improve the sustainability of cements (alternative manufacturing processes, formulations with mineral additives or new binders, improvement of the usage properties); (b) various studies focusing on the concrete components and fabrication methods of such parts (inflatable casting systems, modules in textile reinforced concrete, etc.); and (c) the material optimization at the scale of the building, such as thin shells. While the first group is being developed almost exclusively by cement manufacturers, the two others can be considered as an interdisciplinary exercise more or less intrinsically associated to the first one: changing the properties of materials will tend to modify their usage and proposing new usages and structural/fabrication paradigms will encourage the cement industry to develop more adapted products. Available on the market for fifteen years, Ultra-High Performance Concretes (UHPCs and UHPFCs) are amongst these new materials that could allow a radical evolution in concrete structure morphologies: in addition to their resistance to high compressive strength, their (metallic for structural use and polymers for more decorative ones) fibres potentially remove the need for rebar (for flexure or pure tension), and their low viscosity enables the manufacturing of more and more complex structural components. Nevertheless, in most cases these components either follow the more or less usual scale of concrete construction (for example moving from a 30 cm diameter column to a 20 cm one shall not be considered as a real leap) thanks to the high compressive strength allowing classical optimizations, or constitute classical claddings and shells whose "simple" topological complexity would lead us to place them in the 2D "components family". We believe that the work we are presenting in this paper is based on an entirely different approach, one oriented towards more elaborate topologies made of extremely thin edges joining high connectivity nodes. The density of UHPC being, for example, much lower than steel, under specific conditions a fully stressed UHPC structure might theoretically be as light—or even lighter—than an equivalent steel structure at a given scale. This is the reason why we emit the assumption that, as self-placing rebar-free materials, the use of UHPCs is increasingly a more sustainable solution than regular concrete; given that the fibre repartition and the rheology are well controlled and that the volume of the expensive raw materials present in high performance concrete is minimized (Figs. 1 and 2).

From Nanometric Metamaterials to Lightweight Concrete Lattices

The behaviour of a spatial structure being by its very definition *spatial*, it is more efficient to minimize the use of material to look for solutions that are already optimal in three dimensions rather than maintaining design rules based on planar sections and patterns (2D networks and surfaces, including shells). As observed at the micro and nano-scale in material sciences (especially in the domain of architected materials), the repartition of matter plays a huge role in the characterization of its stiffness. The hypothesis emitted by EZCT Architecture & Design Research, via the "u-Cube: Universal Cube for Discrete Construction" (Fig. 3) and the "Studies in Recursive Lattices" (Figs. 1 and 4) projects, was to consider that observable improvement of architected material assemblies at the mesoscopic scale can be transported to architectural scale concrete structures, thus reducing their density and cement content. To have a better understanding of such an objective, one could take a four-storey building as an example, then consider this object as a solid, itself constituted by smaller ones (i.e. walls,

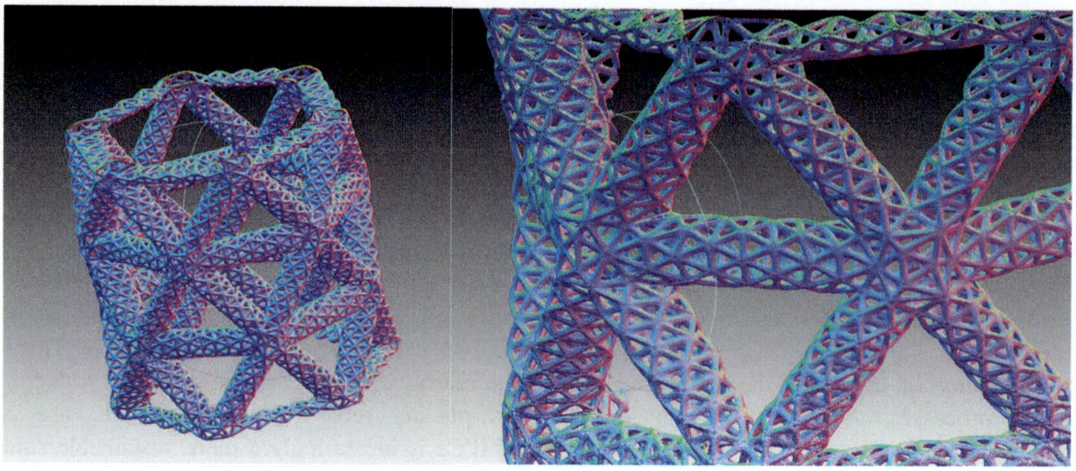

Fig. 1 EZCT Architecture & Design Research, 2012. Recursive Truss structure

slabs, etc.), themselves being divisible to another scale of material arrangement (i.e. concrete blocks). As each component depicted here is a volume, it is natural to focus on the design of spatial structures emulating the mechanical comportment of these multiple assemblies, each of them at a specific scale. In order to obtain spatial patterns that will minimize the density of a structure or a material while maintaining its mechanical properties, material scientists and engineers have been developing a large range of strategies to obtain micro or nanostructured metamaterials. Recently, two main research hypotheses have received a lot of attention by this community: recursive (or fractal) lattices, and topology optimization. The first method, very similar to traditional structural design, is to use patterns that have been isolated for their capacity to modify specific mechanical properties of materials to obtain various performances as elasticity, negative Poisson ratio, etc. The works of the HRL laboratories (Schaedler et al. 2011), Dr. Justin Dirrenberger (Dirrenberger 2012) as well as multiple scientific works dedicated to the microstructure of Euplectella Aspergillum marine sponges were amongst the main inspirations that motivated the range of prototypes that will be presented in the following part of this paper. Since then, much research has been conducted on the power of the recursive structural approach, for example at the California Institute of Technology (Greer Research Group) on multi-scale artificial beams (Montemayor et al. 2014). We now consider this kind of approach as a "new classic" even if some of the most famous examples of structural engineering in history, for example the Eiffel tower, already embedded such a recursive approach (Fig. 2).

The second method consists of using topology optimization with compliance objectives to

Fig. 2 Montemayor et al. (2014). Miniature Truss work

Fig. 3 EZCT Architecture & Design Research, 2012. *U-Cube* project, presentation at Centre Pompidou, during the *Multiversités Créatives* exhibition, Paris

Fig. 4 EZCT Architecture & Design Research, 2013. Studies in recursive lattices, 1:1 scale prototype (approx. 1.9 × 1.6 × 0.9 m. The total volume of concrete is only 34 l). First presentation at Archilab 2013, FRAC Centre, Orléans (this prototype, with other research components, is now part of the Centre Pompidou permanent collection of architecture)

obtain a repartition of matter directly from a numerical model without an underlying predefined pattern other than the discretisation of the solution space. While being theoretically the most optimal solution for the constraints that have been set in the model, those relative to manufacturing techniques are quite difficult to formalize: the morphologies obtained via topology optimization methods are very difficult if not impossible to realize without an appropriate additive manufacturing technology or complex casting systems (Dombernowsky and

Sondergaard 2011). This is one of the reasons why we believe in the necessity of incorporating right into the design space the numerous constraints associated with the crucial issue of fabrication. In any case, these two density minimization approaches are unfortunately not sufficient to guarantee the sustainability of architectural concrete parts, since most of their properties (durability, etc.) and performances (stiffness, etc.) result from the rigor of quality of their manufacturing. The study of microstructures is a good example of this problem: while the properties of the objects that are manufactured have a high scientific value, their economic and environmental costs are not acceptable yet. For the example methods cited above, expensive additive manufacturing techniques (photon polymerization at the atomic scale, laser sintering for higher scale objects) are most often the only eligible solutions which meet the precision constraints. As engineers and architects fully engaged in bridging the gap between engineering, architecture and macro socio-economic issues (for which the cost of building process at the scale of one-to-one is more crucial than ever before), we believe that much more inventive approaches have to be developed. We will now present some of them, aiming to find compromises between optimal geometries of lattices and various manufacturing techniques to introduce them to the architectural scale, developed during the last three years at EZCT Architecture & Design Research office and UCL Bartlett. They will be presented in a chronological fashion in the next paragraphs.

Research by Design: Experiments at EZCT Architecture & Design Research

U-Cube: Universal Cube for Discrete Construction (2009–2012)

U-Cube was a first experimental project conducted by EZCT Architecture & Design Research, aiming at rationalizing networks of UHPFC rods via the use of "hollow voxels" whose smartly connected cavities would create 3D mesh networks to be filled in with self-placing fibre-reinforced concrete. For the first set of experiments, we used pre-cut EPS moulds which were approximating (obviously through discrete approximation) portions of a shell-like structure, when assembled together with specific combinatorial rules. Each EPS block of 100 × 100 × 100 mm (it is important to note that the hollow voxels were from the outset also intended to be made of other materials, for example wood and cellular concrete) was cut with a hot-wire foam cutter mounted on the office 6-axis ABB IRB120 industrial manipulator, controlled with the HAL Robot Programming & Control software firstly initiated at EZCT and developed by Thibault Schwartz within this context before it became the key product of HAL Robotics Ltd (Schwartz 2012). The assembly of multiple families of hollow blocks describing regular void patterns with specified angles led to the creation of connected paths or "pipelines" enabling the casting of self-placing UHPFC lattices while providing a thermal insulation around the concrete to improve the thermal properties of the resulting parts. This experiment was temporarily discontinued, as numerous limitations arose from the complexity and high costs related to the assembly of facing panels or any other finishing solutions on the crenelated resulting walls.

Studies in Recursive Lattices (2012–2013)

As an extension to the previous project, the Studies in Recursive Lattices aimed to bypass the fabrication limitations of subtractive methods. Using large scale 3D printers developed by the German company Voxeljet AG, large sand moulds were produced to cast a 1.9 × 1.6 × 0.9 m lattice structure with three levels of recursion. The final mould, obtained by the assembly of 126 "crosses" each composed of two separate sides (one towards the interior and

Fig. 5 EZCT Architecture & Design Research, 2013. Studies in recursive lattices, detail of the structure with thinnest section of 3 cm diameter

Fig. 6 EZCT Architecture & Design Research, 2013. Studies in Recursive Lattices, detail of the moulds

one towards the exterior of the lattice) was dimensioned to resist the pressure of the concrete during the casting, and its internal faces were coated to obtain a very smooth finish, allowing the metallic fibres of the concrete to flow easily (Figs. 5 and 6).

Research by Design: Experiments at UCL Bartlett (B-Pro, Graduate Architectural Design, Research Cluster 5)

Three Dimensional Fabric Formwork System (Project by Zhe Xing and Danli Yu), 2013–2014

In the context of the 12-month architectural design seminar co-tutored from 2012 to 2014 by Philippe Morel (EZCT), Thibault Schwartz (HAL Robotics Ltd) and Guan Lee (Grymsdyke Farm), numerous student projects focused on the advancement of the various techniques for the fabrication of spatial concrete lattices. As a supplementary hypothesis for the realisation of affordable moulds, fabric formworks have been experimented by two of our students, by using various CNC machines such as routers and laser cutters. This final prototype, a complex fabric formwork measuring 3 × 1.5 × 1.5 m, is based on a tetrahedral grid which is subdivided or fused to finally obtain a demonstrator using three scales of subdivision depending on the original size of the "cells". While the resulting model shows a good precision along rods, the quality of the nodes was less consistent towards the top of the moulds due to the amount of air bubbles emitted by the concrete but also because of the lack of a satisfactory high pressure injection pump. This solution proved to be very labour-intensive for both assembly and disassembly, thereby cancelling the benefits of having a lightweight and cheap moulding system. Yet we believe this could be overcome if the process were industrialized (Fig. 7). As has already been proven (Figs. 8 and 9), more extensive use of computer simulations would enable improved accuracy and a better prediction of the behaviour of the "inflatable" fabric formwork during the injection phase.

Clay Robotics (Project by Jiashuang Sun, Kelvin Ho and Sihan Wang), 2013–2014

A second group from the previously mentioned design seminar at the UCL Bartlett School of Architecture focused on the use of 3D printing temporary clay moulds. This technique allows the production of relatively accurate, cheap and fully recyclable freeform formworks for concrete elements measuring up to 2.5 m high and weighing up to 500 kg. For this project, a KUKA KR150 was used to manipulate a support plate, used as a mobile reference system and on which a manual clay extruder was dispensing material. By managing the water content of the clay, the students were able to obtain a paste with an adequate rheology allowing them to print slightly cantilevered layers, thus enabling freeform 3D printing as long as fillings and layer patterns were respecting this cantilever limit. Multiple

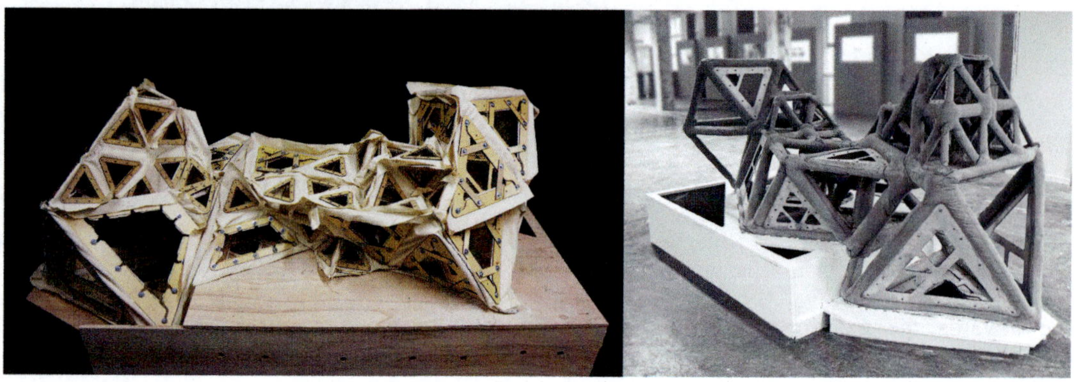

Fig. 7 Xing and Yu (2014). Three dimensional fabric formwork system

Fig. 8 Xing and Yu (2014). Three dimensional fabric formwork, detail and preliminary model (diameter of the section approx. 5 cm)

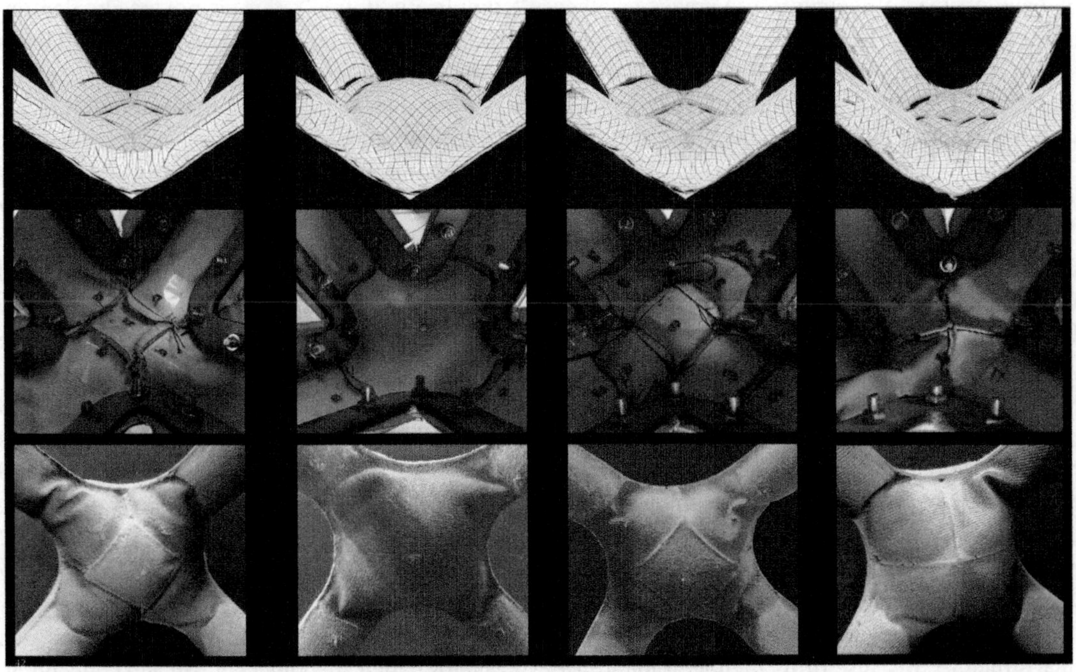

Fig. 9 Xing and Yu (2014). Different sewing strategies and comparisons between the computer simulations and the final results

Fig. 10 Sun et al. (2014). Clay robotics

experiments with varying sizes, wall thicknesses and heights were produced with this system. Finally, it was decided to split the printed elements into 400 × 400 × 400 mm modules to keep the possibility to manipulate and assemble them by hand, instead of having to use cranes or other lifting machinery. The conclusion of this project is that this approach is fully valid and would deserve further investigations at a larger scale within an industrial environment (Fig. 10).

Conclusion: Integrating Multiple Constraints Within the Algorithmic Design

As mentioned, producing accurate, affordable and large scale UHPFC spatial lattices is a difficult task especially because in most cases these parameters are contradictory. Let us take an example: UHPFC is intrinsically and usually more than ten times more expensive than any usual or "traditional" concrete, therefore the quantity of such a "high-tech" concrete should remain low in any final structure. Because of its high-tech nature (it requires extremely precise dosing of additives, etc.), UHPFC seems to be adapted to corresponding high-tech fabrication processes and structural models, for example spatial lattices with complex topologies as those lattices are natural candidates for highly efficient structures in terms of minimization of weight/material quantity. But these structures will also require high definition fabrication methods. Although at the small scale "complexity if free", this is not easily demonstrated when one wants to scale things up. This is especially true for really large scales and even truer when we tend to apply recursive multiple scale approaches, as most of the time—and contrary to what happens with the usual table top 3D printing—the same machine/process cannot be used for these different scales. In brief: if one wants to produce an affordable UHPFC structure, it should be ultra-light, therefore intrinsically spatial as they are natural candidates for structural efficiency, are therefore complex, of a high manufacturing precision, and therefore… expensive. Here one could say that it is easy to produce small components made of UHFPC, like beams, nodes, etc., and assemble them. This would not be wrong at all, except that there is a part of conceptual beauty (and true structural efficiency) in the perfect continuity that one can obtain with a sort of "one time casting" of an entire structure. In order to address such problems which are both technical and conceptual, we believe that as many of these contradictory parameters should be taken into account during the design phase, by making use of algorithmic paradigms which are intrinsically allowing the solving of problems which are either loosely-defined or ill-defined because of the intrinsic contradictions associated with the various parameters (e.g. genetic algorithms or agent-based approaches, which are well known algorithms for the solving of such problems which cannot be "well set"). Such a work is performed at the moment at EZCT Architecture & Design Research by Eng. Romain Duballet (Morel and Duballet 2015).

Acknowledgements Lafarge, Voxeljet AG.

References

Brayer MA, Migayrou F (2013) (ed) ArchiLab 2013: Naturaliser l'architecture. HYX, Orléans

Catsaros CH (ed) Autour d'une structure spatiale légère, Dossier BFUP, Tracés 12, Bulletin technique de la Suisse romande, SIA/Espazium, Lausanne, pp 6–9

Claypool M, Or, the thing itself, in. P.E.A.R, Paper for emerging architectural research, Issue 7 Meaning in Material, pp 42–47

Dombernowsky P, Sondergaard A (2011) Design, analysis and realization of topology optimized concrete structures. Int Assoc Shell Spat Struct 53:209–216

Fellipa CA (2007) Introduction to finite element methods. Department of Aerospace Engineering Sciences and Center for Aerospace Structures, University of Colorado, Boulder, Colorado 80309-0429, USA, Last updated Fall 2004 (notes from the Introduction to Finite Elements Methods (ASEN 2007) given since 1986 in the University of Colorado at Boulder)

Fries-Briggs G (2014) Robotic fabrication in architecture, art and design 2014. Mediating volumetric thresholds. Springer, Wien, New York, pp 211–221

Gerber D, Ibanez M (eds) (2014) Paradigms in computing: making, machines, and models for design agency in architecture. eVolo Press, New York, pp 123–140

Hamda H, Jouve F, Lutton E, Schoenauer M, Sebag M (2002) Représentation non structurées en optimisation topologique de formes par algorithmes évolutionnaires. In: Proceedings ESAIM, actes du 32ème congrès d'analyse numérique: CANUM 2000, (online), vol 11

ISO, ISO/TC 71 (2005) Business plan, concrete, reinforced concrete and pre-stressed concrete, 08/07/2005

Montemayor LC, Meza LR, Greer JR (2014) Design and fabrication of hollow rigid nanolattices via two-photon lithography. Adv Eng Mater 16(2):184–189

Morel PH, Duballet R (2015) Combining 3D-printing and GA-based optimization for the design and manufacturing of UHPC spatial lattices. In: Lecture and paper, presented at 12th international mathematica symposium, 12–14 Jan 2015, Prague

Schaedler TA et al (2011) Ultralight metallic microlattices. Science 334(6058):962–965

Schwartz T (2012) HAL: Extension of a visual programming language to support teaching and research on robotics applied to construction, Rob|Arch 2012. Springer, Wien, New York, pp 92–101

Additive Manufacturing and Multi-Objective Optimization of Graded Polystyrene Aggregate Concrete Structures

Romain Duballet, Clément Gosselin and Philippe Roux

Abstract

We present in this paper a work on multi-material additive manufacturing, linking a GA-based multi-objective optimization form-finding algorithm to a robotic fabrication system for prototyping cement based materials. The genetic algorithm is written entirely in Mathematica® and works on multi-material voxel spaces. It produces optimized material distributions, evaluated both in terms of mechanical performances and thermal insulation abilities. The genetic functions are based on 3D-Voronoi diagram representations and the evolution includes an adaptative penalization strategy (Hamda et al. in Appl Intell 16(2):39–155, 2002) for taking constraints into account (maximum deflections, material resistances, geometric or manufacturing needs). Both thermal and mechanical evaluations, to be combined in a Min-Max method based multi-objective fitness, are conducted thanks to Mathematica® 10 Finite-Element Method library for solving partial differential equations. We have implemented a function to compute volumetric stress distribution in a multi-material voxel space together with structure deformation. The same method is used to evaluate thermal performances: under imposed surface temperatures, the thermal distribution is calculated and the global conductive flow is extracted. The form-finding algorithm is scale-independent, and can work with any list of materials, as long as they can be characterized in terms of Young modulus, Poisson coefficient, and thermal conductivity. We address the prototyping problem by working with concrete mixed with polystyrene beads. Therefore the aim is the making of functionally graded

R. Duballet (✉)
Ponts ParisTech, ENSA Paris Malaquais, Paris, France

C. Gosselin
ENSA Paris Malaquais, Paris, France

P. Roux
Arts et Métiers ParisTech, ENSA Paris Malaquais, Paris, France

M.R. Thomsen et al. (eds.), *Modelling Behaviour*, DOI 10.1007/978-3-319-24208-8_19

material structures, each part (voxel) of them varying from purely structural concrete to almost purely insulating polystyrene. The additive manufacturing system we designed allows us to print different material performance at a large scale (the deposit lace has a diameter of 2 cm). It is composed with an industrial robot, a peristaltic pump and a printing head. The robot controller is given instructions with HAL (Grasshopper plugin) and the printing head controller with Arduino and Firefly® (Grasshopper plugin). We have been able to print prototypes. They were made to calibrate the fabrication process in terms of concrete formulation and deposition reliability. We have also addressed some less straightforward constructive scenarios like printing geometries that include empty spaces to be filled either with flexion-resistant material like UHPFC (as shown in the pictures below), tension cables, or specific architectural parts of the element (pipes, electric wire, etc.). Now that we have a fully operating form finding and manufacturing process for the making of lightweight multi-purpose concrete and polystyrene structures, we are about to link them together. Writing our own algorithm gives a great versatility to our approach, for we can easily simulate any architectural performance or feasibility constraint that our solutions are to respect. Our algorithm also being scale-independent, it is also possible to use it to address multi-scale problems, playing with the search space, the objectives, and the constraints as inputs. Furthermore, because the construction process is completely automated and computer controlled, we can easily adjust each step to ensure our simulation fits with the actual making. Prototyping allows to conduct mechanical and thermal tests, to provide the initial form finding with quantified feedback, and thus to tighten the whole performance oriented (Hensel in Performance oriented architecture rethinking architectural design and the built environment. AD primers. Wiley, New York, 2013) architectural approach.

Multi-Material Additive Manufacturing

Computational design practices are now strongly linked with robotic fabrication, allowing truly integrative approaches and efficient interdisciplinarity.

Our work tries to address large-scale additive manufacturing through the heterogeneous problems of fabrication (robotic process, material behaviour, and geometric control), optimization (multi-material, multi-objective, multi-scale) and use (constructive suitability, architectural pertinence). We have chosen to treat those problems in accordance with their supposed linking, progressively bringing answers and critically considering the possibility and efficiency of their translation from one domain to another.

Several works already exist on large scale additive manufacturing such as Contour Crafting or D-shape. Their work deals with fabrication process of mono-material fabrication prototypes. The particular process we focus on is the design and making of graded polystyrene aggregate mortar structures, that present both structural and insulating properties. Additive manufacturing allows high-resolution placing of customized material, and that resolution directly transfers to a numerical model resolution for optimizing this placing.

We present in this paper the optimization algorithm and the robotic fabrication process. Some prototypes were printed, as well as test cylinders, from which we can measure mechanical performances and compare with poured mortar.

Large Scale Integrative Approach

Multi-material additive manufacturing has mostly focused on smaller scale than architectural design, something that can be attributed to the difficulty of getting productive feedback from architectural objects. Our work tries to focus on the design and making of graded material architectural elements, setting the basis of a generic and integrative approach where digital modelling, robotic manufacturing process and the resulting object are addressed simultaneously, to ensure their efficiency at large scale.

Our aim is to design a fabrication process able to produce graded polystyrene aggregate concrete structures, the purpose of which is to provide both sufficient mechanical resistance and thermal isolation. Additive manufacturing allows us to freely distribute graded material from pure mortar, that shows good resistance but poor insulating properties, to a low resistance but good lagging characteristics polystyrene aggregate mortar. Such a precise organisation of graded material helps us design structurally efficient architectural elements while suppressing thermal bridges.

Our form-finding process is based on a GA-based multi-objective algorithm. As we deal with multi-material search spaces, evolutionary approach is more practical than topological optimization using homogenization (Hiller and Lipson 2009). In addition, we consider this algorithm as a tool, to be enriched with new objectives or design constraints, coming from the diverse feedbacks that the constructive process provides us. We have chosen to implement the algorithm in Mathematica®, for the easy interfacing with other software and languages, on one hand, and for its powerful symbolic calculus functions, on the other hand, which makes the integrative approach very efficient.

Ga-Based Multi-Objective Optimization Form-Finding Algorithm

The evolutionary algorithm works on a three-dimensional voxelized space (Figs. 1, 2, 3 and 4). Each voxel can be attributed either a certain material, with corresponding properties (mechanical and thermal), or void. This representation of material distribution depends on the global dimensions of the modelled object, as well as a resolution. This allows us to be scale-independent in the sense that we can adapt the voxel sizes to fit aimed precision. Typically, we work with a voxel size corresponding to the diameter of the printing drop lace (2 cm for the prototypes shown below), which is the minimum dimension of a measurable material heterogeneity.

The genotype, the encoding of the physical object (phenotype), to be treated by the evolutionary functions (cross-breeding, mutation, selection), is based on a three-dimensional Voronoï representation. Indeed, evolutionary algorithms show better performances with light implicit representation of the phenotype. Between existing possibilities, Voronoï sites allow us to orient the form finding more easily thanks to the evolutionary functions, something more difficult to achieve with encoding based on Fourier transformation for instance.

Each Voronoï site of a genotype is then attributed a certain material -including void- to be transferred to each voxel inside the corresponding Voronoï cell of the phenotype. Each material stands for a series of properties to be included in the evaluation, here a Young modulus, a Poisson coefficient and a thermal conductivity. The evaluations, conducted on the phenotype, feed a multi-objective fitness taking into account both mechanical and thermal performances, combined with a min-max method.

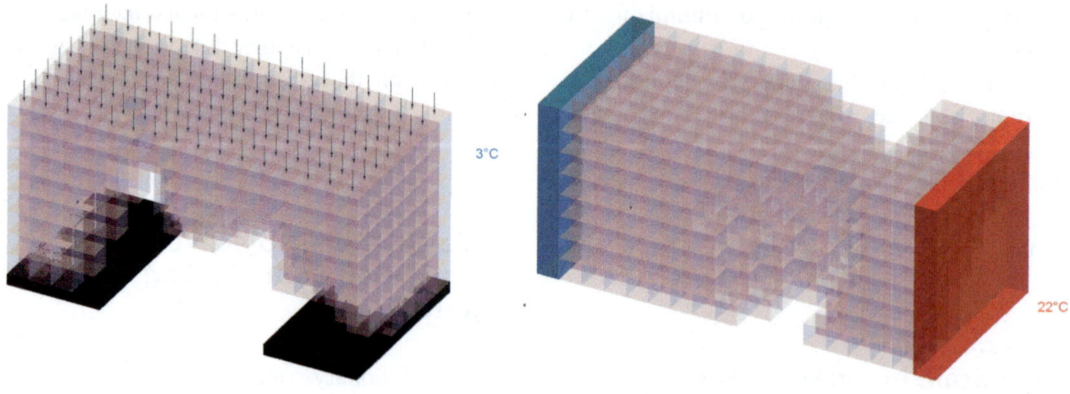

Fig. 1 Boundary problem of the structural and thermal analysis

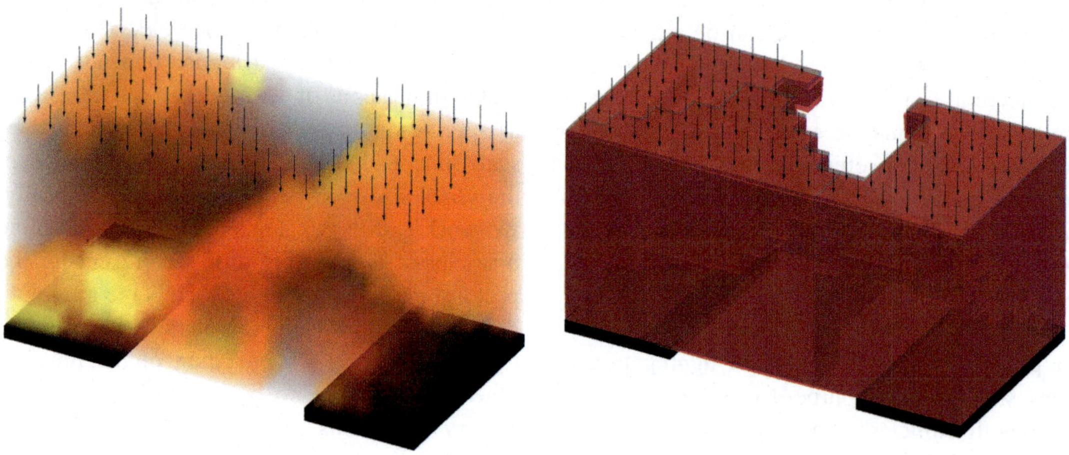

Fig. 2 Stress analysis and deformation analysis

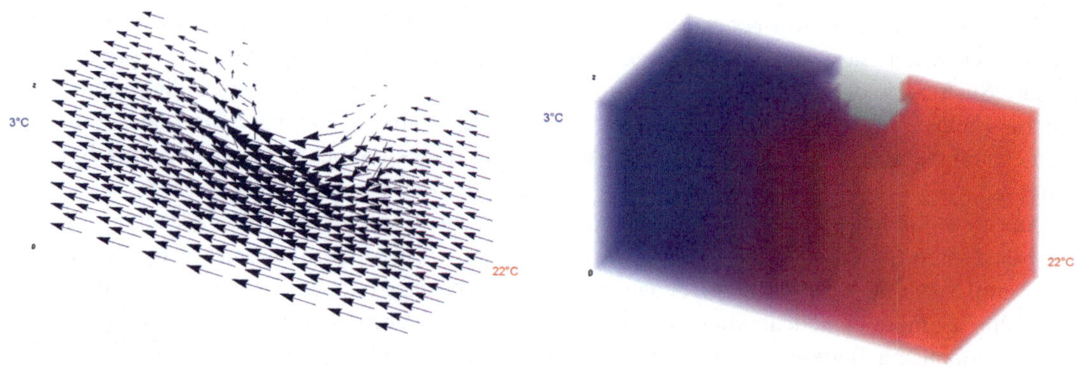

Fig. 3 Heat flow and temperature distribution

The mechanical evaluation uses Direct Stiffness Method to compute the deformation field in the phenotype, due to a certain load case. We use the Mathematica® 10 Finite Element Method native functions for the solving of the equations. We have included them in a mechanical analysis

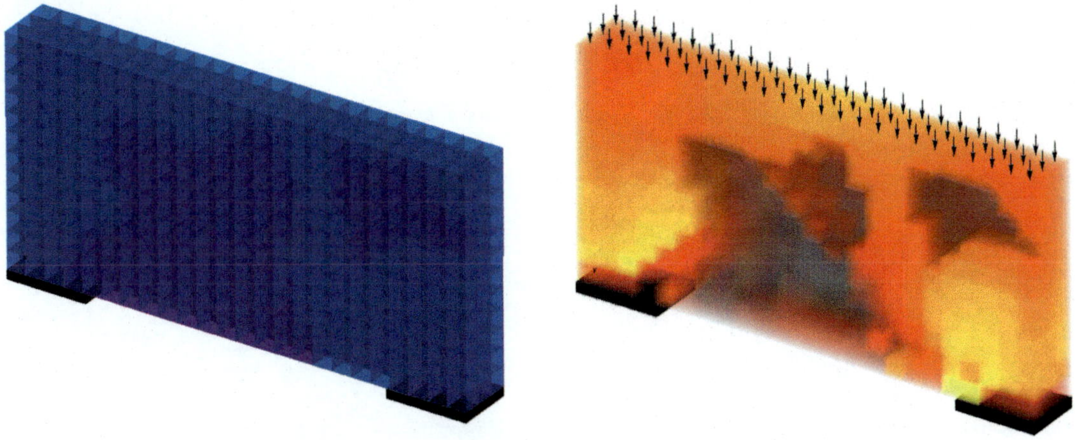

Fig. 4 Wall shaped design space and its stress analysis

function, computing three-dimensional stress distribution in the material. The thermal evaluation follows the same method, but solving a heat equation. The meshing to be taken into account by the FEM directly uses the voxels as hexahedral elements.

Once the selection of the individual has been made, the evolutionary functions act directly on genotypes. The cross-breeding is rather classical; and the mutation can add, move or delete a Voronoï site, or also change its assigned material. The evolution also includes an adaptive penalization strategy (Hamda et al. 2002) for taking design constraints into account, such as maximum deflections, material resistance, geometric or manufacturing needs, directly into the fitness function. That way, the algorithm aims for a certain proportion between feasible (90 %) and not feasible (10 %) solutions (feasible meaning that no constraint is infringed). This strategy allows good convergence, while maintaining diversity.

Robotic Fabrication System

We address the prototyping problem by working with mortar mixed with polystyrene beads. Therefore the aim is the making of functionally graded material structures,each part (voxel) of them varying from purely structural mortar to almost purely insulating polystyrene.The additive manufacturing system we designed allowsus to print different material performances at a large scale (the drop lace has a diameter of 2 cm).

Rheological performances of the mortar were studied in order to achieve its pumpability and its printability. Indeed, the granulometry of the mortar is very small (from 0 to 0.6 mm) to make easier the pumping. Less pressure is needed to achieve it (Chapdelaine 2007). Moreover a small granulometry also facilitates the printing by increasing the mortar plastic yield. Based on the definition that concrete contains gravel that has a diameter of at least 6 mm, we are dealing with mortar. The formulation was thought in two stages related to the printing process. The first one corresponds to the pure mortar pre-mixed pumping. It must be as fluid as possible. This mean it must have a low plastic yield value, be influenced the less possible by thixotropy and not set. The second one corresponds to the printing stage where several additives are added to the mixture. Those additives are used to control material performances and the mortar setting kinetics.

Material performances are controlled by adding polystyrene beads. The proportion between mortar and polystyrene beads gives a range of performance between structural and insulating material. To adjust mortar setting kinetics, an

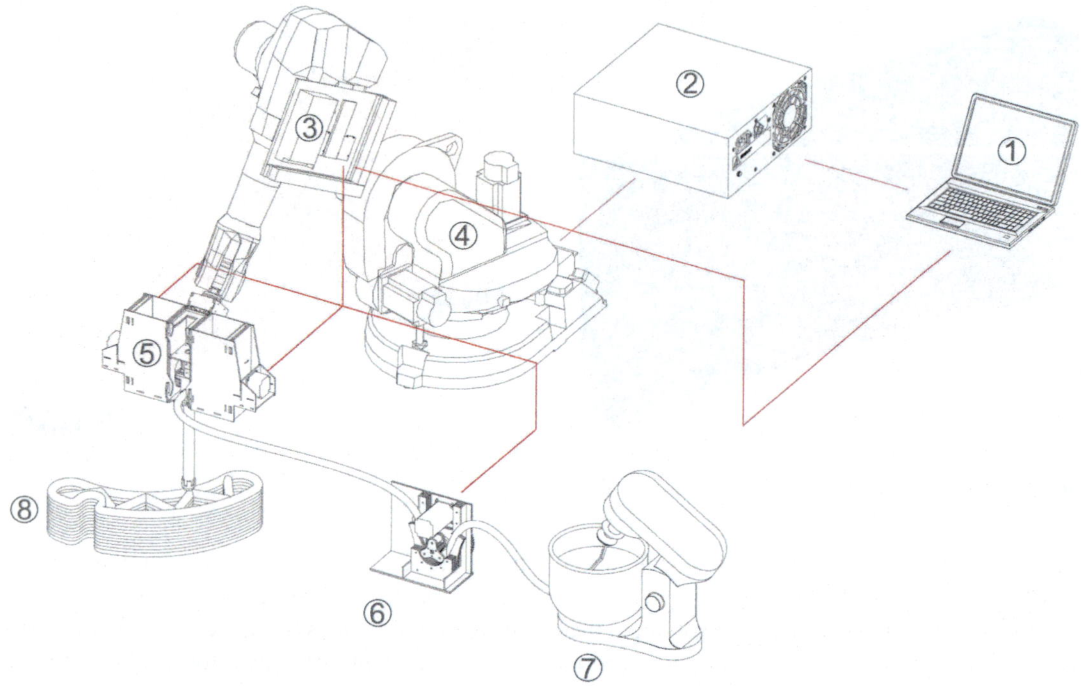

Fig. 5 Diagram of the manufacturing process

additive is added to increase the plastic yield value and accelerate the setting. The printing process is shown on Fig. 5.

1. The additive manufacturing process is controlled by computer. We used Rhinoceros®/ Grasshopper® and Arduino to achieve the communication between the different parts of the printing process.
2. Instructions are given through the robot controller with HAL (Grasshopper® plugin).
3. Instructions are given through the printing head controller with Arduino and Firefly (Grasshopper® plugin). It allows the control of the four motors of the head.
4. An industrial 6-axis robot solves the positioning of the printing head.
5. The printing head is composed of one polystyrene beads screw measure that allows the control of material performances (between structural and insulating), one prompt cement screw measure used to accelerate the mortar setting and increase its plastic yield value, and one mixer to homogenise the blend.
6. A peristaltic pump is used to push the mortar to the printing head.
7. he mortar is mixed to avoid thixotropic issues. Its formulation has been investigated to ensure its pumpability (Chapdelaine 2007).
8. Multi-material object can be printed.

Fabricated Prototypes

We have been able to print prototypes. They were made to calibrate the fabrication process in terms of mortar formulation and lace dropping reliability. The two pictures on Fig. 6 show a piece of wall with a wavy shape in the middle made of structural material and two boundary parts made of insulating material.

The two diagrams on Fig. 7 show the material distribution. There are four material gradients that are numbered from 1 to 4. Material 1 is composed of pure mortar, while materials 2, 3

Fig. 6 Multi-material object printed with our system

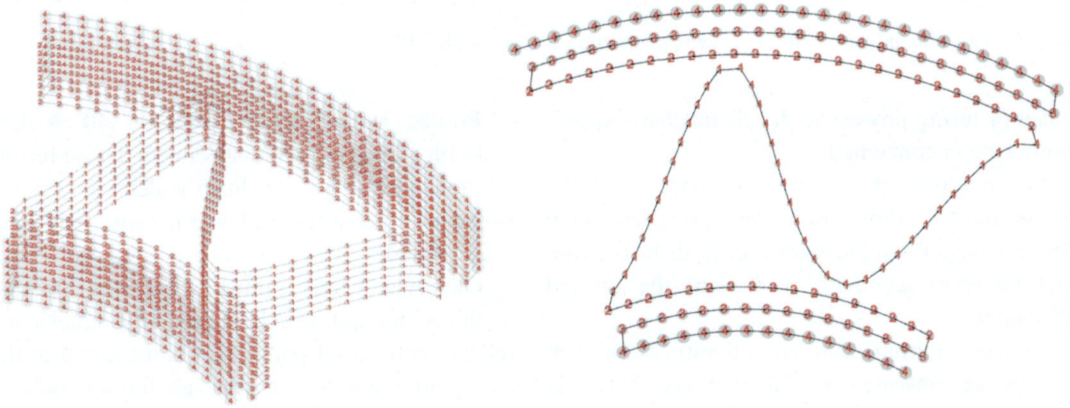

Fig. 7 Multi-material object printed with our system

and 4 are composed of 30, 40 and 50 % volumetric of polystyrene beads respectively. Graded materials are interesting for durability. Indeed, because there is no brutal rupture between the material performances, the stress generated by the distention is spread over several interfaces instead of one. In this way, it could reduce the peeling.

We have also addressed some less straightforward constructive scenarii like printing geometries that include empty spaces to be filled either with flexion-resistant material like UHPFC (as shown on Fig. 7), tension cables, or specific architectural parts of the element (pipes, electric wire, etc.) (Fig. 8).

Mechanical Testing

We made compression tests on 70 by 140 mm cylindrical specimens to compare mechanical properties of poured and printed mortar. Three different types of specimen were made: one poured mortar specimen that was vibrated and two printed mortar specimens, each of them with a different tool path—see Figs. 9, 10 and 11. One was made with an orthogonal tool path and the other one concentric.

This test was not made to prove the viability of printed mortar against poured mortar but to see the issues brought by the additive

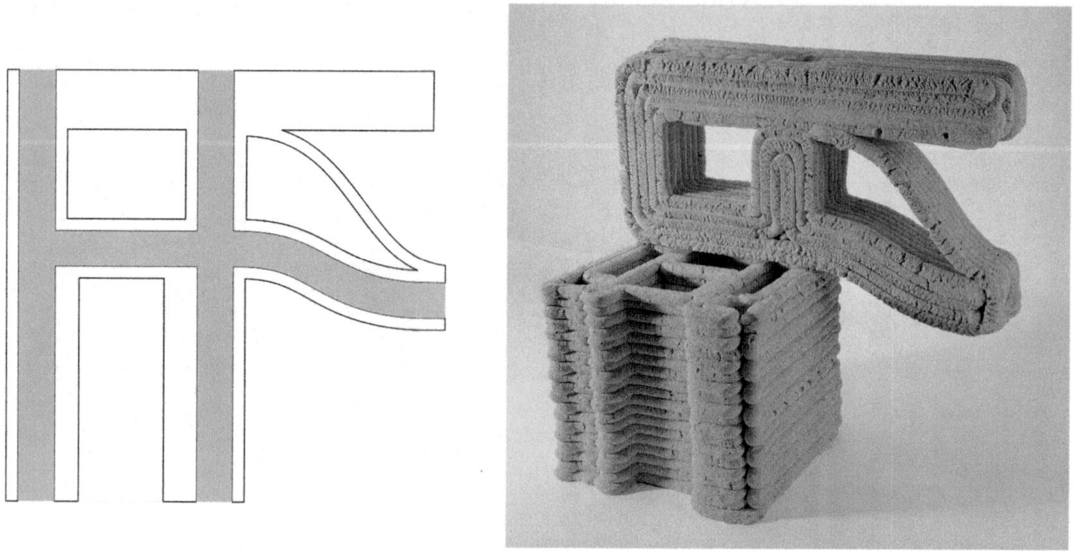

Fig. 8 Assembly printed with our system, connection is made with UHPFC

manufacturing process as far as mechanical performance is concerned.

To compare those specimens we used the same mortar formulation and extracted them from a bigger sample with a core drill to avoid the boundary geometry issues of the printed specimens.

Mortar formulation and compression test results are presented on Tables 1 and 2. Calcia cement (Technocem blanc 42.5R CEM II/B-LL 42.5 R CP2 "SB") was used for all our tests. The sand has a granulometry between 0.1 and 0.6 mm.

Printed specimens are roughly 40 % less resistant than the poured specimen. Those results are related to the specimen's density. Printed specimens' densities are lower because they were not vibrated.

Otherwise, the printed specimens fracture methods are not usual. Indeed, the cracks are related to the tool path. They are located at the laces interfaces as we can see for the orthogonal tool path specimen with the stairs-like cracks and for the concentric specimen with the horizontal cracks. Poured mortar specimen cracks divide the specimen in several distinct parts that are not held together. However, the

Table 1 Mortar formulation

	Water weight /cement weight –	Sand volume /total volume –	Plasticizer weight /cement weight –	Resin weight /water weight –
Mortar	0.35	0.36	0.50	1.20

Table 2 Compression test results

	Density Kg/m^3	Yield stress N/mm^2
Specimen 0 (poured mortar)	2.182	63.194
Specimen 1 (orthogonal)	1.894	36.664
Specimen 2 (concentric)	1.872	35.391

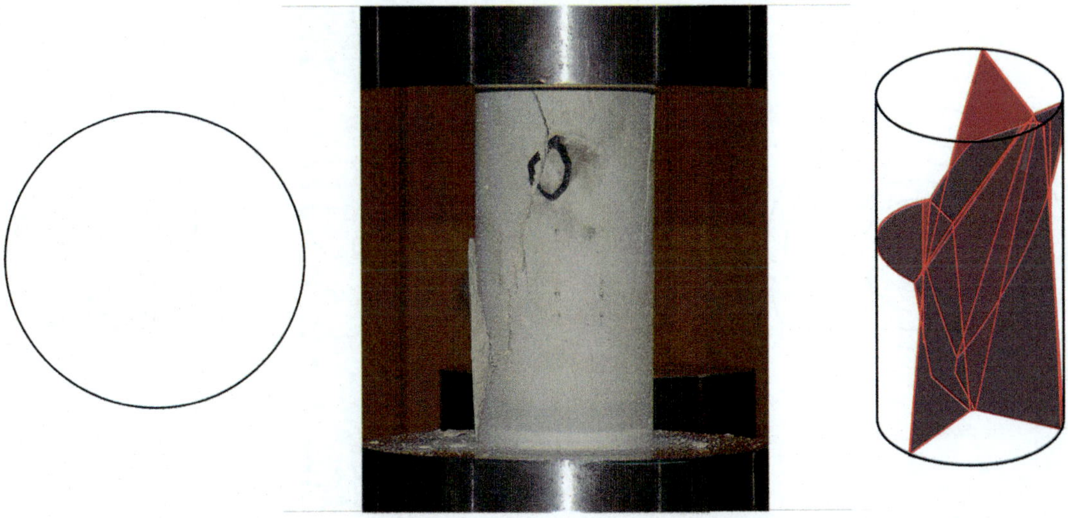

Fig. 9 Poured mortar specimen

Fig. 10 Orthogonally printed specimen

different parts obtained by the cracks for the printed specimens are not divided in several distinct parts, they are still held together. Printed mortar seems to be less brittle hence more ductile than poured mortar. Tools paths, compression test pictures and fracture diagrams are represented on Fig. 9, 10 and 11 for the three specimens.

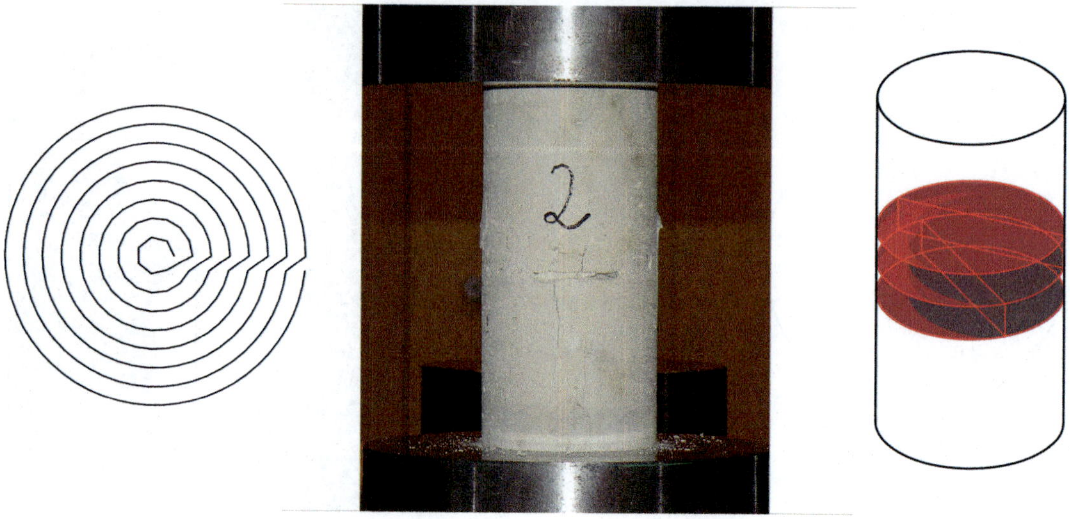

Fig. 11 Concentrically printed specimen

Conclusion

We have now an operating form finding and manufacturing process for the making of light-weight multi-purpose mortarand polystyrene structures. Continuing our work we try to address the various problems we face by modifying the most relevant parts of the process, be it the form-finding one, the pumping one, the printing one, or material properties. The purpose is to maintain a truly integrative approach to be as efficient as possible.

Writing our own algorithm gives a great versatility to our approach, for we can easily simulate any architectural performance or feasibility constraint that our solutions are to respect. Our algorithm also being scale-independent, it is also possible to use it to address multi-scale problems, playing with the search space, the resolution, the objectives, and the constraints as inputs.

The scale of a particular form-finding problem is dictated by the precision of the manufacturing process (printing resolution, geometric control) in relation with the type of constructive piece we are working on (a three meter high wall or a connection nod for instance). For each of them, tests are to be made to evaluate the distance between the real mortarobject and its corresponding virtual solution. Those tests aim less at validating the approach than at providing the whole process (form-finding and manufacturing) with feedback. Everything being computer controlled, we are able to quantify each evaluation to efficiently tighten the approach.

As the tests on cylinders show, the printing process brings a critical modification of the mechanical behaviour of the object. We have identified two main changes that are the global "material" performances, reduced due to lack of density of the pumped mortar, and the failure mode, critically toolpath dependent. This last point shows that a study on material properties of the pumped and printed mortarcan hardly be separated from a geometric and structural approach. Indeed printed architectural elements are not made of a perfectly isotropic material, but are composed by a geometrically organised structure. As a future work we want to address this toolpath problem to be able to implement its evaluation in our form-finding algorithm. In that way, one can imagine that the geometric dependency of the failure mode could be used as a relevant way of controlling the behaviour of a structure.

References

Chapdelaine F (2007) Étude fondamentale et pratique sur le pompage du béton, Université Laval de Montréal

Hamda H, Jouve F, Lutton E, Schoenauer M, Sebag M (2002) Compact unstructured representations for evolutionary design. Appl Intell 16(2):39–155

Hensel M (2013) Performance oriented architecture rethinking architectural design and the built environment, AD primers. Wiley, New York

Hiller JD, Lipson H (2009) Multi material topological optimization of structures and mechanisms. In: GECCO '09 Proceedings of the 11th annual conference on genetic and evolutionary computation, pp 1521–1528

Kwon H (2002) Experimentation and analysis of contour crafting (CC) process using uncured ceramic materials. A Dissertation Presented to the Faculty of the Graduate School University of Southern California in Partial Fulfillment of the Requirements for the Degree Doctor of Philosophy (Industrial and Systems Engineering)

Roussel N, Stefani C, Leroy R (2005) From mini-cone test to Abram cone test: measurement of cement-based material yield stress using slump tests. Cement and Concrete Research

Integrated Design and Fabrication Strategies for Fibrous Structures

Gundula Schieber, Valentin Koslowski, Jan Knippers, Moritz Dörstelmann, Marshall Prado, Lauren Vasey and Achim Menges

Abstract

The aim of the research presented in this paper is to examine the potential of integrated computational design methods for the development of complex fibre-based structures for architectural applications. While in today's building industry structural systems still rely on a set of well known typologies, integrated computational design strategies offer the opportunity to break down traditional strategies and develop new resource efficient architectural systems by synthesizing the material constraints, structural performance and digital production from the very beginning. The approach continues previous integrated design research at the Institute for Computational Design (ICD) and the Institute of Building Structures and Structural Design (ITKE) on fibre-based structures with a mould-less winding technique. This is discussed through the implementation of a full-scale architectural prototype with a pneumatic formwork that reduces the necessary formwork for fibre composite to the minimum and serves after fabrication as envelope. The paper focuses on the project-oriented integrated design strategies, including the challenges and the implementation of new process-specific computational tools.

Introduction

Since the introduction of structural analysis in the second half of the nineteenth century, the paradigm of all structures was based on calculability and predictability. Even though we have today new methods for advanced computational simulation and fabrication, nearly all structures in the building industry still rely on a few well defined structural typologies, which were developed with the help of mechanical theories and the constantly evaluation of construction principles over this time (Knippers 2013, p. 77) (Fig. 1).

G. Schieber (✉) · V. Koslowski · J. Knippers
Institute of Building Structures and Structural Design (ITKE), Stuttgart, Germany
e-mail: g.schieber@itke.uni-stuttgart.de

M. Dörstelmann · M. Prado · L. Vasey · A. Menges
Institute for Computational Design (ICD), Stuttgart, Germany

M.R. Thomsen et al. (eds.), *Modelling Behaviour*, DOI 10.1007/978-3-319-24208-8_20

While these typologies have a limited framework and lead often to standardized building systems with predefined construction elements, previous research on fibre-based structures at the ICD and ITKE has shown that with the help of integrated design and fabrication strategies it is possible to develop novel performance oriented architectural systems which enable an economic use of resources and a wider spectrum of geometric freedom with a new appearance. In comparison to conventional construction materials, like wood or steel, fibre-based structures can be adapted globally and locally through the anisotropic arrangement of fibres to various design and structural requirements that provides the capability for innovative architectural systems through integrated design strategies (Dörstelmann et al. 2014, pp. 219–226).

The production of double curved fibre reinforced composite parts in architectural dimensions typically requires the use of full scale moulds or custom formwork that can be expensive both from a financial and resource vantage point. The two institutes developed in the last years various robotic coreless filament winding techniques with a reduced or reconfigurable formwork. These techniques enable the fabrication of economic and material efficient composite structures in building size with structurally dependent fibre orientations (Waimer et al. 2013; Reichert et al. 2014; Prado et al. 2014).

Based on the previous research, the approach presented in this paper explores load path driven fibre laying technologies to a further reduction of the needed formwork for fibre composites to a 0.2 mm thin inflated membrane that becomes the building envelope after production. To incorporate efficient fibre based load bearing systems, architectural requirements as well as engineering and material science a novel integrated process was developed. While traditionally all geometric decisions are made prior, this process needed a continuously recorded robotic path which is used to place fibres on inflatable structures, allowing the live deflections of the pneumatic to be calculated and re-incorporated into the next iteration of the computational model.

The paper will present this alternative design approach. Instead of going into detail in the individual steps of development the authors will give an overview of the whole process and will

Fig. 1 ICD/ITKE research pavilion 2014–15 during construction in the night

expose the challenges and the continuous and mutual exchange of information.

Investigations and Technical Challenges

Biological Investigation

The diving bell of the water spider (*Argyroneta aquatica*) was identified as a versatile natural role model for the conceptual development of the architectural demonstrator. The water spider, though similar to terrestrial spiders, lives underwater. To do so, it constructs a reinforced web that is used to capture and hold a sub-surface air pocket in a dynamic aquatic environment. Therefore, the water spider has developed an efficient fabrication technique, by systematically reinforcing the air pocket from the inside through various silk laying behaviors resulting in a stabilized airtight composite grid shell-like load bearing system (Neumann and Kureck 2013, pp. 1–5). The water spider has to deal with various underwater constraints and the resulting nest must fulfill a wide variety of functions, which may or may not be suitable for architectural application. The transfer of the biological principles into design and fabrication processes requires a level of abstraction.

The abstraction of underlying principles within the water spiders reinforcement strategy is based on results gathered through an observation of the bubble building and fibre reinforcing behaviour of ten spiders. The analysed photo and video material allowed a comparative study of the spider behaviour in controlled and stepwise altered environments (Fig. 2). Outcome of these investigations is the identification and abstract description of the procedural steps that are employed by the spider for the web building process which resolves in a robust nest under various boundary conditions.

This process was analyzed and transferred into a fabrication strategy for the load adapted fibre reinforcement of an inflated ETFE (Ethylene tetrafluoroethylene) membrane to create a weatherproof fibre composite compression shell without the need of additional formwork.

Bio-inspired Fabrication Strategies for Pneumatic Formwork

The first steps of the water spider, where he builds the outer web for holding the diving bell in place, defining its constantly changing size and geometry doesn`t seem to be suitable for a technical transfer. The focus of the biomimetic investigation was on the building sequences of the inner web (Fig. 2c). The main reinforcement of the diving bell takes place from the inside. While the water spider uses this strategy to have enough air and to be protected from enemies the transfer into a digital fabrication set-up with a six axis robot in its centre to apply extruded fibres onto an inflated membrane from the inside, enable to protect both the robotic fabrication setup as well as the applied fibre composite material from the external environmental conditions (Fig. 3). Throughout these investigations also emerge the concept of using the pneumatic formwork later as a building envelope.

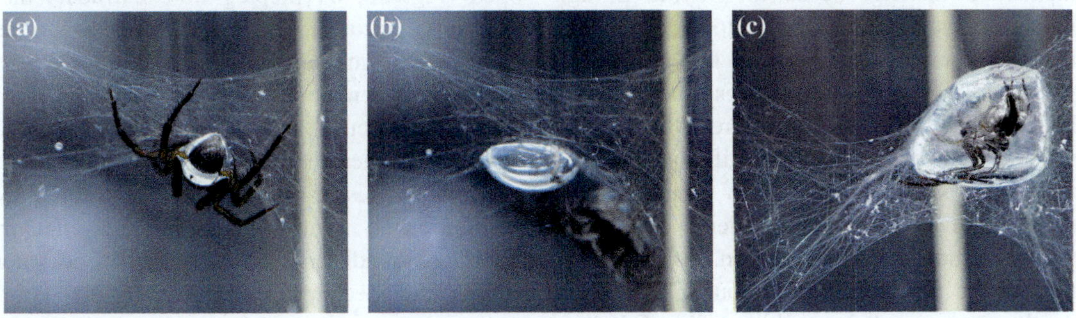

Fig. 2 Water spider (Argyroneta aquatica) during the building process of its diving bell, **a** Water spider building the outer web; **b** Water spider building the diving bell; **c** Water spider building the inner web reinforcing the diving bell

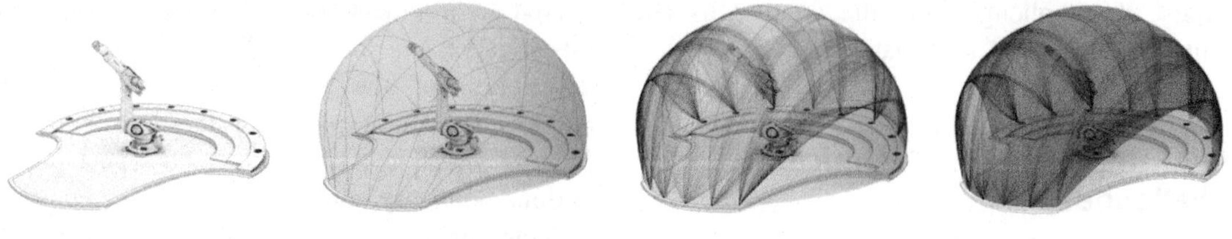

Fig. 3 Bio-inspired Fabrication Concept: 1st Installation of platform and robot, 2nd inflation of the ETFE membrane as pneumatic formwork; 3rd placement of epoxy resin pre-impregnated carbon fibres on the ETFE membrane; 4th deflation of the pneumatic structure, fibres form load bearing structure, ETFE membrane serves as envelope (from *left* to *right*)

Whereas the diving bell constantly changes size and geometry during construction, the inflated formwork needed a predefined geometry with a cutting pattern. This reflects the fact that biomimetic investigations couldn't be seen as a linear transfer of biological principles, but rather as source of inspiration to develop systems beyond preconceived typologies of architecture and engineering.

The water spider has to deal with various underwater constraints and the resulting nest must fulfil a wide variety of functions, which may or may not be suitable for architectural application, the transfer of the biological principles, into design and fabrication processes requires a level of abstraction.

Technical Challenges

The main challenge was to find suitable materials and to develop a new adaptive fabrication strategy to apply fibres from the inside against the direction of gravity on a compliant formwork.

In order to deal with these conditions the membrane material needs to provide a high modulus of elasticity in order to reduce deformations when the robot places fibres on it, as well as a large linear elastic stress strain behaviour that resists the force that results from the fibre placement, without plastifying. Furthermore, transparency, weather UV resistance and on site join ability were considered criteria for the membrane material selection. After various tests of the mechanical behaviour and the production process, a 0.2 mm thick ETFE (Ethylene tetrafluoroethylene) membrane was chosen for the pneumatic formwork and finished building envelope in combination with carbon rovings, consisting of 48,000 filaments that were pre-impregnated with epoxy resin (Fig. 4).

To achieve a load transmitting bond between the ETFE membrane and the carbon fibre prepregs several laying techniques were tested with different epoxy resins, adhesives and foil surface treatments. In order to place the fibres onto the ETFE membrane from below, in other words against gravity's direction, a custom made effector was built that integrates sensor driven online control of various process parameters (Fig. 5). The tension applied to the interface between uncured fibres and the membrane needed to be reduced to a minimum as the placed fibres would otherwise be pulled off. This was an uncommon challenge, as the fibre tension is usually a required factor in fibre placement methods. This was especially different from previous ICD/ITKE projects, where the fibre tension was utilized as a form generating parameter. In order to reduce fibre tension to a minimum, an extruder mechanism was developed (Fig. 5b, 04 prepreg fibre extruder) that unspools the exact roving length in relation to the real time information about robot speed and path length. The remaining challenge was the dead load of the uncured fibre itself that needed to be held by an adhesive. The distribution of the right amount of adhesive was crucial and controlled by an automated spray device on the effector. Therefore an adhesive was selected that does not change the mechanical properties of the composite compound.

Fig. 4 Experimental fabrication set up of primary and secondary fibre bundles on pneumatic formwork

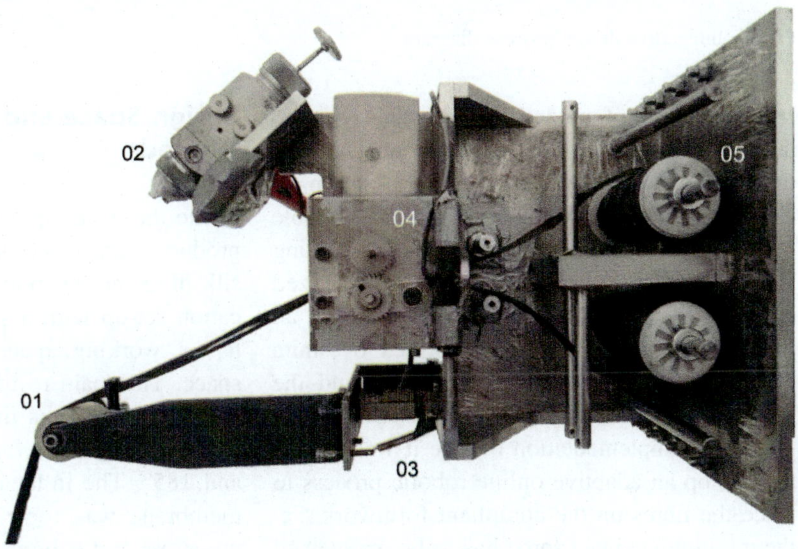

Fig. 5 Custom sensor integrated fibre extruding robotic effector: 01 roller, 02 adhesive spray device, 03 force sensor, 04 prepreg fibre extruder, 05 carbon fibres prepreg spools

Integrated Design Strategies, Methods and Processes

To realize a fibre-based architectural prototype on the campus of the University of Stuttgart it was crucial to develop integrated design strategies that allow formalizing the complex relationships between the various requirements (Fig. 6) and help negotiate them continuously throughout the design process.

Project Specific Integrated Design Strategies

The design space was first defined through the possible workspace of the robot set-up, the

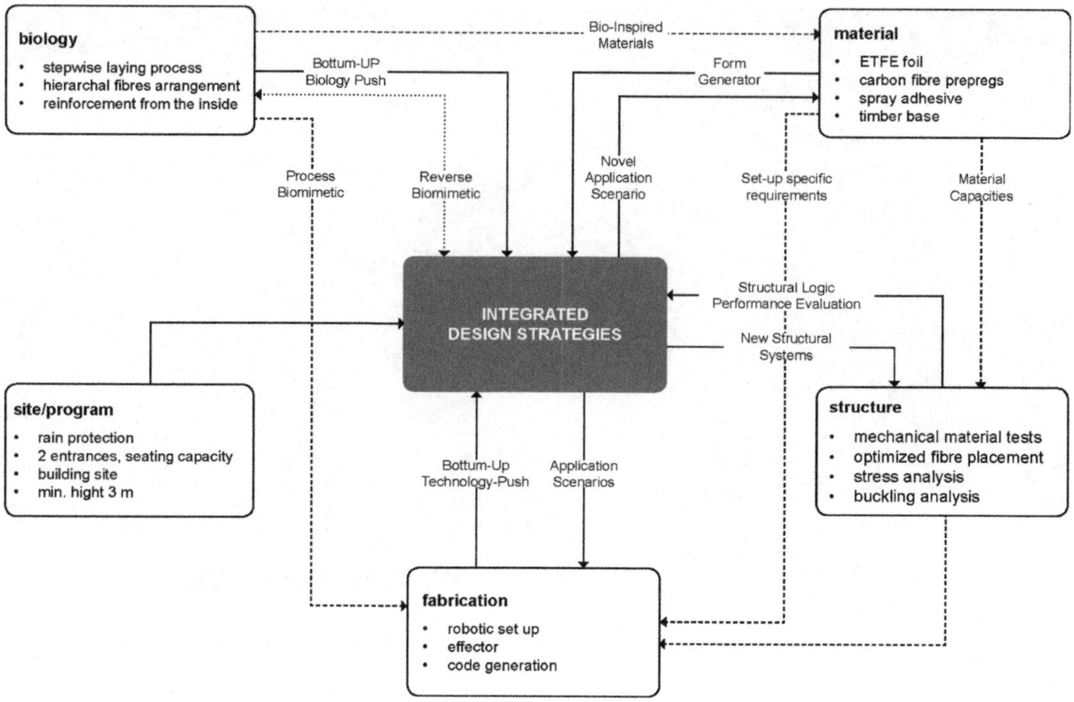

Fig. 6 Integrative design process diagram

solution space of potentially inflatable geometries, the material constraints and the architectural requirements given by site and program. In the next step the geometry of the inflated bubble was defined by the development of a cutting pattern with respect to an enclosed pressurized volume. The fibre layout was generated by an agent based design method informed by finite element analyses, architectural criteria and the working space of the robot.

For the implementation on site it was crucial to develop an adaptive online robotic process to place the fibres on the compliant formwork, i.e. the minimum contact force has to be maintained to establish the compound between the fibres and foil and the maximum contact force to avoid plastic deformations of the ETFE membrane. There a communication strategy that allows the live deflections of the pneumatic formwork to be calculated and re-incorporated into the next iteration of a computational model was developed.

Design Space and Form Finding Methods

While the water spider can easily move around, produce and place simultaneously continuous silk fibres at any point of the bubble, the fabrication set-up with a six axis robot in the centre has a working space that defines the design space. The main restrictions of the robot are its reach as well as its first axis, which is not continuous and allows for rotations between −185° and 185°. The inflatable geometry of the ETFE membrane was form found by predefining the pre-stress and solving for the geometry.

The performance in both the inflated state, which is bearing tensile forces during fabrication, and the deflated state, which is a load bearing system that is mainly subject to compression stresses had to be considered (Fig. 7). A computational tool combining mechanic form finding methods and geometric nonlinear finite element analyses allowed for various design explorations

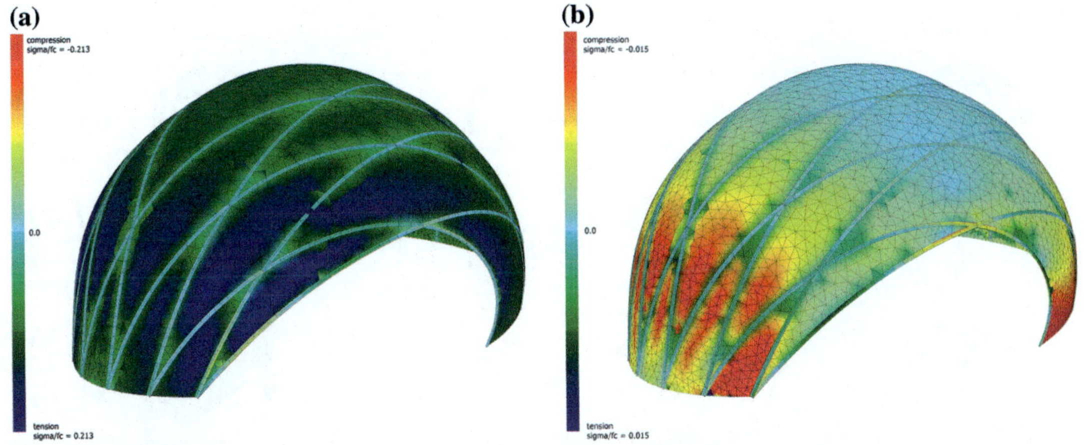

Fig. 7 Finite element analysis (self-weight), primary structure modelled with beam elements, secondary and tertiary elements abstracted to shell elements: **a** Pneumatic structure during production subject to tensile forces; **b** Deflated load bearing system subject to compression

of lightweight potential with respect to self-weight and wind loads.

The final double curved membrane that was structurally simulated in its inflated state was divided into stripes and flattened into a two dimensional cut pattern in order to determine the geometry and to avoid wrinkles. The stripes were welded to form the double curved inflatable bubble. The stress strain relations of the ETFE membrane and the joints were experimentally investigated by mechanical tests. The focus of the investigation was on the elastic stress strain relation in order to avoid permanent plastifications during the robotic fibre placement where a certain force has to be carried in order to allow the bonding between the ETFE and the carbon fibres. The mechanical properties of the carbon fibre composite were experimentally determined by tensile and three point bending tests.

Agent Based Computational Method

The positioning and arrangement of the fibres was derived from an agent based computational model agent in which a single agent acting as the robotic effector directly output approximate robotic tool paths. This agent model took several aspects into consideration including the data of the different structural analyses, the robot's working space, axis limits, and the necessity of path continuity and the location of critical boundary information such as the openings and lights.

The primary structure, which is a system of crossing fibre bundles that structurally act beams, was inspired by the biological role model. Its position and orientation was determined by the working space of the robot so that continuous paths could be laid within a single inverse kinematic configuration. The material system also informed this initial layout, where the main fibre directions were ideally 60 ° relative to each other. Refined finite element models, in which these fibre bundles were modelled as beam elements, considering the experimentally investigated tested material properties of the compound, were used to analyse and optimize the stability of this primary load bearing system, with the aim of achieving maximum stiffness by minimum weight. The desired structural depth of these primary members was revealed through this analysis, and achieved through a behaviour in which the agent would reinforce these paths until a similar structural depth was achieved.

Additional secondary fibre distribution layers were derived from an additional stress analysis in which the membrane, with entrance openings removed, was modelled with shell elements and an isotropic material. This analysis revealed both

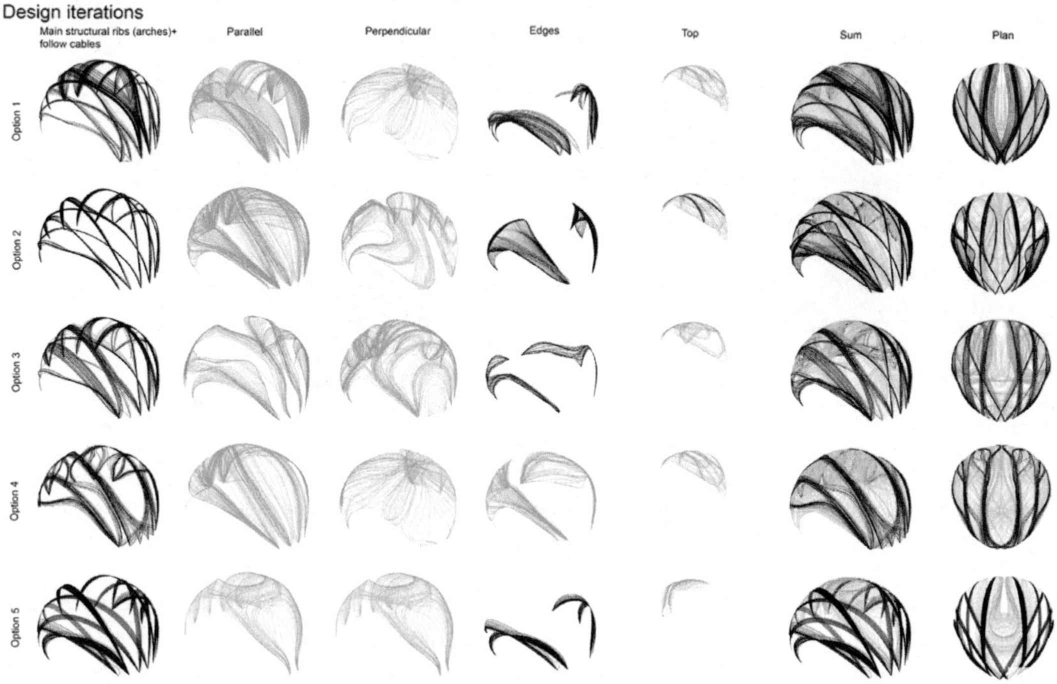

Fig. 8 Design iterations of primary and secondary fibre bundles

local stress directionality and magnitude. The agent was programmed to follow previously laid paths more closely in areas of higher stress, which led to a concentration of material, and to follow the direction of primary stress. Additional geometrically nonlinear finite element analyses were utilized to compare several different schemes for the ideal directionality and location of secondary crossing members for their impact on global stiffness and local buckling behaviour, particularly in the problematic area surrounding axis one.

Collectively these analyses and agent behaviours allow the accumulation, density and directionality of the fibre to be locally adapted to the load bearing behaviour. This approach opened up a novel load bearing system with differentiated levels of structural stiffness and transparency (Fig. 8).

Adaptive Online Robotic Process

Traditional digital fabrication processes are generally programmed off-line through instruction based commands that are static, predetermined, and inflexible. This particular production setup was extremely dynamic due to the nature of inflation and the applied loading during the process of the fibre placement. To achieve the necessary flexibility, online communication between the robot and the secondary devices of the custom robotic effector directly allowed the live modification of an intended fibre path. Online adaptive instructions negotiate critical construction information (i.e. force feedback from integrated sensors) to maintain criteria for fibre extrusion on the active pneumatic membrane.

With each iteration during the construction process, the applied loading force is fed back in the design environment, where a retract vector is recalculated in the current tool coordinate system of the robot. The greater the active force deviated from a target or threshold force, the greater the correction movement is, allowing the path to deviate from the original intended fibre path. Concurrently, the processing environment calculated the current speed and streamed it to the motor, allowing synchronization between the robot and the extrusion, and the pressurized spray device. Additionally, the actual fibre path was continuously recorded, providing the input data that was re-incorporated into the next iteration of the agent based model as was described above.

Result and Conclusions

The integrated design approach allowed to realise a differentiated fibre composite architectural demonstrator. An inflated foil that serves after completion as envelope was used as formwork. The fabrication process was inspired by the biological role model of the water spider. The outlined design and fabrication methods were developed and tested through the construction. The demonstrator covers an area of 40 m^2, encloses a volume of ca. 125 m^3 with a maximum span of 8.4 m and a height of 4 m. The fabrication strategy allowed an automated and simultaneous placement of 9 pre-impregnated, twisted 48 K carbon fibres rovings with a speed of 0.01 m/s respectively 1 kg/h.

The load path adapted placement of the fibre composite material and repercussions of the fabrication strategy are expressed and visible in the architectural outcome. The resulting novel load bearing system and design repertoire is therefore inseparably linked with performative criteria and based on innovation within the design and fabrication processes. Conventional distinctions of pre conceived building elements were eliminated within such an integrated structure as exemplified by the featured gradual transitions in fibre density and the respective correlations between structural capacity and transparency. With that the typical linear process of design and construction of architectural projects is overcome by the integration of all influencing parameters in a closed digital chain.

Potential trajectories of further research could be the development towards a multi layered structure with insulation capacity and functional integration of required building technology like heating and shading. If applied to large scale structures, the fabrication strategy needs to be modified towards a cyber-physical fabrication setup that integrates not only multiple cooperating robots, but links them with large distance material delivery systems.

Acknowledgments Thank you to all students who worked on the research project, in particular Emily Scoones for her heroic effort in coordinating the project. Thank you to the other researchers Ehsan Baharlou, Benjamin Felbrich, Manfred Hammer, Axel Körner and Anja Mader for their effort and time spending on the building site.

References

Dörstelmann M, Parascho S, Prado M, Menges A, Knippers J (2014) Integrative computational design methodologies for modular architectural fibre composite morphologies. In: Gerber D, Huang A, Sanchez J (eds) Design agency, proceedings of the 34th annual conference of the association for computer aided design in architecture (ACADIA), Los Angeles, pp 219–228

Knippers J (2013) From model thinking to process design. Archit Design 02(2013):74–81. doi:10.1002/ad.1558

Neumann D, Kureck A (2013) Composite structure of silken threads and a proteinaceous hydrogel which form the diving bell wall of the water spider Agyroneta aquatica. SpringerPlus 2013. doi:10.1186/2193-1801-2-223

Prado M, Dörstelmann M, Schwinn T, Menges A, Knippers J (2014) Coreless filament winding: robotically fabricated fibre composite building components. In: McGee W, Ponce de Leon M (eds) Robotic fabrication in architecture, art and design 2014. Proceedings of the robots in architecture conference 2014, University of Michigan, Springer, pp 275–289

Reichert S, Schwinn T, La Magna R, Waimer F, Knippers J, Menges A (2014) Fibrous structures: an integrative approach to design computation, simulation and fabrication for lightweight, glass and carbon fibre composite structures in architecture based on biomimetic design principles. CAD J Elsevier 52:27–39. doi:10.1016/j.cad.2014.02.005

Waimer F, La Magna R, Reichert S, Schwinn T, Menges A, Knippers J (2013) Integrated design methods for the simulation of fibre-based structures. In: Gengnagel C, Kilian A, Nembrini J, Scheurer F (eds) Rethinking prototyping, proceedings of the design modelling symposium Berlin 2013, Verlag der Universität der Künste Berlin, pp 277–290

Automated and User Controlled Variation and Optimization of Grid Structures

Eva Pirker

Abstract

This paper discusses newly developed, interactive structural design tools and investigates the capability of design variation, optimization and material reduction of grid structures due to loop procedures. Most existing beam grid structures have same cell sizes and the members have the same cross sections; hence these structures have a large number of member cross sections, which are not adequately utilized. There is potential to optimize material usage in grids. The optimization and variation process in this research results in a material redistribution that reflects the range of internal forces in structural members due to cross sectional or grid density (spacing) variation. The paper presents the development of interactive design tools and uses the case study of rectangular grids as a proof of concept. The aim of the tools is to allow users to integrate structural aspects while maintaining the constructional logic and controlling all spatial design aspects as well. The parametric design tools are developed in Grasshopper using custom made C# scripts. The parametric geometry models are linked to the finite element program RFEM and enable an iterative design and optimization process based on the static analysis as well as on the user controlled parameters.

Introduction

Grid structures are widely used to create architecturally interesting spaces with flexible plan layouts. They can be used to span ranges where plates cannot be used efficiently. Grids, similar to plates, enable biaxial load transfer and also have a two dimensional appearance. The spanned space is not visually divided by single beams as it would be when using slab and beam structures. The supports can be located at any member intersection point. Grid structures are typically built in a highly repetitive manner. The roof of the Neue Nationalgalerie Berlin by Ludwig Mies v. d. Rohe completed in 1968 is a well-known example. Yet, there is great potential for structural

E. Pirker (✉)
Institute of Structural Design,
Graz University of Technology, Graz, Austria
e-mail: eva.pirker@tugraz.at

M.R. Thomsen et al. (eds.), *Modelling Behaviour*, DOI 10.1007/978-3-319-24208-8_21

variation and adaption to individual design concepts as illustrated by the workshop space of the Kanagawa Institute of Technology designed by Junyai Ishigami and Associates. A grid structure, partly filled with glass, is one of its central elements. It covers a big indoor space that gives the impression of being outside (Mehta and MacDonald 2011). The rectangle beam grid is supported by thin columns positioned in an irregular fashion. This is enabled by additional grid members which also enhance irregularity. The big indoor working space is divided into zones only by the large number of columns.

From a structural point of view, grids allow for wider spans and are more material efficient than solid plates. Standard repetitive grid structures with beams of the same cross section, however, are only efficient for small portions of the structure where member cross sections are fully utilised according to internal forces (Fig. 1).

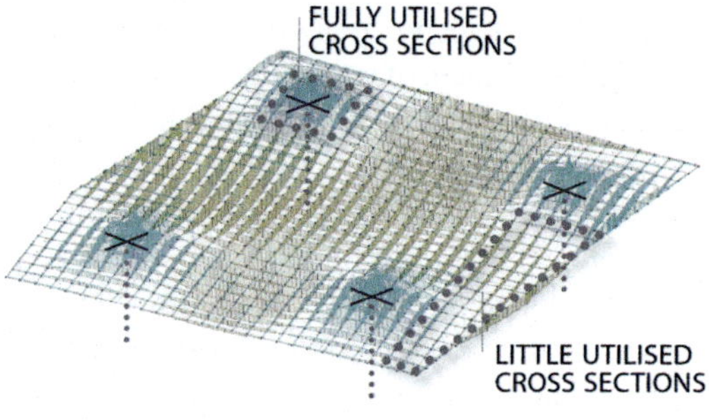

Fig. 1 Repetitive grid—internal moments

Research by Weilandt and Tessmann (2013) and Sundermann et al. 2013 has studied the potential of optimizing material usage in the structure through a variation of the beams' cross sections. The load bearing behaviour can also be influenced by the grid density (spacing of the beams) (Fig. 2). The production of optimized structures, however, often leads to complex and costly fabrication effort due to a large number of individual construction elements. Such nonstandard structures (Dunn 2012; Dimcic 2011) are often designed using genetic algorithms. Graphic statics is another method that can be used for the optimization of beam structures, as demonstrated in (Beghini et al. 2013) for trusses. The optimization resulted in an irregular layout of truss members and, consequently, in a noticeable material reduction. Also topology optimization typically generates solutions which are difficult to construct. Zhu et al. (2014) studied topology optimization of frames incorporating constructability by restricting brace types; however, the presented case study demonstrates a distinct stiffness reduction compared to optimization without these restrictions.

Optimization of grid like geometries can also be maintained by particle swarm optimization (PSO). Research by Nimtawat and Nanakorn (2011) dealt with beam-slab layout design for rectangular floors based on a discrete version of PSO. The objective function was developed considering the efficiency of column utilization. The results were only influenced by geometrical parameters.

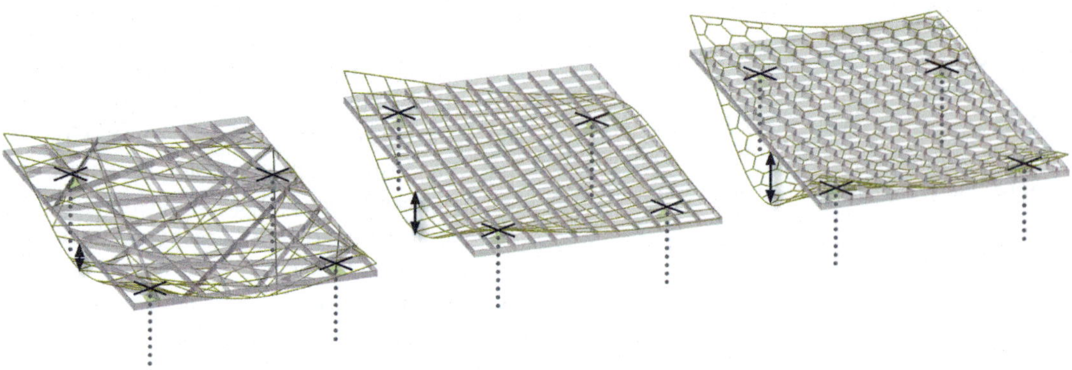

Fig. 2 Deflections—same material usage, different grid geometry/density

The optimization and variation process in this research results in a material redistribution that reflects the range of internal forces in structural members due to cross sectional or grid density variation. Constructional efficiency is maintained, however, by restraining the density variation and maintaining the constructional logic of a small family of beam types and connection geometries. The optimization of building structures often has a visual impact on the space; hence user interaction is crucial (Bülow 2012). Rather than evaluating the range of design options at the end of the design process the newly developed design tools enable influence over the design parameters during the form finding process.

Automated Loop Processes and User Input

The design tools presented here allow designers to control cross-sectional properties as well as grid density in response to the internal moments. The tools can be used in an early design stage. The user influences the design and analysis process by adapting parameters, simultaneously controlling construction relevant dimensions and visual appearance. The parametric design tools were developed in Grasshopper using custom C# scripts, allowing for the resulting geometric models (Rhinoceros/Grasshopper) to connect directly with the finite element program RFEM (Dlubal Software). In the first stage of the research this data exchange is conducted through Excel sheets. The automated integration is a next step in the project. Upon running the analysis the resulting modification of the structure displays automatically through Grasshopper. The design tools support both, the modification of the member cross section and the variation in grid spacing.

The development of the design tools is, among others, based on a consideration of possible grid structures and the resulting grid classification (Fig. 3). Grids can be categorized in structures with continuous or branching members. Generally a cross sectional variation is

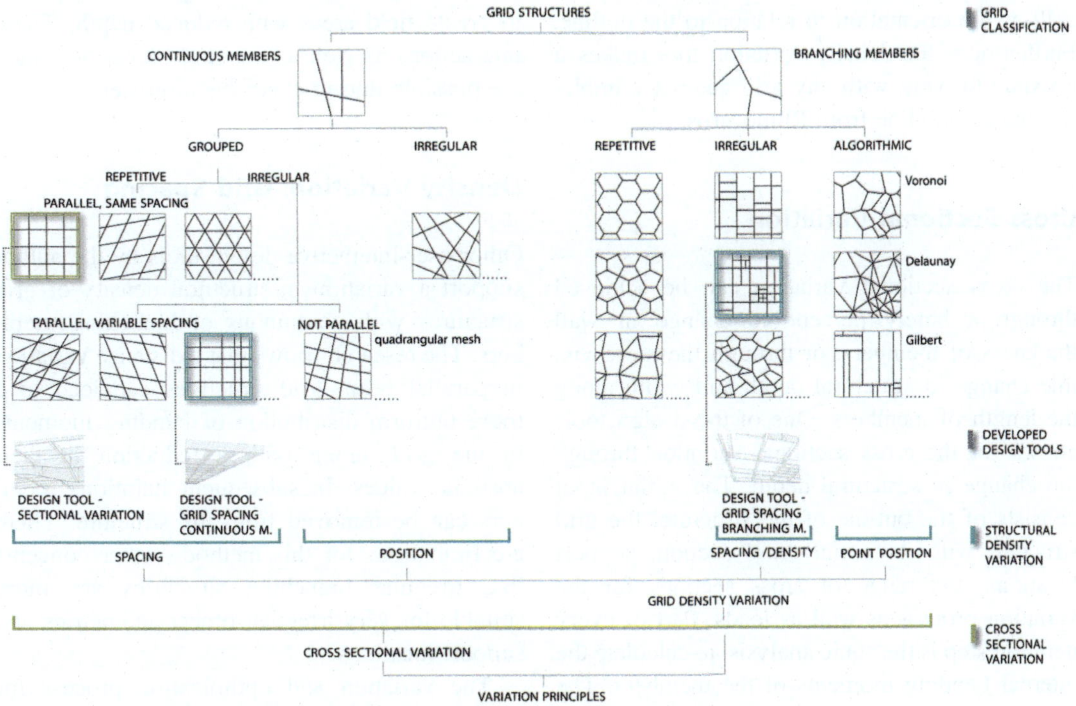

Fig. 3 Grid classification and variation principles

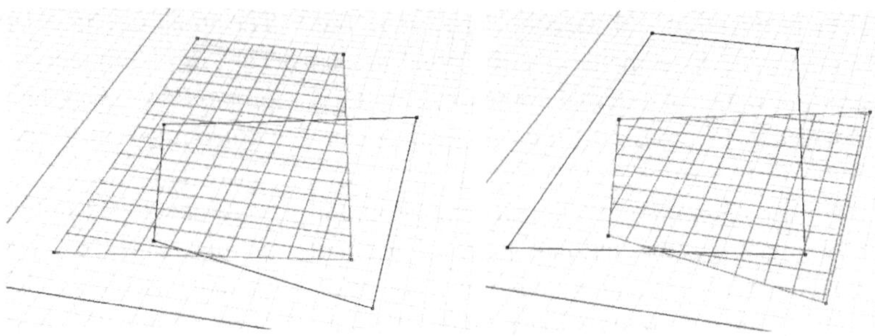

Fig. 4 Any polyline can be used as structure outline

possible for all grid structures. The grid density variation can be achieved for structures with continuous members by adapting the beam spacing and/or orientation. Branching structures can be grouped in repetitive, irregular and algorithmic based (e.g. Voronoi) grids. The density variation of these structures results in irregular geometries. The paper presents the case study of rectangular grids as a proof of concept.

The input data for both design tools is an initial grid. For the first stage the rectangular grid was used. The grid is generated inside of the outline of the structure, which can be imported as a polyline from Rhinoceros (Fig. 4). The spacing of the grid elements can be adjusted in both directions as well as the orientation in relation to the outline. Furthermore the density variation tool makes it possible to work with any grid geometry implemented as polyline from Rhinoceros.

Cross Sectional Variation

The cross sectional variation can be achieved through a barely perceptible change in wall thickness of members, or through the more visible change in structural depth and width along the length of members. One of the design tools allows for the cross sectional variation through the change in structural depth. The initial input consists of the outline of the structure, the grid structure with its initial cross section, support locations, the range of cross sections for the variation process as well as loads. Part of every iteration step is the static analysis, to calculate the internal bending moments of the members. The design tool determines cross sections due to these values and the stress level defined by the user. It is recommended to choose a stress level not too close to the material strength to prevent high deformation. To avoid multiple cross section depths at the member intersections, for each intersection all meeting members are selected and the maximum absolute internal moment is determined. The tool allocates this value to each intersecting member.

The user is able to determine a certain cross section for the members along the edges of the structure. Furthermore, the cross sectional variation can be manipulated by a user controlled parameter. So it is possible to shift stiffness from field to support areas. By doing this it is possible to create field areas with reduced depth. These interactions as well as the stress level definition are possible during every iteration step.

Density Variation: Grid Spacing

Other user-interactive design tools by the author support a variation in structural density of grid structures with continuous or branching members. The research shows that a distance variation in parallel beam grid structures can lead to a more uniform distribution of bending moments in the grid, along with a reduction in their absolute values. In subsequent iterations members can be removed from the structure. There are limitations for this method—other concepts like irregular branching structures are more suitable for very irregular object geometries and support locations.

The variation and optimization process for branching structures (Fig. 5) is based on a grid

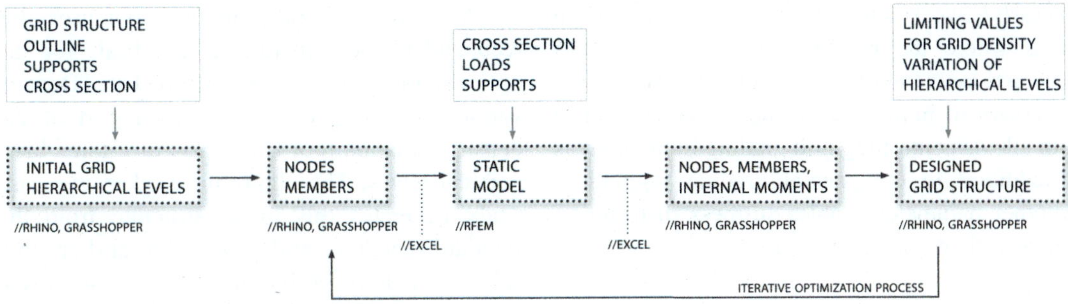

Fig. 5 Variation and optimization process

density variation corresponding to stress levels. Members positioning is automated iteratively based on the initial grid geometry while maintaining the relevant constructional logic. This process can control both the maximum density level as well as the maximum field size of the structure. To ensure a gradual density variation the initial grid has to be dissembled in hierarchical levels. The user is able to determine the grid variation and stiffness distribution by adapting the input values. During the optimization process the member depth can be reduced due to the densification in highly stressed areas.

The first hierarchical grid level defines the fixed members to ensure that also in areas, where the cross sections are not adequately utilised, the members remain on their initial position. The grid spacing of the first hierarchical level limits the maximal spans of the secondary structure. The grids of the further hierarchical levels are graded during the optimization process. All of these grids need to be shaped in a way, that their members divide the fields of grid of the prior level and that no member overlapping occurs. Dependent on the grid structure this can be achieved for example by shifting grids of different levels to each other or by shifting and scaling them. Such a grid composed of at least three hierarchical levels is needed as initial input. Grids can be loaded from Rhinoceros as polylines or an algorithm for generating hierarchical grids can be developed according to the grid structure and implemented into the design tool. In the presented case study of repetitive rectangle grids an algorithm generates a parametric initial grid.

The grading degree of each hierarchical level is controlled by a parameter which can be adapted in every iteration step. If a variation is not appropriate, it is possible to go back to a prior step. During the process the deformation is calculated as part of the analysis and has to be controlled by the user. The change in geometry is always displayed in Rhinoceros. The hierarchical level with the highest degree—the third in the case study—contains the grid with the minimal spacing. This grid is used for additional stiffness in highly stressed areas. The middle hierarchical level ensures sufficient density in average stressed zones. Further user control helps to avoid

Fig. 6 Weakening of zero moment areas; additional grading control in zero moment areas

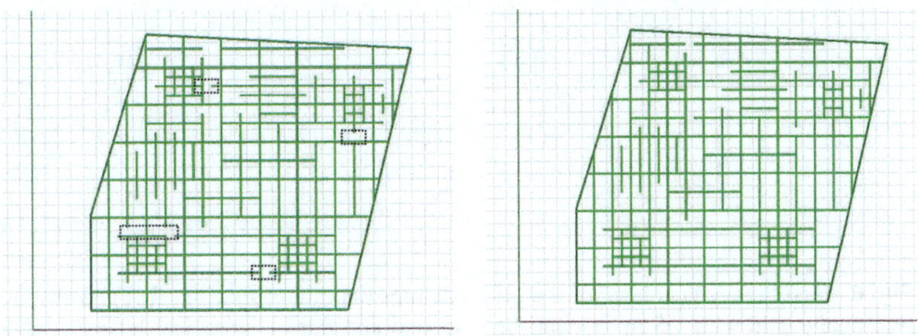

excessive weakening of particular areas, as in areas where the bending moment approaches zero and low stress levels occur (Fig. 6). So it is necessary to implement further parameters. The tool allows controlling grid grading in these areas in addition to the general grading regulation. Another parameter enables stiffness distribution between field and support areas.

Case Study: Repetitive Rectangular Grid

As a proof of concept the author studied the variation and optimization process for a rectangular grid layout within an irregular boundary. The analysis starts from a standard repetitive grid with 1.5 m grid spacing using hollow steel sections. The structure has spans of approximately 15 m and is supported by four columns.

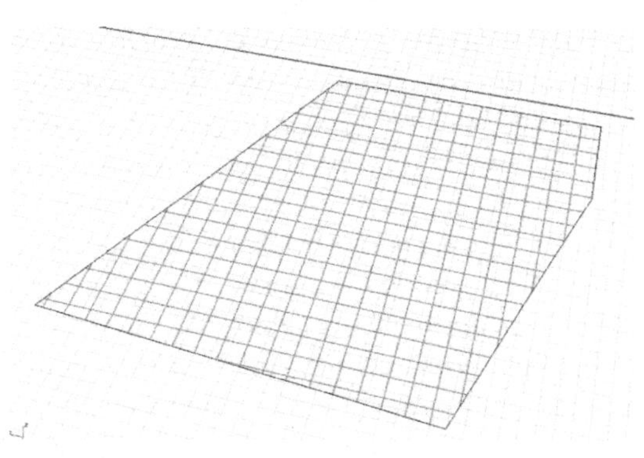

Fig. 7 Case study—rectangular grid

The cross sectional variation tool uses this rectangular grid structure as the basis for the optimization process. As described in "Density variation: grid spacing" the initial grid of the density variation tool is composed of three hierarchical levels. The algorithm used in this case study generates this initial structure from the standard repetitive grid with 1.5 m grid spacing (Fig. 7). This repetitive grid will be optimized. Members will be reduced in areas with low stress and inserted in high utilized zones. Therefore the standard repetitive grid has to correspond to the first and second hierarchical level, thus members of the second level can be reduced from the repetitive grid and members of the third level can be inserted (Fig. 8).

The user input (Figs. 10 and 12) controls the optimization process. Only a few iteration steps (Fig. 9) are needed to enable a material reduction of more than 20 % by either cross sectional (Fig. 11) or grid spacing (Fig. 13) variation compared to the standard repetitive grid. Due to the material redistribution the member depth of the density variation grid could be reduced from 38 to 30 cm.

Depending on the secondary structure (grid covering structure), grid spacing is a crucial factor. As Fig. 9 (right) illustrates many fields of the final grid have an elongated shape. The short span of these fields corresponds to the spacing of the original grid (Fig. 9, left), although the maximum spacing is limited to twice this value (first hierarchical level) (Fig. 8, left). So the standard repetitive grid can be used as reference spacing for the secondary structure and when the distance between two members is too large for a few fields, an additional member can be included.

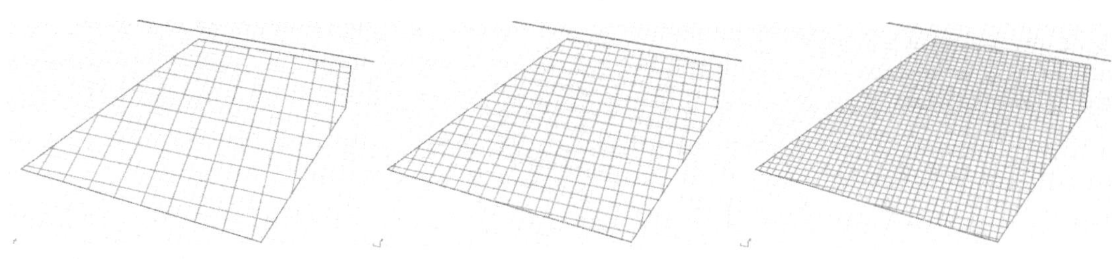

Fig. 8 First hierarchical level of the grid—defines maximum spacing; first and second hierarchical levels–equal to the rectangular grid with 1.5 m spacing; first, second and third hierarchical levels—define highest density

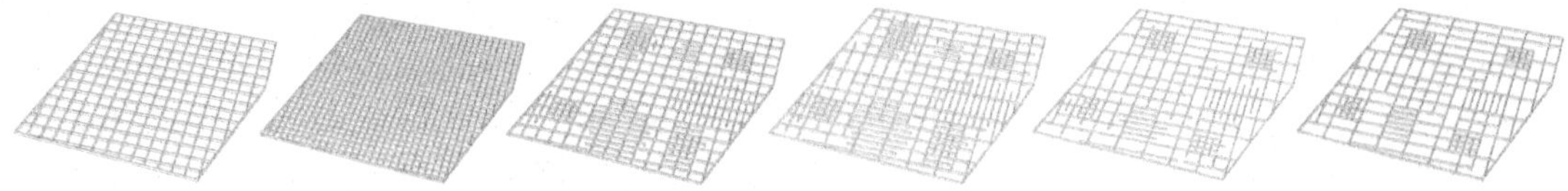

Fig. 9 Grid density variation—iteration steps: standard repetitive grid (first and second hierarchical level), initial grid with three hierarchical levels, further iteration steps, final structure

Fig. 10 Cross sectional variation—input: initial grid, outline, support location, cross section range and stiffness distribution, edge member cross section, loads

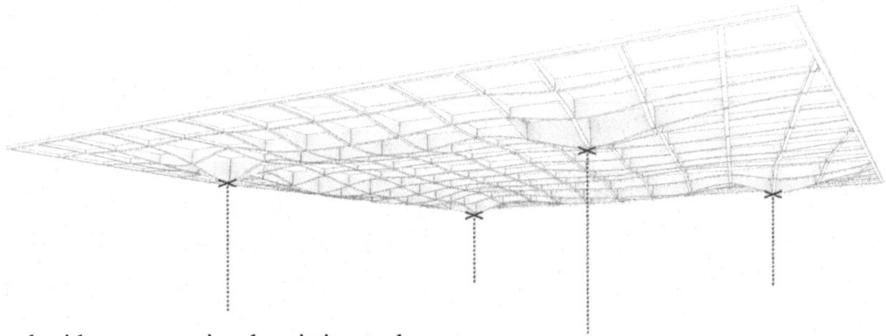

Fig. 11 Designed grid—cross sectional variation tool

Fig. 12 Density variation—input: initial grid with first, second and third hierarchical level, outline, support location, cross section, density levels and stiffness distribution, loads

Fig. 13 Designed grid—density variation tool

A comparison between the standard repetitive grid and the density variation grid shows a comparatively small difference in the number of nodes. The 253 nodes of the initial geometry were reduced by 15 nodes. Although 15 out of 253 nodes is a relatively small number, the amount of connecting members could be reduced significantly. The original structure contains 199 nodes with four intersecting members, whereas the optimized structure only contains 141 of these nodes. Instead there are more nodes with three intersecting members. Nodes along the outline of the grid with members intersecting not perpendicular to each other could be reduced from 60 to 35.

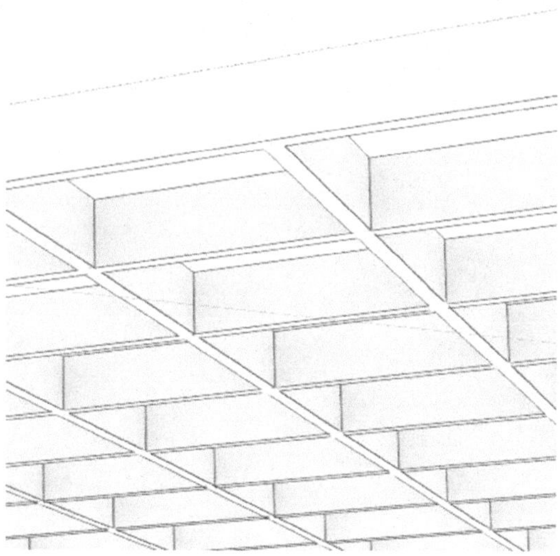

Fig. 14 Detail: initial grid

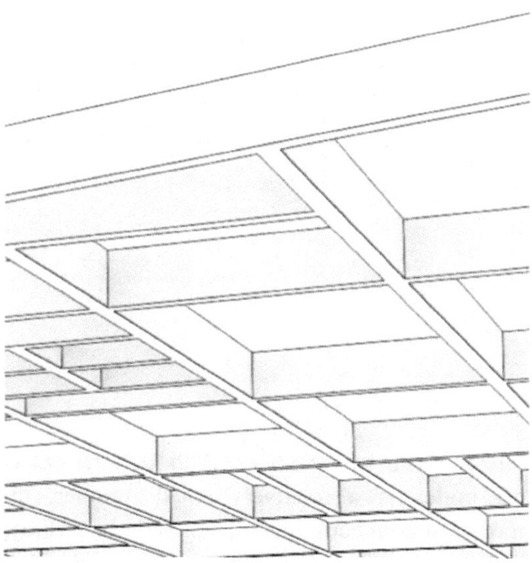

Fig. 15 Detail: optimized grid (density variation tool)

The optimization process influences the perceived visual appearance significantly. As Figs. 14 and 15 illustrate the modified structure appears less dense. This is achieved by the elongated cells and the reduced cross section height. In addition the high density areas are concentrated in the supporting areas and therefore consuming less space in the whole grid.

Conclusions

The research evolves structural form finding concepts based on automated iterative processes according to static analysis and user interaction. This paper presents two approaches for grid structure variation and optimization.

Cross sectional variation is a known principle. This paper describes a method in which the user is able to control certain items, such as the cross section of edge members or the stiffness distribution in field areas. The principle of density variation can be used to design irregular grids with reduced member depth and material, compared to an initial repetitive grid. In the case study the material reduction due to grid density variation corresponds to that of the cross sectional variation. The total number of intersecting nodes could only be reduced very little, however the number of intersecting members could be decreased.

During the density variation the reduction of members in parts of the structure leads to a weakening of the structure in these parts and therefore leading to an increased deformation. When designing the initial grid, the material stress is the limiting factor, in opposition to the optimized grid. In this case the deformation is the crucial factor for the determination of the cross sectional height. In this development stage of the design tool the deformation is monitored by the user during the static analysis. When approaching the deformation limits, the user has to react by adjusting defined parameters. A further evolution of the algorithm enhances the tool as it can optimize the structure also in terms of deformation. Therefore more iteration steps are needed which requires the automated integration of the analysis program RFEM.

Grid structures are more likely to be found as steel structures. The method of density variation could also be applied to the design of ripped concrete slabs and therefore make a contribution to resource efficiency.

References

Beghini L, Carrion J, Beghini A et al (2013) Structural optimization using graphic statics. Struct Multidisc Optim. doi:10.1007/s00158-013-1002-x

Bülow P (2012) ParaGen: performative exploration of generative systems. J Int Assoc Shell Spat Struct 53 (4):271–284

Dimcic M (2011) Structural optimization of grid shells based on genetic algorithms. Dissertation, Stuttgart University

Dunn N (2012) Digital fabrication in architecture. Laurence King Publishing, London

Mehta G, MacDonald D (2011) New Japan architecture: recent works by the world's leading architects. Tuttle, Tokyo

Nimtawat A, Nanakorn P (2011) Simple particle swarm optimization for solving beam-slab layout design problems. In: Proceedings of the twelfth East Asia-Pacific conference on structural engineering and construction, City University of Hong Kong, Hong Kong, 26–28 January 2011

Sundermann W, Eilbracht G, Holzinger C (2013) Anwendungsbeispiele zur Optimierung verformungsempfindlicher Tragwerke. Stahlbau 62:413–420

Weilandt A, Tessmann O, (2013) From structural purity to site specificity—new canopies for the entrance gates of the Messe Frankfurt. In: Gengnagel C, Kilian A, Nembrini J et al (eds) Rethinking prototyping: proceedings of the design modelling symposium, Berlin

Zhu M, Yang Y, Gaynor A et al (2014) Considering constructability in structural topology optimization. In: Proceedings of the structures congress 2014, Structural Engineering Institute of ASCE, Boston, 3–5 April 2014

Simulation Methods for the Erection of Strained Grid Shells Via Pneumatic Falsework

Gregory Quinn and Christoph Gengnagel

Abstract

The practical benefits of strained (a.k.a. 'elastic') grid shells, such as low material usage and fabrication simplicity, are undermined by the methods typically used for their erection. Established erection methods for strained grid shells ('lift up', 'push up' and 'ease down') can be time-consuming, costly and can overstress the system locally (Harris et al. in Build Res Inf 31:427–454, 2003; Quinn and Gengnagel in Mob Rapidly Assem Struct IV 136:129, 2014). The feasibility of and methodology for using inflated pneumatic cushions for the erection of strained grid shells (Otto et al. 1987) is investigated based on geometrically non-linear FE simulations and a scaled physical model for a case study of a dome with a 30 m span, 10 m pitch and constant double curvature. This paper provides a detailed write-up of the scaled physical experiment as well as the developed FE method. A detailed comparison is carried out between different erection methods for strained grid shells in order to evaluate key performance criteria such as bending stresses during erection and the distance between shell nodes and their spatial target geometry. The risk of beam-overstressing for existing erection methods along with challenges caused by modern safety restrictions, scaffolding costs and build duration can be drastically reduced or even eliminated by making use of inflated pneumatic falsework for the erection of strained grid shells. Finally it is argued that the use of pneumatic falsework has the potential to once again facilitate large-span ($L \geq 30\,\text{m}$) strained grid shell structures sch as have not been realised since the likes of the extraordinary "Multihalle Mannheim" (Happold and Liddell in Struct Eng 53:99–135, 1975).

G. Quinn (✉) · C. Gengnagel
Department of Structural Design and Technology (KET), University of Arts Berlin, Berlin, Germany

M.R. Thomsen et al. (eds.), *Modelling Behaviour*, DOI 10.1007/978-3-319-24208-8_22

Dome Case Study

A case study was performed on dome with a 30 m span and a 10 m pitch. The dome has constant synclastic double curvature (i.e. spherical) and a grid member spacing of 1 m. The dome is made of a single layer of timber laths (assumed Young's Modulus, E = 7970 N/mm^2) with a square cross section of 50 × 50 mm (same as the laths in the Multihalle Mannheim). This dome represents a stiff shell shape which was consciously chosen for its simplicity and the subsequent freedom it grants to focus on other test parameters. The mesh grid topology was generated manually using the compass method (Otto et al. 1974) due to the shell's geometric simplicity and lack of need for topological optimisation. The assumed node weight of 400 kg (3.92 N) is based on a slightly modified and lighter version of the steel bolt and plate node from the Downland grid shell (Harris et al. 2003). For the pneumatic falsework being investigated only large spans are of interest financially and practically. A span of 30 m is judged to be the minimum span suitable for when considering erection via pneumatic falsework (Figs. 1 and 2).

Scaled Physical Model

In parallel with FE simulations, a physical model was used to perform a first ever erection of a strained grid shell by means of pneumatic falsework. The scaled model makes use of acrylic beams with an 8 mm square cross section. The scale of the physical model is one tenth (1:10) of the full scale structure, subsequently it has a span of 3 m. This scale was chosen for the following reasons:

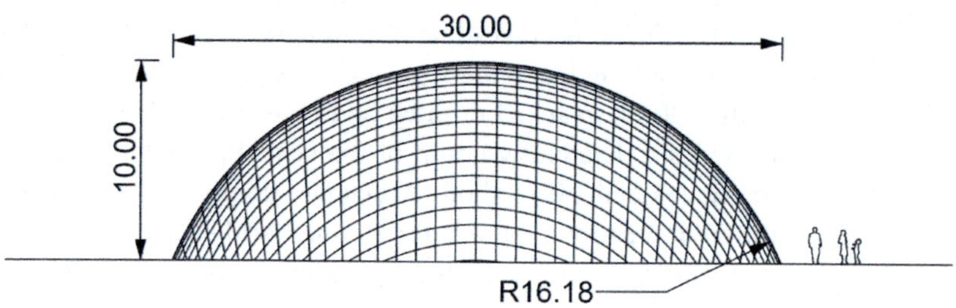

Fig. 1 Span and pitch of case study test dome made from a single layer of 50 × 50 mm timber laths

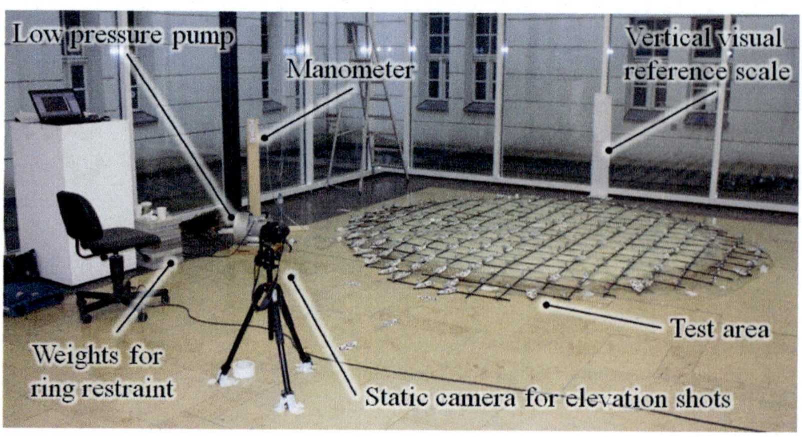

Fig. 2 Experiment set-up for the first test for the pneumatic erection of a (physically scaled) strained grid shell

1. Practicality in terms of model fabrication (e.g. welding the polyamide films for the cushion and suitable sizing for bolted joints between the acrylic beams).
2. Practicality and affordability in terms of the chosen method of photogrammetric 3D scanning technology.
3. To maintain a reasonably large volume in order to appropriate the large volume/low pressure characteristics of the full scale structure.

Dimensional Analysis which makes use of the Buckingham Pi theorem is a method for establishing dimensionless variable groups in order to reproduce physical behaviour in scaled models. This method is well documented in literature (Happold and Liddell 1975; Simitses and Rezaeepazhand 1999, p. 6; Sonin 2001) and so will not be elaborated upon here. The dimensionless group used for the scaling of this model was:

$$\frac{\omega S^3}{\left(\frac{EI}{a}\right)}$$

where ω is the load per unit area, S the span, EI the bending stiffness of the members, and a the member spacing (Adriaenssens et al. 2014, p. 93). The above dimensionless group prioritises bending stiffness which, during the erection, is the dominant behavioural property. Other dimensionless groups can be found for analysing other behavioural properties in the shell such as lateral stiffness from bracing members however this is not of relevance during the erection.

For the 3 m scaled physical model with half the mesh density and the acrylic beams, the self weight of the model needed to be increased by a mass of 104 g per node. This was achieved by means of adding steel weight (nuts, bolts and washers) to each node (Fig. 3). The stiffness of the membrane was not scaled for this model but instead membrane strain is assumed to be negligible at both full and model scales. In total 18 in-line connections were necessary for connecting beams end-to-end. These were made from brass sleeves with single M2 bolt at each beam end (Fig. 4) and their effect on local bending stiffness and global stiffness is considered to be negligible (Figs. 5 and 6).

Young's Modulus of Acrylic Beams

A Young's Modulus for the acrylic bars was determined to be 2783 N/mm^2 by means of a custom 3-point bending test (Fig. 7). Based on Euler-Bernoulli bending theory:

$$E = \frac{L^3 F}{4bd^3 v}$$

where L is the beam span (300 mm in this case), F is the force applied by the turnbuckle and read in the Newton metre, b and d are the cross

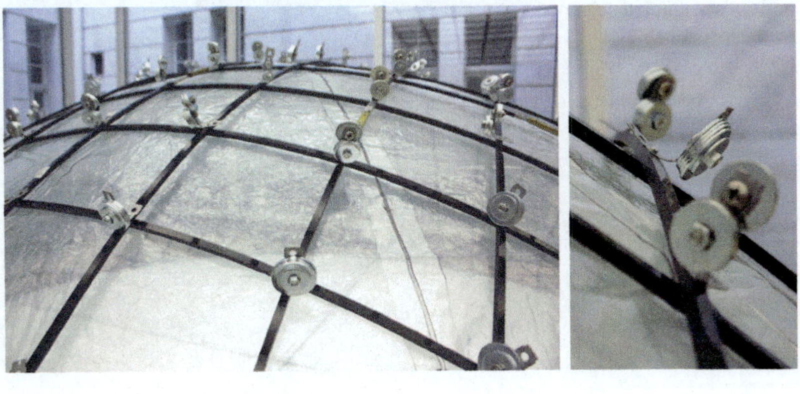

Fig. 3 Increased self weight of the grid shell by means of an additional 104 g at each node to account for scaled physical properties according to dimensional analysis

Fig. 4 One of 18 in-line connections made from a short length of a brass box-section profile one M2 bolt in each beam end creating a semi-rigid joint

Fig. 5 Low pressure blower (Greenco 0.7 kW 50 Hz) used to pump air into the pneumatic falsework using modified bicycle valves

Fig. 6 Comprehensive edge beam detailing which includes clamping strips for membrane lip, support positions with channels to feed support cables and anchorage (screw) for the cable

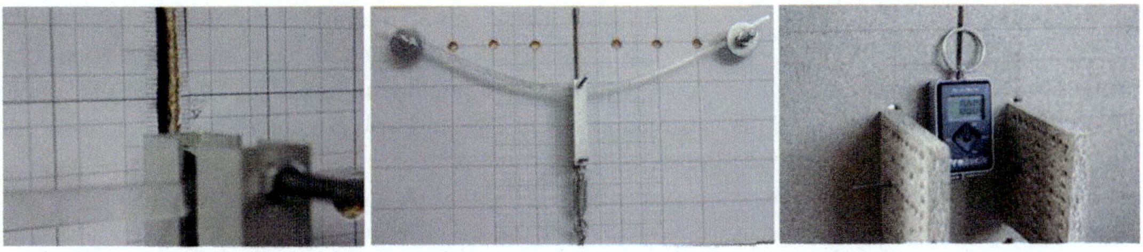

Fig. 7 Custom 3-point bending test to determine the Young's Modulus of the acrylic beams

section dimensions (both 8 mm) and v is the vertical deflection at the centre of the span.

Photogrammetry

A Rhinoceros 3D plugin called Rhinophoto (Photogrammetry plugin for Rhino, n.d.) was used to record the 3D coordinates of 83 potential nodes in the lattice. Using photogrammetry techniques, Rhinophoto calculates the coordinates of printed markers by comparing distances between them in multiple 2D photographs of the same scene. The more photos that are taken, the higher the accuracy of the scan. The software and technique was employed to good effect for the experiment. The biggest handicap in using this technique for the given application was maintaining constant air pressure in the cushion such that the structure and markers would not move during the time it took to take the photos (around 50 per scene). In order to orientate and scale the 3D point cloud during post-processing, fixed markers on the floor are referenced.

Results and Observations

The erection by means of pneumatic falsework for this shell geometry worked extremely well. Both the flat lattice and the uninflated cushion were positioned in the centre of the circular test area before inflation. During inflation the shell remained almost perfectly central with negligible sideways deviation. This is due to the distribution of self weight loading and the inherently stable form of the rotationally symmetric cushion. For asymmetric shapes, the degree of horizontal deviation (and potential need for restraint) during erection it is yet to be investigated. Figure 8 shows a snapshot of the erection process in elevation photos (below) as well as 3D coordinates from the photogrammetric scan (above).

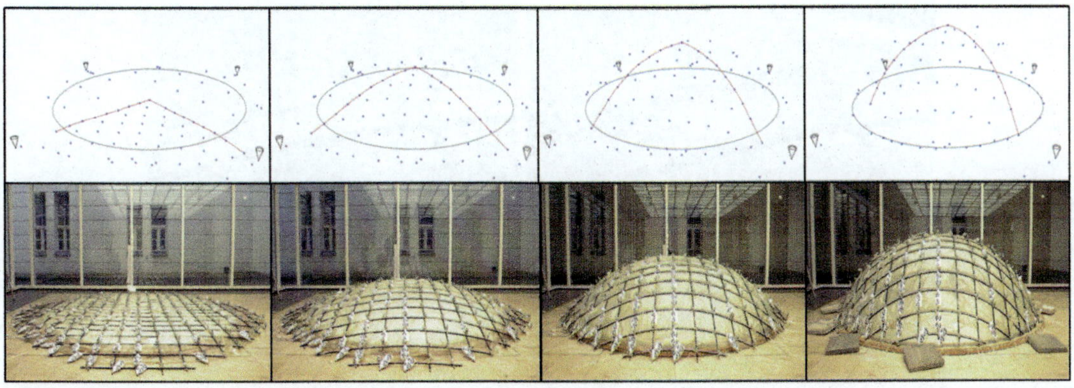

Fig. 8 Four snapshots from the inflation process with elevations photos (*below*) and photogrammetric 3D point cloud data (*above*)

Air Pressure

A total of twelve intermittent steps or "scenes" of inflation were recorded via photogrammetry. Initially, it was planned to subdivide the steps of inflation equally by observing pressure changes in the manometer however the pressure readings proved to be far too imprecise at such low pressures exhibiting variations of around ±50 % between steps. Subsequently a vertical visual reference scale was used to subdivide the inflation process from zero to complete inflation. It cannot be guaranteed therefore that the subdivisions progress with equal intervals of internal pressure change. By making use of the dimensionless scaling group it can be shown that the proportionality of pressure acting on the structure between the real and scaled models is:

$$\omega_{real} = 1.13\omega_{scaled\,model}$$

This means that the internal air pressure required in the scaled physical model is roughly the same as required for the full scale structure. In this case the internal air-pressure ranged from 0 to about 2 mbar (0.2 kN/m^2) when considering self-weight only. This is slightly lower than typical pressures necessary for air-supported membrane structures (0.25–1.0 kN/m^2) and significantly lower than pressures required for air-supported falsework for concrete shells (3.5–10 kN/m^2) (van Hennik and Houtman 2008). These initial findings show that, as long as the shell volume remains large, internal air-pressures for the pneumatic erection of strained grid shells are low and subsequently easily produced and maintained.

Despite the low air pressures in this system, leakages still occurred. In this experiment, as soon as the pump was switched off, gradual loss of air pressure occurred meaning that procedures which took a prolonged period of time (such as fixing the beam ends to their support points) lead to separation between pneumatic falsework and the grid shell (Fig. 9 left). While not critical under self-weight alone, this effect is undesirable because the grid shell has yet to be stabilised and in absence of the air pressure from the falsework, is subsequently prone to large deformations. For concrete shells supported by pneumatic falsework, maintaining a constant internal air pressure while hardening has been shown to be of critical importance due to shells' inherent sensitivity to geometrical deviation from force thrust paths (Levy et al. 2002, p. 39). Over- or under-inflating the cushion can result in global stability failure.

Uplift

During the final stages of the inflation as the air pressure increased, the ring beam began to lift up from the ground due increasing surface tension in the membrane resulting in an uplift force along

Fig. 9 Separation between pneumatic falsework and grid due to gradual deflation in end state (*left*). Re-established contact between cushion and grid after re-inflation (*right*). N.B. wrinkles on surface are from protective film and not the cushion itself

Fig. 10 Uplift at membrane edge-restraint due to internal pressure. *Left* ring beam loaded and restrained against uplift. *Right* Illustrative lifting of edge beam

the perimeter edge. Figure 10 (right) shows the perimeter edge lifting up during erection and then being weighed down with paving slabs to resist uplift (left). In early FE simulations using SOFiSTiK the vertical uplift force along the edge was shown to be around 0.14 kN/m. This uplift will have an effect on the design and detailing of the perimeter foundations which have to perform under a variety of loading situations.

Comparison with Finite Element Simulations

FE Simulations

The distance between the position of the grid shell and its target geometry during the erection in the scaled physical model was compared with three other established erection methods (Fig. 11) which were simulated using FE software SOFiSTiK. While the number of lifting points for cables in the 'lift up' method or spreader beams in the 'push up' method is arbitrary and subject to subjective design choices, a sensible quantity of 12 lifting points were simulated in order to ensure a fair comparison between the erection methods.

From a total of 83 photogrammetric marker points in the physical model, 55 markers were located in a single quarter of the shell. In order to increase computation speed and stability only a quarter model of the grid shell was simulated. Of the 55 marker points available in the quarter, only 42 were recorded successfully in each photogrammetric scene. Subsequently the distance-to-targets was evaluated based on this selection of 42 nodes.

Erection Stages

For both the computational and physical models, the erection was divided into three separate stages: (1) Erection, (2) Beam ends to supports, and (3) Relax. In the second phase the beam ends are pulled to their respective support coordinates in the simulation by means of single-element cables with very low transient stiffness properties such that under increasing load increments of pre-stress they shorten in length; i.e. the elastic cable approach (Lienhard et al. 2014).

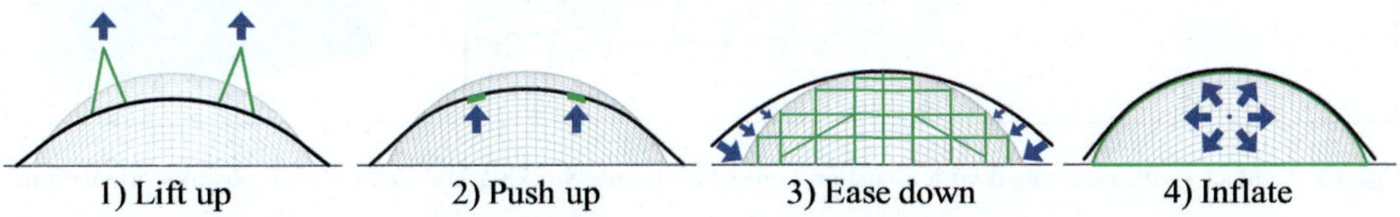

Fig. 11 Schematic representation of three established (*1–3*) and one novel (*4*) methods of erecting strained grid shells

A convergence to null for all cables was not computationally feasible. The average geometric deviation between the beam-ends and the target support points after cable contraction was 77 mm (0.26 % of 30 m span) for the 'lift up' method and 42 mm (0.14 % of span) for the 'push up' method. Such small deviations of the support positions are considered to have negligible effect on the global stiffness of the structure. For 'ease down' the pulling cables are not used and so there was zero geometric error at the support points. The 'relax' phase of the erection procedure allows the grid shell to find a new equilibrium after the beam ends have been fixed and the erection aids have been removed. In order to evaluate larger differences between the simulations, the shell was not stabilized (by cables or a third layer of timber laths) in the final 'relax' stage as would normally be the case for grid shells.

Coupling Elements and Contact Springs

Unlike most structural elements in SOFiSTiK, the geometric orientation of springs and coupling elements does not update after each iteration step. Instead their originally-defined orientation is maintained. Subsequently coupling elements cannot be used to simulate the pinned connection between the beam layers. Instead custom cross section properties (with very high bending stiffness and very low torsional stiffness) are defined for connecting beam elements (Fig. 12).

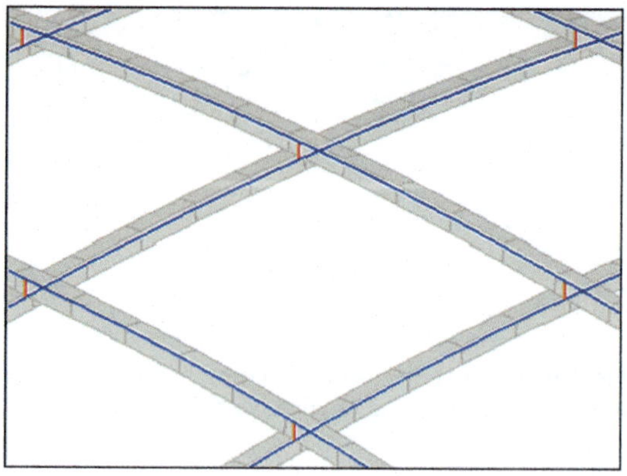

Fig. 12 Custom section properties for 'pin' beam connector elements between timber lath layers

During the simulation of the erection it is important that the laths make contact with the ground in order to accurately recreate the draping effect. Without this effect, incorrect curvatures would be generated for the beams. A custom working law was defined for the contact springs which are very stiff under compression but inactive under tension (Fig. 13, left). The orientation of the floor contact springs is defined vertically (Fig. 13, centre) and, for the contact between grid and pneumatic falsework, a local orientation is defined (Fig. 13, right). Since the orientation of spring elements is not updated after each iteration in SOFiSTiK, the contact between

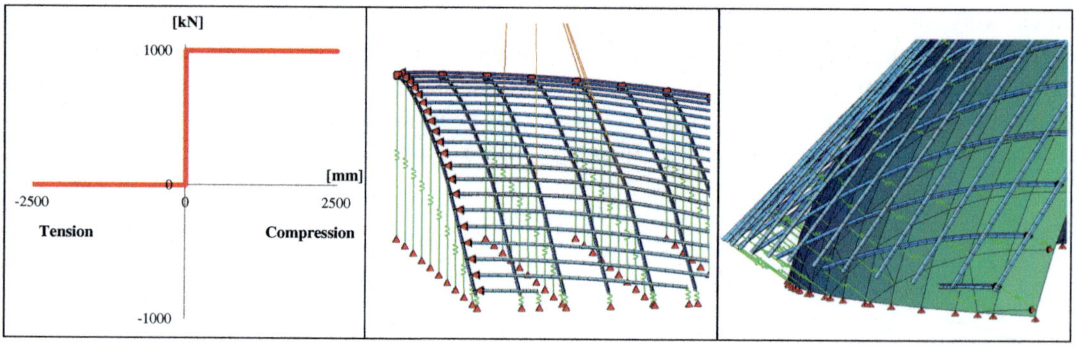

Fig. 13 Contact springs described with a custom compression-only working law (*left*), with a positive orientation

pneumatic falsework and grid shell can be simulated with accuracy only if small deflections occur.

Results

As can be observed in Fig. 14, the simulations show categorically that, of the established erection methods, the 'lift up' method is most likely to result in overstressing of the laths during erection. The maximum stress of the laths for the 'lift up' method is significantly higher than in the shell's final state. So much so, that the ultimate strength of the material is exceeded which can lead to breakages or permanent damages of the laths; a phenomenon that was documented to have occurred in the Essen grid shell (Otto et al. 1974), in the Multihalle Mannheim (Happold and Liddell 1975) and even to a small extent in the Downland and Weald grid shell (Harris et al. 2003). The results also demonstrate clearly that the 'ease down' method is the most controlled, the most precise and also the least strenuous on the laths. The bending stresses do not peak during erection, but instead rise in a controlled manner. Multiple contact points from the scaffold are much more effective at spreading forming loads than singular point loads from lifting cables and pushing platforms.

Figure 15 shows that the 'ease down' method of erection is clearly the most precise, achieving nearly perfect convergence with the target geometry. During the erection (stage 1), the distance to target for the 'ease down' method starts off closer to the target because of the grid being laid out above the shell and not on the floor. Upon closer inspection, Fig. 16 shows that the rate of reduction of distance to target geometry for the "inflate" method is more erratic that the other methods. This is due to the fact that the pneumatic falsework can shift and find multiple intermittent states of equilibrium as it inflates. This is also a result of the lower level of precision in the physical model compared with the digitally created simulation model.

Figure 17 shows that erection by means of pneumatic falsework, at least for this initial experiment, is the least precise in terms of converging with the target geometry; on average 278 mm deviation from the target nodes. However, it should be noted that the FE simulations do not take the effects of friction at the nodes or between beams into account. While some deviation from the target geometry may not be critical

Fig. 14 Equally-scaled stress distribution plots shown at the end of phase 1 (erection) for the 'pull up' (*left*), 'push up' (*middle*) and 'ease down' (*right*) methods. Dominant stress concentrations can be seen in the 'pull up' and 'push up' methods whereas the 'ease down' method shows evenly distributed stresses. Only beams in one of the two directions are show here for graphical clarity

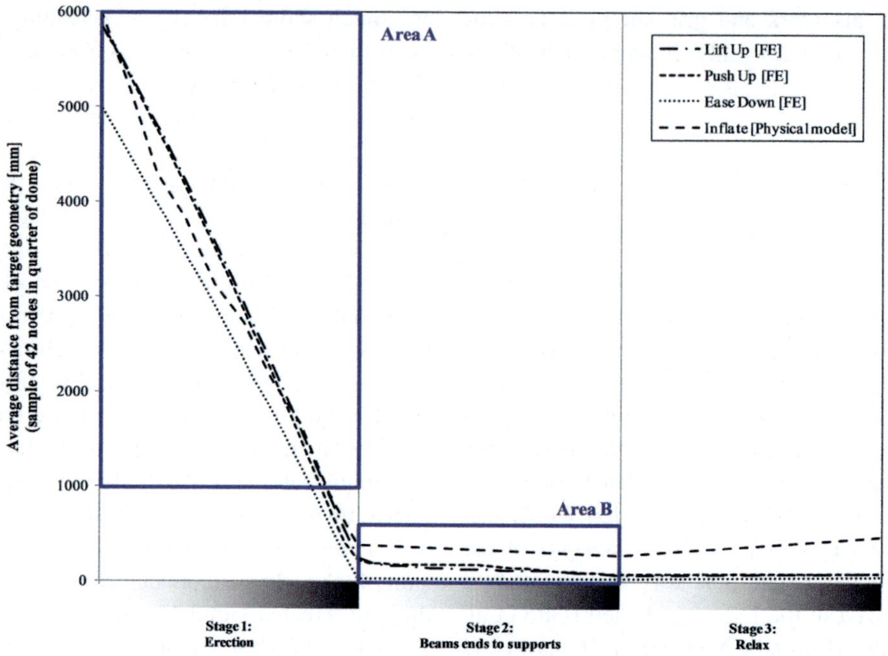

Fig. 15 Average distance from target geometry during erection of a strained grid shell in the shape of a simple rotationally symmetric dome for four different erection methods. Data for 'lift up', 'push up' and 'ease down' from FE simulations whereas data for 'inflate' from scaled physical model

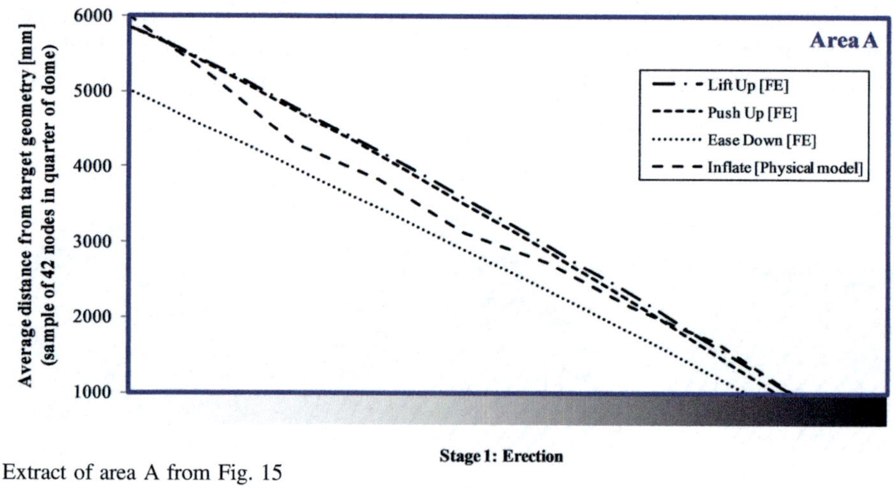

Fig. 16 Extract of area A from Fig. 15

for the shell structure, it can lead to a reduction in global stiffness and as a rule the deviation from target geometry should be minimized as far as possible.

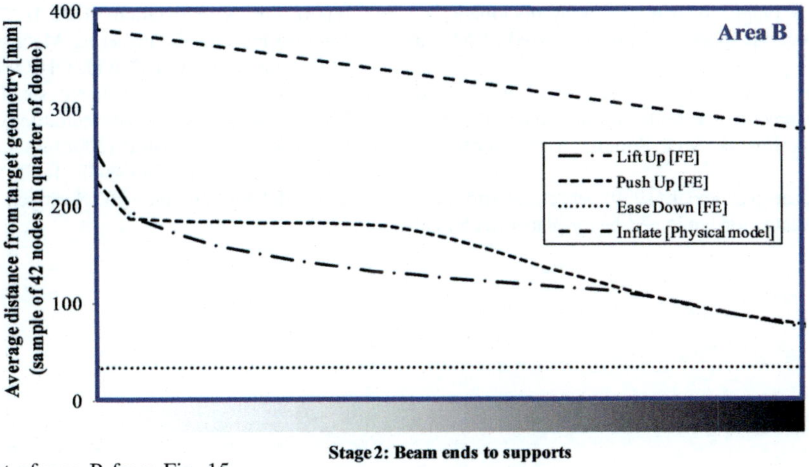

Fig. 17 Extract of area B from Fig. 15

Conclusion

The work presented in this paper demonstrates a proof-of-concept for the novel proposition of using pneumatic falsework for the erection of strained grid shells. Within the context of a geometrically simple strained grid shell, it has been shown that the 'lift up' and 'push up' erection methods are likely to cause overstressing of the laths during erection and that 'ease down' is the most geometrically precise and the least structurally demanding method. However, an alternative method of using pneumatic falsework offers comparable precision but at a fraction of the speed and cost necessary in the 'ease down' method resulting from high scaffolding demands. An accurate and viable simulation approach in SOFiSTiK is detailed and discussed for the established erection methods as well as the contact interaction between a strained grid and underlying pneumatic falsework.

By means of a scaled physical model it was shown that the method of pneumatically erecting a strained grid shell is feasible and practical for simple shell shapes and curvatures. Important behavioural phenomena such as membrane uplift and separation zones are remarked upon. Further investigations are underway in order to better understand which varieties and extremities of shell sizes and shapes are compatible with the pneumatic erection method.

Acknowledgements Special thanks is extended to student assistant Franziskus Rau for his committed and resourceful support for this work.

References

Adriaenssens S, Block P, Veenendaal D, Williams C (eds) (2014) Shell structures for architecture: form finding and optimization. Routledge, New York

Burkhardt B, Otto F (1987) IL 13, Multihalle Mannheim. Krämer, Stuttgart

Happold E, Liddell WI (1975) Timber lattice roof for the Mannheim Bundesgartenschau. Struct Eng 53:99–135

Harris R, Romer J, Kelly O, Johnson S (2003) Design and construction of the Downland Gridshell. Build Res Inf 31:427–454

Levy M, Salvadori M, Woest K (2002) Why buildings fall down: how structures fail. Norton, New York

Lienhard J, La Magna R, Knippers J (2014) Form-finding bending-active structures with temporary ultra-elastic contraction elements. Mob Rapidly Assem Struct IV 136:107

Otto F, Burkhardt B, Hennicke J (1974) IL 10—grid shells (10). Institute for Lightweight Structures, University Stuttgart

Photogrammetry plugin for Rhino [WWW document], n. d. http://www.rhinophoto3d.com/. Accessed 31 March 2015

Quinn G, Gengnagel C (2014) A review of elastic grid shells, their erection methods and the potential use of pneumatic formwork. Mob Rapidly Assem Struct IV 136:129

Simitses G, Rezaeepazhand J (1999) Structural similtude and scaling laws for laminated beam-plates (technical report no. NASA Grant NAG-1-1280). Aerospace Engineering and Engineeering Mechanics, University of Cincinnati, CINCINNATI, OHIO 45221

Sonin AA (2001) Dimensional analysis. Technical report, Massachusetts Institute of Technology

Van Hennik PC, Houtman R (2008) Pneumatic formwork for irregular curved thin shells. In: Textile composites and inflatable structures II. Springer, Berlin, pp 99–116

From 3-Point-Constellations to Self-organizing Folded/Bent Spatial Configurations

Günther H. Filz and Stefan Kainzwaldner

Abstract

Our investigation is based on a seemingly simple phenomenon that we all know from kinking sheets of paper. In this sense it is related to self-organizing processes and forms. The controlled elastic deformation of thin, planar sheet elements into 3 dimensional objects describe above mentioned phenomenon by predefining 3 points (MAB) on the surface with M fixed in space and identified shape characteristics of the short self-organizing curved ridge-line, which is smoothly fading to an elastically bent surface. This paper presents the geometrical and kinematic simulation of controlled elastic deformation of planar sheet elements of any shape into self-organizing folded/bent spatial configurations by predefining any 3-point-constellation (M, A, B) with A, B equidistant and symmetrical to M (Fig. 1). The geometrical description of spatial configurations in all transition phases and its kinematic simulation is based on a microscopic analysis of physical models. From this we receive a line-pattern, which can be applied to the sheet-element as a basis for simulating spatial movements of surface-points. At present simulation and physical models show maximum 5 % deviation only. Several tests and various 3-point-constellations verified that our simulation can be a powerful and reliable tool in exploring design freedoms as well as limits, which will be extended to more complex arrangements.

Context

Our investigation is based on a seemingly simple phenomenon that we all know from kinking sheets of paper. As illustrated in Fig. 2, a characteristic curved ridge-line self-organizing arises when a slightly vaulted sheet of paper kinks in

G.H. Filz (✉) · S. Kainzwaldner
Faculty of Architecture, Institute of Design/unit koge. Structure and Design, University of Innsbruck, Innsbruck, Austria
e-mail: guenther.filz@uibk.ac.at

M.R. Thomsen et al. (eds.), *Modelling Behaviour*, DOI 10.1007/978-3-319-24208-8_23

Fig. 1 Visualization of six simulated, additively arranged sheet elements in fully closed state

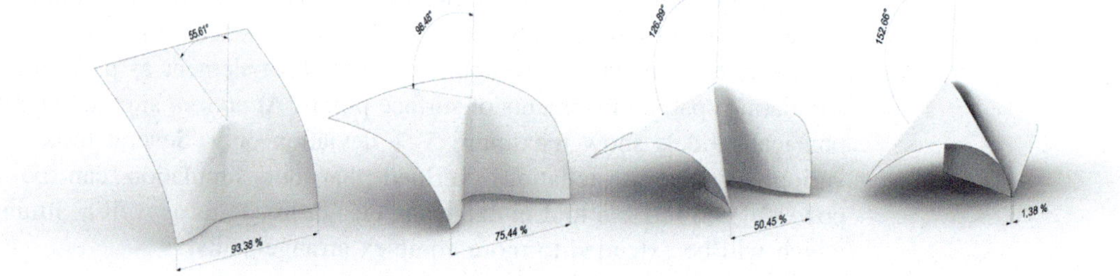

Fig. 2 Simulation of self-organizing folded/bent spatial configurations by predefining any 3-point-constellation—inclination angle αI in M in the range of 0° to maximal 152.66° for 1,0 mm copolyester and angle αMAB = 90°

perpendicular to the vault-direction. With regard to convertibility, minimized material use, structural performance and architectural qualities its basic geometric and structural dependencies as well as formal, kinematic and bending active phenomena can be of great interest for architecture and engineering (Lienhard 2011). The authors described the phenomenon as controlled elastic deformation of thin, planar sheet elements into 3 dimensional objects by predefining 3 points (MAB) (Fig. 1) on the surface with M fixed in space and identified shape characteristics of the short self-organizing curved ridge-line, which is smoothly fading to an elastically bent surface (Fig. 3) (Filz and Kainzwaldner 2014).

General geometric specifications can be summarized as follows (Filz and Kainzwaldner 2014):

- The basic set-up is described by a 3-point-constellation (M, A, B) with A, B equidistant and symmetrical to M, which is fixed in space.
- Kinematic and spatial response of the elastic sheet—change of inclination angle αI in M (Figs. 2 and 4) in non-linear dependence of transition from straight line to a sinus-wave-like, meandering and increasingly spatial curve AB—is given due to counter directional in-plane-movement of A and B. There is need for actuating, but not shape defining impulses, which are triggering controlled elastic deformation.
- The comparison of short self-organizing curved ridge-line shows no perfect match with any other mathematically defined curve (Nettelbladt 2013; Peternell 2009; Pottmann et al. 2010).
- The surface curvature analysis provides a constant relation of convex (87 %) and concave surface area (13 %). Simultaneously the surface curvature analysis can be read as equivalent to a stress-diagram.

In general, self-organizing processes and forms contain physical properties. So specific "architected" designs follow precise rules in the genesis of form—a coalesce consideration of geometrical aspects and structural performance. Finding these rules represents the common basis for tools such as simulation, which is presented with this paper and feeds back designs in meaningful and productive ways—the designer can rely on a certain assurance.

Based on above specified findings this paper presents the geometrical and kinematic simulation of controlled elastic deformation of planar sheet elements of any shape into self-organizing folded/bent spatial configurations by predefining any 3-point-constellation (M, A, B) with A, B equidistant and symmetrical to M, as only point fixed in space. Presented simulation precisely describes any surface point and its spatial movement and therefore a continuously variable transition from plane sheet element to 3 dimensional object in real time. In this context it is possible to define any shape and size of a planar element as well as the geometrical position and orientation of 3-point-constellation (M, A, B) at will. New geometrical insights regarding these 3 point constellations at all phases from plane to object show that A respectively B are part of

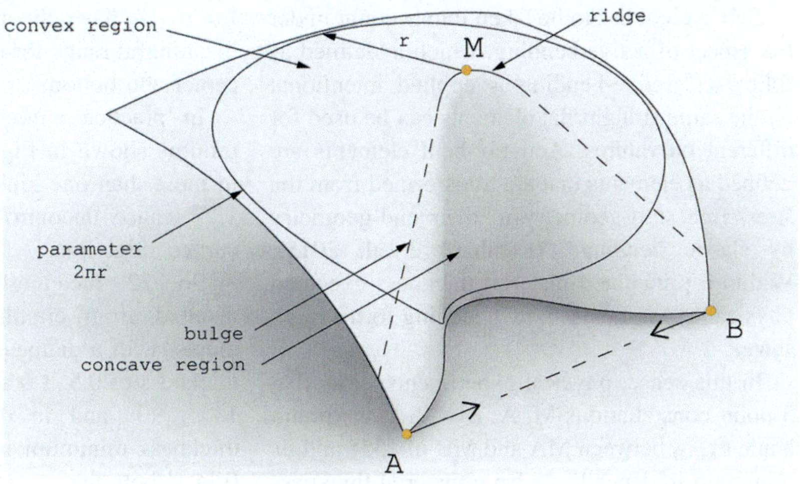

Fig. 3 Geometrical setup of predefined 3 point constellation (A, B, M) generating a self-organized, curved "ridge-line", smoothly fading to an elastically bent surface

straight lined, symmetrical legs starting in M. This simplified the evaluation of different arrangements and their digital simulation. Since M is the only spatially fixed surface point, the planar element can have any orientation in space. Thus, we are able to broadly explore freedoms and limits of these forms within the digital environment (Filz and Kainzwaldner 2015).

Physical Experiments and Resulting Constraints for Geometrical Setup

The self-organizing curved ridge-line is a result of mainly two parameters, firstly a result of physical properties of used material with material thickness to be the most determining factor and secondly a result of applied angle α_{MAB} between MA and MB (Figs. 3 and 4).

Above described phenomenon was tested by numerous physical experiments by means of elastically deformable materials including paper, respectively cardboard, veneer wood, aluminum and copolyester. The size of physical models range from few centimeters to a realized structure of 4 × 4 m shown in Fig. 5. This full-scale object was built in aluminum with a material thickness of 2 mm. For this reason investigated material thicknesses mainly range from 0.5 to 1.5 mm and focus on isotropic material such as copolyester with a Young's Modulus of 2.020 MPa according to ISO 527-2/1B/1.

Self-weight has to be taken into account under the aspect of active bending, which is defined as follows: "Active-bending is applied intentional … the same straight/flat elements can be used for different curvatures. Actively-bent elements are defined as elements that are transformed from the stress-free start-geometry to their end-geometry by elastic bending." (Lienhard et al. 2015). Within a predefined material thickness executed physical experiments show matching forms in all stages.

In this sense, physical experiments show that 3-point-constellations M, A, B with a maximum angle α_{MAB} between MA and MB of 135° and its minimum in dependence from material thickness

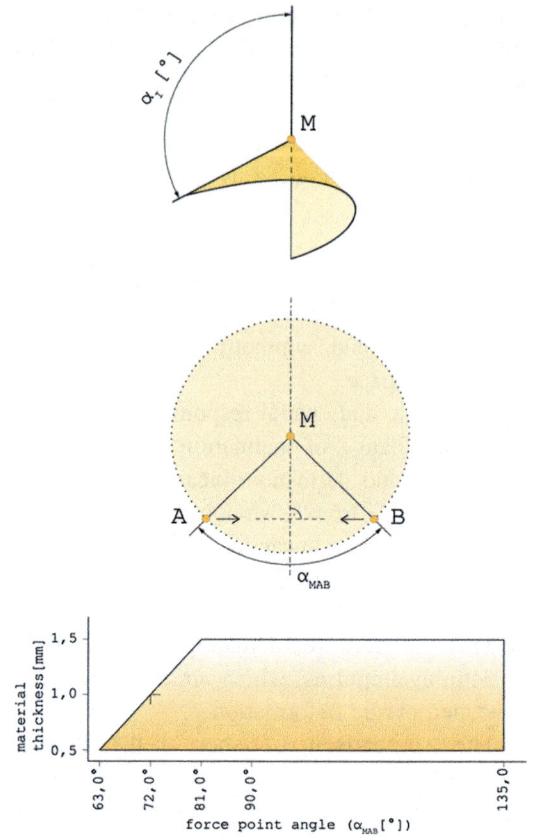

Fig. 4 *Top* Angle of inclination αI. *Center* Angle αMAB between MA and MB. *Bottom* Range of possible angles αMAB in dependence from material thickness

—e.g. 63° for 0.5 mm copolyester—result in a self-organizing curved ridge-line and therefore the objects desired kinematic effect. Other values for α_{MAB} have theoretical meaning only. The meaningful range for α_{MAB} is illustrated in Fig. 4 center and bottom.

In practice angles for α_{MAB} smaller than minima shown in Fig. 4 (bottom) usually result in more than one kink in M, angles bigger than 135° cause uncontrollable lateral spreading of curved ridge-line.

So 72 meaningful physical experiments resulted from circular copolyester sheet elements, with a diameter of 45 cm and a material thickness of 0.5, 1.0 and 1.5 mm and α_{MAB} with 135°, 90° and in dependence from material thickness minimum α_{MAB} of 63°, 72° and 81° (Fig. 4 bottom).

Fig. 5 "Series Curl", realized structure in aluminum, 4 × 4 m with a material thickness of 2 mm

Physical experiments also define following constraints for the geometrical setup of M, A, B (Filz and Kainzwaldner 2015):

- M, A, B must be surface points of predefined planar sheet element
- M, A, B must not have identical position on the sheet element
- Minimum distance of M to A, B is mainly dependent on material thickness
- A, B must not be part of self-organizing curved ridge-line

3D Scans as References

Above mentioned, meaningful physical models were 3D scanned by the means of Kinect (Fig. 6), which originally is a hardware motion sensing input device by Microsoft video game consoles (Filz and Kainzwaldner 2014). Each scanned surface is represented by a mesh, consisting of 224,500 faces in average. This equates to a resolution of about 30.75 dpi respectively a point-cloud with 0.8 mm distance between single

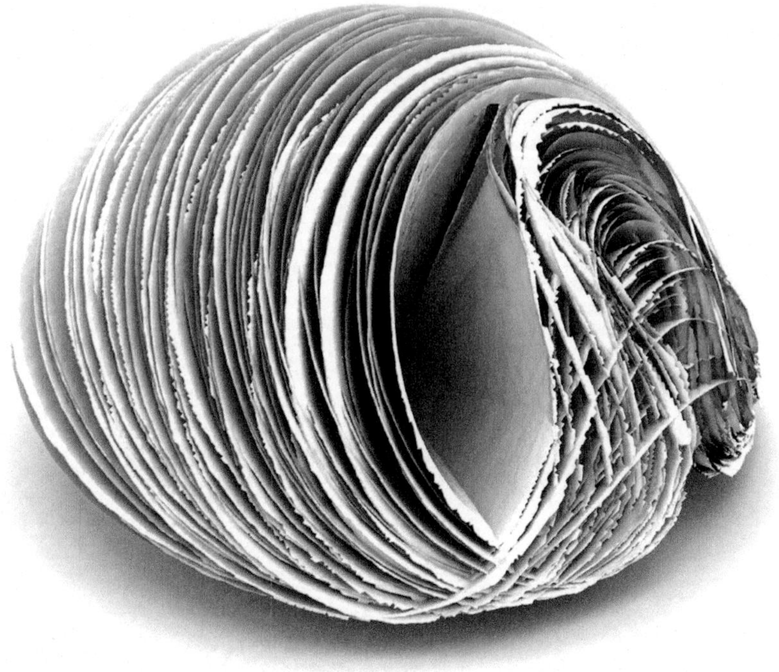

Fig. 6 3D scans by kinect during the kinematic transition-phases

points. Basically Kinect builds on software technology by using an infrared projector and camera. So this 3D scanner system employs a variant of image-based 3D-reconstruction. Due to this fact steep, angles between Kinect and the surface to be scanned impede precise surveying (Fig. 7). Microscribe-3D-Scan provides single points. This is suitable for edge-curves and foldline but less practical for surfaces, which have to be remeshed from point clouds. Therefore only selected surface-points as well as edge curves were scanned by means of Microscribe-3D scanner in order to define boundaries with higher precision and for control purposes.

Curvature Analysis

The geometrical description of spatial configurations in all transition phases and its kinematic simulation is based on a microscopic analysis of physical models (Fig. 8).

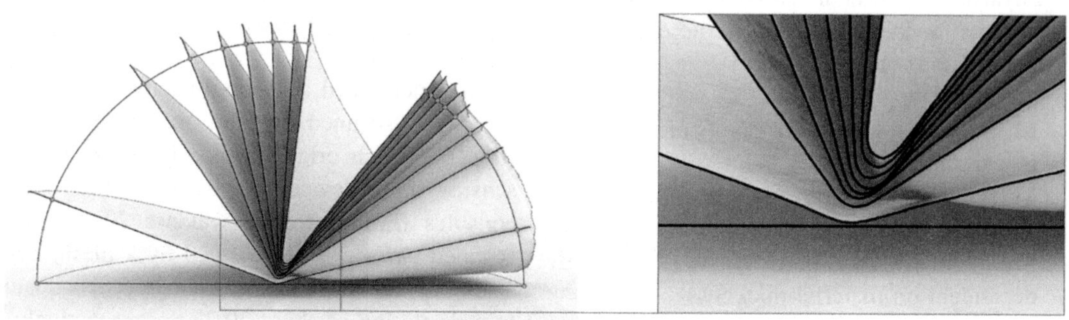

Fig. 7 Steep angles impede precise surveying—comparison of scans coupling kinect and microscribe-3D-scan

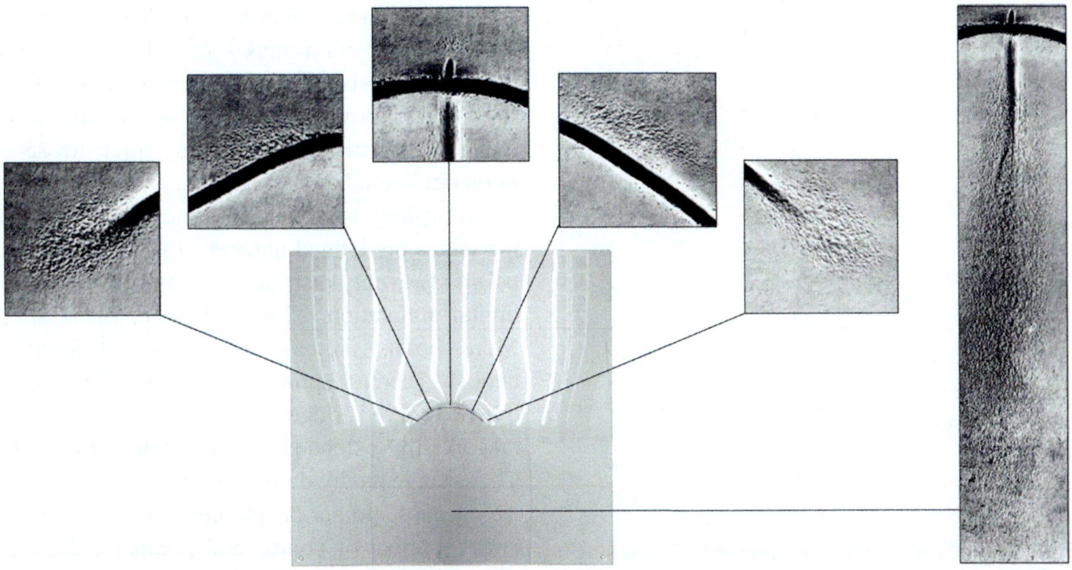

Fig. 8 Formations of microcracks in the region of the self-organizing curved ridge-line in physical model made of copolyester, 0.5 mm

Configurations of e.g. copolyester plates are showing a visual change from transparent to milky in the region of the self-organizing curved ridge-line eccentrically passing with M due to plastic deformation of material, which occurs at an angle of inclination α_I of about 55°. The microscopic analysis shows formations of microcracks with clear orientation, according to local stress peaks. So formations of microcracks match with surface curvature analysis of largest local radii of curvature (Figs. 9 and 10) and stress diagram. From this we receive a line-pattern

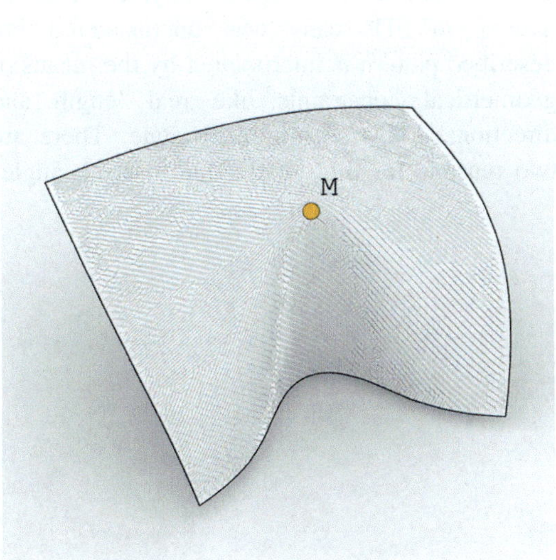

Fig. 9 Surface curvature analysis of largest local radii on 3D scan

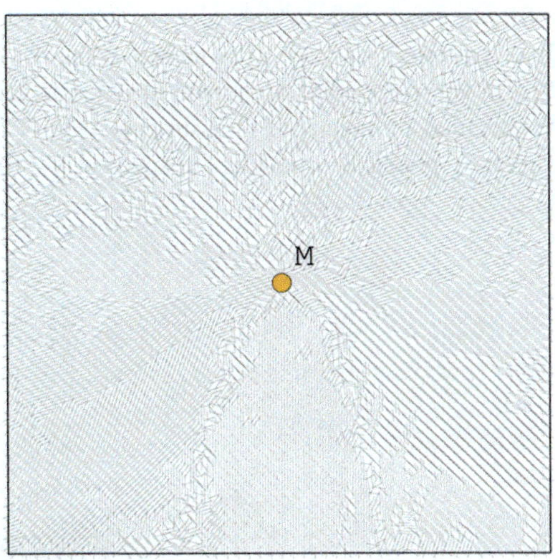

Fig. 10 Planarized line pattern of the 3D surface curvature analysis

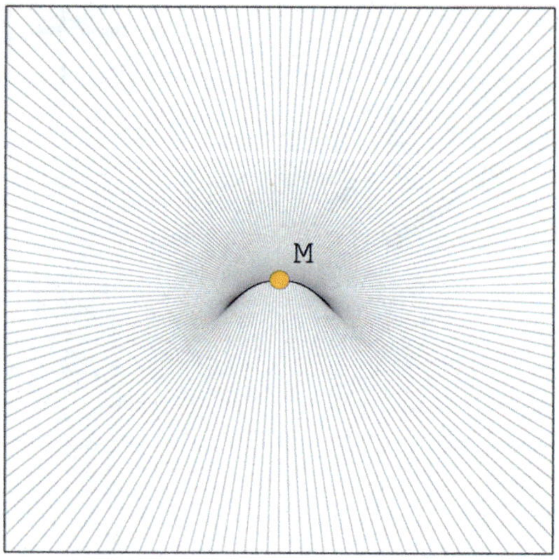

Fig. 11 Application of the curvature analysis resulting in a line pattern to simulate spatial movements of surface-points

(Fig. 11), which can be applied to the sheet-element as a basis for simulating spatial movements of surface-points.

Simulation of Geometry

Since investigated phenomenon results from a self-organized process, the access to its simulation by Origami-techniques or by means of softfolding-software, which operate with predefined patterns, was not appropriate (Zhu et al. 2013). The developed numerical simulation can be executed by means of Grasshopper (Fig. 12), whereas physical models, respectively the comparison and evaluation of 3D scans during all kinematic transition-phases served as references. Surface analysis and line pattern resulting from microcrack formations can be seen as basis for found geometrical solution in order to setup numerical simulation (Fig. 11).

According to circular elements in physical models, a predefined number of points (here 160 points) is evenly distributed along a circle in our digital simulation. This is done in plane state. More points increase precision but simultaneously increase processing time. The characteristic curved ridge-line is implemented and tangentially extended to the boundary circle. These tangents divide points along the circle into two groups, points on the convex side (usually larger number of points) and points on the concave side of curved ridge-line. The convex respectively concave side of curved ridge-line is subdivided accordingly to the number of points of each group. From this procedure we get a line pattern, shown in Fig. 11. Between boundary circle and a circle, which is bigger than the curved ridge-line, a predefined number of circles (in our case 6 circles) are concentrically arranged. This pattern is transferred to all kinematic transition-phases of the element and mapped onto scanned surfaces (Fig. 12 left). In the area of the smallest circle, which is about 9 % of the total area content of investigated physical models, data from 3D scans was unconsidered but described pattern is interpolated by the means of geometrical constraints like real length and directions, known from plane state. There are two reasons for this, firstly due to steep angles

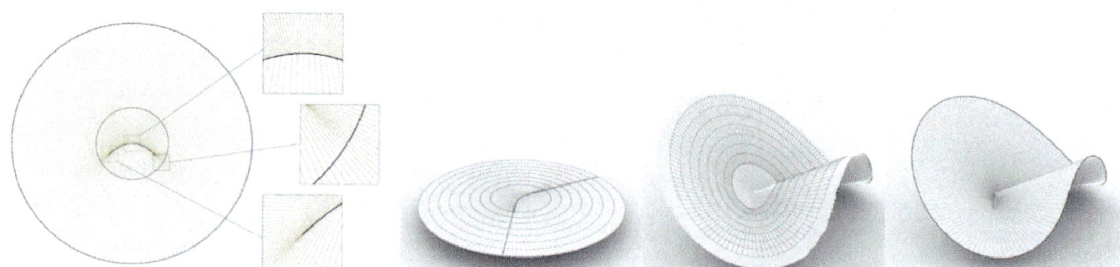

Fig. 12 (Non-radial) distribution of line pattern between *circle* and *curved ridge-line* transferred to all kinematic transition-phases of the element and mapped onto scanned surfaces by the means of Grasshopper

3D scans are quite imprecise for this part of the element (Fig. 7), and secondly physical models show a small area with anticlastic curvature on the concave side of the curved ridge-line driving from plastic deformation. This anticlastic area covers less than 1 % from area content of examined physical models.

Like in physical model, the planar sheet element can be brought into any self-organizing folded/bent spatial configuration with an angle α_I between 0° and in dependence of α_{MAB} and material thickness up to 174.19° (Figs. 2 and 4). Since A respectively B are part of straight lined, symmetrical legs starting in M with a predefined α_{MAB}, they can be equidistantly shifted, which is advantageous for a later physical realization. Since described geometrical dependencies and principles are identical for numerical simulation, the definition of 3-point-constellations in terms of position and orientation as well as the geometrical outline of used sheet material can be chosen by the user and offers a broad variety of constellations and freedom in design. Constraints in design might occur by possible collision/intersection of two or more sheet elements. For these cases, our simulation tool provides several solutions, which are not further described in this paper.

Simulation Tool

Basically our numerical simulation tool consists of 5 components, which can be executed by the means of Grasshopper (Figs. 13, 14, 15, 16, 17 and 18). Component 1 contains the data for the simulation. Component 2 represents the actual simulation. The other components allow analyses and further elaborations of the model.

In general Component 1—Data (Fig. 13) contains all refined data from scans and material data for simulation. This component is separated from other components to make general data

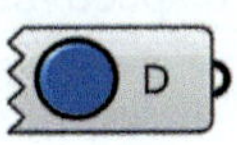

Fig. 13 Component 1—data

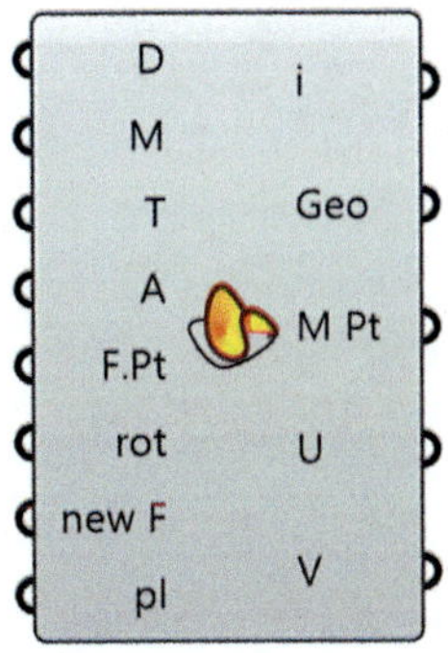

Fig. 14 Component 2—simulation

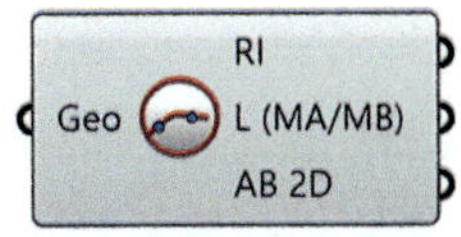

Fig. 15 Component 3—sort geo

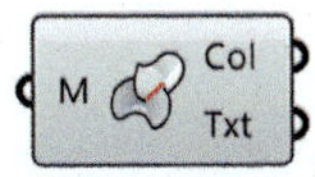

Fig. 16 Component 4—mesh intersection

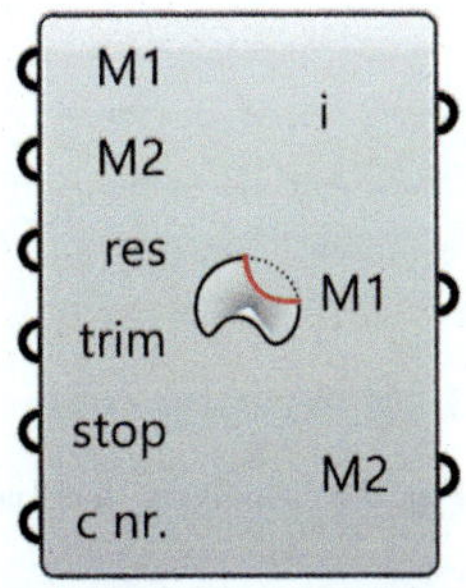

Fig. 17 Component 5—cut mesh

available for simulation of multiple elements, instead of feeding this information each time, which would increase processing time of total simulation and memory capacity.

Component 2—Simulation (Fig. 14) allows for smooth simulation from plane state to 3D object up to defined maximum angle of inclination α_I and user definable angle α_{MAB} as well as continuously variable adjustment of material thickness in the range of 0.5–1.5 mm.

Possible settings for Component 2—Simulation are Component 1—Data, M_Motion, describing non-linear motion of point-cloud from plane state to 3D point-cloud, T_Thickness,

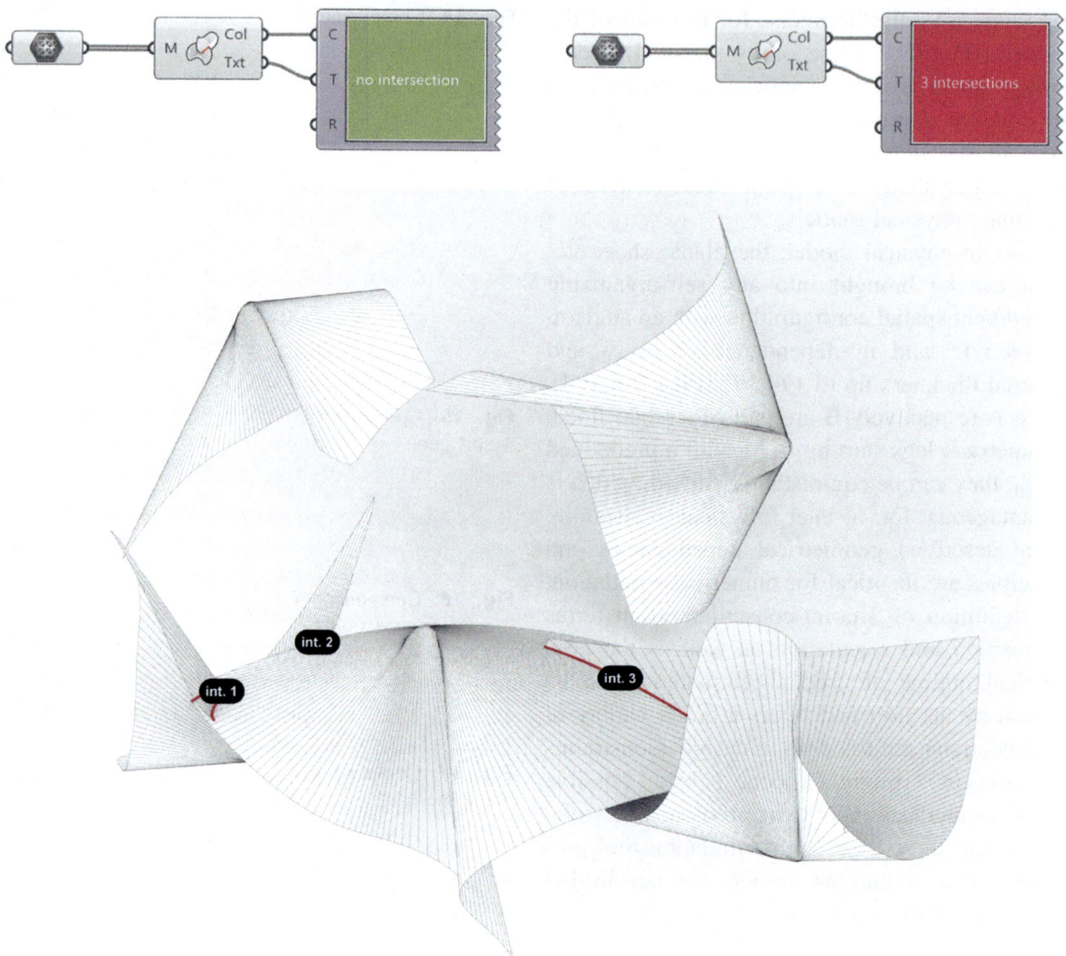

Fig. 18 Analysis, feedback and visualization of mesh intersections

offering continuously variable adjustment of material thickness, A_Angle, setting user definable angle α_{MAB}, F.Pt_Force Point, defining position and orientation of A and B and their spatial path during motion, Rot_Rotation, defining orientation of curved ridge-line in plane state, New F, considering any user definable outline of sheet element, which is different from circular default setup, and Pl_Plane, defining the position of element regarding World XY-plane.

Component 2—Simulation output provides I_Info, information on selected α_{MAB}, selected material thickness and angle of inclination α_I, Geo_Geometry, geometrical information as described below, M Pt_Mesh Points, resulting point-cloud in order to generate and illustrate the resulting surface, U_U-direction, number of points in U-direction and V_V direction, number of points in V-direction (Fig. 14).

Component 3—Sort Geo (Fig. 15) is fed in by Geo_Geometry and provides 2D output for later fabrication. Ri_ridge displays curved ridge-line in plane state, LMA/LMB 2D_Leg MA/MB defines leg MA respectively MB and AB 2D_Points AB 2D defines position of point A respectively point B in plane state.

Component 4—Mesh Intersection (Figs. 16, 17 and 18) is a useful component for the simulation of more than one sheet elements and the detection of possible intersections respectively collisions in realized structures. This component considers parallel, serial and all possible intermediary motion sequences of elements regarding possible intersections/collisions.

Together with Component 5—Cut Mesh (Fig. 17) it is not only meant to be useful in terms of collision detection but also in terms of a creative design process, where elements can be used to "tailor" their neighbor elements. So there are basically three options how to deal with intersections, firstly to search for serial, non-intersecting motion sequences of elements, secondly to trim at a user defined position, which can either be the starting or end-position of elements and thirdly to select elements, which are monitored within a range of α_I, and which are trimmed according to their motion sequence. In this case all select elements (M1_Mesh 1, M2_Mesh 2) are fed into Component 5—Cut Mesh, where all spatial positions are continuously variable simulated within selected range of α_I. Before starting this process it is possible to define that the elements equally trim each other or that there is a hierarchy between elements (Trim; 0 = both elements, 1 = M1 gets trimmed, 2 = M2 gets trimmed). Component 5—Cut Mesh provides output i_info, informing the user in case that trimmed lines or boundary line of elements are intersecting with curved ridge-line, M1_Mesh 1, trimmed geometry of mesh 1, Crv 1_curve 1, generated new boundary line of element 1 and/or M2_Mesh 2, trimmed geometry of mesh 2, Crv 2_curve 2, generated new boundary line of element 2.

Conclusion and Perspective

At present simulation and physical models show a maximum deviation of 5 %. This can mainly be explained by imprecision like asymmetry caused by the fabrication process of the physical model and difficult determination of boundary lines from 3D scans.

Within the digital environment, our simulation delivers a maximum deviation of less than 3 % in surface area and less than 2 % in metrical length of boundary lines between plane state and fully closed 3D state. Like in physical model, the planar sheet element can be brought into any self-organizing folded/bent spatial configuration with an angle α_I between 0° and in dependence of α_{MAB} and material thickness up to 174.19°. So the geometrical precision of our simulation tool can be rated as sufficient and reliable.

At present this is proved for arrangements, respectively naked edges of sheet elements forming a common pattern in plane and arrangements with naked edges of sheet elements forming a common 3 dimensional pattern. In conclusion several tests and various 3-point-constellations in terms of position and orientation on planar sheet elements of various shapes and their transition into self-organizing folded/bent spatial configurations verified that our simulation can be a powerful and reliable tool in exploring design freedoms as well as limits.

Ongoing investigations focus on the wealth of possible forms that can be simulated and accordingly fabricated (Figs. 19 and 20). Therefore, we are looking into constellations and patterns of sheet elements regarding their intersections. Thus, we will develop strategies in order to avoid or using these intersections in the creative design process. Arranging more than one 3-point-constellation on one sheet element is another option to increase formal variety (Filz and Kainzwaldner 2015).

In this context our investigations seem to have close relation to principles of rigid Origami folding (Demaine et al. 2011, pp. 39–50; Jackson 2011). These investigations will be accompanied by architectural design studies and applications to highlight functional, spatial and structural qualities of generated forms and configurations.

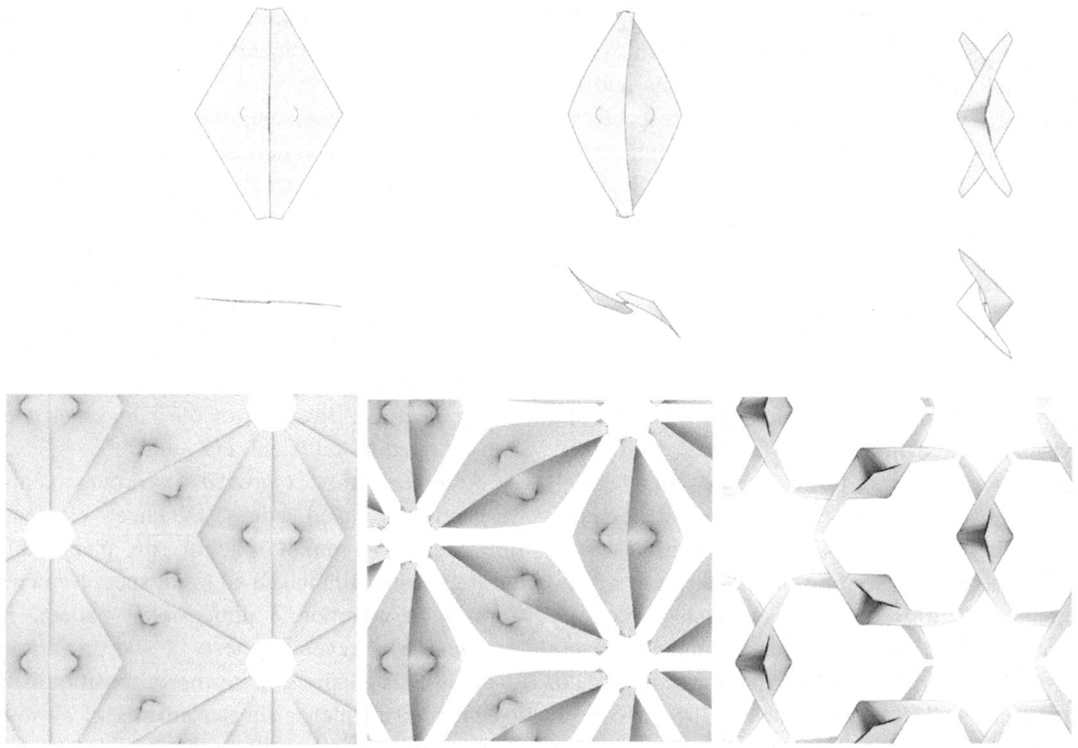

Fig. 19 Simulation of additively arranged sheet elements from planar geometry through continuously variable adjusted states to fully closed 3D-state

Fig. 20 Simulation of surface, randomly subdivided into sheet elements from planar geometry through continuously variable adjusted states to fully closed 3D-state

References

Demaine E, Demaine M, Koschitz D (2011) Reconstructing David Huffman's legacy in curved-crease folding. In: Yim M (eds) Origami 5, fifth international meeting of origami science, mathematics, and education. CRC Press, Boca Raton

Filz GH, Kainzwaldner S (2014) Between bending and folding—buckling from plane to object by 3 predefined points. Paper presented at the VI. Latin-American symposium on tension structures (IASS-SLTE). Shells, membranes and spatial structures: 'Footprints', Brasilia, 2014

Filz GH, Kainzwaldner S (2015) Freedoms and limits in the arrangement of 3 point constellation on planar sheet elements, generating spatial, selforganized folded/bent forms. Paper presented at the international association for shell and spatial structures (IASS), Amsterdam, 2015 (Abstract accepted, full paper in review)

Fuchs D, Tabachnikov S (2007) Mathematical omnibus. American Mathematical Society, USA, pp 185–197, 207–215

Jackson P (2011) Folding techniques for designers—form sheet to form. Laurence King Publishing Ltd, London, pp 176–187

Kilian M, Flöry S, Chen Z, Mitra JN, Sheffer A, Pottmann H (2008) Curved folding. In: Proceedings of ACM transactions on graphics

Leopoldseder S, Pottmann H (2003) Approximation of developable surfaces with cone spline surfaces. Institut für Geometrie, Wien

Lienhard J, Schleicher S, Knippers J (2011) Bending-active structures—research pavilion ICD/ITKE. Paper presented at the international symposium of the IABSE-IASS Symposium, Taller Longer Lighter, London, 2011

Lienhard J, Alpermann H, Gengnagel C, Knippers J (2012) Active bending, a review on structures where bending is used as a self formation process. In: Kim (SD): From Spatial structures to space structures, IASS-APCS 2012. Abstract book. Seoul: Korean association for spatial structures

Nettelbladt M (2013) The geometry of bending. Mårten Nettelbladt, Stockholm, pp 9–11, 17

Peternell M (2009) Unterteilungskurven und—flächen. In: Proceedings of Fortbildungstagung für Geometrie, Wien, 2009

Pottmann H, Asperl A, Hofer M, Kilian A (2010) Architekturgeometrie. Springer, Wien, pp 221, 222, 247–257

Tirapegui E, Martinez J, Tiemann R (2000) Insatbilities and nonequilibrium structures VI. Kluwer Academic Publishers, Dordrecht, pp 269–274

Zhu L, Igarashi T, Mitani J (2013) Soft folding; computer graphics forum, vol 32(7). In: Proceedings of the 21st Pacific conference on computer graphics and applications. Pacific Graphics, Singapore

Simulating Fusion: An Epistemological Analysis of a New Design Tool for an Imminent Multi-material Future

Kostas Grigoriadis

Abstract

Recent research initiatives taking place in architecture indicate that the transfer of graded materials to the construction industry from fields such as aerospace engineering can bring about radical changes in the design and construction of buildings. In this light, the paper addresses the departure from the use of conventional design software that utilise boundary representation geometries to represent volume and the search for alternative methods of designing with gradients. Multi-material design is a process mainly concerned with the assigning of gradients between two or more dissimilar materials, as well as their extents. Particle system elements are by virtue of their computational structure made to simulate natural phenomena and effectively the behaviour of materials. Particle simulations have therefore been utilised as a design tool and in order to allow for the gradients to be computationally calculated based on the physical material properties of the substances to be fused. The main question that the paper effectively addresses, is how can the appropriate materials that will form a larger multi-material entity be selected and how should the relevant simulation environment be set up in terms of the forces that affect material distribution. To address these, two exercises for designing a multi-material panel exposed in high velocity wind conditions are presented in which the agency assigned in the simulation is analysed and critiqued. In the first exercise, the aim is to achieve a similar material distribution to multi-materials resulting from accumulative roll bonding manufacturing techniques. In the second exercise, non-invasive material agency is applied in the simulation, which is informed by the physical forces acting on the panel. Following the initial design experiments, the paper analyses the main parameters that have been set in the tool and presents an epistemological critique of the resulting multi-material design workflow.

K. Grigoriadis (✉)
School of Architecture, Royal College of Art,
London, UK
e-mail: kostas.grigoriadis@network.rca.ac.uk

M.R. Thomsen et al. (eds.), *Modelling Behaviour*, DOI 10.1007/978-3-319-24208-8_24

Introduction

Additive Manufacturing Developments

In a TED article about the future of additive manufacturing published in June 2013 (TED Blog 2013), the objects that we would see 3D-printed soon included food, meat and leather, rocket parts, houses and a moon base. In this rapidly expanding field where technological advancements are taking place exponentially, some of the objects included in the catalogue were already a reality even before the online publication of the article. A 3D-printed handgun was created and tested in May 2013, while research on 3D-printing liquid metal parts was published as "a method to direct-write liquid metal microcomponents at room temperature" (Ladd et al. 2013, p. 1) in July of the same year. A few months later, the breakthrough of building the world's first 3D-printed house was contested between an architectural practice in the Netherlands and a Chinese engineering company.

Further Outlook

A year earlier, in an article by Wired Magazine, an expanded outlook of this additive manufacturing future indicated that there is already research conducted by industrial engineers at Airbus towards 3D-printing by 2050 of a "self-cleaning aircraft with inbuilt neural networks, antioxidant enriched air and body heat harvesting facilities" (Liat 2012). Currently, there are steps being made towards the assimilation of 3D-printing in aerospace, namely in recent instances like the launching by the European Space Agency in 2013 of a program termed "Addictive Manufacturing Aiming towards Zero Waste and Efficient Production of High-Tech Metal Products". The aim is to send in space by 2020 the first satellite printed as a single part, which is something that 'would save 50 % of the costs—(that equates to) millions of euros' (RT 2013).

Multi-material Printing

Expanding on the description of their plans for a 3D-printed airplane, the Airbus engineers suggested that the way an idea like this can remotely be materialised is through *multi-material* printing. In this envisioned future:

> You can dial in the different elasticity of an object, the colour properties, or a continuous piece of material that is different properties over the piece. Certain parts of an airplane need to be strong and flexible [and 3D-printers can create objects] strong just where they needed to be strong, or light where they needed to be light (Liat 2012)

In addition, it is interesting to note that in the article published in 2012 the feasibility of this idea was illustrated by the fact that the company Objet had the technology to 3D-print 107 different materials, which three years later in 2015 have become 140.

Multi-materiality in Architecture and Graded Information CAD

Following up on this principle of exponential technological developments it can be said that *multi-materiality* will also become assimilated in architecture initially confined to design research and small scale manufacturing, which is the case at the moment and eventually fully implemented in building construction (Federal Institute for Research on Building, Urban Affairs and Spatial Development within the Federal Office for Building and Regional Planning 2011). In fact, in material science, multi-materials were invented as far back as the late seventies in the form of *functionally graded materials (FGM)* that at the time were continuous volumes of fused ceramic and steel, utilised in high temperature differential exposure areas of space rocket components. Based on their progressive and increasing use in architecture and design, the main problem when designing with FGM as a built output is the assignment and distribution of gradients between the fused substances. According to Knoppers et al. (2005), the limited CAD techniques that exist for designing with graded material information are confined to finite element analysis, particle system elements, vague discrete and voxel modelling among a few others. Of these techniques, the ones prevalently used by architects and designers is voxel modelling and to a lesser extent finite elements.

Existing Software Critique

A critique that the paper poses, of the procedures that are voxel based (Michalatos, Payne 2013; Oxman 2012; Tsamis 2010) is that although they are highly sophisticated to an extent, at the same time actual material attributes are not taken into account. This critique can firstly relate to the fact that these platforms concentrate primarily on the distribution, positioning and assigning of materials in digital space, with a multi-materiality which nonetheless is of a representational nature as it is attributed as RGB colour values and not as digitised physical properties such as density, viscosity, surface tension etc. Secondly, at the moment the 3D-printed parts that can be built with information generated from these software are similarly representational as the properties of the material that is printed are not fed back and/or simulated into the digital environment. Thirdly, the compatibility between different substances and their chemical capability to bond in order to form a larger multi-material is another primary parameter that is not incorporated. All these can be summarised as the representation of materiality digitally, having as an output a 3D-printed representation of this representational materiality.

Research Question, Methodology and Design Objectives

Framework Definition

What is effectively argued for, is that when the relevant technology eventually allows for direct one-to-one fabrication of the multi-material specified in the digital domain being the multi-material that is 3D-printed in full scale, a direct approach that integrates properties as part of the design process is crucial. Representation should be avoided, for a process where the physical can be simulated in the digital realm, in essence allowing for a closer relationship between the two. Design in this instance should acquire a loose type of control. Its function should be to generate an enclosing framework within which, matter will be allowed to self-structure and different substances fused together into continuously graded topologies. There should therefore be a thin balance between designing these frameworks and enabling self-arrangement.

Particle System Elements and Research Problem

Of the above graded information CAD techniques, particle systems allow for gradients to be computationally calculated based on the physical properties of the substances to be blended. They are by virtue of their computational structure made to simulate natural phenomena and effectively the behaviour of materials in their liquid, malleable state. In this respect, the question that the paper addresses is how can the *a. material type* and *b. affecting agency (or force)* parameters input in the simulation be assigned.

Simulation and Reality

When describing their argument in favour of the epistemological dependence thesis (which suggests that the degree to which simulations can be valid, depends on the degree of their resemblance to experiments), Norton and Suppe suggest that "a valid simulation is one in which certain formal relations hold between a base model, the modelled physical system itself, and the computer running the algorithm" (Winsberg 2003, p. 114). Within the scope of this paper, simulations will be utilised for design rather than scientific purposes and it is therefore beyond the present research to argue for or against particle system modelling bearing a semblance to the chemical attraction behaviour between two or more blending materials. What can be argued on a design level, however, is that these formal relations should be present in the form of simulation parameters assigned in the computer resembling physical material property values as closely as possible and of the simulation forces being informed by practised FGM manufacturing methods and/or by the loads acting on a multi-material segment of a larger overall topology. As it will be discussed, however, in some instances of FGM industrial manufacturing, materials are structured against their natural propensity to arrange themselves and fuse.

Design Brief Definition

To address the research question in 2.2, the exercises that follow demonstrate the use of particle simulations to design a multi-material panel with improved structural performance under tensile loads, as opposed to a conventional singular material. The initial problem, as previously formulated, is the selection of appropriate individual *material types* for designing the multi-material. The argument in response is that in the very large space of possibilities that exist for choosing these, one has to limit the selection to existing, physical, manufactured or found FGM.

Accumulative Roll Bonding Multi-material Manufacturing

In terms of the industrial manufacturing, Schmidt et al. (2012) describe their method of designing "graded materials by particle reinforcement during accumulative roll bonding" according to which, aluminium sheets are sprayed with an aqueous solution consisting of copper particles at 33.3 % concentration with an incremental feed velocity and distance from one end to the other. The sheets are then roll bonded together repeatedly, generating a sandwiched material consisting of copper particles gradually placed in layers within the aluminium body (Fig. 1). The benefit of this technique is that the tensile strength of the aluminium sheet is increased, which is "proven by tensile tests showing a steady and monotonous gradient along the rolling direction" (Schmidt et al. 2012, p. 1009). This also means that the material property gradient can be distributed in a manner that can be "directly opposed to the gradient in loading condition" (Schmidt et al. 2012, p. 1009), therefore having a direct impact on the structural performance and amount of material used in the sheet (Grigoriadis 2014).

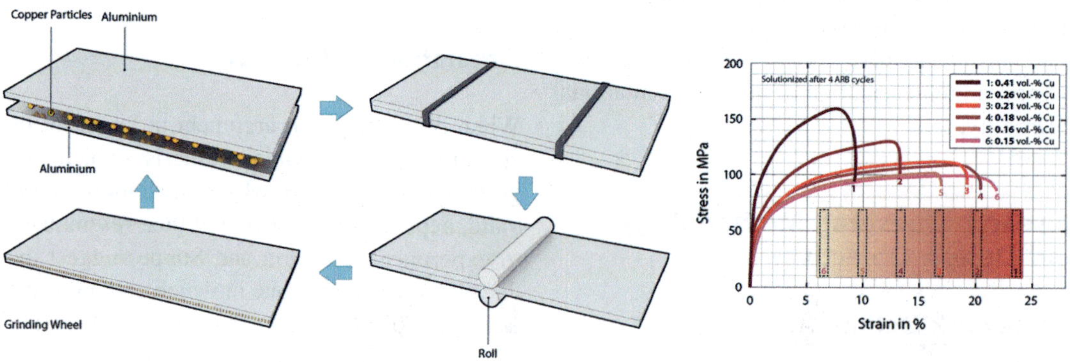

Fig. 1 Diagram of the accumulative roll bonding process, for reinforcing an aluminium sheet with copper particles. A "graded sheet with positions of tensile specimens" is shown on the *right*. Image courtesy by Schmidt et al. (2012)

Material Type Selection

As the intent has been to define a model that simulates real world material gradients, the selection of materials in this instance has been determined by their existence as part of a physical multi-material entity. The technique in 2.5 verified that aluminium and copper bi-materials do exist physically and the two substances have therefore been utilised in order to design a panel hypothetically exposed to a high wind flow velocity condition.

Design Experiments

Experimentation Objective

With this in mind, a first objective has been to decrease the amount of overall material that would otherwise be used for a conventional aluminum panel. A second objective has been to allow for openings on the surface of the sheet that cover approximately 15 % of the total surface area, which would partially allow wind through the element, avoiding that way any turbulence caused as a result of it being completely solid. Primarily, however, the main objective has been to answer research question *b.* in 2.2 and to establish the appropriate *forces* that should be employed in a material blending simulation.

Simulation Global Parameter Set-Out

Initially, a direct frontal wind was simulated for velocities of 90 km/h and a pressure map indicated the load distribution on the surface of the sheet measured in Pascal (Fig. 2). The intent was then to assign the material properties of the two substances in the simulation environment and in their molten form. The liquid temperature for

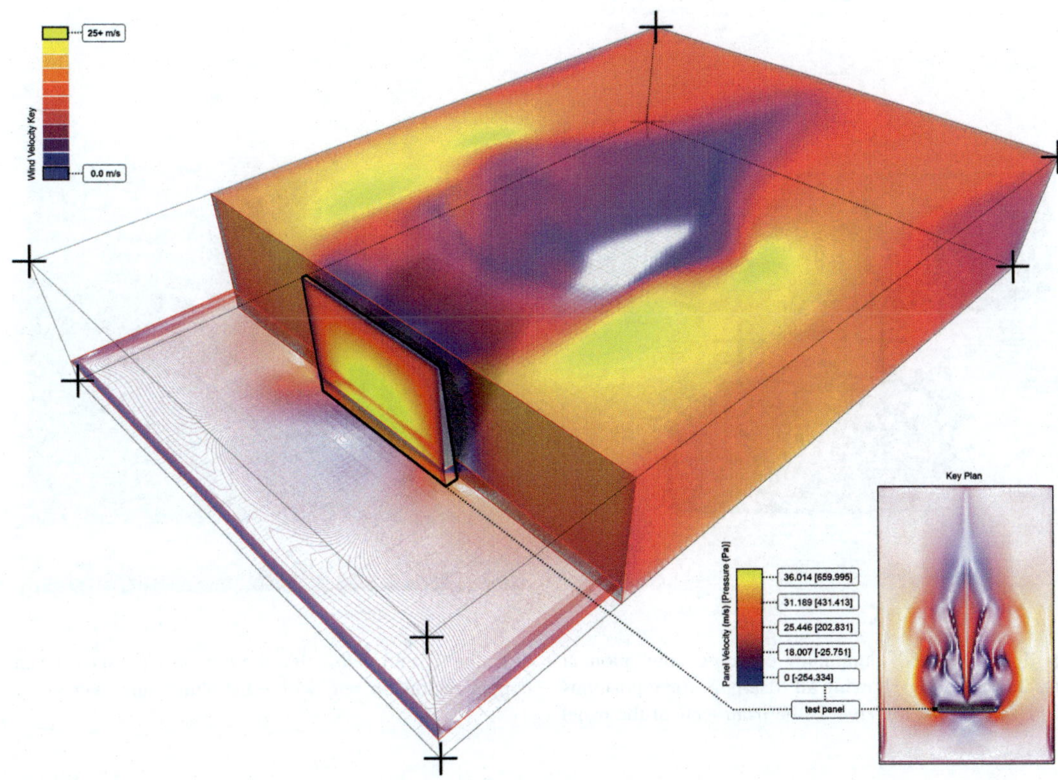

Fig. 2 CFD simulation on the sample panel showing tensile load intensity on the panel's surface

copper was set to 1200 °C (1473 K) and for aluminium to 660 °C (933 K), both slightly above the melting points of the materials. The density and viscosity of the two were also attributed: copper at 1200 °C has a density of 7898 kg/m^3 and its kinematic viscosity at the same temperature is equal to 0.0312 m^2 s^{-1}, while for aluminium the values were at 2375 kg/m^3 for density and 0.01379 m^2 s^{-1} for kinematic viscosity (Grigoriadis 2014).

Continuing Research Delimitation

The ensuing design experiments continue research presented at the eCAADe 2014 conference. Exercise one is further development that has been analysed in conjunction with the previously presented exercise two (Grigoriadis 2014). The simulations are discussed this time with an emphasis on the *affecting forces*, while a new example of an architectural application of the design methodology is presented in 4.1.

Exercise 1: Agency Informed by Accumulative Roll Bonding

In terms of the *affecting agency*, the principle likewise was for it to be informed by forces that would affect the material physically, with one being the standard gravitational force of 9.8 m/s^2. The other forces attributed were 'Limbo daemons' that effectively are two notional planes that constrain the movement of particles in the space between them. This was in order to achieve a more fitting resemblance to the manufactured sheet that had the bulk of copper reinforcement horizontally spread in-between and across the aluminium body. The first simulation was eventually set out with three aluminium and one copper particle emitters placed within a container and allowed to run until the copper accumulated gradually and horizontally across the middle of the aluminium matrix (Fig. 3).

According to research by Schmidt et al. (2011), the tensile strength of commercial purity aluminium ≥99.5 % which is 75 MPa was

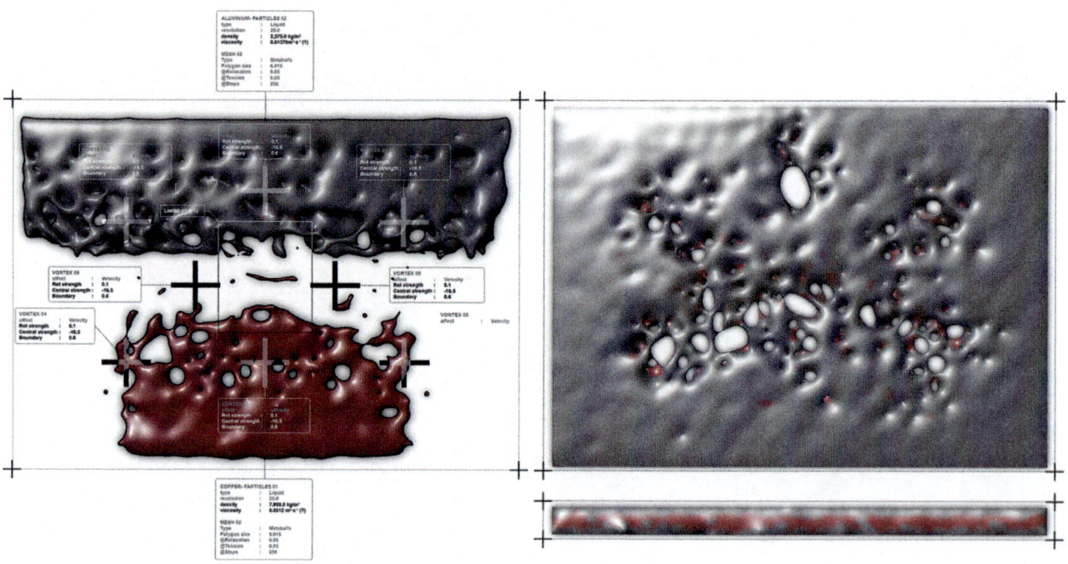

Fig. 3 Copper and aluminium blending simulation at frame 12 on the *left*, showing all 'daemon' force positions and values. On the *top right* is the *front view* of the panel at frame 97 and *below* that the *bottom view* showing the copper region (*in red*) within the aluminium matrix

increased to 160 MPa when copper of volume fraction of up to 2.9 % was added to it, effectively enabling an approximate increase in tensile strength of 53 %. In this case the structural objective would be met as it is envisaged that the tensile strength increase from the bottom of the sheet and towards the top that measured 2000 mm in total, would range in percentages according to copper content, with an average being the aforementioned 53 %.

Critique

A main issue, however, in this first exercise was that the *agency* assigned was not stemming from the physical domain. More specifically, there were forces acting on materials individually, which although can be a possibility (with an example being magnetic forces acting on ferrous materials only), in this instance the forces were software specific and selectively applied rather than reality based. Further research on manufacturing techniques can verify whether the creation of forces of this nature can be attained, but for the moment it is assumed that the use of Limbo forces would be against the objective of bridging the virtual to the actual.

Invasive Versus Non-contact Forces

Additionally, as another take on the problem of material arrangement in a physical environment this time it can be argued that the accumulative roll bonding process described above is contrary to the idea of self-structuring. The treatment of aluminium and copper is forceful, as the energy-consuming severe plastic deformation process that takes place during roll bonding is counter to the principle of self-structuring fusion occurring under *non-invasive forces*. The term *invasive* would equate in this case to a machine, tool, physical part or solid apparatus that coerces form on the two materials. The difference to the *non-invasive* is that industrial mechanic force cancels out the intrinsic chemical-bond capability of materials blending together under centrifugal, gravitational, magnetic and other *non-contact forces*. In addition, when looking into this problem architecturally, the main acceptable force that requires no justification for its application on a material entity would be gravity. For another *non-contact force* to be allowed to act on a material fusion simulation it needs to come from anticipated structural and environmental loading conditions on a particular segment of a larger multi-material entity˙ This additional force can be there to simulate and accelerate these loads in the computer, with the result being a 'natural' arrangement of individual substances in the multi-material space. An example of this can be the aforementioned instance of the wind loads on the aluminium sheet (see Sect. 3.1). The wind applies force at the centre and is then deflected upwards and at the back of the panel, generating a tensile force that is exerted against gravitational pull. This can simply be represented in a digital simulation by an *attractor or vortex force*.

Exercise 2: Agency Informed by Loading Conditions

A second simulation was therefore run in which the gravity force was preserved and the Limbo forces applied in the first prototype simulation removed. In place of these, the second force that would be needed for a gradient to form between the two materials and that was placed in the environment was the aforementioned vortex that generated a centrifugal like force at one end of the container that the materials were poured into. The existence of this in the simulation, as well as its magnitude, were informed (apart from their aforementioned similarity to the wind's tensile force) by industrial casting methods for the creation of continuous gradations in FGM (Watanabe and Sato 2011). This procedure involves a preheated spinning mould that contains one material into which another molten material is poured. The centrifugal force generated during processing, forces the two at the opposite ends of the mould, (which occurs due to the difference in their respective density values) and eventually a gradient is formed between them (Fig. 4).

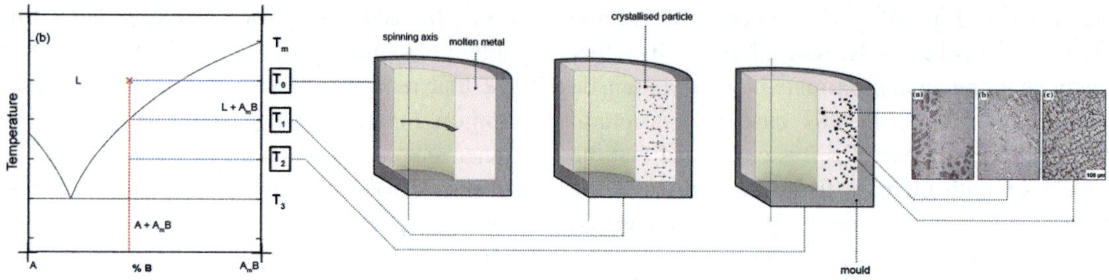

Fig. 4 Centrifugal in situ method for manufacturing FGM. A molten metal is poured in a spinning mould with the centrifugal force separating aluminium (*dark grey*) from aluminium-copper particles (*light grey*), during the solidification process. Image courtesy by Watanabe and Sato (2011)

Result and Critique

With these in place, the simulation run achieved the distribution of copper particles to an area that was approximately the same as the tensile loading area (Fig. 2) output from the wind simulation. The resulting multi-material sheet was solid copper at the lower end and solid aluminium at the top, while the percentage of openings (approximately 13 %) was also close to the objective of 15 % (Fig. 5). What was evident in this case where the materials were given relative freedom to find their own arrangement in space, was that the resulting formation had altogether different qualities from the aforementioned roll bonded sheet (see Sect. 'Accumulative Roll Bonding Multi-material Manufacturing'). This was due to the use of the *attractor force* that resembled much more closely the loading condition on the panel, with its use effectively answering research question *b.* regarding the appropriate *agency* employed in the particular blending simulation.

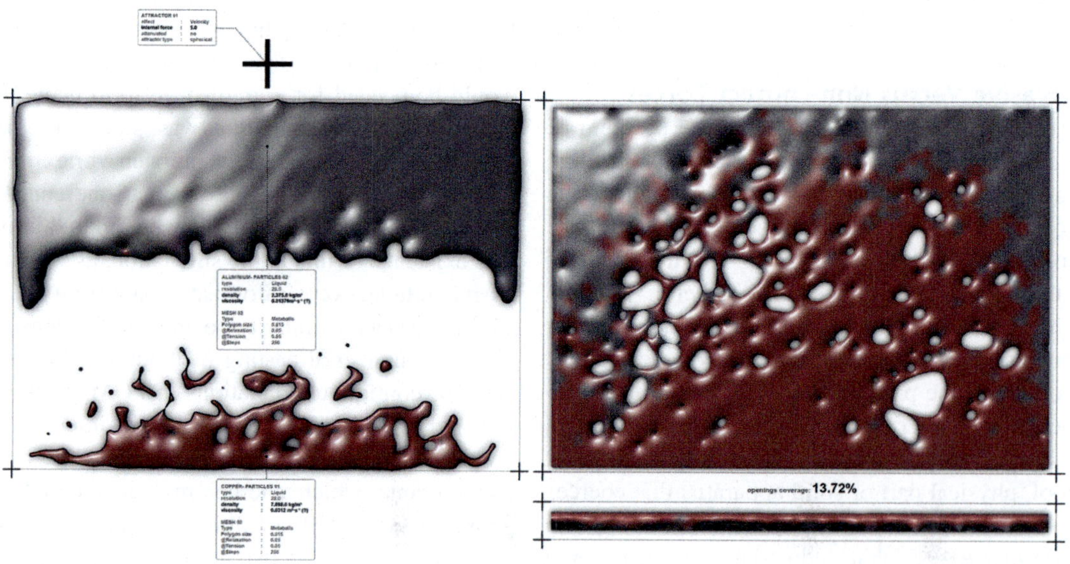

Fig. 5 The second blending simulation at frame 12 on the *left*, showing the non-contact attractor position at the *top*. Shown on the *right* are the *front* and *bottom views* of the panel at frame 79. The base of the panel is now solid copper

Applications and Discussion

Alternative Applications of the Design Methodology

When considering the advantages of multi-material use in the design of architectural elements, an application among others of this aforementioned design methodology can be in conventional curtain wall systems, which similarly to other architectural assemblies are "designed and built by separate organizations, often in a process of chaos and conflict" (Wiscombe 2012, p. 1). The powder coated and anodized aluminium, silicon, EPDM and polyetherimide materials employed in curtain walls (to address the tectonic and material problem of attaching glazing panes to their surrounding aluminium frames) are sourced in remote parts of the world, following messy procurement routes that involve hundreds of different trades and are eventually fabricated through immensely intensive energy-consuming manufacturing methods. Alternatively, in an imminent multi-material future, the messiness of this process could be resolved by the secretion of some of these materials under one print, with aluminium fusing into alumina fusing into glass, forming that way continuously graded material fields the extents and positioning of which can be calculated through material simulations (Figs. 6 and 7).

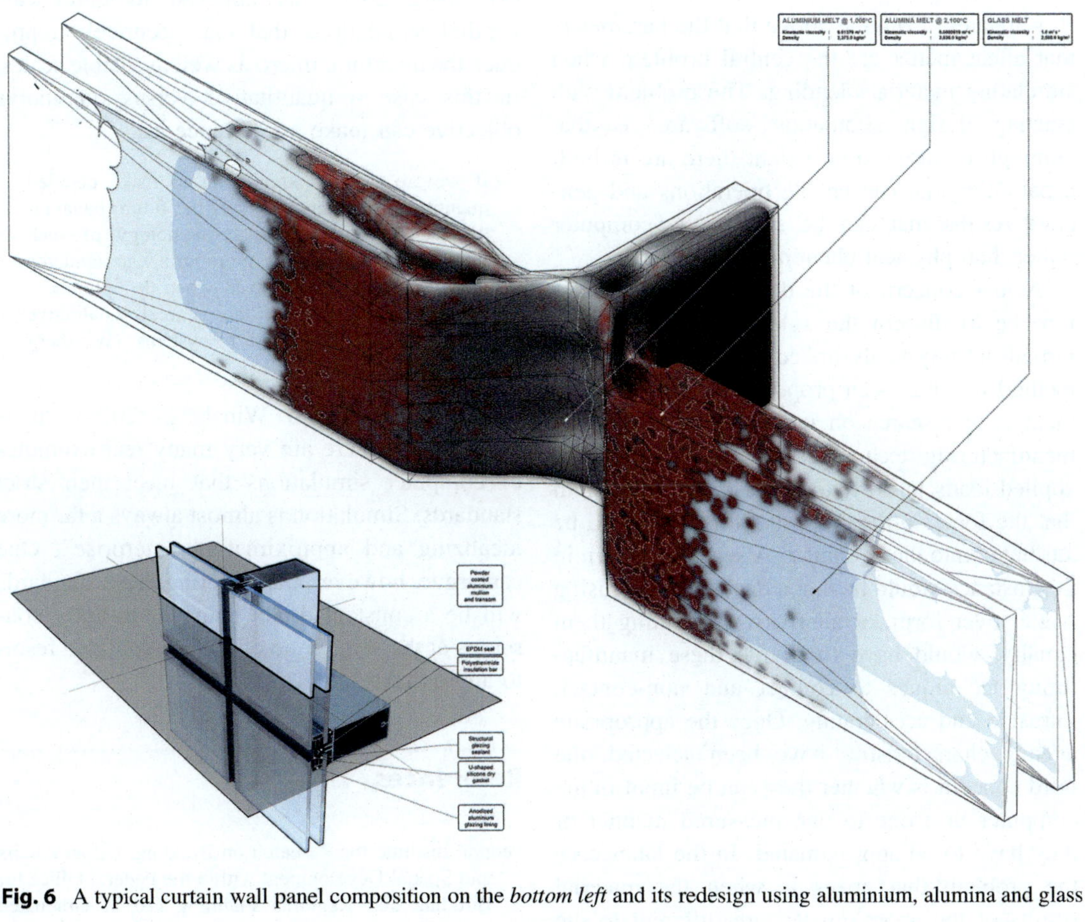

Fig. 6 A typical curtain wall panel composition on the *bottom left* and its redesign using aluminium, alumina and glass

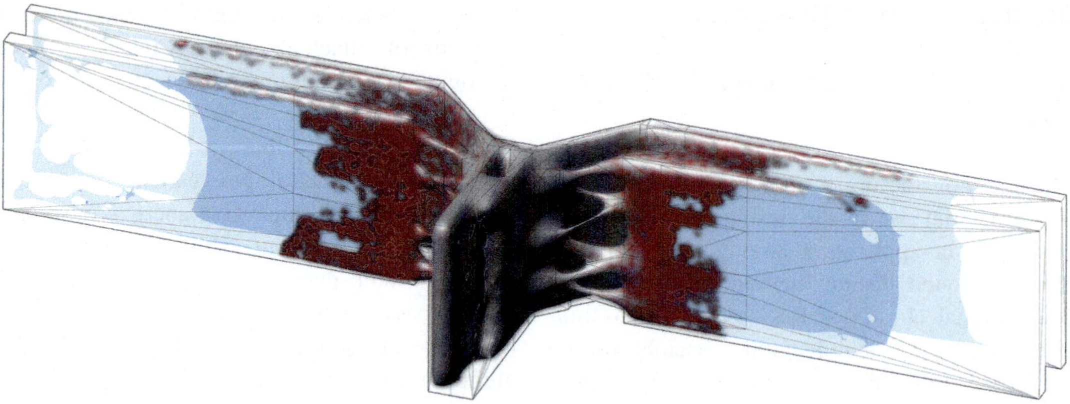

Fig. 7 Internal view of the multi-material panel

Conclusion

It is apparent from the above that the parameters that affect matter are the central problem when simulating material blending. The problem with existing design simulation software, is that although accurate to an extent there are in-built capabilities that can enable operations and generate results that can be said to be computer rather than physical phenomena.

A first concern of the designer would therefore be to discern the solely virtual from the virtual-but-physically-linked capabilities. The method that has been proposed for this discernment is to research on industrial multi-material manufacturing techniques and/or to work out applied loads in order to make the simple claim that the forces utilised/applied there can also be attributed into the computer. A second concern in this instance would be that a designer prioritising matter over form, or alternatively treating them equally, would have to divide these manufacturing techniques to contact and non-contact, intrusive and accelerating. Once the appropriate agency characteristics have been selected, the third concern is whether they can be input in the computer in a one to one measured manner or they have to be approximated. In the latter case the problem that arises is when the material results of the approximated are different to the results of the real. This is especially because of the fact that particle-based design simulations can only simulate a limited range of scales (Winsberg 2014), as opposed to multi-scale parallel simulations that can accurately reproduce the quantum, micro as well as visible scales. In this case a quantitative versus qualitative objective can make a difference as:

> If we are using a simulation to make detailed quantitative predictions about the future behavior of a target system, the epistemology of such inferences might require more stringent standards than those that are involved when the inferences being made are about the general, qualitative behavior of a whole class of systems. (Winsberg 2014)

But according to Winsberg (2009) "it is unlikely that there are very many real examples of computer simulations that meet their strict standards. Simulation is almost always a far more idealizing and approximating enterprise". One can argue, however, that meeting these standards will be a constant drive, being incessantly conscious at the same time of the unavoidable resort to approximation.

References

Federal Institute for Research on Building, Urban Affairs and Spatial Development within the Federal Office for Building and Regional Planning (2011) Graduated building components: production procedures and areas of use for functionally graduated building components

in construction. Federal Ministry of Transport, Building and Urban Development, Berlin
Grigoriadis K (2014) Material fusion: a research into the simulated blending of materials using particle systems. In: Thompson EM (ed) Fusion-data integration at its best. In: 32nd eCAADe conference, Newcastle, Sept 2014. Fusion: proceedings of the 32nd international conference on education and research in computer aided architectural design in Europe 2. Northumbria University, Newcastle, pp 123–130
Knoppers GE, Gunnink JW, Van den Hout J, Van Wliet WP (2005) The reality of functionally graded material products. In: Pham DT, Eldukhri EE, Soroka AJ (eds) Intelligent production machines and systems. First I*PROMS virtual conference, July 2005. Intelligent production machines and systems: First I*PROMS virtual conference. Elsevier Science, Amsterdam, pp 38–43
Ladd C, So JH, Muth J, Dickey MD (2013) 3D printing of free standing liquid metal microstructures. Adv Mater 25(36):5081–5085. doi:10.1002/adma.201301400
Liat C (2012) Airbus designer reveals plans for 3D printed planes by 2050. http://www.wired.co.uk/news/archive/2012-07/12/3d-printed-plane-by-2050. Accessed 11 Apr 2015
Oxman N (2012) Towards a material ecology. In: Johnson J et al (eds) Synthetic digital ecologies. ACADIA 12, San Francisco, Oct 2012. Proceedings of the 32nd annual conference of the association for computer aided design in architecture. California College of the Arts, San Francisco, pp 19–20
Richards D, Amos M (2014) Designing with gradients: bio-inspired computation for digital fabrication. In: Gerber D et al (eds) Design agency. ACADIA 14, Los Angeles, Oct 2014. Proceedings of the 34th annual conference of the association for computer aided design in architecture. Riverside Architectural Press, Los Angeles, pp 101–110
RT (2013) AMAZE: Europe's ambitious aim to 3D print a single-piece satellite. http://rt.com/news/3d-print-satellite-metal-esa-232/. Accessed 02 Sept 2014
Schmidt CW, Knödler P, Höppel HW et al (2011) Particle based alloying by accumulative roll bonding in the system Al–Cu. Metals 1(1):65–78
Schmidt CW, Ruppert M, Höppel HW et al (2012) Design of graded materials by particle reinforcement during accumulative roll bonding. Adv Eng Mater 14 (11):1009–1017. doi:10.1002/adem.201200046
TED Blog (2013) A 3D printed future: 10 surprising things we could see printed soon. http://blog.ted.com/a-3d-printed-future-10-surprising-things-we-could-see-printed-soon/. Accessed 10 Oct 2014
Tsamis A (2010) Go brown: inner-disciplinary conjectures. Archit Design 80:80–85. doi:10.1002/ad.1166
Watanabe Y, Sato H (2011) Review fabrication of functionally graded materials under a centrifugal force. In: Cuppoletti J (ed) Nanocomposites with unique properties and applications in medicine and industry. InTech, Croatia, pp 133–150
Winsberg E (2003) Simulated experiments: methodology for a virtual world. Philos Sci 70(1):105–125
Winsberg E (2009) Computer simulation and the philosophy of science. Philos Compass 4(5):835–845. doi:10.1111/j.1747-9991.2009.00236.x
Winsberg E (2014) Computer simulations in science. In: The Stanford encyclopedia of philosophy. Stanford University. http://plato.stanford.edu/entries/simulations-science/. Accessed 05 Nov 2014
Wiscombe T (2012) Beyond assemblies: system convergence and multi-materiality. Bioinspir Biomim 7 (1):015001. doi:10.1088/1748-3182/7/1/015001

Modelling Behaviour for Distributed Additive Manufacturing

Jorge Duro Royo, Laia Mogas Soldevila, Markus Kayser and Neri Oxman

Abstract

Distributed forms of construction in the biological world are characterized by the ability to generate complex adaptable large-scale structures with tunable properties. In contrast, state-of-the-art digital construction platforms in design lack such abilities. This is mainly due to limitations associated with fixed and inflexible gantry sizes as well as challenges associated with achieving additively manufacturing constructs that are at once structurally sound and materially tunable. To tackle these challenges we propose a multi-nodal distributed construction approach that can enable design and construction of larger-than-gantry-size structures. The system can generate and respond to integrated real-time feedback for parameters such as material curing duration and position awareness. We demonstrate this approach through a software environment designed to control multiple robots operating collaboratively to additively manufacture large-scale structures. We present and report on a novel computational workflow as well as work-in-progress of a digital fabrication environment. The environment combines a centralized system designed to manage top-down design intent given by environmental variables, with a decentralized system designed to compute, in a bottom up manner, parameters such as multi-node rule-based collision, asynchronous motion, multi-nodal construction sequence and variable material deposition properties. The paper reports on a successful first deployment of the system and demonstrates novel features characteristic of fabrication-information modelling such as multi-nodal cooperation, material-based flow and deposition, and environmentally informed digital construction.

J.D. Royo (✉) · L.M. Soldevila · M. Kayser ·
N. Oxman (✉)
Department of Architecture and Urban Planning (SA+P), Media Lab, Mediated Matter Group, Massachusetts Institute of Technology (MIT), Cambridge, MA, USA
e-mail: neri@mit.edu

M.R. Thomsen et al. (eds.), *Modelling Behaviour*, DOI 10.1007/978-3-319-24208-8_25

Introduction

Distributed construction in the biological world enables what is currently lacking in digital design and fabrication: large scale, geometrically complex, materially variable and behaviourally responsive fabrication (Erol 2005, pp. 10–20; Oxman et al. 2015, pp. 1–2). In the biological world, such features are enabled by an organism's ability to respond to environmental cues and adapt to the availability of matter (Wilson 2000; Tan et al. 2013, pp. 18–39). For instance, wasp nests and termite mounts are considered canonical examples for animal-architectures that are coordinated and erected through distributed construction (Wilson 2000). These insects utilize simple communication via sensing stimuli to construct large-scale, highly sophisticated, multi-functional structures that are orders of magnitude larger than themselves (Wilson 2000; Erol 2005, pp. 10–20). In contrast to its biological counterpart, man-made agent-based assembly is generally based on discrete processes, making it challenging to achieve high levels of complexity in agent-to-agent ("node-to-node") communication and material sophistication (Erol 2005, pp. 10–20; Augugliaro et al. 2014, pp. 46–64; Naboni et al. 2014). Higher overlap between material processes, environmental conditioning and fabrication constraints aims to tackle these challenges by increasing the dimensionality of the design space through multifunctional materials, high spatial resolution in manufacturing and sophisticated computational algorithms. In doing so, a holistic vision for design emerges—"Material Ecology"—that considers computation, fabrication, and the material itself as inseparable dimensions of design (Oxman et al. 2015, pp. 1–2) (Fig. 1).

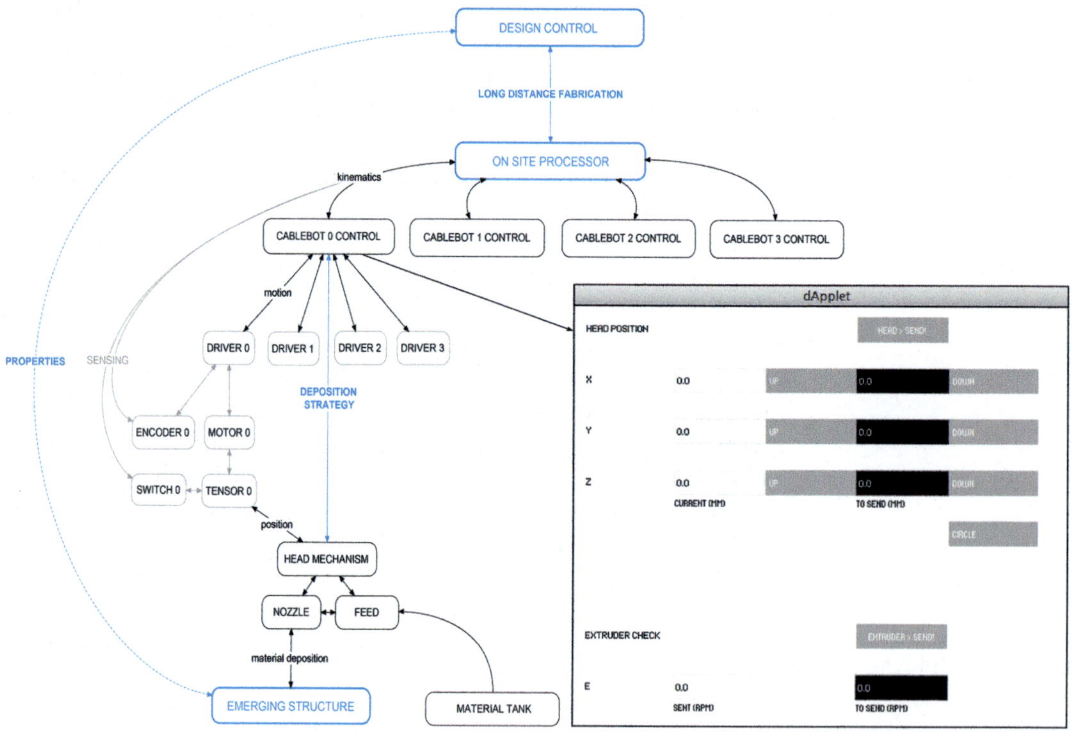

Fig. 1 Flowchart and applet setup mode interface capture of the virtual implementation of a large-scale distributed cable-suspended additive construction platform, Mediated Matter Group—MIT Media Lab

Distributed Construction: Background

The majority of current research efforts in distributed construction focus on the assembly of discrete components (e.g. blocks or beams) held together in ways that are not readily scalable (e.g. magnetism or friction) (Lindsay et al. 2011; Werfel et al. 2014). These systems are typically developed around specific modular or prefabricated components, which limit the range of possible geometries and applications of the resulting structure (Lindsay et al. 2011; Werfel et al. 2014). From a design perspective, such efforts focus either on duplicating existing rectilinear forms as made by conventional construction methods (Lindsay et al. 2011), or on simulation models that fail to be reproduced in physical environments (Tan et al. 2013, pp. 18–39). Few recent projects, such as the one presented here, explore distributed deposition of large-scale structures with tunable material properties (Naboni et al. 2014).

Additive Manufacturing: Background

Current additive fabrication approaches for digital construction are generally limited by 3 major constraints: (1) the typical use of non-structural materials with homogeneous properties; (2) the dependency of product size in the gantry size, and; (3) the typical need for support material throughout the layered deposition process (Augugliaro et al. 2014, pp. 46–64; Naboni et al. 2014; Oxman et al. 2014b, pp. 108–115). A distributed approach to manufacturing carries potential to radically transform digital construction by (1) digitally fabricating structural materials with heterogeneous properties (Oxman et al. 2012, pp. 261–274); (2) generating products and objects larger than their gantry size (Oxman et al. 2014b, pp. 108–115); and (3) supporting non-layered construction by offering novel fabrication processes such as free-form printing and robotic weaving (Oxman et al. 2013b, 2014a, pp. 248–255). Building on our previous work, we propose a work-in-progress distributed multi-robot approach to additive construction at architectural scales.

Behavioural Model for a Cable-Suspended Robotic Construction System: Implementation Strategy

Hardware

Machine control firmware was developed in C and C++ language using micro controller boards (Arduino Mega 2560). The boards distribute serial signals to stepper motors (Gecko 6723-400-4) via the Probotix Bi-polar 7.8 A drivers. The motors are NEMA 23 in size and are rated for a holding torque of 2.83 Nm. The drivers permit a maximum current of 7.8 A and are powered separately from the electronic controls with a 48 V power supply. Constant force spring motor assemblies (Stock Drive Products/Sterling Instruments, ML 2918) are used to spool up excess cable as well to keep tension on the pulleys. The micro controller receives feedback data from incremental rotary encoders (Yumo, A6B2-CWZ3E-1024, 1024 P/R Quadrature) and custom made zero switches comprised of copper contact and a connecting copper element attached to the cable at the right length. Each agent is suspended via four straight centre stainless steel cables which are encased in a helically wound nylon/polyurethane sleeve (Stock Drive Products/Sterling Instruments, Synchromesh, 1.6 mm outer diam.). Each custom-built extrusion head assembly is composed of a stepper motor with a rubber seal and a custom extrusion screw. Lead weights are applied for stabilization of the extruder head; cable fixtures are attached to four incoming cables with machined plastic housing and a material supply inlet Fig. 2. The material feed for each head is composed of a pressure pot containing paste-like material fed to the extrusion heads by narrowing the flexible tubing diameter towards the extrusion head Fig. 2.

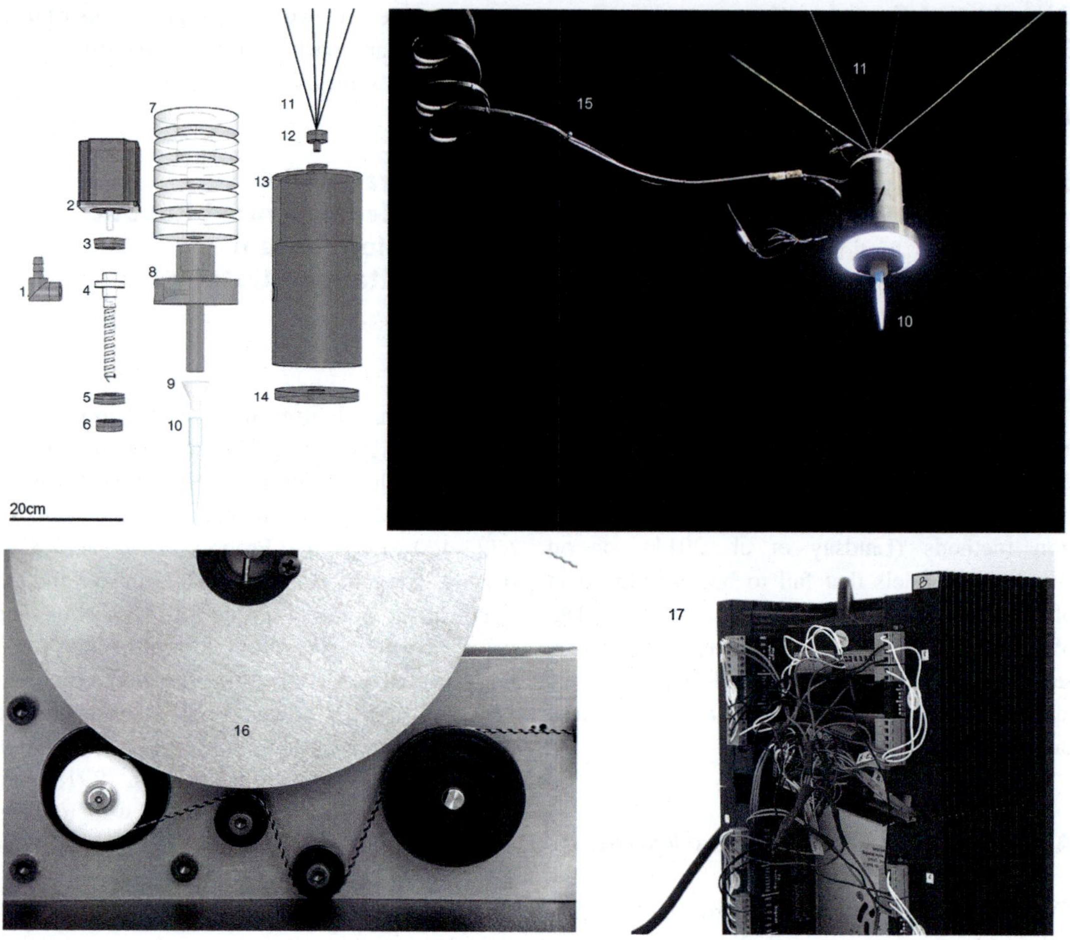

Fig. 2 Electro-mechanical implementation for a cable-suspended construction system. The extruder head assembly is composed of material inlet (*1*), stepper motor (*2*), rubber seal (*3*), custom screw (*4*), rubber seal (*5*), U seal (*6*), lead weights (*7*), screw mixing chamber (*8*), HDPE custom nozzle (*9*), pipette tip (*10*), synchromesh cable (*11*), cable fixture (*12*), delrin housing (*13*), and end cap (*14*). Material is fed via hierarchical tubing (*15*). Spring motor assembly for synchromesh cable (*16*), and electronic control assembly (*17*). Mediated Matter Group —MIT Media Lab

Software

The distributed system is implemented in a Java language customized applet enabling real-time 3D representation of the agents' behaviour. There are two sets of functionalities in the applet relating to two modes, the *setup mode* and the *building mode*.

The *setup mode* takes as input data each agent's envelope dimensions and base. The envelope dimensions of each agent are measured in 3D physical space, where the top corners of each envelope are placed parallel to the construction base plane. The agent's base is an origin point measured in space where the "robot builder" is at rest. With this information at hand the system is able to determine initial cable lengths and to reset motor encoders as a starting point for subsequent construction behaviour. Figure 3 shows the setup interface where initial data is read from motor encoders, and then used to position extruder head as well as to test the extrusion stepper motors.

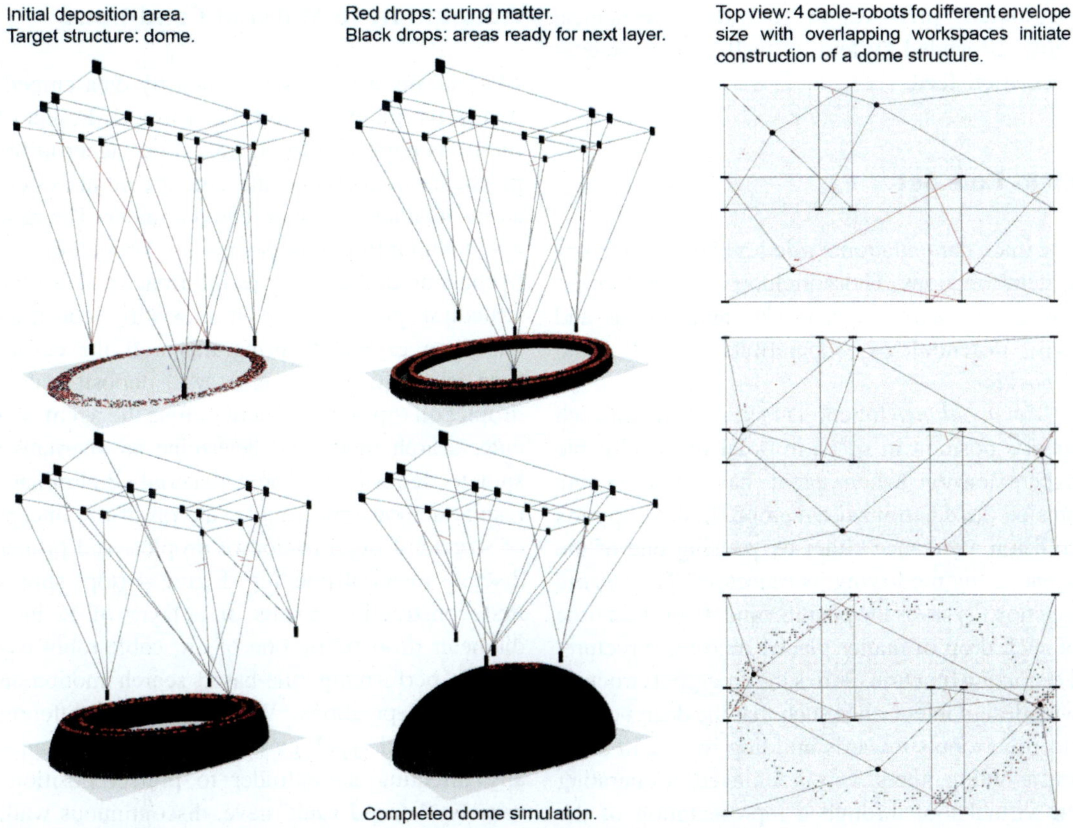

Fig. 3 Virtual simulation of a cable-suspended construction system. Robot envelopes are differently shaped and overlap with each other in order to collaboratively build a structure that is bigger than each individual machine's gantry. Mediated Matter Group—MIT Media Lab

The *building mode* function calculates the trajectories and temporal positions of each agent employing linear trajectories and constant velocity. We use a 3D modification of the Bresenham algorithm (1965, pp. 25–30) to move and track the agents in space via shortening or lengthening one of the four cables that are assigned per agent. The system computes discrete close approximations to linear trajectories from any 3D origin to any 3D target in the agent's envelope. Trajectory corrections are applied to avoid collision although agent envelopes may overlap, allowing for co-construction in specific and designated areas.

Behavioural Model for a Cable-Suspended Robotic Construction System: Rule-Sets and Adaptation

Interesting design opportunities emerge when more than a single material deposition node shares a construction space. The implications of robotic collaboration are vast and must relate to challenges such as agent awareness to boundary condition, envelope sharing across agents, real time multi-agent 3D positioning protocols, means for collision avoidance, and distributed preservation of the mechanical cable-suspended system as well as the material deposition feed.

Main Rule Set

The main computational rule set encodes five key system functions. Those include three centralized operations—*avoidance*, *storing* and *linking*, and two decentralized operations—*search* and *deposition*.

The *avoidance* function (1) keeps track of each robot's position in space in order to avoid cable hyperextension when agent navigation occurs outside the determined envelope. It also supports collision avoidance either by pausing one of the agents or by modifying its trajectory. The *storing* function (2) saves the position and deposition time of each drop of matter placed onto the structure. The *linking* function (3) ties the emergent structure with design intent rules such that the designer can steer the robots towards building in certain areas and avoiding others. This is achieved by operating the virtual tool through a representation of the physical structure as it is being collaboratively built Fig. 3.

Included in the decentralized operations, is the *search* function (4) designed to enable the agents to explore their envelope spaces and determine an adequate deposition location. During search mode, the robots navigate in 3D by employing bouncing trajectories from their maximum envelope until a z-axis threshold is trespassed; then, the agents verify the possibility of depositing material with the central system. Given a 3D position for additional material—if the relative height of the neighbouring structure and the curing time of the underlying drops are adequate —a new drop will be deposited. During the construction phase, the z-axis threshold is modified in order to adapt to the current height of the construction. Finally, the *depositing* function (5) consists simply of depositing a material droplet in a specific position as well as relaying time and coordinates to the central system.

Adaptation to Material Conditions

Material deposition is informed by data embedded in the material itself. Each time a droplet of matter is deposited by a cable-robot agent in the physical environment, the virtual central system stores its data in a clock-based counter. The next agent attempting to deposit a new droplet on top of the stored one, receives information about the structural properties of the existing construct based on expected curing times. If the curing time is adequate, the agent will deposit a new droplet on top of the structure. Else the agent will enter search mode and determine an alternative spot to deposit the stored material. Preliminary results demonstrate small-scale proof-of-concept of structural organization of droplets and proper discrete material bonding. Figure 4 (top) shows first construction results of a layer of ¼ inch diameter droplets by one of our cable-robot extruders performing rule-based search motion in between depositions. We have tested different droplet-based typology configurations by manually directing an extruder to pre-set positions namely; 4-faced vault, nave, discontinuous wall, and cantilevered continuous wall Fig. 4 (bottom). The material system used is industrial soft putty filler paste composed of gypsum plaster from hydrated calcium sulphate and glue. The virtual and physical systems are currently ready for the implementation of real-time feedback using humidity sensors or thermal cameras reporting to the central program.

Adaptation to Design Intent

The construction strategy presented here complies with global electro-mechanical constraints in a rule-based system for the generation of form, while maintaining an adaptation strategy for direct remote intervention. The designer sets up the system to build a structural typology (e.g. a dome, an arch, a column array etc.), and lets the behavioural model initiate its construction in a bottom-up manner. The cable-robots negotiate

Fig. 4 *Top* Initial results of robotic deposition of layered droplets performed by one cable-robot extruder. *Bottom* Preliminary manual experiments in soft plaster deposition including a 4-faced vault, a nave, a discontinuous wall, and a cantilevered continuous wall. Mediated Matter Group—MIT Media Lab

the construction space without top-down specifications for which agent will build what section of the structure. However, in case of local structural instability or in case of design iteration during the building sequence, the designer can steer the agents toward abandoning an area or focus on completing another. This technique enables the emergence of form through robotic node-to-node communication by applying space negotiation rules for each drop deposition. Exploration of this feature is not possible with continuous layering of material employed in traditional 3D-printing extrusion technologies, such as fused deposition modelling (FDM) (Oxman et al. 2014b, pp. 108–115).

Evaluation and Discussion

We designed and built a partly centralized partly decentralized digital fabrication environment of cable-suspended robots. This environment demonstrates the first steps towards the design and construction of a novel fabrication technology made up of multiple fabrication nodes that is designed to support cooperative construction of large-scale structures. The research explores themes of asynchronous motion, multi-nodal fabrication, lightweight additive manufacturing and the emergence of form through fabrication. Importantly, the project points towards a new design *workflow* that is directly informed through fabrication, material and environmental constraints, true to the Material Ecology approach (Oxman et al. 2015, in press) and characteristic of Fabrication Information Modelling (FIM), a novel methodology and framework that we are currently researching (Duro-Royo and Oxman 2015, in press). Although the mechanical hardware requires further development in order to achieve a fully functional large-scale implementation, the computational workflow shows promising results in simulation. Indeed, the successful deployment of a small-scale prototype embodies the benefits of combining a top-down centralized approach to fabrication with a bottom-up decentralized approach. Small size proof-of-concept fabrication was achieved using the described computational workflow and hardware, demonstrating that such an approach and system are feasible. Specifically, improvements to the mechanical system such as higher zero-switch reliability to avoid strain on the hardware at failure must be implemented. In terms of dynamic control, the choice of a 1.6 mm outer diam. Synchromesh cable proved to be challenging in terms of precision and overall dynamic strength. In terms of material supply, continuous feed via a pressure pot could be replaced by a cartridge approach, potentially providing a more stable print head without any physical constraints or interference from the material feed.

Systems such as the one outlined in this paper can be deployed for large-scale construction by attaching a distributed fabrication system to existing objects in the built environment. Cables from each robot can be connected to stable high points, such as large trees or buildings (Oxman 2014bb, pp. 108–115). Such actuation arrangement can enable movement over large distances without the need for conventional linear guides. A cable suspended system is straightforward to set up for mobile projects and affords sufficient printing resolution and build volumes.

In future implementations, sensing feedback can play a key role in the design and development of the agents' rule-based behaviour. By means of 3D scanning and thermal imaging, structural stability and material behaviour can be monitored and fed back into the model. Real-time comparison of virtual and physical built volumes, as well as the aforementioned material property tracking function, can contribute to closing the gap between virtual-to-physical design workflows; as well as opening up new and exciting opportunities for innovative long-distance fabrication environments where the designer provides input from afar.

Acknowledgements This research was conducted by the Mediated Matter Group at the MIT Media Lab. Ideas, methods, products and techniques were developed to support on-going group research focusing on large-scale distributed fabrication systems. The authors would like to thank Mediated Matter alumnus Jared Laucks for his contributions to the project, as well as the MIT Media Lab and the 2013 Lisbon Architecture Triennial for their support.

References

Augugliaro F, Lupashin S, Hamer M, Male C, Hehn M, Mueller MW, Willmann JS, Gramazio F, Kohler M, D'Andrea R (2014) The Flight Assembled Architecture installation: Cooperative construction with flying machines. Control Systems. IEEE 34(4):46–64

Bresenham JE (1965) Algorithm for Computer Control of Digital Plotter. IBM Systems Journal 4:25–30

Duro Royo J, and Oxman N (2015) Towards Fabrication Information Modeling (FIM): Four Case Models to Derive Designs informed by Multi-Scale Trans-Disciplinary Data, Symposium NN: Adaptive Architecture and Programmable Matter: Next Generation Building Skins and Systems from Nano to Macro, 2015 MRS Spring Meeting, MRS Online Proceedings Library, Cambridge University Press

Erol S (2005) Swarm robotics: From sources of inspiration to domains of application. Swarm robotics. Springer, Berlin Heidelberg, pp 10–20

Lindsey Q, Mellinger D, Kumar V (2011) Construction of Cubic Structures with Quadrotor Teams. Robotics Science and Systems, Los Angeles, CA

Naboni R, Paoletti I (2014) Advanced Customization in Architectural Design and Construction. Springer Publishing Company, Incorporated

Oxman N, Ortiz C, Gramazio F, Kohler M (2015) Material ecology. Computer-Aided Design 60. pp 1–2, ISSN 0010-4485

Oxman N, Laucks J, Kayser M, Duro-Royo J, Gonzales-Uribe C (2014a) Silk Pavilion A Case Study in Fiber-base Digital Fabrication. In: Gramazio F, Kohler M, Langenberg S (ed) Proc. Fabricate. pp. 248–255

Oxman N, Duro-Royo J, Keating S, Peters B, Tsai E, (2014b) Towards Swarm Printing. Architectural Design. Special Issue: Made by Robots: Challenging Architecture at a Larger Scale 84(3). pp 108–115

Oxman N, Keating S, Mogas-Soldevila L (2013a) Bio-Beams: Final Project Report. NSF, Cambridge, MA

Oxman N, Laucks J, Kayser M, Tsai E, Firstenberg M (2013b) Green Design, Materials and Manufacturing Processes. Taylor & Francis Pub. ISBN: 978-1-138-00046-9

Oxman N, Firstenberg M, Tsai E (2012) Digital Anisotropy. Journal of Virtual and Physical Prototyping (VPP) 4(7):261–274

Tan Y, Zheng Z (2013) Research Advance in Swarm Robotics. Defence Technology 9(1):18–39

Werfel J, Petersen K, Nagpal R (2014) Designing Collective Behaviour in a Termite-Inspired Robot Construction Team. Science 343(6172)

Wilson E (2000) Sociobiology: the new synthesis. Belknap Press of Harvard University Press

Design Equilibrium of Form, Materiality and Fabrication: A Bacterial-Inspired Multidisciplinary Optimisation Strategy for Free-Form Concrete Structures

Frédéric Waimer and Jan Knippers

Abstract

Geometric complexity of freeform concrete structures in modern architecture hardly allows for an economical realisation anymore and traditional methods for construction, fabrication and planning seem increasingly inadequate. Through a recently developed hybrid CFRP-concrete composite construction technique and an automated high-precision production method, it was possible to develop a new strategy which facilitates providing a new freedom of form in architecture, economically and in high quality. In order to reap the full benefits of the structural properties and to reduce production costs, the authors propose a multidisciplinary design optimisation (MDO) strategy. Through an integrative simulation model, it is possible to bring manufacturing, structural behaviour and geometry in an optimum ratio. As optimisation algorithm, a stochastic method was developed, which is inspired by the foraging of bacteria. The biological role model, the transfer to a numerical algorithm and the practice by means of case studies will be presented in the following paper.

Introduction

For architects, digital design constitutes an increasingly important tool in the conceptual phase, and has certainly been one of the causes for the increase of geometrically complex structures in the past years. Increasingly complex geometric requirements for the building project can hardly be realised economically on site any longer. Thus, with respect to freeform concrete structures for example, the production of the complex formwork is one of the main challenges (Fig. 1). It is predominantly done in manual labour, which can account for up to 60 % of the overall cost of a concrete structure (Nawy 2008). After the formwork has been used once, reutilisation is not possible due to its unique shape.

F. Waimer (✉) · J. Knippers
Institute of Building Structures and Structural Design (itke), Stuttgart, Germany
e-mail: info@itke.uni-stuttgart.de

M.R. Thomsen et al. (eds.), *Modelling Behaviour*, DOI 10.1007/978-3-319-24208-8_26

Fig. 1 Formwork for concrete structures (*Image* B+G, design to production)

Geometric complexity also leads to problems concerning the load bearing behaviour of the structure. Since mainly aesthetics and functionality define shape, the rate of bending is accordingly high.

The implementation of freeform shell structures is therefore limited to only a few construction projects. At this point, the authors pursue a novel approach and believe that they can unlock potential for innovation and development through the use of fibre-reinforced polymers (FRP) in order to resolve the difficulties described. Crucial advantages are the range of different processing techniques and the good mechanical properties of this material that allow the production of components with complex shapes.

Integrative Concept

Through integration of an automated production method and by the help of developed simulation programmes for CAD, CAE and CAM, the authors were able to originate a new concept that provides an economical, high-quality freedom of form.

For that it was necessary to conceive an integrative design and optimisation strategy and hereby generate the ideal overlap between geometry, structural behaviour and production (Fig. 2).

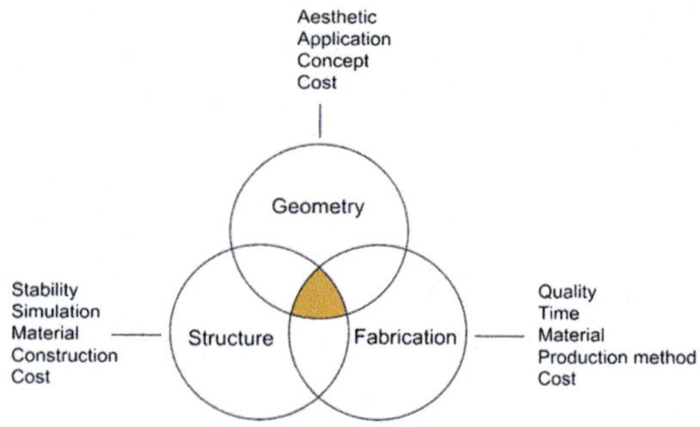

Fig. 2 Ideal overlap after successful optimisation

Developed Production Process

The developed production method to fabricate individual components of batch size 1 is based on the classic filament winding technique. This process was one of the first automated production methods in fibre-reinforced composite technology. The basis of the process is the continuous winding of rovings at a given speed around a rotating positive core. The filament winding method is used for conical and cylindrical components with closed cross-sections. Classically, large numbers of pipes or GFRP tanks are produced with this technique, the core in this case being made of steel. First approaches to waive such a winding core, such as in the so called coreless winding, led to interesting results (La Magna et al. 2014). The filament winding technique qualifies for the new construction for

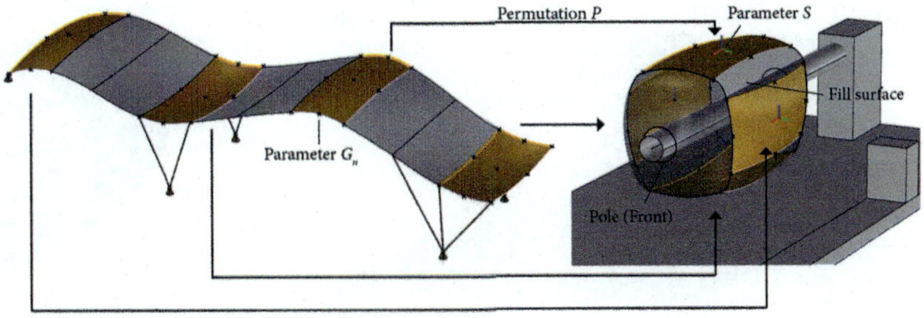

Fig. 3 Concept of the core generation

several reasons. It allows for the production of large components with high filament volume content. Hereby filament orientation and filament layout can be adjusted to the mechanical load. Moreover, the filament winding method is inexpensive in comparison to other production techniques (Bader 2002).

Here, the method of the assembled core is proposed. Components of a global geometry are arranged in such a way that they form a core geometry. Assembly of the components on the core must fulfil the criteria for winding. Figure 3 shows the schematic process of the core generation. A freeform geometry that is subdivided into single components is the starting point. Simulation of the core generation process can be classified into the following four steps:

1. Categorisation of the components
2. Generation of the fill surface and the poles
3. Simulation of the winding process
4. Analysis of the production behaviour

Step 3 and 4 are of high significance for the evaluation of an economical production. It serves, on the one hand, to define an applicable winding pattern and, on the other hand, to ensure winding around the core is flawless. Winding is characterised substantially by the bridging effect. Bridging occurs during the winding process when the roving loses contact with the core because of the core geometry. Consequences are a decrease in pre-stress in the roving, the roving deviates from its desired position, and the cross-section of the roving constricts in a circular shape. This means for the material characteristics of the component being produced that strength and stiffness of the laminate are weakened.

Consequently, the winding process depends on the geometry of the components and the corresponding assembled core (Fig. 4a). Thus, no statement can be made in terms of producibility of the components without a simulation and an analysis of the winding process. As opposed to commercial programmes that work with discretised surface models, the authors pursue a method of analysis based on differential geometry. Thus, the roving can be abstracted into a curve, which makes it possible to identify the exact curvature orientation at any

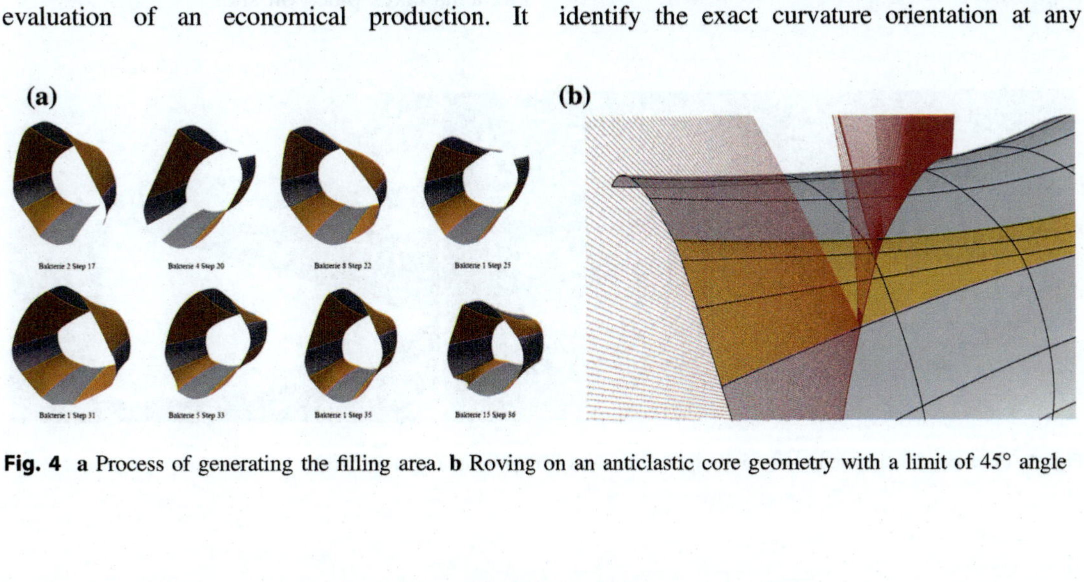

Fig. 4 **a** Process of generating the filling area. **b** Roving on an anticlastic core geometry with a limit of 45° angle

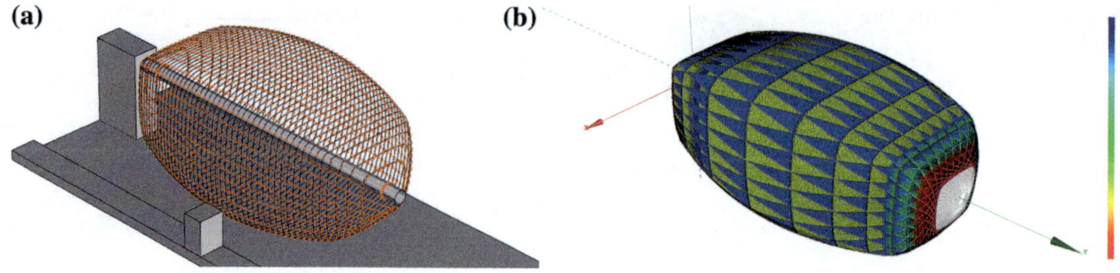

Fig. 5 **a** Data model for the determination of the winding pattern. **b** Winding pattern and slipping analysis

point on the curve. If the curvature is not continuous or if a change in curvature takes place, this means the roving lifts and bridging occurs. Accurate modelling at this point pushes the boundaries to the limits of the production method (Fig. 4b).

In addition to the examination of bridging, determining the winding pattern plays a crucial role for the mechanical properties of the laminate. The data model in Fig. 5a defines the corresponding winding pattern under a certain winding angle. The winding angle to be considered depends on the desired structural properties of the component. After the simulation of the winding process it is investigated whether the laid roving maintains its intended position and does not slip (Fig. 5b). Once planning the production is completed, CNC data for the laying paths are generated from the winding pattern. For the last step in the process—trimming the components off the core—milling paths are generated on the basis of the simulation model (Fig. 6).

Construction Method and Load Bearing Effect

An essential part of the new construction method is the so called sandwich composite formwork (Fig. 7a). The sandwich composite formwork consists of a foamed core with a bilateral, 2 mm thick surface layer of CFRP. The sandwich structure offers sufficient stiffness for the deadweight of the fresh concrete before hardening. In order to guarantee for a form-fit connection between the materials after the concrete has hardened, the sandwich system is perforated with blind holes (Fig. 7b). That allows the concrete to fill in the indentations and to interlock with the sandwich after it has hardened. After production, the formwork is assembled on the construction site, provides the shape during concreting, and ultimately remains in the structure. In case of bending moments in a shell, the CFRP formwork absorbs tensile forces and the concrete compressive forces. Assemblage of the formwork elements takes place on site.

Fig. 6 Process of assembled-core winding and finished SIP-participating formwork

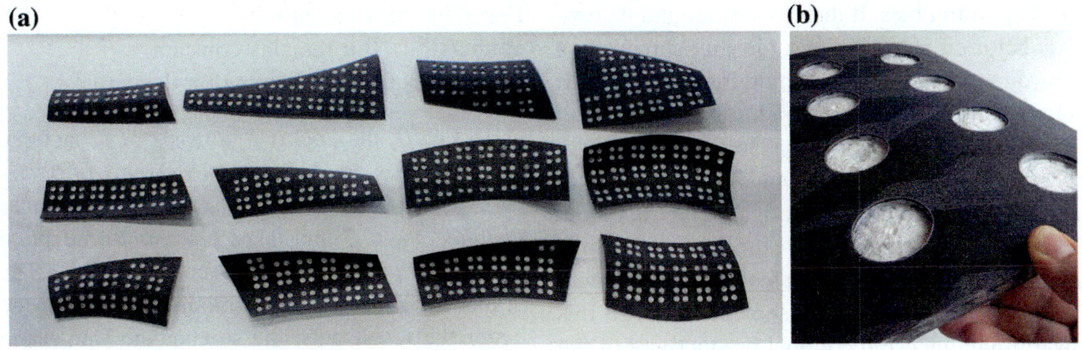

Fig. 7 Sandwich formwork composite with blind hole. **a** Different components. **b** Close up

Efficiency of the new construction method is essentially influenced by its shear characteristics. These have been determined in experiments (Waimer et al. 2015). Besides the determination of the shear characteristics, the properties of the connecting materials (adhesive and rivet connection) and of the winding laminate dependent on the winding angle were determined (Waimer 2015). With the help of these tests the characteristic values for the simulation models have eventually been identified.

Multidisciplinary Optimisation Strategy

Integrating the various requirements of the production and the final structure is a challenging task. The use of an optimisation algorithm promises to create the optimum ratio of form, structural behaviour, material arrangement and production.

Biological Role Model of the Optimisation Algorithm

The manner of food intake and the right conversion are decisive for the survival of any individual. E. coli bacteria belong to the oldest living beings on earth and have been refining their search for nourishment over the millennia. They consist of a plasma membrane, a cell wall and a capsule which in turn is made up of cytoplasm and nuclides. The pilus is used as a means of genetic transfer and the flagella enables them to move (Fig. 8a).

Foraging takes place in a collective of several bacteria. The behaviour of a single bacterium is depicted in Fig. 8b. Each bacterium generally has the ability to move or to tumble. Straight movement allows the bacterium to follow a path with more nourishment in all direction. However, if the bacterium finds itself in an environment with low concentration in nourishment, or if it is

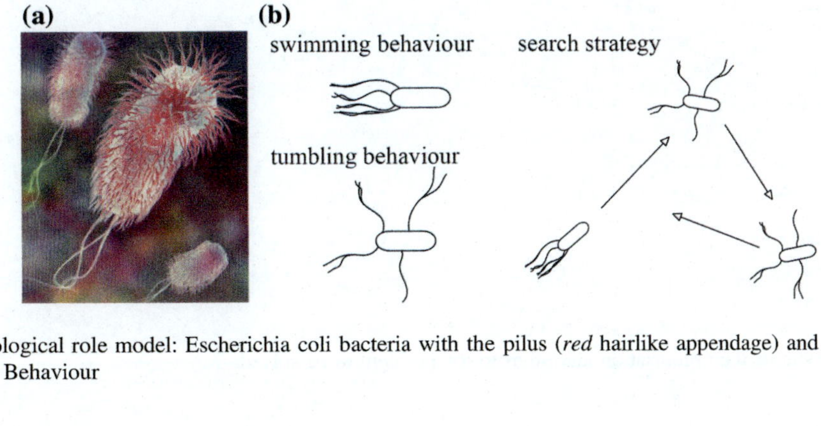

Fig. 8 **a** Biological role model: Escherichia coli bacteria with the pilus (*red* hairlike appendage) and the flagella for protrusion. **b** Behaviour

used up, it tumbles. It then rotates around its own axis before it starts looking for nourishment in a new direction with a swimming move forward. If a bacterium is in a low nourishment environment for a longer period, it dies. However, if a bacterium is in an environment with good sustenance, it multiplies through cell division. Moreover, by releasing an attractant, it informs other bacteria that it is located in a sphere with high concentration of nourishment. Through this swarm intelligence the global productivity of the colony increases. A detailed description of these bacteria and their collective behaviour in the search for nourishment are provided in (Passino 2002).

Transfer and Digital Implementation

The numerical implementation of the bacteria's searching behaviour will be presented by means of an exemplary three-dimensional problem with a global minimum (Fig. 9). The solution space is given in the following function:

$$J(\theta), \theta\, \varepsilon\, \mathbb{R}^p$$

θ represents the bacterium's position. Therefore the concentration changes depending on this variable:

For $J(\theta) < 0$ it increases.
With $J(\theta) = 0$ it remains constant.
For $J(\theta) > 0$ the concentration decreases.

In Fig. 9, the *colony S* consists of 4 bacteria. The colony is divided into two *groups of bacteria* N_g. The value of each bacterium's *swimming distance* N_W is 3 and takes place in an adaptive method dependent on the concentration. The higher the concentration of nourishment is or the better the result gets, the smaller the step range becomes. The *chemotactic step* N_C, which is defined as 2 in this example, describes the live span of a bacterium before it dies or multiplies. The number of *swimming steps* N_S defines how many steps occur in the chemotaxis and how many times. The *reproduction step* N_{re} represents the outmost loop and takes place here one time. It determines how often the die-off of bad bacteria and the cell division of better bacteria take place. The behaviour of the algorithm can be seen most clearly on bacterium 3. After it has run through all chemotactic steps, it jumps back to the position with the best result. Afterwards, reproduction of bacteria sets in. Bacterium 3 finds itself in an environment with lower nourishment concentration in comparison to that of bacterium 2 and consequently dies. Bacterium 2, however, multiplies and goes through the typical slumping and swimming movements again.

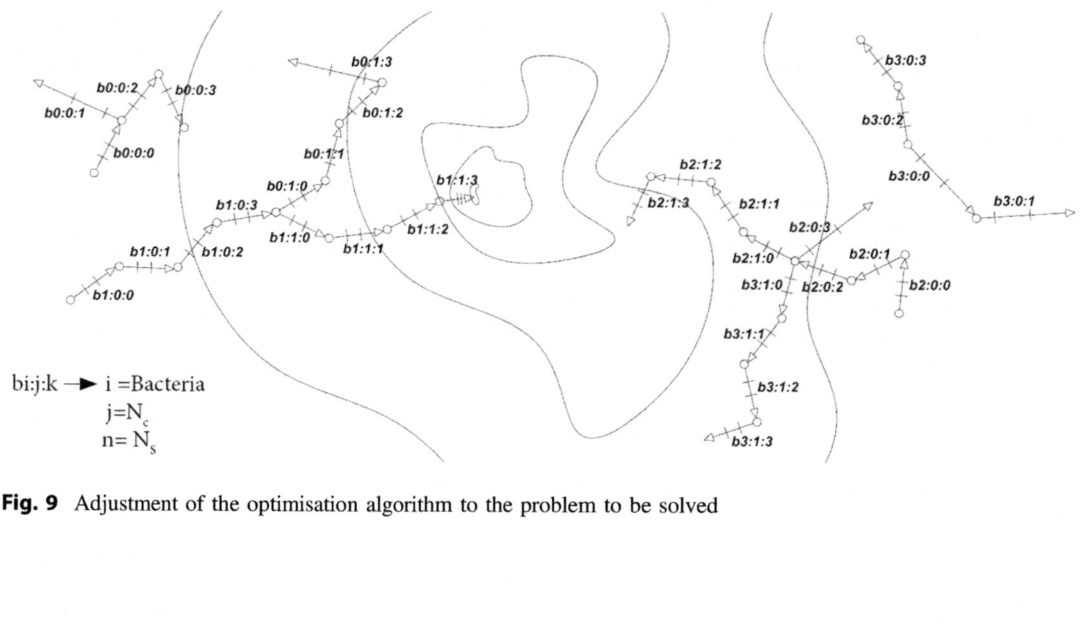

Fig. 9 Adjustment of the optimisation algorithm to the problem to be solved

The authors' implementation of the bacterial searching strategy is different from the one in Passino (2002) in several respects, because the algorithm was tuned to solve the computational task as fast as possible. First, the use of random distribution of bacteria is left aside and second, the concentration is not measured over the entire live span of a bacterium. The bacterium possesses the ability to memorise the location of the highest concentration and to jump back on it. Furthermore the step range of the swimming movement is not constant, but adaptive. The vector change adapts to the level of concentration and is determined through the previous result. A decisive difference lies in the introduction of groups within the colonies. This has the advantage that calculations for bacteria can be run on separate CPU cores through *parallel programming* (Freeman 2010). In this way the efficiency of the algorithm can be enhanced significantly. That led to an increase in the rate of convergence by up to 75 %. With the help of test functions, comparative tests with an evolutionary algorithm were performed too. The computation time of the proposed algorithm was 11 times less.

Digital Framework

The framework developed for this project was created in coding language C#. *Rhino3d* offers an open source platform for C# developers. There the optimisation algorithm as well as the simulation of the production process were implemented as a C# class. The structural behaviour was simulated using the FE software *Ansys*. Control of the programme and administration of model parameters was also done with the use of a C# class. Calculations of structural and production behaviour took place simultaneously so as to reduce the computation time during one iteration of a bacterium (Fig. 10).

Variables of Optimisation and Objective Function

The variables influential for the specific planning models are depicted in Fig. 11. The global geometry is described by the parameters G_n. They manipulate the radii of curvature of each component and therefore influence structural behaviour and fabrication. The parameter of *permutation P* defines the array of the components for each core (Fig. 3). It possesses a strong influence as it makes the solution space increase fivefold. The parameters of the *rigid-body movement S* define the position of each individual component in the core. For each component of a core exist 6 parameters, 3 rotations and 3 translations. Thus, this means that there is a maximum of 24 degrees of freedom available for each core. The parameter for the *fibre angle*, *F*, is essential in the analysis of the core winding. Contrary to the parameters of the core, this parameter also manipulates the structural behaviour of the global structure. Parameter *L* describes the impact of the thickness of the

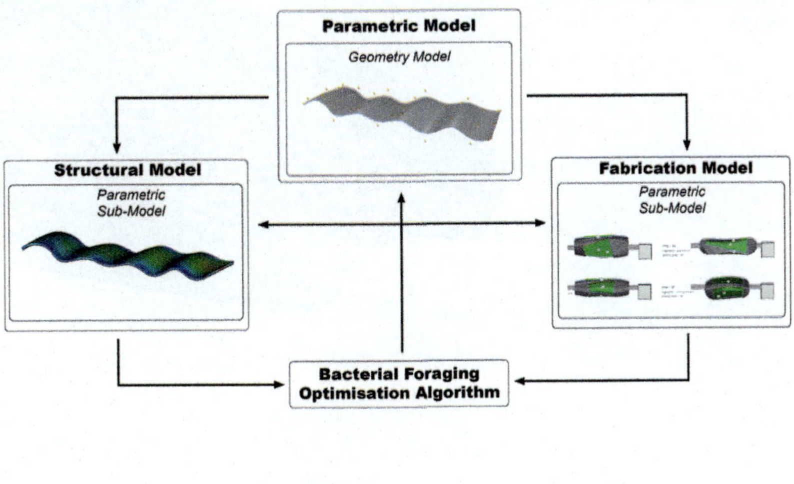

Fig. 10 Multidisciplinary optimisation with simultaneous simulation of structure and production

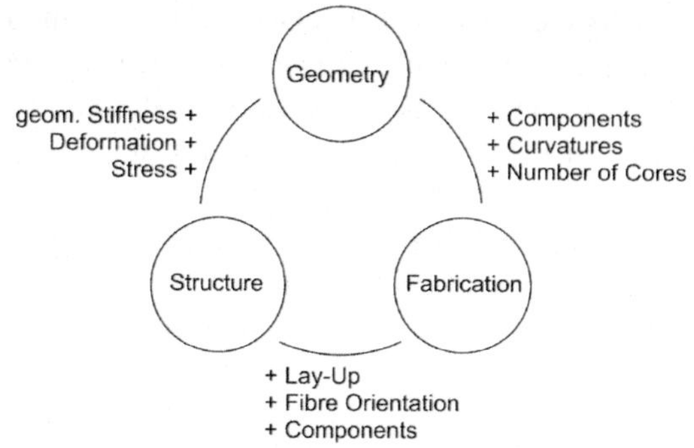

Fig. 11 Interactions in planning and essential parameters

laminate layer. It allows for the stiffness as well as the potential tensile strength of the laminate to be increased. Through parameter of concrete height, *B*, mostly the bending strength can be improved.

The optimisation variables significantly define the size of the solution space and influence the simulation models in various ways. It is possible to determine the priorities within the three planning models individually. However, it is also feasible to balance the priorities of the three models individually in order to focus on aesthetic or economic advantages in the planning process. All this is achieved by the definition of the optimisation variables and the objective function. A further description of the influence of the variables, the objective function and the geometric pattern can be found in Waimer (2015).

Results: Designing and Applications

By means of the following two examples, the potential architectural applications of the construction and production method shall be outlined, and the efficiency of the optimisation strategy presented shall be demonstrated.

Wide Spanning Shell Structure

The shell structure in the first example has a length of approx. 34 m and a maximum width of nearly 18 m (Fig. 12). The structure is divided into 16 components and shall be produced using two synclastic and two anticlastic cores. This leads to 35 parameters, which influence geometry, production and structural behaviour.

The settings of the optimisation algorithm are as follows: S = 40, $N_g = 4$, $N_{re} = 60$, $N_c = 10$ and $N_s = 8$. In order to reduce the computational effort, the static model is not computed until the waste falls below a predetermined percentage (here: 45 %). This is incorporated into the optimisation by means of a *penalty function*. After a configuration of parameters has been established with low waste, a simultaneous calculation of the models is performed. The algorithm then also attempts to optimise the structural properties of the components through the fibre layout. At the same time it is considered that the components that create a common core must also have the

Fig. 12 Wide spanning concrete shell structure with G_n

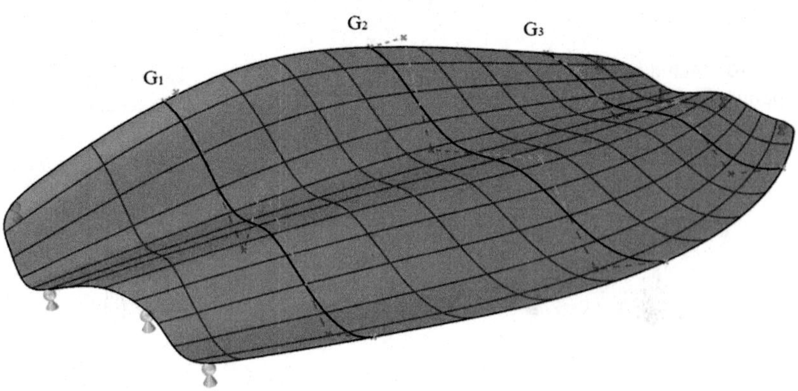

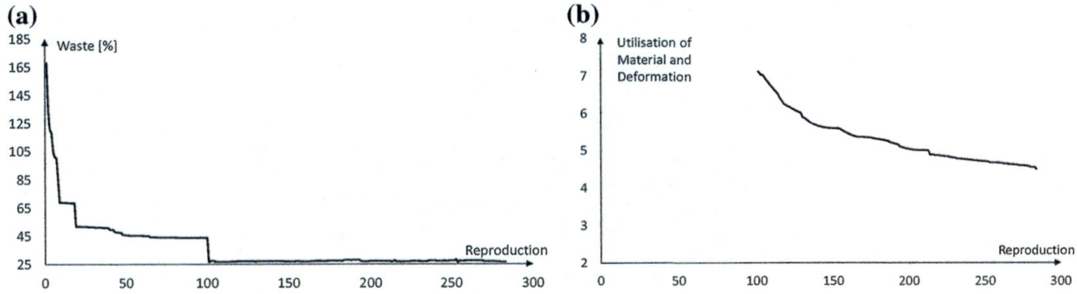

Fig. 13 Optimisation progress of production (**a**), and the structural characteristics (**b**)

same fibre layout. By parallel simulation of the production it is also guaranteed that the corresponding assembled core can be fit with fibre under this winding angle on the one hand and that its production is economically reasonable on the other hand. The optimisation aims to reduce thickness of the laminate layers. That has a positive economic effect on the material consumption of the structure as well as on the production time. One further influence on both models is created by the geometrical parameters, as they regulate the curvature of the surface. For the structural behaviour, curvatures prove advantageous because they generate membrane forces. For the production, however, it is favourable to reduce curvatures so as to minimise the percentage of waste.

The progress of the optimisation is broken down to two graphs and depicted in Fig. 13. Production costs were lowered thanks to the optimisation, while at the same time structural performance could be improved too. In addition to an increased stiffness of the system, material consumption was reduced by 40 %. Material utilisation could be brought down enough so as it does not exceed the specific values from the tests. Thereby the parameters of the global geometry changed only slightly. The divergence of the geometry to be optimised from the one to be initiated can be seen in Fig. 14b. The constellation of the assembled cores is depicted in Fig. 14a.

Physical Prototype

With the help of this example, planning as well as practical aspects of the production of formwork using the winding procedure were tested. In the production of the prototype, a winding machine was used. The core size is limited to a maximum diameter of 750 mm on this machine. The result of the completed integrative planning process is shown in Fig. 15.

Architectural scope of the hybrid construction is visible in Fig. 16. The built prototype, which consists of 14 different components, illustrates the producible components and their high quality. For production of the components, four cores were needed. Besides, the material for the core could be re-used several times. After assembling the formwork elements with an adhesive joint, a 3 cm thick concrete layer was applied manually.

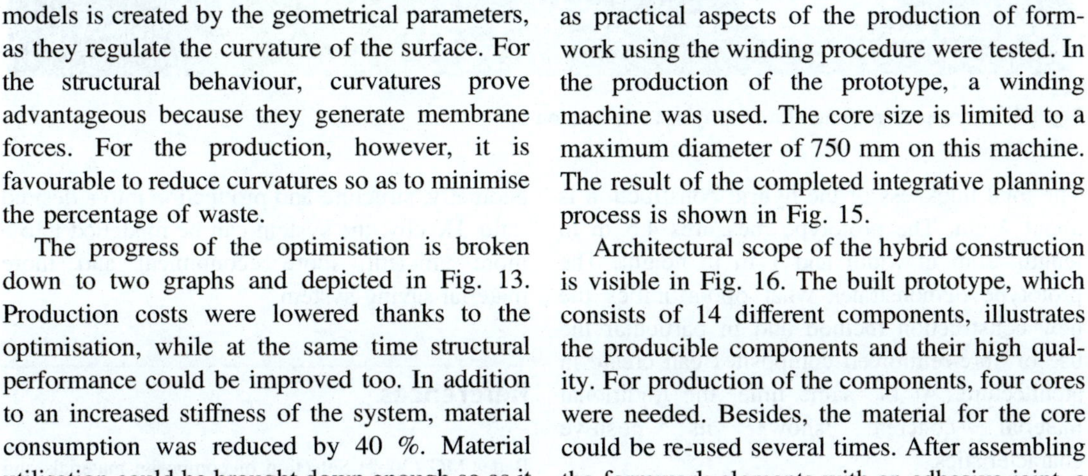

Fig. 14 **a** Assembled cores after the optimisation process. **b** Change of geometry

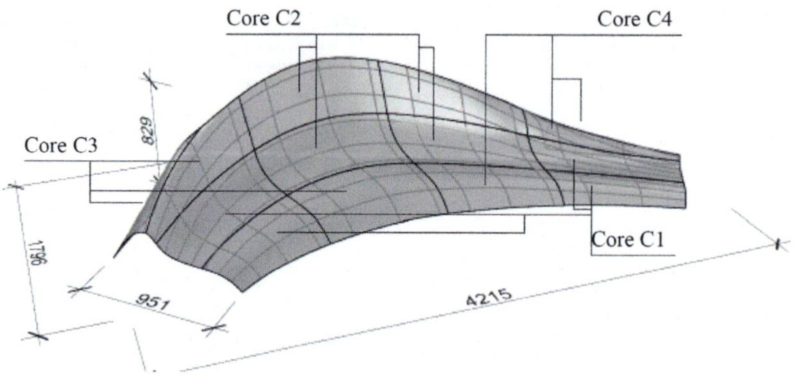

Fig. 15 The component divided according to fibre orientations

Fig. 16 Prototype consisting of a new hybrid construction

The total thickness of the hybrid construction is about 4 cm. The prototype measures 4.5 m in length, 2 m in width and 1 m in height. The prototype demonstrates what opportunities the new construction method and in particular the use of fibre-reinforced composites can create in architecture. At the same time, the traditional material concrete shows its positive characteristics.

Conclusion

In this paper the authors have presented a concept for freeform shell structures, which includes the aspects of structural design, construction method and planning. By means of the developed optimisation strategy it is possible to bring geometry, structure and production into a desired ratio. Hereby, the system can be modelled into a more efficient, more economical and more material-saving system.

References

Bader MG (2002) Selection of composite materials and manufacturing routes for cost-effective performance. Compos Part A Appl Sci Manuf 33(7):913–934

Freeman A (2010) Pro .NET 4 Parallel Programming in C#. Apress, Berkely

La Magna R, Waimer F, Knippers J (2014) Coreless winding—A novel fabrication approach for FRP based components in building construction. In: Proceedings of the international conference on FRP composites in civil engineering, Vancouver, Canada, 2014

Nawy EG (2008) Concrete construction engineering handbook. CRC press, Boca Raton

Passino KM (2002) Biomimicry of bacterial foraging for distributed optimization and control. IEEE control systems magazine

Waimer F (2015) Integrative Planungs- und Optimierungsstrategie von freigeformten Verbundflächentragwerken aus CFK und Beton. Dissertation, University of Stuttgart

Waimer F, Knippers J (2015) A novel hybrid composite construction for complex concrete shells in architecture. In: Proceedings of the international association for IASS Symposium 2015, Amsterdam, Netherlands, 2015

Design with Material Uncertainty: Responsive Design and Fabrication in Architecture

Felix Raspall

Abstract

Inspired by the smartness and creative opportunism of design practices in the informal city and by artists using found objects, this research develops the theoretical framework and enabling technologies for responsive design and fabrication. In this approach, the design project adjusts in real-time to take advantage of the contingencies of the fabrication process. The aim of the investigation is to expand scope of architectural design, embracing design and construction processes, which involve higher levels of uncertainty. As the main case, the research engages the creative challenges of designs that reuse discarded materials, in which the quantity, quality and properties of the available cannot be predicted with precision. Customizable and user-friendly computational systems seamlessly combining design computation, computer vision and robotics are developed and implemented to verify the feasibility. The paper presents theoretical findings, technology developments, and results of empirical experimentations, demonstrating the possibility of reproducing the adaptability of informal practices within the design disciplines.

Introduction

Construction practices in the informal city and in the work of artists who reuse found objects are often praised by architects for their resourcefulness and smartness. Their creative process, *informal production*, gives birth to useful and expressive designs out of what is immediately available, despite and by virtue of the large uncertainties about the materials and construction methods at hand. While the profession of architecture encourages complete designs before construction and requires standard and predictable construction methods, informal production leaves designs open-ended until the very moment of fabrication, extending the decision-making process into the construction process, when the material uncertainties are finally resolved.

F. Raspall (✉)
Harvard University Graduate School of Design, Cambridge, MA, USA
e-mail: raspall.felix@gmail.com

M.R. Thomsen et al. (eds.), *Modelling Behaviour*, DOI 10.1007/978-3-319-24208-8_27

This paper address the following question: Is it possible to create design projects that can automatically adjust themselves during construction, in response to the contingencies of the construction process? By investigating this question, this research aims to expand the scope of architectural design, enabling the use of a wide range of building materials and construction methods that involve uncertainties larger-than-usual and therefore remain incompatible with conventional professional methods.

The research examines design and construction processes under uncertainty from a theoretical and instrumental perspective and proposes *responsive design and fabrication* as a means to encode decision-making behaviours into the design project. The paper is organized into three parts. The first part presents a theoretical framework studying informal production vis-à-vis the well-established professional model, discussing the concepts of informal production and material uncertainty. The second part proposes responsive design and fabrication, developing the strategies, technologies and workflows to enable the design of responsive systems that encode the capacity to adjust the design in response to the contingencies of the construction process. The third and final part of the paper discusses the implications of an informal method for the discipline of architecture and debates the specific challenges and limitations of the developed technologies.

Background

Academic research exploring responsive design and fabrication in architecture presents both a theoretical and an instrumental dimensions. On a theoretical level, research studying the implications for the discipline are scarce. A previous article by the author discusses the consequences of real-time material feedback for architectural design (Raspall 2014) and his unpublished doctoral dissertation addresses provides a more detailed theoretical framework (Raspall 2015). A related study by Dörfler and Rust (2014) studies fabrication processes in which construction involves "vagueness" and investigate opportunities for designers.

On an instrumental level, two lines of research are closely connected to this research effort. The first one is on-site robotics, which studies how industrial robots need to be equipped and programmed to manage the irregularities of construction sites. The projects Endless wall and Stratifications by the Chair of Architecture and Digital Fabrication Gramazio and Kohler show investigate how digital workflows and feedback can be used to receive human instructions and manage construction tolerances using sensors (Helm et al. 2014). The second relevant line of research is the processing of materials of organic precedence, such as sawmill and food processing industries. Dealing with similar yet unique work-pieces, the production system acquire the exact characteristics of each piece using sensors and create custom one-off instructions. Microtec CT.LOG technology is an illustrative example (http://www.microtec.eu/en/solutions/logs/quality/ctlog).

Material Uncertainty Within the Professional Model and Informal Production

Architecture is an anticipatory practice. It brings to the present moment a building not yet exists with sufficient level of detail to secure its spatiality and constructability. The effectiveness of this creative process is supported by a controlled context. Uncertainties related to the construction process are controlled through standardized building materials and construction processes. However, many practices, usually considered outside the professional domain, are able to achieve useful results without a well-defined project. For research purposes, two abstract models represent the extremes of a spectrum: the professional model and informal production.

The Professional Model

The profession of architecture encourages the development of comprehensive projects in which all significant design decision are made before construction. The construction process is, ideally, a translation of a clearly defined project into built reality in which only minor design decisions are made. The conventional use of CAD, CAE and CAM technologies reinforces this structure further increasing the level of control. However, the professional model relies on a reliable infrastructure of standard materials and construction processes.

Informal Production

In real life, several practices defy the professional model. Laymen, artists and unconventional designers often undertake construction processes without a well-defined project and rely on design decisions done during the very moment of fabrication, using building materials and construction processes that involve much larger uncertainties. Informal production is ubiquitous but its creative process is poorly understood. Figure 1 analyses how a very simple shelter could have been conceived and created using estimated dimensions and materials and defined during the assembly process.

Responsive Design and Fabrication

As a means to mediate between (one) the requirement for complete and fixed projects in the professional model and (two) the limited control over construction materials and building processes which requires constant decision-making

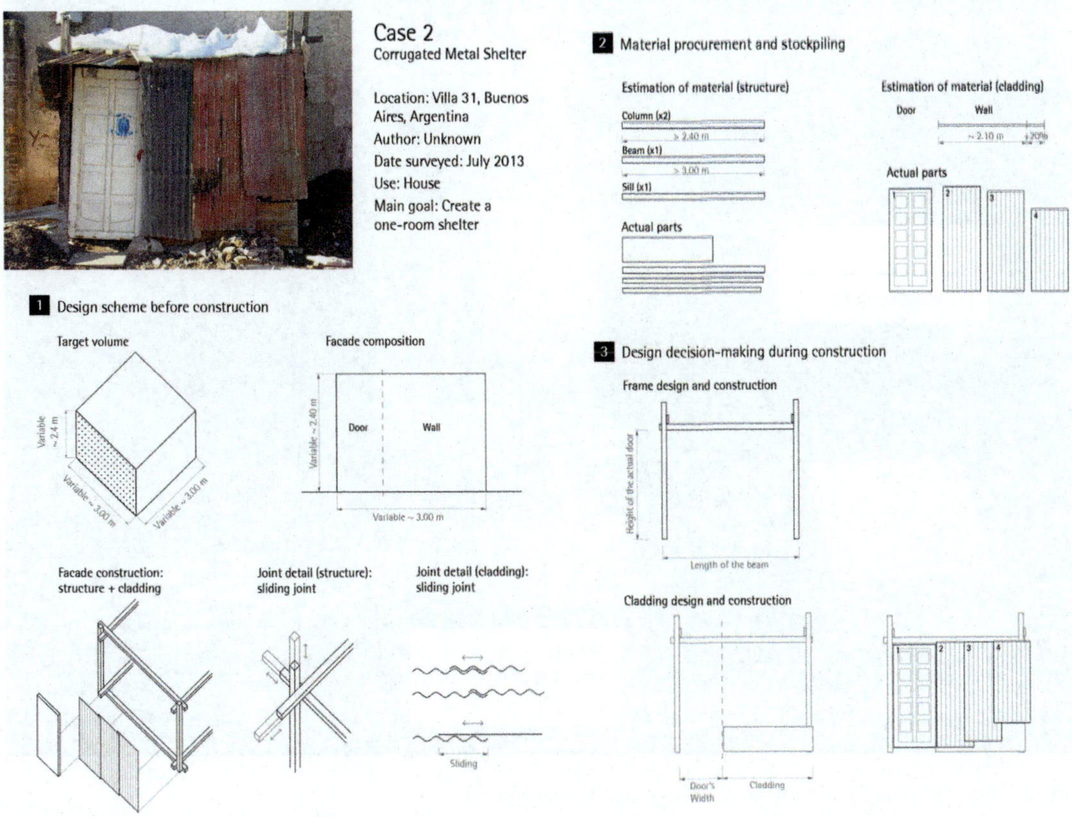

Fig. 1 Analysis of the creative process in the conception and construction of a simple shelter

during construction, the research proposes *responsive design and fabrication*. This approach considers the design project not as a final and fixed product but as a responsive system that encodes the required behaviours to automatically determine how the project should unfold during the construction process.

Such approach to design is enabled by the integration of three technologies: sensing (material scanning), controlling (design computation), and actuation (online robotic fabrication) (Figs. 2 and 3). Material scanning is used to obtain information from the specific construction materials at hand at the time of construction. Design computation encodes design decision-making rules in computer algorithms. And, online fabrication allows to send instruction on real-time as the project is defined during construction.

Although these three technologies are currently available, they remain disconnected from each other and there is no platform that integrates them in a user-friendly interface for designers.

IDR

To implement responsive design and fabrication, the research developed an application that integrates computer vision, online robotic control and design computation a user-friendly and customizable programming environment. In this way, the system is able to recognize the geometrical features of the materials at hand at the specific moment of construction (Fig. 4), and automatically determine how to organize the material to achieve a design goal (Fig. 5).

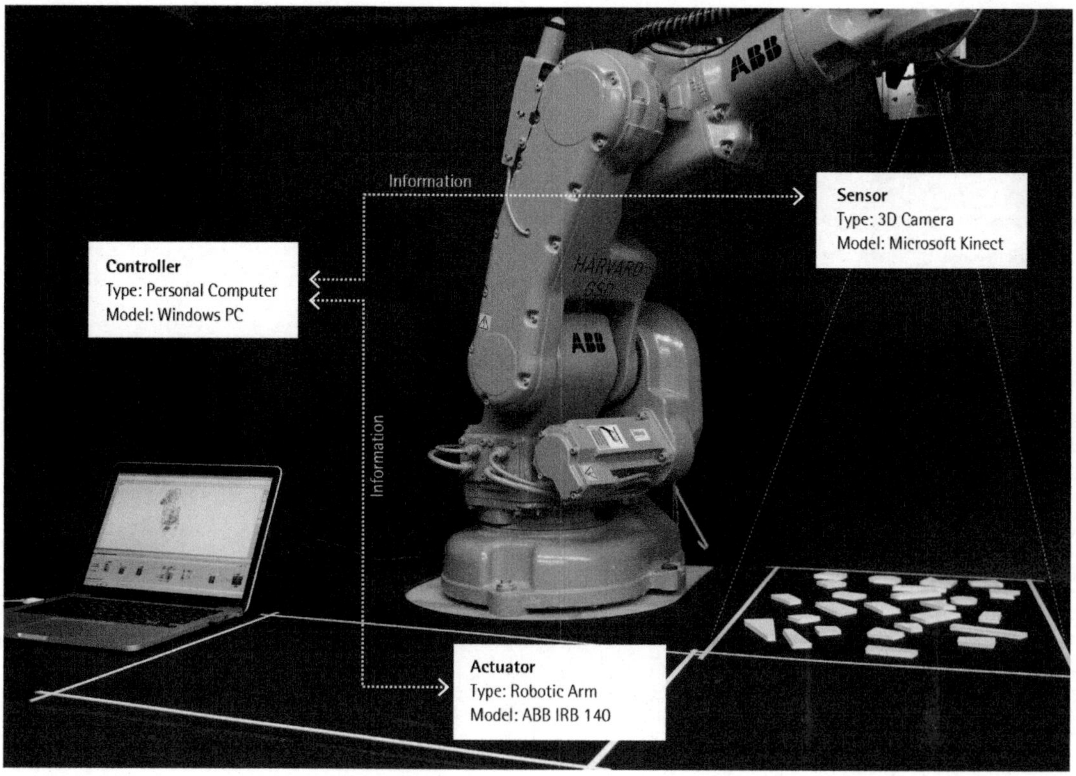

Fig. 2 Basic technologies in responsive design and fabrication

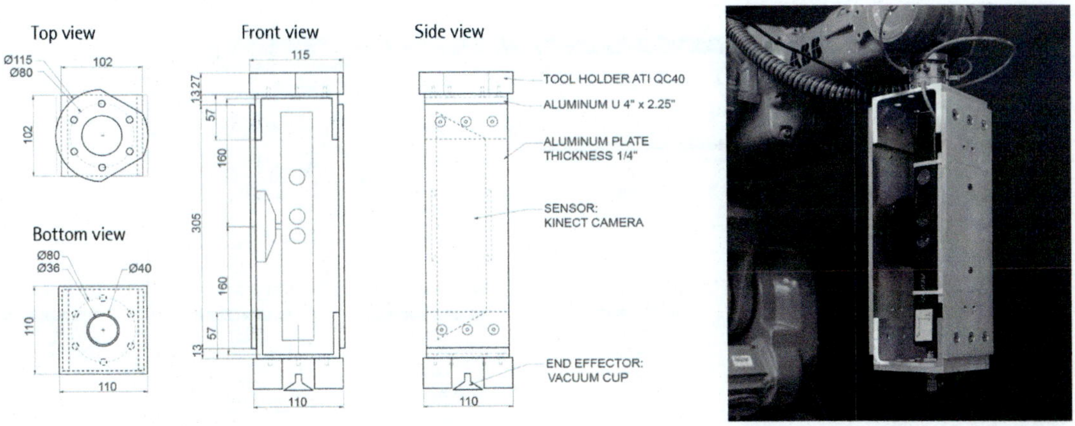

Fig. 3 Detail of the tool, which combines the 3D camera and the end-effector

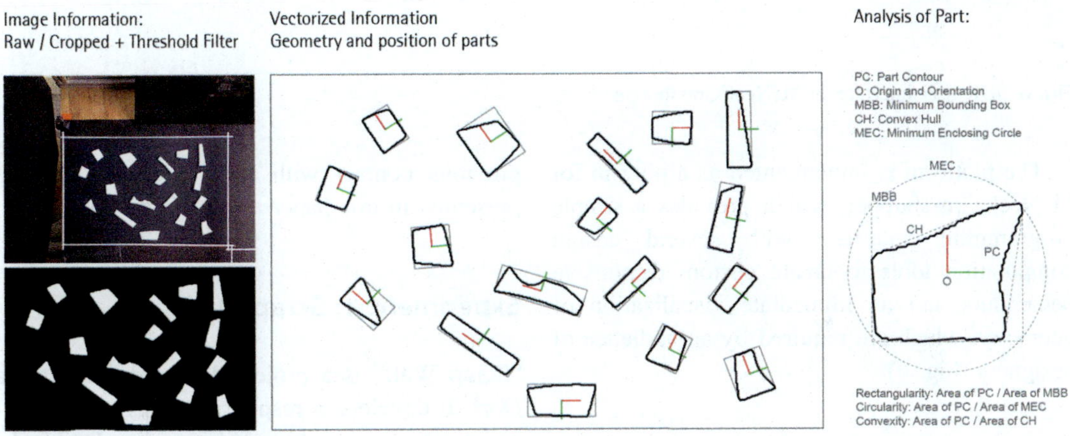

Fig. 4 Robovision functionalities include image trimming and filtering, contour detection and analysis

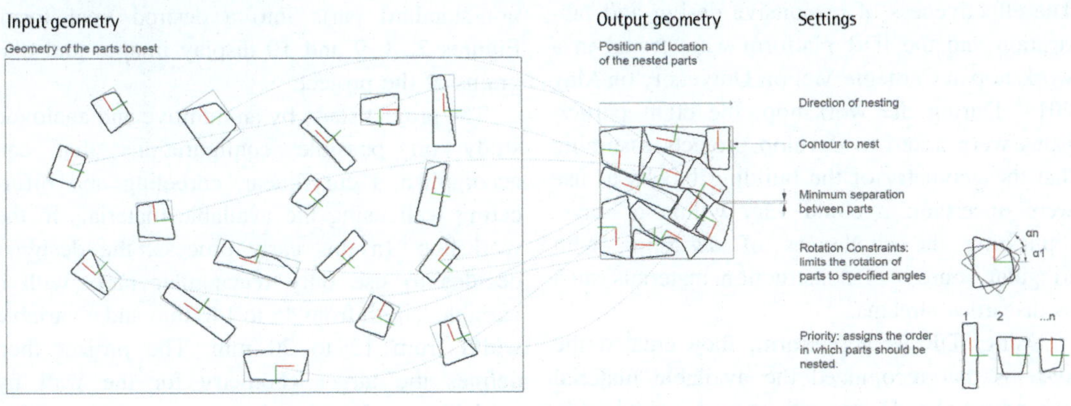

Fig. 5 Nesting process using the implementation of the geometric global NestLib application

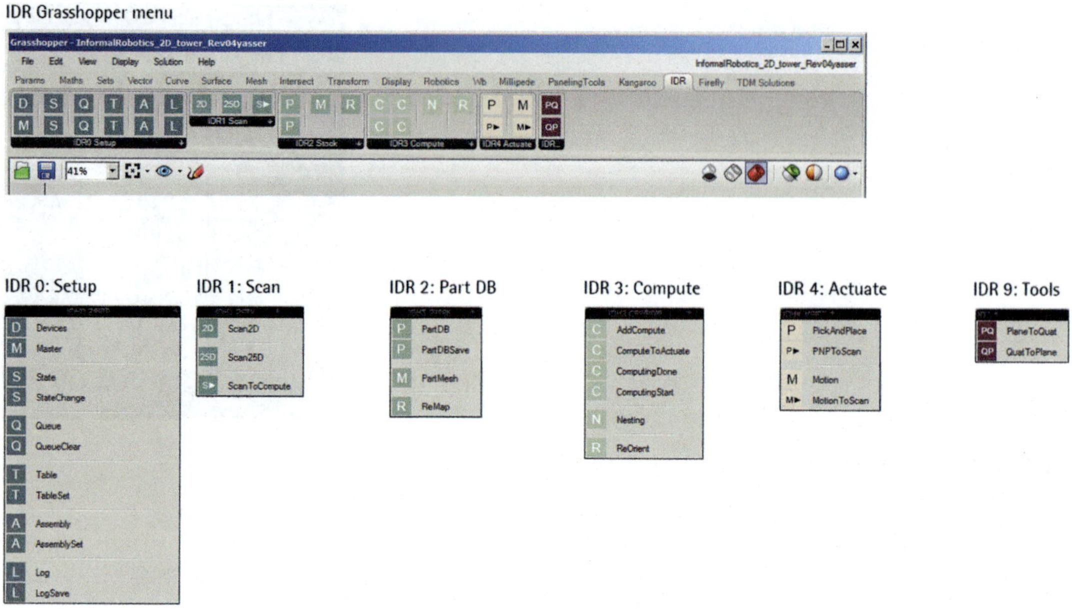

Fig. 6 IDR user interface in McNeel Grasshopper

The platform is implemented as a plug-in for McNeel Grasshopper, which provides a simple programming interface with several design computation tools to create custom responsive behaviours and an immediate visualization of geometry, which are required by an audience of designers (Fig. 6).

Two Design and Fabrication Experiments

The effectiveness of responsive design and fabrication and the IDR platform were tested in a workshop at Carnegie Mellon University on May 2014. During the workshop, the eight participants were asked to develop projects assuming that the geometry of the building blocks to use were uncertain; it could vary within a range, replicating the challenges of designing with irregular sources of construction materials such as discarded streams.

Using IDR as a platform, they created the systems that recognized the available material and adjust the design project and construction process accordingly. The participants had no previous contact with IDR. Two projects are presented in this paper as a proof of concept.

Experiment 1: Scrap Wall

"Scrap Wall" is a project designed by Antonia Dorf. It develops a responsive design and fabrication process capable of using irregular blocks in the creation of a wall. The project deals with a very frequent problem in designs that reuse discarded materials: the arrangement of various non-standard parts into a desired final form. Figures 7, 8, 9 and 10 display images and diagrams of the project.

The project starts by an intuitive and analogue study on possible configurations that can accomplish a curvilinear, corbeling and bifurcating wall using the available materials in the workshop. In this early process, the designer decides to use only rectangular parts with a variable length from 45 to 140 mm and a variable width from 12 to 20 mm. The project then defines the target geometry for the wall by modelling a 3D boundary representation. This geometry is then sectioned with horizontal

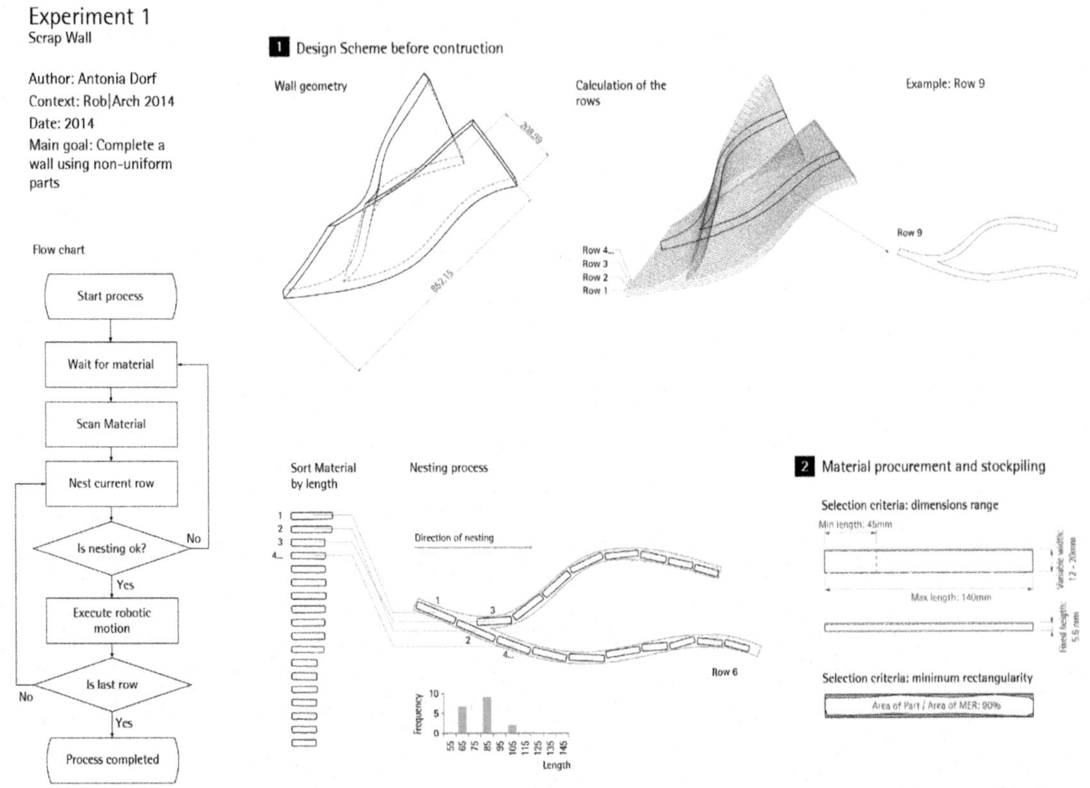

Fig. 7 Experiment 1

planes, resulting in the boundaries for each row in the wall (Fig. 7).

The system operates in the following way. The system scans the parts and recognizes their 2D geometries. Parts that do not meet the selection criteria are filtered out. Then, the system calculates the arrangement of the parts in the current row. The decision-making algorithm is relatively simple: parts are sorted from longer to shorter and then nested inside the row's contour. If the nesting process is successful, the system starts the execution of the pick and place operation. The level of "success" is measured by the ratio of the area of the nested parts to the area of the contour to nest. Figure 10 shows the Grasshopper script. The system repeats the cycle until it runs out of parts. In that case, the system stops and waits. The system follows an incremental decision-making approach, one row at a time.

The experiment 1 produced the following results:

A. Fluidity: The system operated continuously and smoothly. It only required human intervention when the available parts were insufficient to complete the current row.
B. Customizability and user-friendliness: Using IDR's platform, the development of the responsive system was a simple process. The system was designed and programmed by Antonia Dorf with basic knowledge of Grasshopper and minimum knowledge of robotic fabrication, based on a template provided by the workshop.
C. Capacity to achieve the design goal: The system successfully produced the desired wall using irregular sourcing of building parts. The final output matches the predefined form and

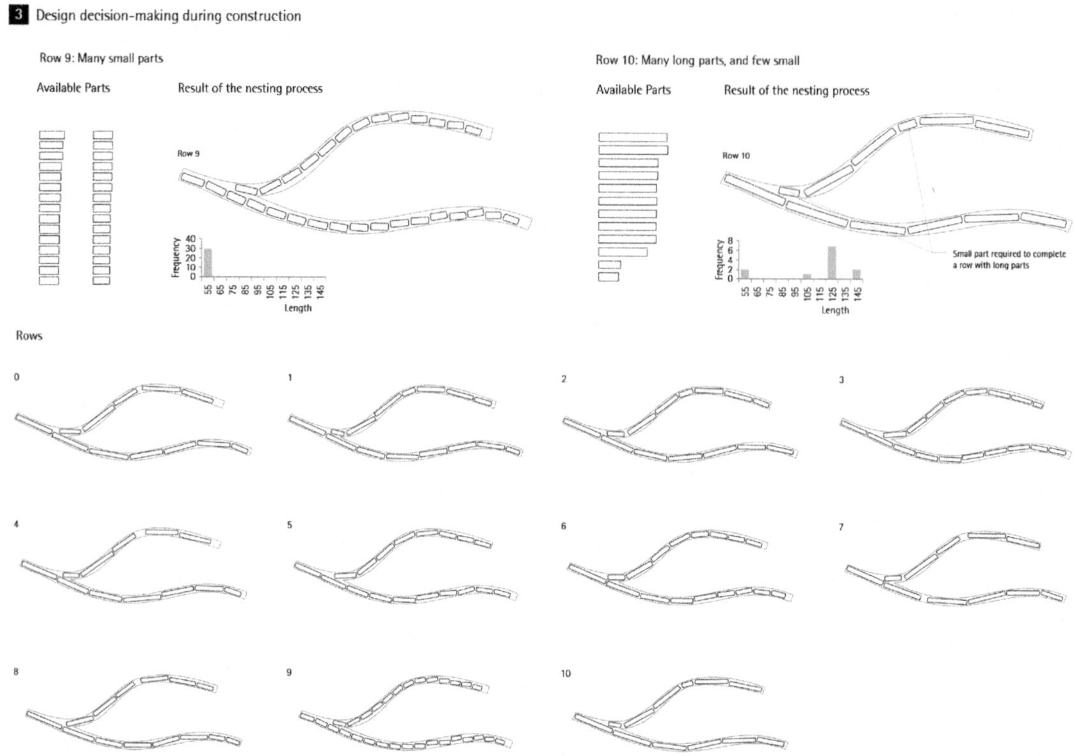

Fig. 8 Experiment 1

Fig. 9 Experiment 1

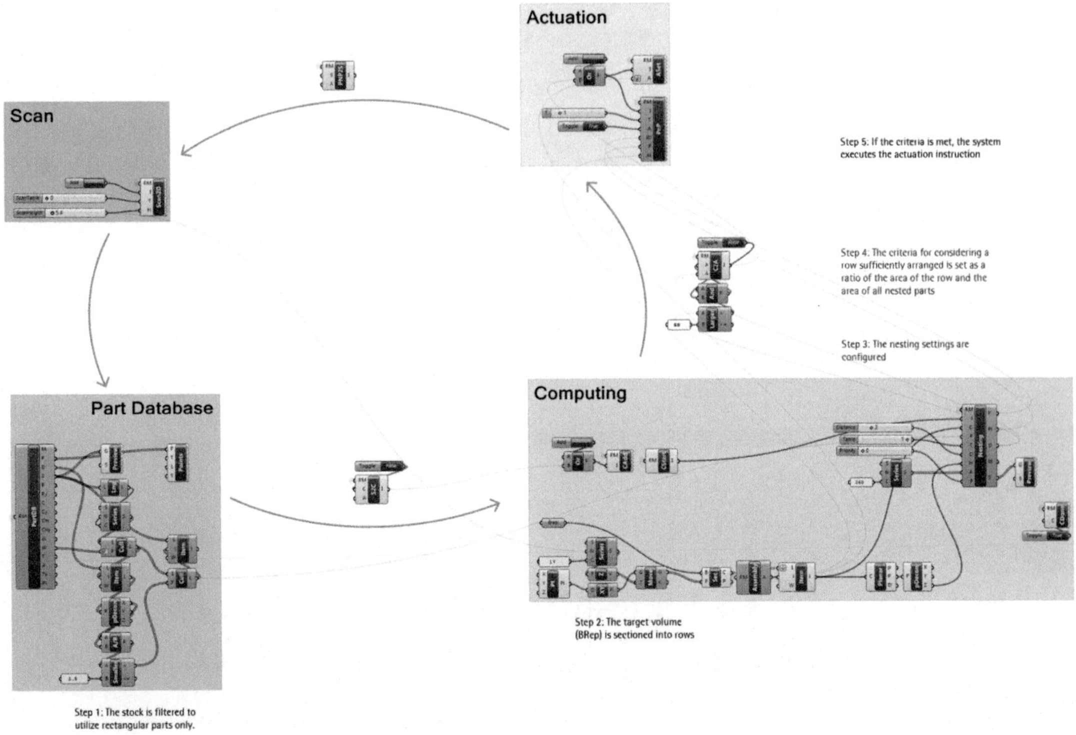

Fig. 10 Experiment 1

achieves a particular irregular texture which is a consequence of the randomness of the material supply (Fig. 9).

D. Sensitivity analysis: The project tested different assortments of parts for each one of the rows. Figure 8 displays the final arrangement for each row. In Row 9, the system operated with only small parts. In Row 6, the system worked only with medium length parts. However, when the system attempted the design of a row with only long parts, it could not find combinations that successfully cover the contour with the available pieces. In Row 10, this issue was solved using few small parts to compensate the spaces left by the large pieces. Other rows tested different combinations of lengths of parts.

Experiment 2: "Scrap Column"

The project "Scrap column" was designed by Pedja Gavrilovic. It develops a responsive system capable of arranging irregular parts in the creation of a column-like organization. In contrast with the Experiment 1, the final form of the product is not explicitly defined in this project; it is determined by the explicit arrangement logic and the available building materials at the time of construction. Leaving the final form open-ended facilitates the use of irregular materials by removing a geometric constrain that may be too demanding and therefore unfeasible. Figures 11, 12, 13 and 14 display images and diagrams of the project.

The design consists of a stack of linear parts, arranged in pairs alternating one pair perpendicular to the next one. It shares the same selection criteria as the Experiment 1: rectangular pieces with a variable length (from 45 to 140 mm), a variable width (from 12 to 20 mm).

The system operates in the following way. It scans the available parts and recognizes their geometries. Using this information, parts that do not meet the selection criteria are filtered out. With the information on the available parts, the

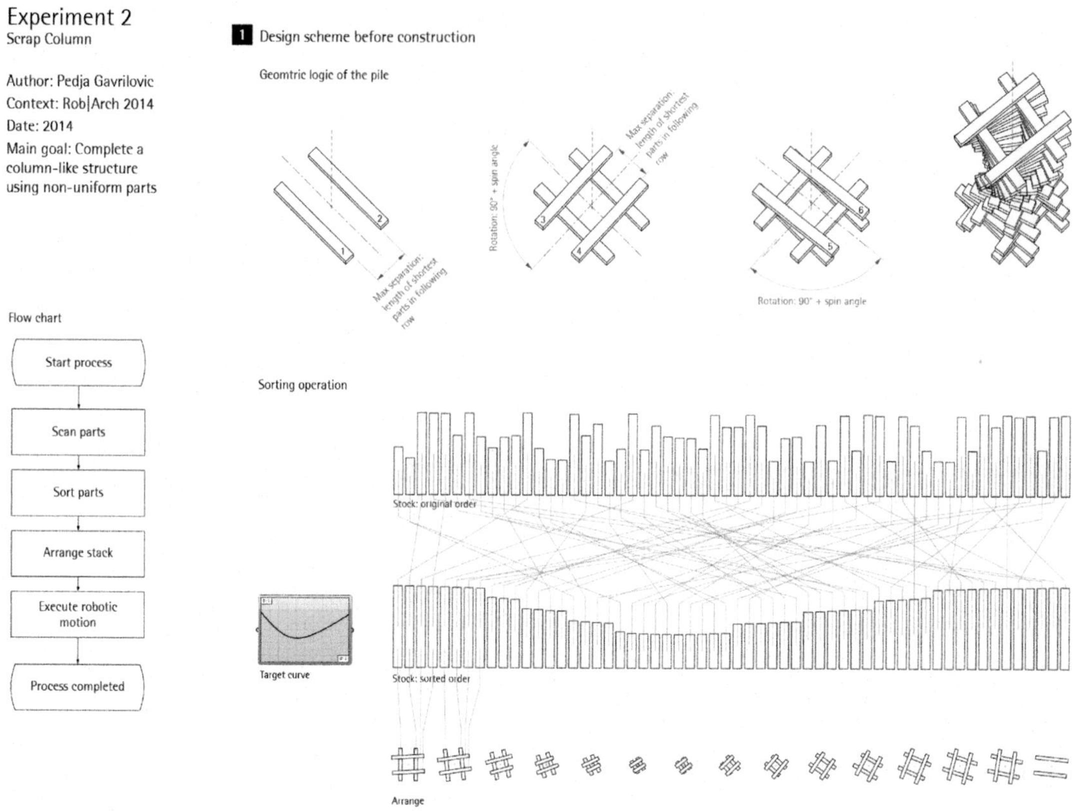

Fig. 11 Experiment 2

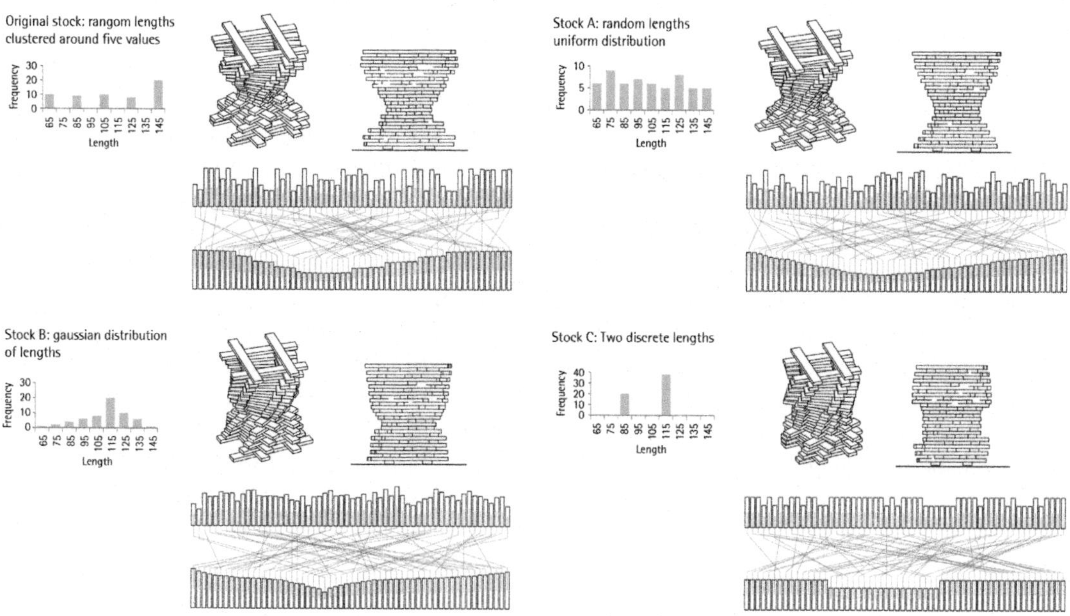

Fig. 12 Experiment 2

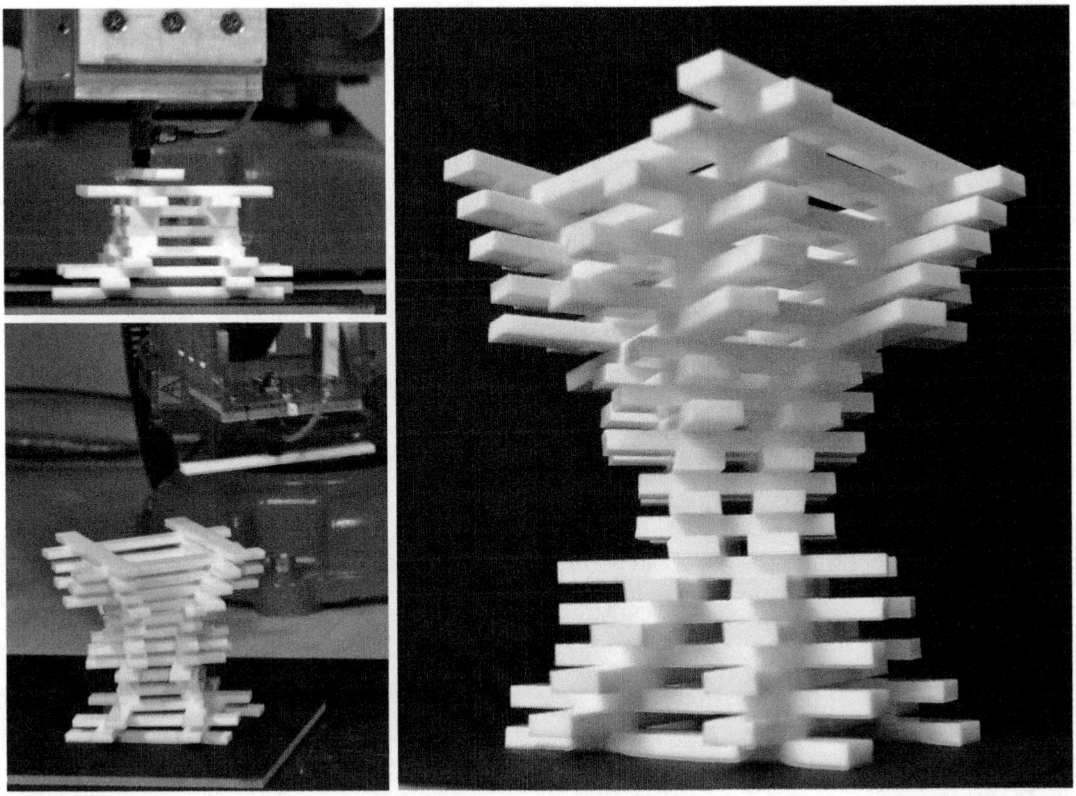

Fig. 13 Experiment 2

system calculates the final design and executes the construction.

The decision-making algorithm is relatively simple. Figure 11 displays diagrams of the decision-making process. Available parts are sorted by length according to a desired profile for the column, defined by a mathematical curve. Then, the sorted parts are grouped in pairs and arranged in a stack, where each row is perpendicular to the previous one. A small angle increment provides a torsion effect to the whole organization. The separation between the two parts in each pair is determined by the length of the pieces that will rest over it, as a way to secure that each part has a place to rest when placed. Although the target curve is defined before construction, the final arrangement heavily depends on the available parts. Then, the system starts and completes the fabrication process.

The experiment 2 produced the following results:

A. Fluidity: The system operated continuously and smoothly. Once the parts are placed on within reach of the robot, the system performs the scanning, computing and assembly reliably without human intervention.
B. Customizability and user-friendliness: Using IDR's platform and a template, the development of the system was a simple process. The system was designed and programmed by Pedja Gavrilovic with intermediate knowledge of Grasshopper and robotic fabrication. Figure 14 displays the Grasshopper script developed for this project.
C. Capacity to achieve the design goal: The system successfully produces columns for different material inputs. The final output accomplishes a stable column-like structure, in which the final form is partially dependent on the available parts. Figure 13 shows the final output of the assembly process.

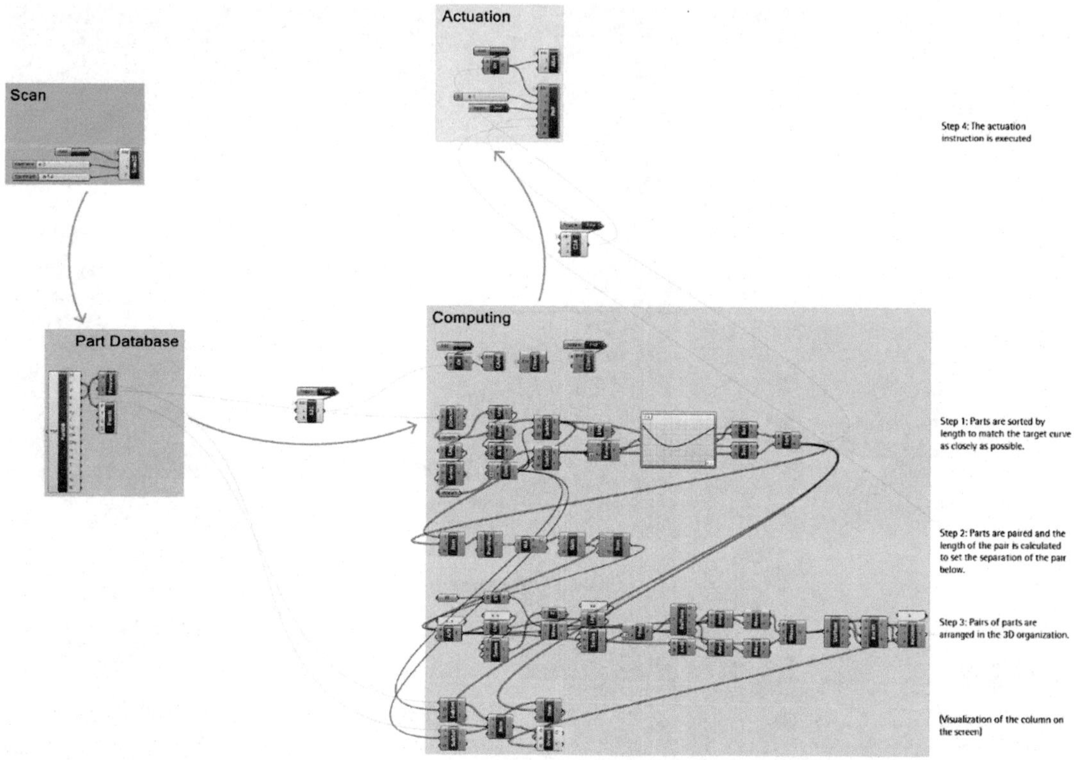

Fig. 14 Experiment 2

D. Sensitivity analysis: After the workshop, the project was tested using different assortments of parts. Figure 12 displays the results for each one of the collections of parts. The original stock, the one used during the workshop experiment, contained parts of different lengths clustered around specific lengths (60, 80, 100, 120 and 140 mm). Stock A, B and C tested the effect of set with uniform distribution, normal distribution (Gaussian) and only two discrete types of parts.

Conclusions

The results of the experiments prove the initial feasibility for responsive design and fabrication. Experiments demonstrate that designers were able to effectively create responsive systems that can embed design decision-making capacity to their projects, so the designed system can refine and adjust their configurations in response to the available materials. The IDR platform was instrumental in the creation of these system and the operation was smooth.

Limitation and Future Research

This research presents a basic framework for responsive design and fabrication and a platform to implement its workflow by designers. However, it is an early research and critical issues remain to be studied to accomplish real-life implementations. The experiments and the IDR platform works well in the controlled environment of the fabrication lab and at a scaled down model.

Three main areas of research should follow this investigation:

- Scaling up: Although the workflow presented in this paper is independent of the scale, real issues will come from the manipulation of real materials which are heavy, bulky and more irregular. Research on robotic manipulation and more precise sensors and computer vision processes are necessary.
- Simplification of programming decision-making algorithms: Throughout the research, it became evident that the creation of algorithms to arrange multiple one-off parts into useful configurations is not a simple task and not addressed by current design computation tools.
- Although IDR is fairly simple to use, the sequential programming style of the McNeel Grasshopper is not the ideal platform and event-based programming alternatives should be studied.
- Design with other sources of uncertainty: This research focuses on the design process in which the geometry and quantity of building materials are uncertain. However, other relevant parameters of building materials such as weight, translucency and strength are not considered in the process. In addition, the problem of uncertainty in architecture extends beyond the material dimension, including rapidly changing urban contexts, and fluctuating construction costs, among others.
- In addition, future research should also cover theoretical questions such as the implications for the design profession.

Acknowledgements The research presented in this paper is part of the doctoral research conducted at Harvard University Graduate School of Design. Special thanks to my thesis committee Professor Martin Bechthold, Professor Panagiotis Michalatos and Professor Sergio Forster for their valuable and generous advice. Thanks to Emgu and Geometric Global for granting access to their libraries EmguCV and NestLib respectively.

References

Dörfler K, Rust R (2014) Moderation of vagueness: experiments on the interconnection between physical and digital processes of form generation. GAM Graz Archit Mag 10:184

Helm V, Willmann J, Gramazio F, Kohler M (2014) In-situ robotic fabrication: advanced digital manufacturing beyond the laboratory. In: Gearing up and accelerating cross-fertilization between academic 63 and industrial robotics research in Europe, pp 63–83

Raspall F (2014) Digital-material feedback: extending intuition into the production process. GAM Graz Archit Mag 10:184

Raspall F (2015) Design with material uncertainty: responsive design and fabrication in architecture. Unpublished doctoral dissertation, Harvard University Graduate School of Design, Cambridge, MA, USA

Harnessing the Informal Processes Around the Computational Design Model

Jonas Runberger and Frans Magnusson

Abstract

This paper presents a strategic framework that facilitates the introduction of computational design techniques into architectural practice. The presented architectural design case, and the strategic framework itself, were developed within the Dsearch, a computational development team part of the R&D at White arkitekter AB. An important aspect of the work within the team is to support the integration of computational processes new to the practice, and promote organisational learning that enables a continuous development. The strategic framework therefore is related to certain concepts within the fields of Sociology and Knowledge Management, such as the notion of boundary objects as first defined by Susan Leigh Star and James R. Griesemer, then later developed by Etienne Wenger as an important factor for collaboration within communities of practice. The strategic framework—an assembly of a number of boundary objects, helps elevate the design model to a design system—a project specific set-up that facilitates design versioning, quality control of processes, and organisational learning. Examples are provided through a case project—the development of a 60 m public bench for Forumtorget (Uppsala, Sweden).

Introduction

This paper is written from the vantage-point of a team of computational designers, Dsearch, founded with the explicit purpose of introducing digital and computational strategies and techniques to White arkitekter AB, a large Scandinavian architectural firm. The chosen approach has been to overlay existing projects with computational development, rather than forming a strictly back-office tooling department. The shift from experimental computational practice, to computational design implemented in conventional practice, has highlighted a need for an office infrastructure, mediating between new and old methods and models. The paper details a strategic

J. Runberger (✉)
White arkitekter AB/KTH, Stockholm, Sweden
e-mail: jonas.runberger@white.se

F. Magnusson
White arkitekter AB/Chalmers, Gothenburg, Sweden

M.R. Thomsen et al. (eds.), *Modelling Behaviour*, DOI 10.1007/978-3-319-24208-8_28

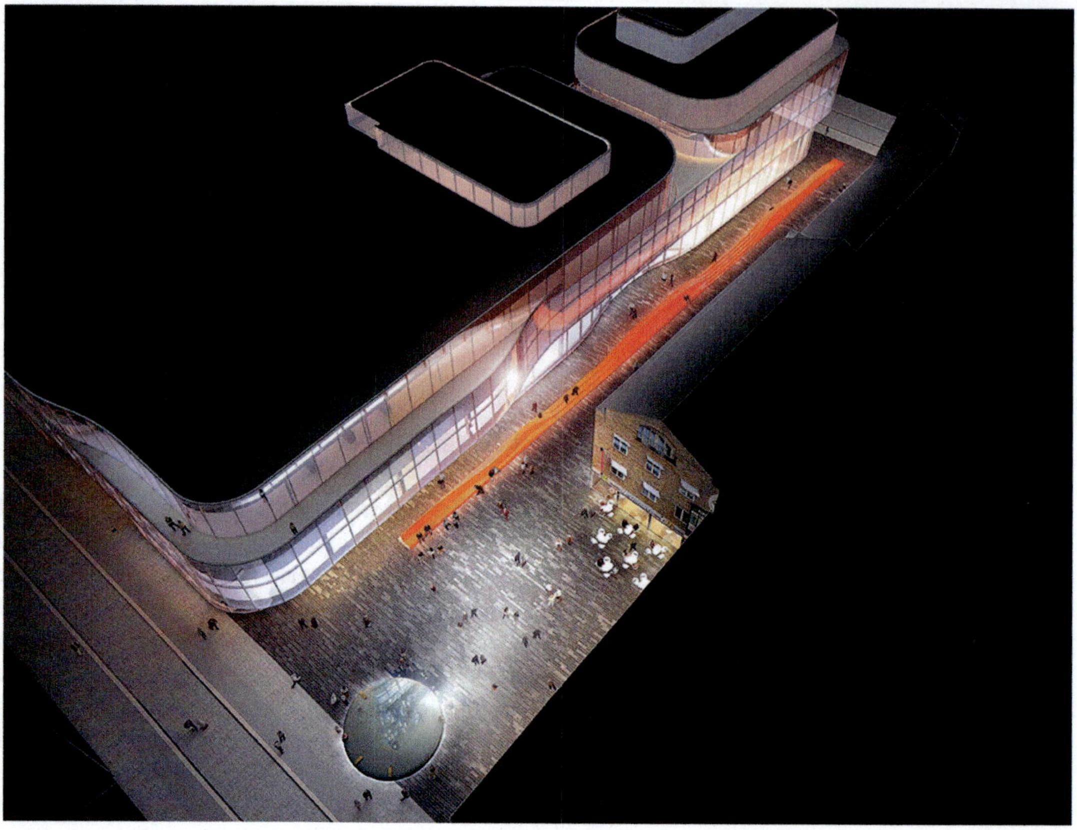

Fig. 1 Representation from the competition entry of the Forumtorget case project—the square and urban furniture

framework, informed by current computational design thinking with a theoretical underpinning from the fields of Sociology and Knowledge Management, and presents how it has been deployed by Dsearch to establish this infrastructure. The framework comprises a collection of so-called boundary objects, where the Design System, expanding the scope of a Design Model, plays a vital role (Fig. 1).

Context

The introduction of computational design into general architectural practice calls for an explicit attention to process and methodology, which in itself can be a provocation to an implicit consensus on methodology—Dsearch has not been preceded by a corresponding "analogue methods team". Together with the approach of active project participation by Dsearch in both design and methodological development, within a context having little previous experience of computational design, this may induce an anxious form of organisational learning—one where the stability of architectural practice is shaken.

The most transformative aspect of this change, is the notion of second order development—to borrow the term introduced within cybernetics, i.e. the process of developing a system that in turn generates, regulates and/or evaluates design (Heylighen and Joslyn 2001). Second order development creates models that are associated with first order design processes and procedures—generated automatically or through the interactions with a designer. In a very literal sense design processes are codified into the model alongside actual proposals. The models produced are at once more powerful, demanding and fragile than traditional design media; they are more open to the receiving context and have embedded agency as long as they are housed within a specific technical and managerial infrastructure.

Second order development also introduces a new specialized culture within architectural practice. Where different participants before could share a drawing, a physical model etc., the design models today also consists of code that requires a new kind of knowledge and technical infrastructure to create, read and use. While graphical programming environments have to some extents lowered the learning curve, there are still distinct boundaries between architects who understand computational design, and those who don't. This is also true for the relation to recently established workflows where BIM software introduces a configuration mode of design. The change in agency of the model media introduced through computational design, together with the cultural and technical implications, often has to be reflected in the mode of operations of second order computational design.

A Strategic Framework of Boundary Objects

Organizational learning—how teams and overall organizations can learn and adapt beyond individual skills and talents, is of great importance to the context of this paper. Most management routines target efficiency and quality in regards to time and delivery, and can thus be related to the idea of single-loop learning—where goals, values and strategies are taken for granted, as opposed to double-loop learning that requires reflection and innovation (Argyris 1999, pp. 68–75). In single-loop learning, each problem is addressed within an existing framework of solutions, with smaller deviations or corrections of errors. Personal experience or already set procedures endure, and a defensive position is established against the unfamiliar. Double-loop learning entails looking at the task at hand with an open mind, and making theories explicit and clear in order to find new solutions. It is a dialogical, or associative approach—employing new modes of operation to achieve new goals. With this in mind, situating experimental computational design processes in conventional practice requires a strategic framework that allows for double-loop as well as single-loop learning.

Susan Leigh Star and James R. Griesemer's work to understand the communication and cooperation in heterogeneous groups of actors has good bearing on the task set for Dsearch. They developed the concept of *Boundary Objects* to explain the successful cooperation of these actors without explicit consensus regarding the aim of their activities. The basis was the use of common "objects which both inhabit several intersecting social worlds… …and satisfy the informational requirements of each of them" (Star and Griesemer 1989, p. 393). Flexibility of interpretation combined with their identity and capacity to structure work and flow of information throughout various social groups are key to boundary objects. Star and Griesemer also identify several different types or categories of boundary objects, such as repositories, ideal types (maps or diagrams) or standardized forms.

Etienne Wenger adds another important aspect to this concept; reification—to make the abstract concrete and legible, a process that could be embodied also in artefacts (Wenger 1998, p. 58). Similar to the way that architectural representations act as boundary objects that reify abstract ideas into important steps in a design process (such as sketching and model-making), the reification of boundary objects in a computational design environment may help both internal development and communication to general design teams. This may be even truer for second order design teams with no formal training, i.e. architects shifting from design to computational design.

The way design steps are formalized and made explicit within second order design follows in itself a process of reification, and in this way the computational design model becomes a boundary object that can support collaboration between several specialists. This notion of process as boundary object is a distinct difference to the conventional practice of end-results (proposals, deliveries) as the boundary objects and clarifies the relation between second and first order of design. In order to facilitate collaboration between first and second order design teams, auxiliary boundary objects such as standards, logs or repository content, are necessary. In this way, the design model can be expanded to a design system—a project specific assembly of boundary objects and actors that supports communication and workflow.

As part of general process development the full set of boundary objects are continuously refined constituents of the strategic framework, in essence a boundary infrastructure (Star 2010, p. 602) that employs specific constellations of boundary objects to establish flows of information and structure work across various environments within White arkitekter, including the internal environment of Dsearch (Fig. 2).

Deployed Boundary Objects

The *Service Matrix* is a two-dimensional diagram targeting a need to specify and communicate the conditions for and expected benefits of [group] participation in a project to potential collaborators—internal and external to White arkitekter (Fig. 3). Specification is carried out as a continuous dialogue with the project principal to establish an awareness of mutual expectations, as well as a reduction of uncertainty in terms of development time and deliveries. In this sense the matrix fits the description of a boundary object of the Ideal type:

> It is abstracted from all domains, and may be fairly vague. However, it is adaptable to a local site precisely because it is fairly vague; it serves as a means of communicating and cooperating symbolically- a 'good enough' road map for all parties. (Star and Griesemer 1989, p. 410)

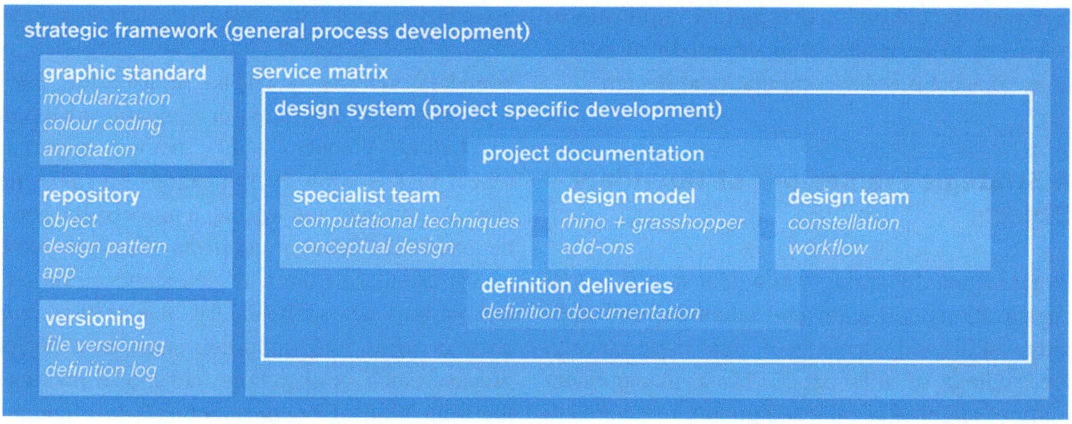

Fig. 2 The strategic framework and deployed boundary objects

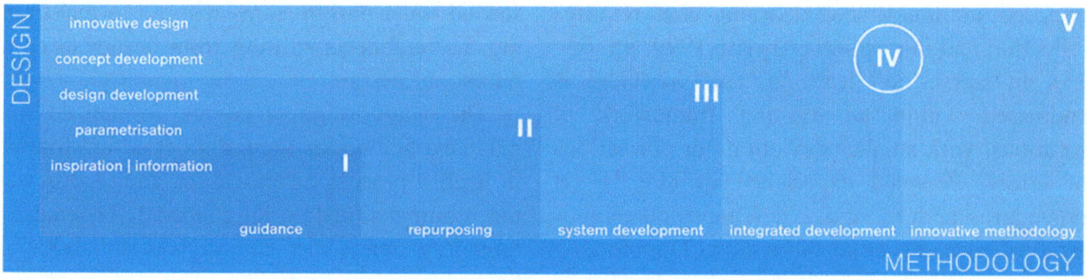

Fig. 3 The service matrix, with the status of the case Forumtorget marked

On the matrix design axis, representing the expected level of design development and architectural innovation, the *Inspiration and Information* level means that the project is discussed in relation to design references and computational methods. In cases where a design concept is well developed and explicitly articulated, a post factum second order implementation is regarded as a *parametrisation.* When Dsearch uses a developed model to deliver a design proposal, it is labelled as *design development* whereas the *concept development* level implies Dsearch taking part in shaping the design concept—introducing computational thinking. *Innovative design* starts with the core value of innovation and is benchmarked against the international design field.

On the methodology axis, representing the level of complexity in computational development, Dsearch can provide *guidance* to the design team regarding relevant methods for the specific project. Design patterns from the repository can be *repurposed* to fit the project needs, but if the necessary modifications are plentiful, complicated or involves new methodology this is considered *system development.* The next level applies when general and computational design methods are intertwined with other specialist methods (energy, daylight, structure etc.) into an *integrated development. Innovative methodology* is needed to address problems where no preceding methods are identified.

The *Design Model* is a first and a second order model, coupled; in our case, this currently means a Rhino/Grasshopper, or a Revit/Dynamo model, plus associated data files. When applied in projects, the design model is expanded to a *Design System,* a curated assemblage continuously formed by the models, but also by the design problem at hand and the human actors of the project in addition to methods and tools—bespoke and conventional. The design system is explicitly situated on a specific coordinate in the service matrix. In this way the matrix mediates the formation of each design system, aiding the specification of what resources from the strategic framework and the larger practice will have to be included in the design system in terms of engagement and expected outcome.

The continuous specification of the design system establishes explicit outer borders for the project—design problem, concept, client intentions, etc.—for both second order development and first order design teams. In that way the strategic framework regulates collaboration and supports the flow of information within and outside of the design system. Star and Griesemer terms this a *Coincident Boundaries Object* where the "…result is that work in different sites and with different perspectives can be conducted autonomously while cooperating parties share a common referent." (Star and Griesemer 1989, p. 410).

Re-usable sections of definitions are routinely extracted, stored and indexed in one of three formats—Objects, Apps or Design Patterns—at a central *Repository,* along with a documentation of key specifications (Fig. 4). The act of editing and documenting parts for re-use also doubles as an explicit annotation to the original definition. Star and Griesemer mean that the index and modularity of a Repository allows that people "…from different worlds can use or borrow from the 'pile' for their own purposes without having directly to negotiate differences in purpose". One intention behind the repository is just that: hopefully objects developed by Dsearch will find their way to unexpected uses within the larger practice (Star and Griesemer 1989, p. 410). *Objects* (User Objects in Grasshopper nomenclature) wrap a section of the definition into a component with the same affordances as standard components (such as search, legibility and commenting) with the difference that the underlying section still is available for later revision. This format is used for generic processes with an expectedly wide use; different projects or multiple instances within the same definition. *Apps* (in the everyday sense) are complete and freestanding definitions with a specific purpose, as close as Dsearch gets to software development. These repository formats can be said to act as Standardized Form boundary objects in the sense that they strongly regulate the flow of information

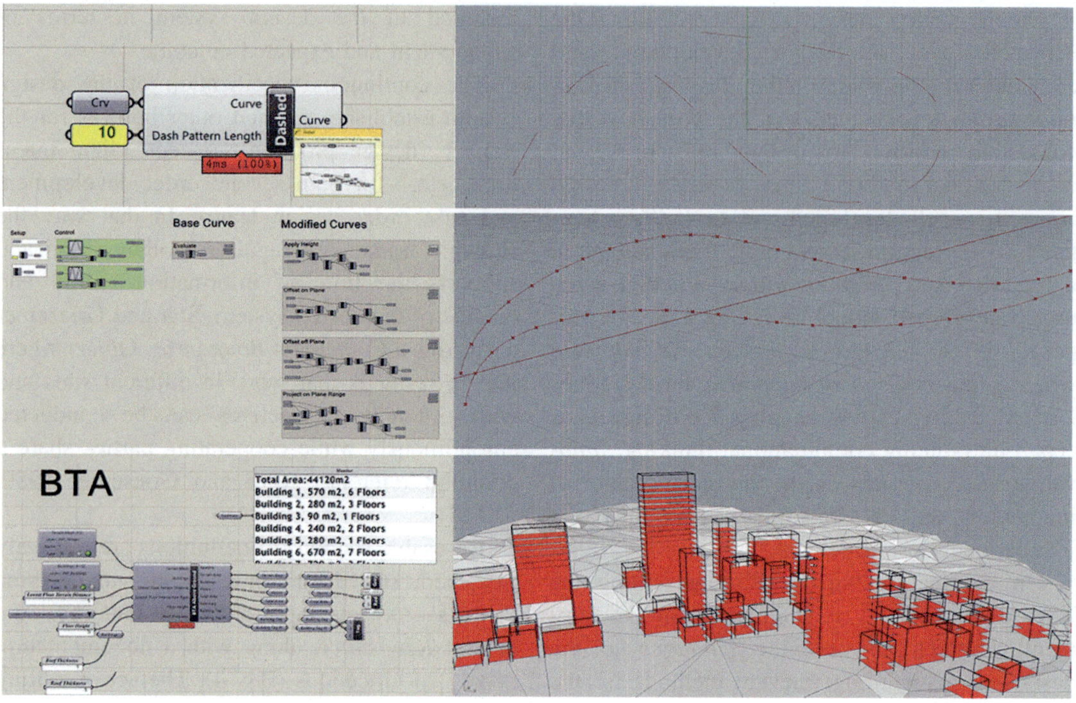

Fig. 4 Examples of objects, design patterns and apps

within a design model in a way that is consistent across all contexts. "The advantages of such objects are that local uncertainties … are deleted" (Star and Griesemer 1989, p. 411).

Design Patterns, "above nodes but below designs" (Woodbury 2010, p. 187), on the other hand, are more complex, specific and often need modification when applied to a new definition context. They may or may not have a distinct geometrical output, but are seen as distinguishable parts of a definition that provides a particular outcome. Preserving ease of editing is prioritized before application, thus patterns are stored as snippets in the standard definition format.

Other prominent boundary objects of the framework are the *Graphic Standard* (Ideal Type, Fig. 5), stipulating annotation and modularization of the definitions for legibility and

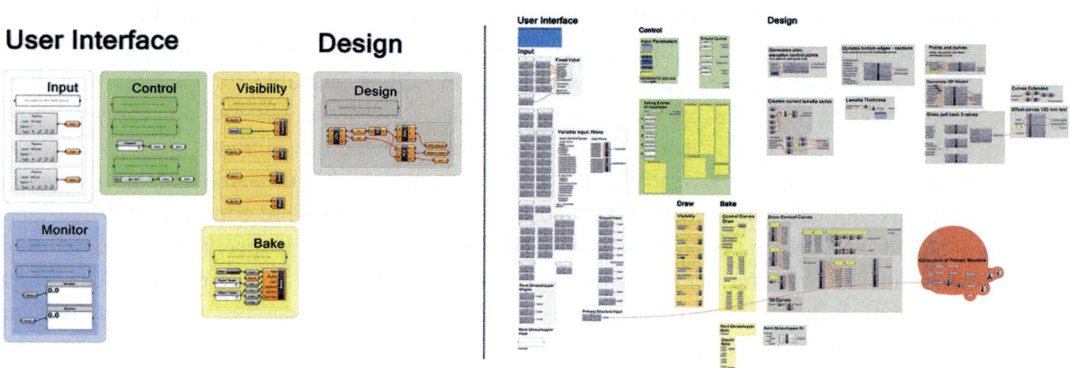

Fig. 5 The graphic standard for grasshopper, as implemented in Forumtorget

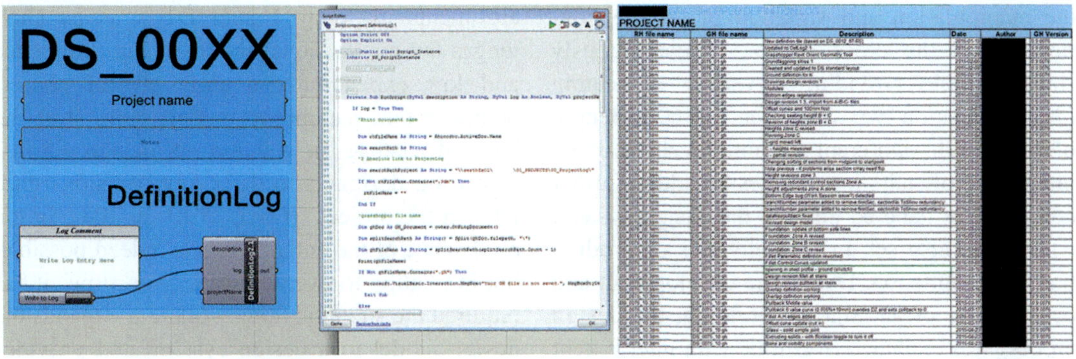

Fig. 6 The definition log object (with script) and definition log

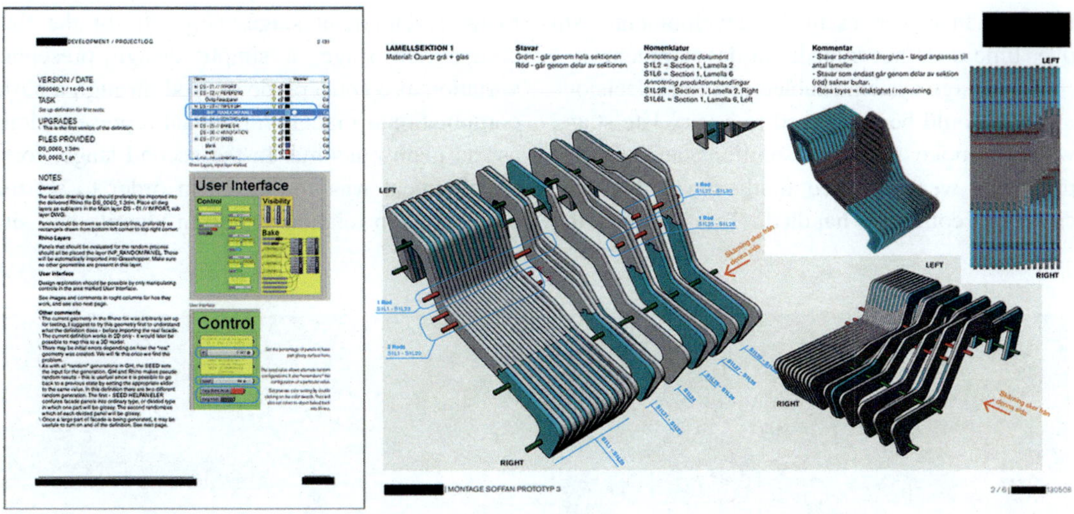

Fig. 7 The definition documentation and project documentation from Forumtorget

reusability (Davis et al. 2011), and different ways of regulating the versioning of design systems—the *Definition Log* object documents brief technical development to a spreadsheet archive, a *Definition Documentation* template is used to communicate changes in functionality between key stages and for deliveries (Fig. 6) and a Project Documentation to communicate design outcome (Fig. 7). As boundary objects they belong to the Standardized Forms type—methods for common communication across dispersed work groups (Star and Griesemer 1989, p. 410). The graphic standard and definition log facilitates easier collaboration within the computational design team, while the definition documentation documents advanced use and provides relevant information for a general practitioner to use a delivered design system. A separate *Project Documentation* template is used to present design outcome to external parties.

Applied Strategies in Project Development

Based on a competition win for Forumtorget, Uppsala, in 2011 (Fig. 1), this urban furniture has been developed through series of computational design models, physical models and prototypes. The programmatic and formal concept includes the manipulation and variation of seating configurations—the cross section is continuously shifting while providing a variety of seating opportunities on two sides over the 60 m length of the bench. Different ground levels and the integration of stairs into the design added further complexities handled through the design system, which was also set up for deliveries of production documentation. Due to the development and construction of an adjacent building, the design process was extended over time with intermittent activity. This allowed a development in distinct phases, dependent on the strategic framework to make sure that past design evaluations are not lost, and that changes in the development team over time does not impede on development.

The design system enabled that final decisions on form could be postponed to a very late stage, while bespoke technical solutions and general principles were resolved to the level that procurement could be handled. This follows the arguments made by Daniel Davis as he challenges the conventional model of high influence and low expenditure in early project phases, claiming that parametric models can "shift the cost of change in relation to design effort", "allowing designers to defer key decisions until later in the project" (Davis 2013, p. 208). In the experience of the authors, this is indeed possible, but in relation to overall project workflows it has required the strategic framework that expands the design model to a design system, and provides boundary objects that supports interactions over time.

The Forumtorget project has undergone four main development stages (Fig. 8). In the first competition stage, a simple design presented variation as a concept, developed through a basic computational model with visual representations as the main outcome. In the second stage, form development was conducted in order to set the boundaries in relation to design identity and site

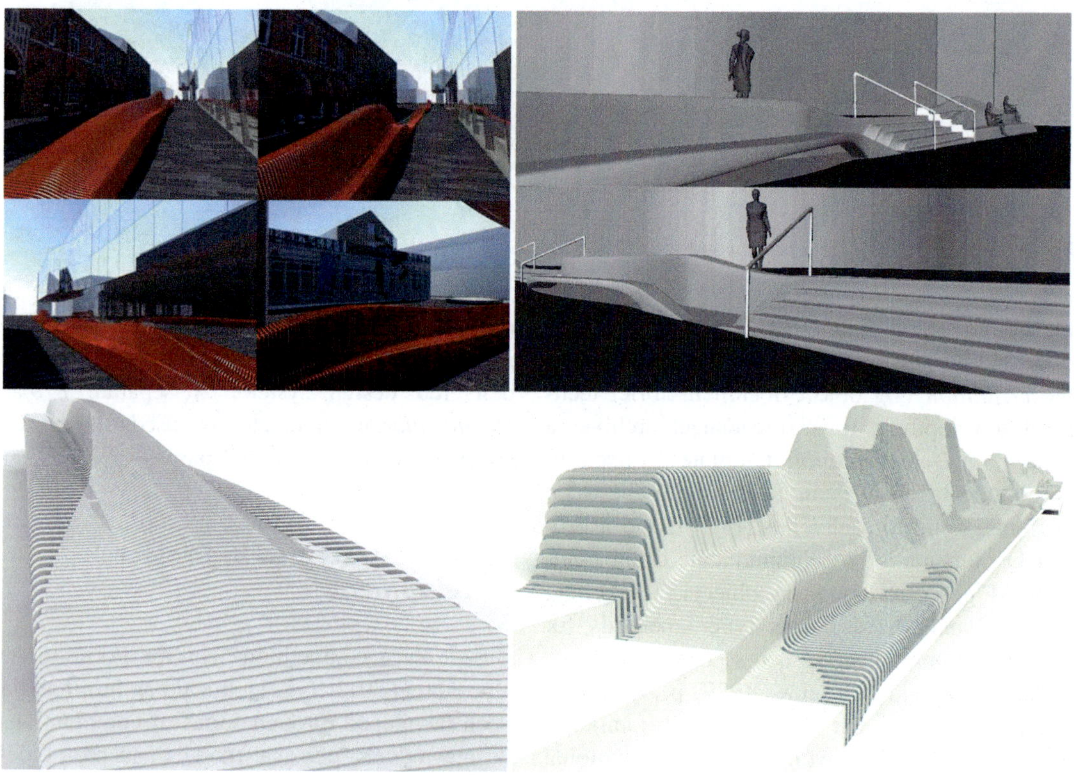

Fig. 8 Representations of Forumtorget from stages 1, 2, 3 and 4

specific conditions such as the two different levels of ground. The third design stage involved an iterative study of the form, structure, manufacture and assembly of the bench, through a continuously refined parametric model, with input from specialists as well as feedback from series of physical models and full-scale prototypes.

During the fourth stage, the technical details and structure were refined, the overall form revised, the design system consolidated to avoid errors in final production, and the generation of documentation for procurement and production initiated. From stage two to four, care was taken to keep exploring the overall spatial principles, yet not fixating any part completely.

A number of early key decisions allowed the continuous development based on the design system. A lamella-based form and fabrication principle that allowed rational fabrication of the "free-form concept" was selected. The cross-section was based on a seven segment polyline with individual control of all corners and their fillets, setting restrictions of possible formal variation, yet handling the zone-specific conditions and providing a formal continuity along the length of the bench as well as an adaptation to differentiated ground conditions. The use of control diagrams; the computational transformation of series of 2d curves to the design model, provided a control interface to overall form as well as detailing (Fig. 9).

Once the premises for the design system were in place, the project was developed through the established versioning approach facilitated through the strategic framework. Supported by a series of design reviews where alternate solutions could be discussed with different specialists, the second order development could operate through iterative loops, allowing technical solutions to be set while the specific form was continuously refined. The use of the definition log also provides a back-log of the development, where the following milestones can be identified:

- Versions 5–8: Setting up basic geometrical generation from control curves (Fig. 9).
- Versions 23: Generating production documentation for first prototype (Fig. 10).
- Version 38: Optimizing definition performance by introducing python scripts.
- Versions 41–58: Developing second prototype (Fig. 10).
- Version 88: Setting up relation to Revit for interface to overall project.
- Version 90: Setting up modularization for production and assembly.
- Version 95: Developing concrete foundations design model.
- Version 97: Developing joint between glass and solid materials.
- Version 100: Overall formal design revisions.
- Version 101: Developing primary steel structure.
- Version 102: Generating procurement documentation (Fig. 11).

The extended development time of the project made documentation very important—a number of different Dsearch members have entered and

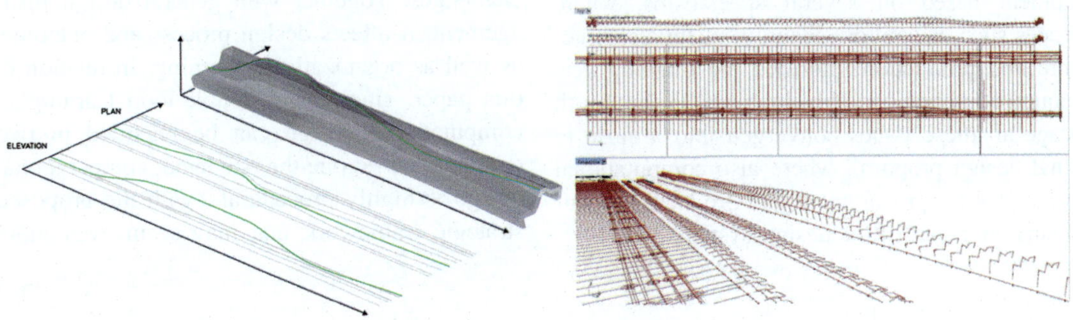

Fig. 9 Control curves representing plan and elevation as a diagram, and view from the design model

Fig. 10 First and second physical prototype, and representations from stage 3

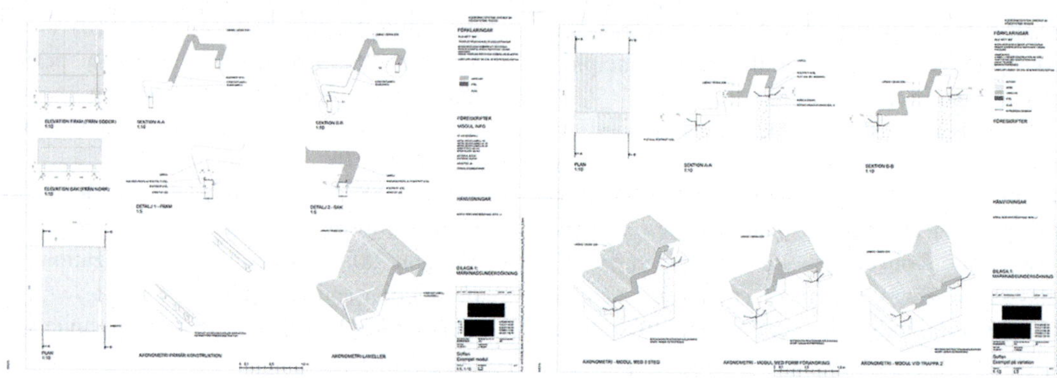

Fig. 11 Procurement documentation from stage 4

exited the project, and the definition log and definition documentation allowed an understanding of previous development. The close collaboration with other specialists and the client also required a continuous use of a project documentation, in particular the development and assembly of full scale prototypes. The overall design process in this way reflected an exploration of two related trajectories; the design development based on several specialisms, where issues such as comfort, identity, materials, structure and production informed the ongoing versioned computational development. In the fourth stage all these issues converged into a close to final design proposal, where also computational principles for procurement and production documents were part of the design system.

Concluding Remarks

Computational and second order design provides architectural practice with new assets, but the deployment in practice may face unprecedented challenges. This paper shows how a strategic framework can be applied in order to establish an infrastructure that in practice mitigates these challenges. Together with general design management, it affects design process and outcome, as well as organizational learning. In relation to this paper, single- and double-loop learning in computational design can be regarded in two opposing ways; one the one hand, computational design is highly procedural—with the proposed strategic framework that may seem even more regulated—suggesting single-loop learning within a given framework; on the other hand, it can be regarded in contrast to existing management models within architectural practice, where the procedures taken are very familiar—we know the modes of representation, (we believe) we know what a client wants, and we understand how the construction industry operates. Given that computational design provides new opportunities in conventional practice, but depend on special skills and a rapid technological advancement—the purpose is to use the strategic framework to enable a continuous exchange between first and second order design, providing double loop learning within both contexts. As the field of computational design continues to develop, the framework enables additional aspects to be introduced, such as iterative analysis, material performance and additive manufacturing.

The strategic framework and its constituent bounding objects is to be seen as a bottom-up approach to managing design processes and flows of information across a heterogeneous environment within a larger firm. Grounded in computational design thinking, it provides an alternative to current trends of explicit process modelling and comprehensive BIM standardization. The associated terminology from Sociology and Knowledge Management has in the development of Dsearch provided useful concepts bridging between first and second order of design and addressing issues beyond computation and design—the interactions between designers of different backgrounds. In essence, it provides a deeper understanding of elements that previously made sense in an intuitive way and articulates them. The terminology opens up these interactions for a wider discussion—by aligning ourselves with recent research on practice cultures elsewhere, we are able to distinguish and evaluate potential assets for future practice and research.

Acknowledgements Forumtorget design team: Gustav Jarlöv, Sam Keshavarz, Jens Modin, Andreas Milsta, Torbjörn Eliasson, Michael Malmborg, Jonas Wannfors. Dsearch team: Jonas Runberger, Frans Magnusson, Hamia Aghaiemeybodi, Pedram Seddighzadeh Yazdi, Vladimir Ondejcik.

References

Argyris C (1999) On organizational learning, 2nd edn. Blackwell, London

Davis D (2013) Modelled on software engineering: flexible parametric models in the practice of architecture. PhD dissertation. RMIT University, Melbourne

Davis D, Burry J, Burry M (2011) Understanding visual scripts: improving collaboration through modular programming. Int J Architectural Comput 9(4):361–376

Heylighen F, Joslyn C (2001) Cybernetics and second-order cybernetics. In: Meyers RA (ed) Encyclopedia of physical science and technology, 3rd edn. Academic Press, New York

Star SL (2010) This is not a boundary object: reflections on the origin of a concept. Sci Technol Hum Values 35:601

Star SL, Griesemer JR (1989) Institutional ecology, translations' and boundary objects: amateurs and professionals in Berkeley's Museum of Vertebrate Zoology 1907–39. Soc Stud Sci 19(3):387–420

Wenger E (1998) Communities of practice, learning meaning and identity. Cambridge University Press, Cambridge

Woodbury R (2010) Elements of parametric design. Routledge, London

A Generic Communication Library for Human-Robot Interaction on Construction Sites

Thibault Schwartz

Abstract

After a first decade of re-discovery of robotic technologies via design-oriented experiments, "digital fabrication" specialists are starting to reach technical limitations common to pre-existing industrial machinery research communities. In their quest for applications of robotics in construction sites and prefabrication plants, their work heavily rely on available programming and communication utilities, and ultimately on the underlying constructor-specific languages and protocols that these interfaces are based on. These last layers, including their documentation, are unfortunately limited by manufacturers for security or intellectual property reasons, resulting in technical limitations which affect naturally the whole research field. One of these bottlenecks is the absence (or lack of abstraction) of unified communication services available between machines themselves, their operators and additional sensing devices which continuously inform the two pre-mentioned types of industrial actors. In order to tackle this problem, this paper proposes a new communication protocol allowing high-frequency structured messaging between robot controllers and external clients over regular TCP/IP networks, and respecting the full abstraction of each robot programming platforms. As a conclusion, we exemplify the use of such service in a couple of application scenarios adapted to construction sites.

Introduction

Over the past decade, the use of robotic arms for research and development of novel prefabrication and building processes has raised numerous technical, economical and sociological questions in the architectural design and engineering communities. Amongst the main technical issues that are being addressed, the improvement of sensor technologies for robot-environment and human-robot interactions is considered as a necessity to allow the deployment of such machines on construction sites. As a result, it is expected that the use of these sensors will overcome manually-operated static calibration procedures that are traditionally employed by the machine operator to "teach" the robot the

T. Schwartz (✉)
HAL Robotics, London, UK
e-mail: t.schwartz@hal-robotics.com

M.R. Thomsen et al. (eds.), *Modelling Behaviour*, DOI 10.1007/978-3-319-24208-8_29

specificities of its environment and/or equipment. In other industries, numerous sensing technologies have been used with success from the beginning of numerical machine control to achieve these objectives: force sensors, proximity sensors, computer vision, etc. These various capture methods proved to be efficient ways to decrease or remove the need for human guidance and supervision in managed environments such as laboratories and factories. Used appropriately in conjunction with robust machines, they also allowed the development of new processes that would not be achievable manually with the same constraints.

Construction sites though, as highly specific, unstructured environments with very different characteristics (presence of mud, dust, approximated dimensions, workers and other machines in close proximity, etc.), seem to be problematic to equip with such devices (Stentz 2001). Not only are these places currently unsuitable for standard robotic arms because of their work envelope limitations or maintenance reasons—these can be resolved in many ways—, but also because of the inherent property of a building site to continuously change at every scale (Warszawski and Navon 1998). As sensing techniques for these complex environments are still to be conceived, construction sites can be theoretical and practical laboratories for the development of intuitive human-machine cooperation at a higher level of abstraction than in automotive industry. This approach however requires that we start to consider on-site workers with their knowledge and/or aptitude to adapt to the changing properties of the building site instead of considering them as a flawed calibration tool (Everett and Slocum 1994).

On a technical level unfortunately, essential "bricks" that would allow engineers and architects to easily implement such novel procedures are still missing or carefully restrained by robot manufacturers (Utz et al. 2002), for security or intellectual property reasons, or for a supposed lack of interest of the consumers for such features. The conservation of obsolete control technologies encouraged by the inertia of the traditional robot markets, in addition to these undocumented/unavailable control layers are limiting factors not only slowing down the emergence of construction automation, but are also critically orienting the solutions in development in our domain to the profit of autonomous, off-site plants using relatively standard prefabrication techniques.

Addressing this problem, this paper introduces a new generic communication library developed for robot controllers. This library, based on TCP/IP socket messaging and using a custom communication protocol, supports all types of movement interpolation, facilitates the streaming and monitoring of any data including system parameters and user-defined structures, while operating with the minimal latency allowed by the controller specifications. In a first part, we expose the characteristics of this communication library, followed by the presentation of the architecture that was chosen in regards of these requirements. Thirdly, the implementation of this library is exemplified on ABB IRC5 robot controllers using the ABB RAPID programming language, in both mono-threaded and multi-threaded applications. We conclude this contribution with application scenarios focusing on the real-time edition of a previously sketched toolpath using different sensor outputs, which could illustrate a strategy of context-dependant assistance to artisanal techniques.

Technical Requirements

Mostly based on proprietary computer hardware and software, industrial robot controllers have dramatically changing specifications depending on their manufacturer, the controller generation, and optional features they are equipped with. The explanation of this variety can be found in the development and marketing cycles of the automation industry: as the conception of a robot controller generation usually takes 10 years, constructors have been unable to follow Moore's law, thus generating a very disparate collection of fixes to keep up with the evolution of industrial needs and standards. This everlasting obsolescence marked by sudden peaks of high-tech patches—worsened by warranty limitations from component suppliers and the troubling technical inertia of the robot user community—have led to the current industrial robot market, where floppy readers are as easy to find than EtherCAT sockets. In order to overcome this disparity of IT-related standards and formats, a low-level solution is traditionally employed: machines are sharing electric signals via networks of Programmable Logic Controller (PLCs), centralizing communication problems by simply suppressing any digital data. This work-around, which now represents a multibillion-dollar market and on which depends any automated plants, is unfortunately too limited for rich applications such as human-machine natural interaction.

Research communities, united in associations such as the ROS Industrial Consortium (Evans and Edwards 2012), have proposed since a few years a series of generic software driver solution (ROS.org 2015) to bypass the lack of abstraction inherent to ageing robot controller technologies and PLC communications. These drivers have a pertinent design: they are enabling the direct control of joints for various industrial machines. This way, the machine controller is virtually "removed" from the control loop, and the motion only results from the messages that are sent via the driver at a high frequency. Unfortunately, this strategy implies that it is the user's responsibility to re-implement numerous basic control features such as linear interpolation and speed management, which usually constitute the value of industrial controllers. This restricting streaming solution is completed with a very partial monitoring of the robot and controller activities. If these features are enough for low-precision movements, they unfortunately are not in adequacy with the manufacturing requirements of the construction industry.

The purpose of the library discussed in this paper is to overcome these limitations, by proposing a robust architecture for the development of truly flexible, generic communication drivers for industrial robot controllers.

API-Independent Communications

Manufacturers' APIs, when provided, usually allow developers to access most of the controller domains such as the robot programs having an impact on the machine movements (*motion tasks*), background programs (*non-motion tasks*), the signal management interfaces (*I/O*), the system variables and configuration (*system*), and the management of files and directories (*files*). While being relatively accessible and easy to use for robot specialists, these APIs can only be compiled in certain languages and formats, and are rarely natively compatible with each other. Similarly, the robot programs which need to run on controllers to feed the robot motion planner are unique for each manufacturer and controller generations. In order to insure its portability/cross platform compatibility, our unified communication library needs to be publicly accessible via a standard communication protocol (Ethernet/IP, DeviceNet, etc.), with all platform-specific code being integrated in the controller-specific driver that will be executed by the machine. As a result, machines will be able to communicate with each other directly as long as they support the same communication protocol, without the need for an additional server or any external computer-based application. Similarly, an external computer will be able to be inserted in the network and communicate with existing machines without the need to rebuild machine-specific code. To ensure performance, the information exchanged by machines will have to be structured adequately to reduce congestion, latency and packet losses over the network.

Abstraction Management via Progressive Enhancement

Robot programming languages specifications widely vary between manufacturers and controllers. As an example, we can analyse the content required in the end-effector declaration instructions of ABB and Universal Robots languages:

ABB: *[OnRobot?,TCPCoordinates,TCPOrientation][Mass,CenterOfMassCoordinates, CenterOfMassOrientation,MomentOfInertia]*

Universal Robots:*[TCPCoordinates, TCPOrientation]*

While depicting the same object—the tool to be used by the robot and its characteristics—, these two instructions are obviously different in terms of abstraction. As a consequence, they differ in terms of size: the ABB instruction is 73 Bytes long, the UR one being 24 Bytes. This difference can affect the design of our message structure, and have a great impact on the network congestion. A Graceful Degradation strategy could help solving this variation: our "tool" message would then take the form of the most limited "tool" representation. As a result unfortunately, most of the advanced features of industrial equipment would become unavailable via our communication library. As a conclusion, we will need to use a Progressive Enhancement design, thus allowing users to have the most complete control over their machines while providing a downgraded solution for the most limited controllers.

Template Metaprogramming Compatibility

Amongst the reasons of the success of hardware PLCs are their inherent compatibility with a large range of equipment, but also their ease of use and reprogramming. In order to be a plausible alternative, our library needs to be accessible, and to only require a small amount of prerequisites and efforts to be integrated and maintained. Most importantly, this library will have to be simple enough to be used by users with basic programming knowledge. For these reasons, the machine-specific implementation of our library will have to be compatible with a template metaprogramming approach, while staying flexible enough to allow highly structured tasks programming for expert uses.

Architecture

In the following paragraphs, we will discuss the various design choices taken to match the aforementioned prerequisites.

Dual Full-Duplex TCP Connection

In our everyday human-to-human conversations, we naturally convey different levels of information via both language and gestures: gestures can then be understood as a way to precise what could sometimes sound ambiguous (Gullberg 1993). In order to explicitly control and monitor machines in real-time, we decided to demultiplex the usual incoming and outcoming communication channels to be used via our library, thus allowing the user to send and receive specifications about the content of the message that he wishes to send or to receive simultaneously (Fig. 1). This solution allows both channels to be multiplexed or demultiplexed without the need to establish any supplementary socket connection, and can also provide a workaround to thread locking when the robot language used for the implementation of the driver does not allows Data Peeking.

Dictionary-Based Nested Messages

As the bandwidth of industrial networks is often restricted, packet sizes have to be minimized in order to guarantee low-latencies and a reduction of the network congestion. At the same time, complex applications need to exchange structured types, thus requiring some form of data nesting to reproduce a similar memory allocation in the client system. The solution used in this communication library is to establish a markup-language-like nesting at the packet structure level (Fig. 2), via the use of a generic 5 Byte-long header packet. This packet (1) explicitly separates the nested value from the message trunk by stating its length (2 Bytes);

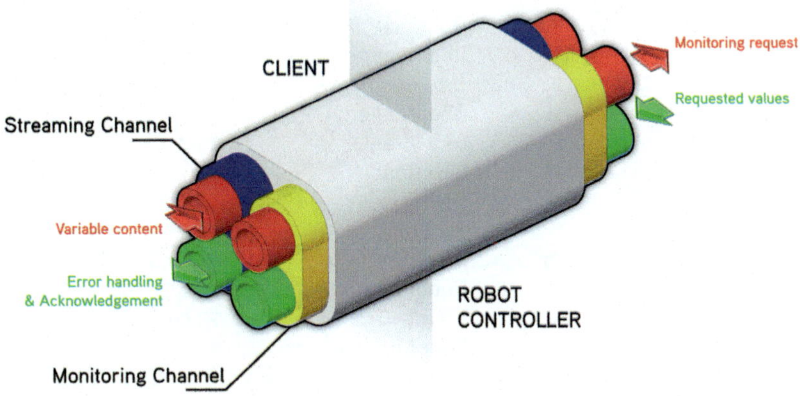

Fig. 1 Dual full-duplex connection, with dedicated channels for both streaming and monitoring data exchange

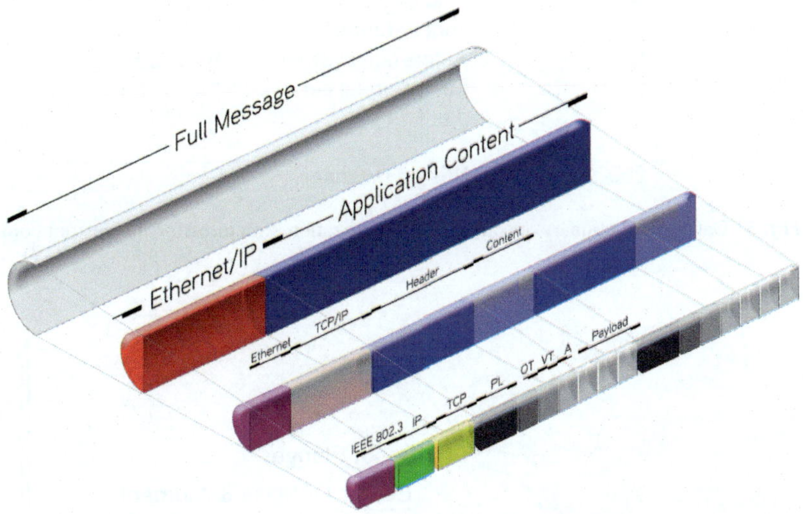

Fig. 2 Packet structure of a typical nested message generated by the library and shared over an Ethernet/IP network

(2) specifies the type of content based on a type dictionary (1 Byte) (Fig. 3); (3) specifies its optional membership to another structured variable using the same type dictionary (1 Byte); and finally (4) specifies its address in this targeted variable (1 Byte).

Changing the X coordinate of the current Tool Centre Point (Fig. 4) to 123.45678 mm will thus be translated to a 9B-long message: *[9,0,15,5,0],[223,233,246,66]*. Similarly, a message containing two nested sub-messages representing the Y and Z coordinates of this same variable will be translated to a 23B-long message:

[23,0,255,2,0],[[9,0,15,5],
[1,223,233,246,66],[9,0,15,5],
[2,223,233,246,66]].

By using this simple addressing system, the user can send 1 kB-long messages composed of any of 255 data types (including machine-specific structured types), which can then be populated by a series of 255 different data types, which can in turn be assigned to 256 local variable arguments. The example above, depicting a simultaneous compensation of the robot end-effector calibration in X and Y coordinates can be sent 113 times in every 1 kB packet. With these specifications, even old controllers equipped with an RS232-to-Ethernet adaptor and communicating over a 60 m shielded cable with a diminished data rate of 2.4 kB/s could constantly receive 250–270 compensation values per second. In more common industrial installations using

Index	Symbol	Description	Structure	Size (Bytes)
0	B	Boolean	B	1
1	I8	8b integer	I8	1
2	I16	16b integer	I16	2
3	I32	32b integer	I32	4
4	I64	64b integer	I64	8
5	F	32b float	F	4
6	St	String	St	1-1019
7	P	3D Point	F F F	12
8	Or	3D Orientation	F F F F	16
9	C	Robot configuration	I8 I8 I8 I8	4
10	Fr	Frame	P Or	28
11	Rj	Robot joint values	F F F F F F	24
12	Ej	External joints values	F F F F F F	24
13	Sp	Movement speed	F F F F	16
14	Z	Interpolation zone	B F F F F F F	25
15	T	Tool definition	B Fr F Fr P	73
16	R	Reference system	B B Fr Fr	58
17	Ct	Cartesian target	Fr C Ej	56
18	Jt	Joint target	Rj Ej	48
...				
255	N	Nested message	–	1

Fig. 3 Data type dictionary specifying the serialization structures for the custom communication protocol

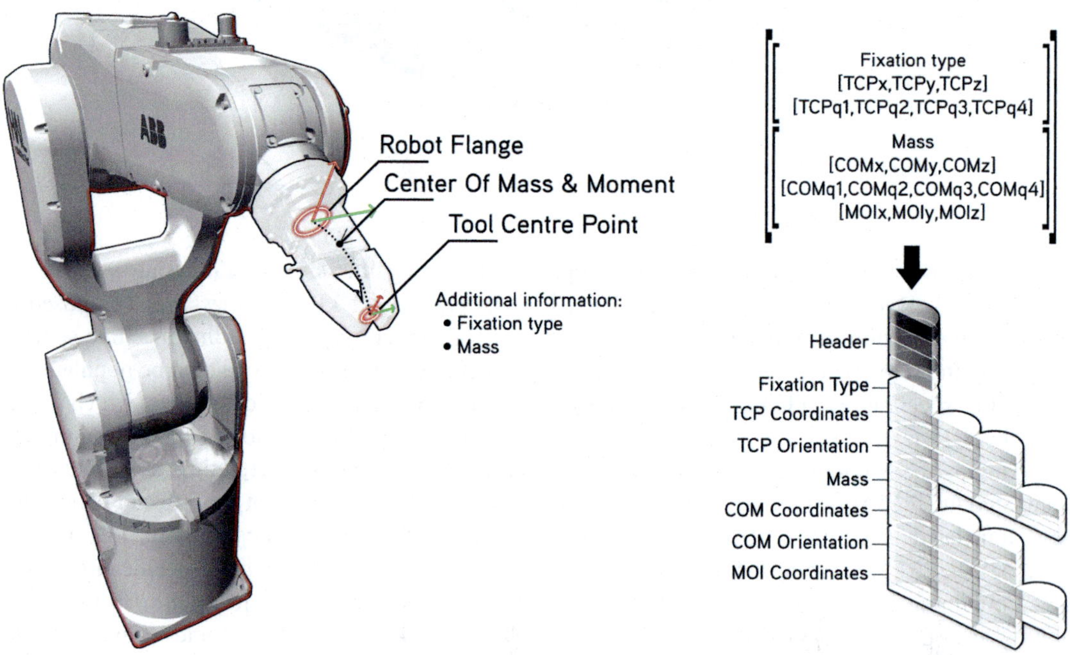

Fig. 4 Description of an end-effector, its properties, and its packet representation in the communication library

10/100/1000 Mbps Ethernet, the network would theoretically be able to handle 1–100 million similar compensation values per second. In real conditions, the computing power of the controllers would limit such transfer frequencies.

Distribution as Robot Languages Extension Packages

In order to ease the distribution and integration of such communication library, we recommend to package it as a single shared system library extending the native robot language. This solution allows to unify the naming of variables shared between tasks, procedures and functions called or populated by the communication system, and to maintain a simple, concise documentation of the library. As a side benefit, such package is easy to integrate in commercial post-processors or other third party robot OLP software like ABB Robot-Studio, and in VPL-based interfaces such as the HAL robot control suite for McNeel's Grasshopper plugin for Rhinoceros (Schwartz 2012), thus enabling its use in various programming environments for any type of robot user.

Implementation

The following implementation examples are performed in the ABB RAPID language used in ABB IRC5 robot controllers, and are using the Socket Messaging (first and second examples) and MultiTasking (second example only) software options. These examples are demonstrating how the specific architecture specifications presented above are met, in respect to the language limitations.

Channel Creation Over a TCP/IP Socket

The connection of incoming and outcoming channels are separated in order to allow the multi-threading of the two communication procedures, but their initialization is similar. Both channels are instantiating a TCP Server socket, waiting for client connections on a specified port (the Incoming and Outcoming ports must be different, and can be specified via a dedicated argument in the channel creation instruction). Once a client is detected, a simple handshaking procedure is triggered: the robot controller sends an acknowledgement message which is completed and sent back by the client. The controller then sends a second acknowledgement with a session ID, and the server socket stays open until the session is terminated by the client (connection aborted) or the robot (execution error). If the client is disconnected, the channel is reinitialized and waits again for new clients (Fig. 5).

Message Reception (Instruction Content Streaming)

The message reception sequence is subdivided in 5 steps, to stay compatible with any usage scenario: firstly, an initial attempt to read messages on the incoming channel during a specified amount of time is triggered. When a message is detected, the two first bytes of the header content are unpacked to verify that the full packet was successfully received, thus discarding invalid packets. After validation, the messages are fully stored into a local queue: this allows messages to be stored and parsed at different times depending on the consumption logic of the program. At consumption time, messages are passed to the library custom byte parser, and the content automatically populates shared variables of the same type than the target data type specified in the third byte of the header.

This way, the user is able to stream and receive various data types at the same time, and to only trigger a manipulation procedure when the validity of these variables is established. Similarly, it is possible to use local buffering of similar values that are sent in the same package, for example for coordinates averaging purposes. If the message reception or parsing fails, the

Fig. 5 TCP socket initialization, usage, and disposing procedures as used in the protocol

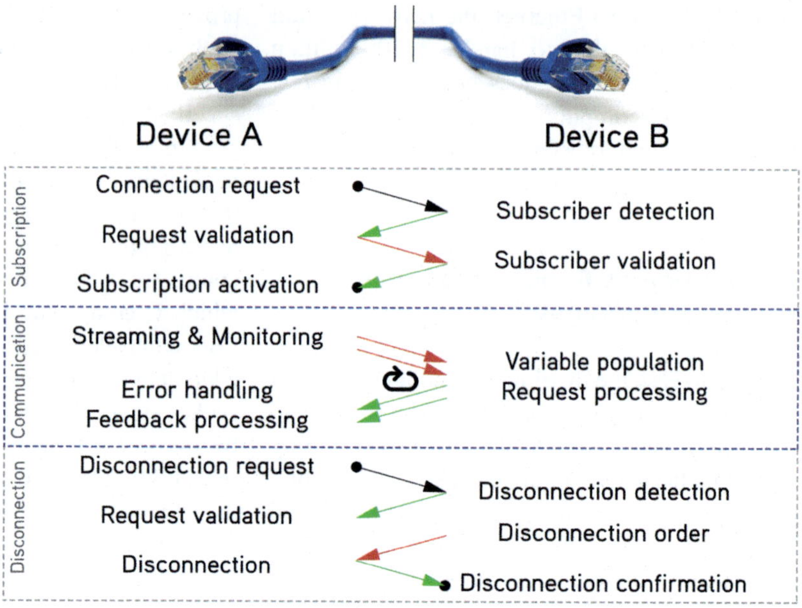

client is automatically notified and can thus trigger a resolution routine before re-emitting the information.

Message Emission (Robot Monitoring)

Monitoring requirements are usually application-specific. For this reason, the message emission system of our communication library allows to systematically rebuild the feedback structure packet to be sent at each iteration, thus bypassing usual packet pre-formatting limitations that users can experience using XML-based configuration files like in ABB RRI option. This dynamic packaging solution is also completed in 5 individual steps: to initialize the packet structure, the oncoming channel listens to the client packets representing the feedback request. This packet, similarly to the reception packet, has two first bytes specifying the packet length for validation purposes. After the processing of this header and the validation of the request packet, the feedback structure is built incrementally based on the other header information: the shared data type to be monitored, an optional specific property of this data type, and the address of this optional argument. Requesting the monitoring of the Tool Centre Point would then be equivalent to sending the header of a Tool Centre Point message, etc. This unified grammar between reception and emission packets are thus facilitating the extension of the data type library, as a single grammar reference file can be used for both packet types. Once the variable packet structure has been received, the outcoming packet content is packaged iteratively until the request structure becomes null. It is then added to the emission message queue, which can then be processed with a specified emission rate to avoid network congestion problems.

Mono-threaded Program Example

A mono-threaded program (Fig. 6) will use the previous features in a single program. The sequence of reception/action/emission will have a direct impact on the machine behaviour and latency, as some operations over the network are locking the thread.

```
module MyCommunicationModule
    !<Declaration & Initialization of the module variables>
    var bool UseCommunicationChannels := true;
    proc Main()
        !< Declaration & Initialization of the procedure variables>
        var num MaxTimeOverride := 10; !Maximum connection time in seconds.
        !============CONNECTION============
        !//The connection order can be changed//
        !-----------------------------------------------
        ! Create an incoming channel using a specified port and a custom maximum connection time
        HAL_CreateServer IncomingSocket, HAL_IncomingPort \HAL_MaxTime:=MaxTimeOverride;
        HAL_ServerStart IncomingSocket; ! Start the incoming channel connection
        ! Create an outcoming channel using a specified port and a custom maximum connection time
        HAL_CreateServer OutcomingSocket, HAL_OutcomingPort \HAL_MaxTime:=MaxTimeOverride;
        HAL_ServerStart OutcomingSocket; ! Start the outcoming channel connection
        !=========CLIENT CONNECTED=========
        ! As long as the robot has to use the communication channel...
        while UseCommunicationChannels do
            !//The message use order can be changed//
            !-----------------------------------------------
            ! Listen to incoming messages. [Thread Locking]
            HAL_UseIncomingMessages IncomingSocket \ParseCompositeMessages:=true;
            ! If a new incoming value is detected...
            if HAL_NewIncomingValue then
                ! Reset the new value event
                HAL_NewIncomingValue:=false;
                !<Do something here using populated data types>
                ! Example: open a gripper based on any incoming boolean value
                ActivateGripper HAL_Bool;
            endif
            ! Listen to outcoming message requests and immediately send back the requested packet.
            HAL_UseOutcomingMessages OutcomingPort \MessageQueueDelay:=0; [Thread Locking]
        endwhile
    endproc
endmodule
```

Fig. 6 Implementation of the developed communication mechanism in a single thread, in ABB RAPID language

Multi-threaded Program Example

In order to have a better control over the communication and execution procedures, the user can implement the various library functionalities is separate threads/tasks. In this case, the recommended multi-tasking structure is: one task for motion-related operations (HAL_Move in our example), one task for incoming messages (HAL_Receive), and one task for outcoming messages (HAL_Send). Application-specific operations can then be set in additional tasks, or embedded in one of the three main ones. The example below shows how tasks can be synchronized while maintaining locking on variables being used. First, the three tasks are initialized and synchronized. Secondly, the HAL_Receive task gets a message that is then shared with the HAL_Move thread as soon as it allocated. Based on the HAL_Bool value (which can be changed by the received message), the HAL_Move thread executes a program, then triggers the HAL_Send thread so the HAL_Bool variable can be sent back in the feedback package. Then the whole program restarts (Fig. 7).

Towards Applications of Real-Time Trajectory Edition with Sensors

Beyond the understanding of the technical foundations of the communication protocol depicted in the previous paragraphs, we propose to exemplify possible application scenarios applicable to the construction industry.

Integration of Machine Localization

If we consider the inability—by design—of industrial robots to move freely in a building site space, there is a high probability that external devices such as cranes or mobile platforms

Fig. 7 Implementation of the developed communication mechanism in three individual threads, in ABB RAPID language

would be used in the future to relocate cells specifically designed for certain construction phases. In order to automatically take this physical repositioning into account, robot tasks need to integrate variables generated by localization devices such as Inertial Measurement Units (IMUs), odometers, SLAM applications etc. If a specific task needs to be operated synchronously during this positioning procedure, the proposed communication library can ease the integration and fusion of sensor data directly in the robot controller. This method presents the advantage of removing the need for an additional "master" computer, while maintaining a real-time network access for safety reasons and/or to enable operator control at a higher priority.

Adaptation to Process-Specific Context Variables

Following the simultaneous development of industrial vision systems, reliable torque sensors and efficient motion planning algorithms, it is becoming common to see industrial robots performing advanced applications relying on object recognition and feature tracking (random bin-picking, object tracking and measurement). In complex environments such as construction sites, robots will similarly have to rely heavily on their ability to adapt to the evolution of their context, including the behaviour of surrounding operators and the inconstancy of dimensions (tolerances of construction components, relative precision of pre-existing manual assemblies). Such versatility will have to be rapidly (re)programmed at each construction phase, based on the requirements of the task to achieve. With this technical situation in mind, it will be vital for workers to be able to quickly interconnect devices together, so they can start sharing context variables impacting machine behaviours and automatic reports used for site management. We can then imagine intuitive collaborative robotics applications being developed with the proposed library, such as pre-programmed assembly tasks for heavy construction parts edited on-the fly by workers using a voice control interface, with torque monitoring being used to verify the fitting of the elements.

Conclusion

This paper proposes a new application-level communication protocol based on the TCP/IP protocol, satisfying the need for generic interconnection solutions for industrial robots in collaborative environments such as construction sites. Following the establishment of specific technical constraints for such system, we discussed the design of this protocol, its dependencies and its flexibility. In its current state of development, this software tool allows to create simplified communications between machines with low bandwidth requirements, and to maintain the abstraction of each local robot programming platform by exchanging structured data types. The implementation of the protocol is exemplified for ABB IRC5 controllers.

Further development of this network service will focus on the implementation of drivers for additional robot controllers from various manufacturers such as KUKA, Universal Robots, Staübli, Fanuc, Comau, Yaskawa/Motoman and Mitsubishi. Additional benchmarking, packaging and implementation utilities for microcontrollers (Arduino, Phidget, RasperryPi, etc.) and computer applications will be released in a second phase. This latter stage of development will allow to obtain performance indices of this protocol, as the measurements obtained solely between equivalent controllers cannot be generalized.

References

ABB Robotics (2014a) Application manual controller software IRC5

ABB Robotics (2014b) Technical reference manual: RAPID instructions, functions and data types

ABB Robotics (2014c) Technical reference manual: RAPID kernel

Everett J, Slocum A (1994) Automation and robotics opportunities: construction versus manufacturing. J Constr Eng Manage 120(2):443–452

Evans PT, Edwards SM (2012) A disruptive community approach to industrial robotics software. Presented at the 2012 RoboBusiness Leadership Summit, Pittsburgh, October 2012

Gullberg M (1993) The role of co-verbal gestures in second language discourse—a case study. Working Papers, Lund University, Department of Linguistics, vol 40 pp 49–70

ROS.org (2015) ROS-industrial robot driver specification (Final). http://wiki.ros.org/Industrial/Industrial_Robot_Driver_Spec. Accessed 15 Jan 2015

Schwartz T (2012) HAL: extension of a visual programming language to support teaching and research on robotics applied to construction, Rob|Arch 2012. Springer, Wien, pp 92–101

Stentz A (2001) Robotic technologies for outdoor industrial vehicles. In Aerospace/defense sensing, simulation, and controls. pp 192–199

Utz H, Sablatnög S, Enderle S, Kraetzschmar G (2002) Miro-middleware for mobile robot applications. Robot Autom, IEEE Trans 18(4):493–497

Warszawski A, Navon R (1998) Implementation of robotics in building: current status and future prospects. J Constr Eng Manage 124(1):31–41

Part III

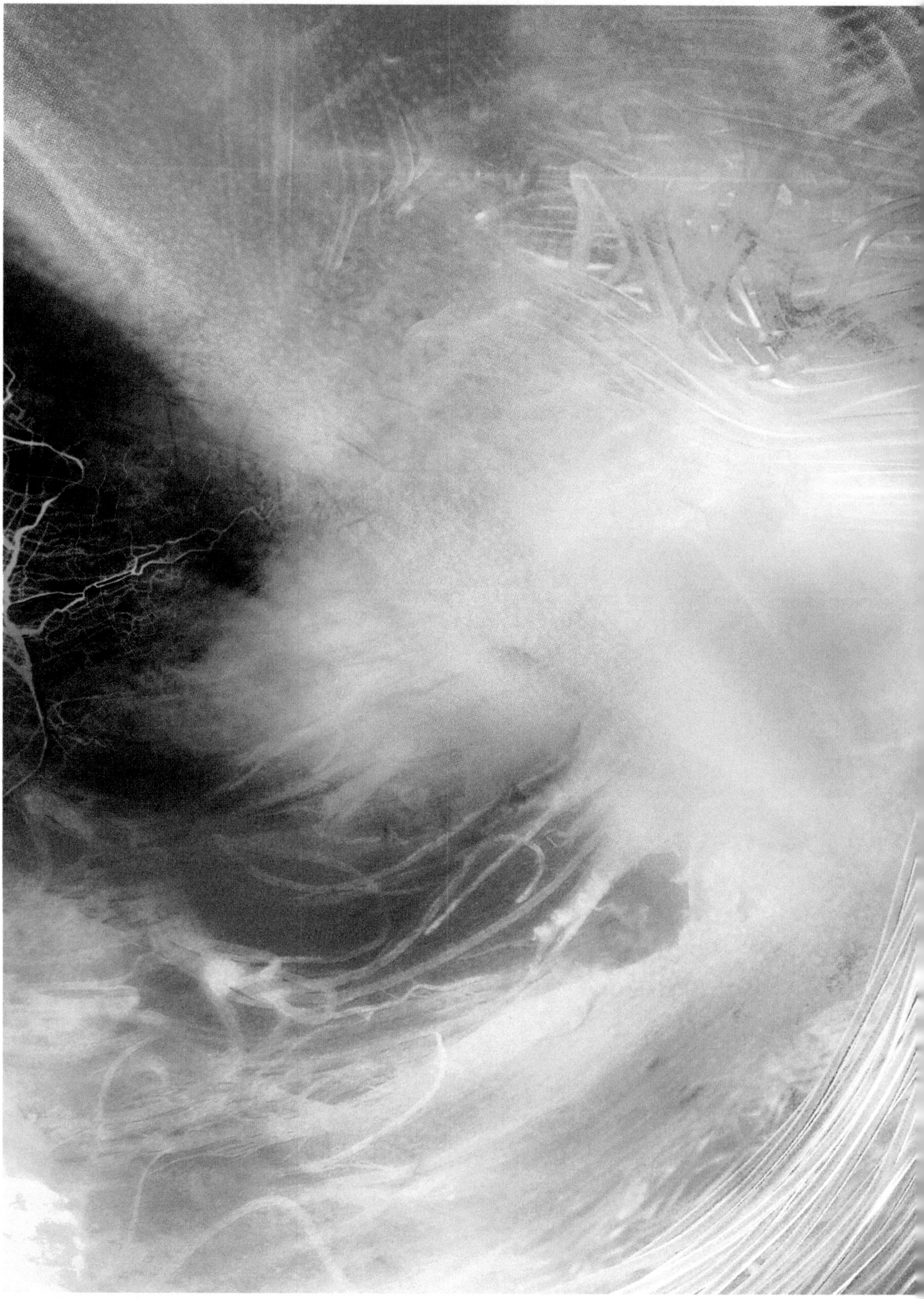

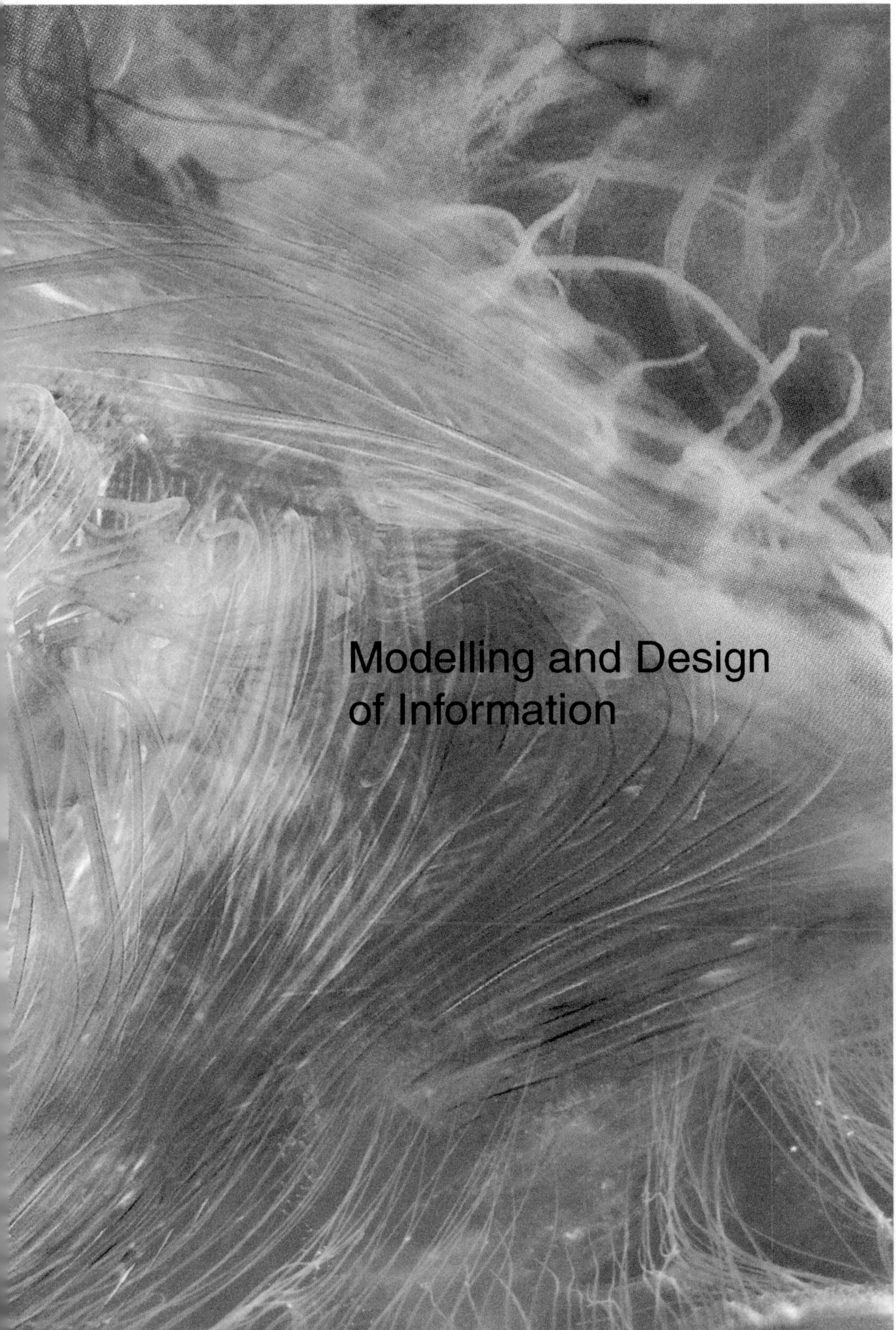

Modelling and Design of Information

Towards AI Drawing Agents

Robert Vierlinger

Abstract

This paper speculates on the applicability of methods from artificial intelligence in generative design. In particular, Cppn-Neat as a recent concept is implemented and tested on an established platform for procedural design. The method evolves artificial neural networks and has proven successful in control of robotic navigation, cognition, and collaboration, therefore is proposed to control the behaviour of drawing agents for architectural optimization tasks. A projected goal is the development of networks which, once evolved, instantly perform specific tasks on generic geometries, to be re-used on different projects and save run-time in the optimization process. Potentials of the method regarding general encoding and novel modes of formal articulation are examined in a first set of parameterization-studies. Further examples are testing the scalability of results from different problem domains and work towards the evolution of large ANNs to be suitable for the synthesis of higher level concepts.

Introduction

Motivation

Borrowing paradigms from nature has proven widely successful in design. Especially architectural engineering optimization can benefit largely from evolutionary strategies (Kicinger et al. 2005) where parallels to the development of artificial intelligence (AI) inherently can be drawn. The field of AI has seen much progress over the last decade (Deng and Yu 2014), where the use of Artificial Neural Networks (ANNs) for e.g. audio-visual pattern recognition are state of the art. Two fundamentally different methods have to be distinguished: Training of an ANN by supplying sample-data, and meta-heuristic evolution of an ANN by evaluation of a fitness function. Many popular results in AI-research recently have been achieved by the former (Quoc et al. 2012; Gregor et al. 2015), where high

R. Vierlinger (✉)
Institute of Architecture, Institute for Structural Design, University of Applied Arts Vienna, Vienna, Austria
e-mail: robert.vierlinger@gmail.com

M.R. Thomsen et al. (eds.), *Modelling Behaviour*, DOI 10.1007/978-3-319-24208-8_30

performance GPU-computing and large sets of sample data allow a great complexity of the tasks. Still the latter method seems to be more adequate for the use in design, as the diversity of tasks requires a flexible high-level implementation of the problem. Also, sufficient amounts of sample data mostly cannot be supplied.

Research on analogue computation (Siegelmann 1995) shows the Turing-Completeness of ANNs, which means that they can simulate and compute any arbitrary program or problem. Gershenfeld (2011) argues that methods of this kind can align soft- and hardware in a more direct way, resulting in new levels of scalability, performance, and simplicity. The problem in the case of ANNs merely is how to 'program' them, i.e. to find the right topology and weight-matrix for their internal configuration. When evolving an ANN, the configuration of the evolutionary process extends this task, to e.g. prevent deception. Further the nature of heuristic optimization requires more runtime to configure a well-performing ANN than to train it with sample data.

Cppn-Neat

To evolve large scale ANNs more efficiently, Stanley (2007) describes a general purpose mapping called Compositional Pattern Producing Network (Cppn). Its original purpose is the indirect encoding of the large ANN's weight-matrix, generally speaking to parameterize complex formations by a few parameters while exploiting geometric regularities and spatial conditions. Cppns are a slightly modified version of ANNs, which can add many other activation functions to the classic Sigmoids and Gaussians (Fig. 1). The choice of functions biases the output toward specific types of patterns and regularities. Developmental biology is a good example for a powerful indirect mapping, producing a precisely defined complex output, the organism, based on a very limited set of instructions, the DNA. As an indirect encoding, the general principle of weighted function-superimposition already proved useful in engineering optimization (Michalatos and Kaijima 2014; Vierlinger and Bollinger 2014), but many forms of indirect encoding also exist in computational design. Cellular Automata (Wolfram 2002), Cartesian Genetic Programming (Miller 2000), or recursive branching systems (Lindenmayer 1968; Dawkins 1986), could be named as just a few related examples.

With Neuro Evolution of Augmented Topologies (Neat), Stanley (2004) also provides a powerful algorithm to evolve and optimize Cppns. A Cppn maps a numeric vector of m dimensions to a vector of n dimensions, where Neat is used to evolve the internal configuration of the Cppn from the simplest possible towards a more complex mapping function by means based

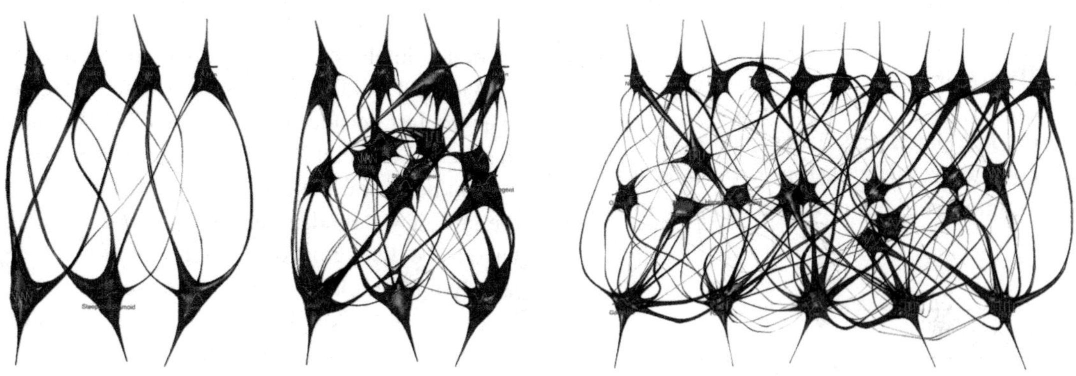

Fig. 1 Visualization of Cppns as organic nervous systems. Their neurons are holding different activation functions to map sensory input values (*top row*) to the output nodes (*bottom row*). *Left* is a typical, simple starting position for Neat, where the *two right* networks already are evolved and complexified

on a genetic algorithm with diversity maintenance through speciation. The implemented version of ES-HyperNeat (Risi and Stanley 2012) additionally features multi-objective optimization and novelty search (Lehman and Stanley 2011).

Problem

This paper works towards an application of Cppns to control the behavior of agents. Research exists on related topics such as collaborative swarm control (D'Ambrosio and Stanley 2008), agent-navigation (Pugh and Stanley 2013), and cognitive tasks (Stanley et al. 2009) and is used as a basis for case studies. Some of the works referenced dedicate tests to the generalization of behavior, which means Cppn-Neat is not just used as a mapping for a single optimization problem, but intended to be trained on one example and then applied to different cases without or with little additional training. One of the main questions is how to use Cppn-Neat within a generative setup, as previous research shows that successful evolution strongly depends on the formulation of the problem around the Cppn.

This work focusses on two topics in the context of Cppn-parameterizations mostly based on agents:

1. Simple ways of parametrically describing complex geometry;
2. Re-usability of a successful optimization result.

Method

An experimental plug-in for the parametric modelling platform Grasshopper® is developed within this work, basing on the implementation of ES-HyperNeat by Sebastian Risi (http://sebastianrisi.com; http://eplex.cs.ucf.edu).

Parameterization

Cppns enable an easy encoding of complex multi-dimensional fields which always exhibit perceptible regularities (Fig. 2). This is exemplified on a recursive branching system which mimics the original Biomorph program by Dawkins (1986). The example shows the maintenance of high causality in complex mappings (Fig. 3). A four-to-seven Cppn is queried with a point's [x|y|z] coordinates and the normalized progress of the Runge-Kutta-4 integration. The seven numbers obtained are [x|y|z] components of the directional vector; three thresholds for termination, branching, and snapping of the direction to a regular grid of vectors; one value for the thickness of the curve segment. A 2D slice through the hyperspace of a mapping is produced by leaving all surplus inputs of the Cppn constant (Fig. 4). The encoded fields further can be easily fostered to resemble features of existent fields such as drawings (Fig. 5) (Clune et al. 2013).

Modern structural design ideally merges with and reacts to spatial requirements and formal expression. In the interface of architectural and structural designer traditional methods of sketching and conceptualization still are necessary, but as the complexity of requirements and projects increases it can be critical to quickly produce manifold alternatives for further decision making. The implementation of a Cppn in this case mostly leverages its flexibility of parameterization. The agents are made aware of their position and distance to different elements of the building such as cores, facades, hulls, slabs, cantilever, open spaces, etc. and alter their directionality and branching behaviour based on this information. Selected results have been used

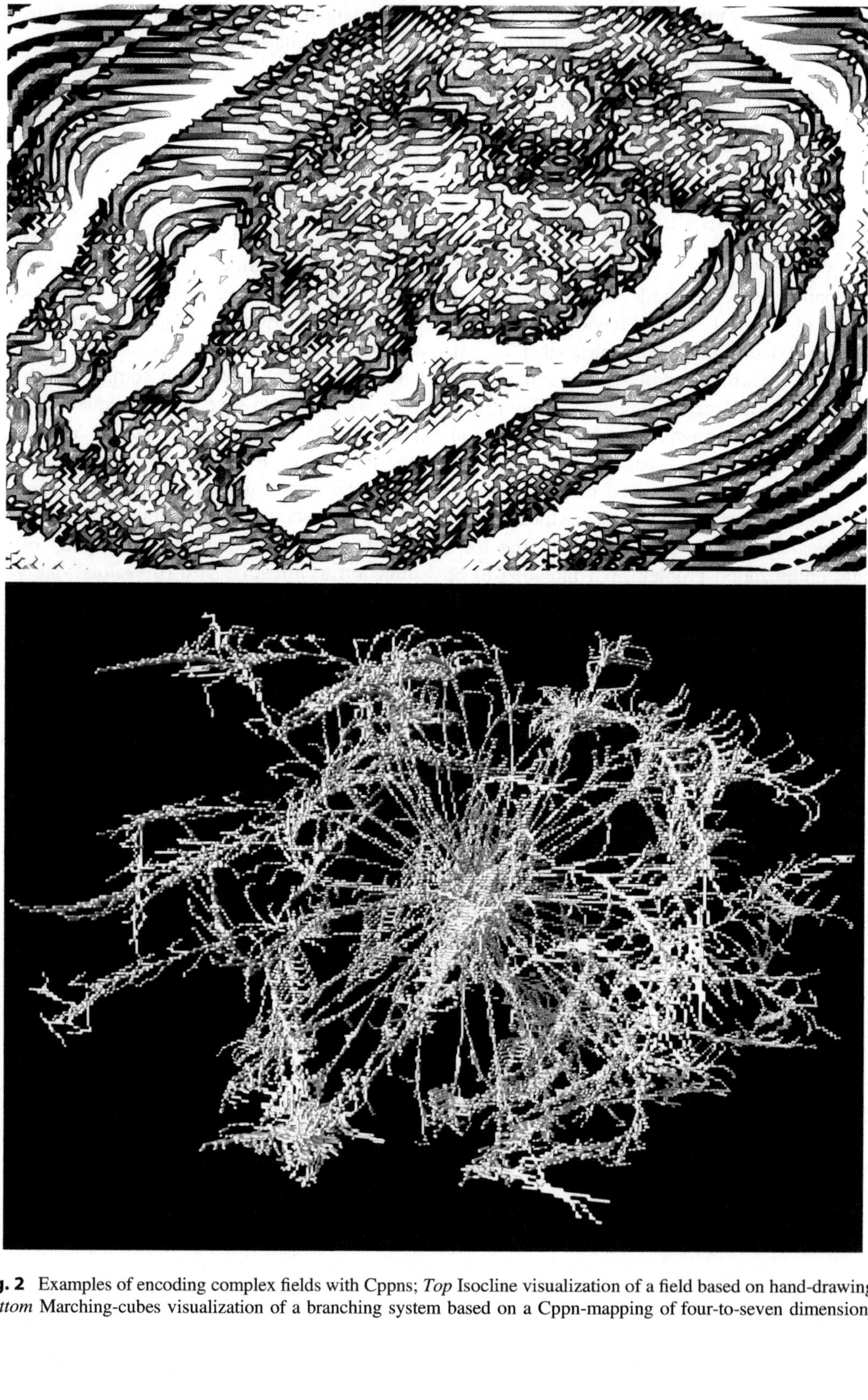

Fig. 2 Examples of encoding complex fields with Cppns; *Top* Isocline visualization of a field based on hand-drawings; *Bottom* Marching-cubes visualization of a branching system based on a Cppn-mapping of four-to-seven dimensions

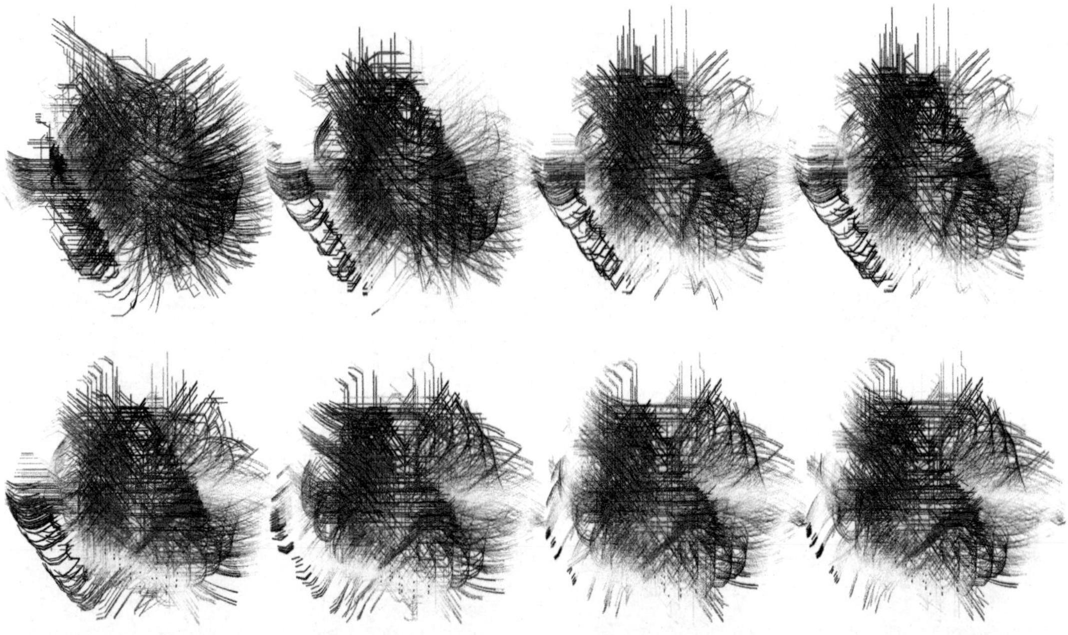

Fig. 3 Cppn-based branching system which is randomly mutated from *top left* to *bottom right* successively. Up to 20 % of the connections are modified by up to ±10 %. The randomly generated Cppn contains 5 hidden nodes and 50 connections

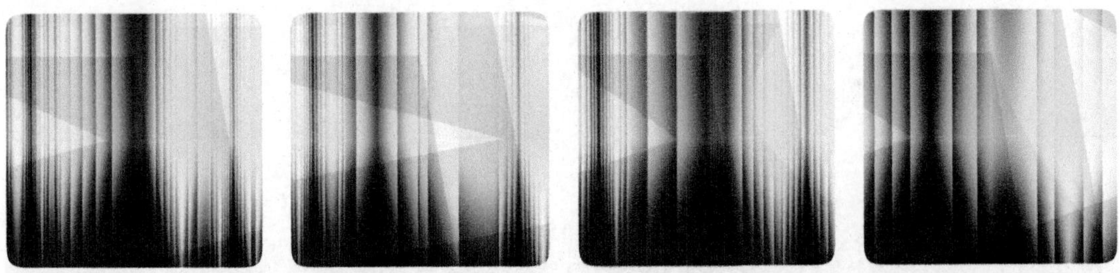

Fig. 4 2D slices through every 2nd Cppn of Fig. 2; sampling just the first two inputs with [x|y] of a pixel, while keeping the other input values constant slices the hyperspace. The first output value is used as the greyscale value of the pixel

as the basis for deciding on the final structural articulation (Fig. 6).

Classic Optimization

Cppn-Neat has proven well suited for the evolution of indirect encodings showing pattern-like regularities (D'Ambrosio and Stanley 2007). However examples of classic optimization often fail with Cppn-Neat. Galapagos (Rutten 2014) and octopus (Vierlinger 2013) both produce better results by adjusting x- and y- coordinates of 20 of control points (Fig. 7).

Connectivity Patterns and Scalability

Connectivity patterns defined by a Cppn can be applied to design spaces of different resolution without losing the basic qualities of the pattern, hence a low-res version could be used to reduce

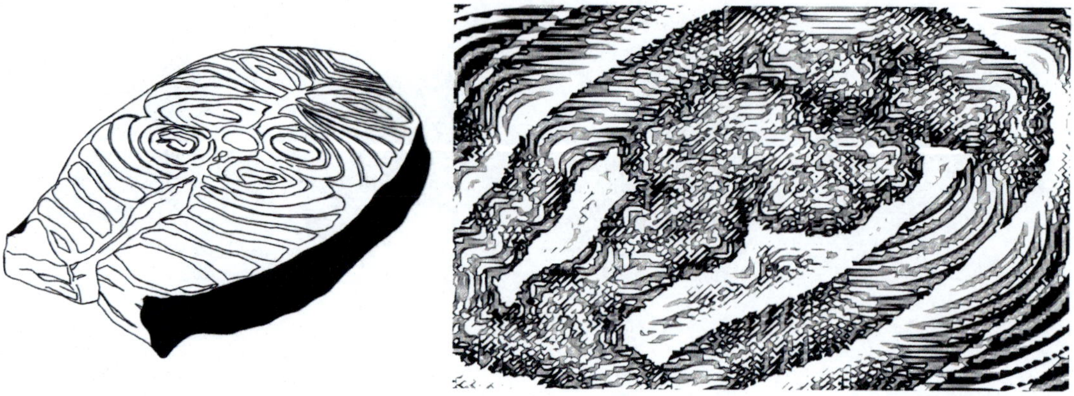

Fig. 5 Field injection by supplying the closest distance to hand drawn curves (*left*) as an input to the Cppn (Clune et al. 2013); resulting field visualized by isoclines (*right*); *drawing left* by Pauline Jocher

Fig. 6 Application of Cppn-neat in a large scale construction project. Given the architectural envelope, slabs, and cores, aesthetically different versions of a branching-stitching-structure are drawn and optimized to support the three cantilevering corners. The false-colour plot visualizes a slice through the multi-dimensional mapping; Image by Melanie Kotz de Acha

computational complexity in the training phase. Further the run-time needed to apply an evolved Cppn to a higher resolution design space is magnitudes lower than for the initial training. The example takes 20 min to evolve while the application to a different resolution takes 0.1 s. However when increasing the resolution, noise in the Cppn can become visible and diminish the quality of the result. Further, higher resolution design spaces might enable better solutions than initially found in the low-res training setup.

Structural topology optimization of a cantilever truss shows an application of the HyperNeat principle (Stanley et al. 2009): links between fixed points are controlled by the Cppn, but are interpreted as structural members instead of connections in a large ANN (Fig. 8). Each element is queried in a 4-to-1 mapping by supplying [x|y] of start- and end-point, then receiving a weight-value to determine presence of the member. Karamba (Preisinger 2013) is the parametric structural modelling software used to

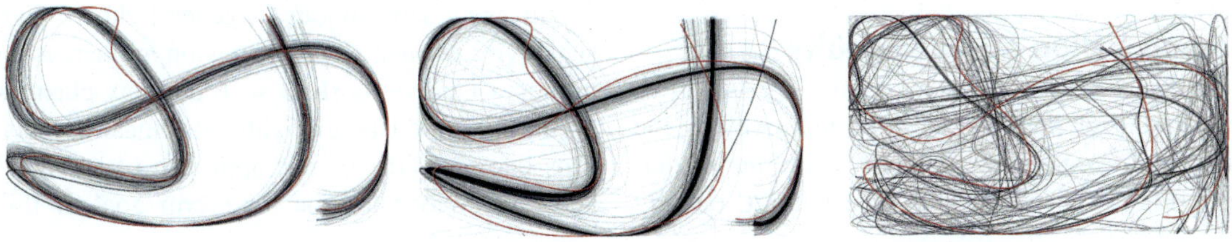

Fig. 7 Results of different evolutionary solvers for Grasshopper® fitting a target curve (*red*). *Left* Set of solutions after 1000 generations of search with octopus' standard settings and minimization of two goals: distance between 20 equally spaced points along both curves and proximity of 20 points on the solution curve to the target curve. *Middle* Last 1500 improving solutions of Galapagos with evolutionary solver and standard settings. *Right* Solutions of generation 1000 produced by Cppn-neat, distorting a regular grid of 20 points to place the control points

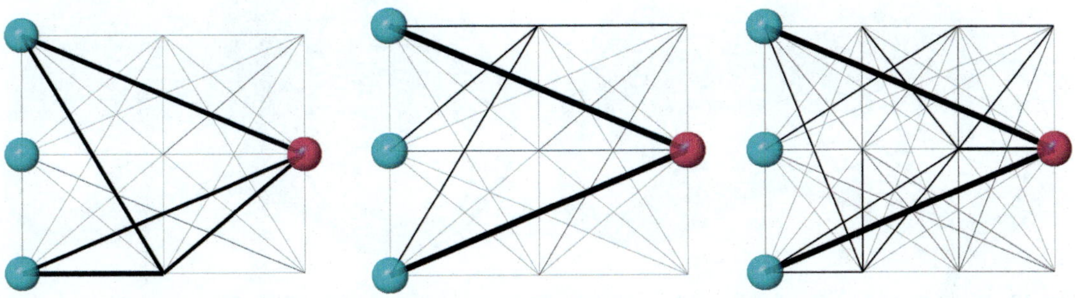

Fig. 8 Topology optimization using hyperneat to determine active structural elements. *Blue dots* are supports, *red dots* are loads. *Thin grey lines* are possible elements, whereas *thick black lines* are activated structural elements. *Left* Result of square beam benchmark (Richards and Amos 2014) considering buckling; *Middle* Result without buckling; *Right* Scaling up the substrate resolution while features of the originally trained example are maintained

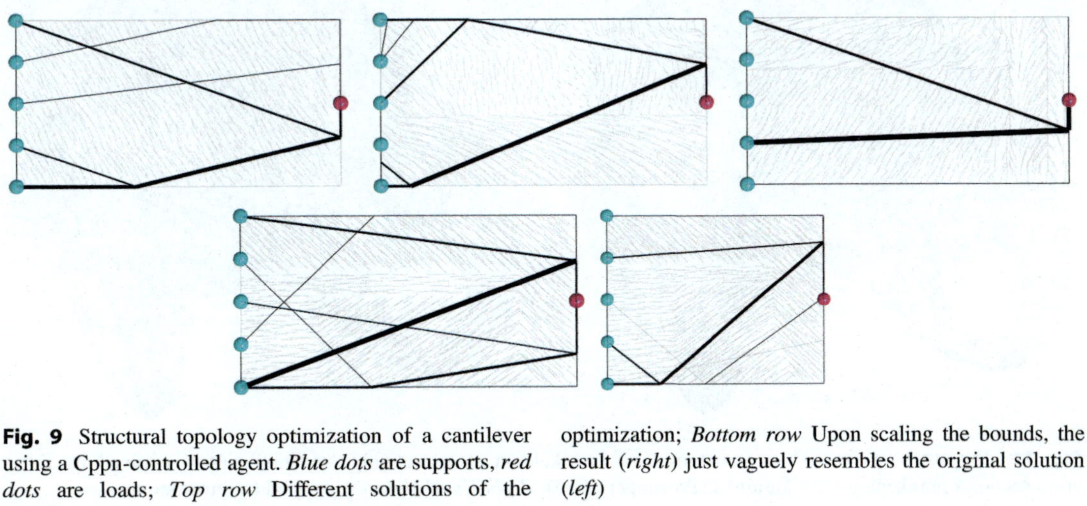

Fig. 9 Structural topology optimization of a cantilever using a Cppn-controlled agent. *Blue dots* are supports, *red dots* are loads; *Top row* Different solutions of the optimization; *Bottom row* Upon scaling the bounds, the result (*right*) just vaguely resembles the original solution (*left*)

analyze the structural system, where the active members are sized to meet a target stiffness. Minimizing the resulting mass of the system is used as the fitness function.

Agent Vision and Scalability

A vierendeel cantilever is to be improved by the traces of agents which bounce inside the rectangle (Fig. 9). They are equipped with 20 radially arranged distance sensors which feed the 20 inputs of the Cppn. The two outputs form the [x| y] components of the agent's next movement direction. If the direction points outwards the agent is stopped. The trails start at the supports.

An agent's behaviour in general shows high sensitivity as each step depends on an increasing number of previous steps such that any changes accumulate and lead to highly non-linear outputs. This questions the classic approach of localized agent behaviour for the long-term goal of this

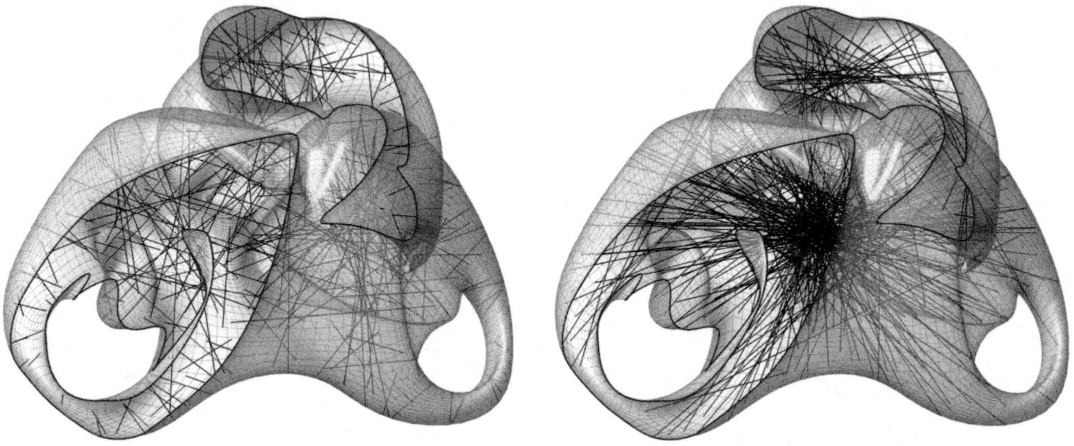

Fig. 10 Directional biases in a free-form shape; *Left* Normal vectors of the surface; *Right* Direction of highest spatial depth

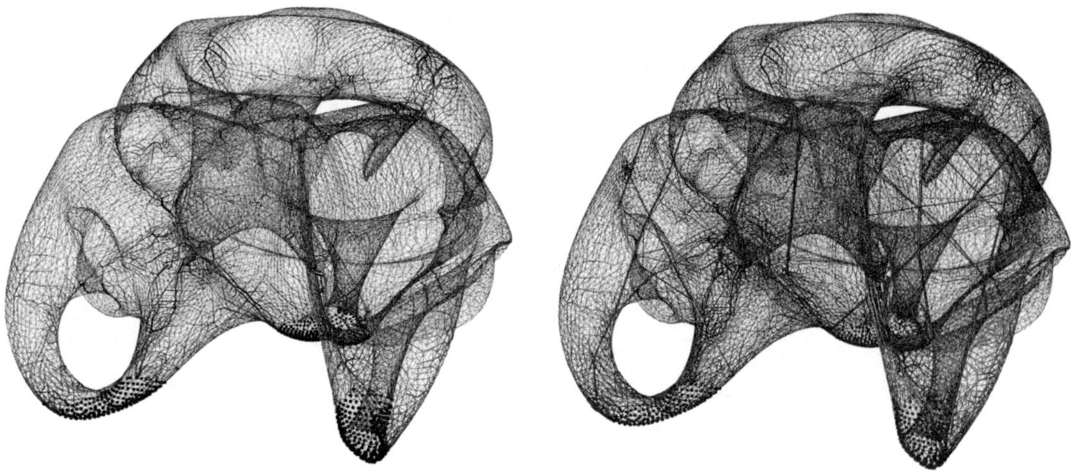

Fig. 11 The stitching is used as structure; *Left* Exoskeleton structure only, venation pattern evolved by linear cross-section optimization with karamba (Preisinger 2013); *Right* Exoskeleton supported by inner struts

work to develop re-useable shape-invariant patterning agents. Working towards this goal, it is speculated that the network must become aware of the global context which it is operating on. Increased complexity of the task is addressed with large scale ANNs in the last two examples.

Biased Agent Navigation

To enhance the discovery of good solutions, general guides for an agent's direction can be implemented to augment the directional decision of an agent's Cppn based on distance-vision (Fig. 12). An additional output of the Cppn

Fig. 12 False colour visualization of the shape's directional field, generated by agent vision and a Cppn; Basic shape and image by Melanie Kotz de Acha

weighs each guidance-direction (Fig. 10) to bias the original direction. A simple fitness function for even coverage of the interior space is formulated as a goal to ultimately form a supportive structural system (Fig. 11).

Complex, Re-usable ANNs

The number of connections in an ANN can reach up to billions (Quoc et al. 2012) for the recognition of high-level concepts. Cppn-HyperNeat can maintain evolve-ability for large ANNs, which is tested on a visual discrimination problem (Stanley et al. 2009). Each possible connection is queried against a Cppn of 8 inputs to obtain one output, feeding a connection's start point's [x1|y1|z1] coordinates, the end point's [x2|y2|z2] coordinates, plus a geometric bias [abs (x1 − x2)| abs(y1 − y2)] as additional information on locality. The value obtained is used to determine the presence and weight of the connection. To create a design-space of possible connections, points have to be positioned representing input-, output-, and hidden neurons which can be a sensitive task (D'Ambrosio and Stanley 2007) but also provides the Cppn with useful geometric information of the problem domain. The automated placement of hidden neurons based on areas of high information density in the Cppn is known as Evolvable Substrate (ES) HyperNeat (Risi and Stanley 2012) and is implemented in the tool developed within this work.

The task for the network is to distinguish a large black dot from a small black dot, by producing an output-image which points at the larger dot (Fig. 13) with its highest value. The large ANN exhibits 100 input-, 100 output-, 36 hidden neurons, and 17,200 possible connections (Fig. 14) and is successfully evolved by the two-step parameterization of Cppn-HyperNeat. In

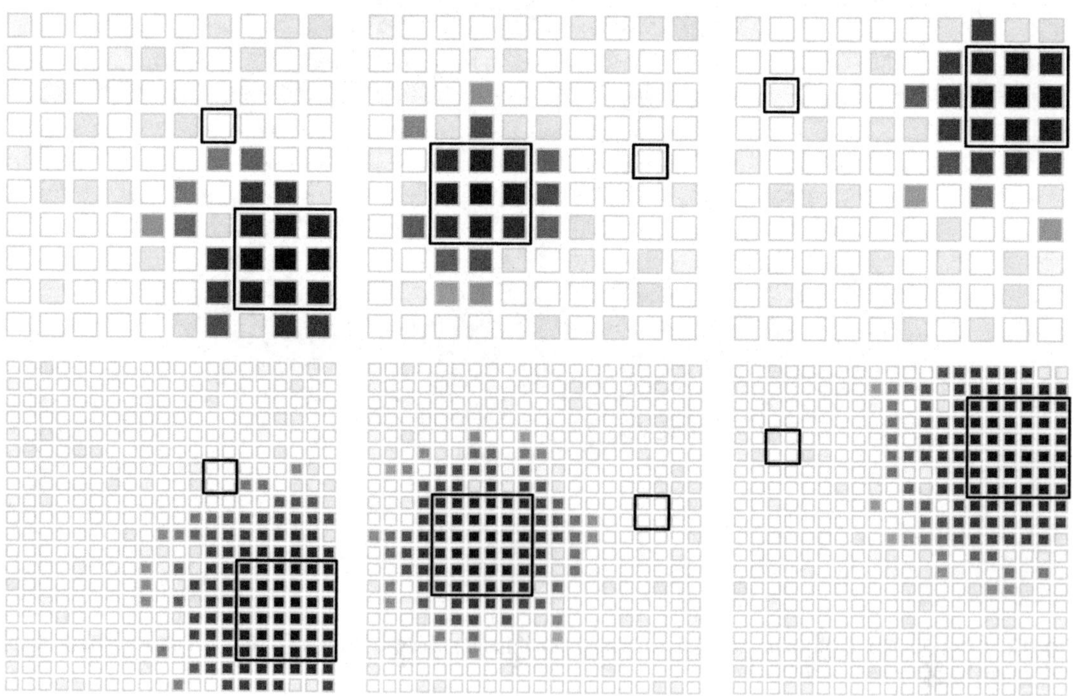

Fig. 13 Validation of trained network; *Top* big and small input-dot (*black quadrangles*) on a 10 × 10 sensory grid; output pattern in *greyscale*, pointing to the *larger dot*; *Bottom* Scaling of the design space to 20 × 20 inputs and outputs

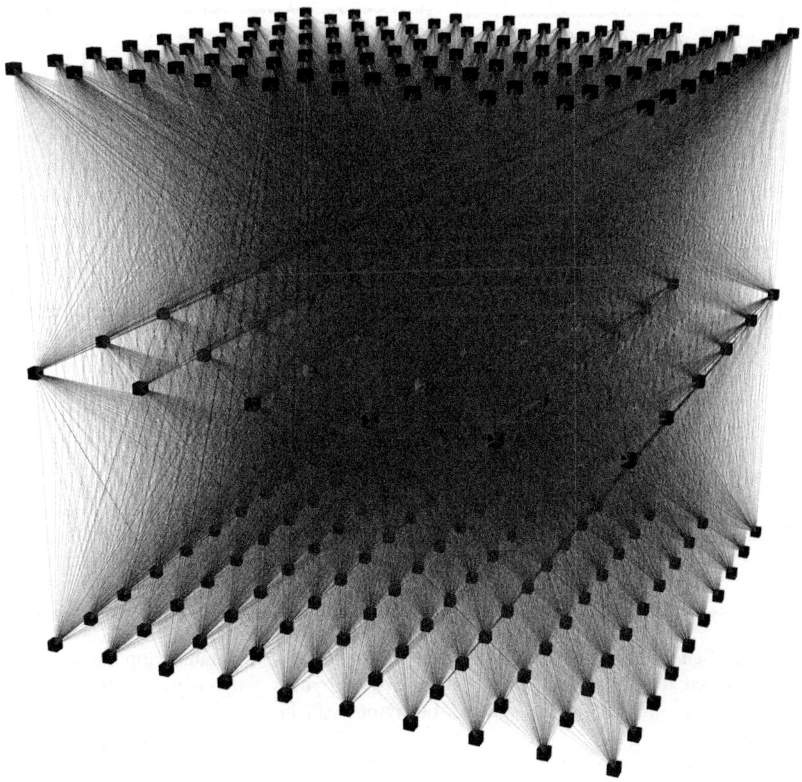

Fig. 14 Large neural network with 100 inputs, 100 outputs, 36 hidden nodes for the visual discrimination task. 17,200 connections are successfully controlled by the Cppn

each evolutionary try, ten different dot-placements are evaluated to achieve re-usability. Scaling of the substrate here works well due to the comparatively simple nature of the task.

Re-useable Navigation Agents

The maze navigation task (Pugh and Stanley 2013) uses agent vision to feed the network and in turn get instructions on movement. When using a Cppn only, training of a single agent in a single setting is successful, but fails for different starting points and mazes. The reason is that a universal behaviour follows patterns, which are hard to be evolved by Neat. Patterns in the network are firstly generated by the Cppn when used to parameterize a second, potentially larger ANN.

For the maze-navigation task novelty search was used (Lehman and Stanley 2011). The novelty is measured as the distance between the solutions' 100-dimensional vectors of [x|y] coordinates of 50 equally distributed points along an agent's trace. The network is trained with five starting positions on one maze only. A good exploratory coverage of the maze is achieved in the training task as well as upon application of the trained agent on other mazes (Fig. 15). The example proves the evolve-ability of re-useable behavioural patterns. By using seeded Cppns for subsequent evolutionary searches, the collision-avoiding behaviour could form the basis for a training of more intricate tasks.

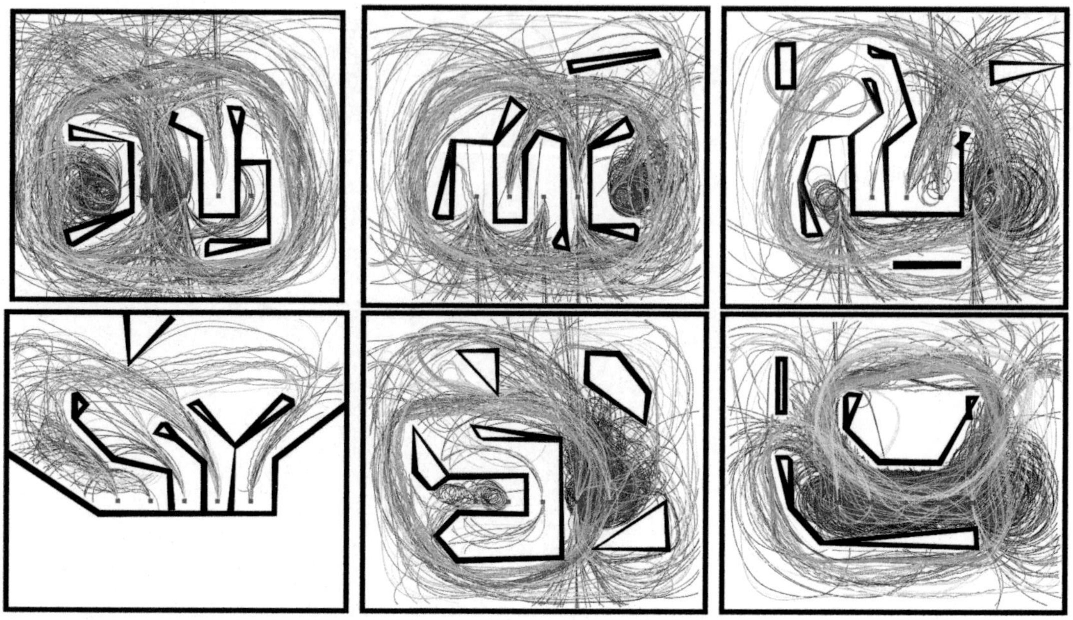

Fig. 15 Sensory agents are trained for the *top left maze*, then applied to the others. Based on behavioural rules learned in a single example, specific behaviour can be instantly applied to different conditions. *Red dots* indicate the starting positions of the agents

Conclusion

A number of studies firstly show ways using Cppns to encode complex field conditions as a basis for formal articulations. The flexibility in dimensionality and the intrinsic correlations among those sampling channels together with the computational efficiency are the most apparent advantages of the method as an aid for novel modes of formal expression. A lack of intuitive control over the resulting function hereby is best compensated with conceptual changes in the mapping itself by addition or interrelation of sampling dimensions.

Evolutionary optimization examples with Cppn-Neat are characterized by a high sensitivity on the formulation of the problem around the Cppn. In most examples it is possible to evolve a good solution for a specific setting, however the setup of a successful optimization might require experience and deepened knowledge on the theoretical implications of the method. This further supports the approach of developing ready-made mappings which are fully or partially trained to successfully perform a specific task on different inputs, as shown in the maze navigation and visual discrimination example. Evolutionary algorithms generally suffer from long runtime, but the training of re-usable, instantly applicable networks is seen as a promising approach to the time-consuming process of heuristic optimization.

It is shown that Cppns in control of large ANNs are able to be evolved for tasks of higher complexity. The process performs particularly well when the exploitation of geometric regularities and patterns is beneficial for the performance of the network. It is to be shown if and how a large ANN can be evolved to master the recognition and synthesis of higher level concepts such as the ad hoc design of rules for efficient structures on different complex geometries. Drawing conclusions from related fields of AI-ANN research could help in designing suitable methods and is subject to ongoing research.

References

Clune J, Chen A, Lipson H (2013) Upload any object and evolve it: injecting complex geometric patterns into CPPNS for further evolution. In: Proceedings of the 2013 IEEE congress on evolutionary computation (CEC), Cancun Mexico. pp 3395–3402, 20–23 June 2013

D'Ambrosio DB, Stanley KO (2007) A novel generative encoding for exploiting neural network sensor and output geometry. In: Proceedings of the genetic and evolutionary computation conference (GECCO-2007). ACM, New York, 7–11 July 2007

D'Ambrosio DB, Stanley KO (2008) Generative encoding for multiagent learning. In: Proceedings of the genetic and evolutionary computation conference (GECCO 2008), ACM, New York, 12–16 July 2008

Dawkins R (1986) The blind watchmaker. Norton, New York

Deng L, Yu D (2014) Deep learning: methods and applications. Found Trends Signal Process 7(3–4):197–387

Gershenfeld N (2011) Aligning the representation and reality of computation with asynchronous logic automata. Comput Spec Issue Bio inspired Comput 93(2–4):91–102

Gregor K, Danihelka I, Graves A, Rezende DJ, Wierstra D (2015) DRAW: a recurrent neural network for image generation. In: Proceedings of the 32nd international conference on machine learning, JMLR: W&CP, vol 37. Lille, France

Kicinger R, Arciszewski T, De Jong K (2005) Evolutionary computation and structural design: a survey of the state of the art. Comput Struct 83(23–24):1943–1978

Lindenmayer A (1968) Mathematical models for cellular interaction in development. Theor Biol 18(3):300–315

Lehman J, Stanley KO (2011) Abandoning objectives: evolution through the search for novelty alone. Evol Comput 19(2):189–223

Michalatos P, Kaijima S (2014) Eigenshells: structural patterns on modal forms. In: Shell structures for architecture: form finding and optimization. Taylor & Francis, Routledge, London, pp 195–210

Miller JF (2000) Cartesian genetic programming. In: Genetic programming. European conference, EuroGP 2000, Edinburgh, April 2000. Lecture notes in computer science, vol 1802. Springer, Berlin, pp 121–132

Preisinger C (2013) Linking structure and parametric geometry. Architectural Des 83(2):132–135

Pugh J, Stanley KO (2013) Evolving multimodal controllers with HyperNEAT. In: Proceedings of the genetic and evolutionary computation conference (GECCO 2013), ACM, New York, 6–10 July 2013

Quoc VL, Ranzato M, Monga R, Devin M, Chen K, Corrado G, Dean J, Ng A (2012) Building high-level features using large scale unsupervised learning. In: Proceedings of the 29th international conference on machine learning, Edinburgh, 26 June–1 July 2012

Richards D, Amos M (2014) Designing with gradients: bio-inspired computation for digital fabrication. In: Proceedings of ACADIA 2014, association for computer-aided design in architecture, University of Southern California, Los Angeles, 23–25 October

Risi S, Stanley KO (2012) An enhanced hypercube-based encoding for evolving the placement, density and connectivity of neurons. Artif Life 18(4):331–363

Siegelmann HT (1995) Computation beyond the turing limit. Science 268(5210):545–548

Stanley KO (2007) Compositional pattern producing networks: a novel abstraction of development. Genet Program Evolvable Mach 8(2):131–162

Stanley KO (2004) Efficient evolution of neural networks through complexification. Dissertation, University of Texas

Stanley KO, D'Ambrosio D, Gauci J (2009) A hypercube-based indirect encoding for evolving large-scale neural networks. Artif Life 15(2):185–212

Vierlinger R, Bollinger K (2014) Accommodating change in parametric design. In: Proceedings of ACADIA 2014, association for computer-aided design in architecture, University of Southern California, Los Angeles, 23–25 October

Wolfram S (2002) A new kind of science. Wolfram Media, Illinois

Agent-Based Decision Control—How to Appreciate Multivariate Optimisation in Architecture

Kristoffer Negendahl, Thomas Perkov and Jakub Kolarik

Abstract

Early stage building performance optimisation as a viable approach is yet to be applied efficiently in building design processes, especially in the early design stages where the design space is open and changes are inexpensive. This article proposes a method of entire building energy optimisation in the early design stage. The main focus is to demonstrate the optimisation method, which is done in two ways. Firstly, the newly developed agent-based optimisation algorithm named Moth is tested on three different single objective search spaces. Here Moth is compared to two evolutionary algorithms. Secondly, the method is applied to a multivariate optimisation problem. The aim is specifically to demonstrate optimisation for entire building energy consumption, daylight distribution and capital cost. Based on the demonstrations Moth's ability to find local minima is discussed. It is concluded that agent-based optimisation algorithms like Moth open up for new uses of optimisation in the early design stage. With Moth the final outcome is less dependent on pre- and post-processing, and Moth allows user intervention during optimisation. Therefore, agent-based models for optimisation such as Moth can be a powerful substitute for traditional stochastic optimisation.

Introduction

Building performance optimisations during early stages of the design process are not only related to risks of high uncertainty (Lin and Gerber 2014), but also require an excessive amount of calculations that are very time consuming (Salminen et al. 2012). The early design stage can be characterized by a limited amount of information about the building's architecture and at the same time a high frequency of design changes. In contrast to that, most optimisation methods rely on clear objectives and well defined boundaries. Such requirements are rarely associated with the early design stage. The process of designing

K. Negendahl (✉) · T. Perkov · J. Kolarik
Technical University of Denmark, Kongens Lyngby, Denmark
e-mail: krnj@byg.dtu.dk

M.R. Thomsen et al. (eds.), *Modelling Behaviour*, DOI 10.1007/978-3-319-24208-8_31

can be far better described as an exploration of boundaries and objectives rather than finding the one solution within a fixed number of boundaries and objectives. For optimisation to be truly appreciated in building design both qualitative and quantitative objectives as well as "real" human control of these need to be part of the exploration process during the building optimisation.

Background

Few researchers address the complicated methodological challenges in the need to combine human evaluation with computational speeds in early stage optimisation. Nonetheless, design optimisation of building performance is increasingly being used in practice. A popular method for application of optimisation in design is based on generative models in combination with stochastic optimisation algorithms. To maintain artistic control of the optimiser (tool) many different attempts have been made. One dominating idea is to let the designer control the objectives rather than the actual geometry, for example (Caldas and Norford 2001; Gaspar-Cunha et al. 2011). Caldas and Norford (2001) argues that weights on objectives give the designer the highest level of control over the optimisation process. Caldas and Norford (2001) also notices that weighting factors to attributes of different objectives as of will to some extent constrain the design process itself. This is partially because changes of weighting factors may be discontinuous and can lead to dramatically different solutions and partially because setting up weighting factors requires a large amount of insight into all factors involved in the decision-making procedure.

Agent-based models (ABMs) used for optimisation are another method for stochastic optimisation. They are not as commonly used in building design as e.g. genetic algorithms, but ABMs find increasing interest in other design areas such as in transportation and manufacturing industries (Barbati et al. 2012). What distinguishes ABM from the more classical heuristic optimisation approaches (including stochastic and deterministic as mentioned above) is their ability to decompose a global problem into a number of smaller "local" problems that may be solved individually and simultaneously (Davidsson et al. 2007). ABMs, it should be noted, are from time to time mixed up with particle swarm algorithms (PSAs) (Liu et al. 2005). In this article we will clearly distinct the two algorithms by defining PSA-agents as individual solutions roaming in a competitive space of solutions, where ABM-agents compete in same-state solutions.

To complicate matters further the Building Performance Simulation (BPS) tools used to feed the optimisation algorithms affect the actual outcome of the optimisation process. Indeed the type, precision and quality of the BPS tool matter to great extent, but what is less obvious is that most BPS solvers approximate solutions due to adaptive variations in solver iterations (Wetter and Polak 2004) and thereby form discontinuous search spaces for the optimisation algorithm. As Wetter (2011) explains, nonlinear programming algorithms (typical stochastic optimisation algorithms) are computationally efficient (over deterministic counterpart), however they often require the cost function to be differentiable in the design parameters. Therefore, there is a conflict of the choice of optimisation algorithms preferred by designers and the tools applied in the process.

Method

The Moth Algorithm

The presented method makes use of a new adaptive, open agent-based optimisation algorithm named Moth; it has been developed by the authors. The algorithm allows parametric geometry and any other parametric variables to be controlled by individual agents. Moth stands for Multivariate Optimisation with Heterogeneous agents. It is a fully scalable algorithm, thus capable of taking any number of quantifiable objectives with any number of boundaries. Moth is a heterogeneous system; this means that every agent actually comes in opposite pairs; a minus- and a plus-agent. Moth is developed for the Rhino-Grasshopper environment (Robert McNeel & Associates 2013), and in this way it is integrated in the fast growing parametric universe supported by many enthusiasts and designers. Moth is open source and built in the IronPython programming language (Viehland et al. 2015) (Fig. 1).

Each Moth-agent has as a goal to improve its own dynamic objective. The Moth-agents can be manipulated by the model operator (designer) during the optimisation process as seen in Fig. 2. This gives the designer an option to focus on selected areas of a building (in the Euclidean space). The idea is to give the designer live decision control over a complex system that continuously seeks to find balance in the multivariate decision space. The proposed method depends on one or more building performance simulation (BPS) tools, a design tool (CAD) that can visualise the results from the BPS tools and a visual programming language (VPL) that facilitates the data shared between the BPS tools and the design tool. These kinds of models are also known as integrated dynamic models (Negendahl 2015).

The paired agents (−agent and a +agent) seek to reach the same objective by inversely changing design variables. These changes are based on the evaluation feedback from a coupled BPS tool. The evaluation of feedbacks is tied to the agent behavior through a *promotion* system. The strongest agent in the pair gets promoted by a positive feedback, thus giving the promoted agent more authority to change its design variables in its own preferred direction (either positive or negative). In this way every agent pair does not need to have any pre-defined preference or knowledge of what it is supposed to do other than what design variables it is allowed to manipulate. More details on how Moth works is found in the results Part 2.

Following Bento and Feijó's taxonomy (Bento and Feijó 1997) agents can be defined in a simple set of attributes and specifications, see Table 1.

Here, structural attributes may be physical (e.g. colour), or behavioural specifications (e.g. temperature = 35 °C, obtained from a thermal analysis). In an optimisation process, functional specifications (e.g. pleasant temperature) tend to be transformed into performance specifications defined in ranges (e.g. 18 °C < temperature

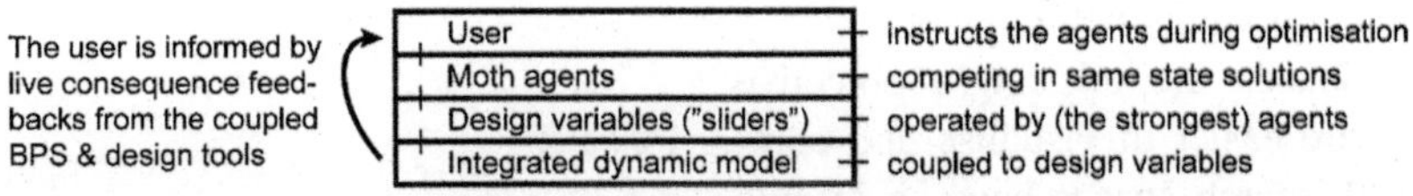

Fig. 1 Method—ABM optimization with Moth. By contrast to most stochastic optimization algorithms Moth allows changes of hyper parameters (agent instructions) during optimization

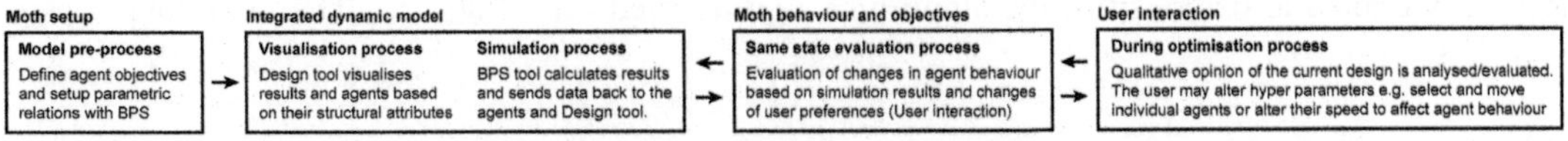

Fig. 2 MOTH—the agent-based optimisation algorithm allows user feedback and feedback from one or more coupled BPS tools facilitated through an integrated dynamic model. No post process is necessary

Table 1 Attributes of an agent-based optimisation algorithm

Type	Attribute	Representation
Structural	Colour, radius, shape	Geometrical representations
Behavioural	Temperature, daylight, energy consumption	Performance specifications

< 26 °C). Finally there are objectives describing the designer's intention for the design optimisation. The objectives may be associated, but not exclusively confined, to the performance specifications. In the following results we have demonstrated how Moth performs. The demonstration is composed in two parts:

Part 1. Moth is demonstrated on three different pre-defined single objective search spaces.
Part 2. Moth is demonstrated in a simple multi-objective case study.

Part 1

The three search spaces were tested by the Moth-algorithm as visualised in Fig. 3. The tests are compared to two other popular algorithms made available to Rhino/Grasshopper, namely Galapagos Evolutionary Solver (Rutten 2010) (the genetic algorithm, GA with default properties is used) and Goat (Simon et al. 2015) [the solver based on an controlled random search with local optimisation CRS2 (Price 1983) implemented in the NLopt nonlinear-optimisation package (Kaelo and Ali 2006)]. The purpose is not to compare different algorithms, but only to show some of the thoughts behind Moth.

The search spaces are all single objective spaces, as they are easier to visualise and compare, however all search spaces have an infinite number of "best performing solutions" thus representing an inner space of optimal solutions within the search space. The search spaces are artificially crafted to demonstrate the algorithms' ability to find optima and avoid sub-optimal solutions. These optima are represented on each search space in Fig. 3 by a dashed line, where all optimal solutions are placed exactly on the dashed line. Search space (a) is the simplest search space. It has a single large and centrally distributed circular optimal space with gradients pointing towards the optimal space in almost any point in the search space. Search space (b) is more complex and does not have any particular global gradients pointing towards the optimal space. Search space (c) is the most complex space of the three. The search space is modelled after Wetter and Polak's (2004) parametric plot of energy consumption of a building based on cooling and lighting as a function of the width of east and west facing windows. According to Wetter and Polak due to the loose solver tolerance, the cost function may exhibit "smooth" regions that are separated from each other by discontinuities. These "smooth" regions are visible as flat plateaus. The optimal spaces are placed on a very steep gradient. This type of search space is quite likely to be the type of space building designers would encounter in practice. However, in cases where multiple objectives are in play, the search space grows in dimensions and is difficult to visualise. Therefore, to demonstrate Part 2, the search space remains unknown.

Part 2

In Part 2 we applied three objectives to a test case (energy use, daylight availability and cost of windows) to a very simple building design as illustrated in Fig. 7. The building design

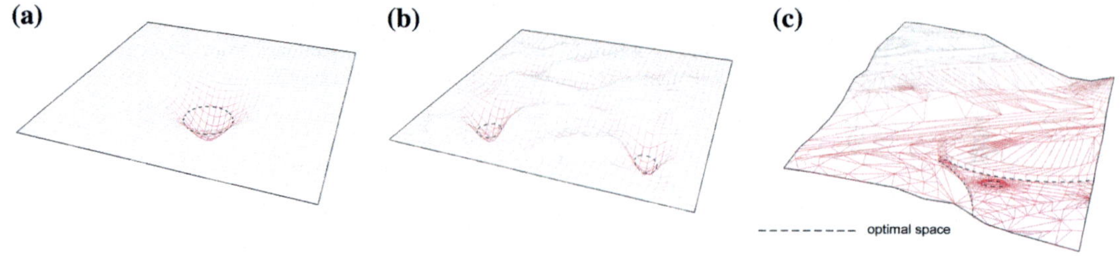

Fig. 3 Search space (**a**), (**b**) and (**c**). The optimal space is infinite solutions *on* the *dashed line*

variation for the demonstration is very limited and was controlled by three design variables. The two windows (widths) are each controlled by individual design variables and a third design variable controls window types (see Table 2). We have set three arbitrary performance specifications; daylight factor (DF %), cost of windows (−) and annual energy consumption (−) to the case. Cost indexes are defined to demonstrate the method, and do not represent real relative cost of window types.

Objective Functions

Three objective functions; $f_{energy}(x)$, $f_{cost}(x)$ and $f_{daylight}(x)$ are minimized by Moth.

$$\min_{x \in \mathbb{R}^s \times \mathbb{Z}^t} f_{daylight}(x) = O_{daylight} - nk / \left(\sum_{k=1}^{nk} (DF_x) \right) \quad (1)$$

where the fixed objective $O_{daylight}$ = 3 (%), k is the measurement point and nk is the number of measurement points in the building. DF_x is simulated by Radiance through Ladybug (Roudsari et al. 2013) for every solution, x.

$$\min_{x \in \mathbb{R}^s \times \mathbb{Z}^t} f_{cost}(x) = O_{cost} - \sum_{j=1}^{nj} (A_{w,j} \cdot C) \quad (2)$$

where the fixed objective O_{cost} = 15, j is the window and nj is the number of windows in the building, $A_{w,j}$ is the window area of the j'th window and C is the cost index shown in Table 2.

$$\min_{x \in \mathbb{R}^s \times \mathbb{Z}^t} f_{energy}(x) = O_{energy} - \sum_{i=1}^{ni} \left(Q_{heating,i} + 2.5 \cdot (Q_{cooling,i} + Q_{vent,i} + Q_{light,i}) \right) \quad (3)$$

where the fixed objective O_{energy} = 215, i is the load condition of the particular condition, ni is the number of load conditions. $f_{energy}(x)$ is simulated by Be10 (SBI 2013) through the Termite (Negendahl 2014) interface for every solution, x. The primary energy factor 2.5 is multiplied with electrical energy uses according to the Danish building regulations (2013), and the primary factor for the heat supply is set to 1.

Results

Part 1

The first comparison of the three algorithms seen in Fig. 4 shows clearly visible differences in the algorithms' approach to find minima. First, there is the "spread" of the solutions. The spread is defined by the way the algorithm distributes its solutions in the search space. One may think of this process as a search space sampling process for possible minima. Moth, compared to the two other algorithms generates less spread after a few iterations. The spread clearly only develops in a diagonally shaped direction from iteration 50 to 1000, so the first 50 iterations have a very large impact on the solutions to be explored. It almost looks like Moth already knows where to find the

Table 2 Window properties

Name	Configuration	Uwin (EN10077-1)	Ug (EN673)	G (EN410)	LT (EN4I0)	Eref (DS418)	Cost index, C
GlzTypel	4-16-4 argon	1.33	1.02	0.52	0.77	−34	1
GlzType2	4-16-4 argon	1.40	1.11	0.62	0.80	−24	1.1
GlzType3	4-16-4 argon	1.53	1.26	0.77	0.82	−11	2.6
GlzType4	4-12-4-12-4 argon	1.11	0.72	0.50	0.72	−17	1.9
GlzType5	4-12-4-12-4 argon	0.79	0.5	0.42	0.63	−2	3.6
GlzType6	4-12-4-12-4 argon	0.82	0.53	0.50	0.72	8	4.7
GlzType7	4-12-4-12-4 argon	0.90	0.62	0.62	0.73	21	6.1

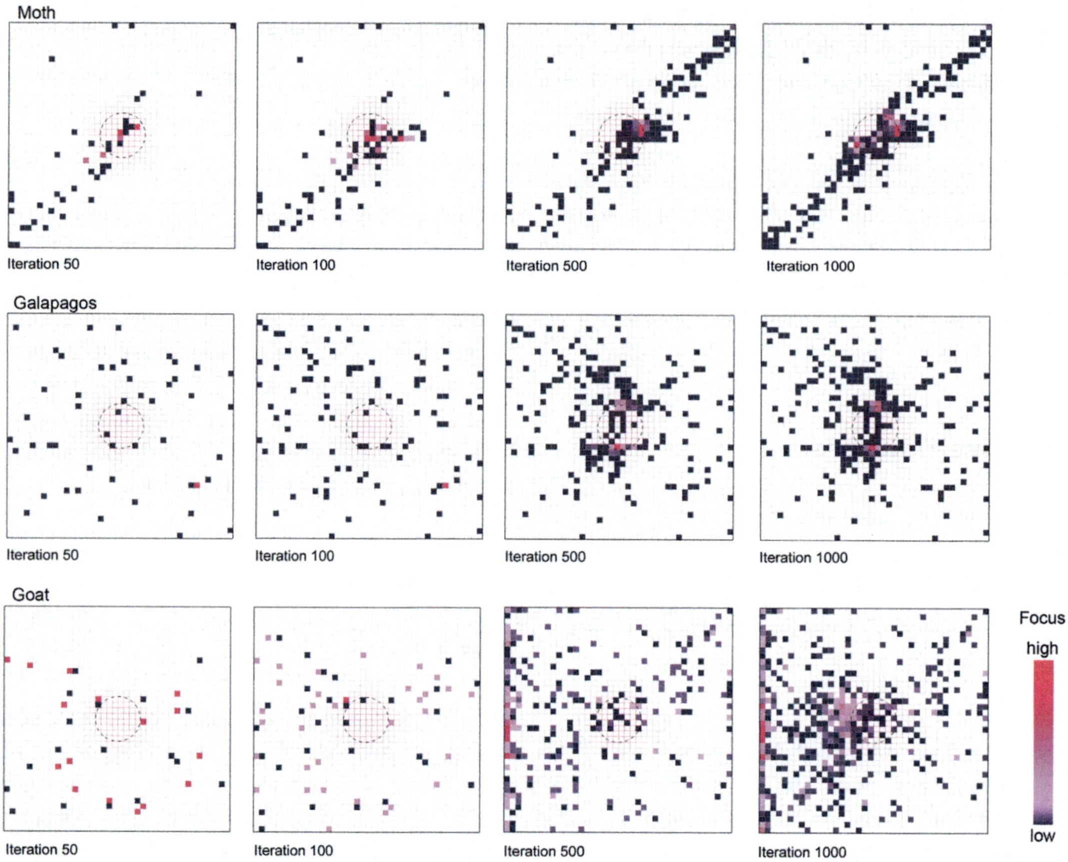

Fig. 4 Search space *A*. *Top* Moth, *middle* Galapagos, *bottom* Goat

minima. There are various explanations to this phenomenon, and this is elaborated on in the discussion.

We can also see a difference in the colouring of the solutions visited during the iterations. Dark coloured solutions have been visited once or very few times, whereas the more magenta the solutions are the more visits the areas have had. One may think of this property as the algorithm's ability to focus its search on specific areas of the search space. Here Moth is distinguished by its attention on actual minima. Around 1/3 of the circle is covered by the attention from the Moth algorithm at iteration 1000. The attention is mainly split into two clusters.

The second search space (Search space B, Fig. 5) contains two clusters of minima. The only algorithm that had success in finding both clusters was Galapagos. What distinguishes Moth from the two other algorithms is its focus and how much of the (discovered) circular optimal space it gives most attention. Moth is also much faster in homing in on one of the two clusters than any of the other algorithms. This, however, may be by chance alone given the very narrow diagonal band of spread. Again Galapagos is most evenly spread out and Goat dedicates much of its searches on the periphery of the search space.

The third search space (Search space C, Fig. 3) is by far the most difficult for optimisation algorithms to navigate in. It contains plateaus and abrupt cliff-like landscapes which arguably (Wetter 2011) are characterised by many BPS tool solver feedbacks. As illustrated in Fig. 6 Moth again finds a minimum quite fast, and again one may argue that this is based on chance, given the fact that the spread is rather limited.

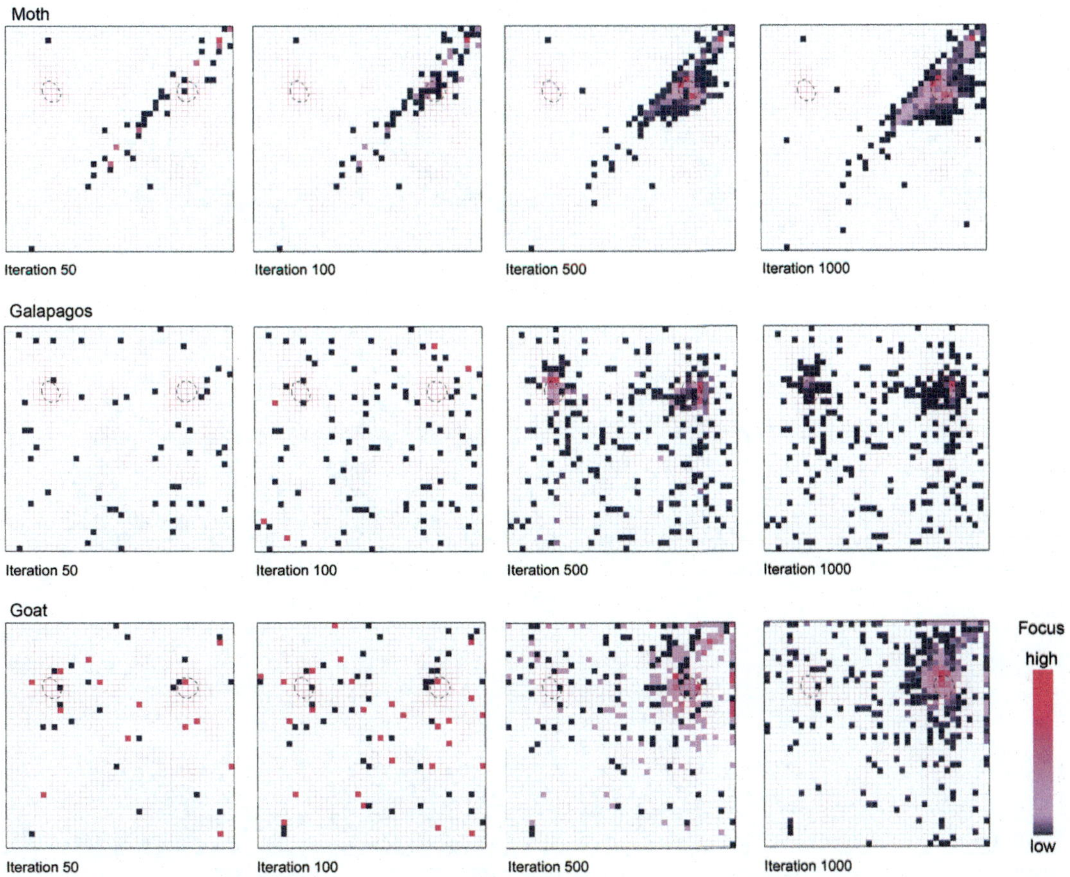

Fig. 5 Search space *B*. *Top* Moth, *middle* Galapagos, *bottom* Goat

What is interesting to notice is that the focus is not fixed upon one area but keeps moving along several minima. The last iteration reveals that the algorithm has explored about half of the optimal space and the focus is concentrated in the area that has the highest concentration of optima. Galapagos starts again with a uniform spread. The algorithm has difficulties in locating the minima and finds a near minimum plateau in the bottom right corner. What is worth noticing is that the algorithm does little to find alternatives to the point on the plateau it has *homed in* on. Goat shares the same fate as Galapagos. It has difficulties in finding the gradients that are particularly steep at the point of the optimal spaces. Interestingly, in the last iteration most of the optimal space has been uncovered by the algorithm. But its focus, similar to Galapagos, is fixed at random points on the *near minimal* plateau. And once again this will generate misleading results. Goat may be seen as the better choice for this type of search space, given its ability to uncover the widest field of near optimal space. However, Goat has not shown interest in any of the real minima spaces, so in this sense the algorithm fails to achieve what it is created to do. Moth is the only algorithm that dedicates its focus to real optima, even though the algorithm seems to settle with the left side "optimal bank" and it has not just started discovering "the island" at the latest iteration.

Part 2

The real excitement in using Moth-agents is introduced when multiple objectives and multiple agents are used in the same model. Here, various

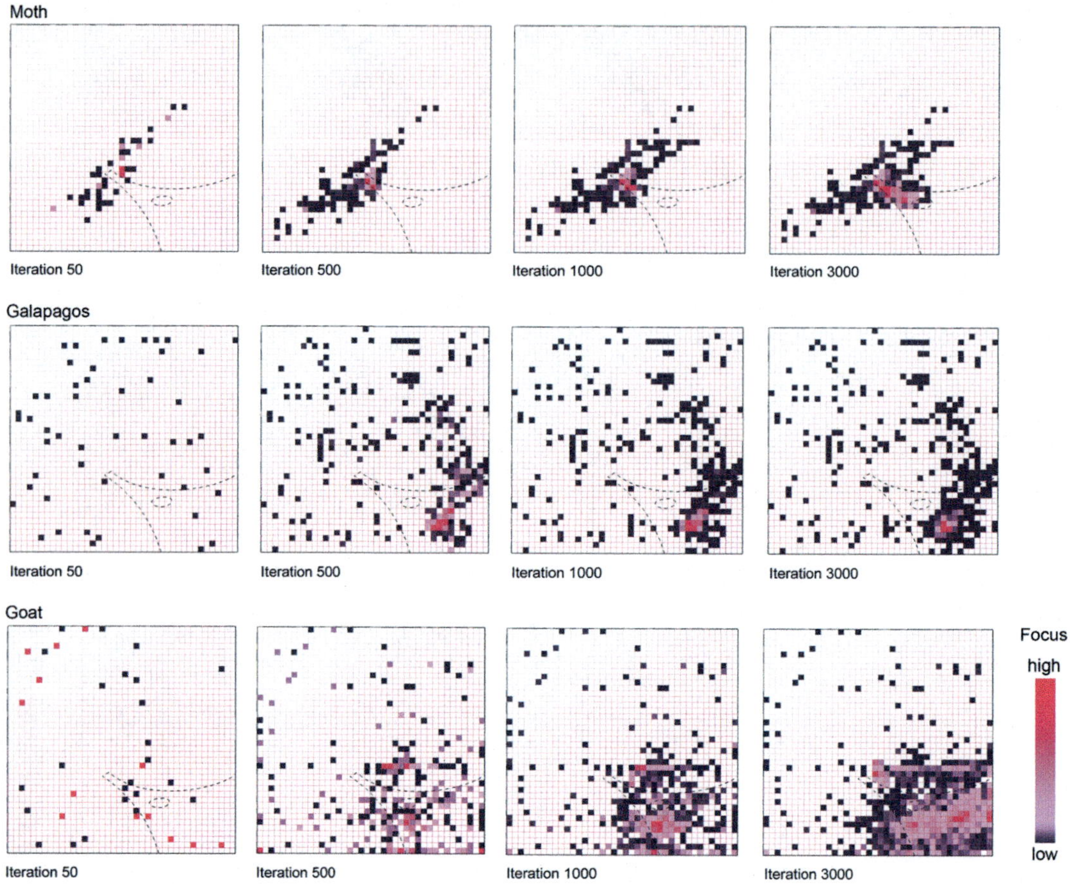

Fig. 6 Search space *C*. *Top* Moth, *middle* Galapagos, *bottom* Goat

agents will seek to disrupt one another's attempts to optimise their own criterion, but over time a multivariate optimum (a convergence) may be found. This is illustrated in Fig. 7. The control of hyper parameters, changing behavioural attributes of agents during optimisation, is unique to ABMs. Moth allows the building designer to change focus and alter agents' search interests simply by altering the agents' behavioural attributes during optimisation. The designer is allowed to manipulate, remove, add, move, and put weights on individual agent-pairs during the optimisation process. This means that the user participates actively in finding a multidimensional balance of the systems. For example, it is possible to change their speed and amount of attraction towards windows or the user may change the objectives without restarting the optimisation process (which is needed when using Galapagos or Goat). This means that the user is able to modify the limits of the search space during the optimisation. In the test case we did not change the objectives at any point. This can be seen in Fig. 8, where the dashed lines represent the constant objectives: $O_{daylight}$, O_{cost} and O_{energy}.

Moth-agents will be attracted or flock around geometrical features in the model representing the design variables. In our case the variables are the windows and a proxy point placed in the centre of the building representing the window type. The number of active agents assigned to a particular design variable will define the focus of the search. In our case we assigned three agent pairs to each of three design variables. As a consequence, we allow each agent-pair to equally

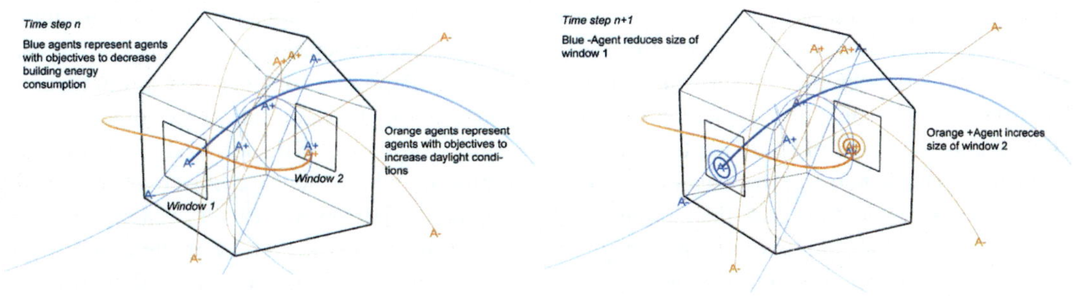

Fig. 7 Example of Moth-agents in action. *Left part* shows time step 1 and *right part* shows time step 2

modify every design variable and we do not distinguish between the objectives.

In Fig. 7 we illustrate how Moth works during optimisation. Here, to make the process simpler, two types of the agent-pairs and two variations of the building geometry are illustrated. The two types of agent-pairs seek to modify the design variables to reach their individual objectives; daylight and energy consumption. The orange Moth-agent objective is to increase daylight by affecting window openings in the building. The Moth-pairs will be attracted towards windows and each +agent will try to increase the window size, while each −agent will do the opposite. Based on the feedback, the +agent will over time get promoted (as larger windows will let more daylight inside the building), and the +agent it will try to further increase the windows sizes.

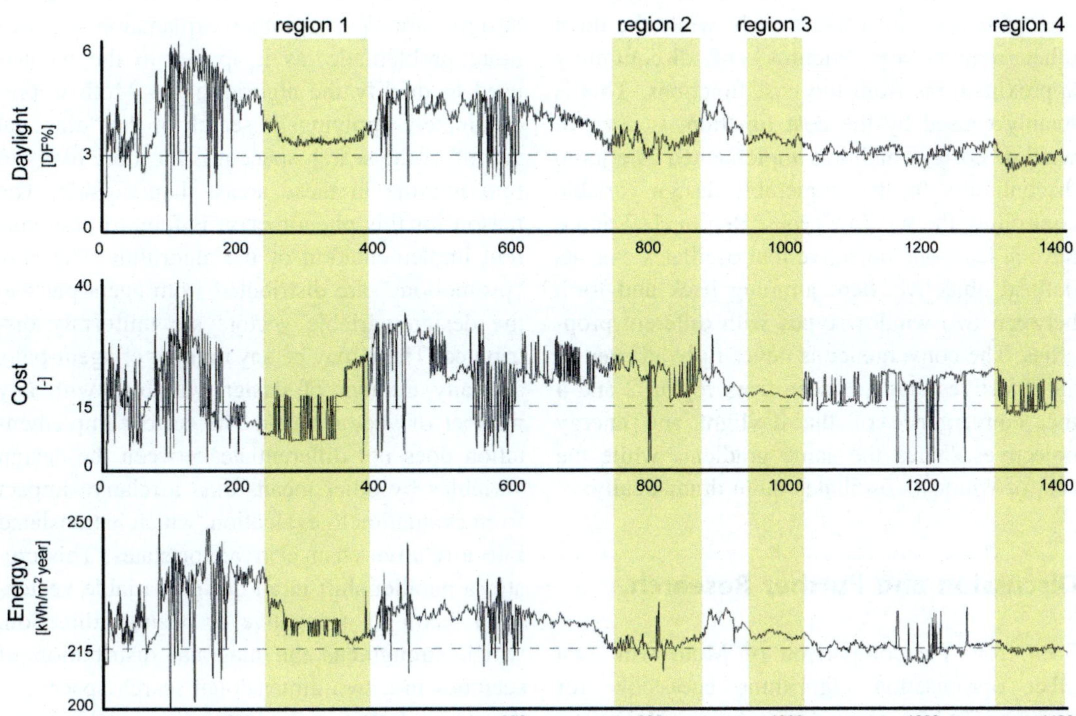

Fig. 8 The multivariate optimisation process using Moth. Regions *1–4* are highlighted to explain certain properties of the Moth algorithm

The −agent will on the other hand have decreasing success in affecting geometry, as it will be gradually demoted during the process. The promotion can be explained as the Moth-agents' 'strength of focus' on the attracting geometry—here windows. The +agent will have more success in circling and thus hitting and affecting the windows that provide most daylight by changing (increasing) the size, whereas the −agent will seem to seek other areas to improve its own goals, that is reducing window sizes, with little success. Over time the +agent will increase windows that make greatest impact on the daylight evaluation, this process is considered a *single objective optimisation*. However, in our case we have three types of agents with three objectives as seen in Fig. 8.

In Fig. 8 four regions are highlighted. The highlighted regions show situations where optima or near optima have been found. What is interesting to notice, is the way Moth concentrates its search in areas where all three objective conditions are close to be fulfilled. Only region 3 maintains a gradient-like search, while the three other regions are "victims" of discontinuity approximations from the cost functions. This is mainly caused by the cost function, $f_{cost}(x)$, of window costs, which are implemented as a jump discontinuity by its enumerable design variable controlling the window types. Region 1, 2 and 4 have at least one objective that oscillates over its defined objective, here jumping back and forth between two window types with different properties. The convergence is never fully adapted to all objectives, but it is seen from region 2 and 4 that convergence of the daylight and energy objectives shares the same gradients while the cost of windows oscillates quite dramatically.

Discussion and Further Research

Seen from the comparison of Moth with two other optimisation algorithms accessible for designers in the early design stage, Moth is a solid choice in terms of accuracy, speed and ability to find multiple optima and near optima in the search space. However, more work is needed to make Moth efficient in finding small minima in larger search spaces. In our multivariate case study in Part 2 we did not find a simultaneous convergence between all three objectives; in other words, we did not find a multivariate minimum, where the three objectives are equally important. However, as the search space is unknown, we do not know if there are any of these minima, and furthermore the demonstration showed that Moth is able to balance two objectives near the optimum while attempting to adapt a third in the process.

From these qualitative tests, the Moth-algorithm seems very efficient in homing in on gradients. There are several explanations why this is happening. First, as the user defines one or several objectives, the algorithm has a measure of "how far" its current performance of a particular solution is from the objective. This is directly translated to the behaviour of the agents, thus making agents far from objectives "more active" and those close to their objective "less active" (and thus less willing to change the design variables). Another explanation is much more problematic, as it applies to the method used to qualify the algorithm. As Moth is predominately applying its search in the "diagonal space" of the search space it is far more likely to find minima in these areas, than outside. The reason for this phenomenon is found in the current implementation of the algorithm. The way "promotions" are distributed from agent-pairs to the design variable vectors are uniformly distributed. There may be any number of agent-pairs and any number of design variables with any number of vector states. The current implementation does not differentiate between the design variables by other means than a relative impact from evaluation to evaluation, which is translated into a relative change in vector states. This creates a parallel shift in all design variable vectors by a factor in a negative or positive direction, which strengthens the diagonal distribution of searches in a two dimensional search space. All three search spaces in part 1 have at least one cluster of minima within the "line of sight" of the algorithm. Therefore it is likely that the performance of Moth is overestimated. Consequently, there are reasons to evaluate ABM-methods like Moth in a more quantitative way. Specifically the comparison of AMBs such as Moth to other methods like Pareto post-balancing methods suggested by e.g. Wang and Zmeureanu (2005) may be necessary to really evaluate Moth as a viable multi objective optimisation method.

A second concern of the ABM method is the oscillation effects. If the agents approach the gradient from a specific angle the search space algorithm can easily get stuck in a multi-dimensional recoil effect. In two dimensions this would look like two very unimaginative or equally skillfull players playing back and forth on a tennis court where they never miss a hand. The reason for this effect is the simplistic nature of the promotion system. Once an evaluation of an agent-pair goes from positive to negative, there is a small chance that the next evaluation again will flip the negative to a positive agent again.

The Moth-agents find only one in infinitely many balances when the optimisation is stopped. But more important to notice is the one solution, is a solution the designer has "moulded" into his or her own personal preference. It can be argued, that this "moulding process" or hyper control of agents is in fact part of the optimisation process, where the qualitative (un-quantifiable) objectives have become an integrated part of the multivariate design domain. We argue that the main reason why ABMs are more relevant for the early design stage, opposed to optimisation methods that depend on post processing, is the direct control of design variable boundaries and control of hyper parameter space during optimisation. ABMs such as Moth allow: (1) the designer to re-evaluate own objectives during the optimisation procedure, (2) the designer to change search focus to dedicated areas of the search space by manipulating behavioural attributes, (3) the optimisation procedure to be part of an exploratory process.

Conclusion

It is concluded that agents may provide valuable design feedback on building performance (e.g. energy, capital cost and indoor environment). Agents may help optimise open ended multivariate design problems, however, the presented system is not particularly efficient in doing so. Given the many fixed and dynamic constraints as well as discontinuities in design variables the algorithm has difficulties in finding true global multivariate optima. However, the presented agent-based system is fully open and adaptable and will allow a high degree of operator intervention during optimisation. For this reason these agent-based optimisation algorithms such as Moth is found better suited in the decision support of early design stages in contrast to traditional stochastic optimisation algorithms e.g. genetic optimisation, which does not support operator intervention during optimisation.

Acknowledgments The work is part of PhD research currently held at the Technical University of Denmark, and is performed in collaboration with Grontmij and Innovationsfonden.

References

Barbati M, Bruno G, Genovese A (2012) Applications of agent-based models for optimization problems: a literature review. Expert Syst Appl 39(5):6020–6028. doi:10.1016/j.eswa.2011.12.015

Bento J, Feijó B (1997) An agent-based paradigm for building intelligent CAD systems. Artif Intell Eng 11 (3):231–244. doi:10.1016/S0954-1810(96)00045-3

Caldas L, Norford L (2001) Architectural constraints in a generative design system: interpreting energy consumption levels. In: Seventh international IBPSA conference, pp 13–15

Danish Building Regulations (2013) Lavenergiramme for kontorer, skoler, institution m.m. Section 7.2.4.2. http://bygningsreglementet.dk/br10_03_id161/0/42

Davidsson P, Persson JA, Holmgren J (2007) On the integration of agent-based and mathematical optimization techniques. In: Proceedings of the 1st KES international symposium on agent and multi-agent systems: technologies and applications, pp 1–10. doi:10.1007/978-3-540-72830-6_1

Gaspar-Cunha A, Loyens D, van Hattum F (2011) Aesthetic design using multi-objective evolutionary algorithms. In: Evolutionary multi-criterion optimization, Springer, pp 374–388

Kaelo P, Ali MM (2006) Some variants of the controlled random search algorithm for global optimization. J Optim Theory Appl 130(2):253–264

Lin S, Gerber DJ (2014) Evolutionary energy performance feedback for design: multidisciplinary design optimization and performance boundaries for design decision support. Energy Build 84:426–441. doi:10.1016/j.enbuild.2014.08.034

Liu B, Wang L, Jin Y-H, Tang F, Huang D-X (2005) Improved particle swarm optimization combined with chaos. Chaos Solitons Fractals 25(5):1261–1271. doi:10.1016/j.chaos.2004.11.095

Negendahl K (2014) Termite. http://cobalab.dk/2014/07/15/termite-1-0/. Retrieved from http://cobalab.dk/2014/07/15/termite-1-0/

Negendahl K (2015) Building performance simulation in the early design stage: an introduction to integrated dynamic models. Autom Constr 54:39–53. doi:10.1016/j.autcon.2015.03.002

Price WL (1983) Global optimization by controlled random search. J Optim Theory Appl 40(3):333–348. doi:10.1007/BF00933504

Robert McNeel & Associates (2013) Rhino. Robert McNeel & Associates. Retrieved from http://www.rhino3d.com

Roudsari MS, Pak M, Smith A, Gill G (2013) Ladybug: a parametric environmental plugin for Grasshopper to help designers to create an environmentally-conscious design. In: 13th Conference of international building performance simulation association

Rutten D (2010) Evolutionary principles applied to problem solving using Galapagos. In: AAG10, Vienna

Salminen M, Palonen M, Sirén K (2012) Combined energy simulation and multi-criteria optimisation of a LEED-certified building. In: Building simulation and optimization conference. Loughborough, UK, pp 372–377

SBI (2013) Be10. SBI. Retrieved from http://www.be10.sbi.dk/

Simon F, Schmiedhofer H, Reis M (2015) Goat. Retrieved from http://www.rechenraum.com/en/references/goat.html

Viehland D, Hardy J, Deville J, Schementi J, Rome K, de Icaza M, Matousek T (2015) IronPython the Python programming language for the .NET Framework. Retrieved 30 April 2015, from http://ironpython.net/

Wang W, Zmeureanu R (2005) Two-phase application of multi-objective genetic algorithms in green building design. In: Ninth international IBPAS conference, pp 1323–1330

Wetter M (2011) A view on future building system modeling and simulation. In: Building performance simulation for design and operation, (i). pp 1–28. doi:10.4324/9780203891612

Wetter M, Polak E (2004) A convergent optimization method using pattern search algorithms with adaptive precision simulation. In: Building services engineering research and technology, vol 25(4), p 327. Retrieved from http://bse.sagepub.com/content/25/4/327.abstract

Implementation of Decentralized Version Control in Collective Design Modelling

Case Study of Web Application Utilizing the Process of GIT

Yasushi Sakai and Daisuke Tsunoda

Abstract

Although the work of Architecture is fully collaboration involving different kinds of specialties with a wide range of consults, Architectural tradition shows that designers still work independently, without clear intension to exchange ideas in design. This view can be derived from what architects uses as tools for design. Attempts to collaborative spatial design have followed since the advent of the internet, yet modern CAD systems still lack a decentralized version control which is crucial to multiuser participation. On the other hand, software development is rapidly changing its process, influenced by the movement of social networking. Today numerous open source programs are developed by the collaboration between vast varieties of people who never meet physically. As a result, the evolution of software resembles the form of nature having different versions and functionality. Moreover, software development can be interpreted as a scientific method where each study increments its knowledge balancing citation and new ideas. Studies achieve reputation from the number of citations as much as the innovativeness. To integrate this accumulation of knowledge in the field of spatial design, it is necessary to measure the difference between alternatives, and clarify what was added and deleted. This study is based on a web application (http://lmnarchitecture.com) which implements the same technology that present software development stands on. The web application was exclusively developed as a browser based 3D modelling site to create simple dwelling like volumes using webGL. Within those plans, the system calculates the resemblance between the models utilizing an algorithm that computes the difference between texts. This algorithm called "Diff" is actually used in a decentralized version control system for software development called GIT. It calculates which line of code was added and deleted throughout the edit.

Y. Sakai (✉) · D. Tsunoda
Nikken Sekkei, Tokyo, Japan
e-mail: mail@yasushisakai.com

M.R. Thomsen et al. (eds.), *Modelling Behaviour*, DOI 10.1007/978-3-319-24208-8_32

In this study, the models are converted in a format that can run this algorithm to gain the similarity ratio. The system requires the users to choose one model that already exists, and calculates the resemblance in real-time. As a result, a tree diagram is collectively achieved with each having the link of inheritance. In GIT, this action is called forking, which copies and edits the existing program and creates a new branch to solve different demands. An evaluation method is concurrently running inside the web application, exchanging points between the "parent" and "child" models accordingly to the calculated similarity ratio gained by the users. The system is distinct to poll systems for achieving collectiveness, involving users to be designers and evaluators at the same time. Along with the anonymous user participation, an automated design generator (bots) was implemented to investigate man-machine collaboration. As a result, 1750 models were submitted to the database between 3 months of experiments. Analysis showed near half of the time users inherited a model different from what they have previously submitted, and 23 % of the submitted models were inherited a machine generated model.

Introduction

Accumulation of Knowledge and Beyond

It is obvious that we humans attained knowledge by learning, which is essentially the stacking of experiences of the past. Comparison with the present models can be interpreted as archives of trial which is a general rule for enhancing architecture. Additionally, ever since science has learned to build theories and experiments on observation upon previous studies by referencing and incrementing steps of novelty, the accumulation of knowledge has bolstered us to discover new ideas. Most CAD systems lack ways to accumulate knowledge and operate simultaneously. Instead, Architects still manually coordinate data using local protocol to avoid interference when collaborating.

This study tries to unleash the power of collaboration and enabling remix of design by presenting a tree diagram. Not only enabling a new way of evaluation by providing the diagram, the achieved data structure has potential for a new subject—automated design intelligence. Supported by today's explosion of data, data mining and machine experiment is to pierce through the process of stacking and automation of intelligence.

Present Tools and Previous Studies

Idea and Methods of Collaborative Modeling

Multi participation in design has long been one goal within architects. The "Open Building" idea will lead in theory, were "users/inhabitants may make design decisions as well" introduced from Habraken. Other methods such as classification of design methods into patterns. Alexander et al. (1977) is another approach for inclusive design.

Internet and Methods from Software Development

The defiance can be observed in the early days of intercommunication progress, starting from the collaboration via internet and video chat.

Attempts followed using a virtual reality atmosphere, such as online conversation using Second Life. As Rosenman et al. (2007) points out, there are still "issues of multi representations of objects, **versioning**, ownership and relationships between objects from different disciplines."

One study from Burry and Dominik (2005) using SVN, a versioning and revision control system that is mainly used from software developers in the process of design, combined with Gehery Technologies' Digital Project, share the same intent. The study argues that direct utilization of SVN was not possible within binary data, and took an approach of a file locking method which simulates the process. To circumvent the binary issue, for this study, the data structure was included in the development of the web application. The scalability to consumer software remains, yet gains control of the version controlling process. A file locking procedure can be seen in modern BIM software giving functionality of multi-user participation. However, the file locking procedure is limited to centralized version control. Only one model can be edited in one time and restricts users to work concurrently in one design. In other words centralized version control creates a hierarchy throughout the versions, this aspect is suitable to later phases of architectural design, where effective accumulation is what is needed to crystalize the proposal. Opposed to SVN, the core concept of Git is the version controlling is "decentralized". Since the proposals have no hierarchy, it is feasible to gain interactions between the versions. One feature that this application lacks is the merging function which the branches are independent once they are forked. However, one type of bot (mentioned below) has the merging feature to combine two models into one proposal.

Functions of Modern CAD and BIM Software

In the world of software for designing, Autodesk's Vault has the capability to track versions in a file level, while the mechanism rely on file locking, essentially a centralized method that hierarchy exists. Moreover, Aish (2000) argues that the concept of "files" affect the project to be fragmented, and negates holistic perspectives to perceive the project. To meet this, modern BIM software has the functionality of multi-user participation within the project, yet Vault and BIM software share the centralization of design documents. Again, centralized methods are said to be effective in the later design phases for sophistication of a single design plan that is ready to be developed. However, such hierarchies if they are too rigidly adhered to, may inhibit the effective flow of information in collaborative design (Cumming 2005).

Another approach, Autodesk's Fusion has partial components that are version tracked and exposed to the internet. Fusion presents the possibility of collaboration not only inside projects, but within projects. The distinction with Fusion and this study is integration of the Diff algorithm which handles the similarity within alternatives (versions). In other words, the system enables us to calculate the distance between given designs, which is crucial to analyse, sort and integrate. There is more than one application that focuses on version tracking and multi-user collaboration, such as the browser based CAD system, OnShape.Version tracking and branching is the cutting edge of next CAD systems and similar approaches is waited to be implemented to consumer software. Fusion and OnShape has implemented this feature adding saving chronological operations. The approach is straight forward for tracking difference, yet the method does not ensure that the output of the model is unique for every archived data. In other words the same geometry may be produced with different operation paths. The study uses the "Diff" approach which compares the final output, which is the state of what is evaluated. Another point is saving chronological operation is essentially the same to what present CAD users are doing. This can be infinite, and the amount of information would not able to be determined before the process. While both approaches have potential for implementing machine learning, comparing geometrical data is finite for this particular application.

System and Flow

The User Work Flow

The system holds two aspects of method; the technical software architecture side and the social architecture side that takes account of the incentives of the participants.

There are two activities for the users, model creation and evaluation. To be a user, the system requires the designer to sign-in and register to this system. After logging-in, the system gets the last time of each activity from the user database. This time interval constrains excessive operation that will destruct the system. The evaluation needs 10 s to recover, while creation requires 3 min.

There are three major views: tree diagram, model evaluation, and model creation. Figure 1 shows the first view, the tree diagram, that any visitor will see when she accesses http://lmnarchitecture.com.

The tree diagram has a link to every single model, directing to the evaluation view (Fig. 2). The evaluation view shows a basic information of the model along with the 3D representation, this view enables users to give point to the model by clicking "add points" (Fig. 3).

The points are affected by the "similarity ratio", which returns a part of the points depending to how similar the model is to its inheritance. This reimbursement of the points is a recursive process until the points are lower than the value of 1.

This view can lead to the third view, model creation by clicking "make plan" (Fig. 4). The system forces the user to select a base model to inherit.

The model creation is the place where users are stack different types of blocks of function and create a new model. The grid space is 12 × 12 × 3 (W × D × H). This limitation lowers the threshold of design participation and sets a finite solution space. The application sets an ideal

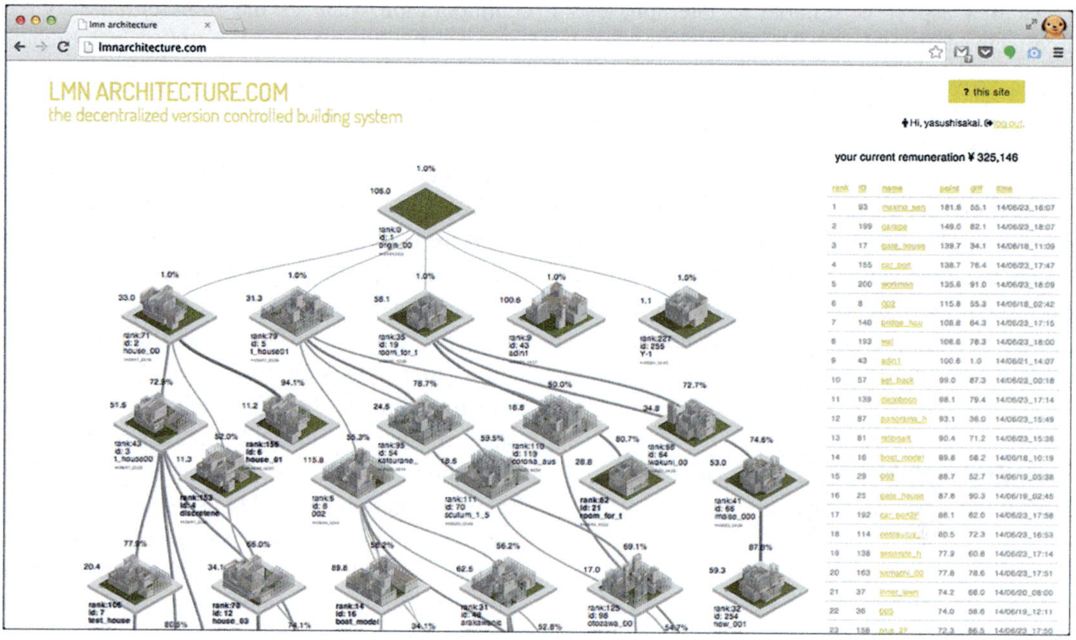

Fig. 1 Tree diagram view, ranking listed in the *right side bar*

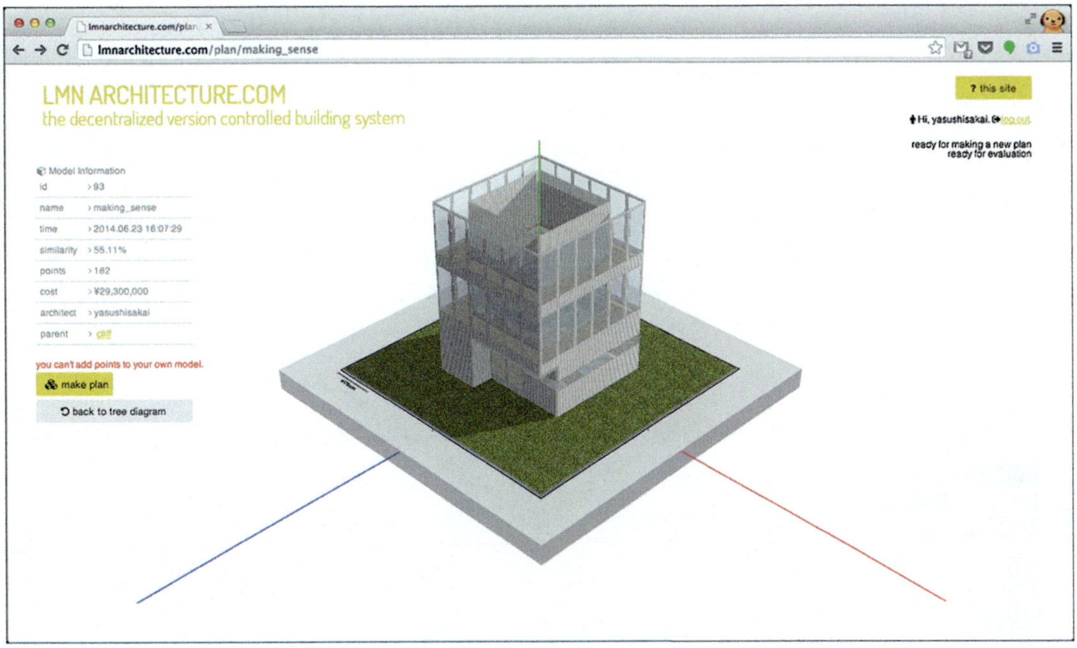

Fig. 2 Evaluation view. 3D representation of the model with basic information in the *left*

situation with the same site region, and without environmental information. In other words, all models share the background and have the same context throughout the system. While modelling, there is one constrain that the total cost of the model should not exceed 30,000,000 yen. Each function has different appearances and cost showing in Table 1.

Additionally, if the model has two or three stories, the total cost is multiplied by 1.2 and 2.0 respectively. The budget constrain moderates the models while controlling the solution space to make popular models that is credited by evaluators. The "similarity ratio" is concurrently calculated, using the "Diff" algorithm, introduced in the next section. After the modelling process, the user gives a unique name to the plan. The model attains points from the parent model corresponding to the similarity ratio, and saves the image and adds the data to the database as shown in Fig. 5. Selecting a model that is present and revising the models is called "branching" in the context of GIT workflow.

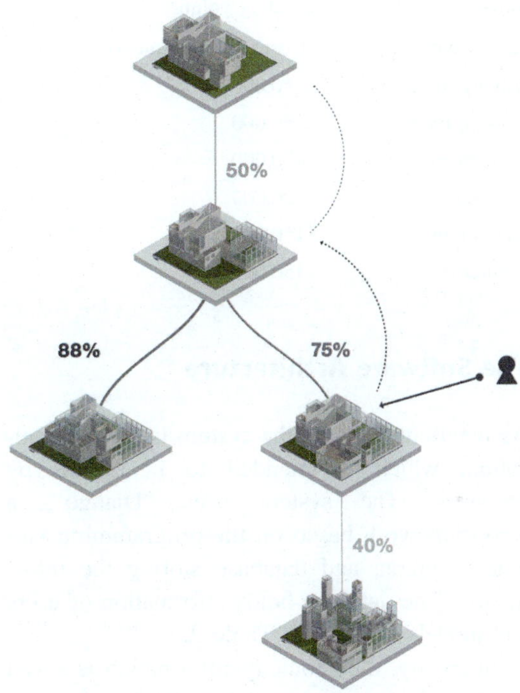

Fig. 3 Reimbursement method. Recursive cash back of points goes to the ancestor models

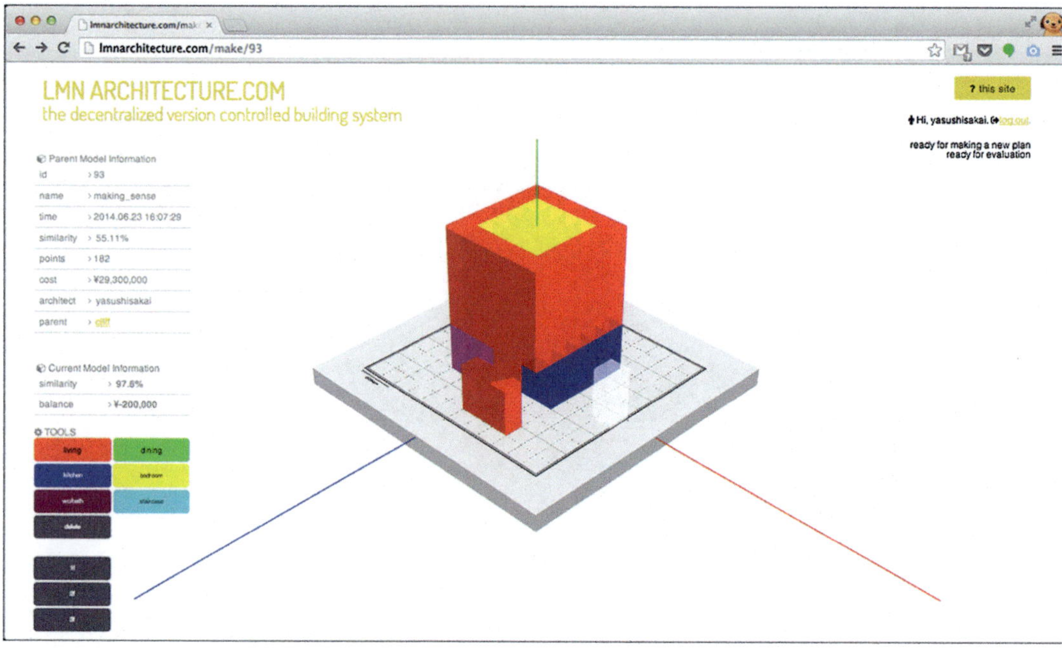

Fig. 4 Plan creation view. *Colour blocks* indicate the function of the volume

Table 1 Cost of each building volume

Block type	Cost for each unit (yen)
Living room	150,000
Dining room	150,000
Bed room	150,000
Kitchen	200,000
Bathroom	150,000
Staircase	150,000

The Software Architecture

As mentioned above, the system is a web application, which is intended to be access by browsers. The system uses "Django", a web-framework based on the programming language Python, and database storing the information. The database holds information of users and models as listed in Table 2.

Internally, the geometry information is saved in an exclusive structured format to make possible of comparing model data. Each model has a name that points to the image stored in the media folder, and holes that userid data it was created. Additionally the model data has the id of the inherited model, the parent. The inborn points is the points that came from the parent when generated, in contrast, the acquired points represents the points gained after generation.

The study utilizes Google's "Diff-Patch-Match"; the Diff algorithm calculates the subtraction within the two models. The algorithm takes two strings (data type for text data) and compares it to get what was added and deleted throughout the revision. The system utilizes this algorithm to achieve the distance by getting ratio of the amount of altered data identical data (Fig. 6).

Non-human User, Automated Generation of Models

Apart from the human user participation, the system holds two types of methods to propagate models automatically using previously submitted models. They can be interpreted as "bots", alike to what we have in the internet to collect data and link information relevance known as search bots.

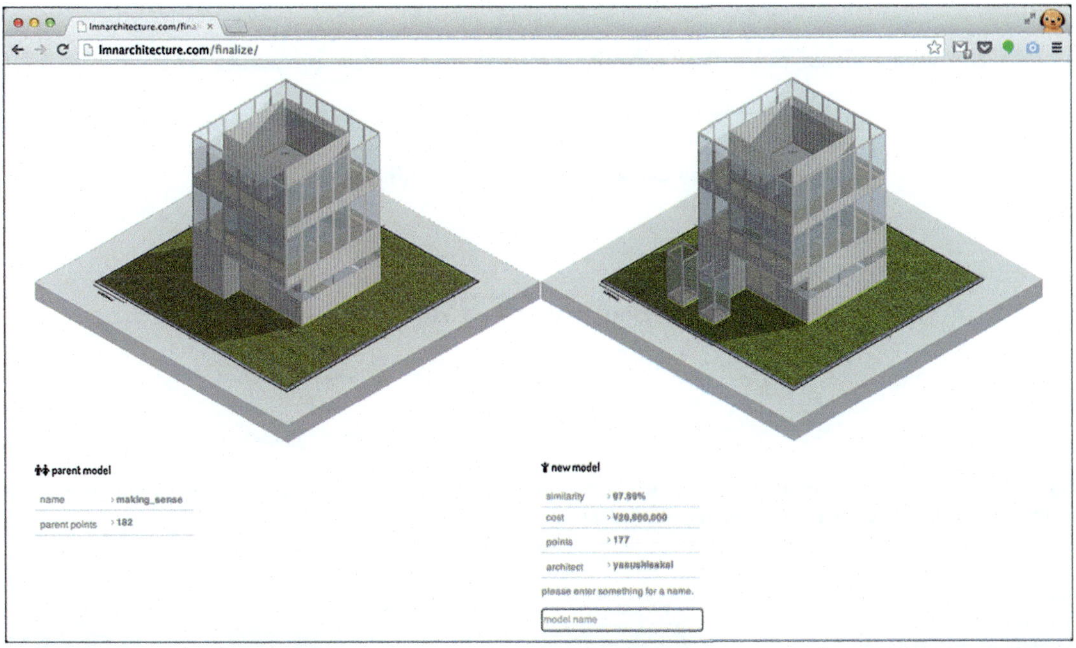

Fig. 5 Model finalization process. User gives unique name to new model

Table 2 Class properties and methods of user and model

User	Model
Name	Id
Email	Name
Password	Geometry
Last evaluation date	Similarity
Last creation date	Parent model id
	Points inborn
	Points acquired

One type of bot is based upon randomness adding and deleting cluster of function blocks (=rooms). This bot has limited intelligence, since there is no strategy in creating a better model, which is wholly dependent in the whole systems natural selection (Fig. 7 left).

Another bot uses the technique of "crossing" known in the field of genetic algorithm, sampling different parts from two models and combining them together (Fig. 7 right). This bot is capable of looking up the top model and the second model to patch work the plan. This way the bot reflects the current trend and bolster what is popular. With the human participants and bots, the entire system collectively searches for the best model.

The Social Architecture

In contrast of the informational technology side, this study has another attempt of designing with the public which includes users that do not have proper training in architecture or design. To include the public in the design process, design of incentives is necessary. To meet this, we have used a competition style to gamify the process. As mentioned in the software architecture, using a browser makes participation easy to everyone; models are created and points are exchanged between the points within generation. Ranking the most granted model and rank the user's contribution to that system delivers a competitive situation. Moreover, we used a real competition that was held in the same time of the examination (Fig. 8).

This way the prize money can be redistributed to the users corresponding the result of the

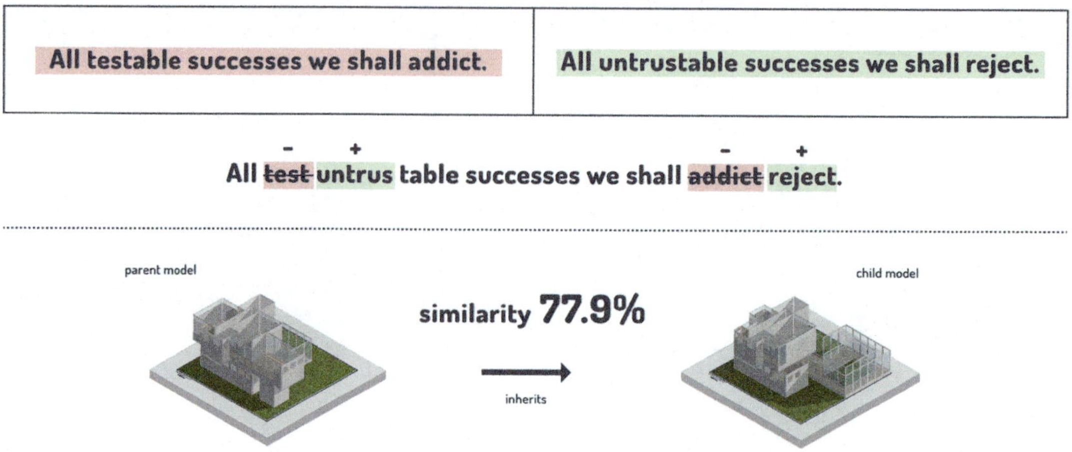

Fig. 6 Diagram of Diff algorithm. By getting the number of characters that was altered, the similarity ratio is achieved by the ratio of the value and the whole length of data

Fig. 7 Two types of automated model generation. (=BOTS) *left* adds clusters of volumes randomly, in the *right* crosses the popular models

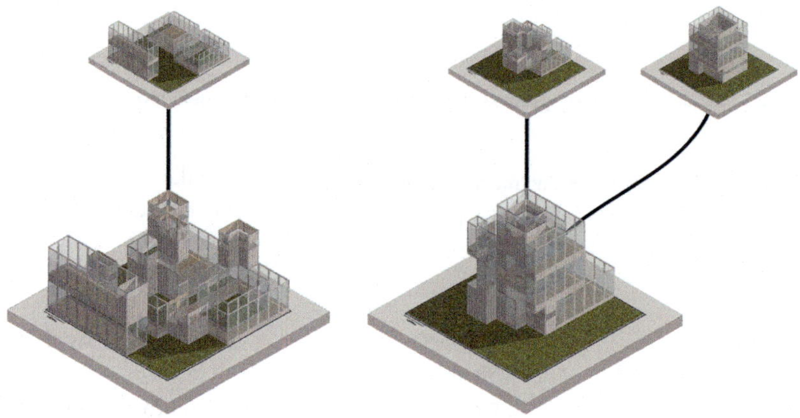

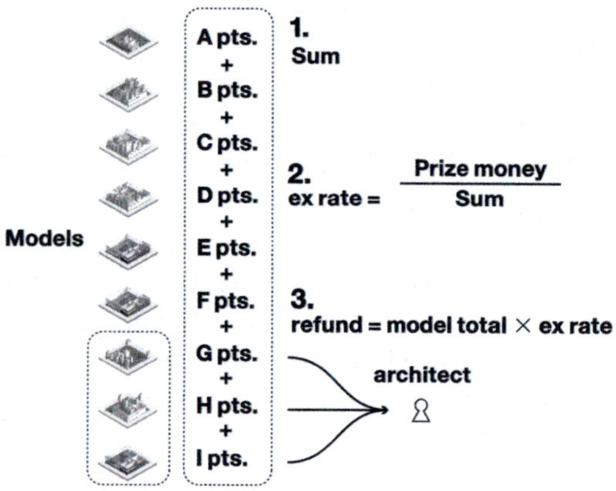

Fig. 8 Diagram showing the distribution of prize money

system. Eventually the system earned prize from the "parent competition" which was able to cashback to the participants.

Results

There were 1750 models submitted to the database within 3 months. Inside the 1750 models, 583 models were submitted by human users. Within the two types of bots, each type generated one models when users added a model (Fig. 9).

Fig. 9 The overall tree diagram that was submitted to the web application [accessed 29 April 2015]

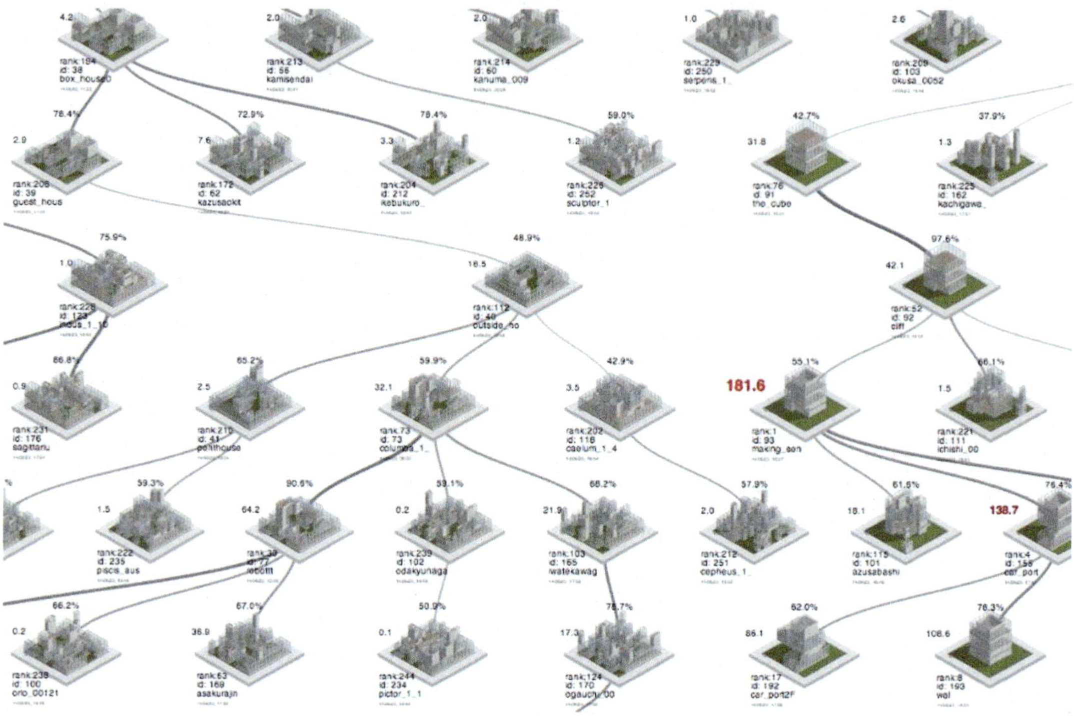

Fig. 10 Magnified view of network of plans

Figure 10 shows a partial image of the tree diagram each connected to their parent (inherited) model. Stroke thickness indicates the similarity ratio: the stronger it resembles the parent.

16 users including one account as BOTs was registered to this system. Figure 11 shows the top rated models.

Table 3 shows the mean of three values: Cost, Similarity, and Point ratio. Point ratio indicates how much acquired points (points achieved after generation) did the model gained within the total points.

Another aspect to investigate is the relationship between the parent and child model. The result shows that within the models that was submitted by human, 49.7 % of the model inherited the same user (by themselves) (Table 4).

Discussion

This situation of nested competitions, questions the present system of competition in architecture practice, which in this case, the prize and honour (while "honour" is not quantitative, the amount of contribution can be measured inside this system.) is distributed in a sequence of contribution of searching the solution space. This study shows that strategy to do so is necessary to multi-user collaboration in design. The tool focused to observe collaborative work within voluntary users by calculating the geometrical difference of model data. Thus the application does not assure technical quality (e.g., structural analysis, and local building code). Yet adding an exhaustive list of simulation tools could be done for complete evaluation of the models, and once implemented, other simulation tools will work as thresholds like in natural selection. The aesthetical side of the models will be evaluated by the anonymous public. The evaluation method will work in a similar intention for narrowing down the possibilities.

Neither of the bots had a learning routine, meaning lacking strategy across plans. Although, this can be said that bots had no intelligence to learn, the result showed a spontaneous strategy to

Fig. 11 Top rated plans, 'jaldabaoth' is responsible of auto-generation of plans

Table 3 Mean values of cost, similarity, and point ratio

	Cost (yen)	Similarity (%)	Point ratio (%)
Human	¥21,664,922	70.33	15.19
Machine	¥28,243,580	63.27	50.27
Total	¥27,525,566	64.04	10.89

Table 4 Inheritance tendency

		Parent model	
		Human (%)	BOT (%)
Child	Human	76.96	23.04
	BOT	13.74	86.26

Design collaboration between humans and bots

Table 5 Ratio of inheritance on models produced by human

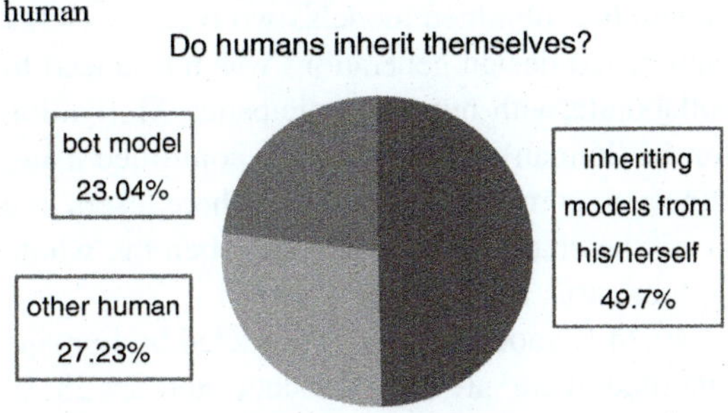

overcome the human competitors. Unlike human participants, the automated process has no limitation in time, in other words the cost for producing one model is nearly zero for machines. The machine has potential to effectively produce accurate models, once the appropriate leaning mechanism implemented.

Table 5 indicates the chart of inheritance within man made models. This can be interpreted the ratio of collaboration. Throughout this experiment, approximately half of the users used models of their own, in other words, modellers are willing to collaborate with different individuals near 50 %. Moreover, the result shows that in a blind situation, humans are open to work together with bots.

It is clear that collaboration in design still needs further exploration, and the result of this study needs constant iteration to measure the will and potential of true collaboration. However, this study constructed one example of a system of collaboration in design. This approach will be one alternative using crowd sourcing and machine learning (using methods that require lots of trials) compared to the top down approach.

Conclusion

Summary

This study conducted an experiment of collective, collaborative design method using a web application that implements a technology poplar in the field of software development. With a renowned algorithm called "Diff" and the simple volume stacking app combined, 16 users signed up for creating 1750 models between three months. An evaluation method was concurrently running, making it possible to add points to the desired model once in an interval. The points are exchanged corresponding to the similarity between the parent and child (inherited) models. Within the submitted models, two types of BOTs (automated design generation) was introduced to collaborate with human participants. Motivation for the (human) participants was controlled using a nested competition system, where users are given the chance to gain money when the whole system earns some prize.

49.74 % models that was modeled by humans inherited there previous models, and 27.23 % inherited models from other humans, and 23.04 % inherited models that was produced by an automated process.

Further Exploration

While not immediate, the method of implementing distributed version control to the process of design has potential to both professional designers and non-proper individuals. The system that was introduced in this study has few elements that are scheduled to progress. From the aspect of human participation, crowd sourcing the generation will be more efficient to collect designs. Services such as Amazon's "Mechanical Turk" will be suitable for distributing work to the mass. As the data produced by humans aggregate, there will be more opportunity for the machine to learn. The possibility of implementing machine learning will be the second field for further investigation. Remaking the method as a mobile application is another strategy to multiply the number of users, since smartphones are getting more and more important when gaining information from users.

Other than including non-proper individuals, application to consumer software for professionals is one path of evolution. Although the data complexity is much higher and dense then the system we have developed, investigation is necessary to implement to tools that is utilized in practice. A relatively straight forward step is to integrate this idea to a visual programming environment, such as the Grasshopper plugin of a 3D modeller called McNeel's Rhinoceros. Specifically, Grasshopper files have an xml-like data structure, which is easy to apply Diff and other operations to analyse the data. BIM software have the same potential, since the data structure is easy to be interpreted as database structure.

References

Aish R (2000) Collaborative design using long transactions and "change merge", eCAADe 18, pp 107–111

Alexander C, Ishikawa S, Silverstein M (1977) A pattern language: towns, buildings, construction. Oxford University Press, United Kingdom

Autodesk Inc (2015) CAD, Modeling and machining features|fusion 360 tools. Accessed 01 May 2015

Burry J, Dominik H (2005) Sharing design space: remote concurrent shared parametric modeling, eCAADe 27, pp 333–339

Cumming M (2005) Distributed and constructed knowledge in design education, eCAADe 21, pp 501–504

Davis D, Burry J, Burry M, (2011) Untangling parametric schemata: enhancing collaboration through modular programming, CAAD futures, pp 55–68

Habraken J (2000) Open Building: brief introduction. http://www.habraken.com/html/introduction.htm. Accessed 04 May 2015

JoeSperrazza (2015) Autodesk Inc. (2015 Autodesk Vault—Wikipedia, the free encyclopaedia. http://en.wikipedia.org/wiki/Autodesk_Vault. Accessed 01 May 2015

Kelly K (2011) What technology wants. Penguin Books, New York

Lima M (2014) The book of trees: visualizing branches of knowledge. Princeton Architectural Press, New York

Martens B, Dokonal W, Schmindinger E, Voigt A (1996) Collaborative teamwork—challenges of the future, eCAADe 12, pp 263–272

OnShape Inc (2015) Future of cad|onshape. https://www.onshape.com/. Accessed 14 June 2015

Rosenman A, Smith G, Maher L, Ding L, Marchant D (2007) Design world: multidisciplinary collaborative design in virtual environments. Autom Constr 16 (1):37–44

Assessing Implicit Knowledge in BIM Models with Machine Learning

Thomas Krijnen and Martin Tamke

Abstract

The promise, which comes along with Building Information Models, is that they are information rich, machine readable and represent the insights of multiple building disciplines within single or linked models. However, this knowledge has to be stated explicitly in order to be understood. Trained architects and engineers are able to deduce non-explicitly explicitly stated information, which is often the core of the transported architectural information. This paper investigates how machine learning approaches allow a computational system to deduce implicit knowledge from a set of BIM models.

Introduction

The adoption of Building Information Modelling (BIM) constitutes a radical shift in the way models in the building and construction industry are described. Traditional representations for architectural knowledge, such as architectural drawings, 3D models, technical descriptions and spreadsheets are transitioning into semantically-rich information models. Building related information no longer exists within discrete entities, but is kept in an interlinked context. BIM authoring tools for Design and Construction and Facility Management systems provide now semantic information about elements and spaces within buildings, their constituting elements, as their interrelation and performance. Currently huge efforts are put in place to create links between buildings, on local, regional and international level, through e.g. standardisation committees (buildingSMART 2014) or the Geospatial communities (Geospatial Media and Communications 2014), but as well through international research projects, such as

T. Krijnen (✉)
Department of the Built Environment, Eindhoven University of Technology, Eindhoven, The Netherlands

M. Tamke
Centre for IT and Architecture, Royal Danish Academy of Fine Art, School of Architecture, Copenhagen, Denmark

M.R. Thomsen et al. (eds.), *Modelling Behaviour*, DOI 10.1007/978-3-319-24208-8_33

DURAARK (Durable Architectural Knowledge) (DURAARK 2015), a three year EU funded project on the creation and maintenance of semantic links between representations of buildings.

Building Information Models, in formats such as a Revit database or IFC, have become the bond that connects disciplines by streamlining data exchange and connecting the construction with the operational phases of a building lifecycle. Building related knowledge is herein represented in an object oriented way, holding building element geometry, properties, and its interrelation to other objects. These objects can be part of the described building, but in addition relate to external objects or other sources of information, including building element libraries. Information can be related to physical entities, like a wall, as well as to intellectual or organisational constructs, for example spaces or organisations. Hence, the model can support many facets of the construction phase, and in addition guide the building's operation with Facility Management tools or the planning of retrofitting tasks.

The new class of information is directly machine-interpretable, as it conforms to a structured schema. The use of BIM models in current practice is however predominantly focussed on explicit information, such as property values, augmented with aggregate functions for the extraction of quantity information and clash detection based on geometrical inference (Tamke et al. 2014a). BIM models hold however information that is not explicitly stated, but lies implicit in the interrelation between the entities within a single model or in the interrelation of a large variety of models. And while years of practice train a building professional to immediately apprehend the functions of a space by means of merely symbolic two-dimensional representations, this information can currently not be assessed by machines. We ask how these implicit second order descriptors can be assessed and whether this approach holds the potential to describe the qualitative aspects of a building.

State of the Art

This paper presents experimental approaches directed at extracting implicit data from building models. It shares its interest with parallel research into the assessment of architectural models. This is for instance concerned with parametric models (Davis 2014), the way these are set up and the complexity they address. The descriptors for this, such as dimensionality or cyclomatic complexity, stem from Computer Science and are produced by algorithms reading the parametric models. Stasiuk and Thomsen (2014) investigate open-ended processes of discovery and categorical description of form-found design models. Rather than looking at the properties that constrain the form-finding process, the study uses machine learning algorithms, such as k-means clustering, to categorize the models by means of emerging properties. Machine learning is particularly interesting to the here-described approach, as it is able to address the great variety that is inherent in architectural designs and identify similarities rather than conformity.

Approaches for explicitly querying building models can be qualified based on their intended use, their underlying technology and expressiveness of the query language, their amount of abstraction from underlying data and various other measures. A language like BimQL (Mazairac and Beetz 2013) provides means to extract entity instances from IFC models using the entity and attribute names defined in the IFC EXPRESS schema, as such it is an effective tool for people to extract explicit information. Other methods are developed for the purpose of conformance checking. Examples of this include Solibri Model Checker (Solibri 2014) which checks models for conformance to BIM standards or clashes between elements using hardwired constraints, or the mvdXML checker (Zhang et al. 2014), which checks for the conformance of a certain model to a Model View Definition, a construct that imposes additional constraints for validity onto an IFC file in addition to the constraints as dictated by the schema.

The expressiveness of the mvdXML language however does not include querying for implicit relations between different instances, for example the spacing between two columns, as it has no notion of binary operators. Other approaches express topological relations, such as containment or adjacency, and implement these in query languages (Daum and Borrmann 2014). Such an approach enables a query for binary relational aspects of the model. Within the DURAARK project, metadata extraction utilities are provided (Beetz et al. 2014) to automatically extract literal values from IFC files according to a metadata schema. This has been extended to determine Level of Development information by assessing calculated attributes based on geometric detail (Tamke et al. 2014b).

Applications of Machine Learning

The querying approaches presented above have in common that they are tailored to specific scenarios with predefined outcomes as they allow for no novelty or discovery. Conversely, the promise of machine learning is the extent to which it is able to make predictions and detect patterns. In order to make such predictions and find such patterns, typically, metrics about actual buildings would be collected, such as the measured energy efficiency. These real-world metrics can then serve as a label to the set of building models and, hence, patterns can be unravelled between the building configurations described in the models and their measured performance.

In this paper both supervised and unsupervised machine learning (Mitchell 1997) of BIM models will be discussed. An unsupervised learning approach will be presented to show anomalies in building models. Unsupervised approaches work without the premise of aforementioned labels. More speculatively, it can be seen as a method to reduce the failure costs in the construction industry by flagging uncommon situations that might need additional checks or coordination. These anomalies could include unusually large overhangs or other situations that constitute an unusual confluence of several building elements.

In addition, a supervised machine learning approach based on a neural network is discussed. It is able to classify floor plans according to its intended function. Such a system can be seen in the light of a large archival framework for building models in which information pertaining to intended function and use is often fragmented and incomplete. This paper argues that supervised learning can be used to complete missing attributes in such a dataset. Such an approach can be extended to classify based on other criteria, such as iconic and exemplary architectural edifices, based on geometrical descriptors. Both these attributes are common in the world of archival, but seldom explicitly documented for newly built artefacts.

Both supervised as well as unsupervised learning algorithms typically assess a sample, in this case a building, by means of a set of features that describe the building. How such features can be extracted automatically from building models is described in the next section.

Implementation

The IFC Machine Learning platform presented in this paper is built on top of the DURAARK IFC metadata extractor (Beetz et al. 2014; Krijnen 2015a, b). This utility is able to extract literal values, aggregates and derived values from IFC SPF files. The extraction utility presents a simplified Domain Specific Language (DSL) that enables users with little programming experience to map query paths pointing to literal values in the IFC file to keys in a metadata schema. Furthermore, the DSL provides functions to compute aggregates, such as to count the extent of a list or sum or concatenate attribute values. The DSL is a subset of the Python scripting language and is in fact executed as a regular Python program. In addition, serialization formats are provides to output the data in a format suitable for further processing. For example, in the context of the DURAARK workbench, the extracted values are written into a linked data RDF graph. This version of the extractor has been published under an open source license (Krijnen 2015a). For the Machine Learning platform, an additional output format has been added to output Comma Separated Value files.

What follows now are short excerpts of the data extraction system to highlight some key aspects of its use. For example, to query the unique identifiers of all wall elements one could invoke the following statement:

The hypothesis in this paper is that geometrical properties of the building elements and their relations can be used to build an implicit architectural knowledge model that can be assessed with Machine Learning. Therefore, in order to extract geometrical quantities that are derived from the body representation of IfcProducts, the DSL has been extended with geometrical operators. In addition, to calculate the minimal distance between these products, the concept of binary functions is introduced that signify a relationship with other products in the file. The geometrical and topological analysis functionality is implemented on top of IfcOpenShell (Krijnen 2015b) and pythonOCC (Paviot 2014).

```
csv_formatter << [
        file.IfcWall.GlobalId >> "ifc_ml:wall_identifiers"
]
```

Listing 1: Code required to output wall identifiers

	wall_identifiers
0	2O2Fr$t4X7Zf8NOew3FL9r
1	2O2Fr$t4X7Zf8NOew3FLIE
2	2O2Fr$t4X7Zf8NOew3FLPP
	...

Listing 2: Output from the program provided in Listing 1

```
csv_formatter << [
        file.IfcWall >> count >> "ifc_ml:wall_count"
]
```

Listing 3: Code required to output the number of walls in an IFC file

wall_count
57

Listing 4: Output from the program provided in Listing 3

```
elems = file.IfcProduct
    >> segment(by_entity)
    >> segment(by_attribute("GlobalId"))

csv_formatter << [
        elems >> shape_area >> "ifc_ml:area",
        elems >> shape_volume >> "ifc_ml:volume",
        elems >> shape_gyradius >> "ifc_ml:gyradius"
]
```

Listing 5: A more elaborate example that lists geometrical attributes for all building elements

segment_0	segment_1	**area**	**gyradius**	**volume**
IfcBeam	2OrWItJ6z...	9.17	1.78	0.04
IfcSlab	1CZILmCaH...	53.89	2.10	3.81
IfcSpace	0BTBFw6f9...	67.50	1.80	33.51
IfcWallSt...	0dxE1Sy6n...	23.64	1.20	1.64
...	...	...	...	...

Listing 6: Output from the program provided in Listing 5

Unsupervised Learning: Anomaly Detection

As an unsupervised method, outlier detection is applied to the geometrical attributes of the elements in a model. Outliers are the samples that deviate from an observed median area. In this experiment the duplex apartment model is used (NIBS 2013). The building element samples are segmented according to their entity type (wall, window, and so on). For each of these element types, an elliptical envelope is fit through the sample data. For this purpose the scikit-learn toolkit (Pedregosa et al. 2011) has been used. For demonstration purposes, a two-dimensional plot of the data is shown below, where the dimensionality of the data is reduced by using the ratio between the geometrical attributes. The plot highlights a clear centre in which most of the samples reside and that there are clear outliers outside of this centre area. The samples are coloured according to their Mahalanobis distance to the centre. This distance metric accounts for the distribution of the data, contrary to how a Euclidian distance would and is a common measure for classification purposes. Contours of the elliptical boundary are shown in dashed lines (Fig. 1).

	Mahalanobis distance
0iEHWY1$XA8eQeeULq4jpl	2.00135121249
0jf0rYHfX3RAB3bSIRjmpw	87.4188172025
1aj$VJZFn2TxepZUBcKphf	2009.27365845
3Y4YRln2r91vflHcHE5ITm	38142.0133829
...	...

Listing 7: An excerpt of the wall GlobalIds with their Mahalanobis distance

Visually, the same information can be represented by colour coding the model elements, as can be seen in Fig. 2. In this overview it can be assessed that the ten elements with the highest distances are in fact miscategorised to be wall elements. If one inspects the definition of a wall in the IFC schema one can see that these elements do not fulfil a role in bounding or

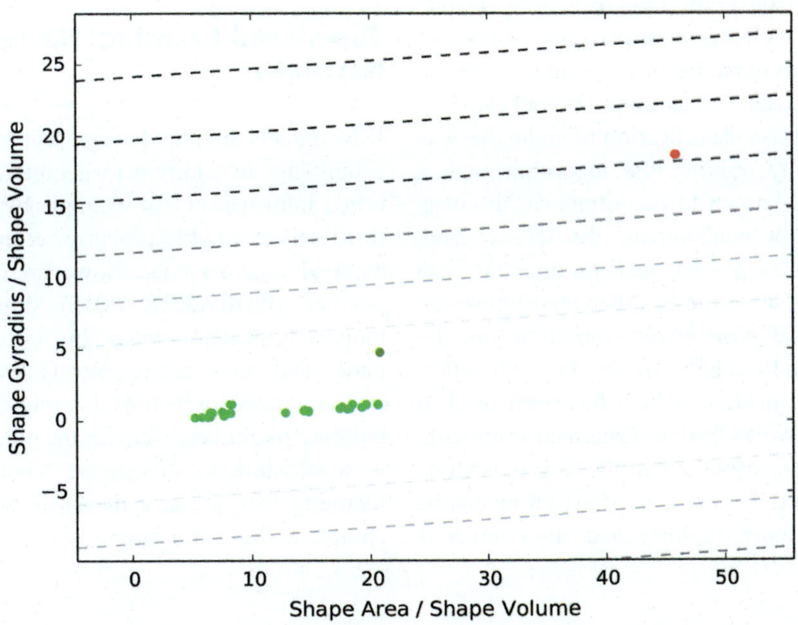

Fig. 1 IfcWallStandardCase anomalies in the duplex apartment model according to their geometrical attributes

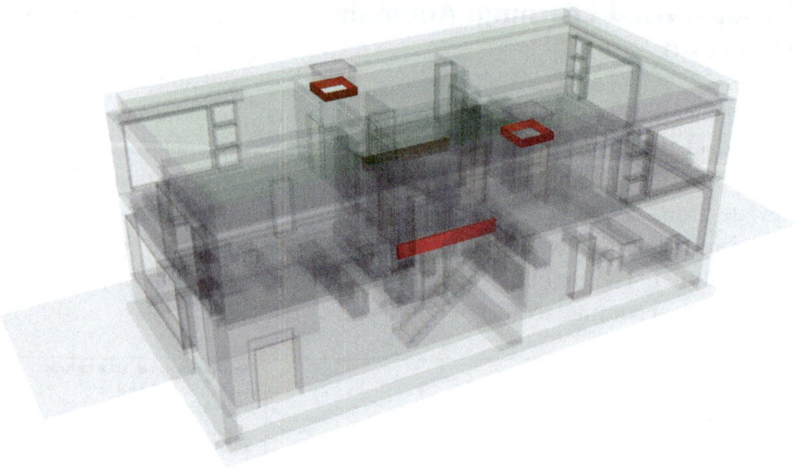

Fig. 2 Ten elements in the duplex model that are misclassified as wall found by anomaly detection

subdividing the construction work. In fact, they appear to be narrow and horizontal structural members and therefore could have been more suitably classified as a beam.

To summarize, by looking at a single model, the algorithm has been able to identify the geometrical essence of a wall and found elements that deviate from this idea. It seems that these are likely misclassified elements. In building projects misclassifications can be problematic as they might result in elements escaping from being verified by the appropriate person, in this case perhaps by a structural engineer, as he might inspect only a subset of the model corresponding to a structural view, to which architectural wall elements do not belong.

A more precise classification of anomalies can be obtained by training the algorithm with a dataset that is known to be correct. In this way outliers do not contaminate the dataset and, therefore, a more precise decision boundary can be obtained. Furthermore, other algorithms can be applied that pose fewer restrictions on the distribution and modality of the data. The elliptical envelope method, which has been used in this example, works best on Gaussian, symmetric and unimodal distributions of the feature vectors.

It is assumed that this line of reasoning can be extended to more sophisticated discoveries if relational parameters to other nearby products are introduced. In such a way, clearance areas, for example which can be found near stairs or doors, can be assessed. Or the load-bearing characteristics of a column can be schematized as an element that directly connects to an element to the top and to the bottom of it. For these advanced relational concepts the definition of orientation of elements is crucial. These experiments are to be done in follow-up research.

Supervised Learning: Neural Networks

BIM models are able to represent many facets of a building, in addition to geometrical and relational information, for example by using predefined and extensible property set or reference to external data sources. However, datasets from practice (DURAARK 2015) show that BIM models typically contain this information only partly and have heterogeneous levels of information. Metadata records that might exist for one building might be absent for the other. As such, it is worthwhile to investigate whether Machine Learning can act as a means to supplement or complete this information. In this experiment,

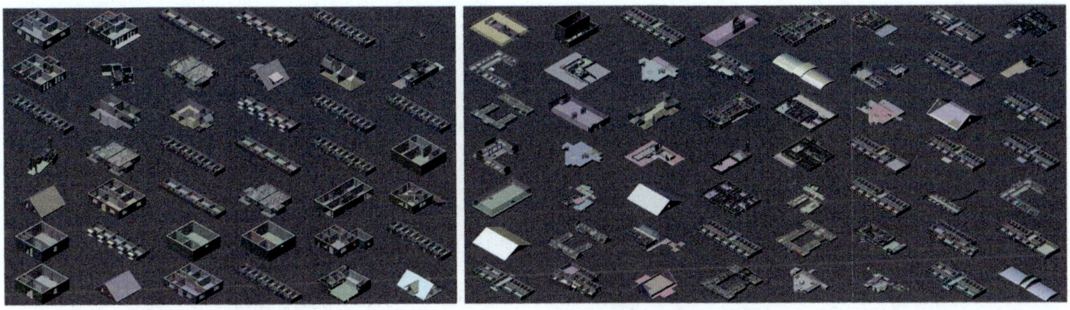

Fig. 3 Residential (*left*) and Non-residential (*right*) building storeys in the dataset

individual building elements are no longer subject of examination, but instead building storeys, a common aggregate structure in building models are assessed in order to classify them according to their function and intended use. This is metadata that is typically not semantically available in an IFC building model, but is very relevant in a digital archive. By looking at the spatial configuration of a floor plan, a neural network is trained to differentiate between residential and institutional facilities.

branching corridors. To be precise, in this specific case, the gyradius refers to the radius around the vertical axis through the centre of mass of the solid volume. It is a measure for the distribution of mass around this vertical axis. This implies that for a shape that describes a network of corridors this measure will be higher than for a cylinder, even if they bear the same volume, the latter being the most compact 2.5D shape. An overview of the gyration radius and centre of mass for two distinct solid shapes can be seen in Fig. 4.

Space - wall volume ratio	Measures the amount of spatial segregation
Space - slab volume ratio	Measures the amount of spatial segregation in vertical direction
Doors per space	Measures the connectivity of spaces
Average space volume	Measures the size of the spaces
Space volume variance	Measures the extent to which spaces vary in size
Average space gyradius	Measures the compactness of the spaces
Space gyradius variance	Measures the extent to which the compactness of the spaces varies
Column wall ratio	Measures whether walls or columns are used for load bearing

Listing 8: Features incorporated in the floor plan assessment

The dataset consists of publicly available models augmented with models that have been aggregated over the course of the DURAARK project. An overview of the models, separated by floor and divided by their residential or non-residential labels is provided in Fig. 3.

The geometrical attributes that are used to assess the floor plans are provided in Listing 8. As a measure for the compactness of a space, the radius of gyration (gyradius) is used. Alternatives to this metric include for example the ratio between surface area and bounding volume (Corney et al. 2002), but it is assumed that for this purpose a high gyradius specifically signifies long

During the training phase of a neural network a (locally) optimal configuration of connections between neurons will be formed that pertains to the situation at hand. This is implemented by means of a gradient-descent based optimization algorithm, called back-propagation. Such an algorithm functions best on normalized features so that the range of values for different features roughly corresponds. Thus, the geometrical attribute values are normalized and scaled so that they all cover, more or less, the same range and their median values are close to zero. An example of two features before and after feature scaling is provided in Fig. 5. Also, notice how

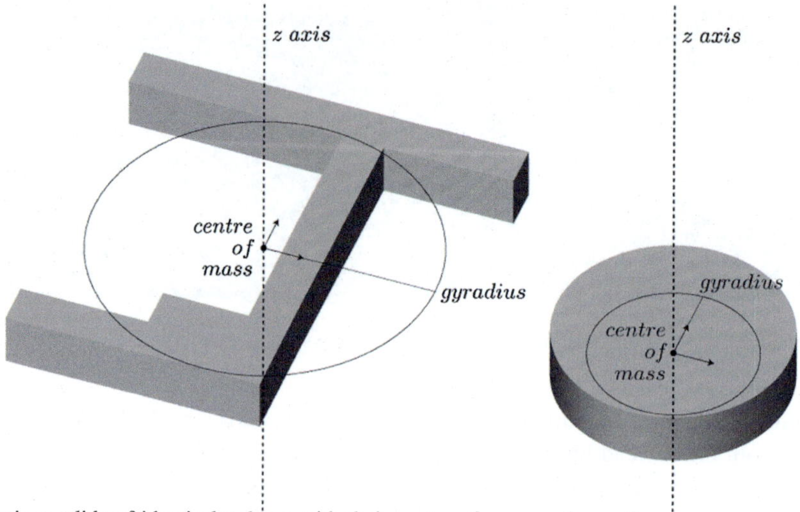

Fig. 4 Two distinct solids of identical volume with their centre of mass and gyradius

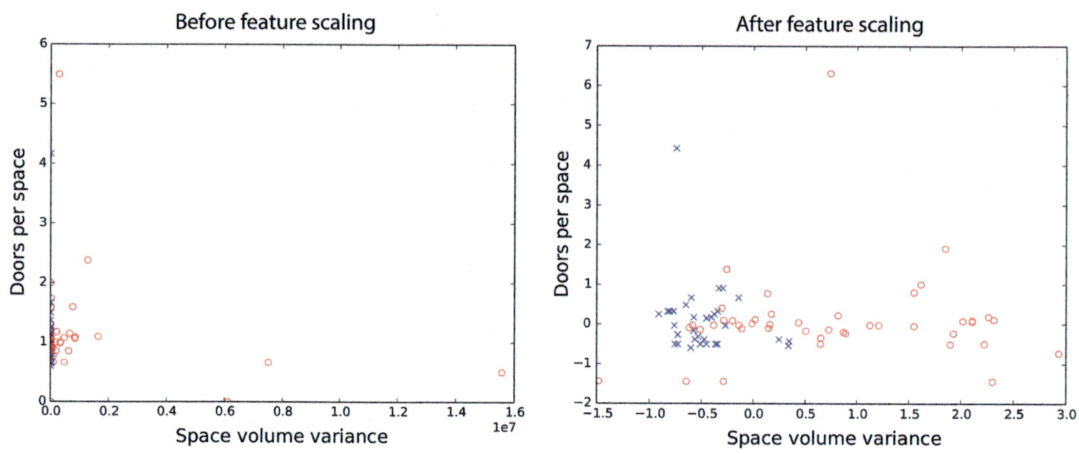

Fig. 5 Two features before and after scaling

these two features already exhibit some grouping in terms of the labels, mostly the variance in space volumes appears to be much higher for the non-residential facilities.

According to best practices, the labelled dataset is divided into three distinct subsets: a training set (70 %), a cross-validation set (20 %) and a test set (10 %). The training set acts as the input for the back-propagation algorithm and determines the weights of the network edges between the neurons. In order to pick a network that pertains to a general solution and not just specifically to the training set input, the optimal network is selected based on its misclassification error on the cross-validation set, which has not been used to build the network. In this particular case a network with bias units and without hidden layers proved to result in the lowest misclassification error on the cross-validation set. Both these concepts in general make the network more capable to fit complicated problems, as the number of neurons increases. The network has

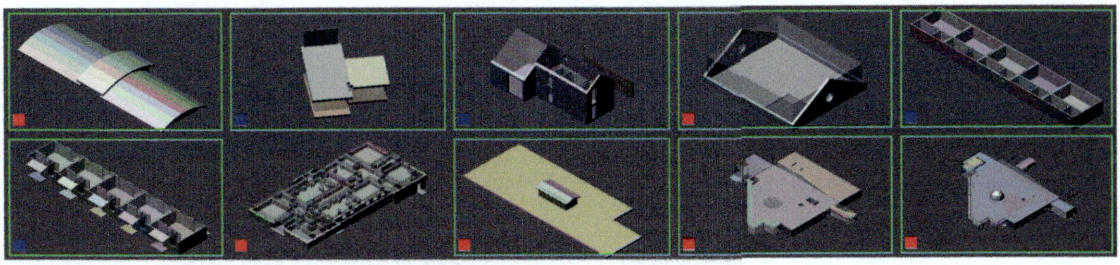

Fig. 6 Results on the test set with all successful classifications marked in *green*

been trained for 4000 iterations. The remaining final 10 % of the models provide an indication of the actual performance of the network on unseen cases. The models are provided in Fig. 6. According to the manual labelling process there is a single model that is misclassified, it is a non-residential floor layout categorized as a residential floor plan. In defence of the algorithm, one might argue that it is a difficult case, as it is the cellar of an historic building for which there are no other samples in the training set. The neural network has been implemented in the Python programming language using PyBrain (Schaul et al. 2010).

To conclude, it has been shown that a neural network has been able to differentiate floor plan layouts into two categories based on their geometrical appearance. The authors assume that such classification can be augmented with metrics from spatial connectivity graphs, including aspects like centrality and clustering coefficients. With a bigger dataset and more features, future research will indicate whether it is possible to distinguish between more categories of building functions. In this example, the neural network has been formed to some extent by trial and error, which is not uncommon in this domain. A more systematic and exhaustive application is of neural networks is beyond the scope of this paper and will have to be extended upon in future research. From both machine learning examples in this paper one can draw that the geometrical nature of the elements that constitute a BIM model provides insight into the nature of these elements on their own and into the assembly they form as a whole.

Conclusion

The paper demonstrates the application of both supervised and unsupervised machine learning methods with Building Information Models. The implicit attributes and qualifications can currently not be found by means of traditional computational approaches in architectural practice. Results from machine learning on architectural datasets provide a relevant alternative view to explicit querying mechanisms and provides useful insights for more informed decisions in the design and management of buildings.

This paper presents initial research into the use of machine learning to create architectural insights. The feasibility of the approach was proven on rather basic examples. Future research will have to indicate how this can be extended and generalized into other areas for instance for the more complete quality assurance of BIM models. The search for misclassified elements is here a starting point into further research, which might even address the assessment and classification of designs and architectural qualities.

In particular relational characteristics on the level of the building element, such as clearance areas and typical confluences of specific element types, might yield a richer comprehension of implicit knowledge and would further exploit the relational nature of an IFC file. On the level of building stories, graph measures relating to spatial connectivity might be vital to develop a more detailed understanding on the exact spatial configuration and therefore a more precise functional classification.

The computational approaches, which are currently used in practice to assess BIM models, build upon explicit rules (Solibri 2015) built by experts. Machine learning might enable less technical users to query complex BIM datasets for highly practice and project specific insights.

References

Beetz J et al (2014) D3.3 semantic digital archive prototype. http://duraark.eu/wp-content/uploads/2014/08/DURAARK_D3_3_3.pdf. Accessed May 2015

buildingSMART (2014) buildingSMART. http://www.buildingsmart.org/. Accessed May 2015

Corney J et al (2002) Coarse filters for shape matching. IEEE Comput Graph Appl 22(3):65–74

Daum S, Borrmann A (2014) Processing of topological BIM queries using boundary representation based methods. Adv Eng Inform 28(4):272–286

Davis D (2014) Quantitatively Analysing Parametric Models. Int J Archit Comput 12(3):307–320

DURAARK (2015) DURAARK Project. http://duraark.eu/. Accessed May 2015

Geospatial Media and Communications (2014) GeoBim. http://geo-bim.org/. Accessed May 2015

Krijnen T (2015a) DURAARK/pyIfcExtract. https://github.com/DURAARK/pyIfcExtract. Accessed May 2015

Krijnen T (2015b) IfcOpenShell. https://ifcopenshell.org. Accessed May 2015

Mazairac W, Beetz J (2013) BIMQL—An open query language for building information models. Adv Eng Inform 27(4):444–456

Mitchell T (1997) Machine learning. McGraw Hill, New York City

NIBS (2013) Buildingsmart alliance common building information model files and tools—national institute of building sciences. http://www.pythonocc.org/. Accessed May 2015

Paviot T (2014) Pythonocc, 3D CAD/CAE/PLM development framework for the python programming language. https://github.com/DURAARK/pyIfcExtract. Accessed May 2015

Pedregosa F et al (2011) Scikit-learn: machine learning in python. J Mach Learn Res 12:2825–2830

Schaul T et al (2010) PyBrain. J Mach Learn Res 11:743–746

Solibri (2015) Solibri model checker. http://www.solibri.com/products/solibri-model-checker/. Accessed May 2015

Stasiuk D, Thomsen MR (2014) Learning to be a vault—implementing learning strategies for design exploration in inter-scalar systems. Newcastle upon Tyne, England, pp 381–390

Tamke M et al (2014a) D7.7.1—Current state of 3D object processing in architectural research and practice. http://duraark.eu/wp-content/uploads/2014/06/duraark_d7.7.1.pdf

Tamke M et al (2014b) Building information deduced—state and potentials for information query in building information modelling. Newcastle upon Tyne, England, pp 375–384

Zhang C, Beetz J, Weise M (2014) Interoperable validation for IFC building models using open standards. ITcon vol 20, Special Issue ECPPM 2014—10th European conference on product and process modelling, pp 24–39

BIM-PIM-CIM: The Challenges of Modelling Urban Design Behaviours Between Building and City Scales

Mark Burry, Justyna Anna Karakiewicz, Dominik Holzer, Marcus White, Gideon D.P.A. Aschwanden and Tom Kvan

Abstract

This paper discusses the challenges that designers face when modelling the anticipated behaviours of people: their movement and transactions around and within precinct scale development. Building Information Modelling (BIM) software philosophy contrasts with that of *City Information Modelling* (CIM)—the route by which we consider how precinct scale development, being somewhere between the two (BIM and CIM), requires a wholly different approach to information and behaviour modelling. The authors offer evidence of the value of augmenting the planners' analytical approach with the architects' synthesis from data leading to meaningful speculations on otherwise unanticipated future scenarios for the precinct far beyond expectation. Novel approaches to modelling behaviour at precinct scale suggest alternative readings of precincts, which require a wider set of approaches to *Precinct Information Modelling* (PIM) software development than simply an expansion of BIM.

PIM: Stuck Between Two Stools

In the most simplistic terms, the urban planner has plentiful computational resources with which they can analyse the vast data-scape that represents the city's actuality, and to a certain extent reliably predict the effect of some relatively simple decisions (as evidenced by http://www.spacesyntax.com/project/trafalgar-square). The architect, in contrast, can use data to synthesise. For them systems theory has peripheral interest, more intellectual in nature than say a builder's practical instincts, for which a systems theory approach may help create more efficient Gantt charts and work flow (Jongeling and Olofsson 2007; Kenley and Seppänen 2009). The precinct designer has several unique challenges in contrast. Firstly, a precinct boundary can be quite arbitrary, and it is not clear whether there is such a thing as a precinct that is too large to be considered as such, or necessarily too small. Secondly, the precinct designer would ideally work

M. Burry (✉) · J.A. Karakiewicz · D. Holzer · M. White · G.D.P.A. Aschwanden · T. Kvan
Faculty of Architecture Building and Planning, University of Melbourne, Melbourne, Australia
e-mail: mark.burry@unimelb.edu.au

M.R. Thomsen et al. (eds.), *Modelling Behaviour*, DOI 10.1007/978-3-319-24208-8_34

with digital tools that can synthesise at many times the scale of a building information model (BIM), yet integrate analytical perspectives at the level of system theory, not least Complex Adaptive Systems (CAS) (Geli-Mann 1994). Thirdly, in many cases the precinct designer works on a precinct that exists already as an evolving urban molecule rather than commencing with a greenfield site as a blank canvas.

From BIM to PIM?

With increase of computing power following Moore's law, is it a case of when rather than can BIM be scaled-up to PIM? Let us takea snapshot of where BIM currently sits from the perspective of a building designer rather than a building deliverer and facilities manager.

Through the digital design revolution the architect as '*Building Information Modelling enabled professional*' has an ever improving set of interoperability opportunities through the protocols developed in partnership with all the principal stakeholders: designers, engineers, cost estimators, constructors, facility managers, and regulatory authorities. BIM detractors and BIM proselytisers share the same platform with regard to the universally acknowledged current deficiencies (Kensek and Noble 2014; Burry 2014):

- Extremely large file sizes hindering collaboration (cloud collaboration tools developing too slowly).
- Relatively undeveloped cross platform exchange (loss of data—particularly between different consultants).
- Limited number of analytical tools beyond individual building assessment (energy rating etc.).
- Limited parametric and computational capabilities (though new tools consistently promise more).
- Little focus given to collaboration across a diverse group of stakeholders.
- Limited functionality for validation across a range of related (yet often not yet interconnected) datasets.
- Too much focus on geometry and not on spatial data.
- Poor capabilities for risk management and risk allocation across the participants.
- Barely any investigation into legal aspects of digital representations of the built environment—in particular when we start considering potential interfaces between human behaviour and its spatial context via e.g. Big Data.

BIM proselytisers point to what BIM is already capable of achieving while its critics tend to focus on the performance gaps as stunting design aspirations. As analysed by Simeone and Kalay (2012) BIM is still too often only utilised to produce 'models of something' instead of 'models for something'; modelling behaviour of a diverse set of occupants is not routinely considered.

Whereas those applying BIM to boost their design or delivery capabilities appear to be satisfied with the benefits it offers them to fulfil their distinctive tasks, research suggests that on a grand scale BIM is very much still in its infancy (Aram and Eastman 2013; Holzer 2011). What happens with a virtual model or the data that is (or can be) associated with it by others up or down the supply chain remains mostly unexplored. Therein lies the conundrum: The way BIM gets applied is still only dictated by immediate, isolated, and often mundane tasks associated to traditional project delivery. The silo-mentality inherent to the construction industry gets mirrored in their engagement with data associated to the BIM produced by individual parties. Zhang et al. (2009) describe how BIM starts to have a positive effect on the connectedness between stakeholders from varying disciplines, but how it currently gets applied in practice does not yet affect the construction industry on a meta-level. This is especially the case when the transformation of project procurement is based on an understanding of the wider information-ecology that extends beyond a project itself. Speaking from an information-management perspective, what is the footprint of a project? How can its functionality, its value or its environmental/societal impact be validated within a wider set of criteria? BIM does not attempt to answer these questions.

It does not help that software applications available for creating and managing information via the BIM process are highly nonintegrated. They predominantly cater for those either authoring BIM (e.g. Revit™, Microstation™ or ArchiCAD™), those coordinating BIM (Solibri™, Navisworks™), those extracting and managing quantitative (e.g. cost) data (CostX™ or Vico™), and those that interface BIM data with tasks for ongoing Operation and Maintenance (Ecodomus™, FM Systems™, Zuuse™, Zutec™). There still exists a disconnect between BIM for design and BIM for construction. Solutions for carrying information all the way through to Facilities Management (FM) are sparse. One might expect that owner/operators of assets push for BIM output to interface directly with their Computer Aided Facility Management (CAFM) systems. Yet in current practice this is barely the case. Even in those cases where there exist regulatory frameworks (such as the Level 2 BIM requirements imposed by the UK government on publicly procured projects from 2016), the information transfer to clients serves mainly the purpose of allowing them to grasp information already inherent to their assets instead of allowing them to learn lessons about future procurement.

Current trends towards the development of Precinct Information Modellers (PIM) run the same risks as the evident shortfalls of BIM from the designers' perspective, namely clients wanting virtual representation of their precincts without necessarily aiming to use it for future planning and testing of scenarios. In that sense Facilities and Asset Management are mere procrastinators of 'activated' PIMs which, it might be argued, hide behind the creative disconnects between BIM and how designers prefer to produce their designs (Fig. 1).

Some Novel Approaches to Modelling Precinct Behaviour

Attempts to model the scale between BIM and CIM exist, but both constructability and feasibility seem to be missing from the PIM discussion, despite being a likely principal driver for urban renewal. The design work featured in this paper draws from a nationally funded initiative to provide an 'Australian Urban Research Infrastructure Network' (AURIN). Our precinct design modelling ambitions include demonstrating the capability and design decision support value of being able to use data and insights from a wide range of sources such as AURIN, building developers, planners, and local government in order to offer visions of possible future precincts.

The foundation of every PIM or CAS is data that allows for the creation of an initial state of the model. Common problems include inconsistent formatting, access to data, or missing documentation of existing datasets. Data exchange

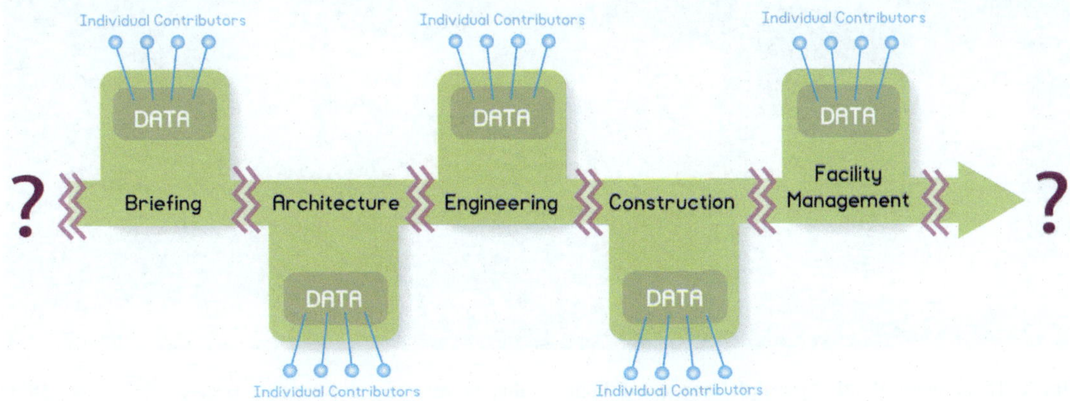

Fig. 1 The disconnects between datasets and disciplines that frustrates some of the BIM PIM and CIM ambitions

platforms such as AURIN are therefore essential for the creation of models. Raw data needs to be processed and transformed to be valuable for individual evaluation programs. For example, the processing of data for a multi agent transport simulation (MATSIM) for Singapore took seven work years (Ordóñez Medina and Erath 2013).

Punggol

Some attempts to deploy a PIM exist but are limited in their application. The Singaporean 'Eco Town' model for Punggol was used to evaluate the impact of new buildings on the wind within the precinct, but neglects the surrounding area (Ming et al. 2010) (Fig. 2). The model used for the 'ETH Science City' campus in Zürich, Switzerland, evaluates the annual energy consumption of the campus as a whole (Goffin et al. 2011). Since the buildings use a bus system to exchange excess heat in combination with ground storage, the sustainability of a new building is measured in the impact it has on the whole system. An attempt that combined historical analysis, building system, agent-based pedestrian and traffic simulation was used to evaluate the impact of new building systems on the use of public space in Singapore (Bruelisauer et al. 2013a, b). The difficulty of going beyond a single purpose of a PIM, for example, is in the limitations of purpose-built datasets and lack of interoperability of evaluation software. An attempt to overcome this is the extension of a GIS platform with shape grammars (Burkhard and Schmitt 2009; Aschwanden et al. 2012). The advantage for this method lies in the availability of the shape grammars to provide data in different forms that can be used by different evaluation tools (Fig. 3).

Drawing from AURIN

The Arden-Macaulay Urban Street Tree Renewal Study involved a tree renewal scenario testing study of an inner suburb within the City of Melbourne municipality. The project investigated the potential of botanically accurate high-polygon proxy object tree modelling at multiple suburb scale involving thousands of trees measuring solar/shade quantity impact (White and Langenheim 2014).

The project integrated textured 3D mesh geometry generated using Google's photogrammetry software Building Maker with datasets from AURIN. The AURIN datasets included Public Sector Mapping Agency (PSMA) road and transport network data, The Census of Land Use

Fig. 2 15_Punggol_Wind: Procedural model of Punggol Singapore generated with CityEngine used to estimate material flow and simulate wind with ANSYS that is reprojected into two planes on 25 and 55 m height. Image courtesy of Gideon D.P.A. Aschwanden 2014

Fig. 3 15_Rochor_Access_PT: accessibility to public transport nodes in Rochor Singapore. Image courtesy of Gideon D.P.A Aschwanden 2014

and Employment (CLUE) datasets, Vicmap Elevation (Contour) datasets, and the City of Melbourne's 'urban forest' dataset which details the location, species and lifespan of over 70,000 of Melbourne's trees in the public domain.

Tree datasets were transposed (projecting from GDA94 to MGA coordinate systems) from GIS to CAD converting point attribute table datasets via a seed file to attributed blocks. The tree block attributes were then used to generate procedurally distributed proxy object trees within Autodesk™ 3ds max™ determining the placement, size and specific tree species over the Arden-Macaulay area. A series of shade analysis experiments were then performed on the existing trees, then on procedurally driven scenarios, testing future potential visual and shade impacts for different tree spacing and alternate street tree species (Figs. 4 and 5).

Rapid Precinct Design—Datasets Used Generatively

The urban design proposition for the future of Australian cities titled '*Implementing the Rhetoric*' was a winning entrant for the Australian Institute of Architects' '*Now and When*' design competition, exhibited at the Australian Pavilion for the 2010 Venice Biennale. The project deprioritising car dominance explored novel approaches to Transit Oriented Development (TOD), daylight amenity preservation for public spaces, urban forest modelling, and Pareto [in]efficient urbanism.

The proposal was produced using PSMA, Vicmap, Open Street Map (OSM) and Cadastral datasets with flexible procedural modelling for rapid design parameter adjustment allowing real-time, tactile design exploration within Autodesk™ 3ds max™. As with most design competitions, speed was critical, the project was produced on weekends over a 4 week period. The modelling was 'light' (the whole project's files were under 120 MB) and well suited cloud server collaboration for this kind of schematic design.

The concept of an 'extreme TOD' was explored by directly linking levels of pedestrian accessibility modelled using agent based pedestrian modelling with RayFire™ thinking particle flow behaviour modelling and urban density

Fig. 4 Aerial perspective rendered view of Arden-Macaulay showing AURIN datasets, 3D photogrammetry building geometry with procedurally generated high-polygon mesh proxy object trees. Image courtesy of White and Langenheim (2014)

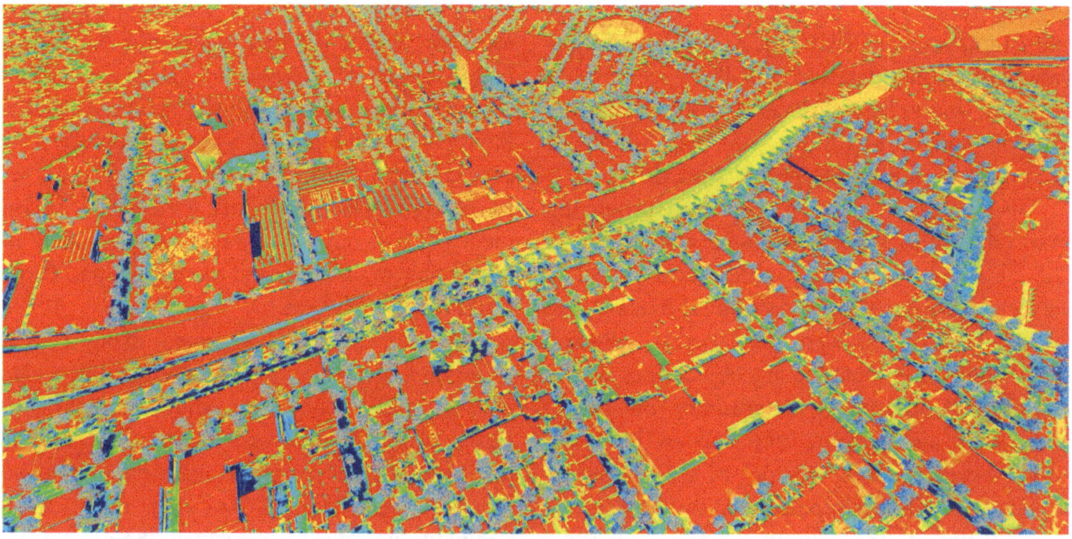

Fig. 5 Solar impact analysis performed on Arden-Macaulay 3D model with future potential trees. Image courtesy of White and Langenheim (2014)

distribution using a displacement and conform "Space Warps" (Fig. 6). Setting up a direct relationship between pedestrian networks and accessibility with density distribution exposed the critical importance of modelling beyond the precinct (Fig. 7).

Creative Disruption: Adding Complex Adaptive Systems to the Mix

In 1995 John Holland suggested that immune systems, cities, and eco-systems, all share certain properties and characteristics that he termed as

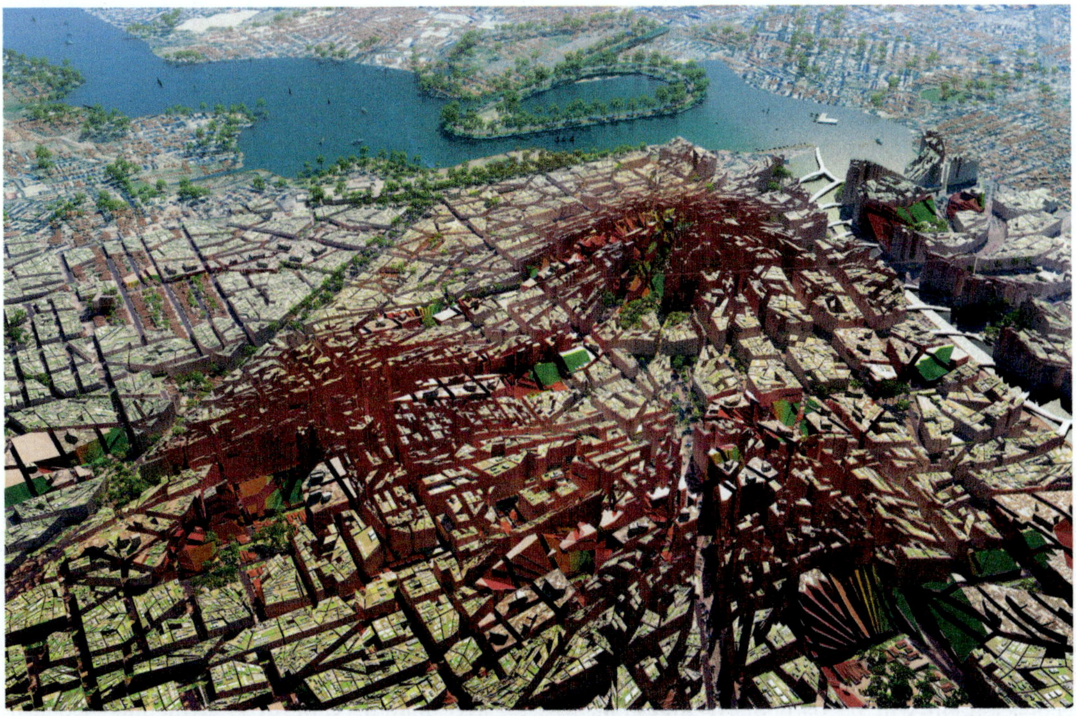

Fig. 6 Aerial image showing density distribution according to level of pedestrian proximity to key transit nodes. Image courtesy of White (2014)

complex adaptive systems (CAS) (Holland 1995 pp. 4–6). Complex adaptive system tools and theories lead to a major shift in perception from static models, based on rational choice to evolution of strategies over time. Therefore, when it came to cities we have realised that cities, just like immune systems—or any other complex adaptive systems, will not be able to settle into permanent structures, but rather remain in a constant state of adaptation and reconfiguration; accordingly we will require new concepts and new techniques to analyse and understand the world around us.

The case studies above identified some of the limitations of current BIM to PIM modelling in

Fig. 7 *LHS* aerial view of 'Subtracto-Sun' solar amenity preservation tool. *RHS* resulting maximum permissible building envelope, any buildings within this envelope will not cast shadows onto the public open space during designated time range. Image courtesy of White (2014)

datasets and analytical systems. As early as 1996 Fritj of Capra had challenged conventional views of the environment coming up with an alternative approach to understand natural processes based on three conceptual dimensions set as questions:

What is it made of?
What is the pattern?
What is the process?

These three conceptual dimensions could be similarly used when analysing cities or parts of the city labelled precincts, for example, Punggol, as discussed above. When we try to apply this method to develop a CAS model for a given precinct, however, we very quickly realise that only one of these three conceptual dimensions relates to material or physical form. Therefore our model will shift its attention from physical objects towards patterns, relationships and networks, understanding that networks with an embedded feedback loop will be particularly important to make sure that the system we are designing remains flexible, adaptable and therefore resilient.

The definition of the precinct as "the area with walls or perceived boundaries of particular building or place" or "an enclosed area or building marked by a fixed boundary such a wall" cannot be considered relevant. When using CAS to analyse a given precinct we also develop very different understanding of data, which we intend to use. We are not able to rely on manipulating independent variables to dependent variables, because our understanding of the world around us has its limits, and what we can observe often misses the most important information. As a consequence producing models for future urban scenarios based on current interpretation of current data, for most of the time, is both quantitatively and qualitatively wrong, they remain "models of" (Torrens 2005).

Not being able to predict urban futures with at least some degree of certainty makes the application of CAS models rather difficult to accept for many. For over 200 years society has believed that "science will yield answers of the simplest kind to a wide range of applicable problems" (Batty and Torrens 2005). Today we are ever more aware that this is not possible.

The quest to build useful theories and models was dominated by the view that we could simplify and distil the essence of things so that we might capture sufficient of the social reality for rudimentary comprehension and decision. If this were possible, we could at least move from "models of" to "models for". Despite recognition that the world was complex, it appeared simple enough to produce robust theory models that might be employed in applications. With increasing uncertainty and growing perception that the systems that we deal with are intrinsically complex, simplicity no longer seemed the watchword in development of techniques and models. Prediction is now couched in qualification, and our science has become less oriented to forecasting, more of an aid to understanding and structuring debate. This is seen nowhere more clearly than in shift to constructing 'what if' scenarios which now dominate most of model building (Batty and Torrens 2005).

Constructing *'what if?'* scenarios is an essential part of the CAS modelling processes. When using CAS models we are suddenly made very much aware of the fact that any action we take as architects, urban designers or planners—whether at the scale of a bus stop, office building or precinct through invoking a CAS we are introducing a major disturbance into an existing system. Knowing that this disturbance will take place we can decide on the role this disturbance might play. Through using CAS models we can elaborate on different scenarios and control, to some degree, that impact of this purposeful disturbance. We can also observe whether introduction of disturbance to the system will allow system operates without major changes, or whether the major change needs to happen in order to change existing system completely and become something new. In this way CAS theory differs from the current direction of proprietary BIM and emerging PIM; instead of offering us predictability and certainty it offers us creative

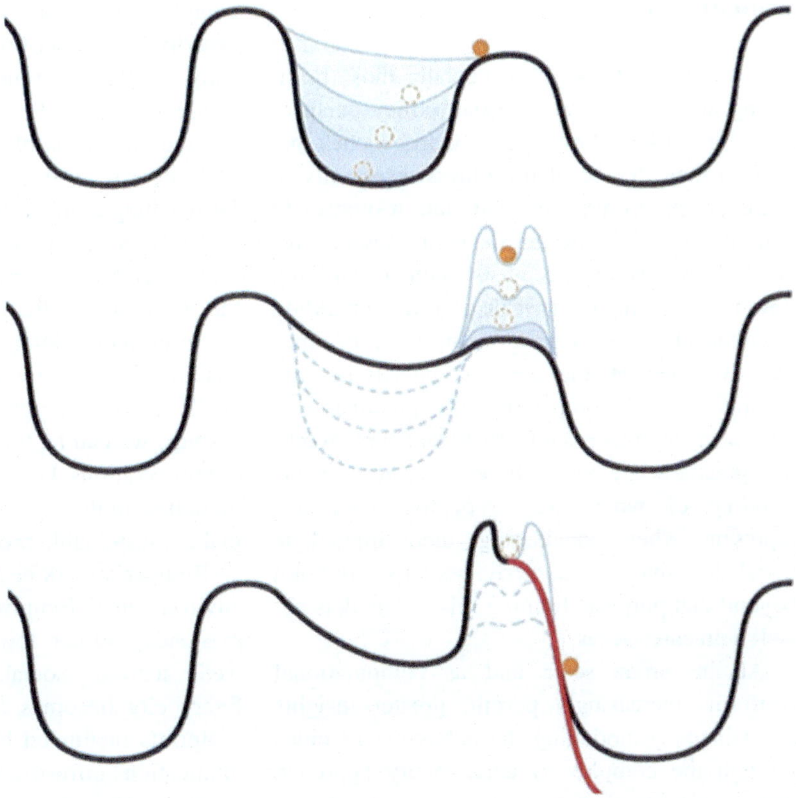

Fig. 8 Possible influence of disturbances within systems illustrating our tendency to actively discourage existing systems to change. We usually do everything in order to maintain the status quo, not creatively disrupt it

uncertainty, and at the same time it offers us a better understanding of the world around us. Through this, we move beyond "models for" and enter into the realm of using "models by which" we explore potentiality (Kvan and Thilakaratne 2003) (Figs. 8 and 9).

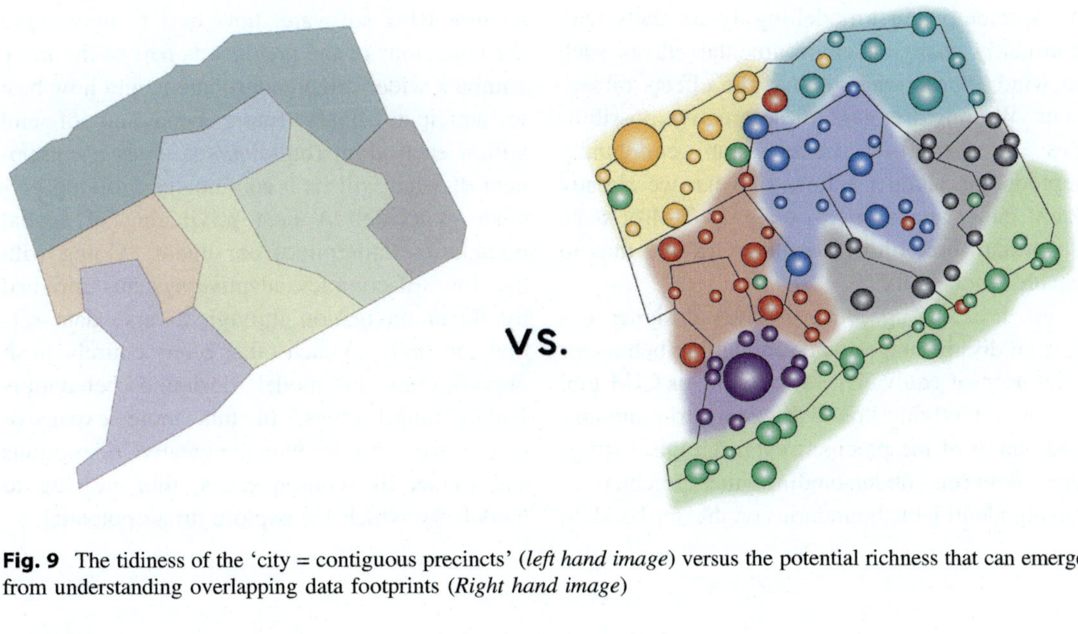

Fig. 9 The tidiness of the 'city = contiguous precincts' (*left hand image*) versus the potential richness that can emerge from understanding overlapping data footprints (*Right hand image*)

Conclusion

Today's urban designers, especially those privileging an architectural or urban policy perspective, are working from digital workbenches that vary in their degree of usefulness, especially in terms of opportunities to assimilate responses to data as part of the design decision-making process. Both sets of specialists such as building designers working to the scale of a single building or urban planners working to the scale of cities have workbenches that are more closely tailored to their needs compared with precinct-scale designers; architects and urban planners' working practices are well aligned to their understanding of what their respective data sets represent when considering their immediate translation into a designerly response, but not beyond that purpose, limiting this to "models of" well delineated areas.

At the urban scale and as computational platforms increasingly permit, greater insights are gained by modelling the behaviour of cities through the complex systems theory approach using software that can assimilate ever-larger datasets and flow, increasingly in real time, and accept the disruptive tendencies of unexpected outcomes. As we have illustrated in the case studies, designers are creating 'work-arounds' for acknowledged current deficiencies allowing them to examine precinct modelling dynamically with constantly changing environmental effects such as wind, diurnal change, and the effects of seasons. We might engage these creative possibilities of disruption rather than continually reproducing along the lines of what we already know either attempting to scale up BIM or scale down and three dimensionalise GIS and thus to enable "models for".

As it is, though, the precinct designer has several disadvantages when modelling behaviour with no neat equivalent to the BIM or CIM tool bench. Underlying this is the geometric unbounded nature of the precinct; even the OED struggles between understanding the precinct as having identifiable boundaries on the one hand, to being an area that is rather more vague in its description. Describing the precinct geometrically is not the issue here; rather it is understanding the complex behaviours that extend from the agglomeration of buildings and urban spaces with diverse uses and occupancies and their far-reaching connections.

As Brenner (2014) has noted, our urban footprint extends across the Sahara Desert and through the Northern Tundra. By applying systems theory to our urban thinking, we are aware that bounded space is necessary for analysis and to constrain the challenge at hand yet, in failing to articulate such bounds, we can be lulled into false confidence in our propositions. For example, a collection of BIM modelled highly sustainable buildings does not make a sustainable precinct or city.

Brenner's work neatly illustrates the extensive environmental footprints of our bounded urban existences. Other footprints can be defined as well, such as social, economic and cultural. Every city becomes a multitude of overlapped footprints facilitated by transportation and communications infrastructures. If we can readily live such overlays through our daily life (ordering items online, commuting to work, consuming distant cultural experiences), how can we then build a precinct model that manages such diverse and discontinuous data boundaries?

It is this that challenges conventional architectural BIM software: how best to understand the behaviour of the precinct as part of the more complex wider urban environment, and how best to anticipate likely future behaviour of and within each atom (building/space) as a component of what will be a continually evolving precinct molecule? A neat privileging of spatial boundaries undermines our intent. 'Going with the flow' of complex adaptive systems enriched by fluent navigation through diverse data sets that can 'talk' to each other offers entirely fresh opportunities to model desirable behaviours within future cities. In this more expansive model, we can introduce the creative disruptions and model the consequences, thus moving to models by which we explore urban potential.

References

Aram S, Eastman C (2013) Integration of PLM solutions and BIM systems for the AEC industry. In: Proceedings of the 30th ISARC, Montréal, Canada, pp 1046–1055

Aschwanden G, Zhong C, Papadopoulou M, Vernay DG, Arisona SM, Schmitt G (2012) System design proposal for an urban information platform: a systems proposal

Batty M, Torrens P (2005) Modelling and prediction in a complex world. Futures (37) pp 745–766. Available online at www.sciencedirect.com

Brenner N (2014) Implosions/explosions: towards a study of planetary urbanization. JOVIS, Berlin

Bruelisauer M, Berthold S, Aschwanden G, Belle I, Ostertag E, Meggers F (2013a) Reclaiming Backlanes–addressing energy efficiency, outdoor comfort and urban space. In: Proceedings of the Sb13 Singap

Bruelisauer M, Meggers F, Engler R, Leibundgut H (2013b) Heat bus for the tropics–exergy analysis of coupling decentralised chillers with central cooling towers. In: Proceedings of the clima 2013 energy efficiency smart healthy buildings

Burkhard R, Schmitt G (2009) Visualising future cities in the Eth value lab. In: Mixed reality in architecture, design and construction. Springer, Netherland, pp 205–218

Burry M (2014) BIM and MetaBIM: design narrative and modeling building information, pp 349–361. In: Kensek KM, Noble DE (2014) Building information modeling: BIM in current and future practice, Wiley, NYC USA

Capra F (1996) The web of life: new scientific understanding of living systems. Anchor Books, Michigan

Geli-Mann M (1994) Complex adaptive systems. Complex metaphors models Reality, 17–45

Goffin P, Ritter V, John V, Baetschmann M, Leibundgut H (2011) Analyzing the potential of low exergy building refurbishment by simulation. In: Proceedings of building simulation

Holland J (1995) Hidden order: how adaptation builds complexity. Addison—Wesley, Reading

Holzer D (2011) BIM's seven deadly sins. Int J Archit Comput 9(4):463

Jongeling R, Olofsson T (2007) A method for planning of work-flow by combined use of location-based scheduling and 4D CAD. Autom Construct 16:189–198

Kenley R, Seppänen O (2009) Location-based management of construction projects: part of a new typology for project scheduling methodologies. In: Winter simulation conference. Winter simulation conference, Austin

Kensek KM, Noble DE (2014) Building information modeling: BIM in current and future practice. Wiley, NYC USA

Kvan T, Thilakaratne R (2003) Models in the design conversation: architecture vs engineering. In: Newton C, Kaji-O'Grady S, Wollan S (eds) Design + research: project based research in architecture, The Association of Architecture Schools of Australasia, Australia

Ming LJ, Suan TP, Toh W (2010) Hdb's next generation of eco-districts at Punggol and eco-modernisation of existing towns. Ies J Part A Civil Struct Eng 3:203–209

Ordóñez Medina SA, Erath A (2013) Estimating dynamic workplace capacities by means of public transport smart card data and household travel survey. In: Transportation research record, pp 20–30

Saurin TA, Rooke J, Koskela L (2013) A complex systems theory perspective of lean production. Int J Prod Res 51:5824–5838

Simeone D, Kalay YE (2012) An event-based model to simulate human behaviour in built environments. In: Digital physicality—proceedings of the 30th eCAADe conference, Czech Technical University in Prague, Faculty of Architecture (Czech Republic) 12–14 September 2012, pp 525–532

Torrens P (2005) Knowledge and complexity. Futures (37):581–584 (Science direct). Available online at www.sciencedirect.com

White M (2010) The subtracto—sun: 4D solar envelope. In: Burry M, Ostwald M, Downton P, Mina A (eds) Homo faber: modelling identity and the post digital, Archadia Press, Sydney, pp 111–124

White M (2014) Preserving open space amenity using subtractive volumetric modelling, global demographic and climate challenges in the city, Aachener Geographische Arbeiten, Heft 50. Department of Geography of RWTH Aachen University, Germany, pp 107–126

White M, Langenheim N (2014) Urban street tree modelling using high polygon 3D models with photometric daylight systems. In: Oliveira V, Pinho P, Batista L, Patatas T, Monteiro C (eds) Our common future in urban morphology. Porto, Portugal, FEUP, pp 256–267

Zhang X, Arayici Y, Wu S, Abbott C, Aouad G (2009) Integrating BIM and GIS for large-scale facilities asset management: a critical review. In: The twelfth international conference on civil, structural and environmental engineering computing, pp 1–4 (2009) at Funchal. Madeira, Portugal

EPIFLOW: Adaptive Analytical Design Framework for Resilient Urban Water Systems

Dana Cupkova, Nicolas Azel and Christine Mondor

Abstract

In a time when digital tools outpace design methods, we see an increasing obligation to provide empirical support for a historically qualitatively-oriented design profession. This condition necessitates a research framework that facilitates a more comprehensive assessment of both context condition and design effect; one that integrates a comprehensive predictive system that allows us to test various scenarios against systemic performance criteria. Motivated by extensive discourse on health and environmental stability surrounding urban hydrologic systems, and the implication of uninformed design in the propagation of health and environmental stability, *EpiFlow* tooling proposes a new way of incorporating consequences of hydrological processes into the early stages of design process. By utilizing computation and visual scripting as a means of embedding information into the modeling environment, *EpiFlow* integrates storm water runoff and water management models directly into the 3d modeling test-bed. The intention is to assimilate direct visual and quantitative feedback on water flow behavior within a single interface, and to provide rapid design feedback that would enable integration of natural systems into holistic design thinking while engaging larger systemic issues of the site. Currently, the algorithms utilized within *EpiFlow* mainly address urban storm water runoff and household waste water systems. These two systems, already addressed within the architectural discipline by standardized metrics, are essential for such analysis, and historically have been central to issues of pollution, health dangers, resource waste, and environmental degradation. This paper explains the *EpiFlow* tool and workflow, as well as explores its implications on design process within an academic setting.

D. Cupkova (✉) · C. Mondor
Carnegie Mellon School of Architecture, Pittsburgh, USA
e-mail: danacupkova@gmail.com

N. Azel
Cornell University, Ithaca, USA

M.R. Thomsen et al. (eds.), *Modelling Behaviour*, DOI 10.1007/978-3-319-24208-8_35

Background

Existing scientific models and engineering formulas related to water behavior within the built environment enable fairly comprehensive predictive systems. However, they are not easily accessible within the design world. They have not yet been integrated into a suitable design interface that would allow for the rapid testing of various scenarios against systemic performance criteria, nor do they inform an intuitive understanding of water behavior by the designer. Currently there are three storm water runoff software packages that can help facilitate design: *Infraworks* from Autodesk, *SUSTAIN* from the EPA, and *SWIMM5* also from the EPA. There are a number of others, but they lack sufficient visual interface or ability of rapid editing to be particularly useful in a fluid design process. *Infraworks* has the significant drawback of only providing numeric feedback, and when facilitating design, inputs are fairly heavily constrained to predefined standard practice solutions. *SUSTAIN* is a robust software but is geared toward optimizing placement and use of green infrastructure within a district seeking to retrofit existing water systems. *SWMM5* is flexible, and can provide robust feedback on water runoff and pipe networks; however, it is grounded in simple 2d representations and a workflow of importing data constructed in GIS. The nature of these exiting tools is indicative of the engineering attitude toward design and management of natural environment, one that suggest that the natural systems are to be controlled, standardized, and optimized by engineers, lacking integration with design systems thinking for the larger built environment.

In order to foster a more environmentally-responsive design framework, *EpiFlow* utilizes contemporary water models and current computational tools to integrate hydrological feedback into the exploratory and intuitive design investigation. Unlike other hydrological models, *EpiFlow* is a design oriented parametric tool that operates with both GIS data and the user's design inputs. This pairing allows for adaptive testing of design scenarios related to water behavior, and offers more comprehensive feedback through the visualization of, and metrics for dynamic water flow patterns. By directly engaging such feedback, we liberate a more radical architectural attitude towards the manipulation of the environment, one that perceives design as the act of redefining a site's nature and casts architecture as an agent of resiliency within the urban network.

Framework and Methodology

Simulation has the ability to represent information flows (environmental, social, performative), to test causality between these systems, and gauge their capacity for adaptation and self-maintenance. For the *EpiFlow* designer, a focus is given on determining capacities and strategies for engaging biotic systems in relationship to urban waste water systems and storm water management; such utility ultimately enables urban ecological recovery in the context of human health and constructed landscapes.

EpiFlow is a sandbox, a type of adaptive test bed where various urban morphological scenarios can be played out, evaluated, and their effect explicitly visualized (Fig. 1). Compared with traditional analysis and mapping techniques that simply produce analysis from data, this framework facilitates creative design operability within the identified parameters of the hydrological model and topographical data sets. Such operability arises out of embedding the water model, geographic data and analytical algorithms directly into "live" digital space; from there, geometry and information can be manipulated in response to, and directed by analytical processes. *EpiFlow* operates within the McNeel Rhinoceros 3D environment and its Grasshopper Plugin, utilizing land-use data, surface permeability, urban features, and topographical data to solicit a predictive response for water behavior in relation to desired urban morphology. Any design changes to the model can be instantly fed through the storm water simulation, resulting in a flow-line interpretation of that particular intervention. Our current version of the simulation and feedback

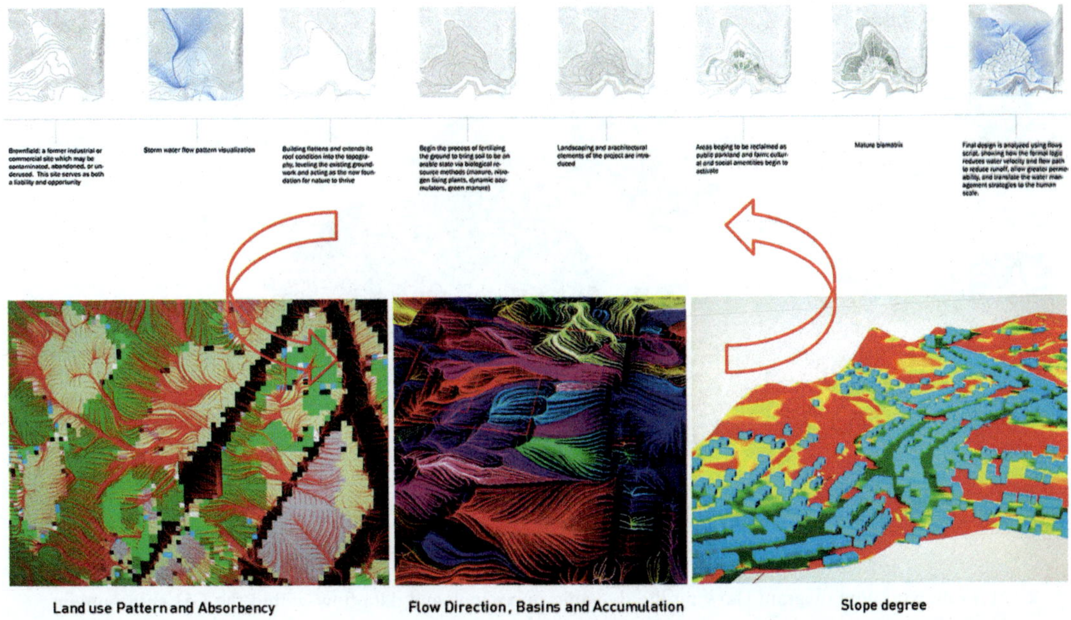

Fig. 1 Sample of design studio workflow utilizing the simulation parameters to determine the land and building form. Student: Thomas Sterling with Prof. Christine Mondor, Flow studio, Fall 2014, CMU School of Architecture

includes a rapid response to the following metrics: water flow, peak flow calculations, bio-swale suitability testing, waste water production from new units, water handling capacity of new infrastructure (based on NFT nutrient film technology), and land-use change over time. This paper further explains the operations within this simulation and shows its pedagogical integration to demonstrate the impact metric-based analytical feedback has on creative design thinking.

Epiflow Design Process

EpiFlow is a dynamic water flow model that responds to changes in urban morphology, topographic manipulation, building placement, and land-use changes. With a bias towards green infrastructure built into the simulation to mitigate systemic water-filtration issues, this test bed provides rapid feedback on water use, water flow visualization and wastewater metrics. Building visualizations of water flow behavior into the McNeel Rhinocerous modeling environment underscores qualitatively interpretable ramifications of formal design moves. Such a workflow supports an iterative design thinking that allows for more intuitive formal negotiation of multiple environmental and programmatic design objectives, while understanding the impact on water-use metrics. *EpiFlow* is a compact Grasshopper plugin (GH) script consisting of eight steps (Fig. 2) and six operations.

GIS Data Formatting and Input

Prior to engaging with the *EpiFlow* simulation script in Grasshopper (GH), it is necessary to collect and format the Geographic Information Systems (GIS) site data that will be used to auto-reconstruct the base site model in the 3D environment. *EpiFlow* uses land-cover data inscribed within a 3D digital mesh. For this purpose, we combine the raster digital elevation model data (DEM) and raster land cover data to define a digital terrain with embedded information. The land-use and land-cover information

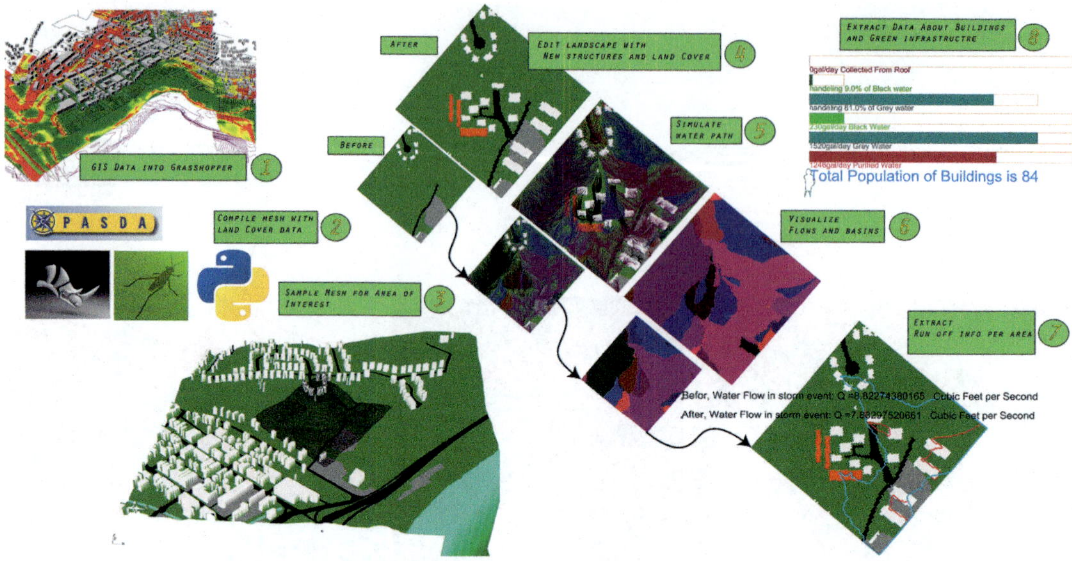

Fig. 2 *EpiFlow* workflow diagram showing the dynamic operations and data flow within the GH environment

affect the degree of water absorbency in future simulation. In any generic GIS-mapper, the DEM and land-cover data can be clipped and aligned to the same resolution and dimensions. The overlay of spatial data is then exported to ASCII format for direct import into Grasshopper. This technique has some limitations with regard to modeling resolution, as it is unusual to find digital elevation model (DEM) data at a higher resolution then 1/9 arc s, or approximately 1 m. (We have had some success developing land-use 3D models using vector data from GIS shape files and non-standard cell-size meshes; however, they are far more inefficient in terms of computation time, and therefore have not been fully implemented in this version.)

The ASCII-formatted site data set is imported into the GH environment using a custom scripted Python component and automatically reconstructed into a standardized 3D mesh. This mesh represents the base site topography, with embedded land coverage, land use, and outlines of urban morphology; it is transformed a second time by importing the geographic elevation vector data, which includes three-dimensional information about buildings and roads. The elevation data is integrated into a single mesh geometry (Fig. 3). This represents the total base geometry for simulation. In the future, we hope to add storm-water infrastructure to this operation.

Fig. 3 Single 3d site information mesh before and after integrating elevation vector data

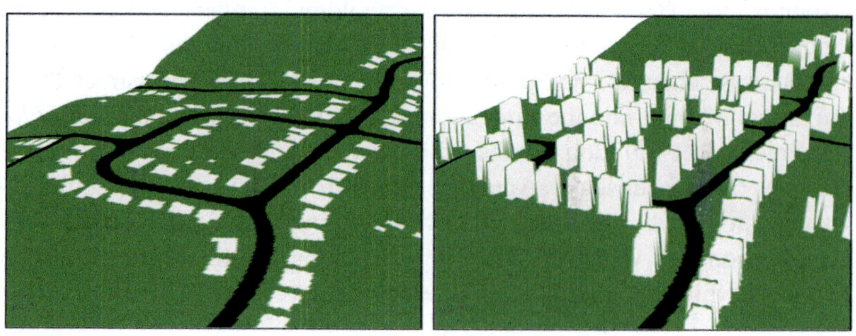

Site Sampling and Simulation Setup

A base 3D model, in the form of a single mesh, serves as a test bed for water simulation and modeling input. It is used as a substrate for a series of interventions, including but not limited to: geometric input in the form of buildings and roads: the change of land use and land coverage, topo-forming, etc. This facilitates an iterative process and allows for a deeper understanding of the consequences particular design decisions have on water-flow patterns and accumulation.

To manipulate the original mesh and test any design changes sequentially, our script incorporates the *Dynamic Pipeline* component available in the Human + TreeFrog Plugin for GH. (Human + TreeFrog GH Plugin has been developed by *Andrew Heumann*, who consulted the integration of the Pipeline Component into the *EpiFlow* simulation and contributed to water basin script development.) Using the input of geometric objects via pre-structured Rhinoceros layer input, this component helps with seamless transfer of information associated with geometries. It allows us to associate variables for water filtration, wastewater output, land cover, surface absorbency, or the differences between multi-story and single-story buildings, roads, bioswales, etc. By using layer-based grouping when importing any new geometry, we are able to assign constant information flow to all similar object types. This assures consistency when determining the final result of the computed water flow metrics. It is a type of BIM management system that adapts to the environmental and performance simulation (Fig. 4). Any new three-dimensional objects, digitally sketched or modeled within the base site mesh geometry, have to be modelled within a particular layer to be associated with the correct information type: i.e. building, road, bioswale, etc. These objects are automatically incorporated into the original single-mesh geometry, and their land-use data is incorporated via the layer structure.

Once a new mesh has been constructed to include any design transformations, the user has the option to redefine the mesh resolution for use in the water flow model. The water simulation uses every vertex within the mesh to determine water flow path. Thus, the mesh resolution greatly impacts the water model runtime. Being realistic about the resolution needs, given the scale of the study area, can help significantly in managing the runtime for modeling water flow. For example, if one wishes to determine the

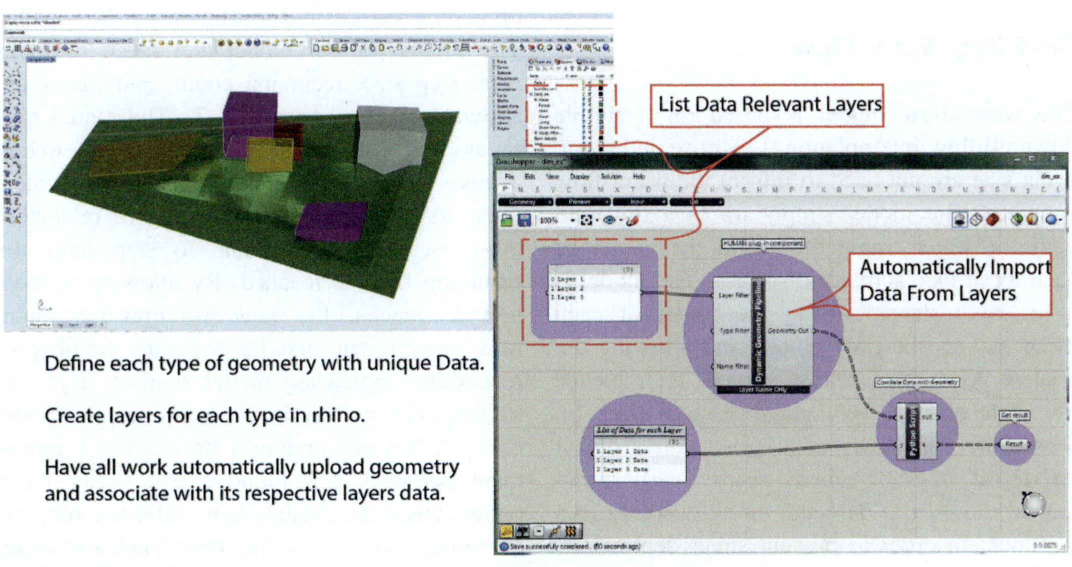

Fig. 4 Streamlined workflow using layer structure to associate data with geometric objects

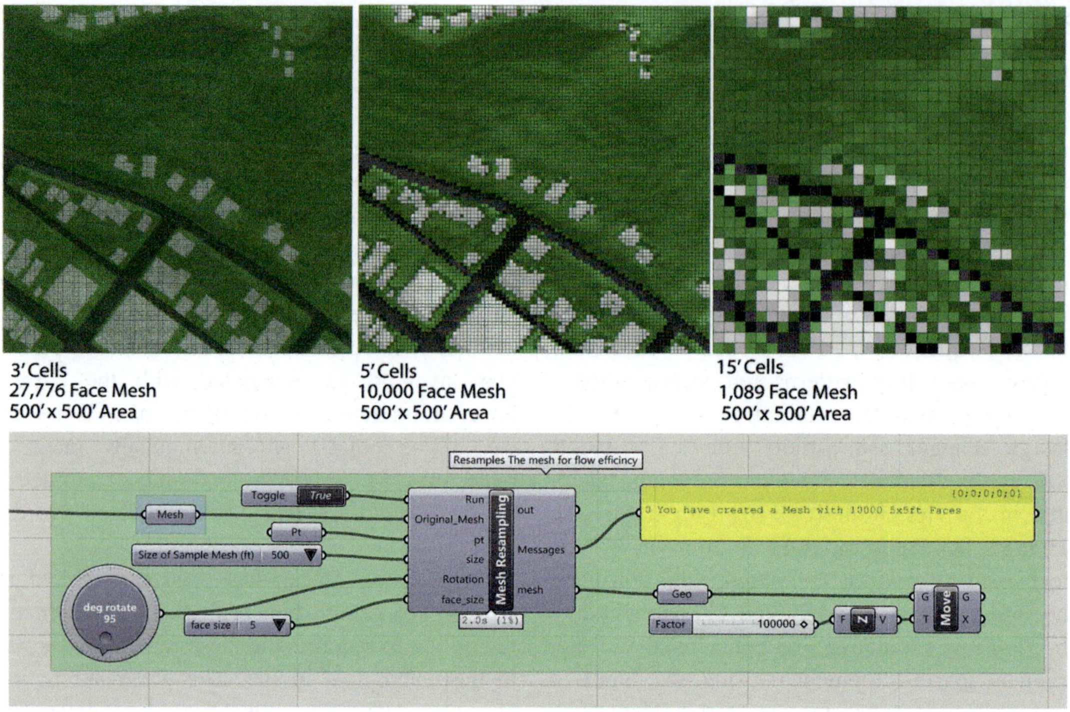

Fig. 5 Mesh sampling comparison and control mechanism within a custom GH component

edges of a large watershed, she would be able to use a 10-m mesh resolution, whereas a small development on a slope may call for a resolution closer to 1/3-m (Fig. 5).

Modeling Water Flow

The water flow model is based on a simple downhill flow interpolation. Land use and storm event size are not used to inform the water flow geometry; the model simply traces a downhill path for water originating from the center of each mesh face across the surface. The model is a recursive stepper, using the center of each mesh face as the flow object, and allowing the relative position of nearby vertices to influence the angle of step, by a given distance (Fig. 6). When the step path of a given flow object arrives at a point where nearby vertices no longer allows it to descend, the flow comes to a terminus. In order to prevent small depressions from obstructing flow, a parameter for vertical tolerance has been added. This allows a flow object to move vertically by a specified maximum, and greatly improves results when defining the limits of a drainage basin. The focus area's overall scale affects the model calibration, producing the desired resolution for flow paths. The inputs for the model are: mesh of focus area, step size, recursion count, and maximum allowable vertical hop (Fig. 7). The model runs by assessing the step of each flow object, one at a time, for each recursion. When a flow object finds its terminus, its recession is relaxed—otherwise, it will continue to step until the recursion limit is reached. By allowing vertical hop (a bounce of the water drop upwards); more flow objects can step for a greater number of recursions, increasing model runtime. By controlling recursion limit, one can decrease model runtime, but risk ending a flow object's movement before it has found a true water basin outlet. Once the models have all been run, the component output is the flow path for water originating from each mesh face.

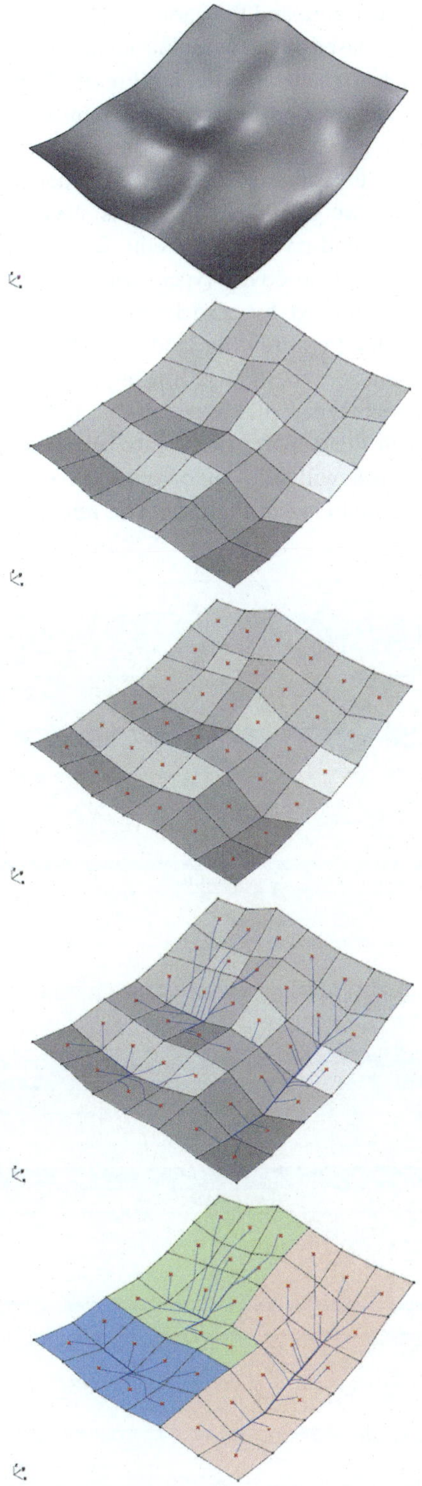

Fig. 6 Diagram of sequence in water flow modeling and delineation of water basins

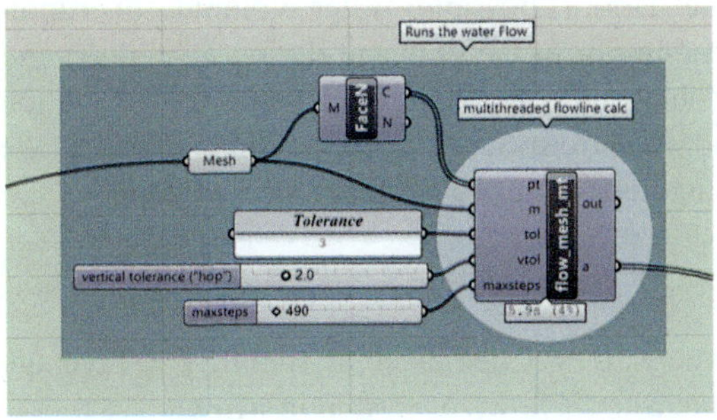

Fig. 7 Water flow component controlling the vertical tolerance and iteration sequence that affects the resolution for the basin outline

Water Basin Delineation

By taking the outputs from the water flow model, all flow paths are put to use assigning each mesh face to their respective water basins. This grouping is based on the proximity of each flow path's terminus. If a set of mesh faces have their flow paths ending within a given distance of one another, then they may be grouped into a basin. Basin delineation is critical for systemic understanding of water flow and outlines the drainage boundaries of a particular body of water in response to topography (Fig. 8).

Compute Storm Event Runoff Using the Rational Method

With the site mesh divided into drainage basins, it is possible to quickly compute peak discharge,

Fig. 8 Detail of water basins delineation

the largest possible water volume, or runoff coming from the drainage during a single event. The formula for the rational method (Calculation adapted from Steven, Nathan, and Woland: Site Engineering for Landscape Architects) is:

$$Q = C * I * A$$

where: C = Coefficient of runoff, I = Average intensity of flow, A = Area in acres.

We have automated the generation of *C* and *A* through a scripted interpretation of the user-specified drainage basin; however, we have left *I* to be a variable. The attainment of *I* via automated means would be possible, but given the traditional use of graphic calculators, to help derive the figure and the variation relative to flow intensity based on geographic location, this figure is left to the user to define (Fig. 9). For *C*, the weighted average of runoff coefficient is derived from the given basin (This formula is described below in the same paragraph). In order to generate the weighted average, it is necessary to input the land-cover coefficients for all the corresponding land-cover types within the basin. This is achieved by associating two data lists: mesh color (representing land cover) and runoff coefficients per mesh face. If using any additional land-cover types that were not initially represented in the imported GIS data set, these additional types, with relevant colors and coefficients, must be included in the lists. The relevant color

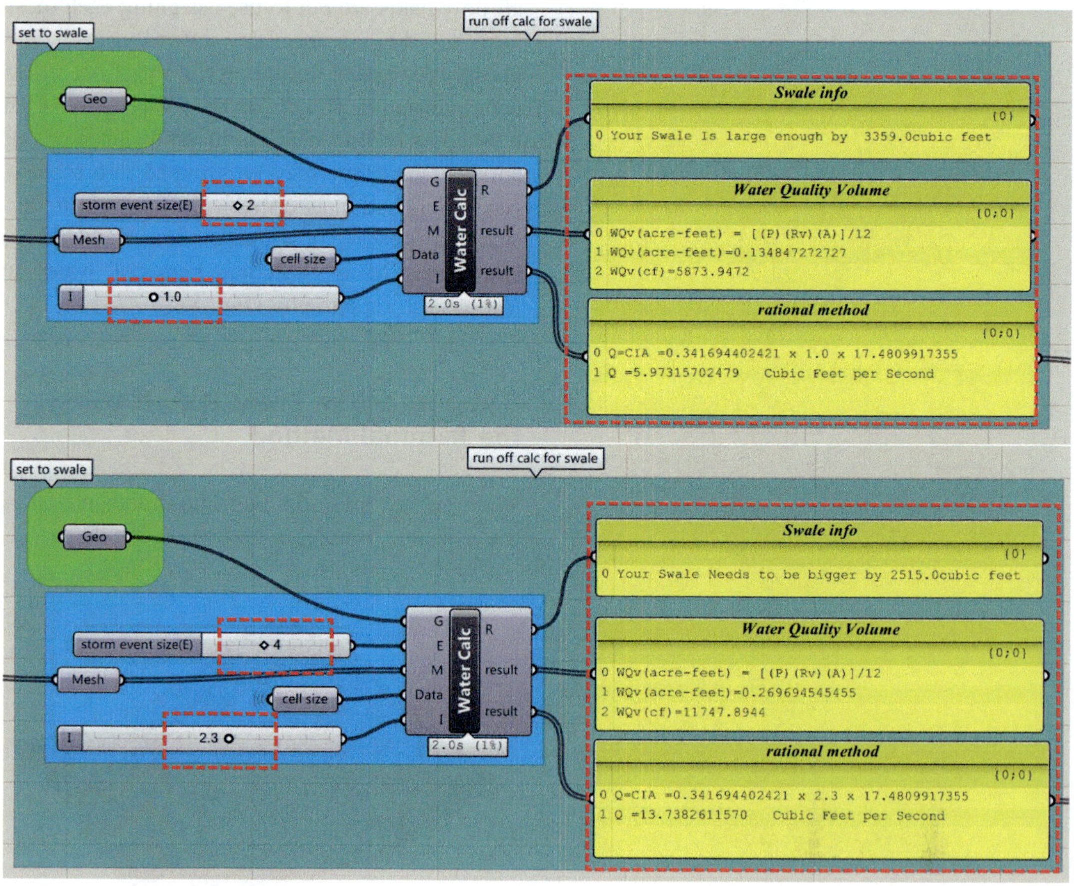

Fig. 9 Showing *I* (Average intensity of flow) variable and *E* (Storm event size) variable for two scenarios in bioswale sizing. As these two variables change the flow calculations adjust. In this case the size of the swale tested is appropriate for a 3″ storm event, but not for a 4″ storm event

must be applied to the given Rhinoceros layer that will contain all geometries of that particular land use.

EpiFlow incorporates a calculation for Water Quality Volume (WQv) adapted from the Center for Watershed Protection, *New York State Stormwater Management Design Manual*. This is a calculation that helps to size runoff detention infrastructure—in this example, a vegetated retention basin referred to as a Bioswale (Fig. 10)—for 24-h pollutant removal before discharge. *EpiFlow* uses the New York State standard for a retention basin. In New York, the standard is 0.9 WQv, or the 90 % rule. This means that a retention basin should be sized to accommodate 90 % of water runoff from impervious surfaces for the purpose of settling particulates and pollution associated with impervious land covers. The formula for acquiring WQv is as follows:

$$WQv = (C * P * A)/12$$

where: C = Runoff coefficient, P = Target rainfall, A = Area draining into the detention basin.

The runoff coefficient is gained by a weighted arithmetic mean for all land cover coefficients within the zone and described as such:

$$C = \left(\sum_{i}^{n} C_i A_{(i)}\right) / \left(\sum_{i}^{n} A_{(i)}\right)$$

where: i = Each land cover, C_i = Coefficient of land cover I, A_i = Area of land cover i.

This calculation is embedded within the *EpiFlow* simulation script and enables the rapid understating of water volumes from the basins' runoff related to land-cover types and various design interventions.

Water Usage and Infrastructural Carrying Capacity

In addition to the water flow model, we have developed a script that provides rapid numerical feedback on green infrastructure's potential for wastewater bio-processing. The wastewater output is based on new housing and infrastructure placement. Using standardized occupancy estimates and water-fixture counts producing maximum flush and flow rates—based on the LEED 4 guidance and building types—this information is imported through the previously-described layer structure. The sizing of the green infrastructure is adapted from the Nutrient Film Technique (NFT) developed by William Jewell as a method

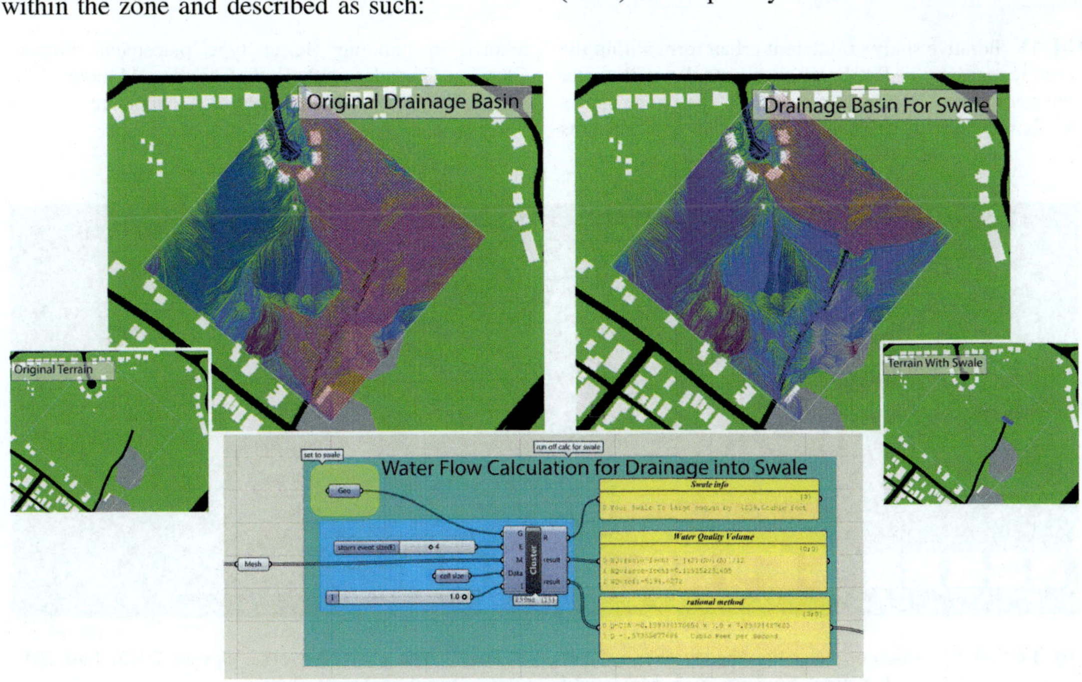

Fig. 10 Bioswale placement and its capacity calculation relative to the relevant drainage basin

for water purification (1992). The simulation is based on the assumption that we want to maximize bio-filtration within the capacity of the landscape. So far, this workflow considers only one type of waste water infrastructure—NFT—divided into two stages: black water and grey water processing. Each of the two treatment areas is evaluated based on the surface area necessary to treat a certain type of waste water per gallon and the time that process takes. Compared to grey water processing, black water requires a longer processing time and requires a larger dedicated surface area per gallon of water processed. Therefore, the metric output designates these two systems separately. By comparing available square feet dedicated to each of the

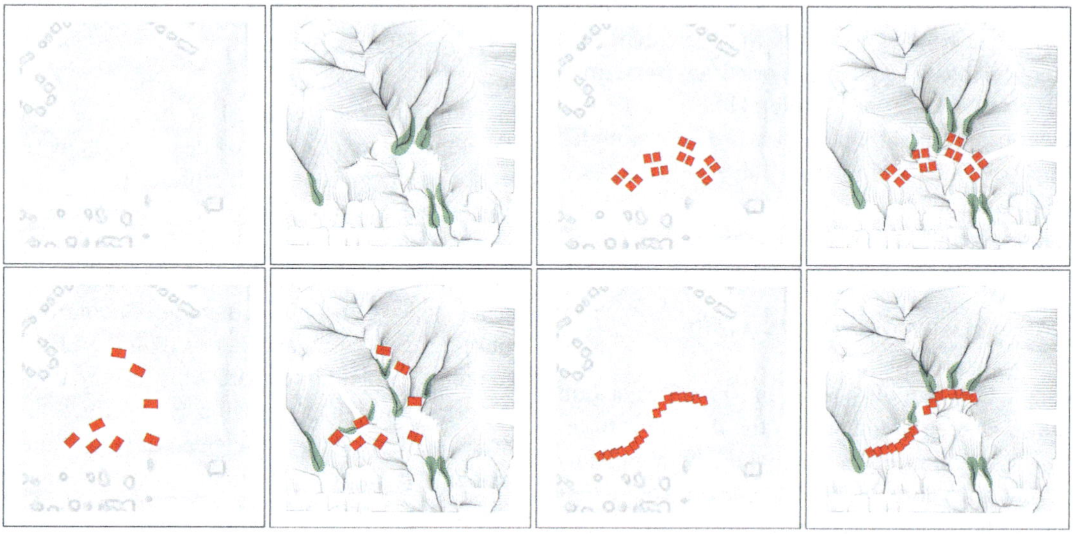

Fig. 11 Iterative study of different urban form within the urban topography of Pittsburgh that struggles with water management and storm water overflow. The image shows the consequence of bioswale generic sizes and locations relative to building cluster type placement. Student: Gloriana Gonzales with Prof. Christine Mondor, Flow studio, Fall 2014, CMU School of Architecture

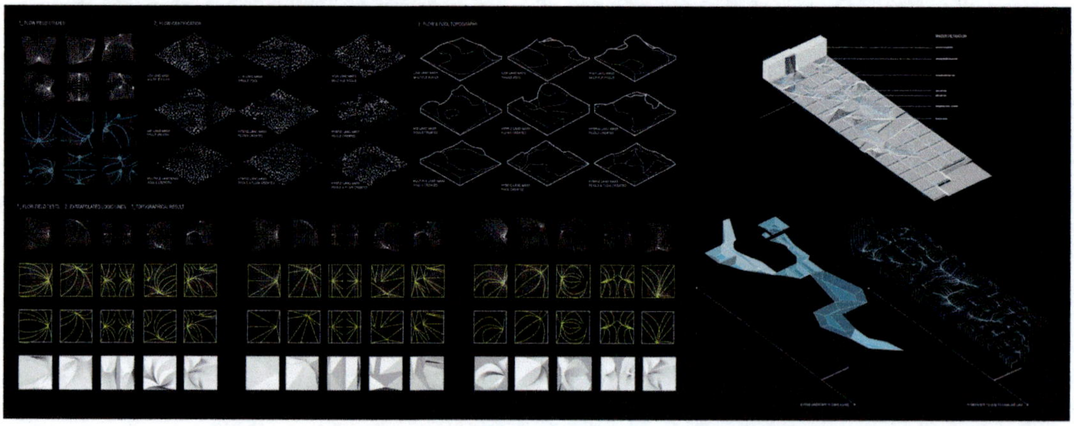

Fig. 12 Iterative study of water flow behavior translated through floodplain adaptation. Student: Kim McDonald with Prof. Dana Cupkova, Epiflow: Towards Resiliency of Post-Soviet City Networks, Spring 2015, Fall 2014, CMU School of Architecture

NFT green infrastructures with the total black and grey water produced by a new development's population in a given day, we are able to quickly compute the percentage handled by this new infrastructure's capacity, and the numeric ratio of gallons treated to gallons produced.

This computation is fully integrated within the visualization and allows the designer to have

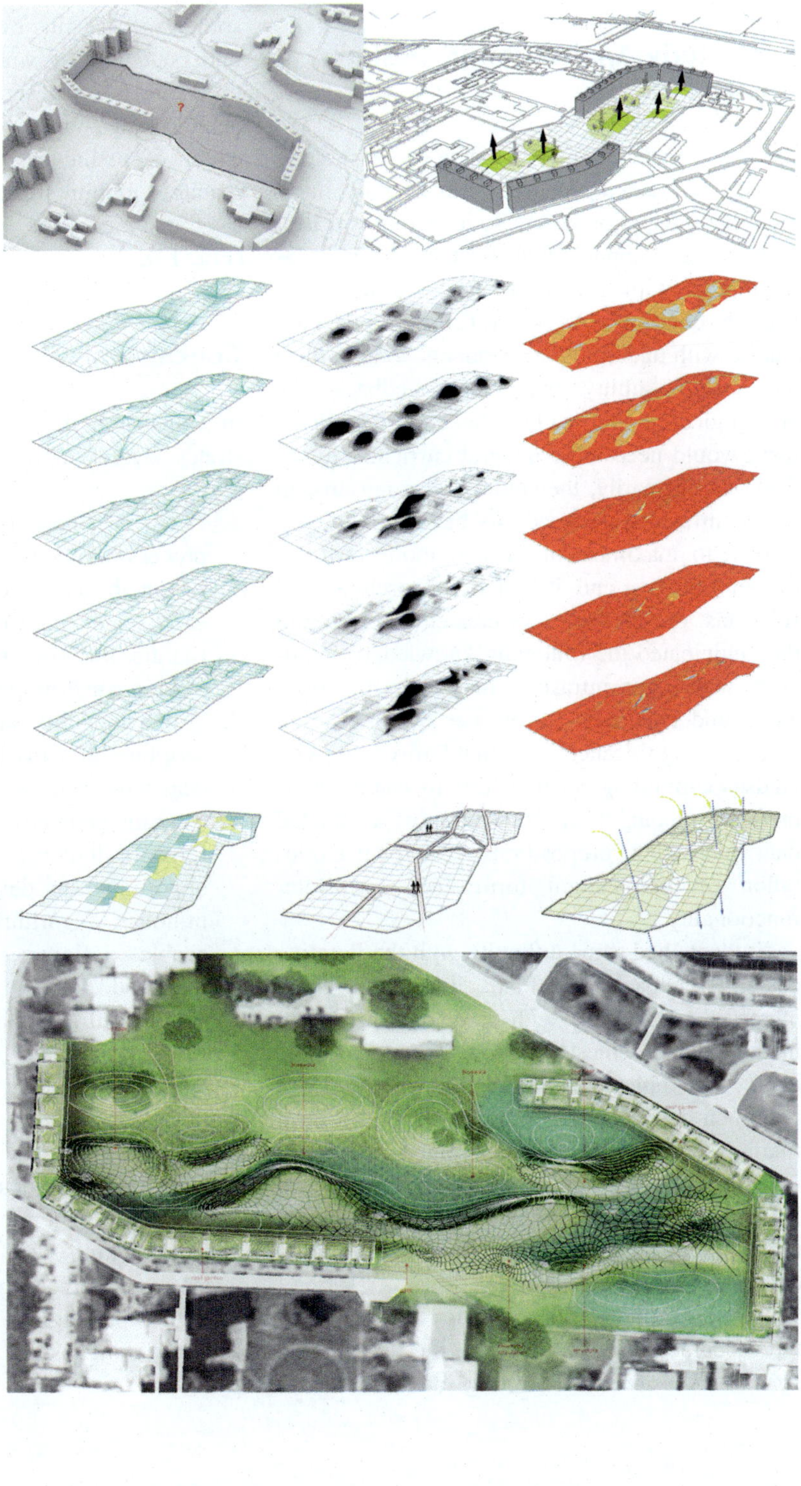

Fig. 13 Design studio workflow sample using waterflow simulation in conjunction with shadow patterns and incident solar radiation data to design a biotic system that integrates human occupancy, landscape, water filtration infrastructure and new programs. Students: Anum Shah + Nouf Aljowaysir, Prof. Dana Cupkova, Epiflow: Towards Resiliency of Post-Soviet City Networks, Spring 2015, Fall 2014, CMU School of Architecture

more freedom in negotiating massing, space allocation, unit distribution, and the total morphology of the development related to the potential natural waste water metabolism of a specific site (Fig. 11).

Case Study Application in Design Studio

EpiFlow simulation, as it is today, leaves more to be desired. Lacking a more user-friendly interface, the navigation of the script requires a user with some computational fluency, as well as familiarity with landscape storm water engineering concepts. As we reflect on our case studies with undergraduate students, to maximize the *EpiFlow* utility and design feedback, the pedagogical model within the studio environment would need more integral curricular modifications. Primarily, the studio pedagogy around green infrastructure needs to be carefully considered to maximize the use of more systemic design thinking and its effect on architectural solutions. For example, we created a curriculum that anticipated the following knowledge: familiarity with green infrastructure and its functionality, understanding stormwater variables and their significance, familiarity with landscape-building relationships to water, interpretation of output and translation of abstracted data into spatial propositions, reciprocal exploration of architectural form and stormwater functionality.

Students had some difficulty with the iterative design feedback at first. They presented increased malleability with computation fluency in the second semester of tool using, and thus some were able to start integrating simulation with their design concepts in early stages of form finding (Fig. 12). Design thinking shifted from understanding the tool at first as purely analytical information, to a later emerging understanding that the tool could directly influence in-parallel design decision making in.

Solving further technical issues that negotiate the accuracy (resolution) of the simulation against the speed of testing are critical to achieving better future results. We intend to address the issues more profoundly given our current understanding of the tool's functionality and our insights into students' capacities to assimilate this information.

While there is still much more to do, we had some success in evolving a more intuitive understanding of water behavior patterns and their effects on design thinking within the students' minds. With architecture positioned as an agent of resiliency within city networks, we managed to think through quantitative analysis as a foundation for creative problem-solving (Fig. 13).

Conclusion

In conclusion this project as a whole demonstrates several important considerations:

1. There is a need for further development of process-oriented computational analysis within the native design environment in the early stages of design. Not as an optimization routine, but to target a creative design thinking informed by performative issues.
2. The availability and accessibility of visual scripting in Grasshopper, and its ability to deal with diverse quantitative computations, dynamic geometric constraints, and nearly all forms of digitized data structures, open up the opportunity to develop unique and custom in-house algorithms and workflows that enable designers to operate in a more informed manner with regard to environmental, infrastructural, and spatial interests, whose complexity may have previously hindered any effective or holistic integration into formal design.
3. Making analytical metrics part of a design workflow is critical to forefront the importance of systemic thinking in integrative form-making. This way of working enables a stronger methodology that reframes the ecological paradigm within the architectural and urban domains.

Acknowledgments This project was funded by *Decade of Design Grant: The AIA Urban and Regional Solutions Challenge*, Targeted Research for Real-World Solutions to Urban and Regional Design Challenges, AIA and ACSA; in consultation with Prof. Nina Baird, CMU.

References

Center for Watershed Protection (2015) New York State stormwater management design manual. New York State Department of Environmental Conservation, Albany

Horner RR, Skupien JJ, Livingston EH, Shaver HE (1994) Fundamentals of urban runoff management: technical and institutional issues. The Terrene Institute, with US EPA, Washington, pp 3–59

Marlow DR, Moglia M, Cook S, Beale DJ (2013) Toward sustainable urban water management: a critical reassessment. Water Res Urban Water Manage Increase Sustain Cities 47(20):7150–7161

Strom S, Nathan K, Woland J (2013) Site engineering for landscape architects, 6th edn. Wiley, Hoboken

United States Environmental Protection Agency (2008) Handbook for developing watershed plans to restore and protect our waters. US EPA, Office of Water, Nonpoint Source Control Branch, Washington

United States Environmental Protection Agency (2012) Urbanization and streams: studies of hydrologic impacts. US EPA, Office of Water. Available via US EPA website: http://water.epa.gov/polwaste/nps/urban/report.cfm. Accessed 26 June 2015

United States Geology Survey (2015) Water resources surface water software. Available via USGS website: http://water.usgs.gov/software/lists/surface_water/. Accessed 10 June 2015

Integrated Forest Biometrics for Landscape-Responsive Coastal Urbanism

Keith Van de Riet and Uta Berger

Abstract

Within this research project, forestry and architecture are merged through an integration of data and modeling protocols for the design of urban landscapes. More specifically, this approach couples simulation of landscape systems, in this case mangrove forest, with 3d modeling software commonly used in design professions. The goals of the work include enabling feedback from living systems in the evolution of urban forms, development of an interdisciplinary design tool for coastal urban areas, and speculation on possible future alternatives for development in ecologically-fragile zones. Although the forms developed herein reduce the challenge to a minimum of variables, they also lay the groundwork for future added complexity in the research. The results of the modeling investigation indicated a higher level of integration would benefit the work, while at the same time maintaining the discrete quantifiable means to extracting values from the system for evaluating the designs. Ultimately, this framework may serve to challenge conventional planning and urban design methodologies that lack ecosystem services as part of the equation in formulating urban landscapes. This more holistic approach may serve to better integrate the dynamics of living systems and to consider real-time feedback as part of the datasets to be utilized in understanding and adapting our habitat as it evolves over time.

Introduction

In order to more effectively manage natural resources in coastal zones, integrated design and modelling frameworks are needed to analyze existing development trends and to speculate on ecological alternatives that may accommodate

K. Van de Riet (✉)
School of Architecture, Florida Atlantic University, Fort Lauderdale, USA
e-mail: keith.vanderiet@gmail.com

U. Berger
Forest Biometrics/Systems Analysis, Dresden University of Technology, Dresden, Germany

M.R. Thomsen et al. (eds.), *Modelling Behaviour*, DOI 10.1007/978-3-319-24208-8_36

the inevitable population growth in these regions. To be effective, these frameworks must account for multiple disciplines in the production of knowledge, while at the same time fostering innovative thinking through design-led research. In this regard, feedback gained through these models should be inclusive of quantitative and qualitative approaches. Of course, this integration of disciplines and model criteria is both a technical and cultural challenge (Ostrom 2007; Carpenter et al. 2009). Models are tailored for data collected by particular disciplines, and the exchange of data across formats, as well as emphasis on the spatial or non-spatial aspects of the simulations, presents additional barriers to fully integrating models (Holdren 2008). As a result, scientific and design communities have identified the need for alternatives to address the many challenges of sustainability in urban areas, and there is a distinct call for new conceptual frameworks that blend areas of knowledge to analyze social-ecological interdependencies (Carpenter et al. 2009; Sukhdev and Economics of Ecosystems and Biodiversity (Project) 2010). In terms of design modeling, there is also a critical need to better understand a building's interaction with the environment to identify more holistic approaches to managing complex systems and to develop more sustainable avenues for the built environment (Olgyay and Herdt 2004; Marsh and Khan 2011). Parallel to this, the frontiers of scientific research have developed simulations that can provide external stimuli for generative architectures that become environmentally-responsive in multiple ways (Hensel et al. 2006).

In the case of analyzing coastal urban landscapes, the challenge to integrate multiple models is compounded by the complexity of dynamic natural systems. In these transitional zones, marine and terrestrial ecosystems merge in hybrid environments linked to the rhythms of diurnal tides. Propagation, in the form of energy waves and offspring dispersal, colonizes areas with ecological feedback loops that more broadly contribute to diversity and balance in these types of environments. Tidal and climatic currents sculpt and mold vegetative, sedimentary and animal layers in the landscape, and the resulting hierarchical structures, as well as the mechanisms responsible for emergent forms in the landscape, present new challenges and opportunities for design integration (Corner 2006; Barnett 2013). Sub-Tropical mangrove forests capture this complexity, and as such, the simulations of these environments must capture the spatial and temporal processes inherent to the geomorphic forms evident in coastal mangrove landscapes (Fig. 1, right).

Mangrove ecosystems perform a number of beneficial ecosystem services that might be integrated through a combined design and planning approach. For example, the wave dissipating characteristics of vegetation, as well as coral reefs and other landscape features, have been attributed critical roles in wave energy absorption

Fig. 1 Equivalent scales of urban development (Miami, Florida, USA) and coastal mangrove forest (Everglades National Park, Florida). Images from Google Earth, Retrieved January 5th, 2015

and erosion control, particularly following the Indian Ocean Tsunami in 2004 (Danielsen 2005; Das and Vincent 2009). In terms of water quality, the EPA reported that wetlands perform to some degree, the biochemical transformations of wastewater treatment common to anthropogenic methods, indicating a potential use for these ecosystems in anthropogenic waste streams (United States, Environmental Protection Agency, and Office of Water 1987). Furthermore, mangrove ecosystems have been shown to have a high capacity to sequester heavy metals, in addition to the surplus nutrients, thereby preventing these hazards from entering the marine environment (Hogarth 2007; Wong et al. 1997). Lastly, coastal ecosystems are responsible for considerable financial contributions to these regions. In Florida, the coastal zone generated $562 billion of the economy in 2006 (Kildow et al. 2006). States like Florida predominantly rely on tourism, food and recreation associated with the coast, and even small increases in ecological productivity may have significant ramifications for a "blue economy" In order to integrate these ecological contributions within coastal communities, design models must account for the environmental requirements of foundation species responsible for these services.

Within this research, forestry and architecture are merged through the integration of design-based and so-called individual-based models (IBMs) for feedback on the growth and development of intertidal mangrove trees alongside coastal urban development. The integration of these two modeling approaches serves to provide a design and visualization tool at the scale of urban and forested landscape, with the goal of developing urban forms that simultaneously support forest structure and an appropriate urban density to underwrite the new construction. The integrated model developed with this work supports data acquisition on the spatial and temporal aspects of urban and forest development, as well as enabling the visualization of large-scale landscapes—conservatively as discreet zones of tree development and more progressively as hybrid coastal landscapes that blend infrastructural and environmental characteristics of the systems. Presented herein are the initial model integration and speculative landscapes, and this modeling groundwork forms the foundation for future investigations to integrate additional design criteria and simulations. Finally, the modeling approach was applied to a test site in Miami, FL to investigate the restoration strategy within a severely degraded mangrove shoreline and to develop design recommendations for future growth within medium to high density urban areas.

Review of Models and Parameters

Two specific models were used in this project: (1) The Kiwi Model, defined by Berger and Hildenbrandt (2000) and (2) McNeel and Associates Rhinoceros 3d Nurbs Modeling (https://www.rhino3d.com/) with Grasshopper Plugin and DIVA solar simulation tool (http://www.grasshopper3d.com/; http://diva4rhino.com/). These models were made interoperable through exchange of image-based information layers, where information on the forest development was used to inform the 3d model form.

Mangrove Forest Model

The Kiwi model parameterizes the three species of mangrove found in Florida (red, black and white) and utilizes the 'field of neighborhood' (FON) approach (Berger and Hildenbrandt 2000) to establish competition for nutrient, light and other resources (Fig. 2). This model describes spatially explicit the interactions of individual trees with their environment. It utilizes information about seed dispersal, individual tree establishment, growth and mortality dependent on the individual tree performance, local abiotic conditions and neighborhood constellation to analyze or forecast the forest dynamics over time (Berger et al. 2008). Each individual tree is described by a zone of influence (ZOI) determining the radius within the tree interacts with its neighbors. The FON includes within this radius, is a scalar

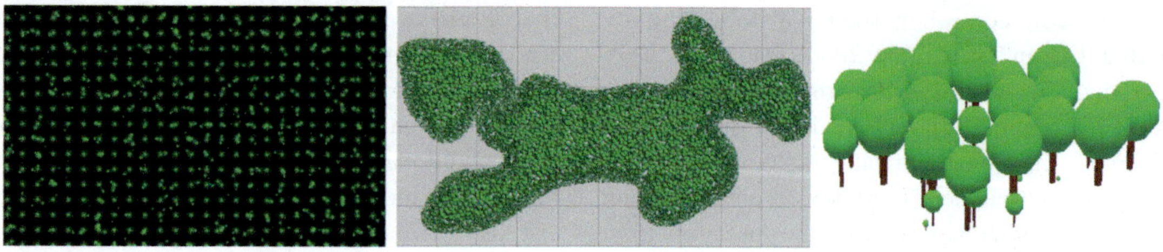

Fig. 2 Computational model of Red Mangrove Forest development—*Left*) plantation origin at 10 years timestep, and *Right* development of an island formation

field diminishing strength of competition moving away from the center of the tree. Hence neighboring trees have overlapping FON; and the amount of overlap defines the strength of competition.

In order to generate useful data with the forest and urban models, parameters for consideration must account for the dynamic interaction of urban and forest with the environment. In the case of forest, biometrics include dimensional aspects (height, diameter, crown radius, etc.), and these rely on environmental factors, such as solar access, hydrological regime, available nutrients, salinity gradients, etc. (Berger and Hildenbrandt 2000). Additionally, the spatial requirements, both in terms of dimensional space to grow and the structural configuration, are critical to assess, and these larger behavioral patterns over time (behaviometrics) are revealed through long-term simulations. The forest model was previously calibrated using live specimens in South Florida, U.S. and through research on specific environmental requirements of the *Rhizophora mangle* (red mangrove) species.

3d Urban Model

In terms of urban form, significant economic pressures exist on major coastal districts to develop large-scale buildings and the concomitant civil landscapes required to support them. As a result, this work focuses on parameters derived from large-scale vertical construction and land reclamation in order to address the existing pressures to develop coastlines with these types of structures. Globally speaking, land reclamation projects are being undertaken at unprecedented scales for residential, commercial and infrastructural uses. Ideally, this constructed land could provide new foundations for beneficial tidal ecosystems, otherwise unable to become established in typical urban edge conditions.

As case study, this project speculates on the redevelopment of Miami's waterfront edge with an engineered-living approach. Miami contains a dense urban core of high rise buildings but quickly becomes low rise residential when leaving the downtown area. A site in the downtown area measuring 400 × 400 m was documented for density, program and population distribution (Fig. 3). The central core has an F.A.R. 24:1 with an incentive zoning option of up to 36:1, and contains a mix of residential, commercial and retail (Miami 21 Code 2012). Historically, mangroves fringed the entire shoreline in this area, with substantial reclamation projects infilling tidal zones in the past century of development (Cantillo, United States, and National Ocean Service 2000). From the site survey, type and number of buildings was ascertained, percentages of lot coverage, street area, surface parking, recreational entities, such as pools and tennis, as well as open green space. Within the present work, the density of existing building structures was used to provide a plausible scale for the speculative modeling.

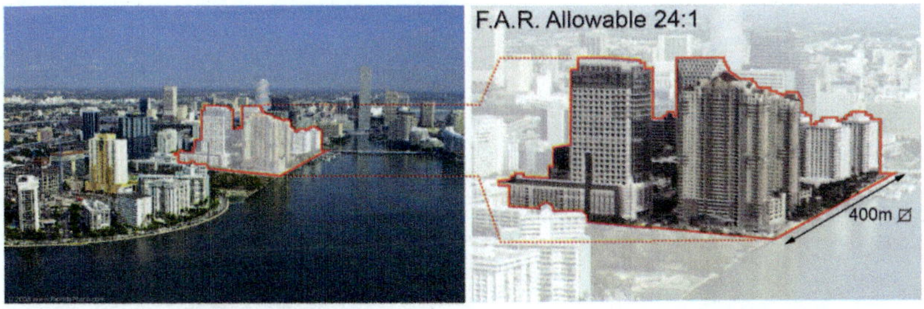

Fig. 3 Extraction of parameters from Miami (adjacent to central business district) for urban model

Methodology for Integration

Multiple methods can be used to integrate the models—through exchanging data in a variety of formats or fully integrated programming. In this case, the approach relied on exchange of data across the two models in the form of bitmaps that translated 3d data for the forest model, and in the other direction, forest data for the 3d model. Zones able to be colonized by trees are defined in the 3d model as objects or landscape surfaces that reside above the tide. These zones are indicated in the bitmap images with black, which is equivalent to areas that receive no sunlight in this study. In terms of forest biometrics, the Kiwi model can output data in the form of spreadsheet or bitmap image to relay patterns of growth or more specific information on individual trees, including diameter at breast height (DBH), age, size, etc. From these extracted values, the results can be further extrapolated to account for habitat area, biomass, and the associated ecological functions.

Because the Kiwi model relies on underlying information to determine forest growth, such as suitable substrate, nutrients, hydrological factors, etc., a hypothetical landscape was used to initiate the model to replace the unknown factors for the site. This landscape was sampled from the west coast of Florida in the 10,000 Islands region of Everglades National Park. Three configurations of mangrove were sampled—inland (swamp), riverine and coastal fringe. These three samples were used to approximate the changing forest conditions at different proximities to the upper limits of the tide and are visible in the model as distinct patches of 400 × 400 m. Without integration of a hydrological and sediment model to supply substrate information, this assumption was necessary for first step integration of the models.

The initial integration was meant to establish an exchange of data across the forest and 3d models using solar access as a means to evaluate the behavior of the model. Both climate and the utilization of solar resource needed to be analyzed prior to the undertaking. Within the Miami climate, an abundance of solar energy reaches the landscape, with approximately 50–60 % of this able to be utilized by plants. Actual capacity (saturation point) for photosynthetic active radiation (PAR) of vegetation varies by species and climate. Solar-driven photosynthesis was reviewed for the particular species of interest—*Rhizophora Mangle* (Fig. 4). This species-relevant information was integrated within the Kiwi model to determine sensitivity to shading by adjacent large-scale structures. Shade tolerant correction used Botkin et al. (1972) and Chen and Twilley (1998) to establish a shading coefficient for the model, which extended the Kiwi model functionality for this research. Utilizing this coefficient with the Kiwi model, the threshold for viable establishment of red mangrove trees was shown to be at levels higher than 30 % of available light in the Miami, FL climate. In the model, areas where trees were unable to become established due to over shading were indicated by darkened zones (Fig. 4, right).

The shading coefficient establishes a minimum level of sunlight required to grow red mangrove trees, whereas the PAR saturation point suggests excess sunlight may be obstructed

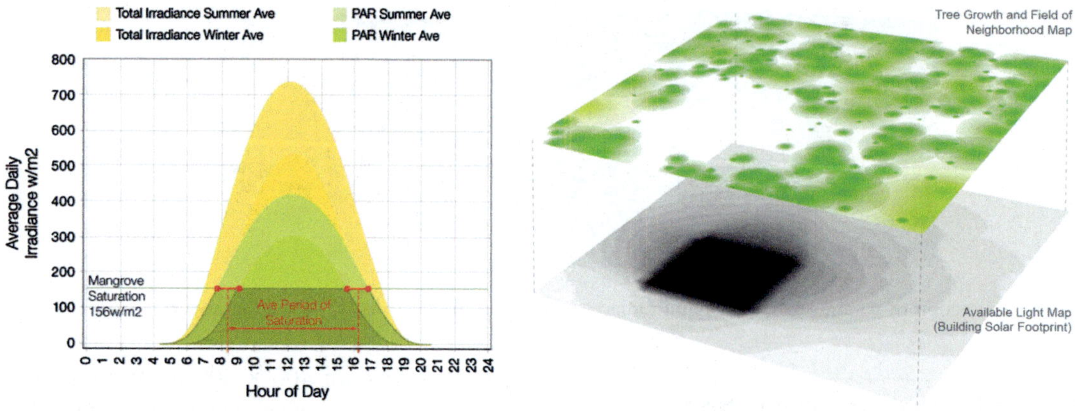

Fig. 4 Photosynthetic saturation point for Rhizophora Mangle (*left*) and solar-based integration of forest and 3d model information (*right*)

without affecting tree growth. Both of these metrics link the environmental requirements of a mangrove forest to the potential impacts of the constructed environment, and this common dimension of solar resource is a metric that can be adapted to other climates and vegetation types.

For this study, the model was applied to two scenarios: (1) a standardized block of buildings with variable F.A.R. and Lot Coverage was simulated for solar access and landscape response, and (2) a larger urban landscape was generated by prioritizing forest parameters in the massing of the constructed elements. The first scenario more closely corresponds to analysis of an existing context with the model, whereas the second scenario projects future development under new zoning regulations that might be partially informed by requirements of the native landscape.

Results

The exchange of information across the models was effective for different purposes at different scales of analysis. At the scale of building or urban block, the simulation provides feedback on growth and development of the forest in response to available sunlight and suitable substrate (Fig. 5). This model established a workflow to evaluate urban forms corresponding to more

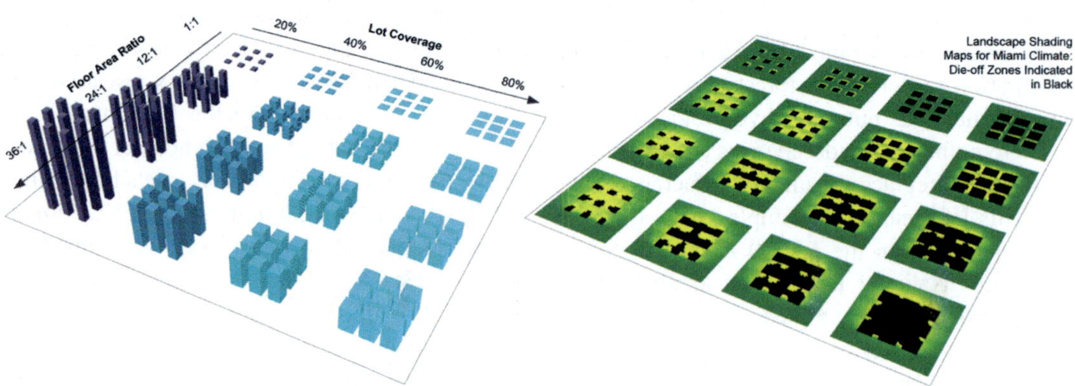

Fig. 5 F.A.R. and lot coverage as variables in building form (*left*) and solar footprint as it affects tree growth (*right*); *black areas* indicate 30 % or less of available sunlight for Miami climate

conventional urban arrangements. The parametric model that defined the urban massing for this study can be adapted for particular waterfront districts to evaluate a restoration strategy or to envision the impact of a changing shoreline in the case of sea level rise.

This first investigation utilized the minimum solar access of 30 % defined in the methodology section, and simple color-coded surfaces were used to indicate likelihood of establishing new mangrove landscapes (Fig. 5, right: green to yellow to black). In this regard, the model gave visual feedback prior to initiating the forest simulation—an advantage in terms of preliminary planning without relying on additional simulations. However, once these landscape maps are generated, the forest model can be initiated to extract further information. This makes it an effective tool for analyzing the impact of urban structures on local environmental conditions, with particular emphasis on vegetation growth and solar resource.

After establishing feedback to visualize the impact of constructed forms on the landscape, the model was inverted to create urban forms from the landscape. Simulated forest growth utilizing the hypothetical tidal landscapes introduced from Everglades National Park created a foundation to an urban development that corresponds to forest morphology. In this scenario, the assumption that

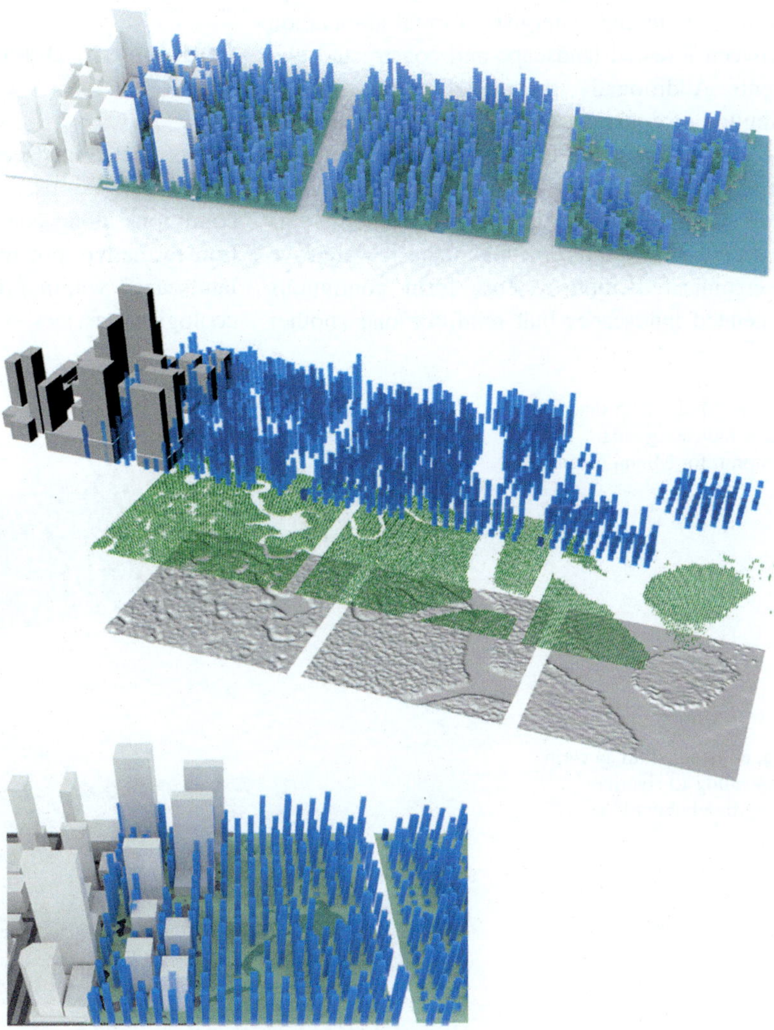

Fig. 6 Visualization of urban morphology based on inland (*left*), riverine (*middle*) and coastal fringe (*right*) mangrove habitats

ecosystem services, reliant on formal parameters of the landscape, take priority in establishing formal characteristics of the urban environment. As a result, forest patterns informed the density and distribution of urban elements, and distinct canopy-like appearances began to emerge due to the underlying informant geometry (Fig. 6).

In these models, the urban form is generated by forest structure, where taller and more established trees correspond to taller and more established urban features. When measuring competition in the forest, more established areas of trees tend to have lower competition due to the increased occupied zones and maturity of the trees. These associated attributes in the model established a zonal correlation among low competition trees and higher density urban development—a potential synergistic formal association between forested landscape and constructed elements. Additionally, the models express a distribution of constructed elements consistent with the continuity of the forested substrate. Interdependencies in mangrove landscapes is critical to their survival, due to the extreme weather events in tropical and subtropical climates. Mangrove interconnected root systems form contiguous vegetated landscapes that reinforce one another against storm surge and regular wave action. These characteristics form the canopy architecture of the forest, which at the scale of urban landscape, informs the model geometry of the constructed elements.

The model can also be used to determine temporal sequences for the urban environment, in terms of spacing and intensity of urban development over time, while simultaneously measuring ecosystem values that may be affected by increasing density or alterations to the urban environment over time. In this regard, the model was able to simulate multiple scenarios that might reflect changes in the constructed environment resulting from shifts in policy within a municipality (Figs. 7 and 8).

Conclusion, Discussion and Future Work

The projected landscapes and datasets within this model can be used to analyze existing built conditions along coastlines and to visualize the transformative potential of integrating natural landscapes within urban areas to capitalize on ecological services. At the larger scale, biomass

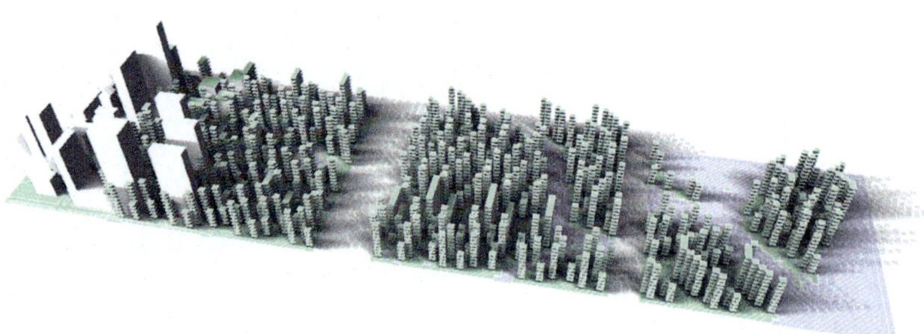

Fig. 7 Model with shadow maps indicating solar footprint for Miami climate

Fig. 8 Isolated urban form responding to riverine mangrove habitat type

and tree density data can be collected for inference of other useful ecological feedback, and the values associated with constructed dimensions, such as floor area and other factors, can be measured alongside these services to determine appropriate tradeoffs between landscape productivity and economics. These large-scale inferences might be used for initial land-use and zoning considerations, but for more specific integration of engineered and natural landscapes, the model platform should be expanded to include more refined spatial information (capacity for multi-scale simulations).

Simulated environments offer an opportunity to project near and long-term human impacts on natural landscapes, and in a more proactive sense, are able to visualize landscape-responsive architecture and urban morphologies based on the resource requirements of multiple species in the models. These "virtual laboratories" enable designers to parameterize ecological systems within spatial and temporal dimensions to inform design and planning decisions, sometimes leading to unconventional forms and structures. To some degree, these models might generate fantasy worlds, but simultaneously, they play out scenarios containing far-reaching implications for how communities live in relation to natural resources and the potential for integrating ecosystem services within urban areas. At a minimum, they provide timely and useful feedback for debate on the management of threatened resources and offer an integrated metric to evaluate alternative methods for conserving and regenerating threatened landscapes.

This work provided a foundation to further develop computational infrastructure to support interdisciplinary work on coastal systems. The models produced in this project contributed to a collaborative forum to speculate radical forms that emerge from a bio-responsive urban model as forest and constructed landscapes interact with the environment over time. By exchanging species and geographical location and accounting

Fig. 9 Layers of simulation and analysis combined in perspective view of building masses, tree and landscape layers

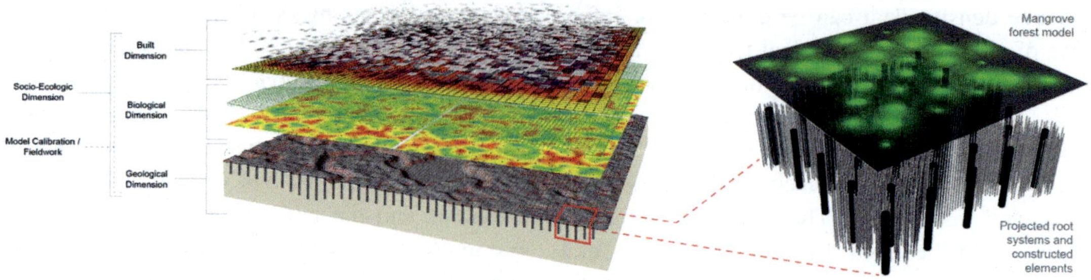

Fig. 10 *Left* Future integration of hydrological and sediment models will support more accurate approximations of landscape interaction with constructed elements (image previously published in Van de Riet et al. 2011); *Right* 3d model of Kiwi mangrove simulation, projected root systems and constructed elements prepared for flow analysis

for various climate change scenarios within the model, the framework can enable planners and designers to visualize strategies for more sustainable and resilient communities in nearly any given context.

Future work will further develop the information exchange across models to better support simulations that include multiple layers of information—from soil, hydrology and climate factors to economic, material and experiential. Additionally, these simulated environments must be developed across a range of scales to ensure appropriate representation of the interacting systems. At certain scales, the datascapes begin to encounter new sets of criteria stemming from urban design, ecological performance and the human perspective (Figs. 9 and 10). Parallel studies, both data-oriented and urban visioning, should be conducted for useful feedback on the performance of these landscapes from qualitative and quantitative perspectives.

Over half the world population currently resides on or near coastlines and is urban, which means models to project landscapes in the realm of "cybernetic ecology" are urgently needed. Although much work has been done on urban forestry and the parameterization of urban vegetation, this information is not readily accessible for design purposes. This limits the potential speculation on the topic—particularly from a spatial and experiential point of departure. Furthermore, these models are needed to play out scenarios that will result from shifting coastlines, community priorities and the types of lifestyles that accompany these environmental adaptations, both negatively and positively. However, methods to measure this coexistence are in many cases severely lacking and require complex interdisciplinary teams to tackle the challenges. Particularly sensitive habitat areas may require even more refined models to determine the impacts of land development on these critical resources. Fortunately, these types of feedback models, specifically within a format accessible for 3d visualization, are now becoming available.

Acknowledgements The authors gratefully acknowledge Anna Dyson at the Center for Architecture, Science and Ecology (CASE) at Rensselaer Polytechnic Institute and Uwe Grueters at The Dresden University of Technology for their support in this research.

References

Barnett R (2013) Emergence in landscape architecture. Routledge, New York

Berger U, Hildenbrandt H (2000) A new approach to spatially explicit modelling of forest dynamics: spacing, ageing and neighbourhood competition of mangrove trees. Ecol Model 132(3):287–302. doi:10.1016/S0304-3800(00)00298-2

Berger U, Rivera-Monroy VH, Doyle TW, Dahdouh-Guebas F, Duke NC, Fontalvo-Herazo ML, Hildenbrandt H et al (2008) Advances and limitations of individual-based models to analyze and predict dynamics of mangrove forests: a review. Aquat Bot 89 (2):260–74. doi:10.1016/j.aquabot.2007.12.015

Botkin DB, Janak JF, Wallis JR (1972) Some ecological consequences of a computer model of forest growth. J Ecol 60(3):849–872. doi:10.2307/2258570

Cantillo AY, United States, and National Ocean Service (2000) Biscayne Bay Environmental History and Annotated Bibliography. Silver Spring, Md.: U.S. Dept. of Commerce, National Oceanic and Atmospheric Administration, National Ocean Service. http://purl.access.gpo.gov/GPO/LPS123392. Accessed 15 June 2015

Carpenter SR, Mooney HA, Agard J, Capistrano D, DeFries RS, Díaz S, Dietz T et al (2009) Science for managing ecosystem services: beyond the millennium ecosystem assessment. Proc Nat Acad Sci 106 (5):1305–1312. doi:10.1073/pnas.0808772106

Chen R, Twilley RR (1998) A gap dynamic model of Mangrove forest development along gradients of soil salinity and nutrient resources. J Ecol 86(1):37–51. doi:10.1046/j.1365-2745.1998.00233.x

Corner J (2006) Terra Fluxus. In: Waldheim C (ed) The landscape urbanism reader. Princeton Architectural Press, New York, pp 21–33

Danielsen F (2005) The Asian Tsunami: a protective role for coastal vegetation. Science 310(5748):643–643. doi:10.1126/science.1118387

Das S, Vincent JR (2009) Mangroves protected villages and reduced death toll during Indian super cyclone. Proc Natl Acad Sci 106(18):7357–7360. doi:10.1073/pnas.0810440106

Hensel M, Menges A, Weinstock M (Guest eds) (2006) Techniques and technologies in morphogenetic design. Architectural Des J. Wiley, West Sussex

Hogarth PJ (2007) The biology of mangroves and Seagrasses. Oxford University Press, New York

Holdren JP (2008) Science and technology for sustainable well-being. Science 319(5862):424–434. doi:10.1126/science.1153386

Kildow JT, Florida, Department of Environmental Protection, National Ocean Economics Program, and Florida Oceans and Coastal Council (2006) Phase I, facts and figures, Florida's Ocean and Coastal Economies Report. Florida Oceans and Coastal Council, Tallahassee, Fla. http://edocs.dlis.state.fl.us/fldocs/dep/focc/phase1Jun2006ff.pdf. Accessed 15 June 2015

Marsh A, Khan A (2011) Simulation and the future of design tools for ecological research. Architectural Des, 1–10. 11 October 2011

Miami 21 Code (2012) The City of Miami. http://www.miami21.org. Accessed 15 May 2012

Olgyay V, Herdt J (2004) The application of ecosystems services criteria for green building assessment. Sol Energy 77(4):389–398

Ostrom E (2007) A diagnostic approach for going beyond Panaceas. Proc Natl Acad Sci 104(39):15181–15187. doi:10.1073/pnas.0702288104

Sukhdev P, United Nations Environment Programme Economics of Ecosystems and Biodiversity (Project) (2010) The economics of ecosystems and biodiversity: mainstreaming the economics of nature : a synthesis of the approach, conclusions and recommendations of TEEB. S.l.: UNEP

United States, Environmental Protection Agency, Office of Water (1987) Report on the use of wetlands for municipal wastewater treatment and disposal. The Agency, Washington, D.C

Van de Riet K, Berger B, Dyson A, Gowdy J, Proffitt E, Zeghal M, Vollen J (2011) Multidisciplinary modeling of coupled mangrove and urban ecosystems. In: Abbate A, Polakit K, Rosemary J (eds) Conference proceedings of subtropical cities 2011, Florida Atlantic University, 8–11 March 2011

Ubiquitous Monitoring and Adaptation of the Tempered Environment

Ryan Welch, Roderick Bates, Christopher Connock and Eric Eisele

Abstract

Recent developments in low-cost, ubiquitous sensing of the built environment offer great potential for improving thermal comfort and reducing building energy consumption. A network of sensors installed during the renovation of a former bottling facility will be used in conjunction with occupant feedback and building automation to enhance thermal comfort while improving building energy efficiency. Adaptive models of thermal comfort such as ASHRAE Standard 55 and its revisions (Nicol and Humphreys in Energy Build 34:563–572, 2002; De Dear and Brager in Energy Build 34:549–561, 2002) characterize thermal comfort in terms of several environmental factors, including air temperature, air velocity, relative humidity, and thermal radiation. While these variables can, in principle, be modeled to any desired fidelity, building energy simulations typically require gross simplifications to address the intractable computational challenges of the Navier-Stokes equations and the radiant heat transfer equation. As a result, air volumes are typically modeled as having uniform properties within a given thermal zone, and each boundary surface is treated as having a uniform temperature. The asymmetries overlooked in such models are the very parameters at the core of many passive and active environmental conditioning strategies. The implications of these exclusions, as well as the performance benefits derived from the provision of this missing data via the deployment of building sensor networks, is the subject that the present study investigates through empirical findings.

Background

Building monitoring as a practice typically makes use of a limited number of sensors to create a coarse understanding of average conditions

R. Welch (✉) · R. Bates · C. Connock · E. Eisele
KieranTimberlake, Philadelphia, USA
e-mail: rwelch@kierantimberlake.com

M.R. Thomsen et al. (eds.), *Modelling Behaviour*, DOI 10.1007/978-3-319-24208-8_37

within a space and at various points within the mechanical system. Such methods, while useful for understanding overall building operation, do not provide thermal information at a resolution that is relevant to the micro-conditions impacting an individual building occupant. The proposed research called for a monitoring platform to quantify boundary layer asymmetries, an investigation that necessitated high density real-time data collection. A review of available off-the-shelf monitoring technology by the research team indicated that such monitoring systems are cost prohibitive when deploying large number of sensors, with cost per data point of $90 + USD for conventional wireless Zigbee, wired serial sensors, or data logging sensors. In response, the team developed their own platform that allows high density data collection at a price point appropriate for experimentation. The cost savings achieved with the technology were ultimately sufficient to offset the cost of development relative to purchasing off-the-shelf technology.

Building Automation System and Passive Controls

The main floor of KieranTimberlake's architecture studio in Philadelphia, Pennsylvania (USA), utilizes a variety of passive and active thermal conditioning systems to optimize user comfort while minimizing energy use in this renovated former bottling facility. An underfloor air delivery system supplies two main "highway" ducts running north-south in the main floor and four separate ducts that supply the perimeter zones, providing a total of 23,000 CFM. Air volume delivered to each occupant is controlled by a manually actuated swirl diffuser located in the floor near each employee's desk. The north, south, and west sides of the building are lined with operable windows, and the high monitor has six remote-operated windows. The system exhaust and building pressure regulation are managed via two fans located on the south and north ends of the monitor; they are capable of extracting a combined 36,000 CFM (Fig. 1).

Two hot-water boilers are used for heating, and cooling is accomplished via a liquid desiccant system, with all systems centrally controlled via an Automatrix JACE. The building has no traditional air-conditioning system, instead relying upon automated MechoSystem shades on the south and west windows, a heavily insulated cool roof, and significant thermal mass.

Depending upon outside temperatures and the building's need for heating and cooling, various modes of operation are engaged. Cooling can be accomplished one of four ways, first by allowing air to enter the building directly through the windows, supplemented by the underfloor system, and then exhausting the air out the monitor windows or fans (Fig. 2). Alternatively, when humidity and dust control is desired, the windows can be closed and air introduced from the underfloor system and exhausted through the central monitor (Fig. 3). The fully active building cooling mode (Fig. 4) consists of dehumidified air supplied via the underfloor system and returned through the two central ducts, where it will be dehumidified again prior to mixing with outside air as appropriate to maintain CO_2 levels. This mode is similar to the heating mode, except the dehumidifier is not active while heating, and the air is reheated prior to recirculation into the occupied space.

Data Collection Method

KieranTimberlake's wireless sensor network is deployed throughout the main office floor. The initial sensor deployment comprises 124 interior surface temperature sensors distributed across the office floor, walls, ceiling, and glazing; 56 air temperature and 56 relative humidity sensors distributed horizontally and vertically throughout the work space; 60 air temperature sensors distributed throughout the underfloor supply air plenum; 120 sensors measuring the surface and core temperature of the structural slab beneath that plenum; and a roof-mounted weather station for measuring ambient temperature, relative humidity, wind velocity, solar irradiance, and rain intensity.

Fig. 1 Main office floor of KieranTimberlake facility in Philadelphia. The historic building was once a bottling plant for a brewery

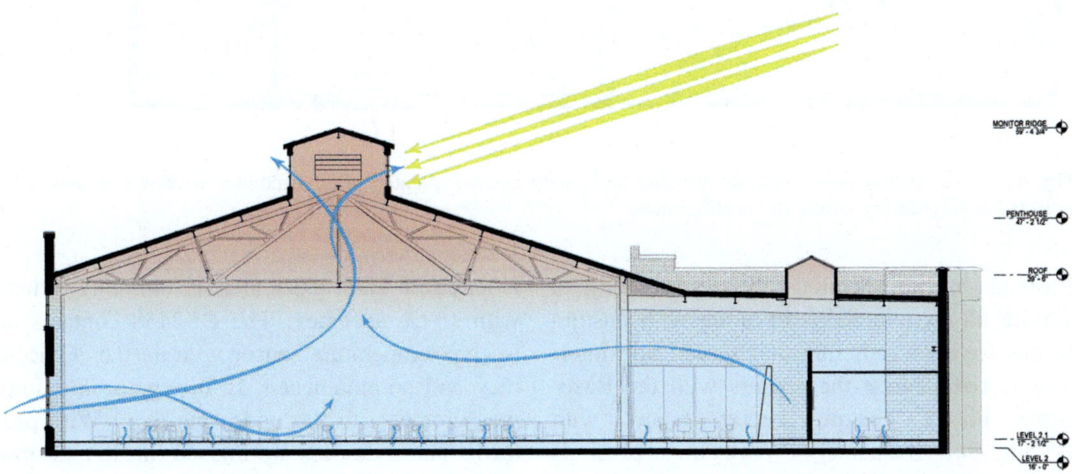

Fig. 2 Combined window and underfloor air delivery, monitor exhausted cooling mode

At the core of the platform is the Maxim 1-Wire™ protocol, which enables communication from sensor to nodes. Each node can interface up to 8 sensor drop cables with a maximum length of 30 m, with each cable capable of interfacing up to 30 1-Wire sensors. This method of connecting multiple sensors to each node limits the marginal cost of a data point to that of only the sensor only. The nodes are Raspberry Pi B+ micro computers with a custom

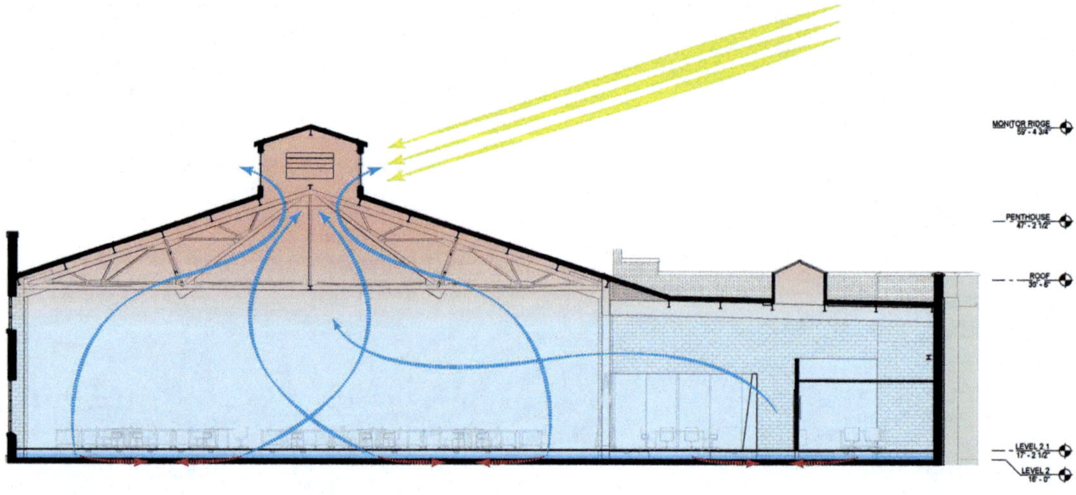

Fig. 3 Underfloor delivered, monitor exhausted cooling mode

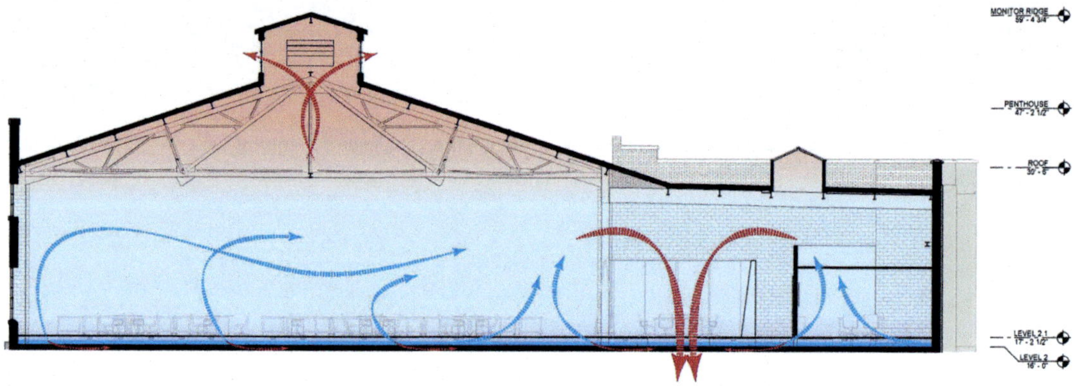

Fig. 4 Active cooling mode with air recycled and excess pressure vented via the monitor windows or fans. Also reflects the air pathway during the heating mode

Raspbian-based operating system running a 1-Wire file system (OWFS) along with custom scripts for data gathering and server side interaction. To integrate the sensors with the Raspberry Pi, a custom circuit board with DS2482-800 8-channel 1-Wire transceiver is employed (Fig. 5). All nodes communicate directly to the Internet via Wi-Fi except for the floor sensors, which use Ethernet for communication due to poor Wi-Fi performance inside the floor system.

The relative humidity sensors deployed in the roof trusses are based on the Maxim Integrated DS2438 1-Wire integrated circuit and the Honeywell HIH-5030 analog humidity sensor with ±3 % accuracy. The DS2438 contains an in-chip temperature sensor with a ±0.5 °C accuracy and communicates 12-bit analog humidity measurements to the nodes via the 1-Wire protocol. These devices are housed inside a custom 3D-printed housing (Figs. 6 and 7), which incorporates magnets to facilitate mounting to steel trusses.

The temperature sensors located in the ceiling and slab core are based on MAX31820 1-Wire temperature sensors on a custom circuit board, encapsulated with silicone in a protective aluminum sheath, measuring 20 mm × 6 mm with

Fig. 5 Sensor node with one drop cable attached

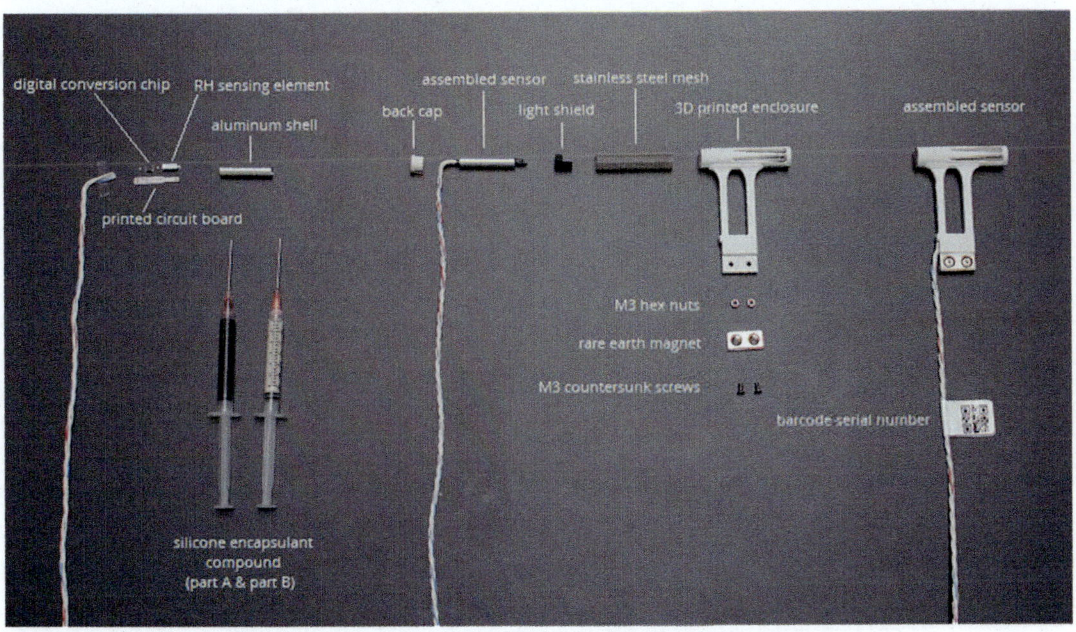

Fig. 6 Relative humidity sensor components

Fig. 7 Relative humidity sensor installed with integral magnets

an accuracy of ±0.5 °C (Fig. 8). The slab core sensors were inserted into 100 mm deep holes and backfilled with DOW 995 structural sealant. The ceiling sensors were inserted into 25 mm deep holes and backfilled with Liquid Nails 950 construction adhesive.

The wall surface, mullion surface, slab surface, floor surface, and underfloor ambient temperature sensors are all based on the MAX31826 1-Wire temperature sensor with a manufacturer stated accuracy of ±0.5 °C. All sensors are mounted to a small custom circuit board and conformal coated, with the surface sensors embedded in a flat block of ABS measuring 8 mm × 25 mm × 37 mm. These blocks were adhered to the slab surface with silicone sealant and insulated with a polyisocyanurate cover to shield from the ambient air temperature. Bare conformal coated temperature sensors attached to a 150 mm height bracket were used for ambient air measurements under the floor.

The glazing sensors are custom fabricated units specifically designed to reduce the amount of heat gain incurred via exposure to direct solar

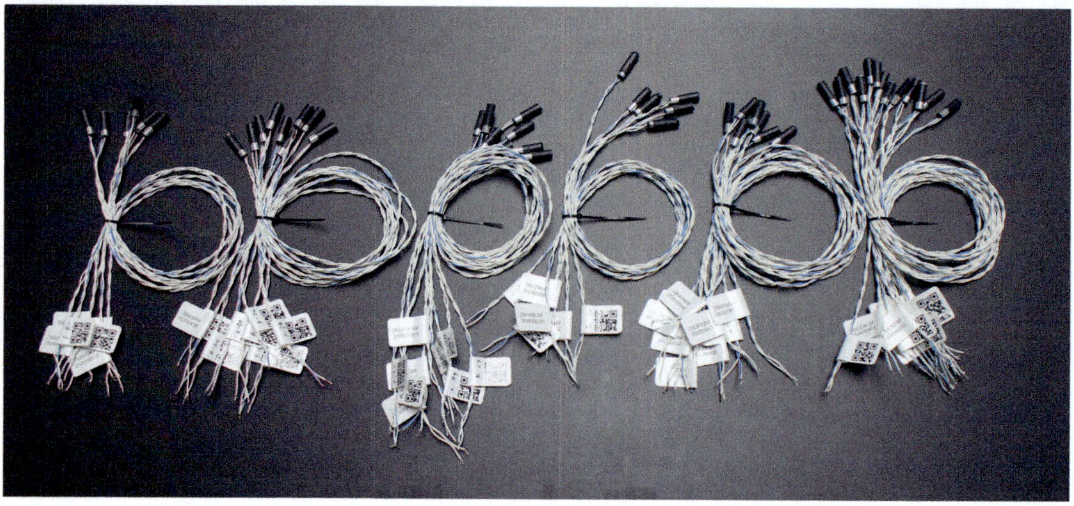

Fig. 8 Ceiling and floor slab sensors

radiation. They are based on the MAX31855 1-Wire thermocouple interface that is conformal coated and protected by a plastic sheath. The thermocouple is assembled from TFE insulated K type thermocouple wire from Omega Engineering with the thermal junction enclosed with 12 mil thick transparent polyurethane film and adhered with a silicone pressure sensitive adhesive (Fig. 9).

The weather station consists of a Vaisala WXT-520 weather transmitter and an Apogee SP-230 pyranometer (with an accuracy of ±5 %) connected to a Raspberry Pi model B+ using a custom board with a RS232 interface and a Texas Instruments ADS1115 16-bit analog to digital converter (Fig. 10). The Vaisala WXT-520 has manufacturer stated accuracy for wind speed and direction of ±3 %, ±0.3 °C for temperature, ±3 % for relative humidity from 0 to 90 % RH and ±5 % from 90 to 100 %, and ±5 % for rainfall.

The data from the sensors is generated continuously and logged in 5-min intervals to a MySQL database. To complement this sensor-derived data, a web app gathers building occupant feedback with regard to comfort and correlates each datum with the effective temperature at the corresponding time and location. These subjective polling results are being considered together with objective data when evaluating the impacts resulting from modifications to building operations.

Fig. 10 Rooftop-mounted weather station

Fig. 9 Installed glazing surface temperature sensor

Experiment 1—Thermodynamics of a Raised Floor System

The raised floor system is designed to counteract the thermal asymmetries that would naturally persist in winter as conductive losses through the building perimeter drive exterior wall surface temperatures well below the interior air temperature. Whereas the bulk of the raised floor system delivers neutral air to meet fresh air requirements, a set of ducted perimeter zones (Fig. 11) provide heated air to elevate the nearby operative air temperature to normal interior levels.

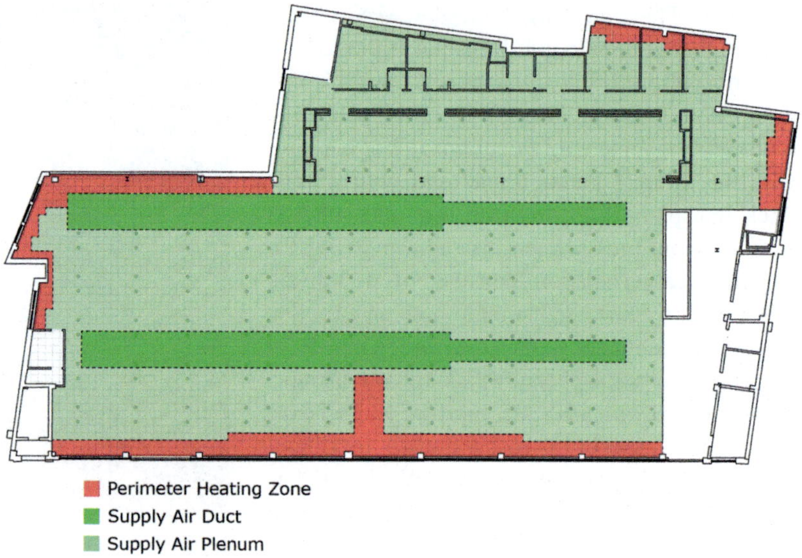

Fig. 11 Plan of raised floor system, indicating perimeter heating zones, supply air ducts, and supply air plenum

Experiment data from March 2015 allows us to draw several initial conclusions about the nature of heat diffusion, which also has implications for summertime conditioning.

Our ability to rely on night-flushing to pre-cool the slab in summertime will depend on the rate of thermal diffusion between the duct air and the slab surface, which also reflects the slab's capacity to reduce the supply air temperature during operation, while the radiative heat exchange between tile surface and slab surface will further support the slab's function as a summertime heat sink. In order to study these coupled thermal reservoirs, we monitored each of the components of the underfloor system (Fig. 12) in 5-min intervals during a typical heating period.

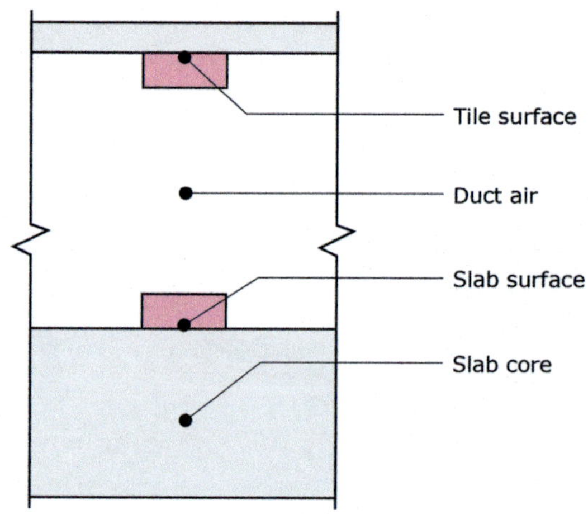

Fig. 12 Schematic of raised floor system monitoring

As the slab core and tile surfaces are in constant heat exchange with the occupied spaces above and below (which are not presently monitored), we treat each of these as a boundary condition. The supply duct air is determined by the state of the mechanical system and is therefore treated as a third boundary condition. As the slab surface has no other thermal reservoirs with which to exchange heat, we would expect its temperature profile to be determined exclusively by these three boundary conditions.

According to the conductive heat transfer equation, we should expect a linear relationship between the rate of temperature change in the slab surface and the instantaneous temperature difference with respect to slab core and duct air. Assuming normally observed temperature deltas (<30 °C), we should also expect the first-order expansion of the radiative heat transfer equation to be a good measure of radiative heat exchange with the floor tile. Therefore, the change in the slab surface temperature over some finite time interval ought to be well-represented by the following linear equation, such that the error term (ε) should be relatively small.

$$\frac{\Delta T}{\Delta t} = \beta_{air}(T_{air} - T) + \beta_{tile}(T_{tile} - T) + \beta_{core}(T_{core} - T) + \varepsilon \quad (1)$$

The terms in parentheses represent the instantaneous temperature difference with respect to thermal reservoirs that are in heat exchange with the slab surface, and the coefficients imply the rapidity that the slab surface will approach equilibrium with each of those bodies. While these parameters will be empirically determined, they are in effect a set of macroscopic analogs to thermal diffusivity. Each term will depend on the heat transfer coefficient between the slab surface and adjacent body as well as on the heat capacity of the slab surface itself.

Determination of the coefficient terms requires that we build up a system of linear Eq. (2) that can be represented in matrix form (3), where each row represents a set of measurements at a particular time.

$$\begin{pmatrix} y_1 \\ \vdots \\ y_n \end{pmatrix} = \begin{pmatrix} x_{air,1} & x_{tile,1} & x_{core,1} \\ \vdots & \vdots & \vdots \\ x_{core,n} & x_{tile,n} & x_{core,n} \end{pmatrix} \begin{pmatrix} \beta_{air} \\ \beta_{tile} \\ \beta_{core} \end{pmatrix} + \begin{pmatrix} \varepsilon_1 \\ \vdots \\ \varepsilon_n \end{pmatrix} \quad (2)$$

$$\boldsymbol{y} = \boldsymbol{X\beta} + \boldsymbol{\varepsilon} \quad (3)$$

Provided the number of observations (p) exceeds the number of parameters (n), we can determine the set of parameters that best fits the data via an ordinary least squares regression, using Eq. 4.

$$\hat{\boldsymbol{\beta}} = (\boldsymbol{X}^T\boldsymbol{X})^{-1}(\boldsymbol{X}^T\boldsymbol{y}) \quad (4)$$

In order to obtain a good fit for Eq. 3, we sampled data over a 36-h period in late March when the heating system was in full operation. Due to the rapid cycling of the perimeter supply air system, in intervals of roughly 25 min, we are able to observe responses in floor tile and slab surface temperature that are clearly independent of the diurnal cycle.

Applying Eq. 4 to the measured data (TestT; TestA;TestS;TestC) shown in Fig. 13, we obtained the following coefficients.

$$\hat{\boldsymbol{\beta}} = \begin{pmatrix} \beta_{air} \\ \beta_{tile} \\ \beta_{core} \end{pmatrix} = \left(\frac{1}{5\,min}\right) \begin{pmatrix} 0.014698 \\ 0.100395 \\ 0.600508 \end{pmatrix} \quad (5)$$

Using the slab surface temperature at the start of the time interval as an initial condition, we applied these coefficients to Eq. 1 in order to plot the predicted slab surface temperatures alongside measured temperature. As Fig. 13 shows, this yielded an exceptionally good correlation, with a root mean square deviation of 0.069 °C, well within published accuracy of the sensors.

These empirically-determined coefficients provide a direct indication of the time scale for two heat thermal reservoirs to tend toward equilibrium, as we find by inverting the coefficients.

$$\begin{pmatrix} {\beta_{air}}^{-1} \\ {\beta_{tile}}^{-1} \\ {\beta_{core}}^{-1} \end{pmatrix} = \begin{pmatrix} 340.2\,min \\ 49.8\,min \\ 8.3\,min \end{pmatrix} \quad (6)$$

It is of little surprise that the slab surface and slab core operate on such a short time scale, as they are in direct thermal contact via a homogenous, uninsulated concrete medium. The long time scale of air-slab heat diffusion places formidable limitations on our ability to pre-cool the slab at night. The ideal time for flushing would be in the late night and early morning hours when temperatures reach their diurnal minimum. However, on humid days, a dehumidification cycle will be required before occupancy to remove the additional latent heat introduced in the night-flush. The remaining time for flushing may not be sufficient to remove enough sensible heat to make effective use of the slab as a heat sink through the late afternoon, when the building is subjected to the greatest thermal loads. The long time scale of the air-slab diffusion also places limits on the temperature drop we could expect to

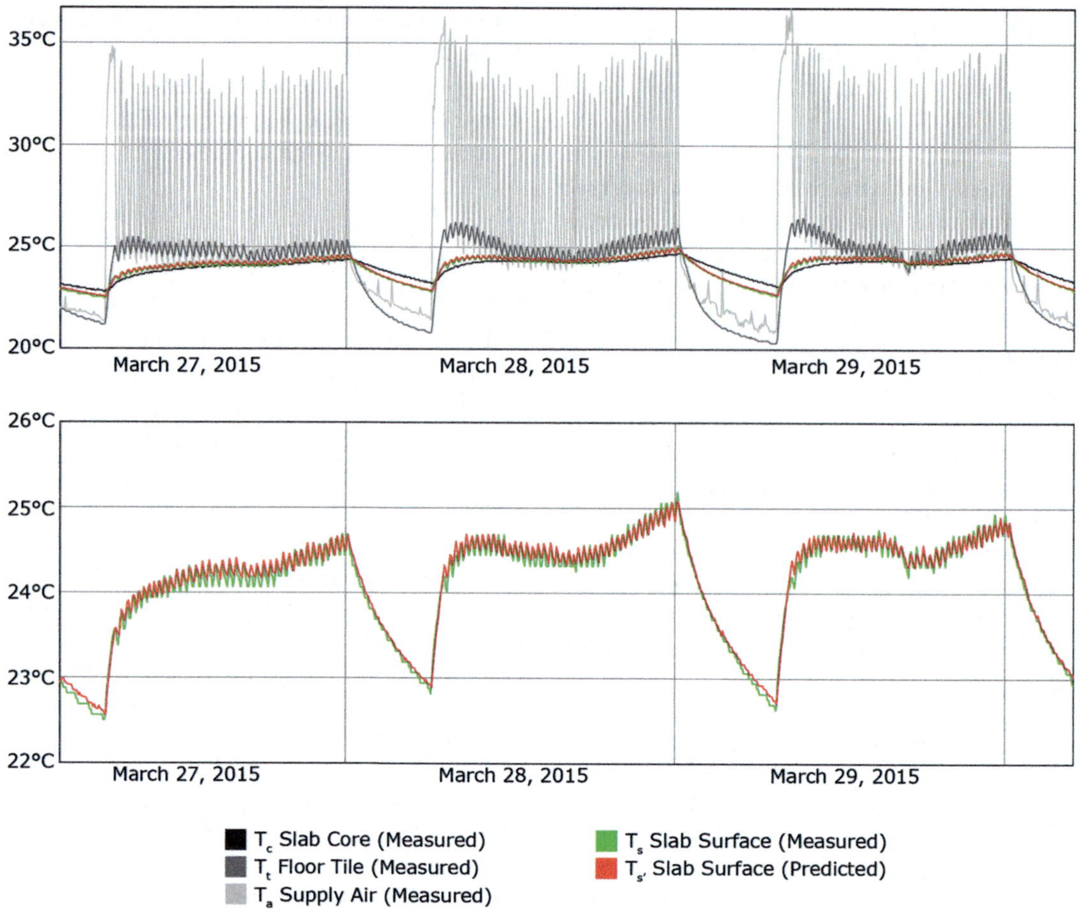

Fig. 13 Temperature profiles for raised floor system measured over a 3-day interval (*above*) and detail showing close correlation of predicted and measured results for slab surface temperature (*below*)

achieve in displacement air as it passes over the floor slab before entering the occupied zone.

Experiment 2—Mechanically Assisted Exhaust and Thermal Comfort

The large volume of the main office level presents opportunities for multiple conditioning modes, as discussed earlier. Under the active cooling mode (Fig. 4), in which monitor venting is limited to overpressure relief, we anticipate that a stratified zone of warmer air will form in the truss space. Over the course of a hot day, the stratified zone will likely reach high enough temperatures to adversely impact the occupied zone beneath. Except under extremely hot, humid conditions, a rapid purge using the monitor fans and operable windows (Fig. 2) may restore the capacity of the upper zone to absorb additional hot air.

In order to test how rapidly the monitor exhaust can vent the stagnation zone, we shut down the building systems to allow warm air to stagnate until the occupied zone reached temperatures that were considered unacceptable. All 28 operable windows were then opened and the monitor exhaust fans set to 100 % power until new equilibrium conditions were met, at which point the previous state was restored and the experiment repeated.

During this time, the ambient air temperature was monitored at four locations across the center of the office. Each location included sensors at 5, 10, 16, and 35 feet above the finish floor to capture the occupied zone and the effects of stratification. Occupants were also asked to report their thermal comfort in hourly intervals through an online survey.

As one would expect, the findings (B-35; B-16; B10; B-05; C-35; C-16; C-10; C-05; D-35; D-16; D-10; D-05; E-35; E-16; E-10; E-05) show positive correlation between temperature and elevation and close correlation among sensors at the same elevation (Fig. 14). Following the change in system state from closed mode to open mode, we see an immediate and rapid decrease in all sensor readings over a 20-min period, followed by a 40-min leveling off as the air volumes approach steady state conditions. The duration of the rapid decline is well within reason given an exhaust rate of 36,000 CFM and an approximate floor area of 20,000 ft^2.

Despite the modest drop in temperature at all elevations, there was not a corresponding convergence of temperature readings across different elevations, indicating that stratification persists despite the rapid change in air volume. This suggests that the internal and envelope loads produce sufficient heat to maintain stratification, which is confirmed by the rapid rise in temperature following the return of the system state to the closed mode. While this study has led us to

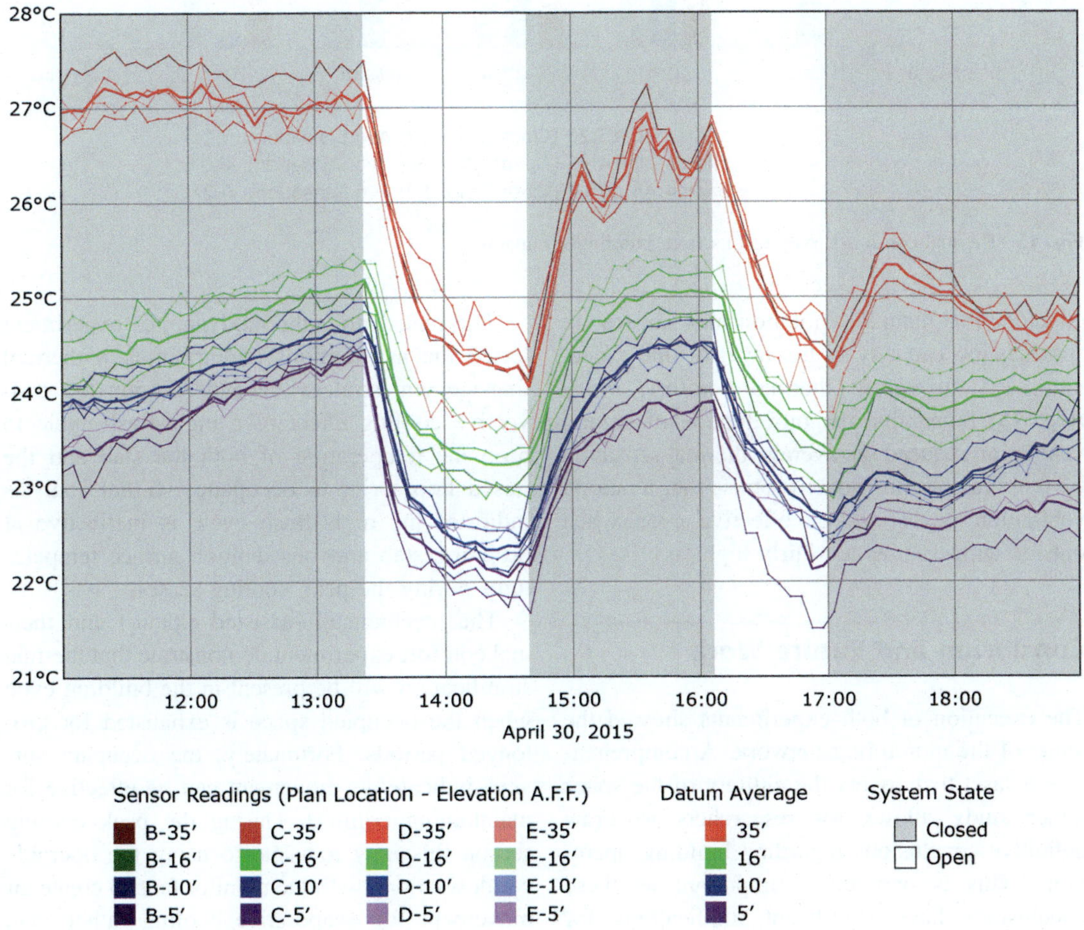

Fig. 14 Temperature profiles for ambient air sensors at multiple elevations

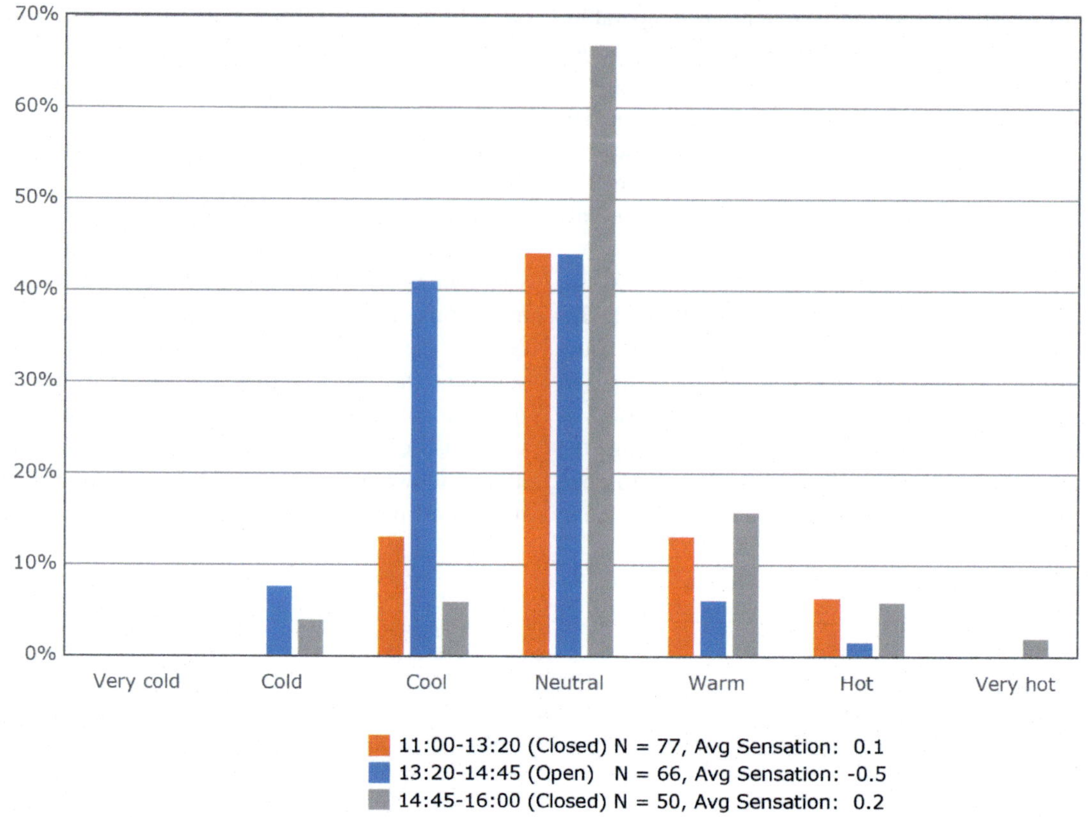

Fig. 15 Thermal comfort survey results from building occupants

conclude that natural ventilation may not restore stratification capacity to the level of our expectations, a concurrent thermal comfort survey (Fig. 15) found that the introduction of natural ventilation reduced the average thermal sensation by 0.6, thus offering some promise that a natural ventilation mode will be effective even when outside temperatures are fairly high.

Conclusion and Future Work

The execution of both experiments showed the value of the monitoring network. A comprehensive dataset that covers the entirety of the space under study allows for researchers to draw definitive conclusions regarding building operations. This is particularly important as these conclusions have significant implications for occupant comfort during the cooling season.

The raised floor thermodynamics experiment tested the assumptions underpinning a thermal management strategy dependent upon the convective cooling effect of a night-flush cycle to drop the temperature of both the slab and the raised floor prior to occupancy. Initial analysis indicates the night-flush cycle is ineffective at dropping slab and raised floor surface temperatures during the peak cooling season.

The mechanically-assisted exhaust and thermal comfort experiment demonstrate that thermal stratification will be present in the building even when the occupied space is exhausted for prolonged periods. Fortunately, the occupant surveys indicate air movement can be effective for maintaining comfort. During the peak cooling season this may translate to using the operable windows along with the monitor fans to create air movement for evaporative cooling rather than sealing the building and relying exclusively on latent cooling through the dehumidification system.

From a building operations perspective, the sensor platform has been incredibly important in framing appropriate expectation of building system performance, particularly the results of the raised floor thermodynamics and the apparent lack of thermal exchange between the supply air and the floor slab, which was intended to be a core component of the summer cooling strategy. The high density monitoring, coupled with surveys, shortened the time necessary to evaluate the effectiveness of the thermal management strategies. For example, the finding that air movement was a central driver for comfort or discomfort led us to alter our modeling and simulation workflows to include this parameter.

Looking toward design applications for monitoring, the data generated from this sensor network will be directly relevant to analyzing the viability of similar passive and mechanical systems in future buildings. The information will also be used to refine the assumptions used in our building energy models, particular with regards to temperature set points, the appropriate weighting of ambient versus mean radiant temperature, and the role air movement plays in occupant comfort.

References

B-35. http://test.kierantimberlake.com/DMS/WSN_SENSOR_11_SAMPLES_1430409600-1430438400.json. Accessed 30 Apr 2015

B-16. http://test.kierantimberlake.com/DMS/WSN_SENSOR_435_SAMPLES_1430409600-1430438400.json. Accessed 30 Apr 2015

B10. http://test.kierantimberlake.com/DMS/WSN_SENSOR_429_SAMPLES_1430409600-1430438400.json. Accessed 30 Apr 2015

B-05. http://test.kierantimberlake.com/DMS/WSN_SENSOR_428_SAMPLES_1430409600-1430438400.json. Accessed 30 Apr 2015

C-35. http://test.kierantimberlake.com/DMS/WSN_SENSOR_19_SAMPLES_1430409600-1430438400.json. Accessed 30 Apr 2015

C-16. http://test.kierantimberlake.com/DMS/WSN_SENSOR_426_SAMPLES_1430409600-1430438400.json. Accessed 30 Apr 2015

C-10. http://test.kierantimberlake.com/DMS/WSN_SENSOR_427_SAMPLES_1430409600-1430438400.json. Accessed 30 Apr 2015

C-05. http://test.kierantimberlake.com/DMS/WSN_SENSOR_425_SAMPLES_1430409600-1430438400.json. Accessed 30 Apr 2015

De Dear RJ, Brager GS (2002) Thermal comfort in naturally ventilated buildings: revisions to ASHRAE standard 55. Energy Build 34:549–561

D-35. http://test.kierantimberlake.com/DMS/WSN_SENSOR_28_SAMPLES_1430409600-1430438400.json. Accessed 30 Apr 2015

D-16. http://test.kierantimberlake.com/DMS/WSN_SENSOR_422_SAMPLES_1430409600-1430438400.json. Accessed 30 Apr 2015

D-10. http://test.kierantimberlake.com/DMS/WSN_SENSOR_423_SAMPLES_1430409600-1430438400.json. Accessed 30 Apr 2015

D-05. http://test.kierantimberlake.com/DMS/WSN_SENSOR_424_SAMPLES_1430409600-1430438400.json. Accessed 30 Apr 2015

E-35. http://test.kierantimberlake.com/DMS/WSN_SENSOR_42_SAMPLES_1430409600-1430438400.json. Accessed 30 Apr 2015

E-16. http://test.kierantimberlake.com/DMS/WSN_SENSOR_421_SAMPLES_1430409600-1430438400.json. Accessed 30 Apr 2015

E-10. http://test.kierantimberlake.com/DMS/WSN_SENSOR_420_SAMPLES_1430409600-1430438400.json. Accessed 30 Apr 2015

E-05. http://test.kierantimberlake.com/DMS/WSN_SENSOR_419_SAMPLES_1430409600-1430438400.json. Accessed 30 Apr 2015

Nicol JF, Humphreys MA (2002) Adaptive thermal comfort and sustainable thermal standards for buildings. Energy Build 34:563–572

TestT. Floor Tile. http://test.kierantimberlake.com/DMS/WSN_SENSOR_154_SAMPLES_1427438414-1427708329.json. Accessed 15 Apr 2015

TestA. Supply Air. http://test.kierantimberlake.com/DMS/WSN_SENSOR_155_SAMPLES_1427438414-1427708329.json. Accessed 15 Apr 2015

TestS. Slab Surface. http://test.kierantimberlake.com/DMS/WSN_SENSOR_156_SAMPLES_1427438414-1427708329.json. Accessed 15 Apr 2015

TestC. Slab Core. http://test.kierantimberlake.com/DMS/WSN_SENSOR_160_SAMPLES_1427438414-1427708329.json. Accessed 15 Apr 2015

Keeping an Eye Out: Real Time, Real World Modeling of Behavior in Health Care Settings

Christopher Beorkrem, Steve Danilowicz, Eric Sauda, Richard Souvenir, Scott Spurlock and Donna Lanclos

Abstract

Imagine a health care facility that is able to track and understand the meaningful behaviors of the patients 24 h a day, 365 days a year, understanding individual variation in behavior over both the short and the long term. Now consider the needs of patients with Alzheimers, who typically have trouble with spatial and visual issues. They are sometimes unable to distinguish between a shadow cast on the floor and a step; they can also spend the entire day at the "front door" anticipating arrivals and departures. The families of these patients, to the best of their ability, want to be able to maintain surveillance and understand the changes in the behavior of their parents or spouses. Continuing on the work of the Computing in Place research group that includes faculty with specialties in architecture, computer vision, ubiquitous computing and anthropology, we propose in this paper a new paradigm for intelligent architectural settings. Health care settings, like most architecture, are generally conceptualized as a spatial volume containing human and technical elements. There is an implicit distinction between the active contents and the passive container. From our research group's expertise in ethnography, we emphasize the importance of meaning to the understanding of behavior, to the idea of place as a construed setting: or, as Clifford Geertz describes it, the difference between "a wink and a blink". Our new paradigm proposes the creation of "intelligent" architectural settings that

C. Beorkrem (✉) · S. Danilowicz · E. Sauda
School of Architecture, University of North Carolina Charlotte, Charlotte, USA
e-mail: cbeorkrem@uncc.edu

R. Souvenir · S. Spurlock
Department of Computer Science, University of North Carolina Charlotte, Charlotte, USA

D. Lanclos
University Library, University of North Carolina Charlotte, Charlotte, USA

M.R. Thomsen et al. (eds.), *Modelling Behaviour*, DOI 10.1007/978-3-319-24208-8_38

capture such meaningful behavior in real time and generate knowledge that is useful both in the real world and in the evaluation of design revisions.

Background

This project began as a collaboration including the development of a computer vision based method for the collection and analysis of ethnographic data. The collaboration included architects, computer scientists and anthropologists. The architects were primarily interested in a deeper understanding of the use of space, the computer scientist with how an understanding of human subjects could be made more sophisticated, and the ethnographers with how their understanding of place could be semi-automated and extended to long time frames. As technology continually shifts so do methods for creating this kind of long-range time scale analysis of spaces and uses. A few recent examples include work AECOM, which started an entire working group called Strategy Plus, which is dedicated to performing this type of analysis. To date they have experimented with a variety of methods using RFID badges as well as intricate room schedules. Additionally, CASE (Davis and Payne 2015) has used Bluetooth Beacons (Estimotes) to analyse the movement of individuals through space. In this case the Beacons function to a certain extent, but require participants to have their phones or devices to be linked to the Beacons during the analysis. In each case the participants are required to be engaged with the analysis to a certain extent and anonymity can not be guaranteed.

Method

First we developed the SENSING Toolkit, framework for storing and providing access to long term, large-scale human behavior data collected from a variety of sensors. Visualization tools generally consider motion lasting only minutes; analysis and visualization of data for extended time scales require new approaches. This required significant advances in the ability to use multiple cameras for the tracking of people in architectural space, as well as the ability to track multiple occupants, including simple pose recognition and orientation. The system can currently track an unlimited number of occupants in a very compact form, allowing us to keep very extended records of human motion lasting weeks, months or even years.

As an extension of the SENSING toolkit, we developed VALSE (Visual Analytics for Large-Scale Ethnography) application, which allows an expert to view avatars representing real-time motion and also visualizations of motion summaries, all controlled using familiar DVR-style interface (Fig. 2). Because VALSE is an interactive, visual analytic approach to the study of human behavior in architectural settings, it requires robust algorithms for human tracking and activity analysis from video and usable interfaces for domain experts to query and understand the data. Our system combines tracking and activity recognition methods into an interactive, real-time system for visual data analysis in the study of motion behavior (Fig. 1).

The current VALSE interface combined a time line visualization with spatial visualizations overlain with the motion of occupants. The visualizations are linked and open to exploration by ethnographers or other domain experts. A critical part of the VALSE interface is the methods of identifying meaningful behavior as the intersection of time, people (one or more) and objects. Once a meaningful event has been identified, the system can search for all incidents of this behavior over very long time periods. This approach leverages the ability of human

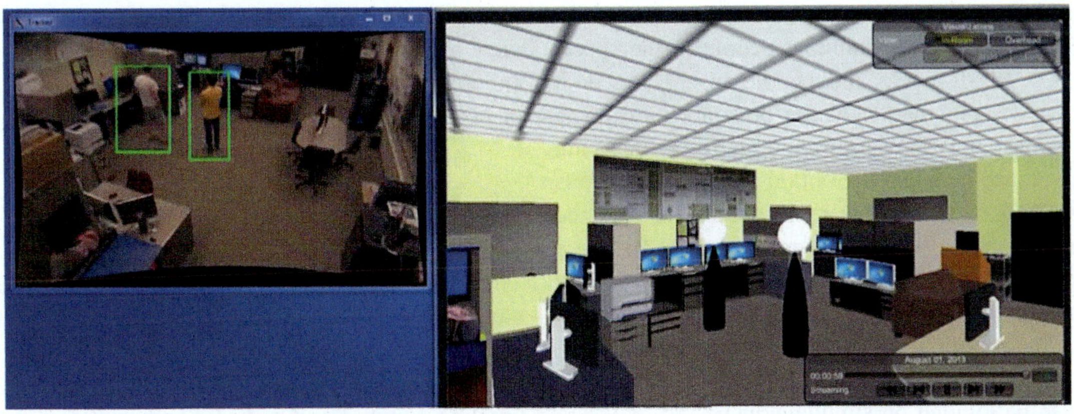

Fig. 1 Camera view of setting and avatar representation. Multiple camera views are resolved into a *single model* for analysis. Note the anonymous nature of the avatar view

ethnographers to recognize meaningful patterns of behavior with the computational ability to apply these insights for very extended periods of time.

VALSE Interface

As a whole, the current VALSE system (Fig. 2) was modelled from a combination of two different types of tools: drawing and annotation software. The drawing software (Adobe Illustrator, Rhinoceros, and AutoCAD) helped to lend methods for organization of elements as well as creation and selection processes. The annotation software (ELAN annotation, VCode, and Morae) lent itself to the control of the timeline and recording events. The expert interacts with VALSE through the Interaction Window with two main portions: the Visualization Panel and the Event Timeline. The Visualization Panel is the region that displays spatial information about the observed area, and the Event Timeline shows and allows the expert to interact with a list of the events that have occurred within the space.

Visualization Panel

The ***Visualization Panel*** is a spatial view, either a floor plan with composite data gathered from the cameras capturing movement or a series of snapshots taken by those same cameras.

The pathways of the people moving through the space can be displayed in multiple ways chosen by the expert. The default display for these paths is the ***dot trail***. This type of representation shows the path of a person as a series of symbols plotted on the floor plan. These symbols are colored "x's", circles, squares, or triangles

Fig. 2 Diagram of the system architecture. The SENSING Toolkit will provide an API for applications to analyze and respond to activity in large spaces. VALSE is built upon the SENSING Toolkit provides tools for visualizing and annotating motions

and can be changed by the expert to help differentiate between multiple people in the space. As VALSE tracks the people in the space, it places a dot marker at the person's location at regular intervals. To provide clarity to that path issue, we included the option to view these trails as lines as well. This representation shows the expert how the person moved through these confusing moments, but does not give the expert an idea of direction or speed. Both of these trail display options may be toggled on or off depending on the experts preference. They may also be toggled on simultaneously.

In addition to the dot trails, ***heat maps*** (Fig. 3) can display the aggregated data of the population in the space over selected time-periods. This data may be shown to the expert as population density, travel velocity, or event hotspots. Population density heat maps shows where people tend to move through or gather and can be useful for large sets of dot trails. Velocity heat maps identify bottlenecks or thoroughfares within the space. Event hotspots heat maps shows the expert the frequency of events within the space over the selected time.

In addition to the two abstracted views presented in the Visualization Panel, there is a third option that shows ***snapshots*** taken by the cameras of the selected location. The Snapshot view show multiple video images to provide the expert with more context than an abstracted trail or heat map.

Event Timeline

The ***Event Timeline*** (Fig. 4) allows the expert to navigate the information presented to the expert from the Visualization Panel. It is divided into two main sections: the control toolbar and the timeline panel. The ***Control Toolbar*** allows the expert to make changes to the selected time as well as playback the recorded information. The ***Datetime*** function on the toolbar allows the expert to specify a precise timeframe. It allows the expert to examine single moments or recurring time spans within the recorded data set; for example, every Monday from 10:00–11:00 am. The toolbar also includes playback controls that allows the expert to navigate the selected time period at a variety of speeds both forward and backwards.

The ***timeline*** panel shows events that are identified by either the system or an expert. The timeline has two sorting options for displaying the events. The first is ***sorting by person*** (Fig. 5).

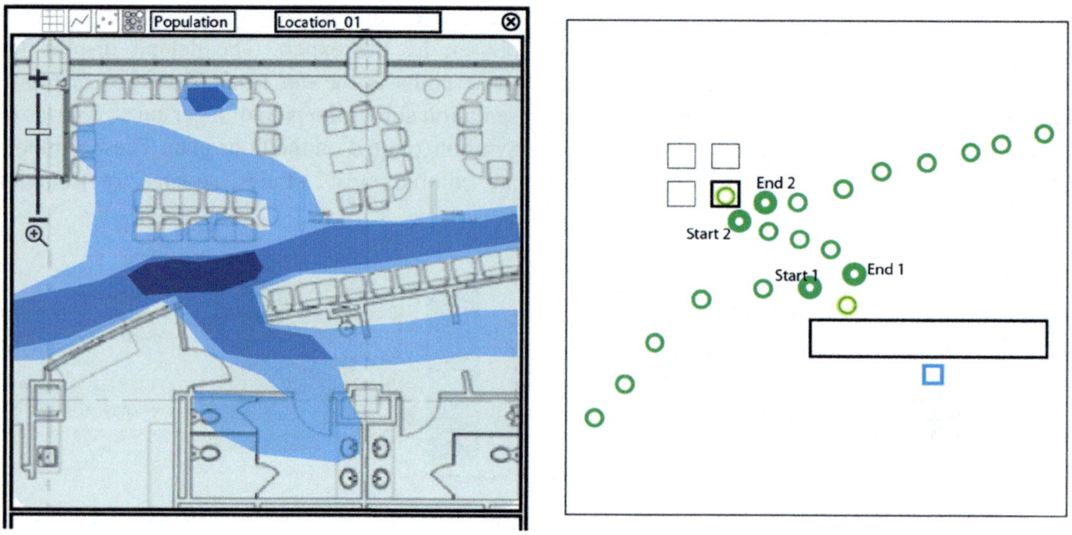

Fig. 3 *Left* Heat map view. *Right* Event selection view

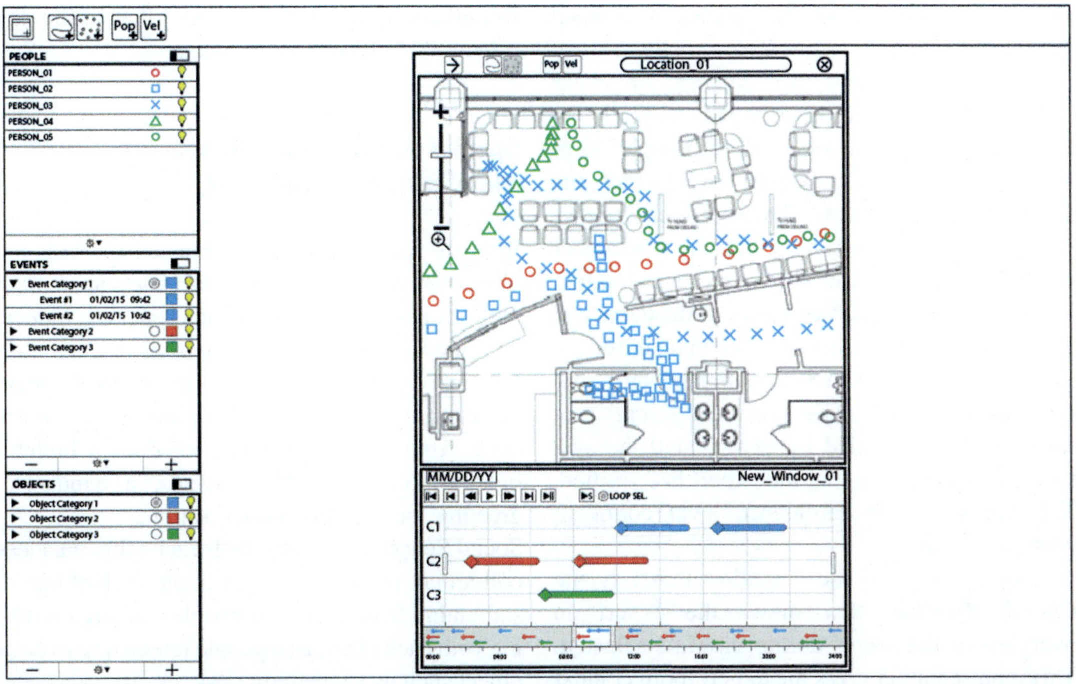

Fig. 4 Prototype of the VALSE interface. Note the use of multiple window which present multiple representations of the data from the setting. This interface is fully interactive and discoverable with the ethnographers studying the setting

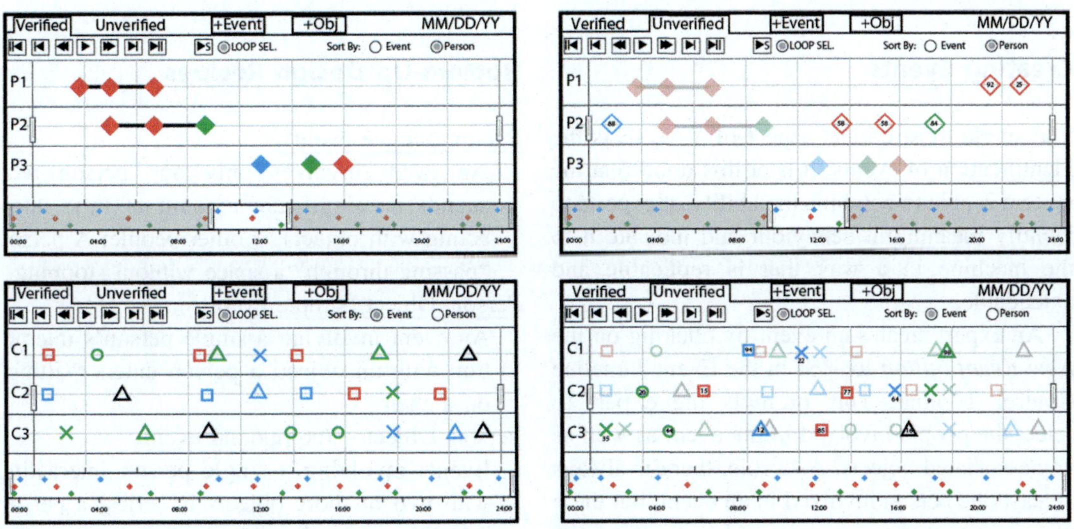

Fig. 5 Timeline sorting options: sorting by event and sorting by person

This method sorts the list of people in the space vertically and displays when an event occurs by placing a colored marker at the appropriate point in the timeline. Events are indicated as either simple or compound. Simple events are independent of other events in the space. Compound events are made of multiple simple events chained together, much like a compound sentence is made of smaller sentences chained with conjunctions.

The second option for ***sorting by event*** (Fig. 5). The timeline is arranged vertically by event category with markers to for people. Representing events in this manner only displays the events as simple events. An expert can understanding how frequently a specific event occurs or when these events occur in relation to each other.

In addition to these two views of the events on the timeline, there are two tabs that allow the expert to switch between ***Verified Events*** and ***Unverified Events***. The Verified Events tab shows events identified by experts and the system generated events that the expert has verified. This tab show the confidence level confident identified events.

The last part of the Timeline Panel is the ***Overall Timeline*** that shows the expert an overview of the events during the time selected. This representation gives the expert more context about the people in the environment, and helps the expert determine which direction the people are moving within the space.

Creating Events

The single most critical aspect of VALSE is the identification of events. It is at this point that the system is able to combine the ability of experts to identify meaningful behaviour and indicate it to the machine in a way that is replicable and extendable.

An expert creates an event by clicking on the ***New Event button*** located in the Event Timeline Control Toolbar. This prompts the expert to select the people involved in the event as well as any associated objects. As a rule, there is always at least one person involved in an event, but there may be no objects involved. Along side of the top-down approach of teaching the system by example through machine learning, we are also looking to approach the design of VALSE from a bottom-up method as well. To do this, we've formulated a series of "recipes" that utilize the principles of spatial proxemics.

Spatial Proxemics in Event Identification (Figs. 6, 7)

Intimate Space: directly interacting with an object or person: sitting in a chair, interacting with a desktop, washing hands, opening a door, hugging or kissing another person.
Personal Space: interacting with another person on a more colloquial level or observing something from a short distance: reading a bulletin board, a couple holding hands, a handshake greeting, or a hushed conversation.
Social Space: interacting with each other on a less colloquial level, or if a person is looking at something from afar: two people walking or talking with each other, two people interacting over an object such as a reception desk.
Public Space: maintaining a distance around a person; a possible indication of no intentions of interaction.

Bottom-Up Design Recipes

Person (simple event)
: An event involves only one person who spends no significant amount of time interacting with objects or other people; A person "passing through" a space without stopping.

Person–Object (simple event)
: An event involving a single person's interaction with an object; a person enters "sitting" on a chair.

Person–Object + (compound event)
: Events involving a single person interacting with two or more objects in series; a person "checking in and waiting" at the receptionist desk and a chair.

Person+–Object (simple event)
: Events involve two or more people and a single object; two people with the receptionist

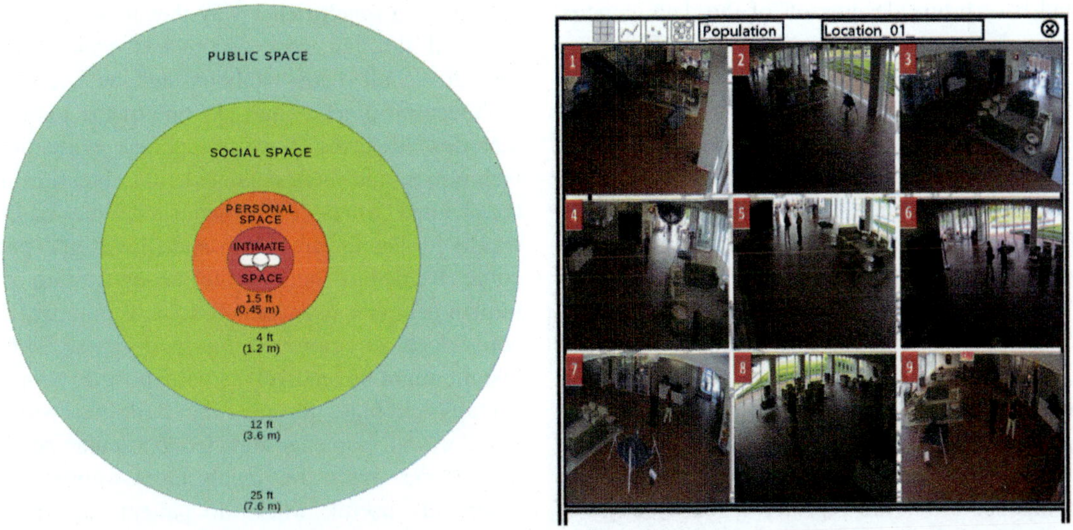

Fig. 6 *Left* Proxemics space diagram used in analysis of possible events. *Right* Snapshot view for closer analysis of meaningful behavior

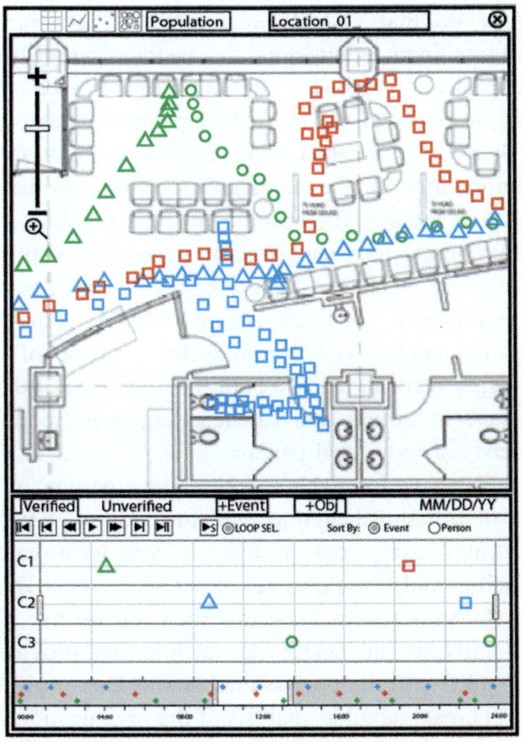

Fig. 7 Spatial view and events mapped on person timelines

desk between them in "conversation". This could be identified as "question".

Person–Person+ (simple event or compound event)

Event involving two or more people interacting in proximity for an extended period of time; a "conversation" or "meeting".

Applications for VALSE System

Our buildings are becoming smarter; imbedded sensors and technology are allowing us to gather data on the way our buildings are functioning. VALSE is an example of the potential for intelligent architecture. Such a system would give architects and building managers a detailed set of observational data about a space that could not otherwise be obtained. Traditionally, observational data collected about a building within a Post Occupancy Evaluation is done through survey and observation. The observation portion may last from between a week to a month, and then is extrapolated over the course of months or years to give architects and companies an idea of how their building is functioning. Architects use this information within Evidence Based Design, and base future designs off of the data gathered from these observations. With VALSE, these observational data points are collected continually, and are not limited to a single week or month. This constant feedback gives architects a more detailed look at the space and can improve the future design decisions made by the designer. Building management can also use this information to help determine ways to improve the functionality of the built environment.

VALSE is best suited to building typologies that use a defined kit of parts in their design. These could be buildings such as schools, hospitals and healthcare facilities, retail environments, and offices. These buildings all have components that are repeated in each design of the building. Schools all have classrooms, offices, and cafeterias; healthcare has waiting rooms, patient rooms, and operating rooms; retail has the sale floor; and offices have a variety of cubicle layouts as well as individual room layouts. These are only a few examples from each, but incorporating VALSE into a building to observe any type of repeatable space can yield observational data that may have otherwise gone uncollected.

Future Work: Health Care Settings

An intelligent architectural setting could be useful in a wide variety of situations, but our focus on health care facilities has several advantages. First, the privacy concerns raised in many situations are less important, as surveillance is precisely what the patient and their family are anticipating and expecting. Second, activities in health care facilities by both the staff and patients are usually highly organized. Whether it is the operation of a surgical suite or the rules and constraints on the patient and their visitors, meaningful patterns of behavior are often very distinctive, repetitive and identifiable in these settings.

We expect two specific outcomes of this research. First it allows for a fine-grained understanding of the specific uses of existing layouts in a much richer way that it is currently gathered through post occupancy evaluations. Indeed health care facilities are an area of architectural practice that has one of the longest histories of using post occupancy evaluation techniques, and there is considerable pressure in the United States to provide outcome based results for health care. Many specific insights can come from this large scale understanding of human behavior over long periods of time which could lead to improved future designs. Some insight might be relatively simple; how do people actually circulate through the building, which architectural elements attracted particular activity? Some might be much more subtle and potentially useful; where do patients and their family gather to console or talk with each other? What kind of spaces attract the most meaningful small scale conversations?

An even more transformative outcome is the potential that VALSE holds to create new form of intelligent architectural that become part of the therapeutic setting, allowing for immediate adjustments to an individual patient's therapy. There is a great deal of literature about the importance of the architectural setting as part of the therapy for specific medical conditions. But we are proposing a more general conceptualization that the ability to create intelligent architectural settings using mixed expert/computational systems holds the promise of a setting that can react to both the dynamics of the usage of groups and can react uniquely to the needs of individual patients.

We have formed a new three-year partnership with a national architectural firm to study two settings using our system; a long-term care facility and a complex set of surgical suites. The architectural firm is using their client list to identify test site health care facilities and provide contact with the professional staffs. This project will create tangible analysis of the organizational systems implemented in clinic or hospital floor designs. Rather than rely on a short-term analysis by an ethnographer or partial semantic data from occupants, the computer vision toolkit creates a comprehensive analysis of the actual use and performance of each hospital floor or clinic. In addition, it will be a pilot study of the possibilities analysis of the behavior of individual patients linked to their medical prognosis.

References

Akdemir U, Turaga P, Chellappa R (2008) An ontology based approach for activity recognition from video. In: 16th ACM international conference on multimedia 2008. ACM

Beorkrem C, Sauda A (2008) Architecture user interface: towards a performative architecture. In: 2008 EAAE/ARCC conference: architectural research and the digital world, Copenhagen

Davis D, Payne A (2015) How beacons will change architecture, http://www.case-inc.com/blog/beacons-indoor-positioning-in-architecture. Accessed 7 Apr 2015

Deng J et al (2009) Imagenet: a large-scale hierarchical image database. IEEE

Ghanem N et al (2004) Representation and recognition of events in surveillance video using petri nets. In: Computer vision and pattern recognition workshop. IEEE

Hakeem A, Shah M (2004) Ontology and taxonomy collaborated framework for meeting classification. In: Pattern recognition. IEEE

Heierman E, Cook DJ (2003) Improving home automation by discovering regularly occurring device usage patterns. In: Proceedings of the international conference on data mining

Miller GA (1995) WordNet: a lexical database for English. Commun ACM 38(11):39–41

Nevatia R, Hobbs J, Bolles B (2004) An ontology for video event representation. In: Computer vision and pattern recognition workshop. IEEE

Noy NF (2004) Semantic integration: a survey of ontology-based approaches. ACM Sigmod Rec 33 (4):65–70

Energy Efficiency Assessment Based on Realistic Occupancy Patterns Obtained Through Stochastic Simulation

Lavinia Chiara Tagliabue, Massimiliano Manfren and Enrico De Angelis

Abstract

Occupant's behaviour determines a significant level of uncertainty in building energy use. The relevant "performance gap" usually encountered between energy performance simulation with standard design conditions and actual energy performance is clearly connected to the biased assumptions made, especially for the parameters that can have a large variability. The paper deals with the variability of energy performance determined by occupants' behaviour. The operation parameters related to occupancy profiles, such as air change rates and internal heat gains, are investigated. A specific analysis of indoor air quality is carried out to verify the internal conditions of an education building, chosen as case study, the Smart Campus Demonstrator of Brescia University in Italy. The parameters related to occupancy and used in energy modelling are derived from realistic dynamic patterns generated stochastically. The data envelopment of the stochastic schedules is used to simulate the potential variability in daily and hourly energy consumption due to changing operational patterns and to highlight the "performance gap" with respect to standard simulation settings. The modelling approach proposed is conceived for an initial energy modelling phase, however it can be extended and validated during real time building operation, by implementing coherent performance monitoring and benchmarking practices. The research work constitutes the starting point of a more general activity aimed at integrating inverse modelling techniques in current design practices for building retrofit. The availability of a calibrated and validated building energy model is fundamental to explore accurately further efficiency intervention for the building.

L.C. Tagliabue (✉)
Angelo Luigi Camillo Ciribini,
University of Brescia, DICATAM, Brescia, Italy
e-mail: chiara.tagliabue@unibs.it; chiara.tagliabue@polimi.it

M. Manfren · E. De Angelis
Politecnico di Milano, DABC, Milan, Italy

M.R. Thomsen et al. (eds.), *Modelling Behaviour*, DOI 10.1007/978-3-319-24208-8_39

Introduction

Modelling realistically building energy behaviour is crucial to optimise energy management practices during the building lifespan. However it meets barriers given by factors such as discrepancy between design and as-built data, simulation settings and real parameters, standard operation schedules and actual users' behaviour, etc. (Burrows et al. 2014). A substantial improvement in building operation management and energy efficiency can be achieved considering the following key issues:

- climate;
- building envelope;
- building services and energy systems;
- building operation and maintenance;
- occupants' behaviour;
- indoor environmental quality.

The determination of the real thermal properties of the envelope as well as the correct modelling of the operation of the technical systems are relevant items to calibrate energy models with monitored data. Further, occupants' behaviour has a significant impact on the discrepancy between actual and simulated energy consumption (Baptista et al. 2014; Fabi et al. 2012). Many researches focused on the statistical relations between energy behaviour and environmental parameters shows that when the goals of thermal comfort and energy savings conflict, the users privilege their own comfort, to the detriment of energy efficiency (Malavazos et al. 2013; O'Brian et al. 2013). Comprehensive studies on the interaction between these factors led to guidelines for energy savings and potential opportunities for efficiency (Parys et al. 2014; Eguaras-Martínez et al. 2014). Occupants' behaviour has been found to be one of the more interesting and variable aspect, difficult to predict and to simulate properly (Jahn et al. 2012). Automation systems and sensing can play an important role in understanding the interaction with occupants by contributing detailed information useful to unveil the dynamic operation patterns (Ioannidis et al. 2014; Krinidis et al. 2014). The present study illustrates the application of a methodological approach to simulate building performance variability due to occupancy patterns for an education building. The performance modelling of this building present relevant technical issues which are generally encountered in existing buildings (i.e. uncertainty in thermal characteristics, efficiency of technical systems, inadequate maintenance, high energy consumption, low level of internal environmental quality and highly variable and uncertain occupants' behaviour).

Case Study

The case study is the Brescia Smart Campus Demonstrator, University of Brescia, in Italy, used for lectures and labs. The building aims to provide insights about smart control and optimized building management by detailed data acquisition and modelling. The building has three floors of classroom spaces and a glazed atrium. An external view of the building is depicted in Fig. 1, including the developed

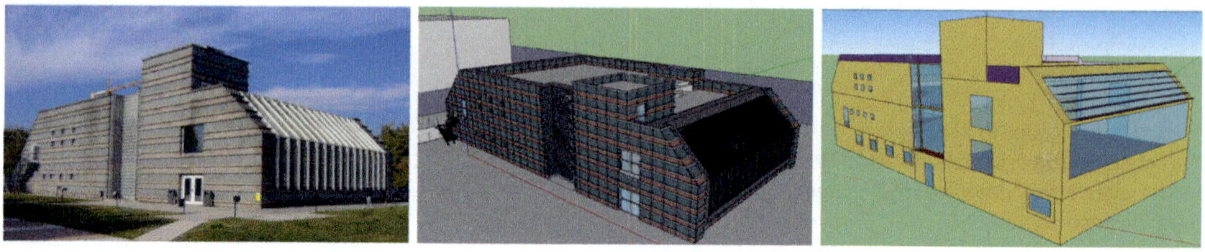

Fig. 1 South-west and south-east façades of the building, BIM model (Autodesk Revit to Sketch-up) and BEM model (Sketch-up to EnergyPlus)

building information model (BIM) and the building energy model (BEM). The BIM model has been generated from geometric data captured with Terrestrial Laser Scanner (TLS). The energy model will be calibrated with real data acquired in the monitoring phase.

Figure 2 shows the internal spaces of the building: the underground floor comprises two computer labs; the ground floor has two classrooms and the first floor consists in a big classroom used for lectures and graduation days. The glazed atrium is used individually by the students.

The building has been analysed to collect information about internal gains showing inadequate conditions in terms of internal air quality (IAQ). At present, only the underground floor has been renovated and the ground and first floors need strong refurbishment (De Angelis et al., in press).

Building Spaces and Ventilation Requirements for Indoor Air Quality

The actual ventilation rate and the ventilation rate to guarantee an adequate IAQ have been calculated for the building spaces (from 8:30 to 19:00). Table 1 resumes the main features of the building spaces, the ventilation requirements and the lacks in terms of IAQ.

Figure 3 shows the indoor air quality gap in the different spaces of the building. The gap is represented by the negative values of outdoor fresh air rates, for two sample days. What appears evident is that the spaces that do not present an air quality gap are the atrium, because of a lower people density and the underground floor labs, that have undergone a retrofit of the AHUs.

Fig. 2 Internal views of the classroom spaces with different uses, occupants' loads, lighting, plugs

Table 1 Features of spaces and classrooms in building and ventilation rates

Location	Definition	Occupancy	Ventilation			
Floor	Name	N. users	AHU flow rate (m^3/h)	Standard air change (m^3/h)	IAQ air change (m^3/h)	ΔV difference standard/IAQ (%)
Underground	MLAB1	56	5000	1639.332	1427.169	13
	MLAB2	82	2000	2245.752	2786.378	−24
Ground	MTA	168	2000	1925.316	4281.507	−122
	MTB	168	2000	1916.784	4281.507	−122
	Atrium	56	5500	1952.424	1902.892	3
First	M1	262	5500	3644.568	6677.113	−83

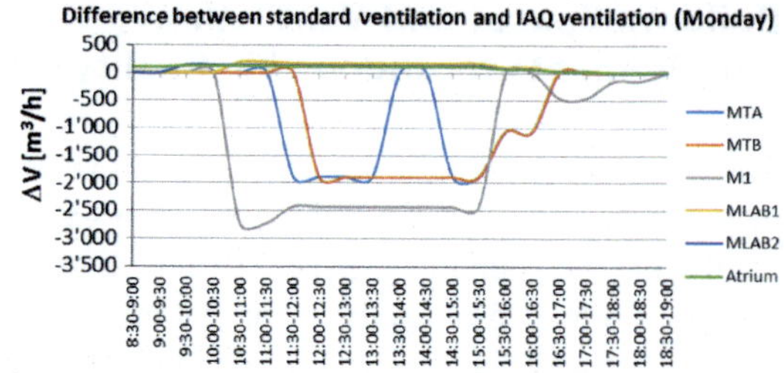

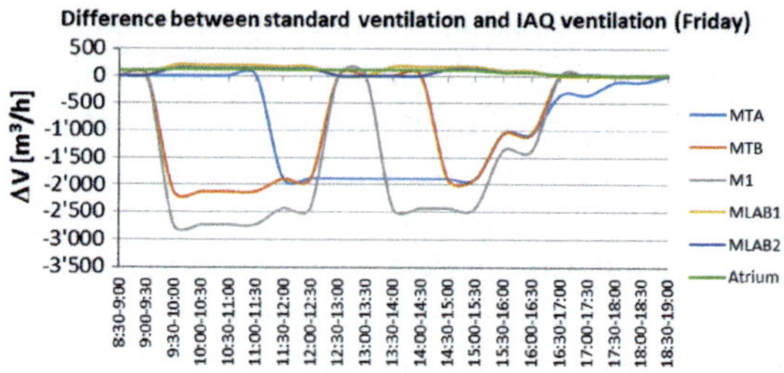

Fig. 3 Difference between actual ventilation rate and ventilation rate to guarantee appropriate indoor air quality on two sample days of operation

Internal Heat Gains Due to People and Equipment

The amount of internal heat gains used in the energy simulation of the building is related to the number of people and to the equipment. Therefore, the calculation of the internal heat gains has been divided into the thermal load due to people and equipment (Table 2).

Occupancy Patterns and Schedule of the Classrooms

In Fig. 4 the weekly occupancy of the classrooms and of the other building spaces is reproduced. The use of the classrooms is intensive but the actual attendance to the lectures has not been directly measured up to now and it will be monitored in the development of the research project.

The use of the spaces during the weekdays is regulated by the schedule of the lectures, thus a 0.1 variable is introduced (Fig. 4). The occupancy patterns have been determined by randomly changing the attendance value with a stochastic simulation process. After that, we considered the envelopment of the data patterns generated, selecting for the performance simulation three profiles for occupancy level:

1. minimum 30 %;
2. average 60 %;
3. maximum 100 %.

The three profiles considered are selected arbitrarily in this initial research phase because of the absence of real metered data. In fact, the methodological approach is conceived to enable a quick assessment of the potential variation of thermal energy demand based on the variation of occupancy patterns, but can be calibrated considering real metered data (Gunay et al. 2014; Aerts et al. 2013). For these reasons, the results obtained for the three profiles are then compared with the ones obtained with standard

Table 2 Specific power for lighting and equipment for the Brescia Smart Campus Demonstrator building spaces

Location	Definition	Lighting and appliances				Equipment	People
Floor	Name	Pls (W/m^2)	Pl (W/m^2)	Pe (W/m^2)	Ps (W/m^2)	Qa (W/m^2)	Qp (W/m^2)
Underground	MLAB1	0.76	15.22	60.95	18.22	95.14	39.84
	MLAB2	0.76	15.22	60.95	18.22	95.14	42.59
Ground	MTA	1.28	25.68	2.50	5.93	35.39	101.78
	MTB	1.28	25.68	2.50	5.93	35.39	102.23
	Atrium	1.40	28.01	1.58	1.98	32.97	33.46
First	M1	2.44	7.34	2.11	8.89	20.79	83.85

Classroom	Week day	8:30-9:00	9:00-9:30	9:30-10:00	10:00-10:30	10:30-11:00	11:00-11:30	11:30-12:00	12:00-12:30	12:30-13:00	13:00-13:30	13:30-14:00	14:00-14:30	14:30-15:00	15:00-15:30	15:30-16:00	16:00-16:30	16:30-17:00	17:00-17:30	17:30-18:00	18:00-18:30	18:30-19:00
MLAB1	M	0	0	0	0	1	1	1	1	1	1	1	1	1	1	1	1	1	1	0	0	0
	T	1	1	1	1	1	1	1	1	1	1	1	1	1	1	1	1	1	1	1	1	0
	W	1	1	1	1	1	1	1	1	1	1	1	1	1	1	1	1	1	1	0	0	0
	T	1	1	1	1	1	1	1	1	1	1	1	1	1	1	1	1	1	0	0	0	0
	F	0	0	1	1	1	1	1	1	0	0	1	1	1	1	1	1	0	0	0	0	0
MLAB2	M	0	0	1	1	1	1	1	1	1	1	1	1	1	1	1	1	1	1	0	0	0
	T	1	1	1	1	0	0	0	0	0	0	0	0	1	1	1	1	1	1	1	1	0
	W	1	1	1	1	1	1	1	1	0	0	0	0	0	0	1	1	1	1	0	0	0
	T	0	0	0	0	0	0	0	0	0	0	0	0	1	1	0	0	0	0	0	0	0
	F	0	0	1	1	1	1	1	1	0	0	0	0	1	1	1	1	1	1	1	1	0
MTA	M	0	0	0	0	0	0	1	1	1	1	0	0	1	1	1	1	0	0	0	0	0
	T	1	1	1	1	1	1	1	1	1	1	0	0	1	1	1	1	1	1	1	1	0
	W	0	0	0	0	1	1	1	1	0	0	0	0	1	1	1	1	1	1	0	0	0
	T	1	1	1	1	1	1	1	1	1	1	0	0	1	1	1	1	1	1	0	0	0
	F	0	0	0	0	0	0	1	1	1	1	1	1	1	1	1	1	1	1	1	1	0
MTB	M	0	0	0	0	0	0	0	1	1	1	1	1	1	1	1	1	0	0	0	0	0
	T	0	0	0	0	1	1	1	1	1	1	1	1	1	1	1	1	1	1	0	0	0
	W	0	0	0	0	1	1	1	1	1	1	0	0	1	1	1	1	0	0	0	0	0
	T	0	0	0	0	1	1	1	1	0	0	1	1	1	1	1	1	1	1	1	1	0
	F	0	0	1	1	1	1	1	1	0	0	0	0	1	1	1	1	0	0	0	0	0
M1	M	0	0	0	0	1	1	1	1	1	1	1	1	1	1	0	0	1	1	1	1	0
	T	1	1	1	1	1	1	0	0	1	1	1	1	1	1	1	1	1	1	1	1	0
	W	1	1	1	1	1	1	1	1	1	1	1	1	1	1	1	1	1	1	1	1	0
	T	0	0	0	0	1	1	1	1	0	0	0	0	0	0	0	0	0	0	0	0	0
	F	0	0	1	1	1	1	1	1	0	0	1	1	1	1	1	1	0	0	0	0	0
Atrium	M	1	1	1	1	1	1	1	1	1	1	1	1	1	1	1	1	1	1	1	1	1
	T	1	1	1	1	1	1	1	1	1	1	1	1	1	1	1	1	1	1	1	1	1
	W	1	1	1	1	1	1	1	1	1	1	1	1	1	1	1	1	1	1	1	1	1
	T	1	1	1	1	1	1	1	1	1	1	1	1	1	1	1	1	1	1	1	1	1
	F	1	1	1	1	1	1	1	1	1	1	1	1	1	1	1	1	1	1	1	1	1

Fig. 4 Classrooms occupancy schedule of the Brescia Smart Campus Demonstrator building during the weekdays from Monday to Friday

assumptions, according to national legislation, used in simulation, which of course neglect the potential variability of conditions (Chapman et al. 2014; Cheng and Gan 2013).

Methodology

The focus of the research is the creation of a methodological continuity in the workflow among design phase, simulation models and techniques to calibrate models with metered data in the operation phase. Therefore, the scope of the research is to overcome the traditional barriers existing in the design and modelling phases (i.e. promoting interoperability between BIM and BEM) and the gaps between simulation models' results and actual building performance (i.e. performance gap, simulation model calibration). The traditional modelling approach describes the occupancy patterns and the end-use of energy from the electric and thermal point of view, in a deterministic way. However, this is far to be realistic and consistent from what happens during real time building operation. The fundamental issues outlined can be partially resolved by adopting a more advanced simulation strategy, coherent with the inherent premises of creating a more realistic model to be used already in the design phase. Further, in order to make the approach more consistent, we conceived it considering the possibility of effectively calibrating the model during real time building operation, by means of inverse modelling techniques. In particular, the simulation approach aims to unveil the interaction of the occupants' behaviour with the building performance, focusing primarily on the fundamental aspects related to thermal comfort, internal air quality, internal heat gains and electricity demand. In Fig. 5 a graphical description of the methodological framework is provided; the research actually described in this paper is outlined with a dotted red line.

In Table 3 the sensors installed for the monitoring phase are described. As explained earlier

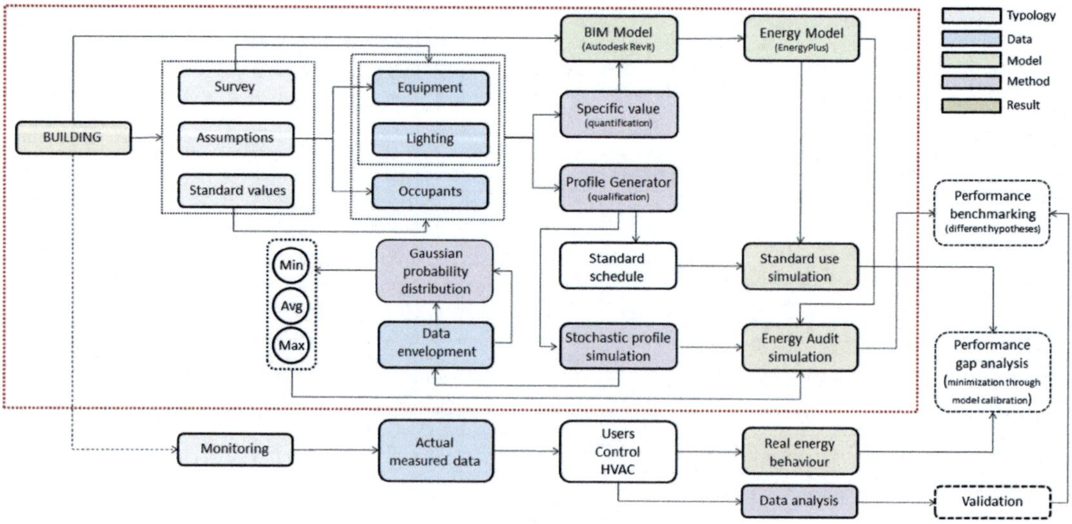

Fig. 5 Methodological framework adopted to evaluate the incidence of users' behaviour and related parameters on whole building performance

Table 3 Typology and number of sensors planned for the installation in the Brescia Smart Campus Demonstrator building

		Sensor	
Area	Function	Typology	Number
Lighting	Presence	Presence	24
	Daylighting control	External luminance	4
Ventilation	Inlet air control	Temperature	15
	Inlet air control	Outdoor air temperature	8
	Humidity control	Relative humidity	9
Heating/cooling	Emission control	Presence, CO_2, ventilation and conditioning dumper	36

the acquired data will enable the calibration of the building energy models. In particular, with respect to the specific case study considered, the presence sensors will be crucial to verify the real occupants' number during the year and the CO_2 sensors will provide detailed information about actual indoor air quality, critical for technical systems' operation management (e.g. modulation of airflow handled by AHUs by means of a stepped or continuous control logic).

With respect to the stochastic behaviour of the occupants' we have to recall some fundamental concepts of "probabilistic design" and "reliability-based design" (Aerts et al. 2014). The "probabilistic design" procedures are primarily aimed at producing safe and reliable design solutions while the stochastic simulation approaches embody a much wider scope, in particular with respect to the evaluation of variability of conditions over time from different perspectives (occupants' comfort, electric and thermal demand, etc.) (Zhang et al. 2011). Therefore, the stochastic simulation approach proposed aims to extend the current design practices for energy simulation, based on deterministic criteria, to more general stochastic based criteria, aimed at giving a more realistic portrait of the building behavioural patterns (Parys et al. 2011). In this research, we assumed a fixed internal distribution of functions, given the characteristics of the chosen case study; however, the approach can be extended, by introducing adequate zoning criteria to account for the variability in end-used from conceptual to scheme or detailed design. In any case, given the design-related uncertainties, the modelling approach becomes relevant if it exploits, as much as possible, the available information. In order to extend the capabilities of the presented modelling approach to real building behaviour, Markovian models can be used. This type of models have been successfully applied in the simulation of occupancy patterns (Widén et al. 2011), lighting system and appliances operation patterns (Richardson and Thomson 2013). As highlighted in Fig. 6, all the aspects are connected among themselves and have a large impact on the overall energy balance.

Other relevant aspects that can be modelled stochastically are natural ventilation driven by windows openings and preferred comfort setting, supplementing existing validated approaches (Gunay et al. 2013). Moreover, as briefly introduced before, the Markovian modelling approach can be used in control systems at the building level (Kun et al. 2010), creating an higher degree of connection between design phase and real time operation.

Results

The variability of occupancy patterns is simulated by means of a stochastic model and the simulation settings data generated are inserted in the main dynamic building simulation models. The possible conditions considered for the comparison are: "minimum 30 %", "average 60 %", "maximum 100 %" and "standard". In Fig. 7 the thermal energy demand (energy need) for heating and cooling on a daily basis is reported, including the daily average outdoor air temperature. The results show the large variability in the daily energy need as a function of the different load profiles. The thermal energy need that will be metered during building operation will be likely included in the envelopment of the results obtained through simulation.

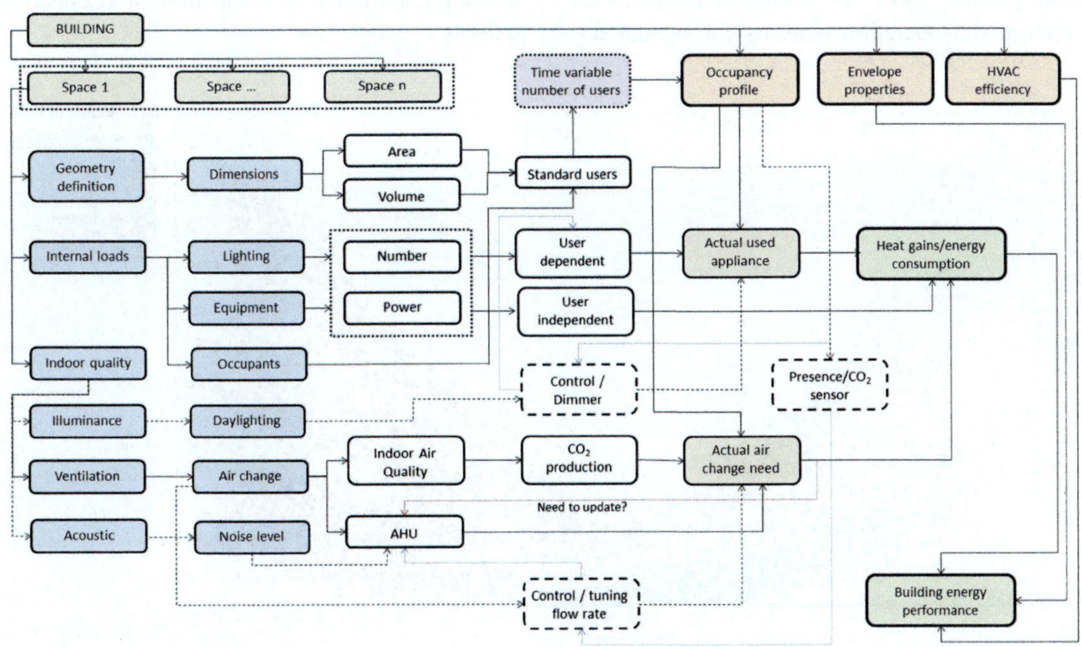

Fig. 6 Correlation among the fundamental aspects of building operation to be considered for energy efficiency and internal environmental quality

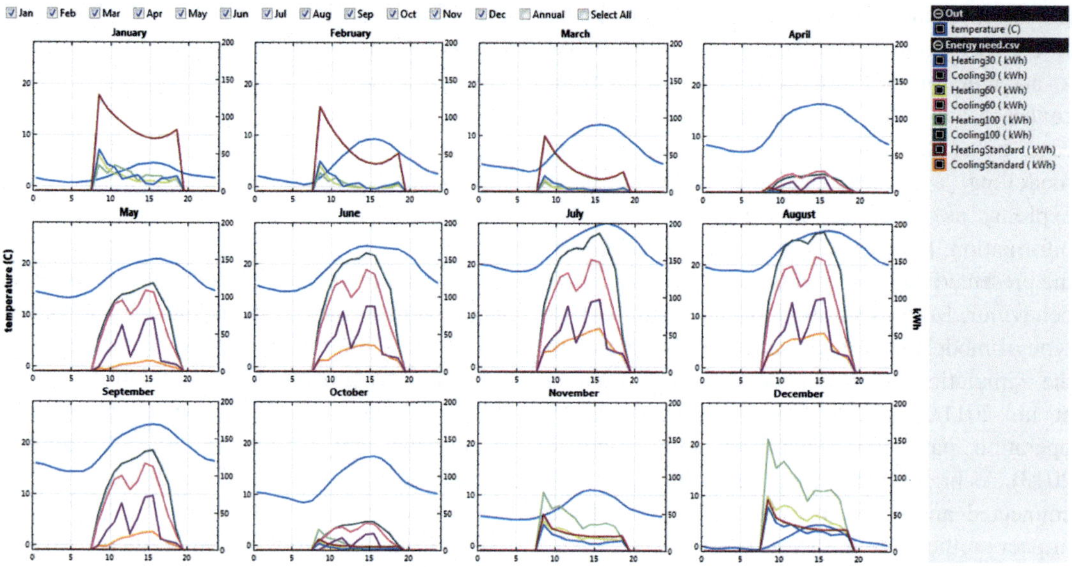

Fig. 7 Hourly load profiles for heating and cooling for the standard case, minimum 30 %, average 60 % and maximum 100 %

The daily control of building energy consumption is not extremely data intensive and has been introduced in several building management systems (Bruton et al. 2015; Lin and Claridge 2015). On the other hand, the hourly load profiles show the potential hourly variations in the operation patterns for typical days of every months. Finally, a scatterplot of hourly thermal energy need with respect to outdoor dry bulb air temperature has been reported in Fig. 8 to highlight the extremely large variability of energy need with respect to outdoor conditions as a function of occupancy patterns.

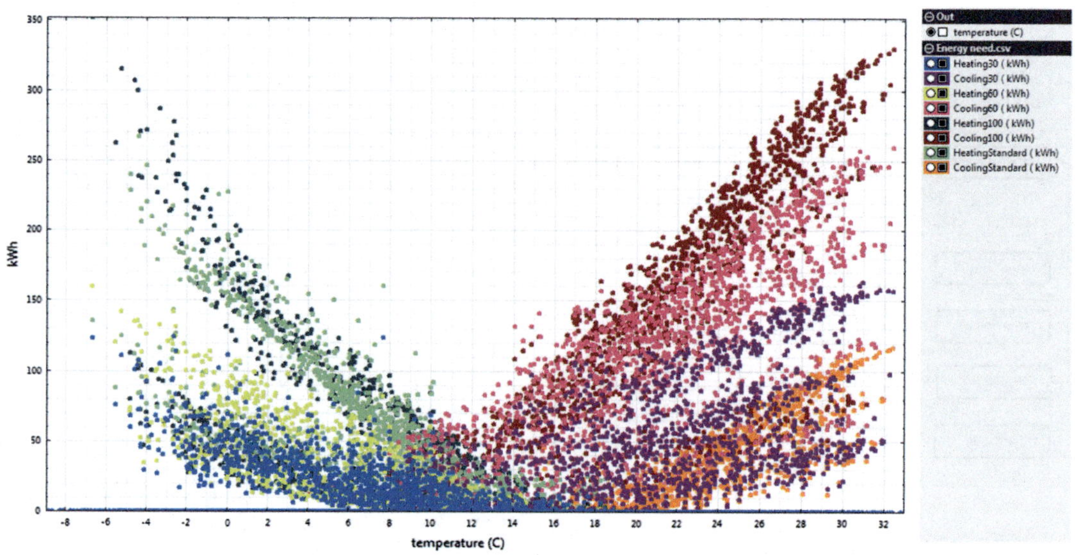

Fig. 8 Energy need for heating and cooling related to outdoor dry bulb air temperature for the standard case, minimum 30 %, average 60 % and maximum 100 %

Conclusions

The paper presents a methodological framework for the assessment of the influence of occupancy patterns on building energy performance by means of a stochastic simulation approach. The approach presented aims to establish a continuity between modelling practices used in the design phase and model calibration techniques used in the building operation phase, for optimal energy management and performance monitoring purposes. In fact, for the specific case study presented, a detailed data acquisition scheme (through appropriately placed sensors) is present and the data will be used to reconstruct the real stochastic occupancy profiles, following a Markovian modelling approach. Modelling techniques, used in building performance simulation, should be considered not merely as a "black-box" approaches, but rather as tools to unveil the relevant probabilistic events happening within the building and their statistical correlation. This represents a fundamental problem if we consider the variability of the performance both in new building and in refurbished ones, especially where techno-economic optimality constraints are present or a clearly defined level of internal environmental quality and comfort are requested. As a conclusion, the generality of the approach proposed aims to fill progressively the gap between real and simulated performance, assessing critically the issue of the "robustness" of the data obtained with conventional simulation approaches, which are based on standard assumptions.

References

Aerts D, Minnen J, Glorieux I, Wouters I, Descamps F (2013) Discrete Occupancy Profiles from Time-use data For user behaviour Modelling in Homes. In: Proceedings of BS2013: 13th conference of international building performance simulation association, Chambery, France, 26–28 August 2013

Aerts D, Minnen J, Glorieux I, Wouters I, Descamps F (2014) A probabilistic activity model to include realistic occupant behaviour in building simulations, IBPSA-Canada eSim 2014, Ottawa, Canada, 7–10 May 2014

Baptista M, Fang A, Prendinger H, Prada R, Yamaguchi Y (2014) Accurate household occupant behavior modeling based on data mining techniques. In: Proceedings of the twenty-eighth AAAI conference on artificial intelligence, 27–31 July 2014, Québec. The AAAI Press, Palo Alto

Bruton K, Coakley D, Raftery P, Cusack DO, Keane M, O'Sullivan DTJ (2015) Comparative analysis of the AHU InFO fault detection and diagnostic expert tool for AHUs with APAR. Energ Effi 8:299–322

Burrows R, Johnson P, Johnson H (2014) Influencing behaviour by modelling user values: energy consumption, second international workshop on behavior change support systems (BCSS 2014) in conjunction with the 9th international conference on Persuasive Technology, Padova, 21–23 May 2014

Chapman J, Siebers P, Robinson D (2014) Coupling multi-agent stochastic simulation of occupants with building simulation, BSO14 building simulation and optimization, second IBPSA—England conference in association with CIBSE, UCL—University College London, London, 23–24 June 2014

Cheng JCP and Gan VJL (2013) Integrating agent-based human behavior simulation with building information modeling for building design. IACSIT Int J Eng Technol 5(4):473

De Angelis E, Ciribini ALC, Tagliabue LC, Paneroni M (2015) The Brescia Smart Campus Demonstrator. Renovation toward a zero energy classroom building. In: Procedia engineering ICSDEC 2015 international conference on sustainable design, engineering and construction, Chicago, 10–13 May 2015

Eguaras-Martínez M, Martín-Gómez C, Vidaurre-Arbizu M, Brennan T, Krinidis S, Ioannidis D, Tzovaras D (2014) Architectural simulation of the integration of building information modelling (BIM) & business process modelling (BPM). In: Mahdavi A, Martens B, Scherer R (eds) eWork and eBusiness in architecture, engineering and construction, ECPPM 2014. CRC Press, Boca Raton, pp 805–811

Fabi V, Vinther Andersen R, Corgnati SP (2012) Window opening behaviour: simulations of occupant behaviour in residential buildings using models based on a field survey. In: Proceedings of 7th Windsor conference: the changing context of comfort in an unpredictable world, Cumberland Lodge, Windsor, 12–15 Apr 2012

Gunay HB, O'Brien W, Beausoleil-Morrison I (2013) A critical review of observation studies, modeling, and simulation of adaptive occupant behaviors in offices. Build Environ 70:31–47

Gunay HB, O'Brien W, Beausoleil-Morrison I, Goldstein R, Breslavc S, Khan A (2014) Coupling stochastic occupant models to building performance simulation using the discrete event system specification formalism. J Build Perform Simul 7(6):457–478

Ioannidis D, Krinidis S, Stavropoulos G, Tzovaras D, Likothanassis S (2014) Full-automated acquisition system for occupancy and energy measurement data extraction. Symposium on simulation for architecture and urban design (SimAUD'14) at the 2014 spring simulation conference (SpringSim'14), 13–16 Apr 2014

Jahn M, Eisenhauer M, Serban R, Salden A, Stam A (2012) Towards a context control model for simulation and optimization of energy performance in buildings. In: 9th European conference on product and process modelling (ECPPM 2012), 3rd workshop on eeBDM, eeBIM, Reykjavik, July 2012

Krinidis S, Stavropoulos G, Ioannidis D, Tzovaras D (2014) A robust and real-time multi-space occupancy extraction system exploiting privacy-preserving sensors. In: International symposium on communications, control, and signal processing (ISCCSP'14), 21–23 May 2014

Kun D, Barooah P, Mehta PG, Meyn SP (2010) Building thermal model reduction via aggregation of states. American control conference (ACC), pp 5118–5123

Lin G, Claridge DE (2015) A temperature-based approach to detect abnormal building energy consumption. Energy Build 93:110–118

Malavazos C, Tzovaras D, Ioannidis D (2013) Activity based and behavioural occupancy modelling for ee building design, Revista de Edificación, pp 41–42

O'Brien W, Kapsis K, Athienitis AK (2013) Manually-operated window shade patterns in office buildings: a critical review. Build Environ 60:319–338

Parys W, Saelens D, Hens H (2011) Coupling of dynamic building simulation with stochastic modelling of occupant behaviour in offices—a review-based integrated methodology. J Build Perform Simul 12 (4):339–358

Parys W, Souyri B, Woloszyn M (2014) Agent-based behavioural models for residential buildings in dynamic building simulation: state-of-art and integrated model assembly, IBPSA conference, Arras, 20th–21st May 2014

Richardson I, Thomson M (2013) Integrated simulation of photovoltaic micro-generation and domestic electricity demand: a one-minute resolution open-source model. Proc Inst Mechan Eng Part A J Power Energy 227:73–81

Widén J, Molin A, Ellegård K (2011) Models of domestic occupancy, activities and energy use based on time-use data: deterministic and stochastic approaches with application to various building-related simulations. J Build Perform Simul 5:27–44

Zhang T, Siebers P, Aickelin U (2011) Modelling Electricity Consumption in Office Buildings: an Agent Based Approach. Energy Build 43(10):2882–2892

Boosting the Efficiency of Architectural Visual Scripts

A Visual Script Structuring Strategy Based on a Set of Paradigms from Programming and Software Engineering

Malgorzata A. Zboinska

Abstract

This research is an initial step towards the advancement of the current routines of architectural parametric modelling, executed using visual programming languages such as Grasshopper. The explored research question is how to effectively organize the infrastructures of visual scripts in order to increase their legibility and hence also their comprehension by designers who work on them. Our proposal advocates the use of an extended visual script structuring tactic, based on well-established solutions developed within software engineering and computer programming. These embrace the notions of design patterns, modular programming, parsing, algorithms, abstraction, enumerative reasoning, static and dynamic data structures, and data hierarchies. The most important outcome of our research is the successful empirical validation of the proposed visual script structuring strategy, achieved through its experimental application to authentic cases of scripts from the architectural practice. A conducted experiment reveals that the structuring principles are valid for various parametric design processes explored using visual scripts, including geometry creation, computational analyses, physics simulations and optimization. Ultimately, the research conclusion is that the application of the structuring principles promises to yield script infrastructures which are highly ordered visually and easier to grasp cognitively.

Introduction

Although recent studies of complex architectural visual scripts show that lack of structure can result in low script legibility and hence difficulty to understand, edit and share, everyday practice shows that most designers still leave their scripts

M.A. Zboinska (✉)
Chalmers University of Technology,
Gothenburg, Sweden
e-mail: malgorzata.zboinska@wp.pl

M.R. Thomsen et al. (eds.), *Modelling Behaviour*, DOI 10.1007/978-3-319-24208-8_40

unstructured. However, since visual programming is now rapidly spreading in the architectural praxis and results in increasingly complex architectural model definitions, it seems that sound script structuring strategies should be developed, so that time and resources are not wasted on deciphering and reworking the scripts, and the promising capabilities of parametric modelling are not compromised.

Since visual scripting classifies as programming, it seems that good practices developed by programmers could benefit the management of architectural scripts. The first steps to adopt such practices in architecture have already been taken by Woodbury et al. (2007), who demonstrated the use of programming patterns, and by Davis et al. (2011), who introduced the notion of modular programming. This study aims to advance these investigations further, by proposing an extended strategy of structuring visual scripts, based on a compound set of paradigms from computer programming and software engineering.

In comparison to prior research, the study brings in three elements of novelty. Firstly, it integrates the application of programming patterns and modular programming within one approach. Secondly, it incorporates the yet unconsidered programming notions: the use of algorithms, parsing, enumerative reasoning, design pattern-based program decomposition and data structures. Finally, it demonstrates how to structure visual scripts not only in the context of generating parametric geometry, but also its computational analysis, dynamic behaviour simulation and optimization. In this way, the strategy interfaces various interdisciplinary aspects, necessarily accompanying advanced parametric modelling.

Programming and Software Engineering Concepts Supporting the Strategy

The following four concepts from software engineering and computer programming are employed to formulate the extended visual script structuring strategy.

Algorithms

The existence of an algorithm, representing the procedures of a computer program in an abstract way alongside the actual code, is believed to positively affect the program's readability, ease of debugging, effectiveness, and user friendliness (Charntaweekhun and Wangsiripitak 2006). Additionally, if the algorithm is represented using graphical means, such as a flowchart or an activity diagram, then other programmers are able to comprehend that algorithm, and consequently the computer program created based on it, much faster and with fewer errors (Scanlan 1989).

It therefore seems worthwhile to consider the use of algorithms and their diagrammatic representations in visual scripting environments. This could potentially help to communicate the design intent and outline the complex design procedures executed to create, modify, analyse and optimise parametric geometry. It appears that this could mitigate a better understanding of the parametric design workflow, both for the creator of the visual script, as well as for other designers who will edit or reuse that script.

Structured Programming

As a remedy to the illegibility of complex computer programs, the paradigm of structured programming emerged. It advocated stepwise program composition, supported by three notions: abstraction (conceptualising program procedures using abstract constructs—algorithms, represented graphically as flowcharts), parsing (decomposition of programs into sub-actions) and enumerative reasoning (stepwise structuring of programs, by placing their sub-actions one after the other, in the actual order of execution, with every sub-action numbered and corresponded to a program's algorithm) (Dijkstra 1970).

Because this strategy proved to be beneficial for classical programming, it is interesting to consider its application in visual programming. Davis et al. (2011) have already implemented one of its notions, i.e. parsing, by advocating the use of grouping of sub-actions into visual script modules. However, this work could potentially be advanced further, by incorporating all three notions of structured programming into the visual script structuring tactic. The notion of abstraction could be implemented through the use of the already mentioned algorithms, representing the design intent graphically. The idea of parsing could be developed by subjecting it to enumerative reasoning. Namely, the program modules could be numbered and labelled based on their order of appearance in the script, and directly corresponded with the particular steps in the algorithm.

Design Patterns

Design patterns are frequently applied in software engineering to facilitate a common understanding of typical programming procedures and to promote their reuse. A design pattern is a conceptual construct which provides an abstract description of a design problem and shows how a general arrangement of program elements solves it (Gamma et al. 1995). Hence, it is a kind of a design template, more specific than an algorithm, but less concrete than the particular object instances which the algorithm creates.

Recent studies suggest that design patterns can also be beneficial for parametric modelling using visual scripts, because patterns foster communication among designers, and they help to learn and to use parametric systems more efficiently (Woodbury et al. 2007). However, although in the current architectural visual scripting practice repeatable design patterns can be detected, in most cases this results not from deliberate pattern application, but rather intuitive use. This can be inferred from the latest survey by Davis (2014), who concluded that most designers leave their scripts in unstructured form, suggesting that design patterns, if present, remain seamlessly blended within the scripts.

Consequently, this study will attempt to make design patterns more visible in visual scripts. Our proposal is to directly link design patterns and modular programming principles. That link can be achieved by using design patterns as frameworks which represent the logic of module grouping in visual scripts.

Modular Programming

Apart from module grouping based on design patterns, entire groups of modules could also be organised into collections. These collections could represent a higher order of modules, i.e. groups of groups, based on the functionality of each collective. As a basis, we propose five main categories of collectives: geometry generation; computational analysis; behaviour simulation; geometry optimization; and helpers. Within the geometry generation collective, additional sub-collectives can be distinguished, such as: geometry manipulation; operations on data; parameter assembly. Also the helpers collective could contain miscellaneous sub-collectives, such as: color-coding; baking; tagging; meshing; nesting; data export etc.

The introduction of this additional higher order of module grouping may be another important step in the management of complex visual script infrastructures. Once connected with the graphically-represented algorithm of the visual script, the bigger collectives of groups could increase script comprehension, allow easier navigation and a better orientation among its particular sub-functionalities.

Data Structures

The latest surveys by Davis (2014) indicate that in the majority of visual scripts the parameters are not strategically managed. Also, they are used sparingly, to avoid overcomplicating the script. Paradoxically, this effort to simplify scripts may prevent the exploitation of 3d model flexibility offered by parametric modelling. It seems that clearer guidelines for parametric data management should be established, so that designers are able to achieve a good balance between the script's complexity and its parametric resilience.

Two concepts related to data management within programming could help to manage parameters in visual scripts: static/dynamic data structures and data hierarchies. A static data structure has a fixed size and is modified by external processes, while a dynamic data structure can grow and shrink, and is modified by internal processes (Pieterse and Black 2007). In a complex visual script, classifying parametric data into static and dynamic types could help to represent its varying roles. When dealing with clearly classified parameter types, the designer is more likely to know which sets of data affect the different properties of the 3d model, and which of them can be modified.

Parameters in visual scripts can further be grouped using two basic data hierarchy categories: fields and records. A field is an individual piece of data holding a single attribute, and a record is a collection of fields (Harrington 2009). In visual scripts, a field can be represented by a single parameter primitive, storing geometry, numbers, colours etc. The collection of these primitives, assembled based on the common function they perform, could then be understood as a data record. The aim of this additional categorization of data is to make it more clear which parameters are responsible for modifying the particular parts of the architectural model. This allows a designer to extract the key controls for the different elements of the model, which can be tampered with without the need to analyse the functionalities of the script.

The Extended Strategy of Visual Script Structuring

In relation to the software engineering and programming concepts outlined above, the following general guidelines for the structuring, management, inspection and optimization of visual scripts were developed:

- **Using abstraction and algorithms**. Conceptualise parametric design workflows as algorithms. Express diagrammatically, enrich with geometry screenshots, indicating which 3d elements of the model a particular step of the algorithm generates or affects.
- **Parsing**. Decompose visual scripts into sub-parts. Parse on several levels: operation class level (module collectives, demarcating major tasks: geometry creation, analysis, simulation, optimization, and helper operations); sub-operation class level (module sub-collectives, nested within main collectives); sub-task level (module groups, performing single tasks).
- **Applying design patterns**. Use as a conceptual base for module grouping, primarily on sub-task level.
- **Enumerative reasoning**. Order, number and describe collections, sub-collections and module groups in sequence of appearance. Refer them to the main algorithm (numbers and descriptions should appear in the visual script twice: in the algorithm and in collection/group headings).
- **Data structure-based parameter management**. Structure the data fundamental for the parametric modification of the 3d model. Categorise that data into static/dynamic types and group into records of fields. Group records according to geometry elements they control (global form, local detailing etc.) and/or additional function (inputs for computational analysis, optimization etc.). Avoid 'tangled spaghetti effect' by hiding wire connections between the main parameters extracted at the beginning of script and the modules buried deeper within the script. Represent numeric variables as sliders and constants as number primitives (parameters not intended for editing should have a fixed form).
- **Modular programming**. Gather parsed script sub-parts into groups. Indicate module inputs and outputs in explicit form, to clearly indicate entry and exit data (by duplication of parameters as primitives). For each sub-task

group visually signal that data, e.g. by bold-style capitalised IN/OUT tagging.

- **Group colouring**. Apply a clear and consequent strategy for color-coding of various group levels and types. Apply special color-coding treatment also to key parameters for manipulating the 3d model, so that users unfamiliar with the script can quickly detect and intuitively use them without having to analyse the entire script.
- **Script optimization**. Upon the end of script development stage, perform script simplification: identify which operations can be carried out using a smaller number of modules or using simpler routines, and restructure these parts of the script accordingly.

The generalised visual script organization scheme, prepared according to these guidelines, is visualised in the popular visual scripting environment Grasshopper (Fig. 1).

Strategy Application: The Script Structuring Experiment

Method

To investigate the process of implementing the structuring guidelines, we chose an empirical investigation method, offering a controlled environment for our inquiry. An experiment was conducted, in which unstructured visual scripts were subjected to structuring using our strategy. Through the experiment, three hypotheses were to be verified: (1) All guidelines of the extended

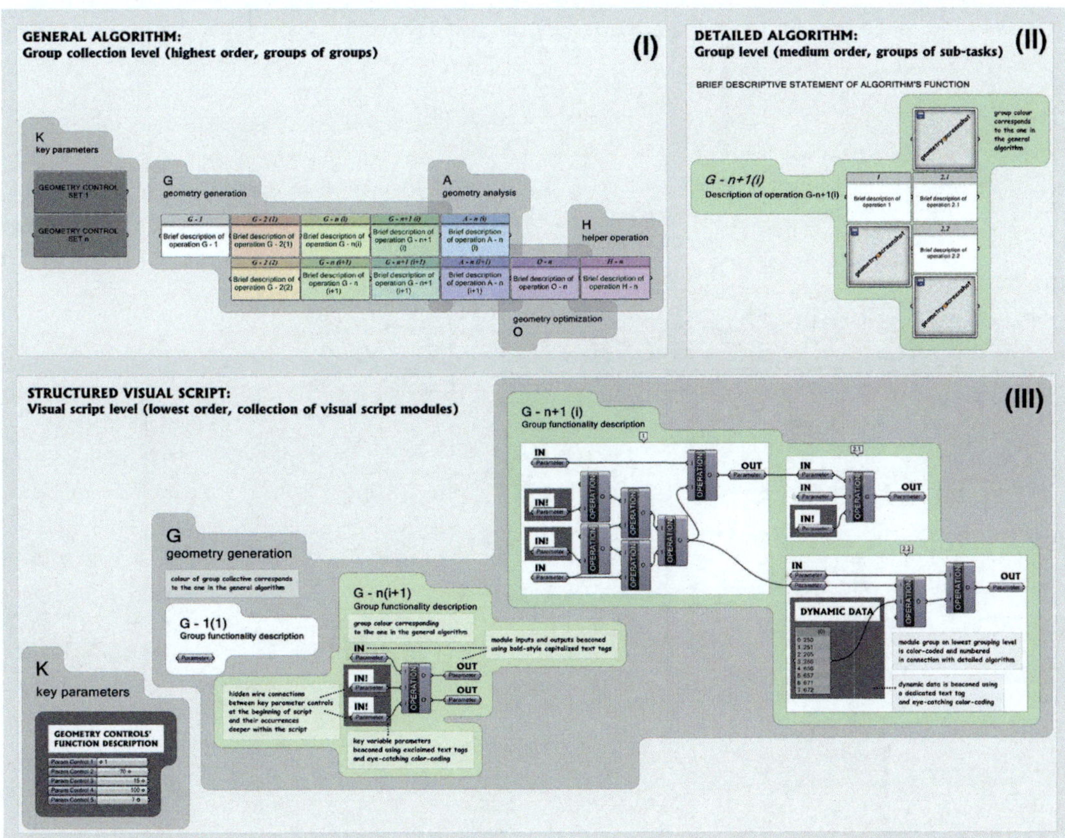

Fig. 1 The generalised visual script structuring scheme

strategy of structuring visual scripts are possible to be applied to visual scripts; (2) The strategy is universal, i.e. it can be applied to miscellaneous visual scripts, having various functions and generating various parametric models; (3) The application of the strategy produces visual scripts that appear highly structured visually, which suggests improved cognitive potential. Additionally, the aims of the experiment were: to obtain practical clues on how exactly to implement the guidelines; to construct a general scheme of visual script organization based on these clues (Fig. 1); to obtain structured script cases, useful for further usability studies.

Study Sample Selection

We used the visual scripts directly available from our parametric modelling practice as experimentation samples. These scripts can be regarded as idiosyncratic, which supports our goal to test the strategy on miscellaneous cases. Ultimately, we chose three visual scripts (Fig. 2), conforming to the following selection criteria: embracing at least three typical parametric design phases (geometry generation, performance analysis, simulation, optimization); having at least: median size (minimum 23 modules), median dimensionality (minimum 6 parametric nodes without

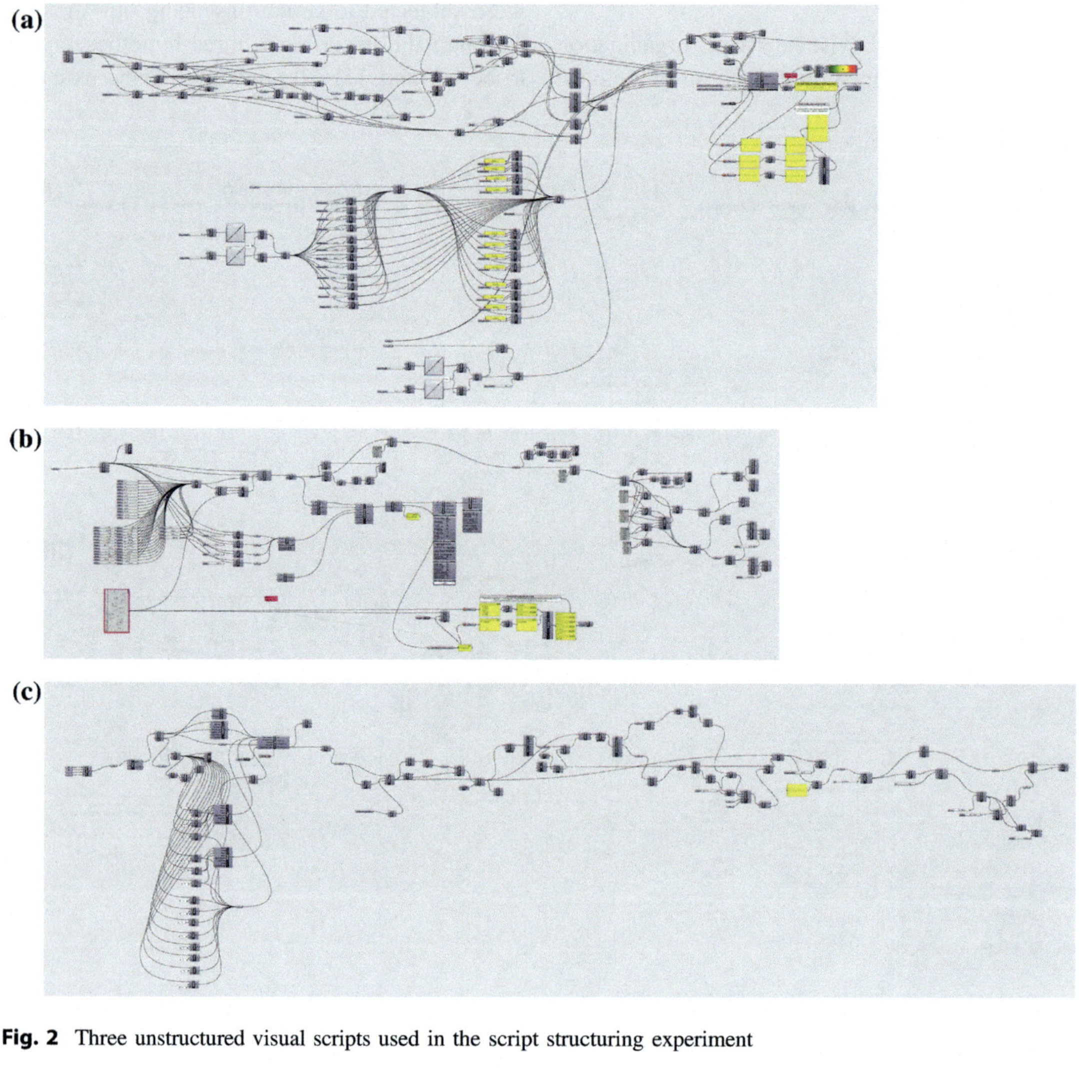

Fig. 2 Three unstructured visual scripts used in the script structuring experiment

parents) and median cyclomatic complexity (minimum 13) (median values adopted from Davis 2014). Moreover, to guarantee sample diversity, we chose scripts which generated various geometries (simple and complex), and which had various purposes (conceptual design, digital fabrication).

The Experiment

In the experiment, we applied our structuring guidelines to each of the three scripts, gradually affecting their morphology. The operations embraced: representation of the design intent using an algorithm diagram (Fig. 3a); numbering,

(a) (b)

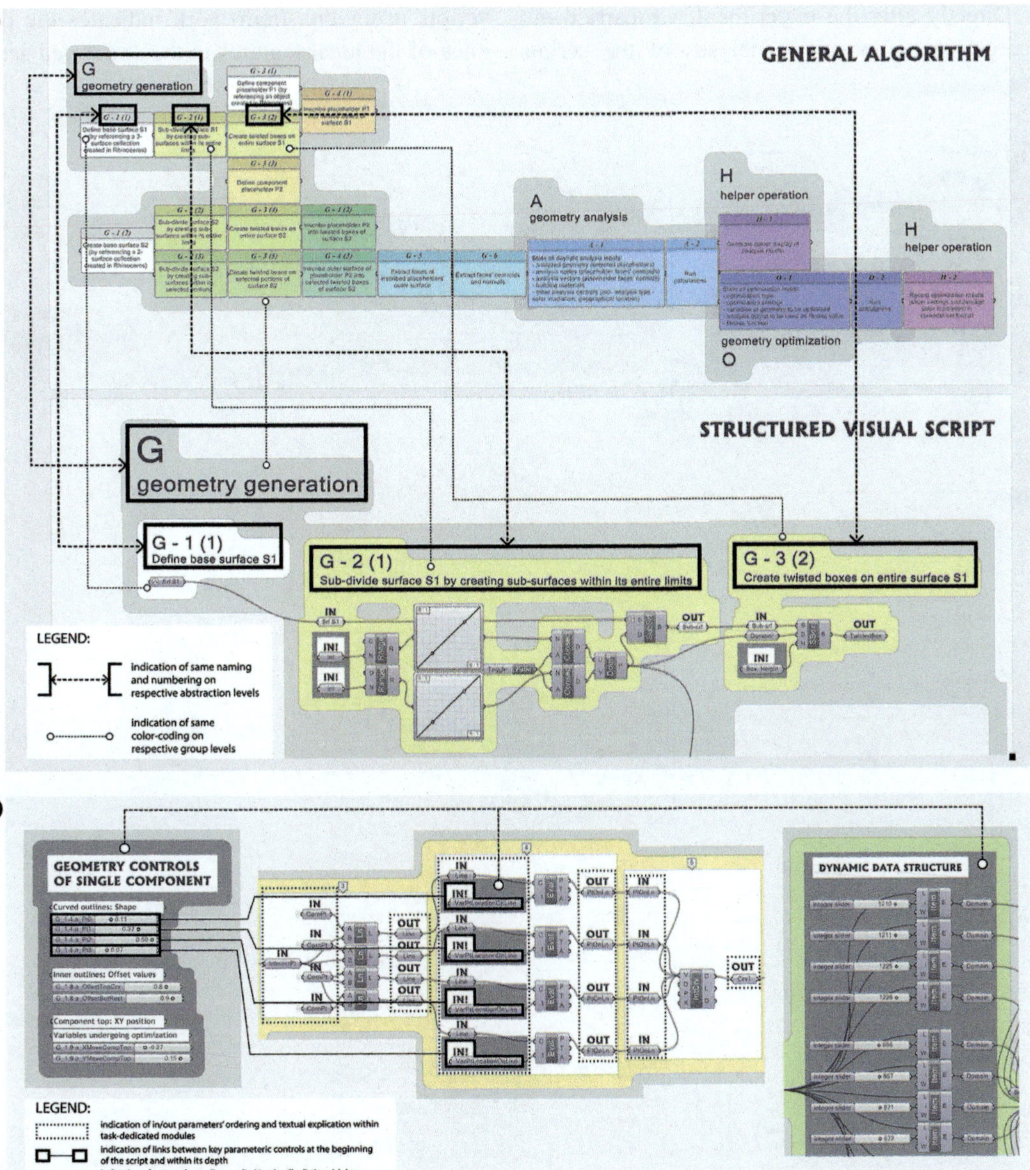

Fig. 3 The application of the script structuring strategy

describing and grouping visual script modules based on the identified operation classes, design patterns and sub-tasks (Fig. 3a); color-coding the groups (Fig. 3a); ordering, explicating and rearranging inputs, outputs and modules in each group (Fig. 3b); establishing and grouping fundamental data structures (Fig. 3b); and performing script optimization. Ultimately, the experiment yielded three organized visual scripts (Fig. 4).

Directly after the experiment, we carried out qualitative comparative analyses of the scripts before and after structuring. For the purpose of that examination, we selected the cognitive dimensions framework of Green and Petre (1996). It is a HCI-based visual programming environment usability evaluation tool, containing a set of criteria which define the morphological features of an ideal visual programming language, user-friendly from the cognitive point of view.

We assumed that if the examination of our scripts using this framework indicates the presence of the ideal features in the structured scripts

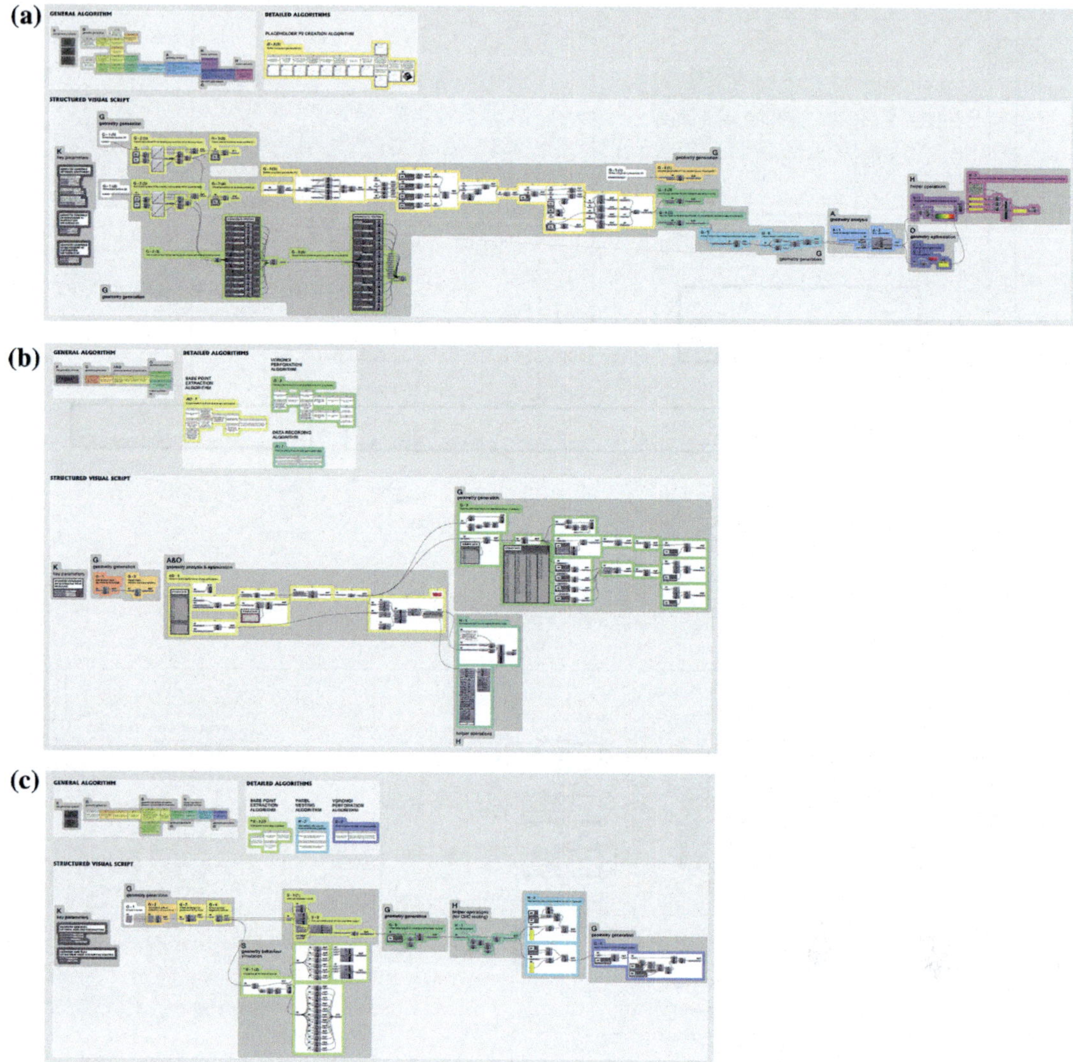

Fig. 4 Three visual scripts after structuring

and their lack in the unstructured ones, then we can infer that the script structuring strategy does promise to increase script comprehension. In our analysis, we focused on identifying the following six ideal features of visual scripts from the framework:

- **Abstraction gradient**: Are varying levels of abstraction present in the script structure? Is each level encapsulated as a whole?
- **Closeness of mapping**: Is each abstraction level related to other levels? Do the relations progress gradually from abstract to concrete ones?
- **Consistency**: Is the script's structure harmonious, i.e. built according to some general structuring rules?
- **Role-expressiveness**: Is it possible to see how each script component relates to the whole?
- **Secondary notation**: Are additional notational means, such as layout, colour and annotations present and do they convey additional meaning beyond the programming language semantics?
- **Visibility**: Is every part of the script visible? Is it possible to know in what order to read the script?

Having established the features we are looking for, we carried out script examinations. The scripts were considered one by one, with each case subjected to visual search for the features outlined above. A positive reply to the inquiries listed above was taken as a confirmation of each respective feature's presence. Once that analysis of the scripts was complete, we compared the results obtained for all six scripts (three unstructured and three structured). This comparison validated the emergence of the following new morphological features of the structured scripts, lacking in their unstructured version:

- **Presence of abstraction gradients and closeness of mapping**. In the structured scripts, abstraction levels are present and they gradually become concretised, starting from the graphically-represented algorithm, through collectives and groups, ending with explicit visual script components. This contributes to the closeness of mapping between abstract design intent, conceptual algorithm, its stepwise procedures, and the singular actions of the program (Fig. 5a). Conversely, in the unstructured script there is only one level of abstraction present (the most concrete one) and hence closeness of mapping is impossible (Fig. 5b).
- **Consistency of the script's conceptual and visual structure**. In the structured scripts, algorithm steps and component groups are consequently numbered, described and mutually associated across the different abstraction levels (Fig. 5c). A common logic of grouping program tasks and data structures is applied across the entire script. In the unstructured version, the morphology of the script arises from the nature of visual programming (components added one after the other) (Fig. 5d). However, there is no additional logic of arranging these components.
- **Presence of role-expressing elements**. In the structured script version, the groupings visually demarcate the different roles played not only by each module, but also by entire bundles of modules (Fig. 5c). The functions of groups, modules and key parameters are also briefly described and present both on the abstract algorithm level as well as the concrete visual script level. In the unstructured version, apart from the default names of the components, there are no role-expressing elements. This makes it difficult to understand how each component relates to the whole (Fig. 5d).
- **Presence of secondary notation**. In the structured scripts operations, tasks and parametric data are classified, bundled and demarcated within color-coded groups. More complicated geometry manipulations are additionally explained using geometry screen-shots. Inputs/outputs for each module are textually and visually beaconed (text tags and duplicated primitive components upon module entry/exit) (Fig. 5c). The unstructured versions have no extra notations which would

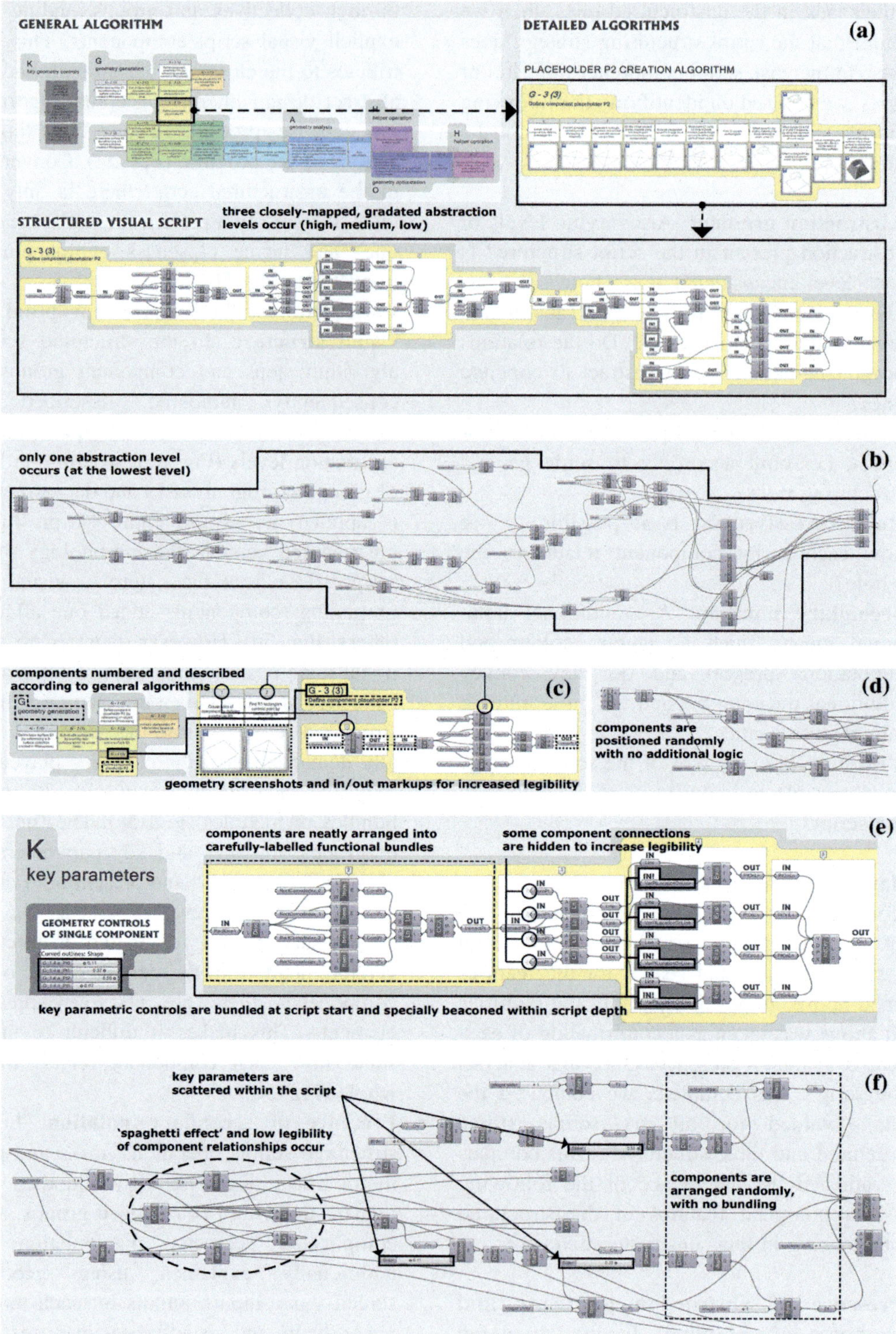

Fig. 5 Comparisons of structured and unstructured script morphologies and indication of new features of the structured scripts

convey additional meaning, going beyond the programming language semantics (Fig. 5d).

- **Increased visibility**. Upon visual examination of the structured scripts, the coloured groups, representing program tasks, are identifiable and clearly visible. The key parametric geometry controls appear at the beginning of the script and are additionally colour-flagged within the script depth. This facilitates their visual identification within the script. Component alignment/centring are used, which prevents visual clutter. Some dependency wires are strategically concealed to diminish the 'spaghetti effect' (Fig. 5e). In the unstructured scripts, visibility is hindered by factors such as: the tangling of wires connecting the components, commonly occurring lack of component alignment and random disposition, and placement of key parameters within the script depth (Fig. 5f).

Conclusions

The key proposition of this study is to advocate the use of an extended visual script structuring strategy, in order to support effective parametric modelling in architecture. The strategy feeds on the notions of design patterns, modular programming, parsing, algorithms, abstraction, enumerative reasoning, static/dynamic data structures and data hierarchies, assimilated from software engineering and programming.

The most important research outcome is the empirical validation of the proposed structuring strategy. The conducted experiment indicated that it is possible to successfully apply the suggested structuring guidelines to authentic cases of visual scripts from practice, regardless of the scripts' purpose and the design phase they concern. The experiment also showed that the structuring principles apply to various processes accompanying parametric design, including geometry analysis, simulation and optimization. The structuring ultimately results in script infrastructures which are highly ordered visually and may hence be easier to grasp cognitively.

By proposing an extended strategy of structuring visual scripts, enriched with fresh concepts acquired from software engineering, this research is the initial step towards the advancement of the current routines in parametric modelling using visual programming. Previous architectural studies indicated that structuring is worth applying, because it can aid visual script and parametric model comprehension, editing, debugging, sharing and collaborative work. Given the effectivity it promises to bring into the design workflow, the extended structuring approach appears to be a good candidate for mainstream practice.

However, because this investigation was limited to retroactive script structuring, the next research step will embrace usability studies in which we will verify whether the structuring strategy could also benefit scripting in an active way, by being applied directly in the course of the design process. In those upcoming studies, we will take a closer look at how different designers approach the problematic of applying the structuring principles, and what are the potential advantages and limitations of this approach.

References

Charntaweekhun K, Wangsiripitak S (2006) Visual programming using flowchart. In: Proceedings of the international symposium on communications and information technologies, Bangkok, 18–20 Oct 2006

Davis D (2014) Quantitatively analysing parametric models. Int J Archit Comput 12(3):307–319

Davis D, Burry J, Burry M (2011) Untangling parametric schemata: Enhancing collaboration through modular programming. In: Leclercq P et al (eds) Proceedings of the 14th international conference on computer aided architectural design, Liège, 2011

Dijkstra EW (1970) Notes on structured programming. In: Dahl O-J et al (eds) Structured programming. Academic Press, London, pp 1–82

Gamma E, Helm R, Johnson R, Vlissides J (eds) (1995) Design patterns: elements of reusable object-oriented software. Addison-Wesley, Massachusetts

Green T, Petre M (1996) Usability analysis of visual programming environments: a 'cognitive dimensions' framework. J Vis Lang Comput 7(2):131–174

Harrington J (2009) Relational database design and implementation: clearly explained. Morgan Kaufmann Publishers, Burlington

Pieterse V, Black P (2007) Active data structure, passive data structure. In: Dictionary of algorithms and data structures. National Institute of Standards and Technology. http://xlinux.nist.gov/dads/. Accessed 24 Apr 2015

Scanlan D (1989) Structured flowcharts outperform pseudocode: an experimental comparison. IEEE Softw 5(5):28–36

Woodbury R, Aish R, Kilian A (2007) Some patterns for parametric modelling. In: Lilley B et al (eds) Proceedings of the 27th annual conference of the association for computer aided design in architecture, Halifax, 2007

Modelling with Forces: Grammar-Based Graphic Statics for Diverse Architectural Structures

Juney Lee, Corentin Fivet and Caitlin Mueller

Abstract

Most architectural modelling software provides the user with geometric freedom in absence of performance, while most engineering software mandates pre-determined forms before it can perform any numerical analysis. This trial-and-error process is not only time intensive, but it also hinders free exploration beyond standard designs. This paper proposes a new structural design methodology that integrates the generative (architectural) and the analytical (engineering) procedures into a simultaneous design process, by combining shape grammars and graphic statics. Design tests presented will demonstrate the applicability of this new methodology to various engineering design problems, and demonstrate how the user can explore diverse and unexpected structural alternatives to conventional solutions.

Introduction

The institutionalized separation of form (architecture), forces (structure), and material (fabrication/construction) has resulted in a geometric-driven contemporary design culture. During early stages of design, an architect tries to control spaces by "finding a form" among countless possible forms, while an engineer tries to control forces by "form-finding" an optimized solution of that particular form. This modern prioritization of expressive form over material and performance is the platform upon which architecture and structural engineering remain divided as schools in academia and as professions in practice today. The development and logic of most Computer-Aided Design (CAD) software are also based on this separation of form and forces. While the rapidly advancing capabilities of computational tools have enabled architects to digitally model almost any form, and engineers to analyse any structure, digital models generated by architects typically must be re-modelled by engineers in a file format that is compatible with numerical analysis software. More meaningful investment of

J. Lee (✉) · C. Fivet
Massachusetts Institute of Technology, Cambridge, USA
e-mail: juney@mit.edu

C. Mueller
Massachusetts Institute of Technology, Department of Architecture, Cambridge, USA

M.R. Thomsen et al. (eds.), *Modelling Behaviour*, DOI 10.1007/978-3-319-24208-8_41

the computational resources that are available today may be in investigating new structural possibilities, rather than developing better ways of analysing what may be inherently bad forms.

There is a need for computational design tools that can not only generate forms, but simultaneously process structural logic and rules, so that the outcome does not need to be constantly remodelled and checked with numerical analysis software. For architects, introducing real-time performance feedback during the modelling process can help in exploring structurally feasible and diverse designs much more efficiently, and improve the general structural intuition of designers. For engineers, adding generative ability to tools that are traditionally meant to only analyse an already defined structure, can potentially expand the creative capacity of the engineer and lead to undiscovered structural possibilities. In order to explore a wide range of diverse design alternatives, computational power combined with controlled randomness can help the designer in unbiasedly exploring alternative solutions that are unexpected, visually interesting and yet performatively adequate.

Background

In conventional parametric modelling paradigm, forms are generated and controlled by parameters or variables. Similarly, in computational design and optimization of structures, the objective function is mathematically formulated and numerical parameters are clearly defined. This means that the design space contains all possible solutions to a given problem. While a sophisticated optimization algorithm may help the user in finding the optimal solution within this parameter-based design space, the result is still limited by the design space itself. During earlier stages of design, a parameter-based design space does not contain the wide range of design possibilities that the designer may want to consider (Mueller 2014, pp. 79–88).

Grammar-Based Design

In order to broaden the design space, a grammar-based approach can be used in place of the conventional parameter-based design paradigm. Grammar-based design, or more commonly known as *shape grammars*, is a set of allowable shape transformations that can be used to define a design language, through which form generations can be automated based on a desired logic, style or objective (Stiny and Gips 1972). It has been used frequently in architectural context to not only analyse design styles and languages, but also generate new ones. William Mitchell hinted at its potential applicability to other fields such as structural design, by incorporating functional attributes and structural criteria to grammar rules in the form of *functional grammars* (Mitchell 1991).

Structural Grammars

Shape grammars have been applied in engineering, most notably by Shea and Cagan, as a method called *shape annealing* (Shea and Cagan 1999). Because this method uses random shape transformations which are entirely geometric, a numerical analysis is required after every operation. In addition, the transformations are guided by a stochastic optimization algorithm, which means that unless the transformation improves the overall performance, it will keep iterating until one is found. While successful in generating unexpected solutions, shape annealing is an optimization driven method and seeks to output one solution. Shape annealing is ultimately resource-intensive, and the resulting diversity is limited. Alternatively, shape grammars can be used to explore trans-typology structures, by randomly mix-and-matching elements of different typologies (Mueller 2014, pp. 79–101). A wide range of unexpected yet structurally equilibrated solutions can be found using a small set of grammar rules. However, previous research in this area was limited to typologies that are specific mainly to bridge structures only.

Graphic Statics

Graphic statics is a graphical method of calculating forces for discrete structures under axial loads (Culmann 1864). It is based on construction of two reciprocal diagrams (Cremona 1890): the form diagram representing the actual geometry of the structure, and the force diagram that represents the internal forces as vectors. Because forces are graphically represented, no numerical analysis is required to calculate the forces. When combined with modern day computation, graphic statics can become a powerful design tool by automating the drawing process, and enabling real-time interaction between the reciprocal diagrams. Most notable examples include Active Statics (Greenwold and Allen 2003), eQuilibrium (Van Mele et al. 2014), and Constraint-based Graphic Statics (Fivet and Zastavni 2013).

Combining Grammars and Graphic Statics

Shape grammars and graphic statics have been explored previously in the field of creative structural design, but never in combination. When shape grammars and graphic statics are combined, several key benefits emerge. First, geometric rules can have direct relationship with corresponding force diagrams so that any geometric transformation results in equilibrium. Because local and global equilibriums are always guaranteed, randomness can be introduced during the generation process to increase diversity of solutions. Second, because force diagrams are constructed for every transformation, there is no need for further numerical analysis. Lastly, the rules have no boundary-specific parameters, which enables the methodology to be applied to a wide range of design problems. By harnessing the intelligent, generative power of shape grammars, and the computational graphic statics that can transform forces into equilibrated forms, architecture and structure can be integrated more seamlessly during conceptual design.

Setup

Conceptual Overview

The proposed methodology automatically and randomly generates designs by iteratively applying a series of rules. The conceptual overview of the computational setup is illustrated in Fig. 1. Generally, the *grammar engine* is responsible for choosing rules, deciding where to apply them and updating the geometry. *Graphical computation engine* functions as the structural blueprint for the all procedures to be performed by the grammar engine.

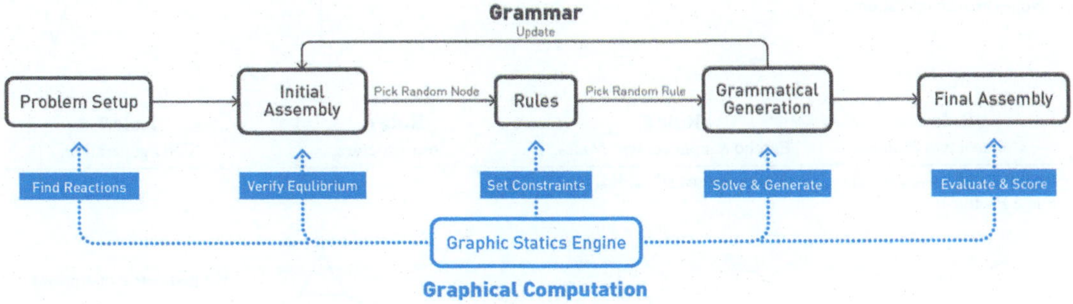

Fig. 1 Conceptual overview of integrating shape grammars and graphic statics

Elements

The proposed methodology operates on three types of computational classes: (1) a *Force* class that is a 2D vector, with a type parameter, direction and amplitude; (2) a *Node* class that includes a coordinate, state parameter (active or not-active), type parameter, and a list of forces; (3) a *Member* class that is a line, with information about its internal forces; and (4) an Assembly class that includes a list of *Node* classes, list of members, the overall system state, and other information about the entire structure. The four elements are summarized in Fig. 2.

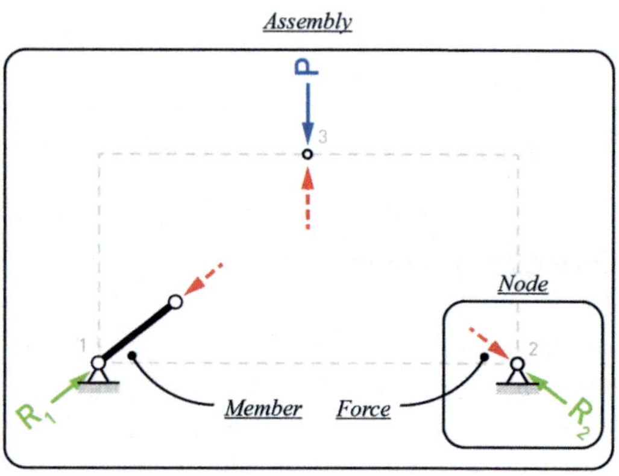

Fig. 2 Four basic elements of the methodology

Rules

The eight rules used for generating designs in this paper, are summarized in Fig. 3. All parameters incorporate structural logic and knowledge, and are always verified by construction of force diagrams. While the parameters are randomly determined, it is constrained by user-defined lower and upper bounds.

Constraints

Generative grammars can be a powerful tool in discovering new structural forms. However, without intelligent constraints, the rules may be too broad and generate forms that have limited practical feasibility. In addition, the grammar rules can potentially be applied recursively, or repeated without an end. The following strategies are used to set constraints.

- *Reasonable range angles for the initial reactions.*
- *Setting reasonable local bounds, such as minimum and maximum angles or lengths.*
- *Global termination conditions, such as generation count and recursion control mechanisms.*

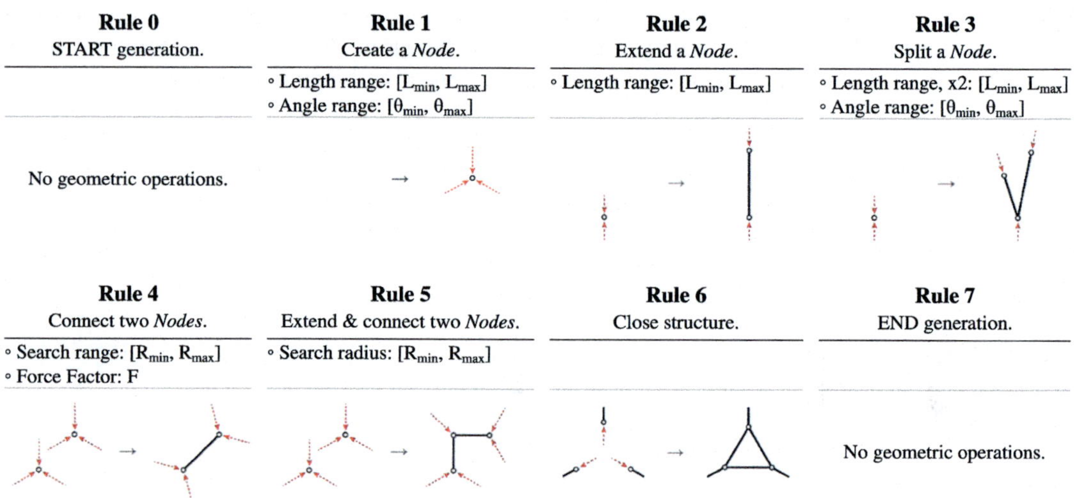

Fig. 3 Summary of rules and parameters

Generation Algorithm

Figure 4 summarizes the automatic random generation algorithm framework. Steps 1a, 1b and 1c are where user input is required to setup the design problem. Then, the initial *Assembly* is constructed. From the parameters defined in steps 1b and 1c, the algorithm randomly chooses a *Node* to apply a random rule. Steps 3 through 8 are repeated until the system reaches a terminating condition defined by the user.

Sample Generation

Figure 5 shows a step-by-step generation sequence of one possible design for a simple problem.

Design Tests

Workflow

Unlike most conventional engineering tools, this methodology begins without a starting geometry. First, the user sets up the problem by defining the applied loads, reactions and the locations of those forces and supports (Fig. 6a). The user then chooses the rules to apply, and weights for each rule which defines how likely it is that a rule will be randomly selected to be applied (Fig. 6b). Finally, the user defines how many options to produce. Results can be further diversified by modifying the following global parameters: (1) minimum number of rule applications for each generation; (2) rule sensitivity towards the beginning, the middle or the end of the generation cycle; (3) termination conditions; and (4) random seed (Fig. 6c).

User interface

The proposed methodology was implemented as an interactive tool, using IronPython, Rhinoceros and Grasshopper. Each iteration instantaneously generates: (1) corresponding force diagrams for every node (Fig. 7a), (2) a form diagram with clear labels (Fig. 7b), and (3) rule history, information and evaluation metrics for the current solution (Fig. 7c). Visual representation of the forces, the evaluation metric, and the rule history which summarizes how the structure was derived, enables clearer understanding of the structure and informs better design decisions more quickly. The rule history, which records all the parameters that were used to generate the current iteration, is an important feature that enables reproducibility of the same iteration

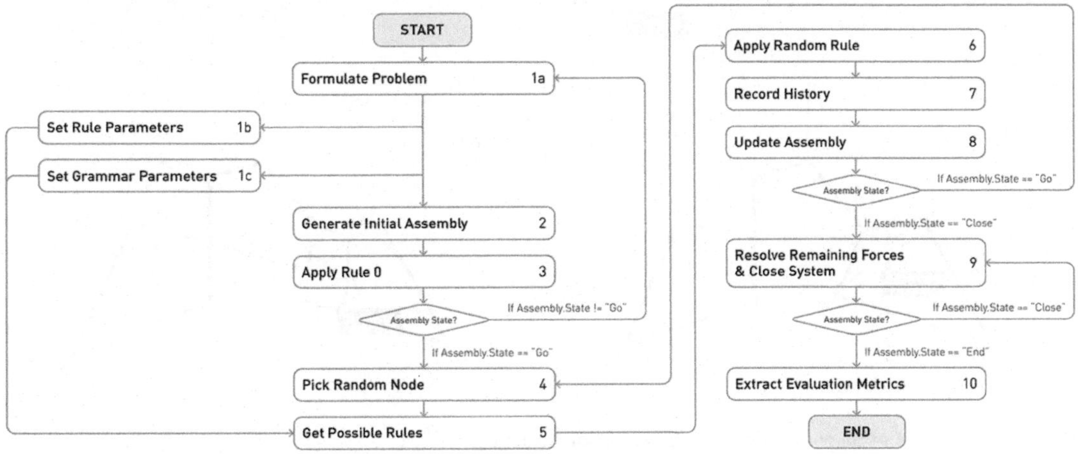

Fig. 4 Framework of the automatic random generation algorithm

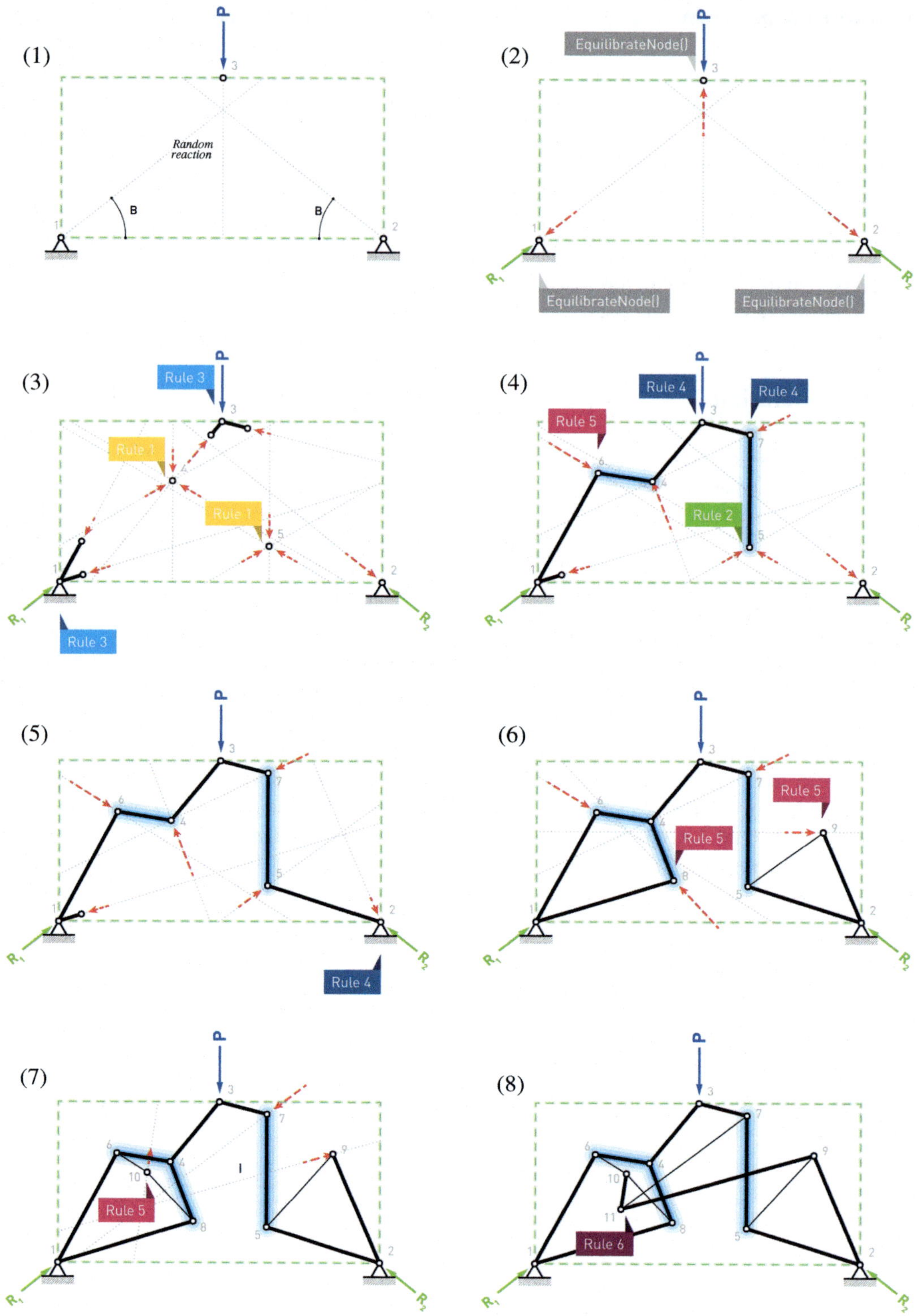

Fig. 5 An example of an automatic random generation sequence

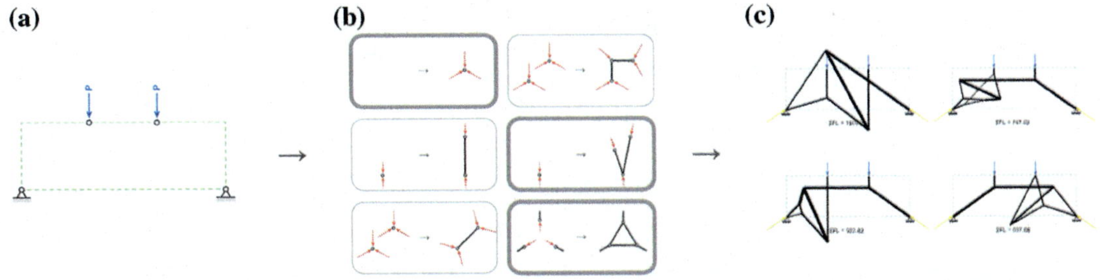

Fig. 6 User workflow of the proposed methodology. **a** Step 1: Setup problem, create initial assembly. **b** Step 2: Pick rules to apply, and set rule weights. **c** Step 3: Generate iterations, and sort based on a desired performance metric

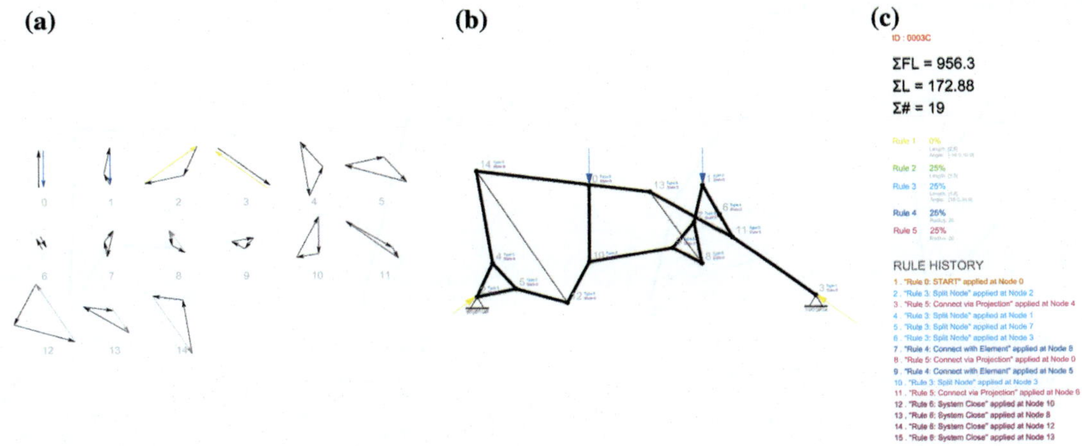

Fig. 7 Screenshot of an example problem in Rhinoceros

during later stages in design, when more information about the boundary conditions and the project in general may be available. This blueprint can also be used to develop more detailed versions of the design, and allow creative breeding using genetic algorithms.

Results

Figures 8, 9, 10, 11, 12 and 13 show the application of the tool on six different design tests. Designs shown are top eight performing solutions out of 80 iterations. All designs have performances that are approximately within 20 % of the solution that can be derived using the fewest number of members, which is given a normalized score of 1.00 in each design test. When a simple, benchmark solution cannot be easily found as in the case of design test 6, the best performing solution out of the 8 designs is used as the benchmark. The performance is based on the total volume of structural material, or equivalently the total load path. Assuming constant internal stress at its optimal or final iteration state, the total volume or load path can be calculated by the simple formula: $\Sigma|F| \cdot L$, where F is the internal force of a member, and L is the length of that member (Baker et al. 2013). Using graphic statics, $\Sigma|F| \cdot L$ can be computed easily by multiplying the length of the member in the form diagram, and the length of the corresponding force vector which is provided by the force diagrams. The designs randomly produced through grammatical exploration exhibit

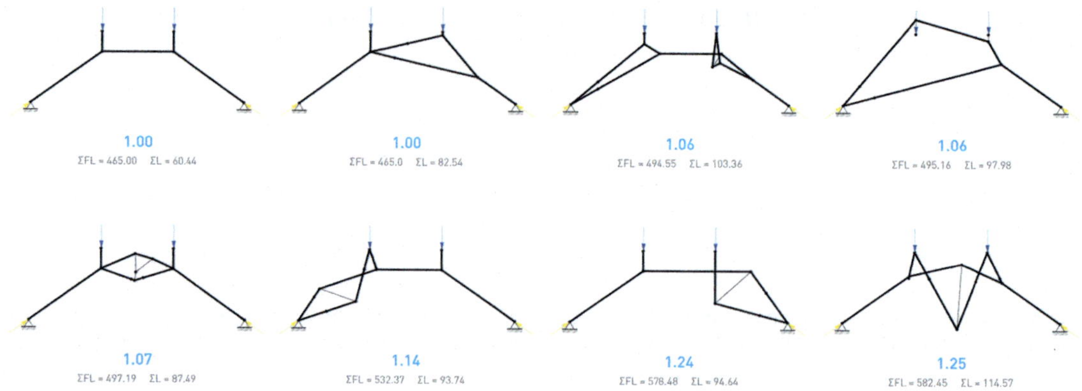

Fig. 8 Design test 1, a span structure

Fig. 9 Design test 2, vertically cantilevered structure

significant diversity, which may often be desirable even at the sacrifice of a small amount of efficiency.

Exploration of Parameters

Several global and local parameters can be modified to explore design alternatives, as well as tradeoffs between various constraints. Parameters described in this section relate specifically to the design test 1 in Fig. 8.

Global Parameter 1: Reaction Angles

Because the support reaction vectors are determined before the automatic random generation begins, a variety of possible solutions with

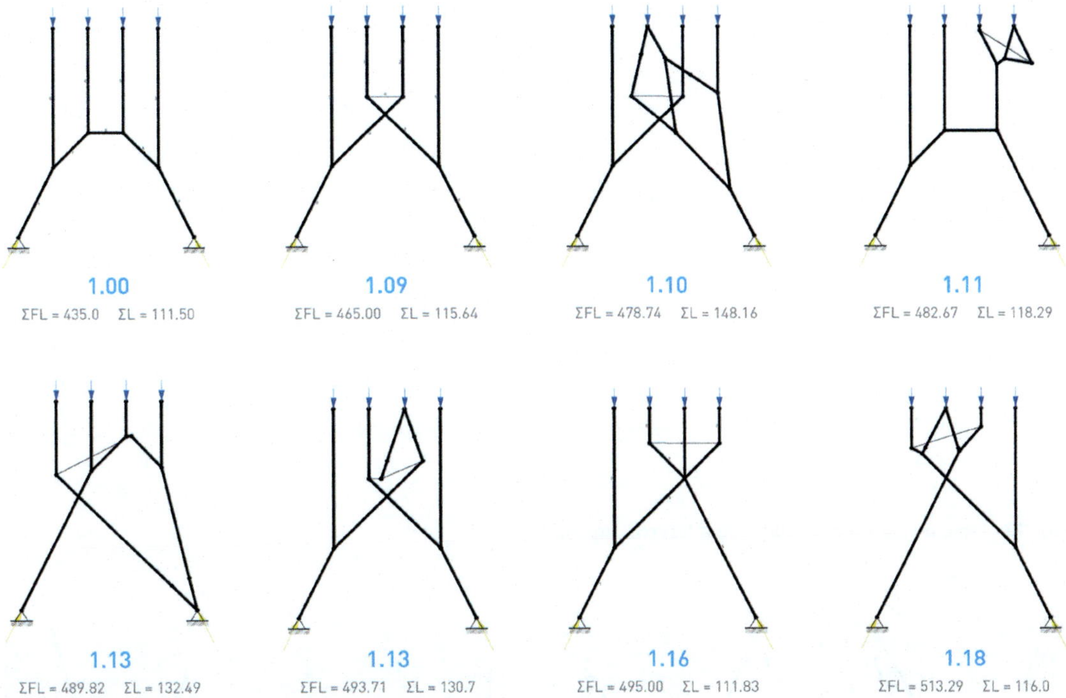

Fig. 10 Design test 3, a wall-like compressive structure

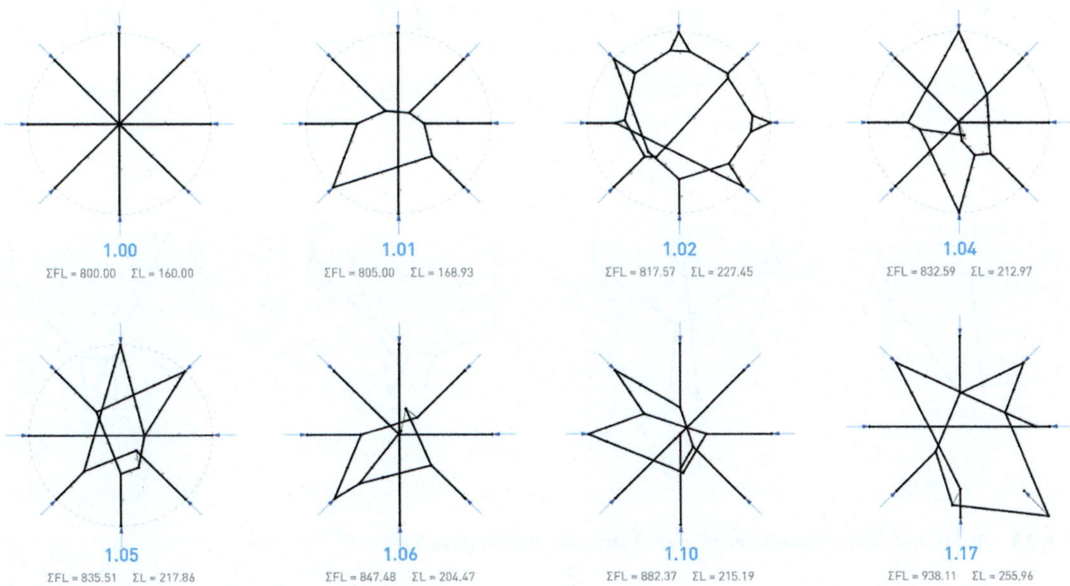

Fig. 11 Design test 4, radial compression structure

varying shapes and performances can be explored by setting a reasonable bound for this angle parameter, as illustrated in Fig. 14. Conversely, the reaction angles can be altered after a design has been chosen to improve the performance. This parameter is most closely

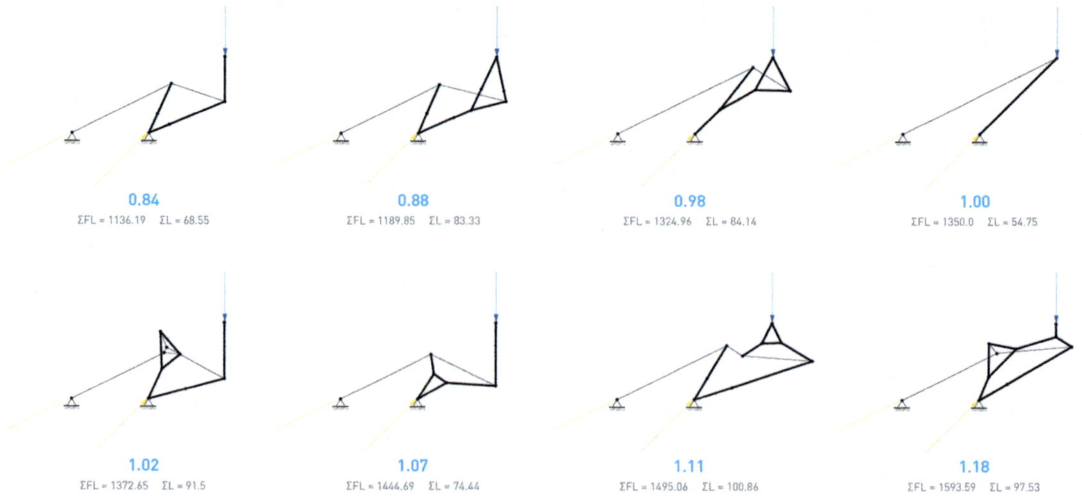

Fig. 12 Design test 5, horizontally cantilevered structure

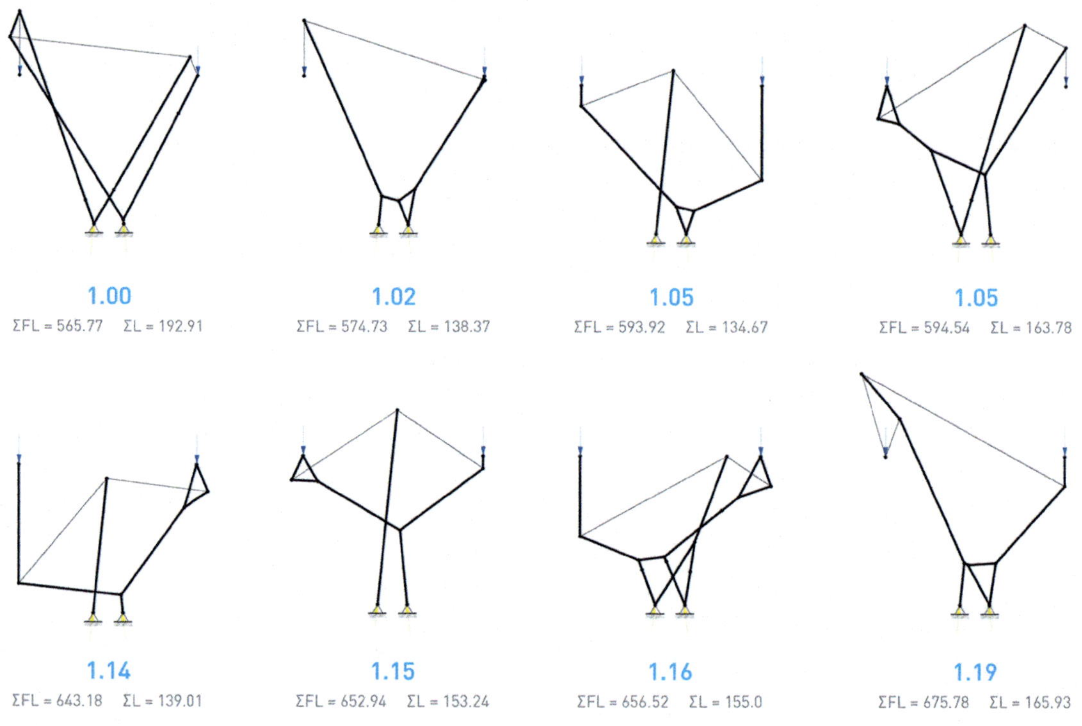

Fig. 13 Design test 6, horizontally cantilevered structure in two directions

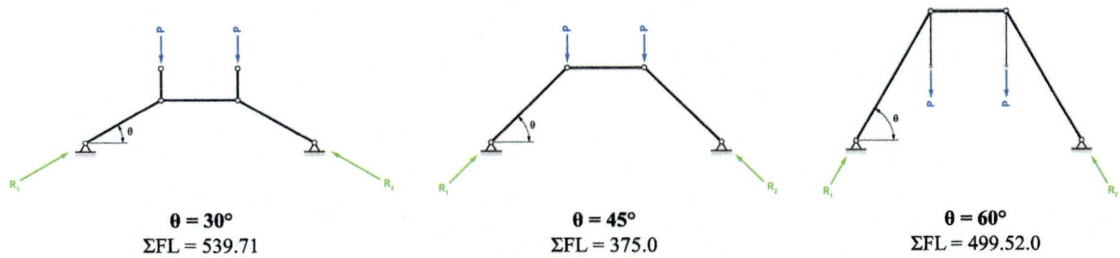

Fig. 14 Changing the angle of the reactions not only changes the shape of the structure, but also its performance

related to the boundary conditions of the supports, and how much horizontal reaction a support can provide.

Global Parameter 2: Generation Count

Generation count defines the maximum number of rules that can be applied in a generation. Figure 15 shows three similar structures with varying generation counts. The generation count can be increased in cases where more redundancy may be required, or more geometric variation and expression within a design are desired.

Rule Parameters

Figure 16 shows the effect that rule parameter variations can have on the results. The only parameter that was changed was the lower and upper bounds for the angle of the split rule. While increasing the range of possible angles does not necessarily improve the performance, the larger angles may be necessary for constructability of joints. Similarly, modifying the parameters for other rules will result in drastically diverse designs.

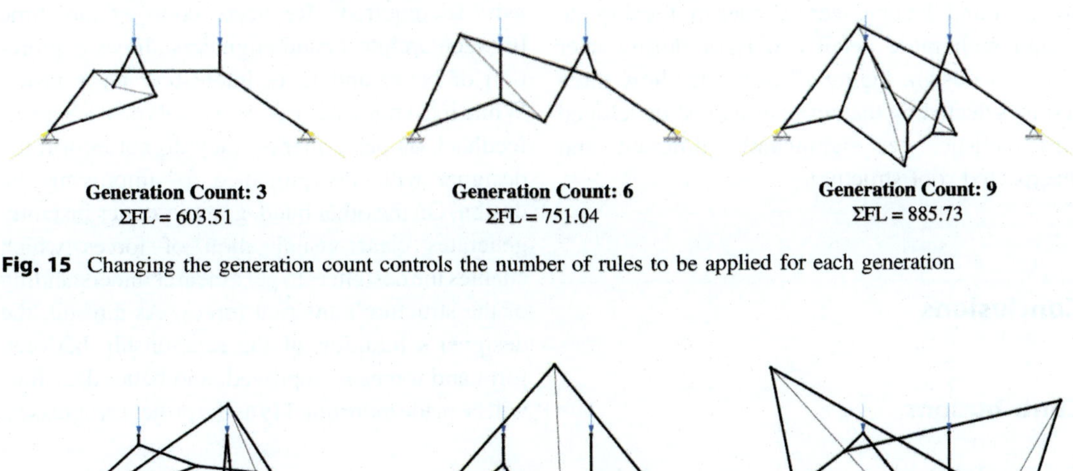

Fig. 15 Changing the generation count controls the number of rules to be applied for each generation

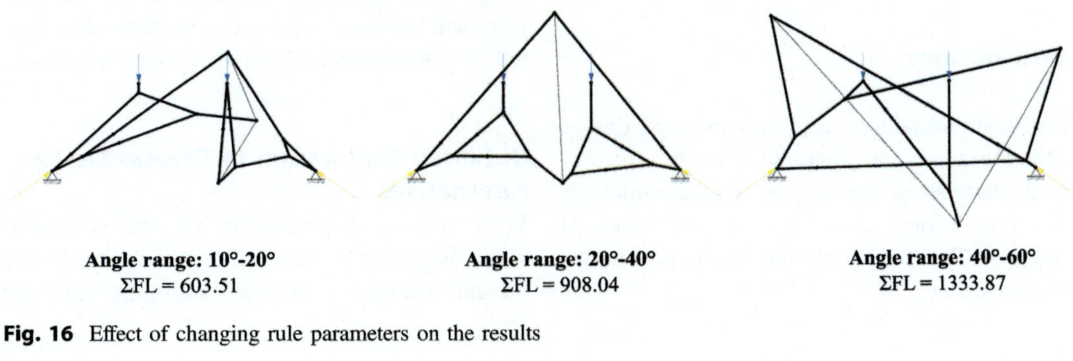

Fig. 16 Effect of changing rule parameters on the results

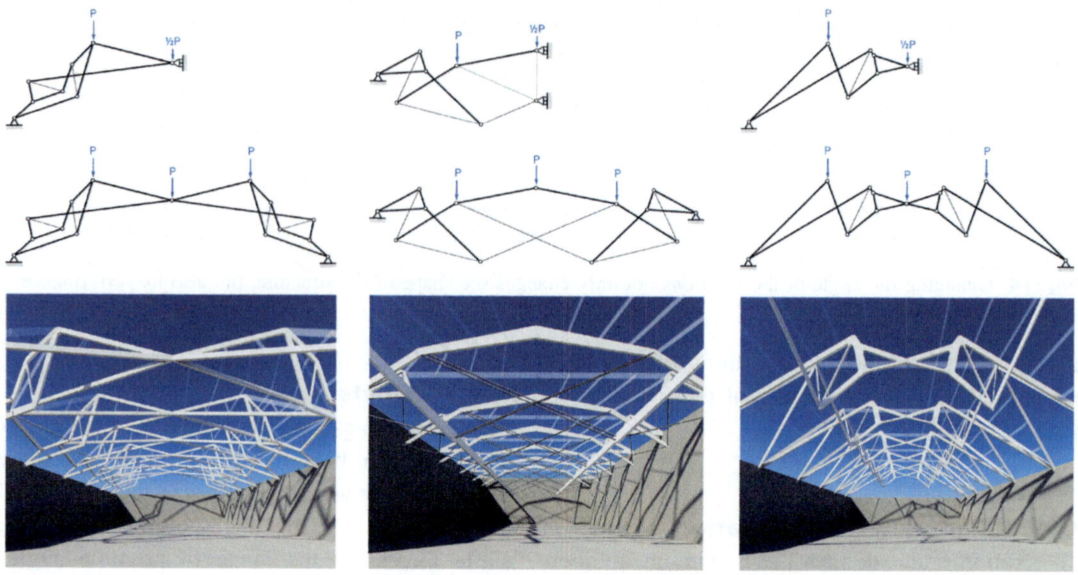

Fig. 17 Sample designs with symmetry enforced, resulting in more practically applicable designs

Practical Applications

While the proposed tool allows fast exploration of design possibilities during conceptual design, the results will need to be interpreted by the architect and the engineer in order to develop the design with more detail and rigor during later stages of design. Figure 17 illustrates how three results selected by the designer can be developed into realistic, yet significantly different and unexpected roof structures.

Conclusions

Contributions

This paper introduced a grammar-based design methodology, as an alternative to the conventional parametric design paradigm, which is limited in variety and often lead to expected solutions. The following specific contributions were presented.

More Trial and Less Error

By incorporating forces during the form generation process, the resulting designs are guaranteed to be in equilibrium. Therefore, no further numerical analysis is required. Reduced coordination time between architects and engineers allows exploration of better and more interesting ideas faster. While most numerical analysis tools provide quick feedback on performance, they do not inform the designer with any guidance for improving the design. On the other hand, graphic statics instantly generates clear visualization of forces which enables the designers to get a clearer understanding of the structure's internal forces. As a result, the designer's intuition of the relationship between form and forces is improved, and better decisions will be made more quickly as the project progresses.

Unbiased Exploration of Diverse Design Alternatives

With automated generation by the computer, which is guided by the design goals input by the human designer, diverse solutions can be generated that simply would not be conceivable manually by a human designer with a pencil or a mouse. In addition, the automated generation of multiple designs at once not only increases the creative capacity of the designer, but also leads to new insights and better understanding of the design problem itself.

Generative Graphic Statics: Beyond Reciprocity

The reciprocal relationship between form and forces in graphic statics, means that one has to be created before the other can be drawn. Therefore, most computational graphic statics tools only work with pre-set problems, functioning mostly as an interactive analysis or visualization tool. By combining graphic statics with shape grammars, the form finding capabilities of graphic statics can be used to *generate* equilibrium structures. Most previous work done on shape grammars require a shape to pre-exist before a rule can be applied. However, the rules presented in this paper are based on *Nodes* or points, and are not dependent on any preceding shapes or conditions. Therefore, the methodology is flexible enough to be applied to a variety of design problems, and is able to generate structures without any prescribed typologies or preferences.

Future Work

Although this paper was a successful first attempt of implementing this new methodology, there are several important directions for future work. First, global parameters could be improved to gain better control of overall generation process, including more intelligent ways in which the rules are chosen and where they are applied. Secondly, more detailed or material-specific constraints, buckling lengths, and minimizing number of overlapping members could be incorporated. Also, because all designs shown in this paper are also statically equilibrated only for the defined load case, it will be important to develop a procedure to check for possible mechanisms and local instabilities. Lastly, while this paper focused on rules based on the form diagram, rules can also be developed for the force diagram (Akbarzadeh et al. 2014), which will further enrich the design possibilities.

Closing Remark

Overall, this new methodology demonstrates the validity in combining and applying shape grammars and graphic statics together to various engineering design problems. The general versatility and customizability of the tool, and the speed at which it can generate unconventional and yet statically equilibrated structures, greatly improves possibilities for creative yet performance-focused explorations during early stages of conceptual structural design.

References

Akbarzadeh M, Van Mele T, Block P (2014) Compression-only form finding through finite subdivision of the force polygon. In: Proceedings of the international association for shell and spatial structures symposium, Brasilia, Brazil, 15–19 Sept 2014

Allen E, Zalewski W (2009) Form and forces: designing efficient, expressive structures. Wiley, New York

Baker W, Beghini L, Arkadiusz M, Carrion J, Beghini A (2013) Maxwell's reciprocal diagrams and discrete Michell frames. In: Rozvany G (ed) Structural and multidisciplinary optimization 48. Springer, Heidelberg, pp 267–277

Cremona L (1890) Graphical statics: two treatises on the graphical calculus and reciprocal figures in graphical statics. Claredon Press, Oxford

Culmann K (1864) Die graphische statik. Verlag Meyer & Zeller, Zurich

Fivet C, Zastavni D (2013) Constraint based graphic statics: new paradigm of computer-aided structural equilibrium design. J Int Assoc Shell Spat Struct 54 (4):271–280

Greenwold S, Allen E (2003) Active statics. http://acg.media.mit.edu/people/simong/statics/Start.html. Accessed 2 Mar 2015

Lee J (2015) Grammatical design with graphic statics: rule-based generation of diverse equilibrium structures. Master thesis, Massachusetts Institute of Technology

Mitchell W (1991) Functional grammars: an introduction. In: Goldman Z (eds) Proceedings of association for computer aided design in architecture: reality and virtual reality, Los Angeles

Mueller CT (2014) Computational exploration of the structural design space. Dissertation, Massachusetts Institute of Technology

Shea K, Cagan J (1999) Languages and semantics of grammatical discrete structures. Artif Intell Eng Des Anal Manuf 13:241–251

Stiny G, Gips J (1972) Shape grammars and generative specification of painting and sculpture. Inf Process 71:1460–1465

Van Mele T, Block P, Ernst C, Ballo L (2014) eQUILIBRIUM: an interactive, graphic statics-based learning platform for structural design. http://block.arch.ethz.ch/equilibrium. Accessed 5 May 2015

ShapeOp—A Robust and Extensible Geometric Modelling Paradigm

Mario Deuss, Anders Holden Deleuran, Sofien Bouaziz, Bailin Deng, Daniel Piker and Mark Pauly

Abstract

We present ShapeOp, a robust and extensible geometric modelling paradigm. ShapeOp builds on top of the state-of-the-art physics solver (Bouaziz et al. in ACM Trans Graph 33:154:1–154:11, 2014). We discuss the main theoretical advantages of the underlying solver and how this influences our modelling paradigm. We provide an efficient open-source C ++ implementation (www.shapeop.org) together with scripting interfaces to enable ShapeOp in Rhino/Grasshopper and potentially other tools. This implementation can also act as a template for future integration of computer graphics research. To evaluate the potential of ShapeOp we present various examples using our implementation and discuss potential implications on the design process.

Introduction

Under the well established geometric modelling paradigms such as constructive solid geometry or spline-based modelling, polygonal mesh modelling yields a good tradeoff between expressibility —its many degrees of freedom allow to approximate an arbitrary design—and computational effort—its inherent linear interpolation reduces mathematical complexity. This has led to the development of various form-finding and modelling tools for the exploration of shape spaces of polygonal meshes. In our context we consider a shape space as a set of all designs that respect given geometric constraints dictated by aesthetic, fabrication and cost requirements. See Fig. 6 in Bouaziz et al. (2012) for an example of a shape space.

Shape space exploration is typically facilitated by an optimization algorithm that negotiates a large number of complex and possibly conflicting constraints to satisfy the design goals. Numerical solvers for constraint satisfaction therefore play a fundamental role in shape exploration environments. A number of requirements on these solvers

M. Deuss (✉) · S. Bouaziz · B. Deng · M. Pauly
École Polytechnique Fédérale de Lausanne, Lausanne, Switzerland
e-mail: mariodeuss@cpfl.ch

A.H. Deleuran
Royal Danish Academy of Fine Arts School of Architecture, CITA, Copenhagen, Denmark

D. Piker
Robert McNeel and Associates, London, UK

M.R. Thomsen et al. (eds.), *Modelling Behaviour*, DOI 10.1007/978-3-319-24208-8_42

are essential for an effective design process, such as numerical robustness, computational efficiency, flexibility to handle a diverse set of design constraints, and extensibility to adapt to new design environments.

Existing shape exploration methods are often restricted by inherent limitations of their optimization approach. They might be tailored to a specific set of constraints, for example planarity of polygons, which can limit design flexibility. Some of them exhibit numerical instabilities or slow convergence, which makes interactive modelling cumbersome. Last but not least, they are often closed, monolithic software, which makes adaptations or extensions in new design tasks difficult. We propose a new computational approach to geometric modelling and design that alleviates these limitations.

We adopt the physics solver proposed in Bouaziz et al. (2014) that integrates a variety of constraints, dynamics and handle-based shape space exploration, and add projective constraints described in Bouaziz et al. (2012). We refer to the combination as ShapeOp. In this paper we evaluate the potential of ShapeOp for design in a number of examples using Rhino/Grasshopper as a graphical user interface. We also discuss and provide our implementation of ShapeOp, which effectively bridges the gap between computer graphics research and practical computational design, and acts as an open-source template for making research available. ShapeOp can also act as a building block for algorithms exploring further aspects of the shape space, e.g. adaptive meshing, evolutionary optimization and automatic constraint selection. The contribution of this paper is three-fold:

1. We propose ShapeOp, a state-of-the-art unified and extensible constraint solver based on the latest research in computer science, and make it accessible to the architectural modelling community.
2. We describe and provide an efficient C++ open-source implementation of ShapeOp and an integration into Rhino/Grasshopper using Python/ctypes.
3. We highlight design applications and demonstrate the extensibility of ShapeOp in various examples.

Related Work

The two papers Bouaziz et al. (2012, 2014) provide a thorough discussion about previous work related to ShapeOp. In the following, we focus on related work in the domain of constraint optimization and form-finding.

For computational design, numerical optimization is a fundamental tool as it allows multiple requirements to be incorporated in the design process. For example, in architectural geometry, design shapes are often optimized according to certain geometric constraints that correspond to fabrication requirements (Pottmann et al. 2015). Typically, a non-linear least squares problem is formulated, with each residual term corresponding to one constraint, and solved using standard solvers such as Gauss-Newton.

Numerical optimizations as discussed above can be time-consuming, due to the need for evaluating the Jacobian and solving a different linear system in each iteration of the solver (Nocedal and Wright 2006). In comparison, each iteration of ShapeOp only involves parallel evaluation of projection operators, as well as the solution of a pre-factorized linear system. Recently, Tang et al. (2014) propose a form-finding technique for polyhedral meshes, with much better performance than classical non-linear least squares formulations. However, their approach still relies on solving different linear systems in each iteration, resulting in poor performance for meshes with more than a few thousand vertices. On the contrary, the fixed linear system in ShapeOp makes it suitable even for large models.

For form-finding, one popular approach is to model the shape as a system of nodes subject to internal and external forces, and to compute the final shape as an equilibrium state of the system (Day 1965; Kilian and Ochsendorf 2005;

Attar et al. 2009; Senatore and Piker 2015). For example, Kilian and Ochsendorf (2005) use particle-spring systems for finding structural forms composing only axial forces. Using an implicit Runge-Kutta solver for computing the equilibrium state, their method allows the user to interact with the simulation while it is running. Such forcebased approach is also adopted in Kangaroo, a live physics engine built on top of Grasshopper (Piker 2013). By modelling geometric constraints as forces, Kangaroo can perform not only form-finding and physics simulation, but also constraint solving and optimization, making it a popular tool among architects.

Although such force systems are intuitive to set up, it is challenging to simulate their behavior in an efficient, accurate, and stable way (Witkin and Baraff 1997). Implicit solvers allow for a large time step and require fewer iterations, but each iteration can be quite costly to compute since it requires solving a system of algebraic equations. Also, adding new forces requires the derivation of a Jacobian, making them more difficult to extend. Explicit solvers involve lower computational cost for each iteration, but require smaller step sizes to produce stable results, which can result in a large number of iterations. For example, one issue of Kangaroo as presented in (Piker 2013) is that the simulation can explode for highly stiff problems, since such problems require a step size much smaller than the default value; as a result, it is difficult to compute a shape that satisfies the given constraints exactly, because this will require large forces for the constraints and lead to very stiff systems.

Unlike such force-based approaches, ShapeOp computes the equilibrium state of a system by minimizing a potential energy that incorporates physical forces as well as geometric constraints. Using the carefully designed numerical solver in ShapeOp, a stable solution can be computed in a small number of iterations with low computational cost, achieving better stability and efficiency than force-based solvers. Another benefit of ShapeOp is that it is fully open-source, with bindings for many languages including C, C ++, C#, Java, and Python. This makes it easy to use ShapeOp from different programming environments, and to extend and adapt the codes according to specific needs.

Solver

ShapeOp is a physics engine as well as an optimization tool, designed for a set of points that are subject to physical and geometric constraints. In dynamic mode ShapeOp simulates physics by preserving momentum. In static mode ShapeOp optimizes for an equilibrium solution, which converges considerably faster due to the absence of oscillations induced by momentum.

For constrained optimization, ShapeOp adopts the iterative solver of (Bouaziz et al. 2014), which models physical potentials as well as geometric constraints including the ones presented in (Bouaziz et al. 2012) in a unified manner. Each iteration of the solver consists of a local step and a global step (Fig. 1):

Local Step

A candidate shape is computed for each set of points that are commonly influenced by a constraint. For a geometric constraint, this amounts to fitting to the points a shape that satisfies the constraint. For a physical constraint, this reduces to finding the closest point positions that have zero physical potential value.

Global Step

The candidate shapes computed in the local step are incompatible. The global step solves for a new set of consistent point positions such that each set of points subject to a common constraint are as close as possible to the corresponding candidate shape (Fig. 1, right).

By repeating the above steps, the overall constraint violation is decreased in each iteration, and the mesh converges to a shape that satisfies all physical and geometric constraints as much as

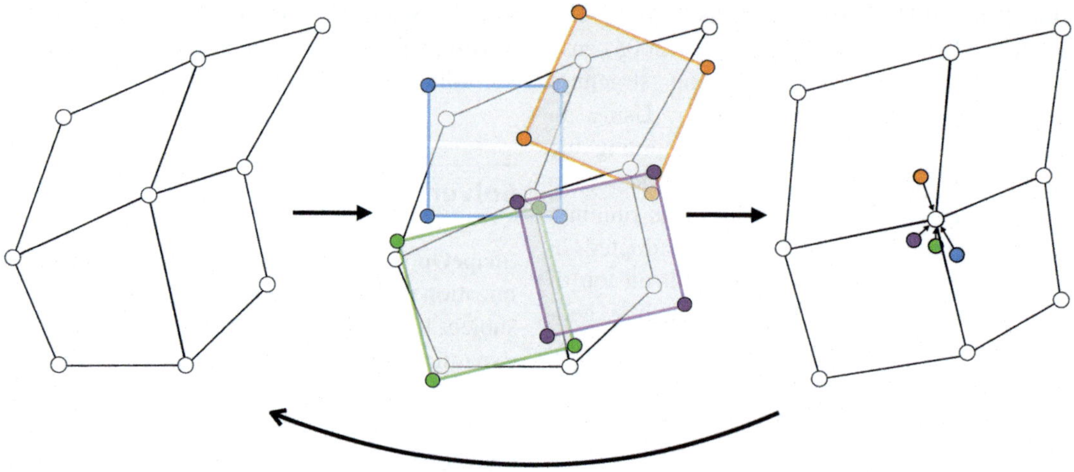

Fig. 1 A quad mesh constrained to consist of squares illustrating the ShapeOp solver. *Left* Initial configuration. *Middle* Local step—projecting each quad onto its closest square. *Right* Global step—joining the individual projections by a global minimization. The resulting mesh is then used as initial configuration and the solver iterates

possible. Moreover, each iteration can be run very efficiently: in the local step different constraints can be handled in parallel, while in the global step we only need to solve a linear system with a fixed matrix.

For dynamics simulation, ShapeOp uses the implicit Euler integration scheme from Bouaziz et al. (2014), where at each integration step the physical and geometric constraints are resolved using the above local-global solver. Thanks to the efficiency of the local-global solver, ShapeOp benefits from the stability of implicit integration, with significantly lower computational cost than traditional implicit Euler solvers. ShapeOp also allows defining external forces such as wind and gravity. For more information, please refer to the paper Bouaziz et al. (2014).

Projections

Central to the constrained optimization solver in ShapeOp are the so-called projection operators, which are used to compute the candidate shapes in the local step. Specifically, given a set of points that are subject to a constraint, the projection operator finds the closest point positions that satisfy the constraint. A new constraint can be added easily to ShapeOp, as long as its projection operator is provided. No changes to global step of the solver are necessary to add a constraint.

A simple example of a constraint is the closeness constraint: It is satisfied if the constrained point x coincides with a prescribed position c. Since the only way to satisfy the closeness constraint is by setting x equal to c, the projection $P(\cdot)$ simply is given by $P(x) = c$. The documentation of ShapeOp also provides a step-by-step tutorial on an orientation constraint, from formulation to implementation. See Table 1 for all constraints implemented in ShapeOp.

While many of the constraints intuitively apply to specific primitives, some of them can be applied to an arbitrary set of points defining novel shape spaces. For example, the circle constraint was often applied to all quads of a mesh because such circular meshes have desirable offset properties (Pottmann et al. 2007). However, it can also be applied to each grid line of a quad mesh, defining an interesting shape space as illustrated in Fig. 5. Also note that the ShapeOp solver has no explicit knowledge of a mesh, but only of a list of points. This allows to apply ShapeOp to any set of points, e.g. mixing different geometric primitives such as splines, tetrahedral meshes, Bezier patches or triangle soups, that are parametrized by point positions.

Table 1 Constraints implemented in ShapeOp

Constraint	Description
Edge, triangle, tetrahedron strain area, volume	Bounds the strain with respect to its initial configuration (Bouaziz et al. 2014, Sect. 5.1)
	Bounds the area (volume) of a triangle (tetrahedron) (Bouaziz et al. 2014, Sect. 5.2)
Bending closeness	Constrains a point to a prescribed position
Line, plane, circle, sphere, rectangle, parallelogram	Constrains points to a lie on a geometric primitive (Bouaziz et al. 2012, Sect. 3).
Uniform laplacian	Constrains a point to the average of its neighbors (Bouaziz et al. 2012, Sect. 4)
Uniform laplacian of deformation	Constrains a deformation vector with respect to the initial position to be the average of its neighboring deformation vectors (Bouaziz et al. 2012, Sect. 4).
Similarity	Constrains points to be similar (related by a rigid motion and uniform scaling) to one of the prescribed set of points. The similarity constraint automatically selects the closest of the prescribed sets of points to project to at each iteration (Bouaziz et al. 2012, Sect. 3.2)
Rigid	This constraint is equivalent to Similarity, only that it does not allow for uniform scaling (Bouaziz et al. 2012, Sect. 3.2)
Angle	Bounds the angle formed by three points (Deng et al. 2015, Sect. 3.3.2)

Implementation

Our implementation of ShapeOp is distributed as a header-only, C++ library. While it is possible to develop C++ plugins for computational design environments this requires a larger and substantially more involved development investment than what is offered by higher level programming languages provided in .NET compliant CAD environments such as Rhino 3D and Revit. Here languages such as C#, VB and Python make development of computational design models fast, responsive, and interchangeable (Fig. 2).

Our implementation can be conceptually divided into two components: The core library *libShapeOp* with various bindings, and *applications* which use *libShapeOp*. The core library contains the abstract C++-classes *Solver*, *Constraint* and *Force*, and many classes deriving from and implementing them. The C-API provides an interface using C only, which simplifies calling *libShapeOp* from other code or programs considerably. ShapeOp also provides everything necessary to use SWIG (www.swig.org), a software development tool that can generate a multitude of bindings for *libShapeOp*. The applications contain the Grasshopper definitions using *libShapeOp*. The definitions use GhPython (www.food4rhino.com/project/ghpython) to enable Python in scripting components. Inside the component we use Python's ctypes to directly call *libShapeOp* (Fig. 3).

There are six ShapeOp Grasshopper components in the current release. The functionality of each is implemented in a Python script. The *ShapeOp Constraint Solver* (*SOSolver*) is the central component and is the only one that calls

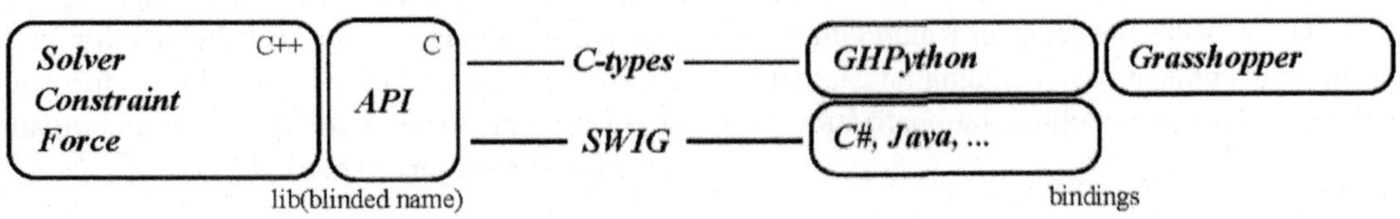

Fig. 2 Schematic overview of our implementation of ShapeOp

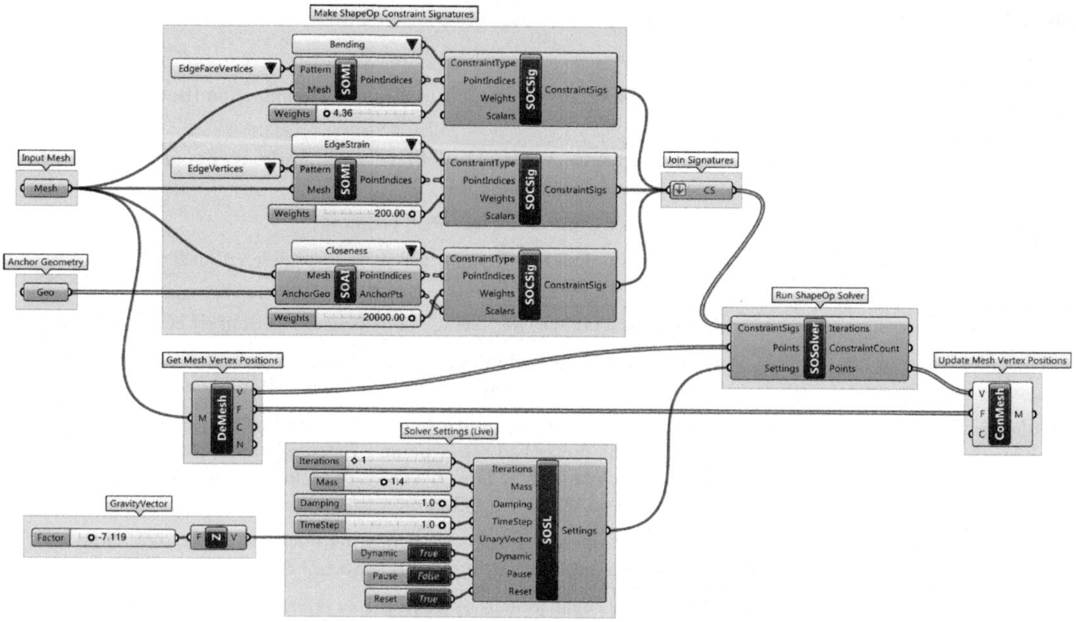

Fig. 3 The Grasshopper definition used for the hanging cloth example seen in Fig. 4

the *libShapeOp* library. It sends points and constraint signatures to the library and retrieves the result. A constraint signature contains all the necessary information to setup a constraint: The constraint type represented by a string; the indices of points to be constrained with respect to the global list of points; the weight of this constraint; the scalars, a list of floating point numbers encoding additional settings of the constraint. *ShapeOp ConstraintSignature* (*SOCSig*) constructs the constraint signatures. Constraint signatures are implemented using a Python dictionary, so they could also be created by custom python components other than *SOCSig*.

Rhino/Ghpython Implementation Examples

In Figs. 4, 5, 6, 7, 8, 9, 10 and 11, we provide some examples of using ShapeOp in Rhino/GhPython for different applications, including physics simulation, constrained modelling, rationalization, and form-finding.

Design Process

We believe that ShapeOp can have a considerable impact on a design process in various ways. In interactive modelling tools the graphical user interface often cost non-negligible fraction of the execution time, in particular on large data. Since ShapeOp is built modularly, it can be run independently of any user interface. Also, due to the C/C++ implementation, ShapeOp runs natively and is heavily optimized by compilers and parallelization with OpenMP (www.openmp.org). ShapeOp can therefore potentially handle huge models.

In ShapeOp the global and local steps are both numerically stable least-squares problems, implying that the overall method is also stable and robust. Also, many constraints such as the plane constraint only concern the relative arrangement of points and stay satisfied after applying a translation to all points. ShapeOp utilizes this in the global step by implicitly solving for the translations for each constraint independently. This allows for constrained points to move arbitrarily far and greatly increases convergence speed.

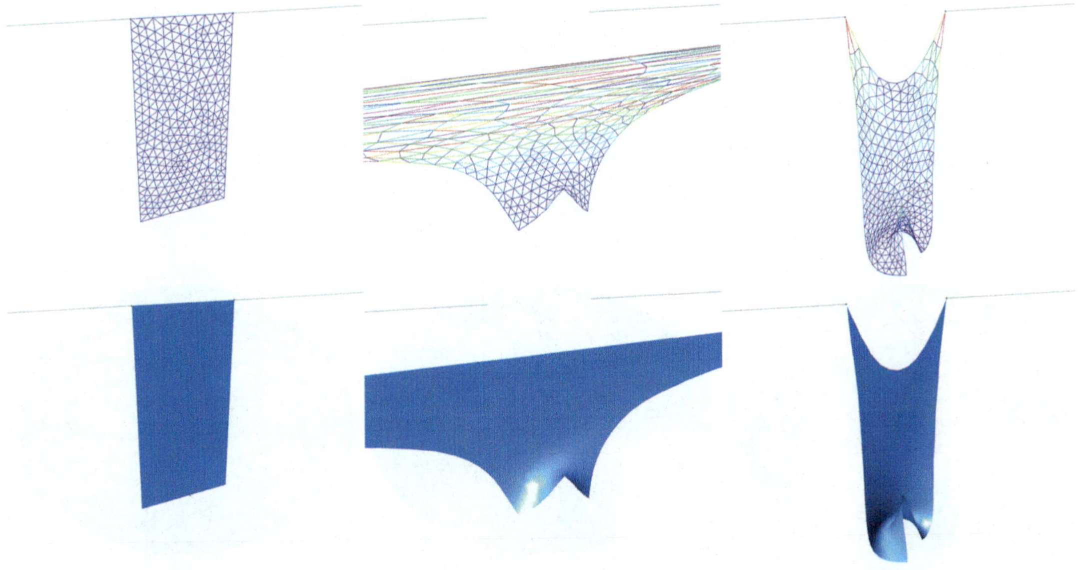

Fig. 4 Use of ShapeOp for physics simulation of elastic materials. A hanging cloth is modelled using edge strain and bending constraints. The three vertices are anchored using closeness constraints and all points are subjected to a gravity load. *Left* The input mesh. *Middle* The constrained mesh at the first solve iteration in which the anchors are immediately moved very far apart. *Right* The constrained mesh after 100 iterations with the anchor point moved back to their starting positions. *Top* Wire-frame rendering with the edges coloured by their strain (*red* high, *blue* low). *Bottom* Shaded rendering. The example demonstrates both the stability and the fast convergence of the solver

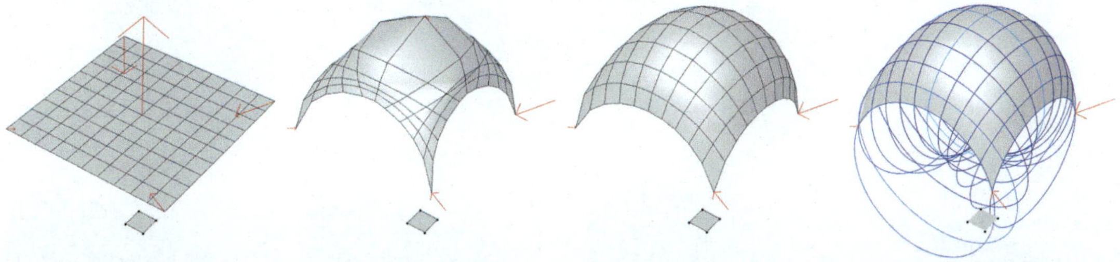

Fig. 5 Use of ShapeOp for constrained modelling of a shell with rational geometric properties. The vertices on the parameter lines of a quad-mesh are constrained to always lie on a circular arc using the circle constraint. Each face is constrained towards being square using the similarity constraint. Five vertices are anchored to different positions than their initial positions, enabling shape exploration. *Left* to *right* (*1*) The input mesh, the face used for similarity and vectors visualizing start/end positions for the anchors. (*2*) The constrained mesh after 10 iterations. (*3*) The constrained mesh after 300 iterations. (*4*) The constrained mesh with circles drawn through each of the parameter line vertices (*red* Line vertices distance to *circle* is large, *blue* Line vertices distance to *circle* is small)

Future Work

As future work we plan to further explore the combination of continuous and discrete constraints. It is a common feature of many design problems to have some design components that can only be selected from a finite set of choices. Unfortunately, such a finite set can be too restrictive to preserve the aesthetic quality of the design. In this case, enhancing the optimization

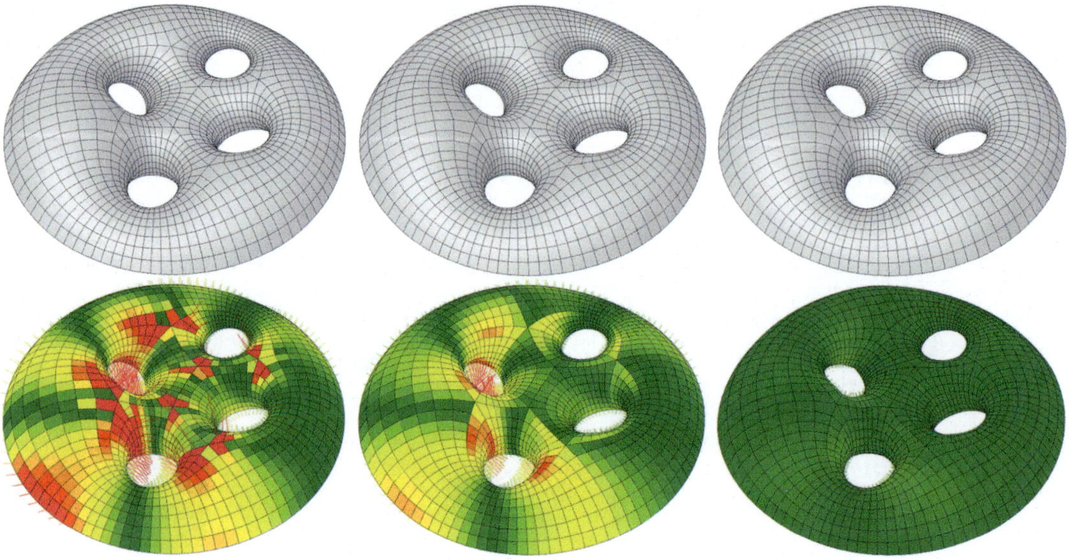

Fig. 6 Use of ShapeOp for rationalizing an existing geometry. Each face of the quad-mesh is constrained towards being planar using the plane constraint. Each vertex is constrained to its initial position using the closeness constraint by a small weight to maintain the shape of the mesh. *Left* Input mesh. *Middle* The constrained mesh after 10 iterations. *Right* The constrained mesh after 200 iterations. *Top* Shaded rendering. *Bottom* Planarity analysis rendering (*Red* Low planarity, *Green* High planarity)

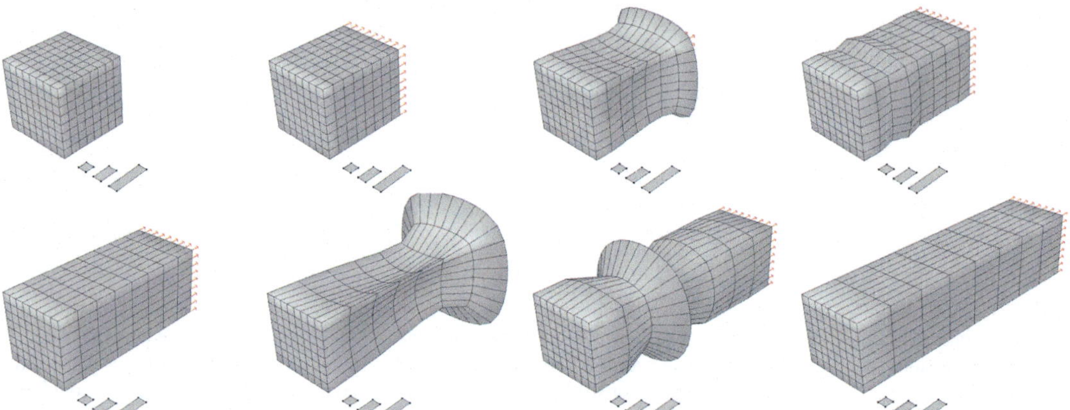

Fig. 7 Use of ShapeOp for constrained modelling of box shape with multiple rigid shape targets. A quad-mesh box is anchored at the vertices on two sides of the box. The image sequence shows the vertices on one side being pulled away over time. As this occurs each mesh face attempts to project itself onto one of the three shape targets below the box. The solver has been initialized using dynamics leading to the rippling effect as the faces switch their projection targets from short to medium to long. This projection type is enabled using the rigid constraint

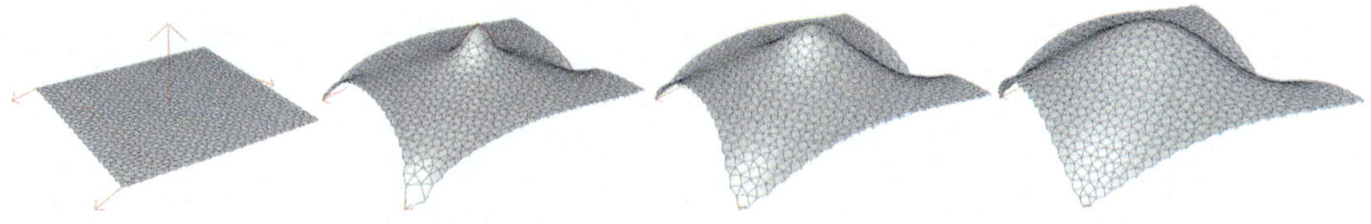

Fig. 8 Use of ShapeOp for constrained modelling of a shell with topologically different shape targets. A planar mesh composed of both triangles and quads is anchored at four vertices using the closeness constraint. Using the similarity constraint, each face is constrained towards their initial shape i.e. an equilateral triangle or a square. (*1*) The input mesh. (*2*) The mesh after 1 iteration. (*3*) After 100 iterations. (*4*) After 500 iterations

Fig. 9 Use of ShapeOp for constrained modelling of a randomly generated quad-mesh with multiple constraints as design drivers. The example demonstrates the effect of applying the same constraints on meshes with different resolutions. It uses three primary constraints: (*1*) Limit the internal angles of each face to be within 80° and 110°, (*2*) The boundaries of the mesh should lie on circles, (*3*) Each face should preserve its area. Additionally, a Laplacian of displacement constraint is added which smoothens out the mesh while maintaining the shape, and a bending constraint is added which ensures that face-face angles do not become too acute. The color code is a based on scoring system: The internal angles for each face are calculated. If an angle is within the desirable range it is scored 0, else 1. The scores are added for each face, best face score is 0 (*dark green*) worst is 4 (*dark red*)

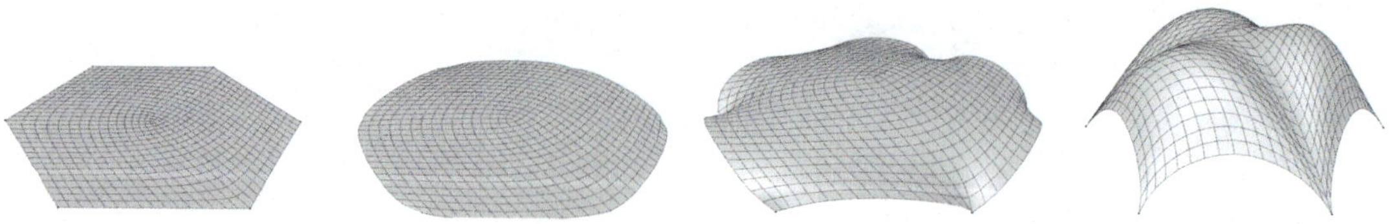

Fig. 10 Use of ShapeOp for funicular form finding. A hexagonal quad-mesh is anchored at each corner and subjected to an inverse gravity load. In this image sequence the only other constraint is that each edge should be 2.0 units long. This is implemented using the edge strain constraint. (*1*) The input mesh. (*2*) The constrained mesh after 1 iteration. (*3*) The constrained mesh after 10 iterations. (*4*) The constrained mesh after reaching equilibrium at iteration 1000

Fig. 11 Use of ShapeOp for funicular form finding under fabrication constraints. Demonstrates the effect of combining different constraints: Desired edge length, planarity of faces and desired range of internal face angles. All images show the constrained mesh at equilibrium. From *left* to *right* (*1*) Shaded rendering. (*2*) Edge length deviation from desired length. (*3*) Face angles deviation from desired angle range. (*4*) Face Planarity Deviation (*red* high deviation, *green* low deviation). *Row 1* The mesh with edge strain constraints. *Row 2* The mesh with edge strain and internal mesh face angles constraints. *Row 3* The mesh with edge strain and face planarity constraints. *Row 4* The mesh with edge strain, internal mesh face angles and face planarity constraints

to automatically detect a sparse set of constraints that needs to be violated in order to better preserve the design intent would lead to a much richer design exploration solution.

Conclusion

We believe that ShapeOp is a promising candidate for state-of-the-art computational geometric design and evaluate its potential in this paper. We explain the theoretical advantages over existing methods, and present the implementation as a simple, fast and extensible C++ library (www.shapeop.org). Our examples use the scripted grasshopper components provided with ShapeOp to highlight its practical importance.

Acknowledgements We thank the reviewers for their valuable comments. This work has been supported by Swiss National Science Foundation (SNSF) grant 200021 137626 and the Danish Council for Independent Research (DFF). This research has received funding from the European Research Council under the European Unions Seventh Framework Programme (FP/2007–2013) /ERC Grant Agreement 257453, ERC Starting Grant COSYM.

References

Attar R, Aish R, Stam J, Brinsmead D, Tessier A, Glueck M, Khan A (2009) Physicsbased generative design. In: CAAD futures conference, pp 231–244

Bouaziz S, Deuss M, Schwartzburg Y, Weise T, Pauly M (2012) Shape-up: Shaping discrete geometry with projections. Comput Graph Forum 31(5):1657–1667

Bouaziz S, Martin S, Liu T, Kavan L, Pauly M (2014) Projective dynamics: fusing constraint projections for fast simulation. ACM Trans Graph 33(4):154:1–154:11

Day AS (1965) An introduction to dynamic relaxation. Engineer 219:218221

Deng B, Bouaziz S, Deuss M, Kaspar A, Schwartzburg Y, Pauly M (2015) Interactive design exploration for constrained meshes. Comput Aided Des 61:13–23

Kilian A, Ochsendorf J (2005) Particle-spring systems for structural form finding. J Int Assoc Shell Spat Struct 46(2):77–84

Nocedal J, Wright S (2006) Numerical optimization, 2nd edn. Springer series in operations research and financial engineering, Springer, New York

Piker D (2013) Kangaroo: form finding with computational physics. Architect Des 83(2):136–137

Pottmann H, Liu Y, Wallner J, Bobenko A, Wang W (2007) Geometry of multi-layer freeform structures for architecture. ACM Trans Graph 26(3):1–11

Pottmann H, Eigensatz M, Vaxman A, Wallner J (2015) Architectural geometry. Comput Graph 47:145–164

Senatore G, Piker D (2015) Interactive real-time physics: an intuitive approach to form-finding and structural analysis for design and education. Comput Aided Des 61:32–41

Tang C, Sun X, Gomes A, Wallner J, Pottmann H (2014) Form-finding with polyhedral meshes made simple. ACM Trans Graph 33(4):70:1–70:9

Witkin A, Baraff D (1997) Physically based modeling: principles and practice. In: Siggraph '97 Course notes

Iterating Towards Affordability

Michail Georgiou, Odysseas Georgiou
and Chris Williams

Abstract

The paper presents ongoing research, aiming towards affordability of complex forms in architecture. It is supported that the above objective can be reached through an enhanced design process integrating materiality and construction logistics in computational feedback models. Within the above framework, the findings of two consecutive case studies are compared, illustrating developments in modelling material behaviour and advances in optimization of construction logistics.

Introduction

This paper presents current developments and outcomes of ongoing research, aiming towards affordability of complex forms in architecture. It is supported that such direction can be approached by re-examining and integrating materiality and construction logistics (focusing on assembly and connectivity) in computational feedback models. As such, an enhanced design process is proposed, incorporating a series of digital simulations verified by parallel physical testing providing input parameters that feed a core computational model. It is illustrated that such model allows *minimization of analytical time*, *optimization of end results* and *simplification of construction processes*.

Materiality is therefore presented as a design tool enabling major cost reductions as opposed to custom, post-designed solutions. Notions such as "active bending" are integrated through finite element analysis in an attempt to fully assess and exploit the behaviour and nature of the materials used. Such behaviour is computationally simulated and physically verified while it is further enhanced by additional linear elements instrumentalized to optimize performance.

In parallel, construction logistics are perceived as proving ground for digital design and fabrication towards affordable complex forms. As such, computational models with built-in construction logic and information allow automation of labour-intensive processes while facilitating assembly of building parts. In

M. Georgiou (✉) · O. Georgiou
HUB Design + Engineering Platform, ARC,
University of Nicosia, Nicosia, Cyprus
e-mail: georgiou.mi@unic.ac.cy

C. Williams
University of Bath, Bath, UK

M.R. Thomsen et al. (eds.), *Modelling Behaviour*, DOI 10.1007/978-3-319-24208-8_43

addition, the use of digital fabrication techniques enables the design of flexible, affordable and adaptable jointing systems, in line with the overall behaviour/materiality of the structure.

Within the above framework this paper expands upon the findings of two consecutive case studies. The comparison between the two examples presents an evolution in modelling material behaviour and optimization of construction logistics with direct measurable benefits.

Both case studies were developed as adaptable structure-membrane interdependent systems. The structure was made out of recyclable PVC electrical conduit pipes, bent in place and secured using custom-made digitally fabricated joints. In both cases, the form was achieved by instrumentalising the bending forces induced on the pipes as defined by a computational model. The case studies illustrate an evolution of the membrane structure to facilitate increased requirements in structural capacity and transportability. Physical computing platforms incorporating programmable interfaces were introduced, to minimize labour-intensive and time-consuming assembly processes.

Prototypes

God's Eye

The Prototype Structure, 'God's Eye' (Fig. 1), was the winning entry of Sukkahville 2013 International Design Competition. The competition called for design proposals on a temporary pavilion pertaining affordable housing. 'God's Eye' was among the six shortlisted entries invited to be constructed in Toronto, Canada on September 2013.

Fig. 1 God's Eye realised Pavilion as it stood in Mel Lastman Square, Toronto Canada

The design consists of a pair of interweaved doubly curved surfaces, forming a central vaulted enclosure (focal point) which corresponds to a small roof opening. The pavilion was developed as a structure-membrane interdependent system. The structure (Primary and Secondary) was made out of recyclable PVC electrical conduit pipes, bent in place and secured using custom-made metal and acrylic digitally fabricated joints. The form was achieved by instrumentalising the bending forces induced on the pipes as defined by a computational model. The membrane, serving structural and sheltering purposes, was realized by interweaving recycled single-sided corrugated cardboard strips on the secondary structural system.

The complete structure was hand-carried from Cyprus to Canada in three sports equipment bags, whereas corrugated cardboard was purchased in Canada. Construction time was less than 16 h over two days by a team of two students and three faculty members.

Halo Sukkah

The Prototype Structure, 'Halo Sukkah' (Fig. 2) was a finalist entry in Sukkahville 2014 International Design Competition. The project was among the eight shortlisted entries invited to be constructed in Toronto, Canada on October 2014.

The pavilion formulates a single conically shaped space emphasizing a central roof opening. Halo Sukkah was also developed as a structure-membrane interdependent system (Fig. 3). The structure was made out of recyclable PVC electrical conduit pipes, bent in place and secured using custom-made metal and 3D printed joints. The form was achieved by instrumentalising the bending forces induced on the pipes assisted by linear tensioning elements as defined by a computational model. The membrane, serving structural and sheltering purposes, was realized by tensioning and interweaving polypropylene twisted twine on the primary structural system.

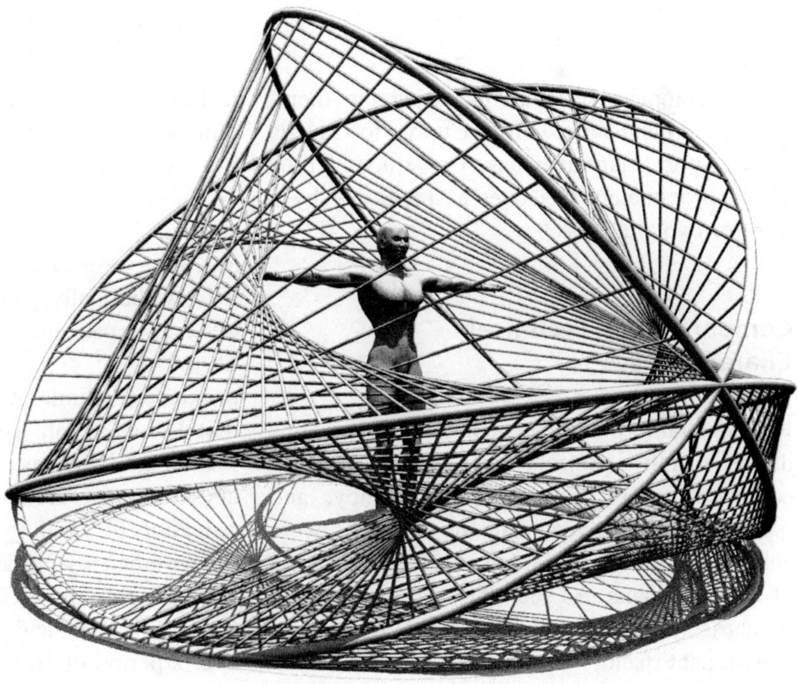

Fig. 2 Halo Sukkah prototype

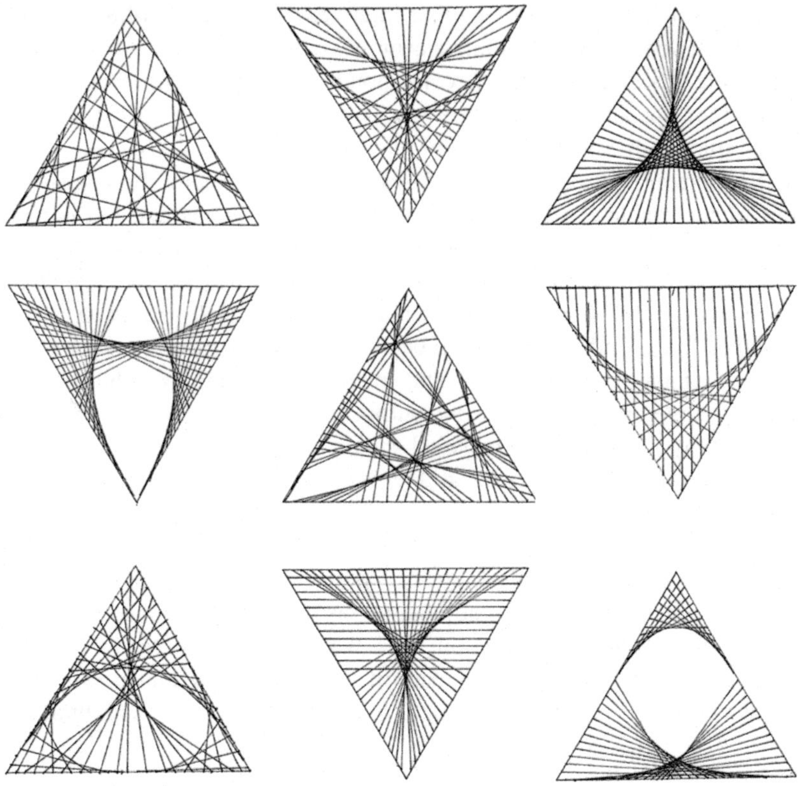

Fig. 3 Halo Sukkah's membrane iterations

The complete structure including the membrane was hand-carried from Cyprus to Canada in one sport equipment bag. Construction time was approximately 6 h over a single day by a team of two students.

Comparison of Technical Characteristics

Key constraints that drove many design decisions during both projects development were cost-effectiveness, construction efficiency, and transportability, given that the projects were aimed to be realized 9000 km away from the base of their designers. As such the characteristics of lightness, elasticity and suppleness have been determinant factors in selecting materials and defining their organizational logic. Given the experience gained from God's Eye, the research team strived for higher levels of construction efficiency and structural capacity during designing and developing Halo Sukkah.

Subsequently, while God's Eye has a larger footprint than Halo Sukkah at a ratio of almost 3:2, their respective weight stands at a ratio of 3:1. Additionally, Halo Sukkah has been more efficient in packaging with an occupying volume of 0.076 m^3 compared to God's Eye with a packaging volume of 0.228 m^3. It is to be noted that Halo Sukkah has been packaged with the membrane whereas God's Eye without. The above figures are coupled with a significant reduction in building material resources. Major improvements were also achieved in manpower and assembly time. God's Eye was erected by a team of five (3 faculty members and 2 students) over a period of 16 h whereas Halo Sukkah was erected in around 6 h by a team of two students. Finally, the aforementioned developments had

Table 1 Pavilion facts' comparison

Case study	Weight (kg)	Footprint (m^2)	Packaging volume (m^3)	Structural members length (m)	Structural joints	Membrane surface area (m^2)	Assembly time (h)	Assembly team	Material cost[a]
God's Eye	100	17.5	0.228	384	302 (4 types)	39.5	16	5 people	350 € 20 €/m^2
Halo Sukkah	30	11.2	0.076	56	13 (2 types)	21.2	6	2 people (Students)	180 € 16 €/m^2

[a]Material cost is calculated based on retail material/fixings prices in Cyprus for 2013/2014

direct benefits in material costs with noteworthy savings of 20 % per m^2 on the already low figure of 20 €/m^2 for God's Eye.

Technical Characteristics—Comparison Table

See Table 1.

Simulating Materiality

A great challenge for both projects was to integrate the physical behaviour of the materials used and their structural and geometrical characteristics in computational models that enable full control over design (Fleischmann and Menges 2011).

In both case studies the geometry was controlled by such computational models, produced using a core parametric definition in Grasshopper 3D (Grasshopper 3D 2015). The models were used throughout the design and construction process, continuously fuelled with information originating from testing parallel physical and digital models from and to which there was a direct transition.

Research on materials was conducted early-on and a selection of locally available resources along with the above properties and performance was documented. The behaviour of these materials was embedded in the computational model, while a series of physical tests verified their ability to be manually shaped as digitally predicted and respond to form geometry and structural efficiency.

Physical tests were conducted with the different pallet of materials used in each case aiming at understanding the behaviour of the structure (Fig. 4). These physical tests, would reveal the "bending active" (Lienhard et al. 2011) behaviour of the material used. During these physical tests, tubes of various lengths, thicknesses and structural properties were manually bent in several chord-length increments. At each increment, the length of the chord and the height corresponding to the deformed arch were measured. The lifting load at each increment was also recorded. An analytical equation (Georgiou et al. 2014) was generated which could approximately relate the vertical deformation of the tube according to its initial length and the horizontal

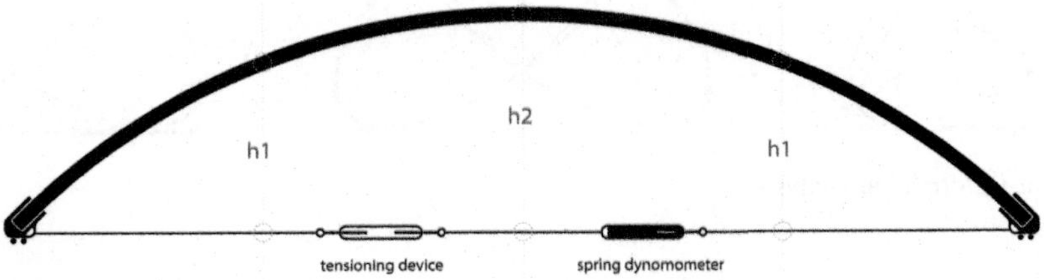

Fig. 4 Bending physical test set-up

Fig. 5 Shapes generated using the parametric model incorporating material properties

displacement. This analytical expression was initially integrated in the parametric models, instantiating a digital form-finding process, embedding within, material properties and approximate performance characteristics (Fig. 5).

The tubes' curvature relationship to the destabilizing load, or the "lift force", was also determined from the physical tests. This relationship specified the actual material stiffness to be used in more advanced numerical analysis. In order to understand the stress distribution of the structure, numerous structural analyses were performed, in which the initial force required to deflect the tube was used as input parameter (Happold and Liddel 1975). The results of the method were integrated in and compared against the parametric model informing and improving its behaviour.

As the shape of the pavilion evolved through digital prototyping and physical testing, its structural capacity was verified at all steps. In order to achieve interoperability between the parametric model and the numerical simulation software, a specialised plugin (Georgiou et al. 2011) was developed.

Departing from God's Eye to Halo Sukkah, the research team aimed for optimising performance and construction efficiency. An advancement towards that direction was the introduction of pre-stressed linear elements. These tensioning members (also forming the membrane) connected and stiffened the primary structure were needed (Dörstelmann et al. 2014).

Given the above, the original assumptions and models proved inadequate for describing the new pavilion. It was realised, that the previously defined analytical relationships responding to in-plane deformations only, could not be used to simulate interaction between members of the primary structure of "Halo Sukkah". Moreover, since there was a need to explore a series of solutions in a short period of time, non-linear structural analysis models couldn't be applied as they involved long computational processing time.

It was therefore decided to utilize the physics engine incorporated in Kangaroo (Kangaroo Physics 2015) for Grasshopper 3d to define the central computational model. Kangaroo solvers were able to simulate the bending active-behaviour structure (Andriaenssens and Barnes 1999). The material properties calculated previously were incorporated in the updated parametric model in order to simulate the behaviour of the new pavilion. Furthermore, at areas where the tensioning elements would act as a diaphragm to enhance stability of the structure, the model enabled monitoring of undesirable deformations (Fig. 6).

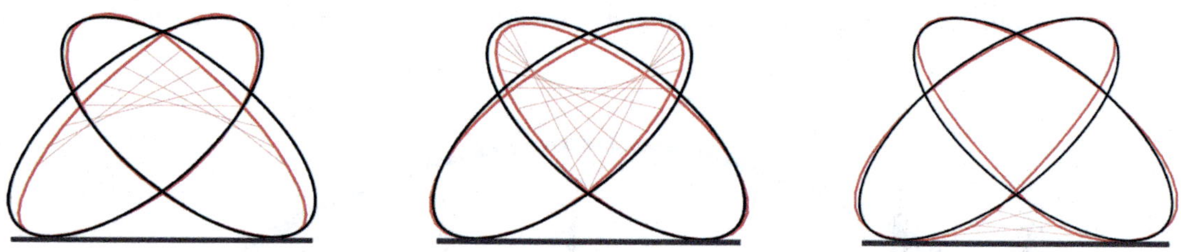

Fig. 6 Structural Deformation analysis

Construction and Fabrication Logistics

Construction and Fabrication Logistics were determinant factors in achieving the desired degree of complexity while maintaining affordability and constructability. Designing systems to streamline construction logistics was therefore considered early on and different strategies were repeatedly revisited during the design and testing of both projects. Specifically, the early design development stages (jointing and materiality), prefabrication and on-site assembly were identified as prominent areas for logistics overload and were therefore addressed accordingly.

During the early design development of God's Eye, UPVC pipes were chosen in order to, apart from their structural advantages, facilitate construction in terms of simplifying the connectivity of constituent parts. UPVC pipes presented an advantage as their in-build socket enabled the formation of continuous members, as well as closed loops. Similarly, among the reasons for selecting corrugated cardboard was its availability in long rolls, enabling the production of continuous cardboard strips virtually eliminating intermediate joints. Weaving the cardboard on the secondary structure removed the need of extra joints as connectivity was based on frictional forces.

While the above strategies had effectively reduced the number of possible joints, the proposal required connectivity solutions between the structural members of the pavilion. Given the tight development timeframe, precise analytical expressions of such connectivity wouldn't have been possible. Additionally, fixed joints would have altered the pavilion's shape depending on their rigidity and would impose a far more rigorous design process. As a result, rotational freedom was accounted for in the analytical models. Such an assumption would also require its physical counterpart; a joint enabling rotation on 2-axis. This was resolved by combining pairs of plumbing pipe clamps using set screws, found in local hardware stores to form a two way 360° joint.

The notion of joint rotational freedom was also adopted for the secondary structure. As revealed by the computational model, the complex shape yielded a large number of different connection angles which would be impossible to design and fabricate within the available time and budget. A universal, innovative joint combining commercial UPVC Tee joints and digitally fabricated acrylic parts, was designed and evolved to both meet the single and double axis rotation requirements.

During prefabrication stage of God's Eye, cutting patterns for the cardboard strips were managed by the computational model providing the length and width of each cardboard strip. Throughout the development of the project, it was ensured that the membrane could be formed out of zero Gaussian curvature strips which could be unrolled to flat, straight pieces. To facilitate construction, the strips were grouped in five sets of different widths, greatly simplifying fabrication logistics without compromising the overall appearance of the pavilion (Fig. 7). A similar process was applied for the PVC pipes in both projects, where cutting and marking schedules were generated for all the primary and secondary structure parts. This enabled waste elimination and simplified the construction process.

Despite the considerable benefits of issuing cutting schedules directly from the central computational model, lots of effort was still devoted in measuring, marking and physically cutting all constituent members (cardboard and PVC pipes). This matter was significantly addressed in Halo Sukkah with the creation of a custom made apparatus to mechanically process the 248 variable length twine lines forming the membrane. The device (Fig. 8) was continually fed by a twine spool and enabled automatic measurement of each different line as generated by the computational model. The electronics were built using an Arduino and a Stepper Driver Board which were real-time driven by Firefly (Firefly Experiments 2015) a plugin of grasshopper. The hardware was inspired and based on a 3d Printing Extruder assembly (Prusa i3 Rework Extruder 2015). Along with the above device, a standard end-loop attachment for all lines was developed and 3d printed to ensure the required levels of precision.

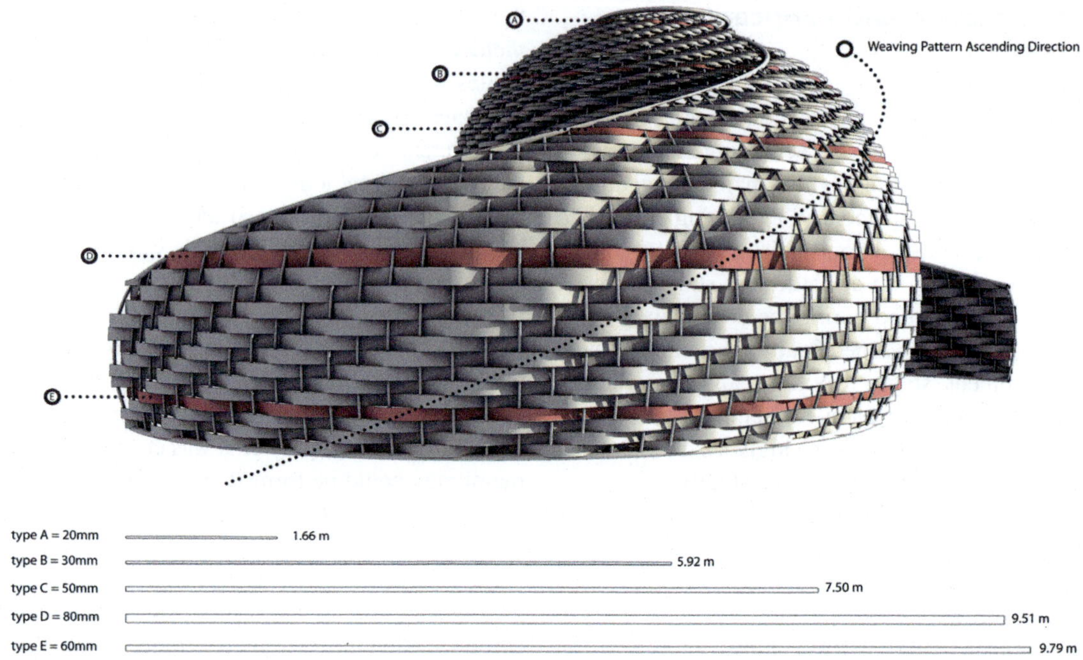

Fig. 7 Weaving Pattern rationalisation for God's Eye pavilion

Fig. 8 Custom made apparatus using Arduino and Firefly experiments

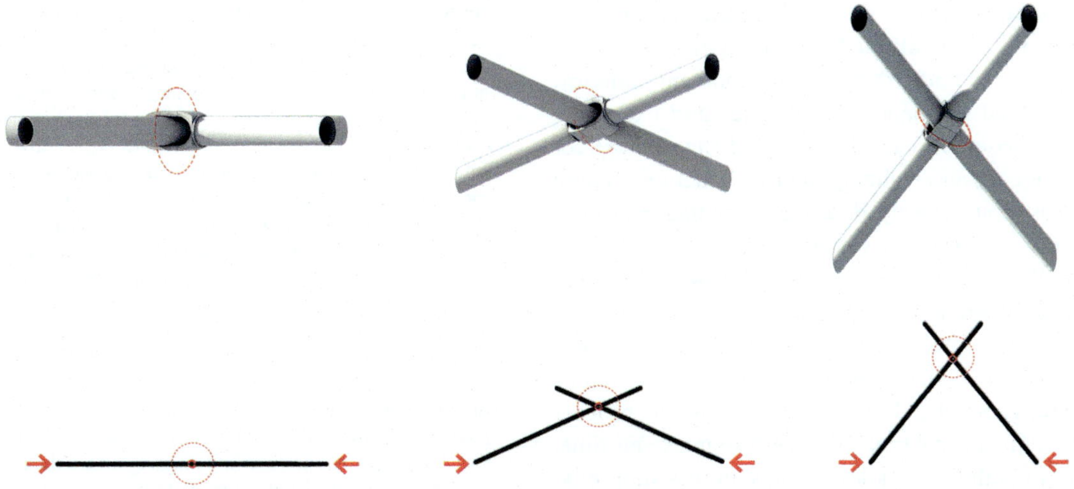

Fig. 9 *Top* rotational capacity of 3d printed joints responding to assembly requirements. *Bottom* assembly process of main structural elements for Halo Sukkah pavilion

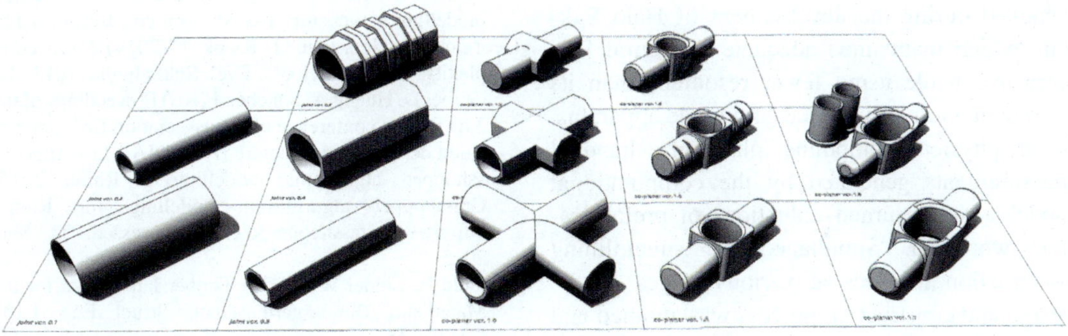

Fig. 10 Evolution of 3d printed joints for Halo Sukkah

Both approaches enabled significant time-savings during pre-fabrication phase and increased the overall accuracy and behaviour of the structure.

On site assembly has been an integral part in the design of both projects especially due to the limited construction time imposed by the competition rules. Capitalizing on God's Eye experience, and enforced by the reduced manpower to erect Halo Sukkah, the research team invested in developing adaptable joints that would facilitate the assembly of the structure similar to gridshell structure precedents (Williams 2014) (Fig. 9, top and bottom). As such, a *co-planar joint*, maintaining continuity and facilitating erection of the main structural members of the pavilion was developed using 3D printing fabrication techniques (Crolla and Williams 2014). The joint was tested and evolved through a series of trials as to withstand considerable forces exerted on it by the UPVC pipes while it would rotate and adapt to fit the desired geometry of the structure (Fig. 10).

Conclusions

The research presented in this paper sets new grounds for affordable complex structures. The two subsequently realized competition pavilions were designed under tight constrains and limited resources which provided fertile ground for testing such affordable paradigm. Restrained budget, short design development phases, unskilled personnel, remote sites and extremely tight assembly schedule defined the context for designing and realizing the two structures. These constraints closely describe designing and constructing for the developing world.

Within the above framework, it was illustrated that integration of materiality in computational feedback models and rationalization of construction logistics enables significant savings throughout the design and construction process in terms of materials, human resources and time.

In both case studies, the design was supported by a central computational model able to simulate the physical behaviour of the structures. Increased levels of understanding and control over the physical response of the structure were achieved during the development of Halo Sukkah, which maintained adequate Structural Performance while using fewer resources than its precedent, God's Eye. A custom made apparatus, using physical computing platforms, handled measurements generated by the computational model which returned reductions of prefabrication workload. Simultaneously, streamlining construction logistics at various stages of the design and construction process was attained and illustrated. In line with the overall behaviour of the structure, a flexible and adaptable jointing system minimized analytical time, fabrication and construction logistics while accelerating the assembly process. Further adaptation of the joints to required performance (assembly optimisation) using rapid prototyping techniques allowed for additional efficiency in construction.

References

Adriaenssens SML, Barnes MR (1999) Tensegrity spline beam and grid shell structures S.M. Eng Struct 23 (2001):29–36

Crolla K, Williams N (2014) Smart Nodes A system for variable structural Frames with 3d Metal—Printed Joints. In: Gerber D (ed) Proceedings of the 34th annual conference of the association for computer aided design in architecture (ACADIA), Los Angeles

Dörstelmann M, Parascho S, Prado M. et al (2014) Integrative computational design methodologies for modular architectural fiber composite morphologies in design agency. In: Gerber D, Huang A, Sanchez J (eds) Proceedings of the 34th annual conference of the association for computer aided design in architecture (ACADIA), Los Angeles

Firefly Experiments (2015) Design Interactively. http://www.fireflyexperiments.com/#home. Accessed 03 May 2015

Fleishmann M, Menges A (2011) ICD/ITKE research pavilion: a case study of multi-disciplinary collaborative computational design. In: Gengnagel C, Kilian A, Palz N, Scheurer F (eds) Proceedings of the design modelling symposium Berlin 2011. Springer, Berlin, p 239

Georgiou O, Richens P, Shepherd P (2011) Performance based interactive analysis. In: Gengnagel C, Kilian A, Palz N, Scheurer F (eds) Proceedings of the design modelling symposium Berlin. Springer, Berlin, p 115

Georgiou M, Georgiou O, Kwok T (2014) Form complexity—Rewind|'God's Eye' Sukkahville 2013. In: Gerber D, Huang A, Sanchez J (eds) Proceedings of the 34th annual conference of the association for computer aided design in architecture (ACADIA), Los Angeles

Grasshopper—algorithmic modelling for Rhino (2015) Grasshopper—algorithmic modelling for Rhino. http://www.grasshopper3d.com. Accessed 03 May 2015

Happold E, Liddel WI (1975) Timber lattice roof for the Mannheim Bundesgartenschau. Struct Eng J 53 (3):99–135

Kangaroo Physics|Grasshopper (2015) Kangaroo Physics| Grasshopper. http://www.food4rhino.com/project/kangaroo?etx. Accessed 03 May 2015

Lienhard J, Schleicher S, Knippers J (2011) Bending-active structures—research Pavilion ICD/ITKE. In: Proceedings of the international symposium of the IABSE-IASS symposium, Taller Longer Lighter, London

Prusa i3 Rework Extruder assembly—RepRapWiki (2015) Prusa i3 Rework Extruder assembly—RepRapWiki. http://reprap.org/wiki/Prusa_i3_Rework_Extruder_assembly. Accessed 03 May 2015

Williams C (2014) The Multihalle and the British Museum: a comparison of two gridshells. In: Adriaenssens S, Block P, Veenendaal D, Williams C (eds) Shell structures for architecture. Routledge, London, pp 239–245

3dj: 3d Sampling Haptic and Optically Performative Textures Remixed from 3d Scans

Sayjel V. Patel and Caitlin T. Mueller

Abstract

3D Sampling is presented as a new method for comprehensive rearrangement and composition of 3D scan data, allowing multi-scale manipulation of surface texturing and subsequent fabrication and re-evaluation using 3D printing. Analogous to sample-based music, 3D Sampling contributes new frameworks, tools, and compositional strategies for the digital remixing, hacking, and appropriating of material qualities, performances, and behaviours directly from physical samples to produce new designs. Case studies are presented which demonstrate *3DJ*: a prototype 3D modelling tool for synthesizing haptic and optically performing textures from 3D scan-generated source material, which can be applied in the context of other 3D modelling techniques.

Introduction

3D Sampling is part of a revolution in sensibilities in the field of design modelling. For the first time, hand-held 3D scanning technologies, such as photogrammetry, enable anyone with a smart-phone to digitally capture real world objects as point-cloud data. In the near future, a treasure trove of 3D materials sampled from real world objects and environments will be available to designers. Presently, the rules of computer modelling software precede and constrain the form of the objects generated. However, 3D scans allow a designer to reverse this methodology, making detailed digital replicas of physical objects the starting point for design (Fig. 1). By contributing methods to interject and collide the complex or structural irregular geometries of 3D scans deliberately into playfield of digital design, 3D Sampling facilitates a new solution space that would otherwise be inconceivable through conventional optimized, rule-based, parametric, or procedural modelling practices. In contrast with approaches taking inspiration from observed phenomena to infer design strategies, such as bio-mimicry and material computation (Oxman 2009), 3D Sampling allows designers to appropriate, hack, and remix the material behaviour of macro-scale texture geometry such as structural corrugation, water shedding, grip, acoustics,

S.V. Patel (✉)
Massachusetts Institute of Technology, MIT-SUTD Collaboration, Cambridge, USA
e-mail: sayjel.v.patel@gmail.com

C.T. Mueller
Massachusetts Institute of Technology, Department of Architecture, Cambridge, USA

M.R. Thomsen et al. (eds.), *Modelling Behaviour*, DOI 10.1007/978-3-319-24208-8_44

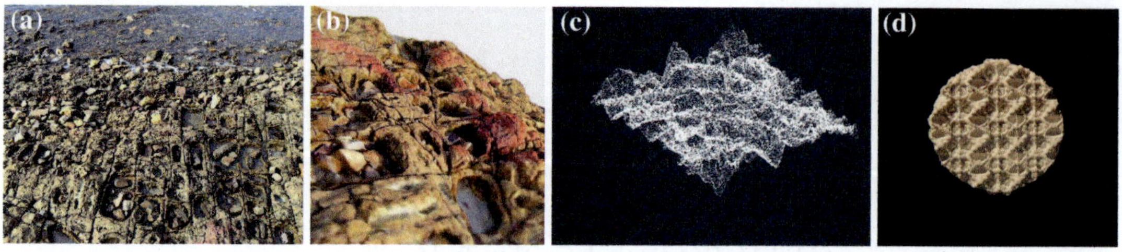

Fig. 1 3D Sampling overview: **a** Photograph, rock formation with convex surface features. **b** Detail, photogrammetric model generated using AutoDesk 123d Catch (http://www.123dapp.com/catch). **c** 3DJ: prototype design tool for textured synthesis from 3D scan samples. **d** 3D printed texture sample

optics, and camouflage (Gage 2013). This paper presents the conceptual framework and technological implementation of 3D Sampling, demonstrated through case studies presenting designs for architectural surfaces textures.

Related Work

Akin to sample-based music, 3D Sampling combines existing modelling tools and methods with concepts and practices drawn from external examples. This section describes: (1) How sampling provides the conceptual inspiration for applying 3D scanners as design generators; (2) Recent work in 3D mesh sculpting and cut and paste interfaces; (3) The technological opportunities and challenges related to 3D scanning; (4) Research efforts in the computer graphics community exploring texture synthesis techniques; and (5) The functional and cultural significance of texture design in contemporary architectural practice.

Sampling

In music, sampling is the act of taking a portion of a sound or recording and reusing it to create original compositions. Walker (2008) provides a comprehensive overview of the field of sample-based music; its motivations, compositional strategies, and history. Some of the central manipulations applied by artists in this field, including digital signal processing, convolution, algorithmic filtering, looping, and mixing, serve as compelling metaphors for 3D Sampling. In architecture, 3D scanning (Sheil 2013) and textural sampling (Gage et al. 2013) have garnered recent attention; however, hybrid applications, as proposed in this paper are scarce.

Personal 3D Scanning

3D scanners are instruments that collect data about the geometry of physical objects, outputting lists of geo-spatial coordinates called point-clouds. Photogrammetry is one such 3D scanning technology that generates textured 3D meshes from photographs. However, because of computational requirements, there is a significant time lag between scanning and model generation. In contrast, low-cost, real-time scanning can be achieved through depth camera technology. KinectFusion applies this technology to 'paint' 3D reconstructions through the motion of Kinect device (Azadi et al. 2011). One inherent disadvantage of this technology is the trade-off between the resolution of reconstructions and the immediacy of the interaction. As discussed in the literature (Boulassai et al. 2009; Arachchige and Perera 2014; Truong-Hong and Laefer 2011), a fast and reliable method for real-time segmentation and conversion of point-clouds into discrete 3D surfaces for computer modelling remains in absentia.

Procedural Texture Generation

Image-based texture synthesis is a topic of continued interest in the field of computer graphics, whereby 2D samples from photographic images can be used to synthesize entire worlds (Efros and Freeman 2001). Texture synthesis algorithms have two primary goals: to sample an existing texture to generate an unlimited amount of new image data that perceived to be the same to humans, and to transfer textures between objects. Texture synthesis works by extracting patches of an existing texture and applying the samples in consistent way. The literature provides a range of possible technical approaches to 2D textural synthesis (Ashikhmin 2001; Ruiters et al. 2010; Efros and Leung 1999; Wei and Levoy 2000). While typical applications of textural synthesis in computer graphics are guided by performance-driven criteria, 3D sampling applies textural synthesis as a new technique for design exploration.

Virtual Clay

Evocative of traditional practices of physical clay modelling, virtual clay has garnered recent attention as an alternative to conventional 3D modelling styles. A review of scholarly work reveals *3D Sculpting* (Dipen et al. 2013; Cit et al. 2013; Cingi and Oghan 2012; Madugalla et al. 2013) and *Cut and Paste* modelling (Sharf et al. 2006; Schmidt and Singh 2010; Bierman et al. 2002) as two inter-related applications of virtual clay. This body of work has leveraged a variety of rapid mesh editing techniques to produce innovative new compositional operations for fusion of discrete parts and the transfer of surface features (Schmidt and Singh 2010). Bringing together the processes of concept visualization and final design virtual clay offers a way to quickly explore a large space of alternative solutions by extending the functionality of existing 3D objects and simplifying user interaction and manual manipulation (Sharf et al. 2006). When used in conjunction with virtual clay, 3D Sampling contributes novel remixing operations which allow a designer extend and exploit a new range possibilities beyond the mere surface of an input 3D objects.

Textures, Patterns, Behaviours

Architects have always relied on textures and ornamentation as a compelling way of resolving the cultural and pragmatic requirements of building details (Fig. 4a). Considering digital technologies and networks, the status of architectural texture and ornament has shifted in response to parametric tools, (e.g. Grasshopper or Maya); now designers can create highly complex and dynamic patterns predicated on relatively simple rules. The emerging interest in texture and ornament is discussed extensively in the literature (Moussavi and Kubo 2006; Lynn 2004; Picon 2013; Gleiniger and Vrachliotis 2009; Levit 2008).

Research Question

The framework and tools presented here address some of the major limitations inherent in currently available computer modelling approaches—specifically, how to define the rules of a parametric system which produce a desired form or material behaviour. Without a clear understanding of the rules from the onset, parametric and procedural modelling approaches inefficiently reach acceptable solutions requiring significant iteration through trial and error. Demonstrating a new field of design and research inquiry, this paper explores several broadly defined questions: What advantages does 3D Sampling offer as a technique for design exploration? What are some possible applications? What is the relationship (if any) between the sampled material and the subsequent textures? How does one evaluate or analyze results? What are the boundaries of such an approach? We postulate that it is possible to 3D scan specific

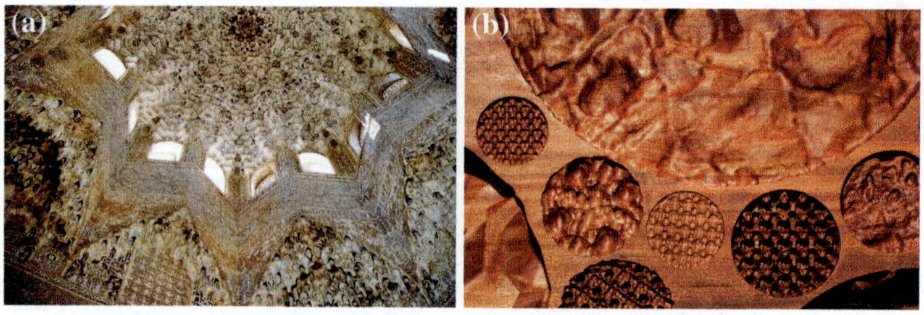

Fig. 2 Performative textures: **a** Macabres, Alhambra (*source* http://en.wikipedia.org/wiki/Alhambra#/media/File:Ceiling_in_Alhambra.JPG, user Liam 987). **b** Haptic texture study, a range of surface textures were generated using the 3DJ modelling tool and CNC milled into a slab of sapele wood

qualities of an object, remix the material digitally, and produce a range of new material qualities from the original scan (Fig. 2).

Methodology

This section outlines the details of 3D Sampling, a novel technique for synthesizing textures with material qualities one or more 3D scans as the foundation. The function and rationale of the new 3DJ 3D Sampling software tool is elucidated.

Overview

Figure 3 outlines the six operations 3D Sampling contributes to traditional 3D modelling. Figure 4 explains conceptual framework of 3D Sampling. The key processes described are: (1) Scanning—the digitization of a physical sample. (2) Synthesis—remixing a digital sample using conventional (e.g. Rhino) and novel (3DJ) digital modelling. (3) Printing—physical materialization of a digital sample, and (4) Analysis—evaluating the behaviour or performance of a physical or digital sample. The framework also tracks the format of information as it translates mediums throughout the sampling process: (1) physical samples—physical objects; point-cloud samples—non-hierarchical lists of coordinates, and 3D mesh samples—hierarchical lists of coordinates.

Technical Framework and Implementation

This section describes the technical implementation of 3D sampling further.

3D Synthesis

3D Synthesis involves the processes of sampling, modifying, and remixing digital samples using conventional and novel software approaches. This section describes *3DJ*, a custom DJ tool developed for manipulating 3D scan point-clouds in the course of this research.

3DJ

Inspired by music sampling tools and methods, 3DJ is a prototype computer modelling application developed by the authors for 3D Sampling (Figs. 5, 6 and 7). Written in Processing, the inputs and outputs of 3DJ are point-clouds. 3DJ provides an interface and additional features allowing designers to manipulate point-clouds in new ways. 3DJ recognizes the practices, methods, and procedures of design established in sample-based music, and organizes the experience of designing with 3D scans into six operations—sample, cut, mix, filter, analyze, and loop—to produce new texture variations The 3DJ interface allows a user to crop regions from

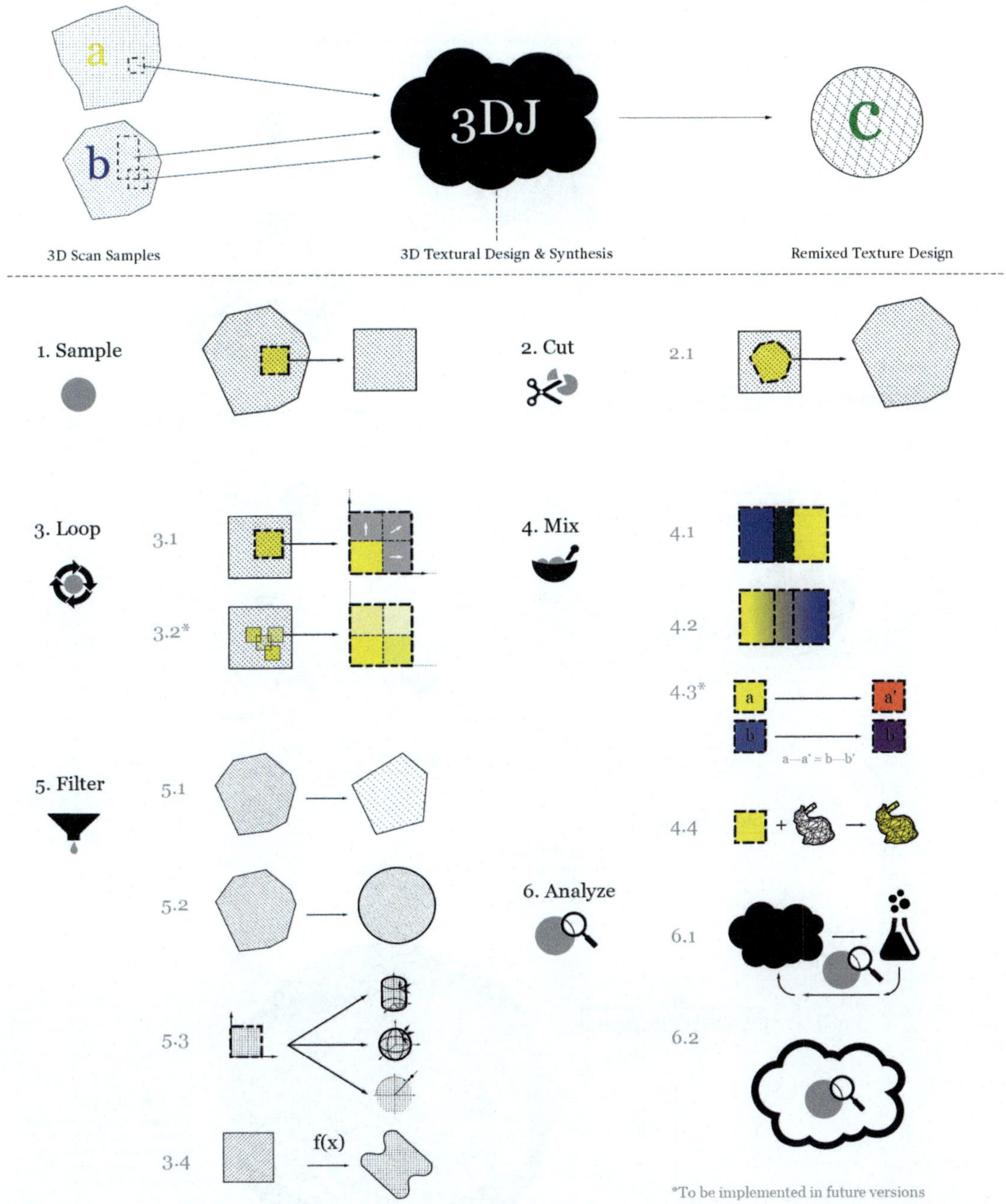

Fig. 3 Summary of 3DJ operations and functions: (*1*) Sample—a set of point cloud data, (*2*) Cut—extracting a set of point cloud data, (*3*) Loop—repetition of geometric features, using a process of arraying tiles (*3.1*) or pixel quilting (*3.2*) as described in (Efros and Freeman 2001), (*4*) Mix—textural blending, superposition, and juxtaposition. Techniques include superposition and averaging of samples (*4.1*), linear blending using distributed averaging of pixel positions (*4.2*), transforms using image analogies (Hertzmann et al. 2001), and 3D mapping of textures using techniques such as projective texture mapping (Segal et al. 1992), (*5*) Filter—including, mesh reduction through localized averaging of clusters (*5.1*), mesh smoothing, coordinate transforms, particularly useful as a precursor to mapping operations (*5.2*), (*6*) Analysis—evaluation and identification of desirable features of a sample, either through empirical testing methods (*6.1*), or computational simulation (*6.2*)

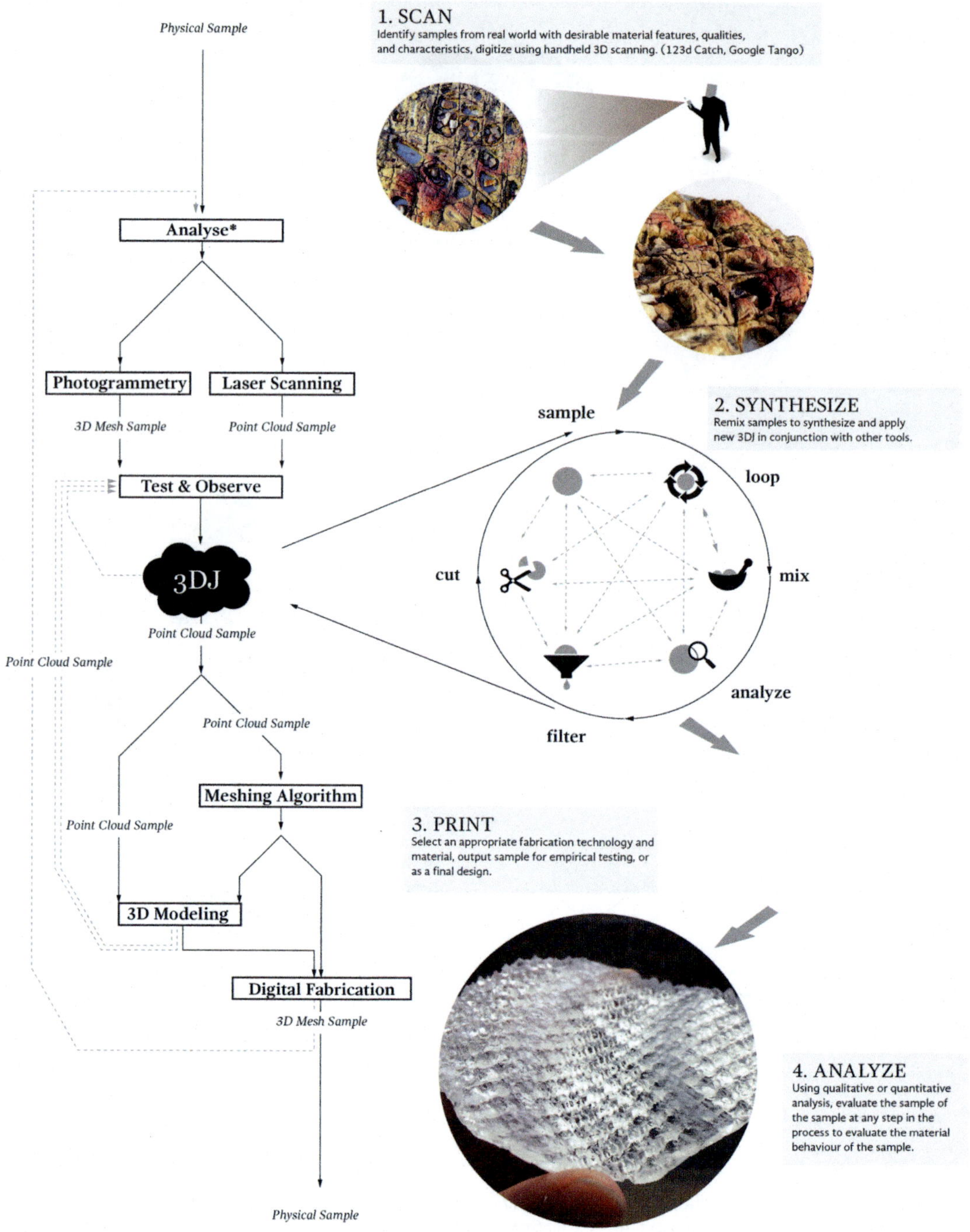

Fig. 4 3D Sampling network diagram: flows, processes and tools

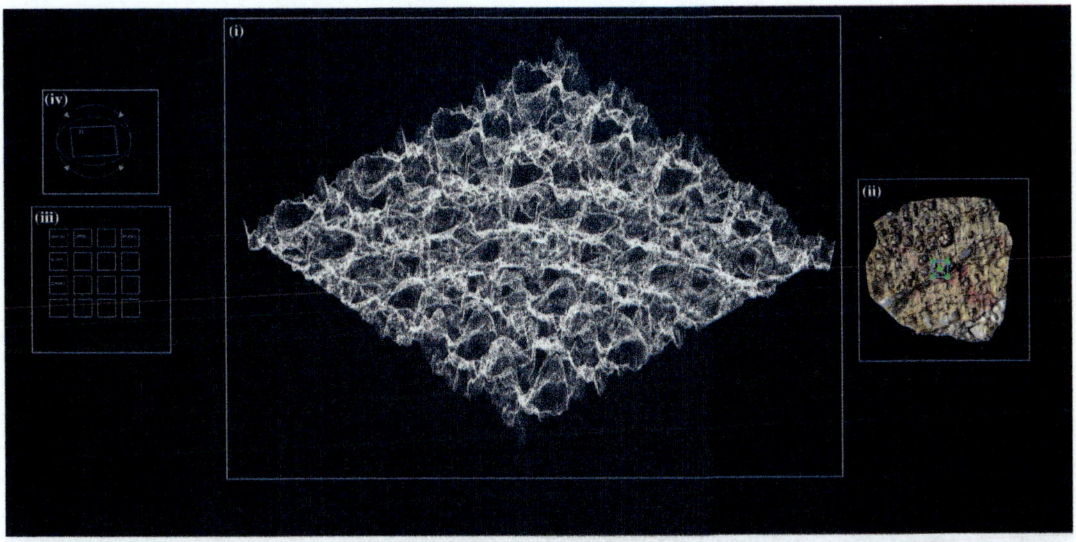

Fig. 5 Screenshot of 3DJ interface: **a** 3D view port. **b** 2D interface, here a user crops of a sub-sample from the input 3D scan data. **c** Customizable buttons for user manipulation, toggles a looping operation, mathematical filters, point cloud reduction and coordinate mapping. **d** 3D view selector—top, bottom, left, right, axo—southwest, southeast, northwest, northeast

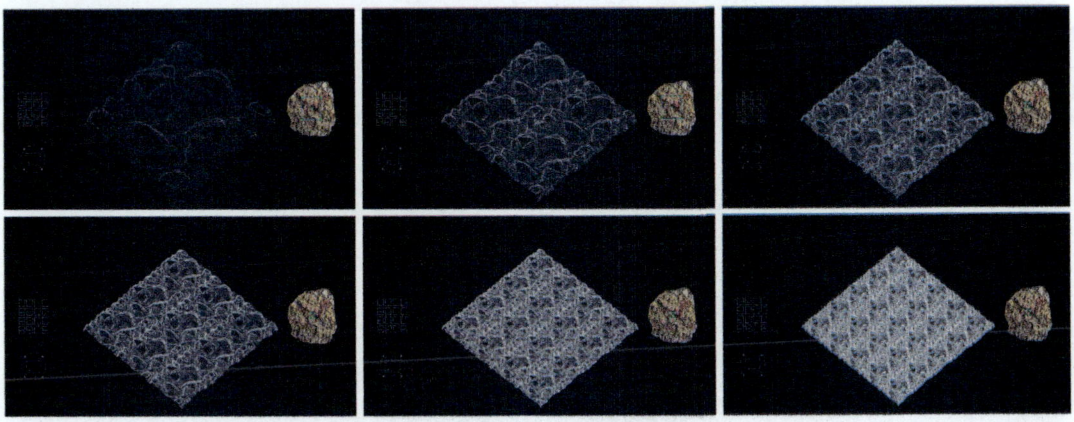

Fig. 6 Looping operation applied to sub-sample cropped from a 3D scan

imported 3D scan point-clouds or meshes (cut), and apply looping, mixing, or filtering operations to synthesize a new point-cloud texture which can is output as a PLY file. While more sophisticated textural synthesis algorithms can be incorporated in the future, the 'looping' operation involves a simple mirroring and tiling of sub-samples. 'Mixing' involves a simple linear blending algorithm to combine two or more scans (Fig. 8). The 'Filter' in this version of 3DJ involves operations to apply global mathematical distortions across the textural field, as well as mesh reduction algorithms; either involving a random re-sampling of the input point cloud within 3DJ, as well as the use of the built-in mesh reduction function in rhino prior to the import of a sample being imported into 3DJ. While future versions of 3DJ could include real-time mesh re-construction; efforts to do so using within 3DJ stalled the tool, undermining

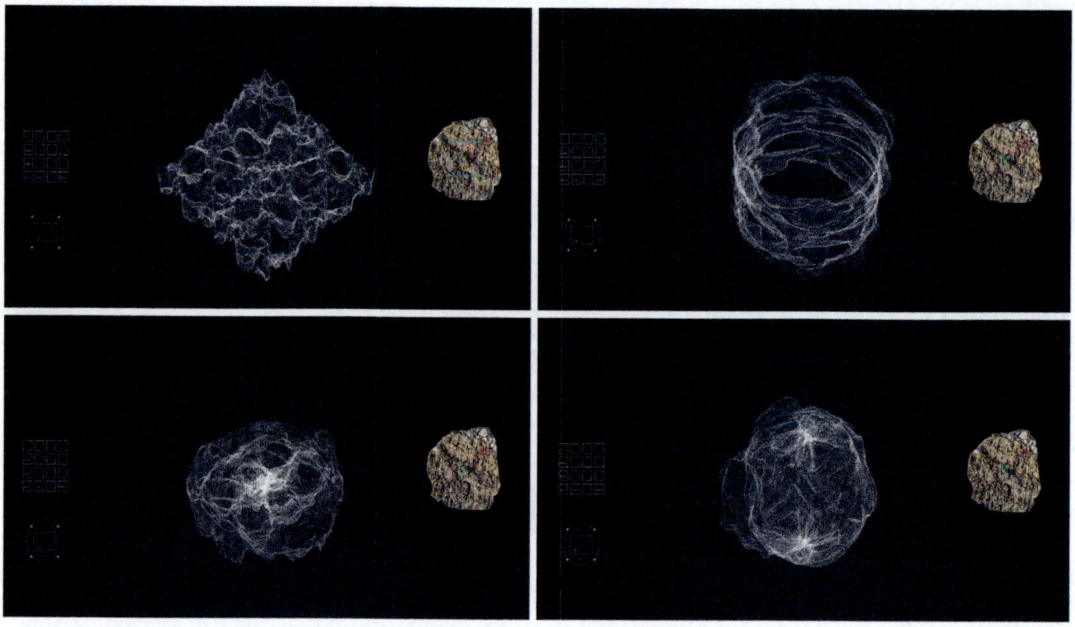

Fig. 7 Coordinate transformations (clockwise starting from *top left*): x-y-z texture, cylindrical coordinate system, polar coordinates, spherical coordinates

Fig. 8 Linear blending—contour drawing of a texture combining four 3D sampled textures

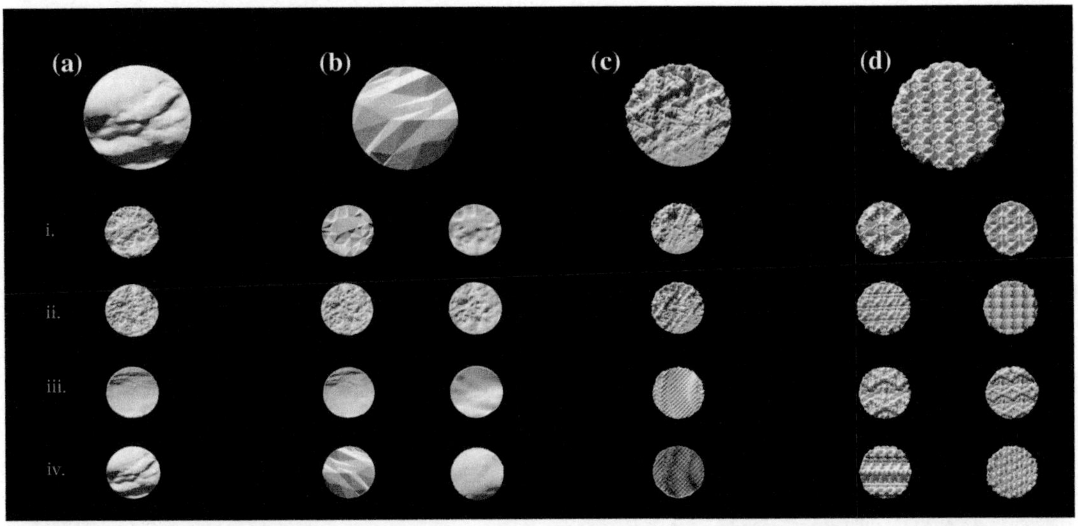

Fig. 9 Design exploration: variations created from prototype 3D texture design tool: **a** Sample: a cropped extract of an oyster shell. **b** Filter: a mesh-reduce operation and mesh smoothing operation. **c** Mix: a mathematical z-distortions applied to textures. **d** Loop: tiling of sample textures achieves scalar-manipulations

Fig. 10 **a** Additive manufacturing: powder printed models were brittle and did not have useful mechanical properties. **b** Additive manufacturing optical qualities—resin stereo lithography on a form-1 3D printer. **c** CNC milling

interaction rate; ultimately a Poisson surface reconstruction algorithm and the built-in Delaunay mesh triangulation were applied using Mesh Lab to produce satisfactory meshes for subsequent milling and 3D printing operations. While similar techniques could be achieved by employing a variety of existing software in concert, the aim of integrating these operations within a single tool was to facilitate greater design exploration (Fig. 9) by allowing the user to actively intervene within the process of guided textural synthesis.

Analysis

Current technological limitations constrain 3D sampling to macro-scale material properties, attributes and characteristics. For the initial tests, high definition starch-based powder 3D printing (Fig. 10a) was used. However the brittle, water soluble, nature of this medium degraded performative qualities, thus other alternatives were explored. 3D printing mediums like resin-based stereo lithography (Fig. 10b) and ABS allow optically and mechanically performative parts.

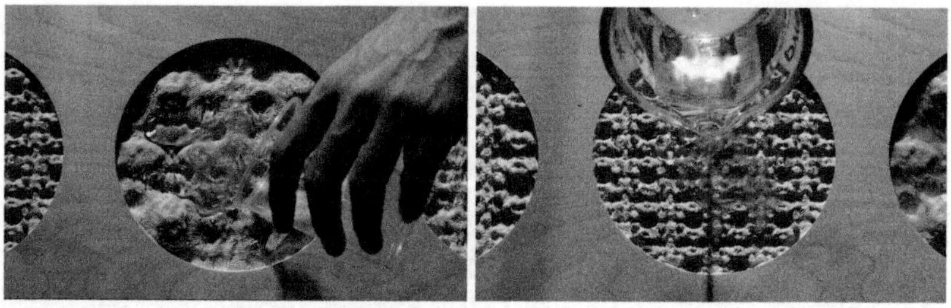

Fig. 11 Initially analysis, exploring how texture designs dispersed fluid across a surface

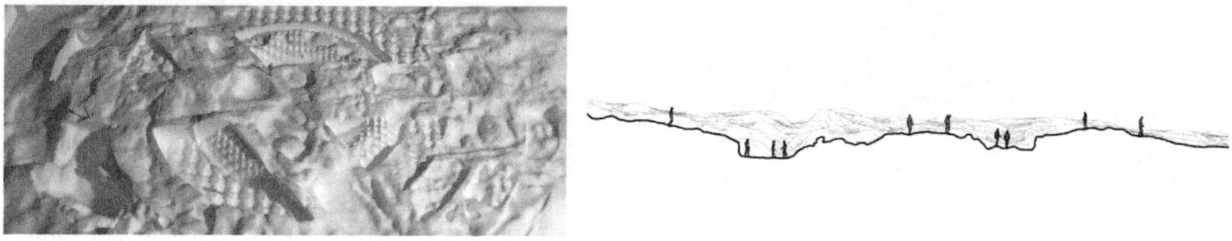

Fig. 12 Landscape scale

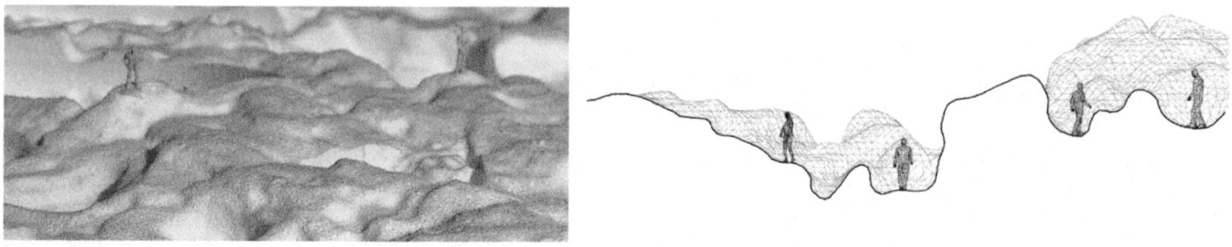

Fig. 13 Proto-architectural scale

However, economies of scale and material flexibility made CNC milling the desired fabrication medium here.

Analogous to how musicians must have the ability to listen to a progressing mix to properly judge their compositions, 3D Sampling requires feedback to evaluate how mixing operations manifest as the sample's physical characteristics. While the initial test applied qualitative analysis and evaluation in the production of texture design (Fig. 11), other possible modes of empirical, and real-time digital analysis exist, including thermal, optical, and haptic performances, and tests quantifying surface qualities (e.g. aerodynamics, roughness, irradiance, etc.) While the physical analysis methods provide immediate, tangible results and verification, real-time digital analysis methods within 3DJ would facilitate a stronger relationship between digital manipulations the final texture designs.

Case Studies

This section describes initial design experiments demonstrating the application of the 3D sampling framework to the design of camouflage, haptic, and optical, qualities using a single 3D scan.

Landscape and Camouflage Textures

Figures 12, 13 and 14 describe the first design experiment. Imagining possible applications to landscape design, this test demonstrates how

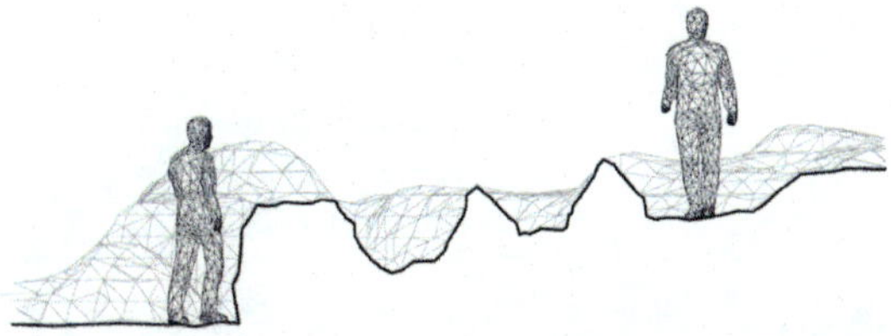

Fig. 14 Human-scale

Fig. 15 Material selection and the translation of material qualities (*touch*) from the original sample

terrestrial features can be 3D scanned and blended at multiple scales of occupation, generating a landscape. Textures representing a range of scales using sampling and looping operations were synthesized in 3DJ and combined through a technique of superposition, to produce a hypothetical landscape providing numerous scales of human habitation. Considering the human body as a datum, we observe three scales in the model: A landscape scale, suggestive of a connection to a broader context, a proto-architectural scale, suggestive of more enclosed and intimate spaces, and finally a human-scale, suggestive of ways in which a person might directly engage the landscape.

Haptic Textures

The second series of experiments (Figs. 15 and 16) explored applications at product design scale by demonstrating possibilities to synthesize textures with tactile qualities (e.g. grip, comfort etc.). Multiple sampling, looping, mixing, and filtering operations were applied using 3DJ to

Fig. 16 Haptic case study: qualitative evaluation

generate a wide range of material textures: from slippery, to sharp, to rough. A key finding revealed that the material selection itself was a design operation, transforming the qualities of the raw 3D scan. For example, an otherwise un-altered geometry sampled from the 3D scan becomes smooth like a piece of cloth when milled into wood. However, applying a filter produces a jagged texture. Applying a loop increases the roughness of the sample. One opportunity demonstrated in this case study is using 3D Sampling as a means to short-scale surface heterogeneity (Fig. 17).

Fig. 17 Textural gradients

Optical Textures

Figures 18, 19 and 20 highlight the third experiment, exploring the synthesis of optically performative textures at the building envelope scale. The methods were applied to demonstrate how material qualities related to light transmission across a CNC milled acrylic surface such as opacity, transparency, and reflectivity could be controlled using the 3DJ tool. From the preliminary results, it was observed that filtering, which removes redundant points, produces greater transparency (Fig. 17), while looping, which increases point density creates a texture with greater diffusion and specular noise (Fig. 18). The unaltered sample geometry (Fig. 19), results in textural chimera.

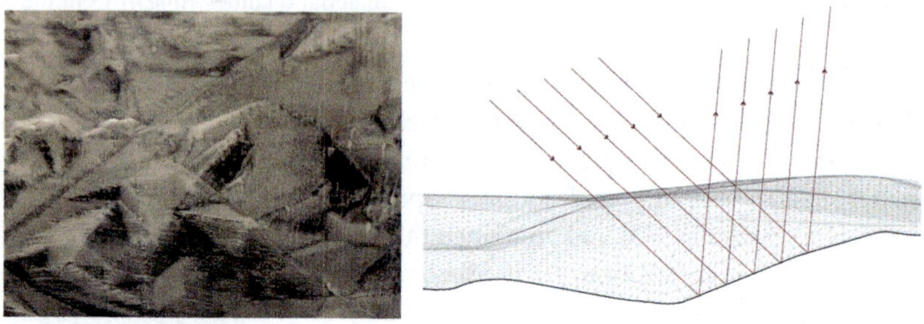

Fig. 18 Optical example 1, mesh reduction filter achieves more directed light

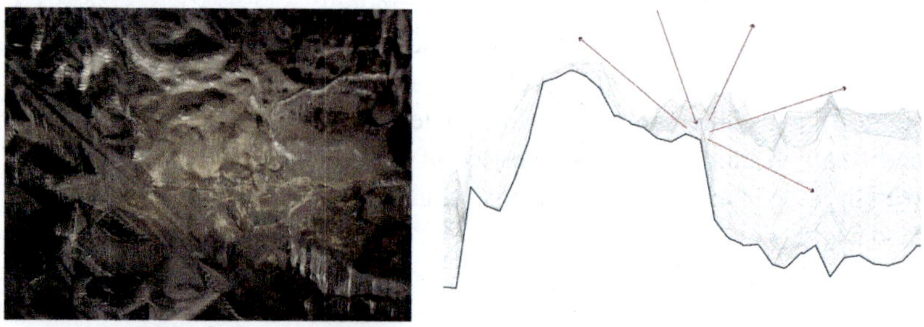

Fig. 19 Looping texture produces more dispersed light and opaqueness

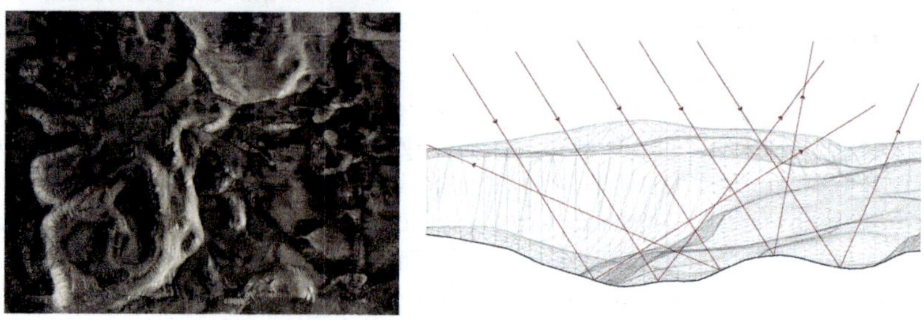

Fig. 20 Unaltered geometry texture from the original 3D scan

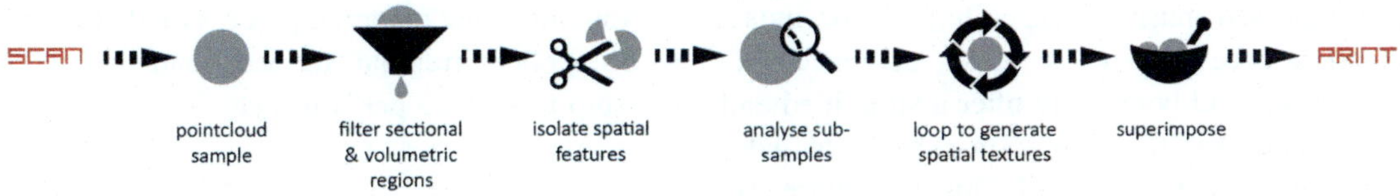

Fig. 21 Camouflage case study

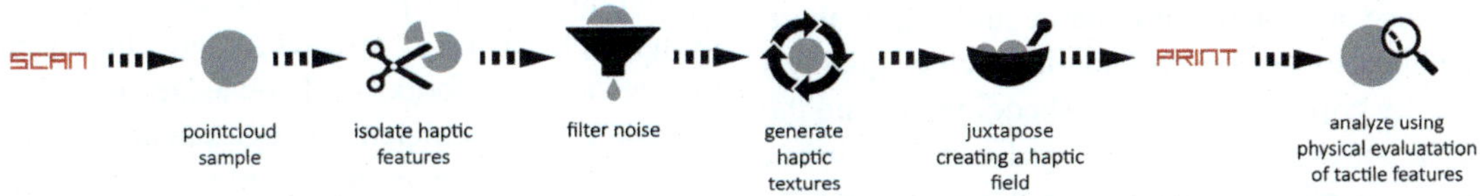

Fig. 22 Haptic case study

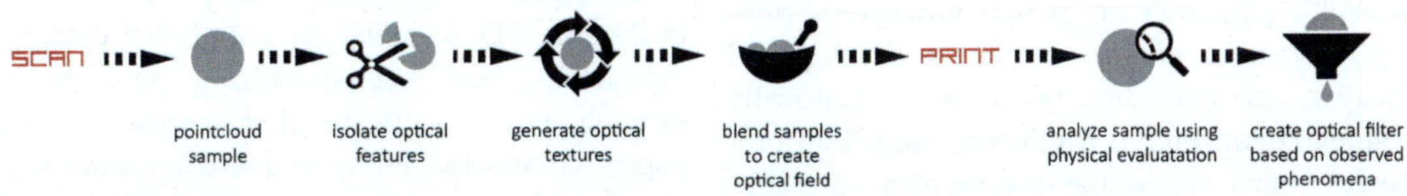

Fig. 23 Optical case study

Case Study Summary

The motivation for the case studies was to demonstrate morphogenic potentials of sampled material. Part of this involved the development of syntax to link operations of the tool and the resultant material performances and behaviours produced. Figures 21, 22 and 23 summarize the application of 3D Sampling operations, and compositional strategies used to generate performative textures. The case studies provide initial results, demonstrating a single iteration of the 3D Sampling framework.

Conclusion

Summary

3D Sampling imagines a future where a designer could instantly sample and remix the geometry and material qualities of objects from the real world, producing new designs. The research presented in this paper contributes: (1) The concept of 3D Sampling as a creative process; (2) A definition for conceptual and technical framework for 3D Sampling; (3) 3DJ, a 3D modelling software tool allowing users to generate new 3D textures from 3D scans; and (4) Three case studies implementing the framework, initial results and evaluation. The speculative tools, frameworks, and experiments presented and assessed in the paper, represent first steps which exemplify the potential future of a new design and research field which re-contextualizes 3D scanning, once conceived as a process of measurement, as a method for creative production.

Potential Impact

While the solutions developed within the three case studies—spatial, haptic, and optical—were framed within an architectural discourse of texture, the implications of this work reach far outside the discipline. Considering how sample-based music practices have made music creation accessible to a wider audience, 3D sampling could potentially offer a simplified and more intuitive way for non-experts to edit, manipulate, and create 3D objects. Future 3D textural synthesis tools, suggested by the 3DJ prototype software, imply a wide range of possible applications in light of different users. For instance a product designer could apply these methods to design ergonomic, tactile, qualities such as grip, comfort, water shedding, around the sense of touch, while a landscape or urban designer might sample of natural features to integrate a structure into a landscape through camouflage. Conversely, a materials engineer could use 3D Sampling to apply a texture with desirable properties but poorly understood principles such as gecko's feet to a develop and analyze solutions directly through empirically testing, contrasting to parametric modelling processes which require the development of a conceptual model of a gecko's foot prior to application within a design space.

Future Work

To further explore the connection between the operations and performance, future work may incorporate additional iterations between synthesis and analysis in order to further knowledge. Quantitative analysis and experimental data would buttress this. Greater interrogation and classification of the initial samples, as well the combination of multiple samples would expand on this initial exploration, such as the possibility of creating material hybrids (e.g. mixing gecko's feet with fur). A promising avenue to advance methods regarding mixing within 3DJ would be to adapt seminal work on 2D texture synthesis in computer graphics (Efros and Freeman 2001; Ruiters et al. 2011; Hertzmann et al. 2001) as a way to conceive of new mixing, looping, and filtering operations. Bump mapping techniques could provide a quick way to test how these techniques for 2D textural synthesis translate to 3D point-clouds. Another possible topic for exploration is the mapping process to 3D textures to be applied irregular surfaces, and their relationship to surface performance.

Concluding Remarks

Within architectural practice, the use of external design precedents, research, and first-hand observations as forms of knowledge are often considered more efficient strategies in developing designs than initiating projects from a tabula rasa (Zarzar 2003). 3D scanning generates a wealth of new digital data that architects can access freely to produce new material, spatial, and even cultural qualities, properties, and effects. By re-framing 3D scanning as a mode of creative production and operationalizing this mode through novel software demonstrations, this paper explores the ability to transcend subjective or experiential qualities of physical reality, and empowers designers with new tools to experience and remix the world.

Acknowledgments Special thanks to: Skylar Tibbits, Terry Knight, Marc Downie, and Mark Goulthorpe for their mentorship, timely ideas, and guidance, Inés Ariza, for being an excellent colleague, and her generous contribution towards the photographic documentation presented here, Alexander Charidis, for sharing his insights and knowledge of related research efforts in the field of computer graphics, Ben Goldstein, for his help editing and clarifying earlier drafts of this paper, and the MIT-SUTD Collaboration and the MIT Structural Design Lab for supporting the work.

References

Ashikhmin M (2001) Synthesizing natural textures. In: Proceedings of the 2001 symposium on interactive 3D graphics—SI3D '01

Biermann H, Martin I, Bernardini F, Zorin D (2002) Cut-and-paste editing of multiresolution surfaces. ACM Trans Graph 21(3):312–321

Boulaassal G, Landes T (2009) Automatic extraction of planar clusters and their contours on building facades recorded by terrestrial laser scanner. Int J Archit Comput 07(1):1–20

Cingi C, Oghan F (2012) Teaching 3D sculpting to facial plastic surgeons. Facil Plastic Surg Clin Am 19 (4):603–616

Efros AA, Leung TK (1999) Texture synthesis by non-parametric sampling. Proc Seventh IEEE Int Conf Comput Vis 2:1033–1038

Efros AA, Freeman WT (2001) Image quilting for texture synthesis and transfer. In: Proceedings of the 28th annual conference on computer graphics and interactive techniques—SIGGRAPH '01

Gage M, Friddle T, Ayas T, Burry M, Cannavino B, Connolly R, Kehagias N, McPhail L, Oncescu C, Sheridan W, Stranix K, Tripodi RJ, Utting B, Wiskup E (2013) Disheveled geometries: the digital stone project. Yale School of Architecture, New Haven

Gleiniger A, Vrachliotis G (2009) Pattern: ornament, structure, and behavior. Birkhäuser, Basel

Hertzmann A, Jacobs CE, Oliver N, Curless, Salesin DH (2001) Image Analogies. In: Proceedings of the 28th annual conference on computer graphics and interactive techniques—SIGGRAPH '01

Izadi S, Davison A, Fitzgibbon A, Kim D, Hilliges O, Molyneaux D, Newcombe R, Kohli P, Shotton J, Hodges S, Freeman D (2011) KinectFusion. In: Proceedings of the 24th annual ACM symposium on user interface software and technology—UIST '11

Levit R (2008) Contemporary "ornament": the return of the symbolic repressed. Harvard Des Mag 28:70–85

Lynn G (2004) The structure of ornament. In: Leach N, Turnbull D, Williams C (eds) Digital tectonics. Wiley-Academy, Chichester, pp 62–68

Madugalla AK (2013) FaceID: a 3D computer graphic application for forensic medicine: a novel semi-automated muscle based digital sculpting initiative for forensic facial reconstruction in Sri Lanka. In: 2013 International conference on computer medical applications (ICCMA). pp 1–6

Moussavi F, Kubo M (2006) The function of ornament. Actar, Barcelona

Menges A (2012) Material computation: higher integration in morphogenetic design. Wiley, Hoboken

Oxman N, Rosenberg JL (2007) Material-based design computation: an inquiry into digital simulation of physical material properties as design generators. Int J Archit Comput 5(1):26–44

Segal M, Korobkin C, Van Widenfelt R, Foran JJ, Haeberlie P (1992) Fast shadows and lighting effects using texture mapping. In: Proceedings of SIGGRAPH '92. pp 249–52

Schmidt R, Singh K (2010) Meshmixer. In: ACM SIGGRAPH 2010 Talks on—SIGGRAPH '10

Sharf A, Blumenkrants M, Shamir A, Cohen-Or D (2006) SnapPaste: an interactive technique for easy mesh composition. Vis Comput Vis Comput 22(9–11):835–844

Ruiters R, Schnabel R, Klein R (2010) Patch-based texture interpolation. Comput Graph Forum 29 (4):1421–1429

Picon A (2013) Ornament: the politics of architecture and subjectivity. Wiley, Hoboken

Miller PD (2008) Sound unbound: sampling digital music and culture. MIT, Cambridge

Wei LY, Levoy M (2000) Fast texture synthesis using tree-structured vector quantization. In: Proceedings of the 27th annual conference on computer graphics and interactive techniques—SIGGRAPH '00

Zarzar KM (2003) Use and adaptation of precedents in architectural design: toward an evolutionary design model: proeschrift

Author Index

M.R. Thomsen et al. (eds.), *Modelling Behaviour*, DOI 10.1007/978-3-319-24208-8

Printed by Printforce, the Netherlands